T0189197

Lecture Notes in Computer Science 10637

Commenced Publication in 1973
Founding and Former Series Editors:
Gerhard Goos, Juris Hartmanis, and Jan van Leeuwen

More information about this series at http://www.springer.com/series/7407

Derong Liu · Shengli Xie
Yuanqing Li · Dongbin Zhao
El-Sayed M. El-Alfy (Eds.)

Neural
Information Processing

24th International Conference, ICONIP 2017
Guangzhou, China, November 14–18, 2017
Proceedings, Part IV

 Springer

Editors
Derong Liu
Guangdong University of Technology
Guangzhou
China

Shengli Xie
Guangdong University of Technology
Guangzhou
China

Yuanqing Li
South China University of Technology
Guangzhou
China

Dongbin Zhao
Institute of Automation
Chinese Academy of Sciences
Beijing
China

El-Sayed M. El-Alfy
King Fahd University of Petroleum
    and Minerals
Dhahran
Saudi Arabia

ISSN 0302-9743   ISSN 1611-3349 (electronic)
Lecture Notes in Computer Science
ISBN 978-3-319-70092-2   ISBN 978-3-319-70093-9 (eBook)
https://doi.org/10.1007/978-3-319-70093-9

Library of Congress Control Number: 2017957558

LNCS Sublibrary: SL1 – Theoretical Computer Science and General Issues

Printed on acid-free paper

This Springer imprint is published by Springer Nature
The registered company is Springer International Publishing AG
The registered company address is: Gewerbestrasse 11, 6330 Cham, Switzerland

Preface

ICONIP 2017 – the 24th International Conference on Neural Information Processing – was held in Guangzhou, China, continuing the ICONIP conference series, which started in 1994 in Seoul, South Korea. Over the past 24 years, ICONIP has been held in Australia, China, India, Japan, Korea, Malaysia, New Zealand, Qatar, Singapore, Thailand, and Turkey. ICONIP has now become a well-established, popular and high-quality conference series on neural information processing in the region and around the world. With the growing popularity of neural networks in recent years, we have witnessed an increase in the number of submissions and in the quality of papers. Guangzhou, Romanized as Canton in the past, is the capital and largest city of southern China's Guangdong Province. It is also one of the five National Central Cities at the core of the Pearl River Delta. It is a key national transportation hub and trading port. November is the best month in the year to visit Guangzhou with comfortable weather. All participants of ICONIP 2017 had a technically rewarding experience as well as a memorable stay in this great city.

A neural network is an information processing structure inspired by biological nervous systems, such as the brain. It consists of a large number of highly interconnected processing elements, called neurons. It has the capability of learning from example. The field of neural networks has evolved rapidly in recent years. It has become a fusion of a number of research areas in engineering, computer science, mathematics, artificial intelligence, operations research, systems theory, biology, and neuroscience. Neural networks have been widely applied for control, optimization, pattern recognition, image processing, signal processing, etc.

ICONIP 2017 aimed to provide a high-level international forum for scientists, researchers, educators, industrial professionals, and students worldwide to present state-of-the-art research results, address new challenges, and discuss trends in neural information processing and applications. ICONIP 2017 invited scholars in all areas of neural network theory and applications, computational neuroscience, machine learning, and others.

The conference received 856 submissions from 3,255 authors in 56 countries and regions across all six continents. Based on rigorous reviews by the Program Committee members and reviewers, 563 high-quality papers were selected for publication in the conference proceedings. We would like to express our sincere gratitude to all the reviewers for the time and effort they generously gave to the conference. We are very grateful to the Institute of Automation of the Chinese Academy of Sciences, Guangdong University of Technology, South China University of Technology, Springer's *Lecture Notes in Computer Science* (LNCS), IEEE/CAA *Journal of Automatica Sinica* (JAS), and the Asia Pacific Neural Network Society (APNNS) for their financial support. We would also like to thank the publisher, Springer, for their cooperation in publishing the proceedings in the prestigious LNCS series and for sponsoring the best paper awards at ICONIP 2017.

September 2017

Derong Liu
Shengli Xie
Yuanqing Li
Dongbin Zhao
El-Sayed M. El-Alfy

ICONIP 2017 Organization

Asia **P**acific **N**eural **N**etwork **S**ociety

General Chair

Derong Liu — Chinese Academy of Sciences and Guangdong University of Technology, China

Advisory Committee

Sabri Arik	Istanbul University, Turkey
Tamer Basar	University of Illinois, USA
Dimitri Bertsekas	Massachusetts Institute of Technology, USA
Jonathan Chan	King Mongkut's University of Technology, Thailand
C.L. Philip Chen	The University of Macau, SAR China
Kenji Doya	Okinawa Institute of Science and Technology, Japan
Minyue Fu	The University of Newcastle, Australia
Tom Gedeon	Australian National University, Australia
Akira Hirose	The University of Tokyo, Japan
Zeng-Guang Hou	Chinese Academy of Sciences, China
Nikola Kasabov	Auckland University of Technology, New Zealand
Irwin King	Chinese University of Hong Kong, SAR China
Robert Kozma	University of Memphis, USA
Soo-Young Lee	Korea Advanced Institute of Science and Technology, South Korea
Frank L. Lewis	University of Texas at Arlington, USA
Chu Kiong Loo	University of Malaya, Malaysia
Baoliang Lu	Shanghai Jiao Tong University, China
Seiichi Ozawa	Kobe University, Japan
Marios Polycarpou	University of Cyprus, Cyprus
Danil Prokhorov	Toyota Technical Center, USA
DeLiang Wang	The Ohio State University, USA
Jun Wang	City University of Hong Kong, SAR China
Jin Xu	Peking University, China
Gary G. Yen	Oklahoma State University, USA
Paul J. Werbos	Retired from the National Science Foundation, USA

Program Chairs

Shengli Xie	Guangdong University of Technology, China
Yuanqing Li	South China University of Technology, China
Dongbin Zhao	Chinese Academy of Sciences, China
El-Sayed M. El-Alfy	King Fahd University of Petroleum and Minerals, Saudi Arabia

Program Co-chairs

Shukai Duan	Southwest University, China
Kazushi Ikeda	Nara Institute of Science and Technology, Japan
Weng Kin Lai	Tunku Abdul Rahman University College, Malaysia
Shiliang Sun	East China Normal University, China
Qinglai Wei	Chinese Academy of Sciences, China
Wei Xing Zheng	University of Western Sydney, Australia

Regional Chairs

Cesare Alippi	Politecnico di Milano, Italy
Tingwen Huang	Texas A&M University at Qatar, Qatar
Dianhui Wang	La Trobe University, Australia

Invited Session Chairs

Wei He	University of Science and Technology Beijing, China
Dianwei Qian	North China Electric Power University, China
Manuel Roveri	Politecnico di Milano, Italy
Dong Yue	Nanjing University of Posts and Telecommunications, China

Poster Session Chairs

Sung Bae Cho	Yonsei University, South Korea
Ping Guo	Beijing Normal University, China
Yifei Pu	Sichuan University, China
Bin Xu	Northwestern Polytechnical University, China
Zhigang Zeng	Huazhong University of Science and Technology, China

Tutorial and Workshop Chairs

Long Cheng	Chinese Academy of Sciences, China
Kaizhu Huang	Xi'an Jiaotong-Liverpool University, China
Amir Hussain	University of Stirling, UK

| James Kwok | Hong Kong University of Science and Technology, SAR China |
| Huajin Tang | Sichuan University, China |

Panel Discussion Chairs

Lei Guo	Beihang University, China
Hongyi Li	Bohai University, China
Hye Young Park	Kyungpook National University, South Korea
Lipo Wang	Nanyang Technological University, Singapore

Award Committee Chairs

Haibo He	University of Rhode Island, USA
Zhong-Ping Jiang	New York University, USA
Minho Lee	Kyungpook National University, South Korea
Andrew Leung	City University of Hong Kong, SAR China
Tieshan Li	Dalian Maritime University, China
Lidan Wang	Southwest University, China
Jun Zhang	South China University of Technology, China

Publicity Chairs

Jun Fu	Northeastern University, China
Min Han	Dalian University of Technology, China
Yanjun Liu	Liaoning University of Technology, China
Stefano Squartini	Università Politecnica delle Marche, Italy
Kay Chen Tan	National University of Singapore, Singapore
Kevin Wong	Murdoch University, Australia
Simon X. Yang	University of Guelph, Canada

Local Arrangements Chair

| Renquan Lu | Guangdong University of Technology, China |

Publication Chairs

| Ding Wang | Chinese Academy of Sciences, China |
| Jian Wang | China University of Petroleum, China |

Finance Chair

| Xinping Guan | Shanghai Jiao Tong University, China |

Registration Chair

Qinmin Yang   Zhejiang University, China

Conference Secretariat

Biao Luo   Chinese Academy of Sciences, China
Bo Zhao   Chinese Academy of Sciences, China

Contents

Computational Intelligence

Biomedical Engineering

Emotion and Bayesian Networks

Computational Intelligence

Multi-Robot Task Allocation Based on Cloud Ant Colony Algorithm

Xu Li[(⊠)], Zhengyan Liu, and Fuxiao Tan

School of Computer and Information Engineering, Fuyang Teachers College,
Fuyang, Anhui Province, China
16556793@qq.com

Abstract. In this paper, an improved ant colony algorithm based on cloud model is proposed to study the multi-robot task allocation problem. The improvement of the proposed algorithm mainly includes the construction of adaptive control mechanism, pheromone updating mechanism and task point selection mechanism. Some important optimization operators are designed such as evaluation of pheromone distribution, determination of suboptimal solution and selection of task point. Simulation results show that the proposed algorithm can obtain high-quality solution and fast convergence, the effect is significant.

Keywords: Multi-robot task allocation · Ant colony algorithm · Cloud model

1 Introduction

Multi-robot task allocation (multi-robot task allocation, MRTA) is a fundamental problem in multi-robot system research. With the increase of robot and task difficulty in the system, the problem of task allocation becomes more and more important [1]. In recent years, some intelligent optimization algorithms have been increasingly applied to solve the MRTA problem, such as genetic algorithm, ant colony algorithm, immune algorithm, particle swarm algorithm. In solving the MRTA problem, ant colony algorithm is considered to be a typical algorithm because of its positive feedback characteristics, parallel distributed computation and simple realization, and many improved ant colony algorithms are proposed [2–4]. However, there are still many disadvantages such as of poor diversity, easy to fall into local optimum and slow convergence. The fundamental reason is that the algorithm lacks the mechanism which can adaptively adjust the degree of randomness and make a dynamic balance between "exploration" and "development".

As an uncertain transformation model between qualitative and quantitative, the cloud model has a good randomness and stability characteristics. In the early stage, the ant colony algorithm has been improved by using the cloud model theory for traveling salesman problem. Based on the previous research and the characteristics of MRTA problem, a Cloud Ant Colony Algorithm (CACA) is proposed in this paper. By building the adaptive mechanism, task point selection mechanism and pheromone update mechanism, the quality and efficiency for solving MRTA problem are improved.

© Springer International Publishing AG 2017
D. Liu et al. (Eds.): ICONIP 2017, Part IV, LNCS 10637, pp. 3–10, 2017.
https://doi.org/10.1007/978-3-319-70093-9_1

2   Multi-Robot Task Allocation

2.1   Problem Description

Multi-robot task allocation is to assign all sub-tasks in the application system to multiple robots to perform. In the process of completing the task, robots need to pay a certain price, such as time, cost, distance and so on. The ultimate goal of multi-robot task allocation is to minimize the cost of the robots when the task is completed [5]. Assuming there are m robots represented as $R = \{R_1, R_2, \cdots, R_m\}$ and n tasks represented as $T = \{T_1, T_2, \cdots, T_n\}$ in the system. Any robot R_k has n kinds of capabilities represented as $B_{R_k} = \{b_{k_1}, b_{k_2}, \cdots, b_{k_n}\}$. The capacity requirement of task T_i is expressed as $B_{T_i} = \{b_{i_1}, b_{i_2}, \cdots b_{i_n}\}$. If robot R_k is able to complete task T_i, then $B_{R_k} \geq B_{T_i}$.

2.2   Mathematical Model

First, the following two variables are defined using (1) and (2).

$$x_{ijk} = \begin{cases} 1, & \text{if robot } k \text{ chooses from task } i \text{ to } j \\ 0 & \text{otherwise} \end{cases} \tag{1}$$

$$y_{ik} = \begin{cases} 1, & \text{if task } i \text{ is executed by robot } k \\ 0, & \text{otherwise} \end{cases} \tag{2}$$

The ultimate goal of task allocation is to minimize the cost of the robots, so the objective function can be described using (3).

$$\min f = \sum_{i=1}^{n} \sum_{k=1}^{m} \cos t_k(T_i) \times y_{ik} \tag{3}$$

where $\cos t_k(T_i)$ represents the cost of robot R_k to complete the task T_i. The cost in this paper represents the distance. Constraints are described using (4)–(7).

$$\sum_i B_{T_i} \times y_{ik} \leq B_{R_k}, k = 1, 2, \cdots, m \tag{4}$$

$$\sum_k y_{ik} = 1, i = 1, 2, \cdots, n \tag{5}$$

$$\sum_i x_{ijk} = y_{jk}, j = 0, 1, \cdots, n; k = 1, 2, \cdots, m \tag{6}$$

$$\sum_j x_{ijk} = y_{ik}, i = 0, 1, \cdots, n; k = 1, 2, \cdots, m \tag{7}$$

Equation (4) indicates that the sum of the capabilities of all the tasks performed by the robot can not exceed the capabilities of the robot. Equation (5) indicates that a task can only be executed once. Equations (6) and (7) show that a task can only be executed by a robot [6].

3 Cloud Ant Colony Algorithm (CACA)

3.1 Cloud Model

Definition: Let U be a quantitative domain with exact values, C is the qualitative concept on U, the quantitative value $x(x \in U)$ is a stochastic realization of qualitative concept C. If x satisfies: $x \sim N(Ex, En'^2)$, where $En' \sim N(En, He^2)$, and the membership degree μ satisfies $\mu = \exp\left(-(x - Ex)^2 \big/ \left(2(En')^2\right)\right)$, then the distribution of x on U is called the normal cloud [7].

The cloud model has three digital characteristics. Expectation Ex indicates the expectation of the spatial distribution in the domain, that is, the point that best represents the qualitative concept. Entropy En represents the uncertainty measure of qualitative concept, reflecting the range that can be accepted in the domain space. Hyper-entropy He is uncertainty measure of the entropy En, reflecting the cohesion of the uncertainty.

3.2 Setting of Cloud Model Parameters

In CACA, Ex is the global optimal solution found by the algorithm. According to the "$3En$" rule, let $Ex + 3En$ be the global worst solution $Worst$, $En = (Worst - Ex)/3$. He is determined by En and the pheromone state value Avg, $He = En/Avg$. Avg is obtained from evaluation algorithm of pheromone distribution (see 3.3).

3.3 Evaluation Algorithm of Pheromone Distribution

Learning from the idea of average node branch, pheromone distribution is evaluated after each iteration by using the algorithm given as follows.

> For each task point i
> > Calculate average pheromone between the point and other points;
> > Count the number of larger than the average, marked as $B(i)$;
> End for

Calculate the pheromone state value Avg using (8)

$$Avg = \sum_{i=1}^{n} B(i)/n \tag{8}$$

where n is the number of task points.

3.4 Determination of Suboptimal Solution

Define a membership threshold q, $q = 1 - 1/Avg$. According to *Ex*, *En* and *He*, calculate the membership degree μ of feasible solution found by the algorithm in each iteration, and then determine the solution of $\mu > q$ as the suboptimal solution.

From the above analysis, we can see that if the pheromone distribution is concentrated, *Avg* and q are smaller, the range of the suboptimal solution will increase, which can improve the algorithm's exploration ability and increase the diversity. Conversely, if the pheromone distribution is uniform, *Avg* and q are larger, the range of the suboptimal solution will be reduced, which can improve the algorithm's development ability and accelerate the convergence.

3.5 Selection Mechanism of Task Point

Since the robot is required to return to the starting point after performing the task, it is necessary to consider not only the distance between the current task point and the candidate task point when choosing the next task point, but also the distance between the last task point and the starting point. If not considered, it is likely to make the final total distance increases.

The optimization information μ_{ij} is introduced using (9), and the selection of the next task point is represented using (10).

$$\mu_{ij} = d_{i0} + d_{0j} - d_{ij} \tag{9}$$

$$\begin{cases} j = \arg \max\limits_{s \in allowed_k} \left\{ [\tau_{is}(t)]^{\alpha} [\eta_{is}]^{\beta} [\mu_{is}]^{\gamma} \right\}, & rdm \leq q_0 \\[3mm] P_{ij}^k(t) = \dfrac{[\tau_{ij}(t)]^{\alpha} [\eta_{ij}]^{\beta} [\mu_{ij}]^{\gamma}}{\sum\limits_{s \in allowed_k} [\tau_{is}(t)]^{\alpha} [\eta_{is}]^{\beta} [\mu_{is}]^{\gamma}}, & rdm > q_0 \end{cases} \tag{10}$$

where, $allowed_k$ is the set of task points satisfying the capability condition of the robot k; $P_{ij}^k(t)$ is the probability that the robot k chooses the task point j from the task point i at time t; $\tau_{ij}(t)$ is the pheromone value between the task point i and j; The degree of expectation η_{ij} of i to j is defined as $\eta_{ij} = 1/d_{ij}$, α, β and r represent the weighting factors; *rdm* is a random number between (0,1), q_0 is the set threshold.

3.6 Pheromone Updating Mechanism

Local pheromone updating. When the robot accesses a task point, the pheromone on the current path will be updated using (11).

$$\tau_{ij} = (1 - \xi)\tau_{ij} + \xi \Delta \tau_0 \tag{11}$$

where ξ is the volatilization coefficient of pheromone satisfying $0 < \xi < 1$, τ_0 is the initial value of pheromone.

Global pheromone updating. The global pheromone update not only updates the global optimal path, but also updates the suboptimal path.

At the end of an iteration, the pheromone on the paths determined as global optimal and suboptimal solutions will be updated using (12).

$$\tau_{ij} = (1 - \rho)\tau_{ij} + \rho\Delta\tau_{ij},$$
$$\Delta\tau_{ij} = \mu/L(T'), \ \forall(i,j) \in T' \tag{12}$$

where T' are the global optimal and suboptimal paths, $L(T')$ are the global optimal and suboptimal solutions, that is, the total length of path T'. The pheromone increment $\Delta\tau_{ij}$ is determined by the solution $L(T')$ and the corresponding membership μ.

If the quality of the solution is better, the corresponding membership will be larger, pheromone increment will also be larger, which can promote the pheromone positive feedback. At the same time, due to the uncertainty of membership value, there will be uncertainty in the pheromone increment, which can also increase the diversity of the algorithm.

3.7 Framework of CACA

The basic framework of the CACA is illustrated as follows.

Step 1: Given the task coordinate (x_i, y_i), calculate the distance d_{ij} and optimization information μ_{ij};

Step 2: Add a starting point, all the robot start from the starting point, and then return to the starting point after performing the task;

Step 3: Select the next task point;

Step 4: Local pheromone updating;

Step 5: Add the selected task point to the taboo table and update the set of task points that meet the robotic capability requirements. If the set is not empty, return Step3 to proceed with the task selection; otherwise replace the robot to select the task until all tasks are visited;

Step 6: Evaluate the pheromone distribution state, determine the global optimal and suboptimal solutions and adjust the cloud model parameters;

Step 7: Global pheromone updating;

Step 8: If the end condition is not satisfied, clear taboo table and jump to Step 2 to continue until the algorithm ends.

4 Experiments

4.1 Experimental Data

Assuming that the robots are homogeneous robots, the starting point of the robot is represented by 0, the position coordinates are (5, 5) and the capability is 5. There are 10 tasks (1–10) distributed in the $10 \times 10 \ m^2$ two-dimensional space. The position coordinates and capacity requirements of 10 tasks are shown in Table 1. The parameters used in experiments are shown in Table 2.

Table 1. Position coordinates and capacity requirements.

Task point	Position coordinates (x, y)	Capacity requirements
1	7.50, 8.86	1.2
2	3.38, 2.46	1.6
3	2.71, 6.31	0.3
4	2.76, 1.44	1.9
5	3.69, 9.31	1.2
6	7.94, 1.25	0.8
7	5.02, 1.95	0.8
8	5.23, 3.99	1.5
9	8.96, 8.43	0.9
10	5.26, 9.71	1.6

Table 2. The parameters.

Number of ants	Number of iterations	α	β	r	ρ	ξ	q_0
10	500	1	3	2	0.3	0.2	0.9

4.2 Experimental Results

In order to verify the validity of the proposed algorithm, we compare CACA with the basic ant colony algorithm (BACA) [8] by 50 independent experiments. The experimental results are shown in Table 3.

Table 3. The experimental results.

Algorithm	Optimal solution	Worst solution	Average solution	Number of optimal solution
BACA	39.8498	40.7322	40.4061	38
CACA	38.0430	38.0430	38.0430	50

Seen from Table 3, whether the optimal solution, the worst solution or the average solution, the quality of solution found by CACA is significantly higher than BACA. In addition, the optimal solution can be found by CACA in each experiment, but found by BACA only 38 times in 50 experiments. It shows that CACA has better stability and global search capability than BACA.

In order to more clearly see the task allocation and algorithm performance, we plot the optimal assignment figure and the evolutionary comparison figure for the two algorithms which are shown in Figs. 1 and 2.

Seen from Fig. 1, the optimal task allocation of BACA requires three robots, the tasks and procedures performed by the robots are as follows: {0-8-6-7-2-3-0}, {0-9-1-10-5-0},{0-4-0}. The optimal task allocation of CACA also requires three robots, the tasks and procedures performed by the robots are as follows: {0-5-10-1-9-0},{0-3-2-4-7-0},{0-8-6-0}. It can also be seen from Fig. 2 that the optimal solution and convergence speed of CACA are obviously better than BACA.

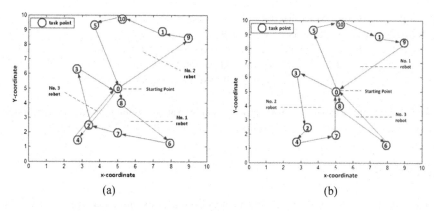

Fig. 1. The optimal assignment figure. (a) BACA; (b) CACA.

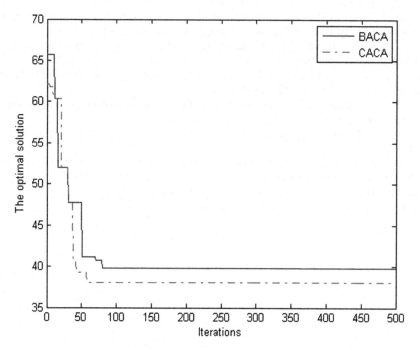

Fig. 2. The evolutionary comparison of BACA and CACA.

5 Conclusions and Further Work

In the study of the MRTA problem, an ant colony algorithm based on cloud model is proposed to improve the quality and efficiency of solving the problem. By evaluating the pheromone distribution status, the cloud parameters such as expectation, entropy, super entropy and membership threshold are adjusted adaptively, the global optimal and suboptimal paths are selected for pheromone updating. At the same time, a new selection method of task point is designed. The simulation results show that the proposed algorithm is efficient. For the further work, the adaptive control mechanism will be further studied by improving the optimization operator and parameter setting. The theoretical analysis and proof of the algorithm convergence will also be further studied [9].

Acknowledgment. This work is partially supported by National Natural Science Foundation of China (No. 61673117) and two other research grants (Nos. rcxm201713 and 2017FSKJ11).

References

1. Fang, B., Li, Y., Wang, H.: Research on emotional robot task allocation algorithm based on emotional contagion. J. Chin. Comput. Syst. **37**(8), 1731–1734 (2016)
2. Peng, Y., Yuan, Q.: A method of task allocation and automated negotiation for multi-robot. J. Changchun Inst. Tech. **17**(1), 53–56 (2016)
3. Duan, J., Zhu, Y., Huang, S.: A new multi-robot task allocation algorithm based on multi-modality synthesis. J. Northwest. Polytech. Univ. **31**(6), 975–978 (2013)
4. Cao, Z., Wu, B., Huang, Y.: The multi-robot task allocation study based on improved ant colony algorithm. Modular Mach. Tool Autom. Manuf. Tech. **2**, 35–37 (2013)
5. Liu, L., Dylan, A.: Large-scale multi-robot task allocation via dynamic partitioning and distribution. Auton. Robots **33**(3), 291–307 (2012)
6. Su, L., Yao, L., Li, X.: Modeling and solution for assignment problem of multiple robots system. J. Cent. South Univ. **44**(2), 123–125 (2013)
7. Liu, C., Li, D., Du, Y.: Some statistical analysis of the normal cloud model. Inf. Control **34**(2), 237–239 (2005)
8. Dorigo, M., Gambardella, L.: Ant colony system: a cooperative learning approach to the traveling salesman problem. IEEE Trans. Evol. Comput. **1**(1), 53–66 (1997)
9. Xu, X., Yan, S., Cheng, R.: Dynamic differential evolution algorithm for swarm robots search path planning. J. Electron. Measur. Instrum. **30**(2), 274–282 (2016)

Firefly Algorithm for Demand Estimation of Water Resources

Hui Wang[1,2(✉)], Zhihua Cui[3], Wenjun Wang[4], Xinyu Zhou[5],
Jia Zhao[1,2], Li Lv[1,2], and Hui Sun[1,2]

[1] Jiangxi Province Key Laboratory of Water Information Cooperative Sensing
and Intelligent Processing, Nanchang Institute of Technology,
Nanchang 330099, China
huiwang@whu.edu.cn
[2] School of Information Engineering, Nanchang Institute of Technology,
Nanchang 330099, China
[3] Complex System and Computational Intelligence Laboratory,
Taiyuan University of Science and Technology, Taiyuan 030024, China
[4] School of Business Administration, Nanchang Institute of Technology,
Nanchang 330099, China
[5] College of Computer and Information Engineering, Jiangxi Normal University,
Nanchang 330022, China

Abstract. Firefly algorithm (FA) is an efficient swarm intelligence optimization technique, which has been used to solve many engineering optimization problems. In this paper, we present a new FA (called NFA) variant for demand estimation of water resources in Nanchang city of China. The performance of the standard FA highly depends on its control parameters. To tackle this issue, a dynamic step factor strategy is proposed. In NFA, the step factor is not fixed and it is dynamically updated during the search process. Three models in different forms (linear, exponential and hybrid) are developed based on the structure of social and economic conditions. Water demand in Nanchang city from 2003 to 2015 is considered as a case study. The data from 2003 to 2012 is used for finding the optimal weights, and the rest data (2013–2015) is for testing the models. Simulation results show that three FA variants can achieve promising performance. Our proposed NFA outperforms the standard FA and memetic FA (MFA), and the prediction accuracy is up to 97.91%.

Keywords: Firefly algorithm · Swarm intelligence · Water demand estimation · Water demand forecasting · Optimization

1 Introduction

Water is a valuable resource for human survival and social economic development. It is an irreplaceable basic natural resource and strategic economic resource. With the speeding up of urbanization process, the demand of water resources is increasing. However, the amount of water resources is limited in nature. Therefore, the optimal allocation of water resources is important to the sustainable utilization of water resources [1].

© Springer International Publishing AG 2017
D. Liu et al. (Eds.): ICONIP 2017, Part IV, LNCS 10637, pp. 11–20, 2017.
https://doi.org/10.1007/978-3-319-70093-9_2

Demand estimation of water resources is a significant step in the allocation of water resources. The estimation of water demand aims to infer the water demand in the future according to the historical water consumption, current situation, and environment changes. As the basic means of water resources planning and management, water demand estimation is the prerequisite for the optimal allocation of water resources. Water demand is related to population, economy, social policy, ecological environment and other factors. Due to some uncertain factors, the estimation error exists. How to exactly estimate the water demand is worthy to be investigated.

Traditional estimation methods for water demand includes time series, regression analysis, gray predication, artificial neural network (ANN), quota method, and so on. For these methods, how to choose weighting parameters is a difficult task because of the random behaviors of water consumptions. In the past several years, some intelligent algorithms have been used to estimate the water demand [1–5]. In [1], a hybrid model based on soft computing techniques was used to improve the demand estimation of irrigation water. Fuzzy logic and genetic algorithm (GA) were combined with computational neural network (CNN). Experimental results show that the hybrid model outperforms the single CNN model. Do et al. [2] used GA to estimate the water demand in water distribution system. Simulation results show that GA can achieve good solutions on a 24-h period case study. In [3], harmony search (HS) was applied to short term water demand estimation, in which HS aims to search the parameter of a double seasonal ARIMA model. Romano and Kapelan [4] combined evolutionary algorithms (EAs) and ANN to construct a smart estimation model. Reported results show that the mean error is about 5%. Bai et al. [5] proposed a multi-scale method for urban water demand estimation. In the approach, an adaptive chaotic particle swarm optimization (PSO) was used to search the optimal parameters of the relevance vector regression model.

Swarm intelligence is a new computational paradigm inspired by the social behaviors from the nature [6]. In recent years, some efficient swarm intelligence algorithms have been proposed, such as artificial bee colony (ABC) [7–10], firefly algorithm (FA) [11–16], cuckoo search (CS) [17, 18] and bat algorithm (BA) [19]. FA is inspired by the mating behaviors of flashing fireflies [11]. Some recent studies show that FA can achieve promising results on many benchmark functions and real-world problems [20]. In this paper, we present an application of FA on demand estimation of water resources in Nanchang city of China. To reduce the dependency of FA on its parameters, a dynamic step factor strategy is proposed. In our approach, the step factor is not fixed and it can be dynamically updated during the search process. Three models based on linear, exponential and hybrid forms are developed based on the structure of social and economic conditions of Nanchang city. In the experiments, the performance of proposed NFA is compared with the standard FA and memetic FA (MFA) [21].

The rest of the paper is organized as follows. In Sect. 2, the standard FA is briefly described. Estimation models are developed in Sect. 3. Our approach is proposed in Sect. 4. Results and discussions are presented in Sect. 5. Finally, this work is concluded in Sect. 6.

2 Firefly Algorithm

Like PSO, FA is also a population-based random search algorithm. Each individual (firefly) in the population represents a candidate solution. The search of FA is inspired by the mating behavior of flashing fireflies. When a firefly is attracted by other brighter ones, it can move toward other new positions and find potential solutions. To construct the search model, Yang [11] proposed three assumptions: (1) one firefly is attracted to other all brighter ones; (2) the attractiveness is determined by the brightness; and (3) the brightness is affected by the given objective function. The assumptions mean that a brighter firefly has a better fitness value.

Let $X_i = (x_{i1}, x_{i1}, \ldots, x_{iD})$ be the ith firefly in the population, where $i = 1, 2, \ldots, N$, N is the population size, and D is the dimensional size. For any two different fireflies X_i and X_j, their attractiveness can be calculated as follows [11].

$$\beta(r_{ij}) = \beta_0 e^{-\gamma r_{ij}^2} \tag{1}$$

where β_0 is the attractiveness at $r = 0$, γ is the light absorption coefficient, and r_{ij} is the distance between X_i and X_j. The distance r_{ij} is defined by [11]

$$r_{ij} = \|X_i - X_j\| = \sqrt{\sum_{d=1}^{D} (x_{id} - x_{jd})^2} \tag{2}$$

when X_j is brighter (better) than X_i, X_i will move toward X_j due to the attraction. In the standard FA, this movement is defined as follows [11].

$$x_{id}(t+1) = x_{id}(t) + \beta(r_{ij}) \cdot \left(x_{jd}(t) - x_{id}(t)\right) + \alpha \left(rand - \frac{1}{2}\right) \tag{3}$$

where x_{id} and x_{jd} are the dth dimensions of X_i and X_j, respectively, $\alpha \in [0, 1]$ is called step factor, and $rand$ is a random value within $[0, 1]$.

3 Estimation Models

In this paper, we focus on an application of FA to estimate the water demand in Nanchang city of China. The water demand is related to the social and economic conditions. Table 1 shows the historical water use in Nanchang city from 2003 to 2015 [22, 23]. It can be seen that the water use of Nanchang is distributed in three departments, agriculture, industry and residents. The average proportion of agricultural water use is up to 57%. It demonstrates that agriculture is the main department of water use. Industrial and residential water use also take large proportions. The ecological water use is only 3%. So, three factors related to agricultural, industrial, and residential water use are utilized to construct the estimation model, while the ecological factor is ignored.

Table 1. Historical water use in Nanchang city from 2013 to 2015 (10^8 m^3).

Year	Total water use	Industrial water use	Agricultural water use	Residential water use	Ecological water use
2003	24.21	9.81	11.55	2.53	0.32
2004	26.22	8.72	14.47	2.75	0.28
2005	28.14	8.30	16.92	2.60	0.32
2006	27.71	8.11	16.73	2.52	0.35
2007	32.55	7.51	21.27	2.92	0.85
2008	30.42	6.90	19.73	2.94	0.85
2009	33.42	6.57	20.15	3.21	3.49
2010	30.87	7.51	17.37	3.49	2.50
2011	31.26	8.97	17.70	4.03	0.56
2012	28.82	9.20	14.68	4.36	0.58
2013	32.62	9.35	18.23	4.45	0.59
2014	31.42	8.92	17.35	4.54	0.61
2015	30.64	9.17	16.21	4.64	0.62
Average	29.87	**8.39 (28%)**	**17.10 (57%)**	**3.46 (12%)**	**0.92 (3%)**

From the above analysis, we use gross agricultural production, gross industrial production, and population to associate with agricultural, industrial, and residential water use, respectively. Table 2 lists the total water use, population, gross industrial production, and gross agricultural production in Nanchang city from 2003 to 2015 [22, 23]. By the suggestions of [24], both linear and exponential forms of models for water demand estimation are defined as follows.

Linear estimation model:

$$Y_l = x_1 \cdot W_1 + x_2 \cdot W_2 + x_3 \cdot W_3 + x_4 \tag{4}$$

Exponential estimation model:

$$Y_e = x_1 \cdot W_1^{x_2} + x_3 \cdot W_2^{x_4} + x_5 \cdot W_3^{x_6} + x_7 \tag{5}$$

where W_1, W_2, and W_3 are the population, gross industrial production, and gross agricultural production, respectively, and $x_i \in [0, 1]$ are the corresponding weights.

In this paper, we propose a hybrid model, which is a middle phase between linear and exponential models. The new model is defined by

$$Y_h = x_1 \cdot Y_l + (1 - x_1) \cdot Y_e \tag{6}$$

where Y_l and Y_e are linear and exponential models, respectively, and $x_1 \in [0, 1]$ is the weighting factor. The hybrid model can be written as follows.

$$\begin{aligned} Y_h = {} & x_1 \cdot (x_2 \cdot W_1 + x_3 \cdot W_2 + x_4 \cdot W_3 + x_5) \\ & + (1 - x_1)\left(x_6 \cdot W_1^{x_7} + x_8 \cdot W_2^{x_9} + x_{10} \cdot W_3^{x_{11}} + x_{12}\right) \end{aligned} \tag{7}$$

Table 2. The total water use, population, gross industrial production, and gross agricultural production in Nanchang city from 2003 to 2015.

Year	Total water use (10^8 m^3)	Population	Gross industrial production (10^8 yuan)	Gross agricultural production (10^8 yuan)
2003	24.21	4437476	250.95	51.29
2004	26.22	4469671	306.08	99.11
2005	28.14	4500672	374.93	115.76
2006	27.71	4530776	448.15	124.58
2007	32.55	4563025	532.75	142.84
2008	30.42	4597936	676.61	171.14
2009	33.42	4632067	753.20	187.20
2010	30.87	5042567	952.75	204.66
2011	31.26	5088996	1223.72	229.70
2012	28.82	5131564	1290.93	249.35
2013	32.62	5184231	1398.63	266.12
2014	31.42	5240179	1500.70	283.63
2015	30.64	5302914	1619.50	296.92

4 Proposed Approach

4.1 Dynamic Parameter Strategy

The performance of FA is seriously affected by its control parameters α and β. In our previous study [25], we analyzed the relations between the step factor α and convergence. If FA is convergent, α should satisfy the following condition.

$$\lim_{t \to \infty} \alpha = 0 \qquad (8)$$

where t is the index of iteration.

According to Eq. 8, a dynamic step factor strategy is designed to automatically adjust the parameter α as follows.

$$\alpha(t+1) = \alpha(t) \cdot \exp\left(-k \cdot \frac{t}{T_{max}}\right) \qquad (9)$$

where T_{max} is the maximum number of iterations. $k \in (0, 1]$ is called decreasing rate, which can adjust the decreasing speed of α. In this paper, $k = 0.2$ is used based on empirical studies. In some recent literature, the parameter α was limited in the range [0, 1]. So, the initial $\alpha(0)$ is set to 0.5, which is the midpoint of the range.

4.2 Normalization

In this paper, the historical data from 2003 to 2015 listed in Table 2 is used for training and testing the estimation models for water demand. To eliminate the influences of different units of data, the normalization method is used. In Table 2, the total water use, population, gross industrial production, and gross agricultural production are normalized as follows.

$$W^* = \frac{W - W_{min}}{W_{max} - W_{min}} \tag{10}$$

where W^* is the normalized value, W is the value to be normalized, W_{min} and W_{max} are the minimum and maximal values for the corresponding variable, respectively.

4.3   Fitness Evaluation Function

The data from 2003 to 2012 is used to optimize the weighting factors of the estimation models, and the rest data (2013–2015) is applied to test the models. To evaluate the quality of obtained weighting factors, sum of squared errors (SSE) is employed to construct the fitness evaluation function.

$$f(X) = \sum_{i=1}^{m} \left(Y_{pre} - Y_{act}\right)^2 \tag{11}$$

where Y_{act} and Y_{pre} are the actual and predicted water demand, respectively, and m is the number of training samples.

5   Simulation Experiments

5.1   Experimental Setup

In the experiments, the proposed NFA is applied to estimate the water demand in Nanchang city. The performance of NFA is compared with standard FA and memetic FA (MFA) [21]. To have a fair comparison, the same parameter settings are used. The population size N and *MaxFEs* are set to 30 and 1.0E+05, respectively. In the standard FA, α and β_0 are set to 0.5 and 1.0, respectively. For MFA, the initial α, γ, β_0, and β_{min} are set to 0.5, 1.0, 1.0, and 0.2, respectively. In NFA, $\alpha(0)$ and $\beta_0(0)$ are equal to 0.5 and 1.0, respectively. The γ is set to $1/\Gamma^2$, where Γ is the length of search range.

Data from 2003 to 2012 listed in Table 2 is used to optimize the weighting factors of the estimation models, and the rest data (2013–2015) is applied to test the models. For each model, each algorithm is run 20 times and mean results are recorded. In the experiments, we use relative error (RE) and mean relative error (MRE) to measure the performance of FA.

$$RE = \left| \frac{Y_{pre} - Y_{act}}{Y_{act}} \right| \tag{12}$$

$$MRE = \frac{1}{n} \cdot \sum_{i=1}^{n} \left| \frac{Y_{pre}(i) - Y_{act}(i)}{Y_{act}(i)} \right| \tag{13}$$

where $Y_{pre}(i)$ and $Y_{pre}(i)$ are the predicted and actual water demand on the ith test sample, respectively, and n is the number of test samples.

5.2   Results

Tables 3, 4 and 5 present the results for the linear, exponential, and hybrid estimation models, respectively. As seen, FA, MFA, and NFA can achieve promising results on three estimation models. The best MRE is 2.09% and the worst one is only 5.76%. It means that the prediction accuracy is between 94.24% and 97.91%. For each model, NFA achieves better results than FA and MFA, and FA obtains the worst performance among three algorithms. The exponential model is better than the linear one. Results on the hybrid model are in line with our idea, which aims to provide a middle phase between the linear and exponential models. For all FA variants, the mean MRE on the hybrid model is better than the linear one, but worse than the exponential one. It is surprised that the best MRE on the hybrid model is better than other two models.

Table 3. Results for the linear estimation model.

Algorithm	Best MRE	Mean MRE	Std	Worst MRE
FA	4.94%	4.97%	3.60E−04	5.05%
MFA	4.92%	4.96%	**2.05E−04**	4.98%
NFA	**4.89%**	**4.95%**	2.33E−04	**4.96%**

Table 4. Results for the exponential estimation model.

Algorithm	Best MRE	Mean MRE	Std	Worst MRE
FA	2.42%	2.53%	1.95E−03	3.00%
MFA	2.36%	2.40%	3.70E−04	2.47%
NFA	**2.27%**	**2.35%**	**3.46E−04**	**2.38%**

Table 5. Results for the hybrid estimation model.

Algorithm	Best MRE	Mean MRE	Std	Worst MRE
FA	2.25%	3.55%	1.39E−02	5.76%
MFA	2.22%	2.82%	7.04E−03	4.25%
NFA	**2.09%**	**2.79%**	**6.98E−03**	**4.06%**

Tables 6, 7 and 8 show the best relative errors for the linear, exponential, and hybrid estimation models, respectively. It can be seen that the linear model achieves good fitting for year 2013, and the exponential model is suitable for year 2014. Results on the hybrid model show that the combination of the linear and exponential models can provide more chances of finding better solutions. Due to the space limitation, some figures and forecasting results from 2018 to 2020 are not presented.

Table 6. The best relative errors (RE) for the linear model.

Year	FA RE	MFA RE	NFA RE
2013	**1.04%**	1.07%	1.14%
2014	4.75%	4.71%	**4.64%**
2015	9.02%	8.99%	**8.90%**
Average	4.94%	4.92%	**4.89%**

Table 7. The best relative errors (RE) for the exponential model.

Year	FA RE	MFA RE	NFA RE
2013	**4.23%**	4.32%	4.62%
2014	0.14%	**0.01%**	0.69%
2015	2.90%	2.75%	**1.50%**
Average	2.42%	2.36%	**2.27%**

Table 8. The best relative errors (RE) for the hybrid model.

Year	FA RE	MFA RE	NFA RE
2013	4.41%	4.31%	**3.90%**
2014	0.51%	0.13%	**0.000037%**
2015	**1.82%**	2.23%	2.36%
Average	2.25%	2.22%	**2.09%**

6 Conclusions

In this paper, we present a new FA (NFA) to estimate the water demand in Nanchang city of China. To improve the performance of the original FA, a dynamic strategy is proposed to adjust the step factor during the search process. By analyzing the historical water use of Nanchang, three estimation models (linear, exponential and hybrid) are developed. Moreover, the normalization method is employed to eliminate the effects of different units of test data.

Data from 2003 to 2012 is used to optimize the weighting factors of the estimation models, and the rest data (2013–2015) is applied to test the models. Simulation results show that FA, MFA and NFA can achieve promising performance. NFA outperforms FA and MFA and the prediction accuracy is up to 97.91%.

This paper only uses three factors (population, gross industrial production, and gross agricultural production) to construct the estimation model. However, there are some uncertain factors, such as social policy and climate change, which may affect the water demand. This will be further investigated in the future work.

Acknowledgement. This work was supported by the National Natural Science Foundation of China (No. 61663028), the Distinguished Young Talents Plan of Jiangxi Province (No. 20171BCB23075), the Natural Science Foundation of Jiangxi Province (No. 20171BAB202035), and the Open Research Fund of Jiangxi Province Key Laboratory of Water Information Cooperative Sensing and Intelligent Processing (No. 2016WICSIP015).

References

1. Pulido-Calvo, I., Gutiérrez-Estrada, J.C.: Improved irrigation water demand forecasting using a soft-computing hybrid model. Biosyst. Eng. **102**(2), 202–218 (2009)
2. Do, N., Simpson, A., Deuerlein, J., Piller, O.: Demand estimation in water distribution systems: solving underdetermined problems using genetic algorithms. Procedia Eng. **186**, 193–201 (2017)
3. Oliveira, P.J., Steffen, J.L., Cheung, P.: Parameter estimation of seasonal Arima models for water demand forecasting using the harmony search algorithm. Procedia Eng. **186**, 177–185 (2017)
4. Romano, M., Kapelan, Z.: Adaptive water demand forecasting for near real-time management of smart water distribution systems. Environ. Model Softw. **60**, 265–276 (2014)
5. Bai, Y., Wang, P., Li, C., Xie, J.J., Wang, Y.: A multi-scale relevance vector regression approach for daily urban water demand forecasting. J. Hydrol. **517**, 236–245 (2014)
6. Torres-Treviño, L.M.: Let the swarm be: an implicit elitism in swarm intelligence. Int. J. Bio-Inspired Comput. **9**(2), 65–76 (2017)
7. Sun, H., Wang, K., Zhao, J., Yu, X.: Artificial bee colony algorithm with improved special centre. Int. J. Comput. Sci. Math. **7**(6), 548–553 (2016)
8. Yu, G.: A new multi-population-based artificial bee colony for numerical optimization. Int. J. Comput. Sci. Math. **7**(6), 509–515 (2016)
9. Lv, L., Wu, L.Y., Zhao, J., Wang, H., Wu, R.X., Fan, T.H., Hu, M., Xie, Z.F.: Improved multi-strategy artificial bee colony algorithm. Int. J. Comput. Sci. Math. **7**(5), 467–475 (2016)
10. Lu, Y., Li, R.X., Li, S.M.: Artificial bee colony with bidirectional search. Int. J. Comput. Sci. Math. **7**(6), 586–593 (2016)
11. Yang, X.S.: Nature-Inspired Metaheuristic Algorithms. Luniver Press, Beckington (2008)
12. Marichelvam, M.K., Geetha, M.: A hybrid discrete firefly algorithm to solve flow shop scheduling problems to minimise total flow time. Int. J. Bio-Inspired Comput. **8**(5), 318–325 (2016)
13. Wang, H., Wang, W., Sun, H., Rahnamayan, S.: Firefly algorithm with random attraction. Int. J. Bio-Inspired Comput. **8**(1), 33–41 (2016)
14. Kaur, M., Sharma, P.K.: On solving partition driven standard cell placement problem using firefly-based metaheuristic approach. Int. J. Bio-Inspired Comput. **9**(2), 121–127 (2017)
15. Yu, G.: An improved firefly algorithm based on probabilistic attraction. Int. J. Comput. Sci. Math. **7**(6), 530–536 (2016)
16. Wang, H., Wang, W.J., Zhou, X.Y., Sun, H., Zhao, J., Yu, X., Cui, Z.: Firefly algorithm with neighborhood attraction. Inf. Sci. **382–383**, 374–387 (2017)
17. Cui, Z., Sun, B., Wang, G., Xue, Y., Chen, J.: A novel oriented cuckoo search algorithm to improve DV-Hop performance for cyber-physical systems. J. Parallel Distrib. Comput. **103**, 42–52 (2017)

18. Zhang, M., Wang, H., Cui, Z., Chen, J.: Hybrid multi-objective cuckoo search with dynamical local search. Memet. Comput. (2017, in press). doi:10.1007/s12293-017-0237-2

19. Cai, X., Gao, X.Z., Xue, Y.: Improved bat algorithm with optimal forage strategy and random disturbance strategy. Int. J. Bio-Inspired Comput. 8(4), 205–214 (2016)

20. Fister, I., Fister Jr., I., Yang, X.S., Brest, J.: A comprehensive review of firefly algorithms. Swarm Evol. Comput. 13, 34–46 (2013)

21. Fister Jr., I., Yang, X.S., Fister, I., Brest, J., Memetic firefly algorithm for combinatorial optimization, arXiv preprint arXiv:1204.5165 (2012)

22. Statistic Bureau of Jiangxi: Jiangxi Statistical Yearbook, Chinese Statistical Press, Beijing (2004–2016)

23. Statistics Bureau of Nanchang: Statistical bulletin of national economic and social development of Nanchang, Nanchang (2003–2015)

24. Assareh, E., Behrang, M.A., Assari, M.R., Ghanbarzadeh, A.: Application of PSO (particle swarm optimization) and GA (genetic algorithm) techniques on demand estimation of oil in Iran. Energy 35, 5223–5229 (2010)

25. Wang, H., Zhou, X.Y., Sun, H., Yu, X., Zhao, J., Zhang, H., Cui, L.Z.: Firefly algorithm with adaptive control parameters. Soft. Comput. (2016, in press). doi:10.1007/s00500-016-2104-3

Using Hidden Markov Model to Predict Human Actions with Swarm Intelligence

Zhicheng Lu[1]($\boxtimes$), Yuk Ying Chung[1], Henry Wing Fung Yeung[1],
Seid Miad Zandavi[1], Weiming Zhi[2], and Wei-Chang Yeh[3]

[1] School of Information Technologies, The University of Sydney,
Sydney, NSW 2006, Australia
zhlu2106@uni.sydney.edu.au
[2] Department of Engineering Science, University of Auckland,
Auckland, New Zealand 1010
[3] Department of Industrial Engineering and Engineering Management,
National Tsing Hua University, P.O. Box 24-60,
Hsinchu 300, Taiwan, Republic of China

Abstract. This paper proposed a novel algorithm which named Randomized Particle Swarm Optimization (RPSO) to optimize HMM for human activity prediction. The experiments designed in this paper are the classification of human activity using two data sets. The first testing data is from the TUM Kitchen Data Set and the other is the Human Activity Recognition using the Smartphone Data Set from UCI Machine Learning Repository. Based on the comparison of the accuracies for the conventional HMM and optimized HMM, a conclusion can be drawn that the proposed RPSO can help HMM to achieve higher accuracy for human action recognition. Our results show that RPSO-HMM can improve 15% accuracy in human activity recognition and prediction when compared to the traditional HMM.

Keywords: Hidden Markov Model · Particle Swarm Optimization · Human activity prediction

1 Introduction

Hidden Markov Model (HMM) was first introduced by Baum in the late 1960s and early 1970s. Being a statistical regression or classification model, HMM still remains popular nowadays because of the following reasons: (1) it has been proven that the mathematical structure of the HMM can be applied to real-world applications; (2) HMM produces accurate results for suitable data sets [1]. HMM and its variations have been widely applied in many fields, most notably, handwriting recognition [2] and speech recognition [3].

HMM consists of three important parameters which are $\lambda = (A, B, \pi)$, which are traditionally optimized by the Baum-Welch method [4]. However, it is generally believed that despite its fast speed, the Baum-Welch method suffers greatly from stagnation in local optima. Particle Swarm Optimization (PSO) was first

© Springer International Publishing AG 2017
D. Liu et al. (Eds.): ICONIP 2017, Part IV, LNCS 10637, pp. 21–30, 2017.
https://doi.org/10.1007/978-3-319-70093-9_3

proposed by Kennedy and Eberhart as an optimization algorithm based on swarm behavior such as fish schooling and bird flocking in 1995. PSO, without the use of gradient, can successfully optimize problems in a high dimension space. Due to its easy to implement property and low computational cost, PSO becomes a popular choice for solving a wide range of problems [5].

Few approaches have been proposed to use PSO based algorithms as an alternative to the Baum-Welch method in optimizing the HMM parameters. Ramussen and Krink proposed a hybrid algorithm of PSO and breeding particles in optimizing HMM parameters [6]. Xue et al. used the traditional PSO algorithm to replace the Baum-Welch method in training HMM parameters [7]. Aupetit et al. developed a hybrid model using both PSO and the Baum-Welch method with HMM space transformation to optimize HMM parameters [8]. Sun et al. used an improved quantum-behaved PSO for HMM training [9]. The main objective of PSO it to optimize the mentioned model parameters $\lambda = (A, B, \pi)$, and thus to improve the performance of the HMM. Experimental results show that all proposed PSO based algorithms outperforms the conventional Baum-Welch method.

Although PSO has the merit of low computational cost and is easy to implement, it is known to contain the drawbacks of pre-mature convergence and loss in particle diversities. This paper aims to further improve the quality of optimization of the existing PSO based HMM (HMM-PSO) by introducing an optimization algorithm, namely Randomized Particle Swarm Optimization (RPSO). In the experiments, the proposed HMM-RPSO will be compared to the conventional HMM, HMM-PSO and HMM-SSO by predicting the state sequence according to the observation sequence and the models. The results will capture the accuracies of the algorithm by documenting the percentage that the predicted states match the labeled ground-truth.

This paper is organized as follows. Section 2 explains the basic concepts of the HMM. Section 3 illustrates the basic concepts of the PSO-based algorithms. Section 4 introduces our proposed RPSO, and their applications on HMM as HMM-RPSO. Sections 5 and 6 presents the experiments and their corresponding experimental results. Finally, Sect. 7 gives a summary along with the future work.

2   Hidden Markov Model (HMM)

2.1   Components of HMM

There are two types of HMM differ in the data type of the output which are discrete HMM and continuous HMM [3]. Discrete HMM refers to a case that the underlying observation of the hidden states occurs in a discrete manner and thus can be characterized by a discrete probability distribution. Continuous HMM, on the other hand, refers to a spectrum of continuous underlying observations which is characterized by a continuous probability distribution such as Gaussian. There are several key elements in both types of the HMM [10]:

1. A finite number of hidden states which are not observed: $X = \{S_1, S_2, \ldots S_N\}$.

2. A set of observation that could be observed in every state: $O = \{V_1, V_2, \ldots V_M\}$. The set O is finite and discrete in the case of discrete HMM and is infinite and continuous in the case of continuous HMM.
3. A set of initial state probabilities that determine the starting state for the first instance: $\pi = \{\pi_i\}$ where $pi_i = Pr(X_1 = S_i)$
4. A set of transition probabilities that gives the likelihood of achieving the next states based on the state of this instance: $A = \{a_{ij}\}$ where $a_{ij} = Pr(X_{t+1} = S_j | X_t = S_i)$.
5. An Observation probability distribution to determine the probability that a certain state emit a certain observation: $B = \{b_j(k)\}$ where $b_j(k) = Pr(O_t = V_k | X_t = S_j)$ for discrete HMM and $b_j(k) = f(k|\mu_j, \sigma_j)$ for continuous HMM. The distribution for $f(k|\mu_j, \sigma_j)$ is generally given by a Gaussian Density Distribution:

$$f(k|\mu_j, \sigma_j) = \frac{1}{\sigma\sqrt{2\pi}}e^{-\frac{(x-\mu)^2}{2\sigma^2}} \tag{1}$$

Among all the notations, model parameters $\lambda = (A, B, \pi)$ is used to denote a HMM [4]. Moreover, there are properties for model parameters since they are all probabilities:

1. $a_{ij} \geq 0$, $b_j(k) \geq 0$, $\pi_i \geq 0$ for $\forall i, j, k$.
2. $\sum_i \pi_i = 1$.
3. $\sum_j a_{ij} = 1$ for $\forall i$.
4. $\sum_k b_j(k) = 1$ for $\forall j$.

2.2 Essential Problems for HMM

With an HMM model, there are three problems that we are interested to solve [11]:

1. Compute the probability of the observation sequence $O = o_1, o_2, \ldots o_T$ given the model parameters $\lambda = (A, B, \pi)$ and the observation sequence O.
2. Compute the optimal sequence of the hidden states $X = x_1, x_2, \ldots x_T$, given the observation sequence $O = o_1, o_2, \ldots o_T$ and model parameters of the HMM.
3. Maximize the probability $Pr(O|\lambda)$ by choosing the model parameters.

In general, problem 1 can be solved by applying forward-backward procedure. For problem 2, the optimal state sequence can be computed by the Viterbi algorithm [10]. Problem 3 is conventionally solved by applying Baum-Welch method. In Sect. 4, alternative approach using swarm intelligence and gravitational search algorithm is proposed to give more accurate results.

2.3 Optimization of HMM Model

The optimization of the model parameters $\lambda = A, B, \pi$ is considered to be the most difficult problem of HMM [10] since there is no known method to solve the optimization problem of HMM analytically. An iterative procedure known as the Baum-Welch method is developed as a conventional way of optimizing the model parameters [10]. The aim of the Baum-Welch method is to locally maximize the value of $Pr(O|\lambda)$ for the given observation sequence by choosing the parameters $\lambda = (A, B, \pi)$. There are two main steps of the Baum-Welch method which are listed below:

1. Transform $Pr(O|\lambda)$ into a new function $Q(\lambda', \lambda)$ where λ' is the new model. Function Q is able to measure the new model because $Q(\lambda', \lambda) \geq Q(\lambda', \lambda')$ implies $Pr(O|\lambda) \geq Pr(O|\lambda')$.
2. Maximize the Q function by continuously replacing λ' with λ.

3   Particle Swarm Optimization (PSO)

3.1   Particle Swarm Optimization (PSO)

PSO, as an optimization algorithm for continuous high-dimension values, derives from methodologies such as artificial life (A-life), bird flocking and swarm theory [12]. In PSO algorithm, particles are placed in the space and move based on their fitness values, particles will gather together after a number of iterations which can be seen as an optimal solution [13].

Formally, the PSO algorithm can be defined as [12,13]:

1. Initialize n particles in a D-dimensional search space with velocity 0. Set the position of each particle as its *Pbest*; compare the fitness values of all the particles, set the position of the particle with the highest fitness value as *Gbest*. *Pbest* refers to the position that, among all present and past positions of a particular particle, gives the highest fitness value. *Gbest* refers to the position that, among all present and past position of all particles, gives the highest fitness value.
2. For each dimension d of particle i at iteration t, update velocity $v_i^d(t+1)$ and position $x_i^d(t+1)$ according to following equations:

$$v_i^d(t+1) = w \times v_i^d(t) + r_1 \times (Pbest_i^d - x_i^d(t)) + r_2 \times (Gbest^d - x_i^d(t)) \quad (2)$$

$$x_i^d(t+1) = x_i^d(t) + v_i^d(t+1) \quad (3)$$

here, w is fixed to some value between 0 and 1. r_1 and r_2 are random values between 0 and 1, they will be randomized in each time when Eq. 2 is applied. The component $r_1 \times (Pbest_i^d - x_i^d(t))$ is referred to as the cognitive part, reflecting the individual thinking of a particle whereas the component $r_2 \times (Gbest^d - x_i^d(t))$ is the social part, reflecting the collective thinking as a swarm.
3. For each particle i, re-evaluate its fitness value and recompute $Pbest_i$.
4. Compare all the fitness values and current *Gbest*, to decide if *Gbest* needs to be updated.

If the iteration reaches the pre-defined number, then stop the algorithm and return the position of *Gbest* as the solution; otherwise, go back to step 2.

3.2   Simplified Swarm Optimization (SSO)

Traditional PSO does have its own drawbacks. Firstly, PSO cannot work on discrete values. Secondly, PSO particles are likely to converge pre-maturely to some local optimums and thirdly PSO often suffers from loss in particle diversities. More complex variations of PSO have been introduced to overcome the above-mentioned drawbacks of PSO [14,15]. SSO is proposed as a simplified version of PSO [14]. The variation of the SSO algorithm can mainly be captured by the equation:

$$x_i^d(t+1) = \begin{cases} x_i^d(t) & R_i^d(t+1) \in [0, C_w) \\ Pbest_i^d & R_i^d(t+1) \in [C_w, C_p) \\ Gbest^d & R_i^d(t+1) \in [C_p, C_g) \\ x & R_i^d(t+1) \in [C_g, 1] \end{cases} \tag{4}$$

The position update of the particles in PSO by Eq. 2 and Eq. 3 will be replaced by a single Eq. 4. In Eq. 4, particles are randomly separated into four categories by the probabilities given by the pre-defined parameters C_w, C_p, C_g. The four categories lead particles to (1) remains unchanged in the same position (2) update to the position of its *Pbest* (3) update its position to *Gbest* and (4) re-initialized the particle in the pre-defined search space. SSO has preserved the exploitation property of PSO by allowing the particles to fully utilize past and present positions as stated in (2) and (3). Moreover, it increases the exploration property by allowing particles to randomly re-distributed in the search space as stated in (4). This enables SSO to provide a potential higher accuracy than PSO.

4   The Proposed Method

4.1   Randomized Particle Swarm Optimization (RPSO)

Although SSO has stronger ability to search the space than PSO theoretically because of its re-initialization, PSO still holds advantages against SSO. One of the main advantages of the PSO is that PSO is capable of searching the area around the *Gbest* more precisely, because instead of jumping to *Gbest* or *Pbest*, particles can move along the intermediate intervals. As a consequence, we do want to merge the advantages of PSO and the advantages of SSO together and the proposed RPSO to do the job.

PSO algorithm can be easily stuck at local optimums. Therefore, we propose Randomized Particle Swarm Optimization (RPSO) to overcome this drawback. RPSO will first check if the position of *Gbest* stays for a pre-defined k iterations. If this is true, 50% of the particles with higher fitness values will be randomly re-initialized to a new position in the search space. By using this approach, our proposed RPSO algorithm can more efficiently in the same solution space.

4.2   HMM-RPSO

Despite its low computational cost, the Baum-Welch method has the drawback of stagnation in the local optima. This induces the introduction of PSO and its variations to optimize HMM parameters [7–9]. However, PSO is also known for its pre-mature convergence to local optimal and the loss of particle diversity. Therefore, in this paper, we propose RPSO to replace the traditional Baum-Welch method.

In order to apply the said algorithm, we need to first put it in the context of HMM. In the HMM model, suppose there are N hidden states and M possible outcomes for an observation. Then the transition matrix A will be of $N \times N$ dimensions, the emission probability matrix will be of $2 \times M \times N$ dimensions for continuous HMM since we need to estimate both the mean and the variance of each emission, the starting probability matrix will be of $1 \times N$ dimensions. Therefore, the total number of dimension D is given by:

$$D = N^2 + 2MN + N \tag{5}$$

The RPSO algorithm is then applied to this D dimensional search space, which will be shown as Algorithm 1.

Algorithm 1. HMM-RPSO

1: $k_0 \leftarrow 0$
2: **while** stopping criteria has not been met **do**
3:     **for** each particle p **do**
4:         **for** each dimension d **do**
5:             update the the position according to Eqs. 2 and 3
6:             normalize the probabilities so that the sum is 1
7:         **end for**
8:         **if** $fitness(p) > fitness(p.Pbest)$ **then**
9:             $p.Pbest \leftarrow p$
10:        **end if**
11:        **if** $fitness(p) > fitness(Gbest)$ **then**
12:            $Gbest \leftarrow p$
13:        **end if**
14:     **end for**
15:     **if** $gebst$ stays **then**
16:         $k_0 \leftarrow k_0 + 1$
17:         **if** $k_0 \geq k$ **then** re-initialized the positions of 50% of the particles
18:         **end if**
19:     **else**
20:         $k_0 \leftarrow 0$
21:     **end if**
22: **end while**

---

5   TUM Kitchen Data

TUM Kitchen Data Set is a comprehensive collection of sequences of human activities recored by multiple complementary sensors located in a kitchen environment [16]. The activities performed are basic manipulation tasks encountered in everyday activities of human life, for instance, setting a table in the kitchen. The motion of the test subjects is recorded by a markerless motion tracker, RFID tags and magnetic sensors for the opening of doors and drawers. The recorded data includes 25 Hz video data, motion capture data, RFID tag readings and magnetic sensors reading. In each video frame, there will be 84 sensors data to describe the poses of the man. In addition to the provided data in the TUM Kitchen Data Set, we have manually labeled class for left hand for use in our experiments.

5.1   Curve for Training Accuracies

Since optimization of HMM is a problem that resides in a 84 dimensional space, it is not uncommon that large number of local optimums exist. If the PSO discovers one of the local optimum in early iterations, all the particles will converge pre-maturely to that local optimum and remain there for the rest of the iterations. This shortcoming of PSO is widely discussed in literature in the field of object tracking, where swarm intelligence is also applied. In the experiment, the training accuracies of HMM-PSO are shown in the Fig. 1(a). In fact, RPSO is the algorithm that is developed to explicitly address such drawback of PSO. After the improvement, the training accuracies can be found in the Fig. 1(b). It is clear that in the training accuracies graph of HMM-PSO, there is a tendency that the algorithms seem to converge after 30 iterations. To improve that, as Fig. 1(c) shows, training accuracies continuously grow with the iteration bound and reach a higher level.

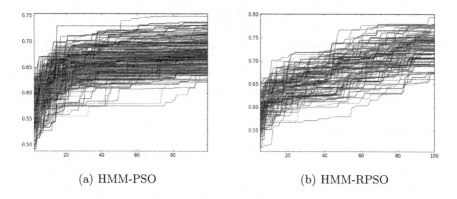

(a) HMM-PSO   (b) HMM-RPSO

Fig. 1. Curve of training accuracies

5.2   Accuracy Tests

In this experiment, there will be 2000 iterations for the conventional HMM, 100 iterations and 20 particles for other optimized HMM. Besides, particles will be re-initialized when *Gbest* stays for 10 iterations. Because the HMM records the relationship between consecutive frames, the observation sequence should not be rearranged. Since the labels are consistent between TUM Kitchen Data Sets, one of the dataset can be used to train models by different optimization methods while others can be used to test the performance of the models. As random values are involved in PSO-based algorithms, accuracies will vary significantly among each time of classification. As a result, the classification will run 100 times and the average will be used as accuracies in the Table 1. From the Table 1, We can concluded that the proposed HMM-RPSO can obtain the best performance with highest human activity prediction accuracy.

Table 1. Results for TUM Kitchen Data Set with four different HMMs

Dataset	Conventional HMM	HMM-PSO	HMM-SSO	HMM-RPSO
1-1	67.33%	71.41%	**76.98%**	76.01%
1-2	66.42%	72.10%	77.59%	**78.88%**
1-3	74.15%	78.05%	77.37%	**78.76%**
1-4	70.09%	76.46%	78.19%	**80.25%**
Average	69.50%	74.51%	77.53%	**78.47%**

6 Human Activity Recognition Using Smartphone Dataset

Human Activity Recognition using Smartphone Dataset [17] contains the data collected from 30 volunteers aging from 19 to 48 who perform 6 activities (Walking, Walking Upstairs, Walking Downstairs, Sitting, Standing, Lying) equipped with a smartphone. By using the accelerometer and gyroscope inside the smartphone, acceleration and angular velocity of human actions can be captured every 0.02 second. As the actions have been recorded, the data for each frame are then manually classified into those 6 classes. 70% of the volunteers are chosen to contribute to the training data, while the rest 30% are for the testing data.

6.1 Accuracy Tests

In this experiment, we adopt the testing strategy of 9 fold cross validation. The whole dataset is reshuffled based on individual volunteers and is then divided into 9 subsets of data. Training is conducted using 8 of the 9 subset and the remaining one is used for testing. All 9 subsets are used as testing data once in our experiment. The experiment results can be found in the Table 2. Similar to the tests in Sect. 5.2, there are 2000 iterations for Baum-Welch method, 20 particles with 100 iterations for other optimized algorithms, the particles will be re-initialized when *Gbest* stays for 10 iterations in RGSO algorithm; the testing results are the average of 50 times of running since random value has a deep influence on the PSO algorithm.

The testing results in the Table 2 illustrates that the original Baum-Welch method yield a 63.69% accuracy on average. PSO is slightly better, yielding

64.23% of average accuracy. SSO performs better than PSO, resulting in an average accuracy of 78.53% while RPSO has the best performance with human activity prediction accuracy higher than 79% which can improve 15% when compared to HMM. Our experiment results suggest that swarm intelligence based algorithms are able to improve the performance of the HMM, and the proposed RPSO is very suitable to optimize the HMM model to predict the human actions.

Table 2. Results for Human Activity Recognition using Smartphone Dataset using four different HMM methods

Testing validation	Conventional HMM	HMM-PSO	HMM-SSO	HMM-RPSO
1	63.76%	64.02%	78.10%	**80.44%**
2	63.88%	64.34%	78.02%	**79.16%**
3	63.86%	64.41%	77.91%	**81.00%**
4	63.69%	64.29%	**78.86%**	77.79%
5	63.71%	64.17%	**78.76%**	78.66%
6	62.99%	63.90%	77.96%	**78.98%**
7	63.67%	64.07%	79.14%	**80.49%**
8	64.03%	64.66%	79.61%	**80.07%**
9	63.59%	64.25%	78.43%	**79.25%**
Average	63.69%	64.23%	78.53%	**79.54%**

7   Conclusion

In this paper, we have proposed a novel HMM-RPSO for human activity prediction. From Table 2, we find that the proposed HMM-RPSO can outperform the traditional HMM by having 15% improvement in human activities prediction. Also, it can have better performance than other swarm intelligence algorithms such as PSO or SSO. Therefore, HMM-RPSO algorithm has been proved to be suitable for the prediction of human action or human activity and apply to the real time video surveillance systems.

Future work can be narrowed in following aspects. Firstly, we can further modify RPSO by parameters to obtain better performance. Secondly, we can try to find a general balance between the number of particles and the number of iterations. Thirdly, we should study the influence of the number of iterations that leads to re-initialization.

References

1. Rabiner, L.: A tutorial on hidden Markov models and selected applications in speech recognition. Proc. IEEE **77**(2), 257–286 (1989)
2. Hu, J., Brown, M., Turin, W.: HMM based online handwriting recognition. IEEE Trans. Pattern Anal. Mach. Intell. **18**(10), 1039–1045 (1996)

3. Juang, B., Rabiner, L.: Hidden Markov models for speech recognition. Technometrics **33**(3), 251–272 (1991)
4. Singh, B., Kapur, N., Kaur, P.: Speech recognition with hidden Markov model a review. Int. J. Adv. Res. Comput. Sci. Softw. Eng. **2**(3) (2012)
5. Shi, Y.: Particle swarm optimization: development, applications and resources. In: 2001 Congress on Evolutionary Computation, Seoul, South Korea (2001)
6. Ramussen, T., Krink, T.: Improved hidden Markov model training for multiple sequence alignment by a particle swarm optimization - evolutionary algorithm hybrid. Biosystems **72**(1), 5–17 (2003)
7. Xue, L., Yin, J., Ji, Z., Jiang, L.: A particle swarm optimization for hidden Markov model training. In: 2006 8th International Conference on Signal Processing, Beijing, China (2006)
8. Aupetit, S., Monmarche, N., Slimane, M.: Hidden Markov models training by a particle swarm optimization algorithm. J. Math. Model. Algorithms **6**(2), 175–193 (2007)
9. Sun, J., Wu, X., Fang, W., Ding, Y., Long, H., Xu, W.: Multiple sequence alignment using the hidden Markov model trained by an improved quantum-behaved particles swarm optimization. Inf. Sci. **182**(1), 93–114 (2012)
10. Rabiner, L., Juang, B.: An introduction to hidden Markov models. IEEE ASSP Mag. **3**(1), 4–16 (1986)
11. Kohlschein, C.: An introduction to hidden Markov models. In: Probability and Randomization in Computer Science Seminar in Winter Semester, vol. 2007 (2006)
12. Kennedy, J., Eberhart, R.: Particle swarm optimization. In: IEEE International Conference on Neural Networks, Perth, Australia, vol. IV, pp. 1942–1948 (1995)
13. Poli, R., Kennedy, J., Blackwell, T.: Particle swarm optimization. Swarm Intell. **1**(1), 33–57 (2007)
14. Bae, C., Yeh, W.C., Wahid, Y.Y., Chung, Y.Y., Liu, Y.: A new simplified swarm optimization (SSO) using exchange local search scheme. Int. J. Innov. Comput. Inf. Control **8**(6), 4391–4406 (2012)
15. Kang, K., Bae, C., Moon, J., Park, J., Chung, Y.Y., Sha, F., Zhao, X.: Invariant-feature based object tracking using discrete dynamic swarm optimization. ETRI J. **39**(2), 151–162 (2017)
16. Tenorth, M., Bandouch, J., Beetz, M.: The TUM kitchen data set for everyday manipulation activities for motion tracking and action recognition. In: IEEE 12th International Conference on Computer Vision Workshops (ICCV Workshops), Kyoto, Japan (2009)
17. Anguita, D., Ghio, A., Oneto, L., Parra, X., Reyes-Ortiz, J.L.: A public domain dataset for human activity recognition using smartphones. In: ESANN (2013)

Outintsys - A Novel Method for the Detection of the Most Intelligent Cooperative Multiagent Systems

Sabri Arik[1], Laszlo-Barna Iantovics[2($\boxtimes$)], and Sandor-Miklos Szilagyi[2]

[1] Istanbul University, Istanbul, Turkey
[2] Petru Maior University, Tirgu Mures, Romania
ibarna@science.upm.ro

Abstract. The increased intelligence of a computing system could allow more efficient and/or flexible and/or accurate solving of problems with different difficulties like: NP-hard problems, problems that have missing or erroneous data etc. We consider that even if there is no unanimous definition of the systems' intelligence, the machine intelligence could be measured. In our research, we will understand by intelligent systems the intelligent cooperative multiagent systems (*CMASs*). Even in a *CMAS* composed of simple agents an increased intelligence emerges many times at the system's level. We propose a novel method called *OutIntSys* for the detection of the systems which has a statistically extremely low and extremely high intelligence, called systems with outlier intelligence, from a set of intelligent systems that solves the same type(s) of problems. The proposed method has practical applicability in choosing of the most intelligent *CMASs* from a set of *CMASs* in solving difficult problems. To prove the effectiveness of the *OutIntSys* method we realized a study that included six intelligent *CMASs* with similar type of operation, composed of simple computing agents specialized in solving a difficult NP-hard problem. *OutIntSys* does not detect any outlier intelligence. It detected just *CMASs* whose *MIQ* is further from the rest but that cannot be considered as outliers. This was expectable based on the fact that the *CMASs* operation was very similar. We performed a comparison with two recent metrics for measuring the machine intelligence presented in the scientific literature.

Keywords: Intelligent cooperative multiagent system · Machine intelligence quotient · Measuring outlier intelligence

1 Introduction

Solving of many difficult problems is based on agent-based intelligent systems, intelligent agents and intelligent cooperative multiagent systems (*CMASs*) [1]. Even in *CMASs* composed of simple agents, an increased intelligence at the systems' level could emerge if the agents cooperate efficiently and flexibly [1].

Very few developed metrics at worldwide are able to make some kind of measuring of the machine intelligence [2–10]. A study related to the design of metrics able to measure the machine intelligence performed at the US National Institute of Standards

© Springer International Publishing AG 2017
D. Liu et al. (Eds.): ICONIP 2017, Part IV, LNCS 10637, pp. 31–40, 2017.
https://doi.org/10.1007/978-3-319-70093-9_4

and Technology was presented by Schreiner [2]. Hibbard [9] proposed a metric for intelligence measuring based on a hierarchy of sets of increasingly difficult environments. Anthon and Jannett [6] propose an intelligence measure that is based on the ability to compare alternatives with different complexity. An universal anytime intelligence test is proposed by Hernández-Orallo and Dowe [3]. Park et al. [5] studied the measuring of machine intelligence of human-machine cooperative systems using a modeling based on so-called intelligence task graph. Legg and Hutter [10] defined a measure presuming performance in easy and difficult environments.

We identified that an intelligent system has a variability in intelligence. As examples of the disadvantages of metrics presented in the literature, we mention the limitation in the: universality, treating the variability and the outlier intelligence values. In our study, we considered the *Machine Intelligence Quotient (MIQ)* of the systems' based on the intelligence in solving difficult problems. The *MIQ* of an intelligent system is obtained by measuring how intelligently a set of difficult problems are solved, and based on that to make some calculus, that will give the value of *MIQ*. We propose a method called *OutIntSys* for the detection of *CMASs* with statistically extremely low and high *MIQ* in solving problems from a set of considered *CMASs*. We call such *CMASs*, *CMASs* with outlier intelligence.

The upcoming part of the paper is organized as follows: Sect. 2 presents our proposed *OutIntSys* method. The performed case study for the validation of the *OutIntSys* method is presented in Sect. 3. Section 4 presents the conclusions.

2   The Proposed *OutIntSys* Method

We propose a novel method called *OutIntSys* for the detection of the *CMASs* with outlier intelligence from a set of *CMASs*, denoted $CM = \{Cm_1, Cm_2, ..., Cm_z\}$, $|CM|$, $|CM| = z$ (the number of studied *CMASs*). In order for the proposed method to be applied, the condition $z \geq 3$ should be satisfied. *OutIntSys* is able to detect at most $z - 3$ *CMASs* with outlier intelligence or intelligence that is not outlier but it is further from the rest, from a set of z studied *CMASs*.

We consider a set of difficult problems denoted $Prob = \{Prob_1, Prob_2, ..., Prob_m\}$ used for the evaluation of the problems-solving machine intelligence denoted MIQ_k of a *CMAS* denoted Cm_k. $|Prob|$, $|Prob| = m$ (the number of problems). $Intellig_k = \{Intellig_{k,1}, Intellig_{k,2}, ..., Intellig_{k,m}\}$ denotes the obtained *IntInd* (intelligence indicators) as the result of the *Prob* problems-solving intelligence evaluations.

A calculated *IntInd* gives a quantitative measure for the system's intelligence, that corresponds to an evaluated problem-solving. In the case of a particular *CMAS*, the researcher who wishes to make an evaluation of the *CMAS*'s machine intelligence should decide on the most appropriate type of indicator of the intelligence. If necessary, the *Int* intelligence indicator value can be calculated as the weighted sum of q type of *intelligence components*, which measure different aspects of the *CMAS* intelligence. $Int = imp_1 \times ms_1 + imp_2 \times ms_2 + ... + imp_q \times ms_q.\ imp_1 + imp_2 + ... + imp_q = 1$. $ms_1, ms_2, ..., ms_q$ represent the considered intelligence components measure at a specific problem-solving evaluation, and $imp_1, imp_2, ..., imp_q$ represent their weights/importance.

For the establishment of $|Prob|$ and $Prob$ ($MPSI$ algorithm), the *Human Evaluator* (*HU*) who would like to analyze/measure the intelligence is responsible. We determined that the $|Prob|$ value should be at least 5 ($|Prob| \geq 5$). *MPSI* should be applied z times for the calculation of the *MIQ* of the each studied *CMAS*, $MIQS = \{MIQ_1, MIQ_2, ..., MIQ_z\}$. MIQ_k denotes the machine intelligence quotient of the Cm_k that should indicate its central intelligence tendency.

MPSI: Measuring the Problem-Solving Intelligence
IN: $Cm_k = \{Agent_1, Agent_2, ..., Agent_z\}$;
OUT: $|Prob|$; $Prob = \{Prob_1, Prob_2, ..., Prob_m\}$;
$Intellig_k = \{Intellig_{k,1}, Intellig_{k,2}, ..., Intellig_{k,m}\}$;
Step 1. *HU decide on the type of System's Intelligence*
@*HU* decide on the type of intelligence that would like to detect and how to measure/calculate it.
//*HU* establish $|Prob|$ and $Prob$;
$|Prob| := m$; $Prob := \{Prob_1, Prob_2, . . . , Prob_m\}$;
Step2. *Obtaining the sample of intelligence indicators*
@Problem-solving intelligence evaluations
$Intellig_k := \{Intellig_{k,1}, Intellig_{k,2}, . . . , Intellig_{k,m}\}$;
EndMeasuringProblemSolvingIntelligence

MIQM algorithm presents the calculation of *MIQS* based on the provided intelligence data obtained for Cm_1, Cm_2, ..., Cm_z. We consider that the most appropriate indicator of the central intelligence tendency of a *CMAS* is: the *mean* in the case of normally distributed data and the *median* in the case of intelligence indicator data that is not normally distributed. The median is more robust than the mean. A very high or very low value influences the median value less than the mean.

There are z sample intelligence data sets based on the fact that there are studied z *CMASs*. In the case of all the sample intelligence data, $Intellig_1$, $Intellig_2$, ..., $Intellig_z$ are verified as a first step, to see if they pass the normality test. If there is at least one *CMAS* that does not pass the normality test, then calculation of *MIQ* of all the *CMASs* as the intelligence indicators median (*TypeMIQ* = *"Median"*) is decided automatically, otherwise the calculation of *MIQ* of all the *CMASs* as the mean (*TypeMIQ* = *"Mean"*) is decided. The human evaluator has the right to make a final change in choosing the mean or the median if he/she have domain and problem specific knowledge.

By applying the *MPSI* algorithm in the case of each analyzed *CMAS*, the problem-solving intelligence indicator is calculated for each evaluated problem. *MIQM* calculates the machine intelligence quotient of Cm_1, Cm_2, ..., Cm_z.

OutIntSys method describes the selection of the *CMASs* with outlier intelligence from the set *CM* of studied *CMASs*. $IdOut^L$ is the set of identified *CMASs* with low outlier intelligence. $IdOut^H$ is the set of identified *CMASs* with high outlier intelligence values. *IdOut*, $IdOut = IdOut^L \cup IdOut^H$ is the whole set of identified *CMASs* with outlier intelligence. Depending on the type of studied cooperative multiagent systems, it could be decided even on the consideration of *CMASs* that do not have outlier intelligence, just intelligence that is further from the rest (much different than expected by the chance).

```
MIQM Method - Measuring the Intelligence Quotient
IN: //the intelligence indicators sample of CM
Intellig={Intellig₁, Intellig₂,…, Intellig_z};
OUT: TypeMIQ; MIQS;
Step1. Calculation of the MIQ
NormPassed:="YES"; TypeMIQ :="Mean";
For (i:=1 to z) Do
  @Verify if Intellig_i is normally distributed;
  If (Intellig_i is NOT normally distributed) Then
    NormPassed :="NO"; TypeMIQ :="Median";
  EndIf
EndFor
//HE has the right to change TypeMIQ if detain some
problem and domain specific knowledge
@HU makes the final decision on the type TypeMIQ
calculation as the mean or the median;
If (NormPassed = "YES") then
  MIQ_i :=MEAN(Intellig₁, Intellig₂,…, Intellig_z);
    Else
    MIQ_i :=MEDIAN(Intellig₁, Intellig₂,…, Intellig_z);
EndIF
EndMeasuringIntelligenceQuotient
```

The *Step 1* of the *OutIntSys* method describes the characterization of the *MIQS* = {MIQ_1, MIQ_2, …, MIQ_z} data. *Standard Deviation* (*SD*) is a measure that is used to quantify the amount of variation in intelligence level of the *MIQS*. A high value of *SD* indicates intelligence level spread out over a wider range of values. The *CV* (*Coefficient of Variation*), $CV = 100 \times (SD/Mean)$ value characterize the homogeneity-heterogeneity of the *MIQS*. We consider the most appropriate homogeneity characterization as follows: $CV \in [0,10)$ - homogeneous variation of intelligence level of *CM*; $CV \in [10, 30)$ - relatively homogeneous variation of intelligence level of *CM*; $CV \geq 30$ - heterogeneous variation of intelligence level of *CM*. *Skewness* is a measure of the lack of symmetry of *MIQS* data. A distribution is symmetric if it looks the same to the left and right of the center point. *Kurtosis* is a measure of whether the *MIQS* data are heavy-tailed or light-tailed relative to a normal distribution.

In the *MIQM* and *OutIntSys* algorithms, the application of a statistical test for the verification of the normality of the intelligence indicators data is indicated. We propose for the normality testing the *One-Sample Kolmogorov-Smirnov Goodness-of-Fit test* [11, 12]. As other alternative options of tests for the verification of the normality we mention the followings [11]: *Shapiro-Wilk*, *Lilliefors* and *Anderson-Darling*. There are many statistical tests for the statistical outlier values detection, like: *Peirce's criterion* [13] and *ROUT test* [14]. In the frame of the *OutIntSys* method we chose the Grubbs test [15–18] for outliers' detection, with the significance level α_out = 0.05. This value signifies that there is a 5% chance to mistakenly identify an outlier in a sample.

```
OutIntSys: Detection of CMASs with Outlier Intelligence
IN: //sample of the machine intelligence quotients
MIQS={MIQ₁, MIQ₂,…, MIQ_z}
OUT: IdOut;//MIQ of the CMASs with outlier intelligence
IdOut^L; IdOut^H;//MIQ of the identified CMASs with low and
high outlier intelligence
Step 1. Characterization of the MIQS
IdOut:=IdOut^L:= IdOut^H :=∅; @Calculates the MIQS: Mean,
Median, SD, CV, Skewness, Kurtosis;
Step 2. Analyze the MIQS data normality
@Verify the MIQS data normality;
@Set the value of normality variable;
If (Normality = "No") Then
   @Ask the HE if he/she decide for the application of a
   transformation;
EndIf
Step 3. Detection of the outlier intelligence
MeanVal:=MeanVal{MIQ₁, MIQ₂,…, MIQ_z}; OutFound :="YES";
While (OutFound = "YES") Execute
   OutFound :="NO"; @Apply the Grubbs Test;
   If (an outlier denoted OUT was found) then
      OutFound :="YES"; IdOut :=IdOut∪{OUT};
      If (OUT< MeanVal) then IdOut^L:= IdOut^L∪{OUT};
         Else IdOut^H:= IdOut^H∪{OUT};
      EndIF
   EndIf
EndWhile
EndOutlierIntelligenceDetection
```

OutIntSys method is able to detect if a *MIQ* value is significantly/statistically different from those others. At the first application, it is able to identify a single outlier. If is identified an outlier, then it can be concluded that this is the most significantly different from those others. If is identified an outlier, then a decision whether the outliers detection test should be applied again may be considered. *Tietjen-Moore test* [19] is a generalization of the *Grubbs' test*. Tietjen-Moore test is appropriate to be applied when the suspected number of outliers is known. In the case of a single outlier, the *Tietjen-Moore test* is equivalent to the *Grubbs' test*.

3   Evaluation of the OutIntSys Method. A Case Study

A set of *CMASs* denoted *CM* specialized in solving a NP-hard problem, the *TSP* (*Travelling Salesman Problem*) [20] is considered. *TSP* asks the following question: Considering a list of cities and the distances between each pair of cities, what is the shortest possible route that visits each city exactly once and returns to the origin city? $CM = \{Cm_1, Cm_2, Cm_3, Cm_4, Cm_5, Cm_6\}$; where Cm_1 operated as an *Ant System* [21–24]; Cm_2 operated as an *Elitist Ant System* [25]; Cm_3 operated as a *Ranked Ant System* [26, 27]; Cm_4 operated as a Best-*Worst Ant System* [28, 29]; Cm_5 operated as a *Min-Max Ant System* [26, 30]; Cm_6 operated as an *Ant Colony System* [23, 24, 31].

Each studied *CMAS* is formed by simple computing agents (artificial ants) that operated by mimicking the search for food of the biological ants. Initially, each agent is placed on a randomly chosen node of the graph. An agent k currently at node i chooses to move to node j by applying the following probabilistic transition rule (1). After each agent completes its tour, the pheromone amount on each path will be adjusted according to (2), (3) and (4). α is the power of the pheromone. β controls the relative weights of the heuristic visibility of the pheromone trail. Q is an arbitrary constant. $d_{k,h}$ is the distance between the nodes (k and h); η_{kh}, $\eta_{kh} = 1/d_{k,h}$ is the heuristic visibility of the edge (k, h). $0 < \rho < 1$, is the trail evaporation factor when the agent chooses a node where it decides to move. m is the number of agents. L_k is the length of the tour performed by the agent k. *HE* established the intelligence consideration based on the best problem solution found by the *CMAS*. The best solution found represented the intelligence indicator. Based on the fact that the intelligence indicator is considered as the global best, lower value indicates higher intelligence. The used parameter values established experimentally were: maps with $nr = 100$ randomly placed cities on the map; $|Probl| = 6$; *CMASs* formed by 10 agents; *number of tests* = 1000; $\alpha = 1.2$; $\beta = 1.3$; $\rho = 0.4$.

$$p_{ij}^k(t) = \left\{ \begin{array}{ll} \dfrac{\tau_{ij}(t)^\alpha \cdot \eta_{ij}(t)^\beta}{\sum_{l \in J_{k(i)}} \tau_{il}(t)^\alpha \cdot \eta_{il}(t)^\beta} & if\ j \in J_{k(i)} \\ 0 & otherwise \end{array} \right\} \tag{1}$$

$$\tau_{ij}(t+1) = (1 - \rho) \cdot \tau_{ij}(t) + \Delta\tau_{ij}(t) \tag{2}$$

$$\Delta\tau_{ij}(t) = \sum_{k=1}^{k=m} \Delta\tau_{ij}^k(t) \tag{3}$$

$$\Delta\tau_{ij}^k(t) = \left\{ \begin{array}{ll} \frac{Q}{L_k}, & if\ (i,j) \in tour\ done\ by\ agent\ k \\ 0 & otherwise \end{array} \right. \tag{4}$$

Table 1 presents the obtained intelligence indicators evaluation results. Table 2 presents the results of the intelligence characterization realized based on the *Intellig* = {*Intellig*$_1$, *Intellig*$_2$, ..., *Intellig*$_6$}. Figure 1 graphically presents the obtained intelligence indicators. Table 3 presents the result of Kolmogorov-Smirnov normality test at $\alpha\_norm = 0.05$ significance level applied to $Intellig_1$, $Intellig_2$, ..., $Intellig_6$. This was made in order to establish the calculation of the central intelligence tendency as the mean or the median. Not all the intelligence indicator data passed the normality tests, in situations when the $P\text{-}value \leq \alpha\_norm$. The decision for opting the mean as the central intelligence was made by HU. The last line of Table 1 presents the obtained $MIQS$. Table 4 presents a characterization of the $MIQS$ data.

Table 1. *Intellig* the intelligence evaluation results.

	Cm_1	Cm_2	Cm_3	Cm_4	Cm_5	Cm_6
$Prob_1/Prob_2$	1137/1110	1112/1118	1296/1383	1845/3038	1755/1808	1289/1296
$Prob_3/Prob_4$	1102/998.9	1035/1011	1311/1304	1949/1888	1813/1663	1265/1276
$Prob_5/Prob_6$	955.1/1108	1103/1026	1271/1381	2018/2011	1563/1655	1293/1256
MIQ	1105	1069	1307.5	1980	1709	1282.5

Table 2. Characterization of the intelligence of the studied *CMASs*.

	Cm_1	Cm_2	Cm_3	Cm_4	Cm_5	Cm_6
Mean/median	1068.5/1105	1067.5/1069	1324.33/1307.5	2124.83/1980	1709.5/1709	1279.17/1282.5
SD/SEM	73.21/29.89	48.5/19.8	46.67/19.05	452.45/184.71	99.07/40.45	16.24/6.63
Kurtosis/skewness	−0.95/−0.996	−2.92/−0.064	−1.66/0.59	5.54/2.32	−1.26/−0.38	−1.68/−0.49
Min/max	955.1/1137	1011/1118	1271/1383	1845/3038	1563/1813	1256/1296
CV	6.85/Hom.	4.54/Hom.	3.52/Hom.	21.29/Rel. Hom.	5.8/Hom.	1.27/Hom.

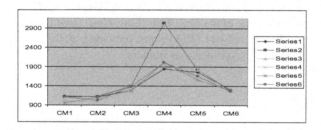

Fig. 1. Graphical representation of the $MIQS$ data.

Table 3. Kolmogorov-Smirnov normality test, at $\alpha\_norm = 0.05$ significance level.

	Cm_1	Cm_2	Cm_3	Cm_4	Cm_5	Cm_6
Statistic/	0.343/	0.268/	0.279/	0.427/	0.18/	0.2276/
P-value/	0.026/	>0.1/	>0.1/	0.001/	>0.1/	>0.1/
Normality passed	No	Yes	Yes	No	Yes	No

The characterization of the $MIQS$ data that allows the formulation of different conclusions related to the intelligence level of the studied intelligent systems, like: amount of variation, homogeneity-heterogeneity, so on.

Table 4. Characterization of the *MIQS* data.

Mean/SD/SEM/Median	1408.83/360.66/147.27/1295
CV/Homogeneity	25.599/(*CV < 30*) Rel. homogeneous *MIQS* data
K-S Normality/P-value/Norm. Passed	0.2773/>0.1/Yes

Table 5. Result of application of OutIntSys method, significance level α_int = 0.05.

CM	MIQ	Out.[*]	Suspicious[#]	No.Id[@]	Type[&]
Cm_1/Cm_2	1105/1069	/	Yes/Yes	4/3	High/High
Cm_3/Cm_4	1307.5/1980	/	/Yes	/1	/Low
Cm_5/Cm_6	1709/1282.5	/	Yes/	2/	Low/

*a *CMAS* is identified as having Outlier Intelligence; # a *CMAS* intelligence is not a significant outlier, but it is furthest from the rest of *CMASs*; @ the application number of the outliers detection test; & indicates the type of intelligence "High" or "Low" intelligence.

Table 5 presents the results obtained by applying the *OutIntSys* method on the *MIQS* data. The *Grubbs test* was applied on the *MIQS* data, with the two-sided significance level α_out = 0.05. The two-sided test was applied based on the consideration that it will be able to detect low (very low and extremely low) and high (very high and extremely high) intelligence values at the same time. At the first application of the outliers' detection test on *MIQS* data, any outlier intelligence (extremely low or high) was detected. This was expectable based on the fact that the data was relatively homogeneous (see Table 4) *calculated CV* = 25.599, (*CV < 30*). Based on this result, it can be concluded that none of the Cm_1, Cm_2, Cm_3, Cm_4, Cm_5, Cm_6 have low or high outlier intelligence. It detected just *MIQ* values, presented in Table 5, that are further from the rest but cannot be considered as outliers.

4 Conclusions

Based on a comprehensive study of the scientific literature, we have not found methods applied for the detection of *CMASs* with high and low outlier intelligence or intelligence further from the rest but that cannot be considered as an outlier. Based on this fact we proposed such a method called *OutIntSys*. The highest/smallest intelligence value does not necessarily mean statistical outlier or statistical further from the rest. *OutIntSys* has practical application in choosing the most intelligent *CMAS* in solving difficult problems. For the calculation of *MIQ* of a *CMAS* that indicate the central intelligence tendency, we proposed a calculus established based on some statistical analysis.

For the validation of our *OutIntSys* method, a case study was performed in which the intelligence of six *CMASs* was analyzed. The *OutIntSys* method does not detect any *CMAS* with outlier intelligence, just *CMASs* with low and high intelligence that is further from the rest but that cannot be considered an outlier. The performed case study proved the fact that the studied type (mimic the operation of natural ants) of *CMASs* machine intelligence, like the human's intelligence, followed the Gaussian distribution.

The main advantage of the *OutIntSys* method consists in the fact that it is universal, it does not depend on details like the studied *CMASs* architecture, the architecture of the systems composing agents. It can be applied for intelligent systems generally.

Metrics for measuring the machine intelligence presented in the literature are based on different principles of measuring the machine intelligence. The metrics for measuring machine intelligence presented in papers [4, 8] are based on a similar principle of considering the ability of machine intelligence for solving difficult problems as the *OutIntSys* method. The *MetrIntComp* metric [8] verifies if a studied intelligent system has the same intelligence with a reference intelligence. The *MetrIntPair* metric [4] compares if the intelligence of two systems is the same taking into consideration the variability. The *MetrIntComp* and *MetrIntPair* metrics are very appropriate to make accurate classification by taking into consideration the variability in intelligence, but they are not able to detect systems with outlier (low/high) intelligence.

Acknowledgment. Sandor-Miklos Szilagyi and Laszlo Barna Iantovics acknowledge the support of the COROFLOW project PN-IIIP2- 2.1-BG-2016-0343, contract 114BG/2016.

References

1. Iantovics, L.B., Zamfirescu, C.B.: ERMS: an evolutionary reorganizing multiagent system. Innov. Comput. Inf. Control **9**(3), 1171–1188 (2013)
2. Schreiner, K.: Measuring IS: toward a US standard. IEEE Intell. Syst. Appl. **15**(5), 19–21 (2000)
3. Hernández-Orallo, J., Dowe, D.L.: Measuring universal intelligence: towards an anytime intelligence test. Artif. Intell. **174**(18), 1508–1539 (2010)
4. Iantovics, L.B., Rotar, C., Niazi, M.A.: MetrIntPair - a novel accurate metric for the comparison of two cooperative multiagent systems intelligence based on paired intelligence measurements. Int. J. Intell. Syst. (2017). doi:10.1002/int.21903
5. Park, H.J., Kim, B.K., Lim, K.Y.: Measuring the machine intelligence quotient (MIQ) of human-machine cooperative systems. IEEE Trans. Syst. Man Cybern. - Part A Syst. Hum. **31**(2), 89–96 (2001)
6. Anthon, A., Jannett, T.C.: Measuring machine intelligence of an agent-based distributed sensor network system. In: Elleithy, K. (ed.) Advances and Innovations in Systems, pp. 531–535. Springer, Computing Sciences and Software Engineering (2007). doi:10.1007/978-1-4020-6264-3_92
7. Besold, T., Hernandez-Orallo, J., Schmid, U.: Can machine intelligence be measured in the same way as human intelligence? KI - Künstliche Intelligenz **29**(3), 291–297 (2015)
8. Iantovics, L.B., Emmert-Streib, F., Arik, S.: MetrIntMeas a novel metric for measuring the intelligence of a swarm of cooperating agents. Cogn. Syst. Res. **45**, 17–29 (2017)
9. Hibbard, B.: Measuring agent intelligence via hierarchies of environments. In: Schmidhuber, J., Thórisson, Kristinn R., Looks, M. (eds.) AGI 2011. LNCS, vol. 6830, pp. 303–308. Springer, Heidelberg (2011). doi:10.1007/978-3-642-22887-2_34
10. Legg, S., Hutter, M.: A formal measure of machine intelligence. In: 15th Annual Machine Learning Conference of Belgium and The Netherlands, Ghent, pp. 73–80 (2006)

11. Razali, N., Wah, Y.B.: Power comparisons of Shapiro-Wilk, Kolmogorov-Smirnov, Lilliefors and Anderson-Darling tests. J. Stat. Model. Anal. **2**(1), 21–33 (2011)
12. Lilliefors, H.: On the Kolmogorov-Smirnov test for the exponential distribution with mean unknown. J. Am. Stat. Assoc. **64**, 387–389 (1969)
13. Ross, S.M.: Peirce's criterion for the elimination of suspect experimental data. J. Eng. Technol. **2**(2), 1–12 (2003)
14. Motulsky, H.J., Brown, R.E.: Detecting outliers when fitting data with nonlinear regression: a new method based on robust nonlinear regression and the false discovery rate. BMC Bioinform. **7**, 123 (2006)
15. Grubbs, F.E.: Sample criteria for testing outlying observations. Ann. Math. Stat. **21**(1), 27–58 (1950)
16. Barnett, V., Lewis, T.: Outliers in Statistical Data, 3rd edn. Wiley, Hoboken (1994). Evolution by gene duplication
17. Grubbs, F.E.: Procedures for Detecting Outlying Observations in Samples. Technometrics **11**(1), 1–21 (1969)
18. Stefansky, W.: Rejecting outliers in factorial designs. Technometrics **14**(2), 469–479 (1972)
19. Tietjen, G., Moore, R.: Some Grubbs-Type statistics for the detection of several outliers. Technometrics **14**(3), 583–597 (1972)
20. Niendorf, M., Kabamba, P.T., Girard, A.R.: Stability of solutions to classes of traveling salesman problems. IEEE Trans. Cybern. **46**(4), 973–985 (2016)
21. Dorigo, M., Maniezzo, V., Colorni, A.: Positive Feedback as a Search Strategy. Dipartimento di Elettronica, Politecnico di Milano (1991)
22. Dorigo, M., Maniezzo, V., Colorni, A.: The ant system: optimization by a colony of cooperating agents. IEEE Trans. Syst. Man Cybern.-Part B **26**(1), 1–13 (1996)
23. Colorni, A., Dorigo, M., Maniezzo, V.: Distributed optimization by ant colonies. In: Actes de la premiere conference europeenne sur la vie artificielle, Paris, pp. 134–142. Elsevier Publishing, Paris (1991)
24. Dorigo, M.: Optimization, learning and natural algorithms. Ph.D. thesis, Politecnico di Milano, Italy (1992)
25. Jaradat, G.M., Ayob, M.: An elitist-ant system for solving the post-enrolment course timetabling problem. In: Zhang, Y., Cuzzocrea, A., Ma, J., Chung, K., Arslan, T., Song, X. (eds.) FGIT 2010. CCIS, vol. 118, pp. 167–176. Springer, Heidelberg (2010). doi:10.1007/978-3-642-17622-7_17
26. Prakasam, A., Savarimuthu, N.: Metaheuristic algorithms and probabilistic behaviour: a comprehensive analysis of ant colony optimization and its variants. Artif. Intell. Rev. **45**(1), 97–130 (2016)
27. Bullnheimer, B., Hartl, R.F., Strauss, C.: A new rank based version of the ant system. A computational study. Cent. Eur. J. Oper. Res. **7**(1), 25–38 (1999)
28. Zhang, Y., Wang, H., Zhang, Y., Chen, Y.: Best-worst ant system. In: Proceedings of the 3rd International Conference on Advanced Computer Control (ICACC), pp. 392–395 (2011)
29. Cordón, O., de Viana, I.F., Herrera, F.: Analysis of the best-worst ant system and its variants on the QAP. In: Dorigo, M., Di Caro, G., Sampels, M. (eds.) ANTS 2002. LNCS, vol. 2463, pp. 228–234. Springer, Heidelberg (2002). doi:10.1007/3-540-45724-0_20. Turning a hobby into a job: how duplicated genes find new functions
30. Stutzle, T., Hoos, H.H.: Max-min ant system. Future Gener. Comput. Syst. **16**, 889–914 (2000)
31. Dorigo, M., Stützle, T.: Ant Colony Optimization. MIT Press, Cambridge (2004)

H-PSO-LSTM: Hybrid LSTM Trained by PSO for Online Handwriter Identification

Hounaïda Moalla[✉], Walid Elloumi, and Adel M. Alimi

REGIM-Lab: Research Groups on Intelligent Machines,
National Engineering School of Sfax (ENIS),
University of Sfax, BP 1173, 3038 Sfax, Tunisia
{hounaida.moalla,walid.elloumi,adel.alimi}@ieee.org

Abstract. The automatic writer's recognition from his manuscript is a topical issue handling online writing. Recurrent neural networks (RNNs) are an effective means of solving such problem. More specifically, RNN networks with Long and Short Term Memory (LSTM) represent an ideal mean for writer's recognition. Intuitively, LSTM networks are based on the gradient method for their learning processes. In addition, an LSTM node presents a complex data processing machine.

Our hybrid approach combining LSTM and PSO (H-PSO-LSTM) presents the purpose of this paper and increases the performance of the network.

Experiments were carried out on a Biometrics Ideal Test (BIT) bilingual database (Chinese and English). The BIT deals with a large number of writers (between 130 and 188). With H-PSO-LSTM, we were able to improve the learning performance accuracy to 91.9% instead of 81.2%.

Keywords: Recurrent Neural Network · Long-Short Term Memory · Particle Swarm Optimization · Biometrics Ideal Test · Random Hybrid Strokes

1 Introduction

Until now, the evolution of technology has offered us a multitude of intelligent devices and fantastic ergonomics of use. Among these devices, we mention the tablets, touch screens, etc. The field of artificial intelligence has largely exploited these devices through advanced research axes such as series prediction [1–3], pattern recognition [4–9], approximation function [3], etc. The recognition can concern fingerprints, speech, handwriting, and so on. The handwriting problem can affect the recognition of fonts [6], words [4], phrases [10, 11], writer [12], etc. In this context, several systems have been established for the recognition of the writer [12] in the English and Chinese languages and the recognition of fonts in Arabic [5].

The writer identification which is based on handwriting data analysis is an efficient and effective strategy for biometrics [12]. Neural networks have proved their effectiveness in the recognition of handwriting in both offline [13] and online [8] modes. Handwriting can be analyzed by stroke segmentation [7] and then construed according to specific recognition characteristics [9] such as segment direction, length of the stroke, etc. Online writing [8] and writer identifications [12] deal with temporal and spatial information recorded in the writing process such as speed, angle, pressure, and so on.

© Springer International Publishing AG 2017
D. Liu et al. (Eds.): ICONIP 2017, Part IV, LNCS 10637, pp. 41–50, 2017.
https://doi.org/10.1007/978-3-319-70093-9_5

In this paper, we won't focus on the process of features extraction. Rather, we will set our attention in analyzing these features to recognize the handwriter. Recurrent Neural Networks (RNN) have made proven a better performance for offline handwriting identification [12, 13]. The online handwriter identification has also been successfully processed for known sentences but for unknown sentences, the handwriter identification is in progress. RNN are a kind of artificial neural networks that contain cyclic connections [10]. RNN appeared in the 1980s with the networks of Elman and Jordan. The Elman ones have information feedback from the hidden layer whereas the Jordan networks use feedback from the output layer [14]. Through these connections the model can retain information from the past, enabling it to discover temporal correlations between events those who are far away from each other in the data [15].

In order to optimize the training time, many algorithms have proved their performance in solving complex problem by optimization [16–19]. In our case, we propose to include the Particle Swarm Optimization (PSO) technique to develop the initial condition of LSTM and boost it to the better initial status. In this context, we propose to preprocess the learning of the LSTM by PSO. This optimization will make it possible to reconcile the values of the initial condition of the LSTM network with optimal values instead of the random ones.

The objective of our work is to compare the obtained results applied to the database BIT [12]. The rest of this paper is structured as follows. An overview of the LSTM architecture and training method is reviewed in the second section as described in the literature. Section 3 will define PSO approach. In the Sect. 4, we will present our new hybrid approach which preprocesses LSTM by PSO. The experimental results details are given in Sect. 4, and conclusion is drawn up in Sect. 5.

2 LSTM

RNN's provide a very elegant way to deal with sequential data through time that embodies correlations between sequential data [20]. During the two last decades, researchers had focused on LSTM networks use for features recognitions. This was motivated by its ability to learn and to process spatio-temporal data.

Figure 1 provides a simplified representation of a LSTM memory block with a single memory cell.

An LSTM node is recurrent and is composed by a memory block [11]. The block contains one or more self-connected memory cells and three multiplicative gates (Input, Output and Forgetting Gates). These gates leave the memory block to write, read and reset the cells [21]. An LSTM network contains an input layer and an output layer having simple nodes whereas the hidden layer contains recursive and fully connected LSTM nodes.

The LSTM node-based learning algorithm uses the gradient method for training. The Feed Forward step of the training is based on the following equations [11]:

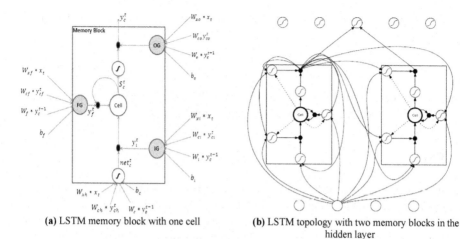

(a) LSTM memory block with one cell   (b) LSTM topology with two memory blocks in the
                                           hidden layer

Fig. 1. LSTM Structure

Block input : $\mathrm{net}_c^t = \delta\left(W_{xh} * x_t + W_{ch} * y_{ch}^{t-1} + W_c * y_c^{t-1} + b_c\right)$ (1)

Input gate : $y_i^t = \delta\left(W_{xi} * x_t + W_{ci} * y_{ci}^t + W_i * y_c^{t-1} + b_i\right)$ (2)

Forget gate : $y_f^t = \delta\left(W_{xf} * x_t + W_{cf} * y_{cf}^t + W_f * y_c^{t-1} + b_f\right)$ (3)

Output gate : $y_o^t = \delta\left(W_{xo} * x_t + W_{co} y_{co}^t + W_o * y_c^{t-1} + b_o\right)$ (4)

Cell state : $S_c^t = y_f^t * S_c^{t-1} + y_i^t * \mathrm{net}_c^t$ (5)

Output memory block : $y_c^t = \delta\left(S_c^t\right) * y_o^t$ (6)

where:

v_c^t: computed output of the memory block at time t;
y_i^t, y_f^t and y_o^t: computed values of gates at time t;
W_{xi}, W_{ci}, and W_i: different weights of input gate;
W_{xf}, W_{cf}, and W_f: different weights of forget gate;
W_{xo}, W_{co}, and W_o: different weights of output gate;
δ: a sigmoid function;
b_c, b_i, b_o and b_f: ifferent bias used in different inputs to the block memory.

LSTM's backward pass is an efficient fusion of truncated Back Propagation through Time (BPTT) and truncated Real Time Recurrent Learning (RTRL) [1, 21, 22]. The BPTT has been truncated after each time phases because, using the forget gate, the memory block will forget the past data which is no longer relevant. This truncation will allow the algorithm, at any given moment, to maintain only the relevant data.

The squared error objective function is represented by the following equation:

$$E(t) = \frac{1}{2}\sum_k e_k(t)^2 \tag{7}$$

where e_k denotes the externally injected error. E will be minimized via gradient descent by adding weight changes to the weight using learning rate.

3   PSO

The PSO algorithm is an adaptive algorithm based on the psychological behavior of birds [23–25]. In PSO learning algorithms, a particle is a matrix having a displacement velocities and positions. A new position is calculated depending on the previous position and velocity. The particles are randomly initialized and then adapted during the learning process to be allowed to approach a desired position by optimizing the value of the fitness function. As described in the literature, every particle has its best position.

S defines a cluster of n particles and can be described as:

$$S = \{x_1, x_2, \ldots, x_n\} \tag{8}$$

where x_i: the particle i among the society of particles.

The ith particle has many positions in the space:

$$P_i = \{p_{i1}, p_{i2}, \ldots, p_{in}\} \tag{9}$$

and its velocity can vary through time as:

$$V_i = \{v_{i1}, v_{i2}, \ldots, v_{in}\} \tag{10}$$

Among all P_i, there is one position that can be considered as the best.

V_i can be updated with the following equation:

$$v_{ij}(t+1) = wv_{ij}(t) + c_1.r_1.\left(p_{il}(t) - x_{ij}(t)\right) + c_2.r_2.\left(p_{ig}(t) - x_{ij}(t)\right) \tag{11}$$

and the particle P_i moves according to:

$$x_{ij}(t+1) = x_{ij}(t) + v_{ij}(t+1) \quad \text{with } 1 \leq i, \ j \leq N \tag{12}$$

where

r_1, r_2:   random variables in the range from 0 to 1;

$p_{il}(t)$:   the best local solution of the ith particle for the iteration number up to the ith iteration;

$p_{ig}(t)$:   the best global solution of all particles;

$x_{ij}(t+1)$:   the movement of the particle at time t + 1.

The best local solution and the best global solution are chosen depending on a fitness function. Several iterations will be applied until a final condition is reached.

4 Hybrid LSTM Preprocessed by PSO

Searchers have proved that training based gradient encounters the problem of long-term dependencies, i.e. c stable back-propagation of error signals over long periods of time is hardly existing [26]. The performance of the PSO algorithm can be improved by using hybrid techniques. The literature is rich by the integration of the PSO with several other learning algorithms (RNN, PSO-ESN, PSO-TSP, PSO-BP, PSO-SA, etc.) [2, 17, 27–30]. All the experiments showed that PSO facilitates the learning of the basic algorithm which brings the error to a stable situation. Nowadays, the LSTM has become largely used since it has proved a very high learning efficiency among neural networks. In this paper, we propose a hybrid method that uses the PSO and the LSTM algorithms applied on database BIT. This is done by running PSO before running the LSTM and will bring the LSTM closer to the global minimum given by PSO. We called the new model H-PSO-LSTM and it can be explained by a simple diagram as shown in Fig. 2.

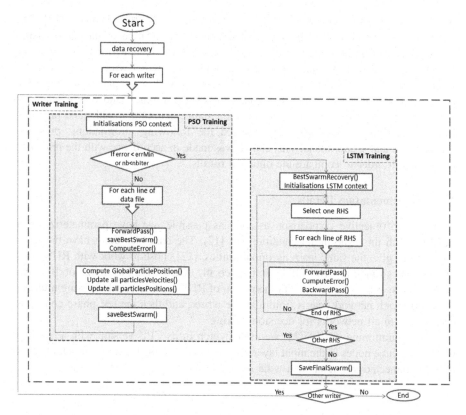

Fig. 2. H-PSO-LSTM diagram

For each writer from data base, H-PSO-LSTM will begin with PSO training before executing the LSTM's. The PSO trains all writer data until reaching a minimum error (Eq. (7)) or ending a *nbIter* iterations. At each iteration, the system computes velocity and a new position for each particle (Eqs. (11) and (12)) and saves the best swarm. Once the PSO processing is complete, the system launches the LSTM's. LSTM training begins by restoring the final best swarm given by PSO as initial condition. LSTM trains many RHSs to ensure good learning using Eqs. (1)–(7). At the end of its processing, LSTM records the final swarm. The saved one is the final learning result of one writer.

5 Experimental Results

In this section, we will present the database on which we have worked since it contains some particularities that deserves to be highlighted. Next, we will present some implementation details that merit to be clarified. Finally, a comparative study of the obtained results will be carried out to show the contribution of the PSO in the LSTM.

5.1 Data Base

We have used some sets from the Biometrics Ideal Test (BIT) database [12]. The database that has been used belongs to the writer identification of the online hand-writing. The BIT online handwriting database contains 1074 handwritten texts in online format from 188 writers applied into two sessions. Each writer wrote a set of texts in English and Chinese language. Each writer's text is saved in a separate file. Each file contains a set of information that characterizes the online handwriting such as: x-coordinate, y-coordinate, time stamp, button status, azimuth, altitude, and pressure. In our experiments, we have only adopted the following informations: pressure, x-coordinate and y-coordinate. This choice was made in accordance with the reference paper [12] in order to compare the obtained results.

5.2 Implementation Details

To make an objective comparison as soon as possible, we have parameterized our program with the same initial conditions as [12]. The data files don't have the same size, and to give the same learning opportunities, [12] chose to work with RHS. This method takes randomly 1000 RHS from each file. Each RHS has a size of 100 successive lines. We also worked with the notion of RHS according to [12] using multiple networks. Each network learns from a single writer, and then the test phase confronts each writer to all networks for decision-making.

For programming our LSTM, we used Python in its CPU version with a learning rate 0.001, three nodes in the input layer, and five nodes in the hidden layer. Our LSTM is fully connected to guarantee a wide use of all the information at any time.

5.3 Results

We conducted two learnings: the first with Chinese data and the other with English data. The tests we have realized are also done in both languages. As indicated in Table 1, we succeeded in reaching the level of performance given by [12] when learning was done with English. Once the learning is done with Chinese language, we recognized the writer from its English text 91.9% compared to 81.2% found by [12].

Table 1. Comparison of obtained results

References	RNN	Training time	English training English test	Chinese training Chinese test	Chinese training English test
[12]	BLSTM GPU	20 h	100%	99.46%	81.2%
Our system	H-PSO-LSTM CPU	36 h	100%	99.52%	91.9%

It should be mentioned that these results were obtained with H-PSO-LSTM in CPU quite simply (compared to BLSTM in GPU of [12]). In the other hand, in spite of the slowness of the training (36 h instead of 20 h), H-PSO-LSTM has proved a better performance in the English test after a Chinese training (91.9% instead of 81.2%). So, we affirm that our system is efficient.

For more details, we can show the tests results and the learnings in the following section.

As noted above, each individual network was trained for one writer. Figure shows error curves for some writers. All chosen curves have the same decreasing appearance, but some of them present sometimes instabilities. This may explain the presence or the absence of discontinuity in the handwriting and the monotony of the writer's writing.

The three curves in Fig. 3 describe the training error of three different writers. All curves show perfectly the decrease of learning patterns. From these curves, we can say that the first one corresponds to a simpler learning compared to the second and the third which presents the greatest difficulty in learning terms.

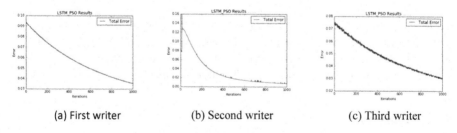

(a) First writer   (b) Second writer   (c) Third writer

Fig. 3. Examples of error curves

Table 2 shows the test results obtained from the two writers presented in Fig. 3. The test is done respectively over different five networks. Each writer is tested with these networks containing the learning results. At each execution, the test gives recognition rates for the five writers according to the used network.

Table 2. Test results of writers (a) and (c) using five networks

Writer	Test results
(a)	[58.6% 13.2% 13.8% 11% 3.4%]
(c)	[28.6% 8.8% 22.4% 36.8% 3.4%]

It is clear that the test of the writer (a), which obtained 58.6%, is much better than the 36.8% for the writer (c). This can be explained by the quality of the training. In both cases, the concerned writer dominates all others.

For more details, the obtained results for the writer (a) can be explained as follows (Fig. 4):

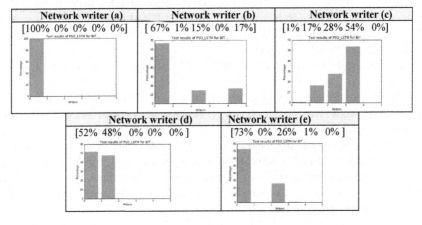

Fig. 4. Test results of writer (a) using five networks

The final result [58.6% 13.2% 13.8% 11% 3.4%] of the writer (a) is obtained by computing averages of each networks results (exp. Avrg (100, 67, 1, 52, 73) = 58.6). If all the networks gave 58.6% that means it was recognized in majority.

The same calculation method is used for all experiments.

6   Conclusion

In this paper, H-PSO-LSTM was proposed firstly for online handwriter identification data base, and can be applied on any other one. The system has achieved high accuracy. The proposed method adds performance in the learning with Chinese language and in the testing with English. We have shown that our approach has increased the accuracy to 91.9% from 81.2% in [12]. We can also mention that such results were obtained using only PSO and LSTM instead of BLSTM.

Acknowledgment. The research leading to these results has received funding from the Ministry of Higher Education and Scientific Research of Tunisia under the grant agreement number LR11ES48.

References

1. Graves, A., Schmidhuber, J.: Framewise phoneme classification with bidirectional LSTM and other neural network architectures. Neural Netw. **18**(5), 602–610 (2005)
2. Cai, X., Zhang, N., Venayagamoorthy, G.K., Wunsch, D.C.: Time series prediction with recurrent neural networks using a hybrid PSO-EA algorithm. Neurocomputing **70**(13), 2342–2353 (2007)
3. Dhahri, H., Alimi, M.A.: The modified differential evolution and the RBF (MDE-RBF) neural network for time series prediction. In: IJCNN 2006, International Joint Conference on Neural Network, pp. 2938–2943. IEEE, Vancouver (2006)
4. Boubaker, H., Kherallah, M., Alimi, M.A.: New algorithm of straight or curved baseline detection for short Arabic handwritten writing. In: ICDAR 2009, 10th International Conference on Document Analysis and Recognition, pp. 778–782. IEEE, Barcelona (2009)
5. Slimane, F., Kanoun, S., Hennebert, J., Alimi, M.A., Ingold, R.: A study on font-family and font-size recognition applied to Arabic word images at ultra-low resolution. Pattern Recogn. Lett. **34**(2), 209–218 (2013)
6. Moussa, S.B., Zahour, A., Benabdelhafid, A., Alimi, M.A.: New features using fractal multi-dimensions for generalized Arabic font recognition. Pattern Recogn. Lett. **31**(5), 361–371 (2010)
7. Bezine, H., Alimi, M.A., Derbel, N.: Handwriting trajectory movements controlled by a beta-elliptic model. In: ICDAR 2003, Proceedings of the International Conference on Document Analysis and Recognition, Scotland, pp. 1228–1232 (2003)
8. Alimi, M.A.: Evolutionary computation for the recognition of on-line cursive handwriting. IETE J. Res. **48**(5), 385–396 (2002)
9. Baccour, L., Alimi, M.A., John, R.I.: Similarity measures for intuitionistic fuzzy sets: state of the art. J. Intell. Fuzzy Syst. **24**(1), 37–49 (2013)
10. Chen, K., Yan, Z., Huo, Q.: A context-sensitive-chunk BPTT approach to training deep LSTM/BLSTM recurrent neural networks for offline handwriting recognition. In: ICDAR 2015, 13th International Conference on Document Analysis and Recognition, France, pp. 411–415. IEEE (2015)
11. Gers, F.A., Schmidhuber, E.: LSTM recurrent networks learn simple context-free and context-sensitive languages. IEEE Trans. Neural Netw. **12**(6), 1333–1340 (2001)
12. Zhang, X.Y., Xie, G.S., Liu, C.L., Bengio, Y.: End-to-end online writer identification with recurrent neural network. IEEE Trans. Hum.-Mach. Syst. **74**(2), 285–292 (2017)
13. Elbaati, A., Boubaker, H., Kherallah, M., Ennaji, A., El Abed, H., Alimi, M.A.: Arabic handwriting recognition using restored stroke chronology. In: ICDAR 2009, 10th International Conference on Document Analysis and Recognition, Beijing, China, pp. 411–415. IEEE (2009)

14. Huang, T.Y., Li, C.J., Hsu, T.W.: Structure and parameter learning algorithm of Jordan type recurrent neural networks. In: IJCNN 2007, International Joint Conference Neural Networks, pp. 1819–1824. IEEE, Barcelona (2007)
15. Pascanu, R., Mikolov, T., Bengio, Y.: On the difficulty of training recurrent neural networks. In: ICML 2013, International Conference on Machine Learning, Atlanta, pp. 1310–1318 (2013)
16. Bouaziz, S., Dhahri, H., Alimi, M.A., Abraham, A.: A hybrid learning algorithm for evolving flexible beta basis function neural tree model. Neurocomputing **117**, 107–117 (2013)
17. Zhang, J.R., Zhang, J., Lok, T.M., Lyu, M.R.: A hybrid particle swarm optimization–back-propagation algorithm for feedforward neural network training. Appl. Math. Comput. **185**(2), 1026–1037 (2007)
18. Elloumi, W., Baklouti, N., Abraham, A., Alimi, M.A.: The multi-objective hybridization of Particle Swarm Optimization and fuzzy ant colony optimization. J. Intell. Fuzzy Syst. **27**(1), 515–525 (2014)
19. Elloumi, W., Alimi, M.A.: A more efficient MOPSO for optimization. In: AICCSA 2010, ACS/IEEE International Conference on Computer System and Applications, Tunisia, pp. 1–7 (2010)
20. Greff, K., Srivastava, R.K., Koutník, J., Steunebrink, B.R., Schmidhuber, J.: LSTM: a search space odyssey. IEEE Trans. Neural Netw. Learn. Syst. (2016)
21. Graves, A.: Supervised Sequence Labelling with Recurrent Neural Networks. Springer, Heidelberg (2012)
22. Hochreiter, S., Schmidhuber, J.: Long short-term memory. Neural Comput. **9**(8), 1735–1780 (1997)
23. Poli, R., Kennedy, J., Blackwell, T.: Particle Swarm Optimization. Swarm Intell. **1**(1), 33–57 (2007)
24. Bali, O., Elloumi, W., Abraham, A., Alimi, M.A.: GPU PSO and ACO applied to TSP for vehicle security tracking. J. Inf. Assur. Secur. **11**(6), 369–384 (2016)
25. Elloumi, W., Alimi, M.A.: Combinatory optimization of ACO and PSO. In: META 2008, Second International Conference on Metaheuristics and Nature Inspired Computing, Tunisia, pp. 1–8 (2008)
26. Feng, M., Pan, H.: A modified PSO algorithm based on cache replacement algorithm. In: CIS 2014, Computational Intelligence and Security, China, pp. 558–562 (2014)
27. Elloumi, W., El Abed, H., Abraham, A., Alimi, M.A.: A comparative study of the improvement of performance using a PSO modified by ACO applied to TSP. Appl. Soft Comput. **25**, 234–241 (2014)
28. Chouikhi, N., Ammar, B., Rokbani, N., Alimi, M.A.: PSO-based analysis of Echo State Network parameters for time series forecasting. Appl. Soft Comput. **55**, 211–225 (2017)
29. Sanjeevi, S.G., Nikhila, A.N., Khan, T., Sumathi, G.: Hybrid PSO-SA algorithm for training a neural network for classification. Int. J. Comput. Sci. Eng. Appl. (IJCSEA 2011) **1**(6), 73–83 (2011)
30. Dhahri, H., Alimi, M.A.: The modified differential evolution and the RBF (MDE-RBF) neural network for time series prediction. In: IJCNN 2006, International Joint Conference on Neural Networks, pp. 2938–2943. IEEE, Vancouver (2006)

A Randomized Algorithm for Prediction Interval Using RVFL Networks Ensemble

Bara Miskony and Dianhui Wang[(✉)]

Department of Computer Science and Information Technology,
La Trobe University, Melbourne, VIC 3086, Australia
dh.wang@latrobe.edu.au

Abstract. Prediction Intervals (PIs) can specify the level of uncertainty related to point-based prediction. Most Neural Network (NN)-based approaches for constructing PIs suffer from computational expense and some restrictive assumptions on data distribution. This paper develops a randomized algorithm for PIs building with good performance in terms of both effectiveness and efficiency. To achieve this goal, a neural network ensemble with random weights is employed as a learner model, and a novel algorithm for generating teacher signals is proposed. Our proposed Randomized Algorithm for Prediction Intervals (RAPI) constructs an NN ensemble with two outputs, representing the lower and upper bounds of PIs, respectively. Experimental results with comparisons over nine benchmark datasets indicate that RAPI performs favourably in terms of coverage rate, specificity and efficiency.

Keywords: Prediction Intervals · Randomized algorithms · Neural networks with random weights · Ensemble learning

1 Introduction

Neural Networks (NNs) are powerful tools for data modelling due to their approximation capability for nonlinear maps. There have been a lot of successful applications of NNs over the last decades [1]. NN-based prediction models are good candidates for modelling the nonlinear relationship between the input and output variables, given a collection of samples. NN-based point forecasting has a limitation in that it lacks a logical interpretation of the prediction results. This could be caused by data uncertainty, which can negatively affect the prediction performance. Insufficient data, noise and outliers can be the main sources of uncertainties. Thus, data-driven point forecasting systems are weak in terms of prediction accuracy. Electricity load prediction [2], [3], load forecasting [4], travel time estimation [5] and financial applications [6] are examples which illustrate this difficulty from various aspects.

A prediction interval consists of two bounds which constrain the uncertainty in prediction results. The goal for the generation of PIs is to specify the level of uncertainty related to point prediction. Also, PIs improve the ability of planners

© Springer International Publishing AG 2017
D. Liu et al. (Eds.): ICONIP 2017, Part IV, LNCS 10637, pp. 51–60, 2017.
https://doi.org/10.1007/978-3-319-70093-9_6

and decision-makers to make decisions by effectively specifying the likelihood of uncertainty related to point prediction. Wider PIs indicate a low level of certainty in decision-making. This could help decision-makers to reverse a decision which may be risky in uncertain cases. Conversely, narrow PIs provide more confidence in decision-making. In the literature, Delta, Bayesian, Bootstrap, Mean-Variance Estimation (MVE) and Lower-Upper Bound Estimation (LUBE) techniques have been proposed for generating PIs [1].

This paper presents a randomized algorithm for constructing PIs using a neural network ensemble with random weights. Our objective is to develop a simple, direct and fast solution for problem solving under relax assumption on data distribution. Simulation results over nine datasets indicate that the proposed algorithm performs favourably in comparison to those generated by the Delta, Bayesian, Bootstrap and LUBE methods. The rest of this paper is structured as follows: Sect. 2 details our proposed algorithm for the construction of PIs. Section 3 reports system performance with result comparisons and analyses. Section 4 concludes this paper.

2   Evaluation Metrics

Coverage rate is an important index which is applied to assess the quality of the constructed PI. Prediction Interval Coverage Probability (PICP) illustrates the probability rate where the true output values (targets) are included in the two bounds of the PI. PICP is described as follows [3,4]:

$$PICP = \frac{1}{N} \sum_{n=1}^{N} \sigma_n \tag{1}$$

$$\sigma_n = \begin{cases} 1, & y_n \in [y^-(n), y^+(n)] \\ 0, & \text{Otherwise,} \end{cases} \tag{2}$$

where N refers to the number of samples in the testing dataset, and $y^-(n)$ and $y^+(n)$ represent the lower and the upper bounds for the n samples of PI, respectively. σ_n represents a Boolean value that illustrates the coverage behaviour of the constructed PIs.

Moreover, measuring the specificity of PIs is important. The Mean Prediction Interval Width (MPIW) is formulated as follows [4]:

$$MPIW = \frac{1}{N} \sum_{n=1}^{N} (y^+(n) - y^-(n)), \tag{3}$$

where $y^+(n)$ and $y^-(n)$ are the upper and lower bounds of the output y_n, respectively. MPIW demonstrates the average width of PIs. This can be normalised by the difference between the highest and lowest values of the underlying target. The mathematical term for the Normalized MPIW (NMPIW) can be defined as follows [3,4]

$$NMPIW = \frac{MPIW}{y_{max} - y_{min}}. \tag{4}$$

Constructing PIs with a narrow width and high coverage rate is significant. From a theoretical view, these two perspectives could lead to a conflict. Increasing the width of PIs usually leads to an increase in the coverage probability and vice versa. As a result, a combinational index of Coverage Width-based Criterion (CWC) is required to assess the quality of the generated PIs using the two indexes, PICP and NMPIW. CWC can be defined as [7]:

$$CWC = NMPIW(1 + \gamma(PICP)\exp{-\eta(PICP - \mu)}), \qquad (5)$$

where η and μ are the two constants and the controlling parameters which are used to measure the penalty that is assigned to PIs with a low coverage rate. μ represents the confidence level related to PIs.

3   Proposed Algorithm

Given a data set $D = \{(x_n, y_n)\}, n = 1, \ldots, N$, where $x_n \in \mathbb{R}^d$ is the input vector, $y_n \in \mathbb{R}$ denotes the output. In this paper, the randomly initialized input weights and bias of Random Vector Functional Link (RVFL) networks [8] are taken from the uniform distribution over a symmetric interval $[-\alpha, \alpha]$, ϕ is the sigmoid activation function $\phi(z) = 1/(1 + e^{-z})$ for the hidden nodes. The output weights $\beta = [\beta_1, \ldots, \beta_L]^T \in \mathbb{R}^L$ can be estimated through minimizing the following cost function:

$$e = \frac{1}{2}\sum_{n=1}^{N}(G(x_n; \beta) - y_n)^2 = \frac{1}{2}\|H\beta - y\|^2, \qquad (6)$$

where $G(x_n; \beta) = \sum_{i=1}^{L}\beta_i\phi(w_i^T x_n + b_i)$, denoted by $\phi(x_n) = \phi(w_i^T x_n + b_i))$, and

$$H = \begin{bmatrix} \phi_1(x_1) & \cdots & \phi_L(x_1) \\ \vdots & \ddots & \vdots \\ \phi_1(x_N) & \cdots & \phi_L(x_N) \end{bmatrix}_{N \times L}, \quad y = \begin{bmatrix} y_1 \\ \vdots \\ y_N \end{bmatrix}. \qquad (7)$$

If the H does not have a full column rank, with the pseudo-inverse calculation, we have

$$\beta^* = arg\min_{\beta} e = H^\dagger y, \qquad (8)$$

otherwise, if H has a full column rank, $\beta^* = (H^T H)^{-1}H^T y$.

To improve the modelling reliability of RVFL networks, we employ a population of M RVFL base models $F = \{G_m, m = 1, \cdots, M\}$ to predict the output. The RVFL network ensemble is expressed as

$$\bar{G}(x) = \sum_{m=1}^{M} a_m G_m(x), \qquad (9)$$

where $\{a_m, m = 1, \ldots, M\}$ is a set of weights, satisfying $\sum_{m=1}^{M} a_m = 1$ and $0 \leq a_m < 1$. For the sake of simplicity, average weights are frequently used, i.e., $a_m = 1/M$. The training error of the whole ensemble model can be defined as

$$E_{ens} = \frac{1}{N} \sum_{n=1}^{N} (\bar{G}(x_n) - y_n)^2. \tag{10}$$

In this paper, we utilize the well-known negative correlation learning technique to evaluate the output weights of the RVFL ensemble model, as done in [9]. Mathematically, the learning error e_m of the $m-$th base model is given with a penalty term p_m as follows:

$$e_m = \sum_{n=1}^{N} \frac{1}{2} \left[(G_m(x_n) - y_n)^2 + \lambda p_m(x_n) \right], \tag{11}$$

where

$$p_m(x_n) = (G_m(x_n) - \bar{G}(x_n)) \sum_{q \neq m}^{M} (G_q - \bar{G}(x_n)), \tag{12}$$

and $\lambda > 0$ is the regularizing factor, p_m is the penalty term for preserving the decorrelation among base models and it allows models to be trained simultaneously. Note that

$$\sum_{q \neq m}^{M} (G_q(x_n) - \bar{G}(x_n)) = -(G_m(x_n) - \bar{G}(x_n)). \tag{13}$$

Thus, (11) can be rewritten as

$$e_m = \sum_{n=1}^{N} \frac{1}{2} \left[(G_m(x_n) - y_n)^2 - \lambda (G_m(x_n) - \bar{G}(x_n))^2 \right]. \tag{14}$$

Therefore, the output weights $\{\beta_m\}$ of RVFL models in the ensemble can be obtained through minimizing all $\{e_m\}$, that is,

$$\{\beta_1^*, \ldots, \beta_M^*\} = \arg \min_{\beta_m^*} \{e_m\}, m = 1, \ldots, M. \tag{15}$$

Remark 1: Random initialization of the basis function could affect the model's capacity and reliability. The scope setting for the random weights and biases of the RVFL networks is critical to ensure the feasibility of resulting randomized learners [10]. Note that a default scope setting (i.e., $[-1,1]$ which is quite misleading indeed) used in Decorrelated Neural Network Ensembles (DNNE) [9] must be carefully chosen in practice.

The main idea behind our proposed algorithm can be summarized as follows. The whole dataset is split randomly into two parts: the training set (Tr) for training and building the model and the testing set (Te) for evaluating the model. The implementation of our algorithm can be accomplished by two phases.

In Phase 1, Tr is further randomly split into two sets: the training set (Trn) and the cross-validation set (Cr). Then, RVFL networks with two outputs are employed as base learner models, and an DNNE can be trained by using proper teacher signals (i.e., upper-bound and lower-bound derived from the original outputs) in Phase 2. The proposed RAPI algorithm for PIs construction is described in the following pseudo-code.

Algorithm 1. RAPI

Input: Training set (Tr), testing set (Te), a sigmoid activation function ϕ, the scope of random parameters $\alpha > 0$, the range of hidden nodes L and penalty coefficient λ, and the number of base models M, Set an initial value of δ as $\delta_0 = 0.9$ and select $\Delta = 0.1$;

Output: Trained PI model;

   Phase 1:
 1: Split Tr into Trn and Cr for 10 cross-validation, corresponding to $Trn^{(k)}$ and $Cr^{(k)}$ represent the trial k=1, 2,..., 10;
 2: **for** $k = 1, 2, \ldots, 10$ **do**
 3:     **while** $\delta_0 \geq 0.1$ **do**
 4:         Set the upper and lower bounds $y^+ = y_n + \delta_0|y_n|$, $y^- = y_n - \delta_0|y_n|$, respectively, $n = 1, 2, \ldots, \#\{Trn\}$;
 5:         Randomly assign w_l and b_l from $[-\alpha, \alpha]^d$ and $[-\alpha, \alpha]$, respectively;
 6:         Train a learner model by using RVFL network;
 7:         Obtain PIs according to the trained model's outputs;
 8:         Calculate PICP on $Cr^{(k)}$;
 9:         **if** PICP < Confidence level **then**
10:             $\delta_k = \delta_0 + \Delta$;
11:             Break(go to **step 2**)
12:         **else**
13:             $\delta_0 = \delta_0 - \Delta$;
14:             Go to **step 4**
15:         **end if**(corresponds to **step 9**)
16:     **end while**(corresponds to **step 3**)
17: **end for**(corresponds to **step 2**)
18: **return** $\delta_{average} = \frac{1}{10} \sum_{k=1}^{10} \delta_k$.
   Phase 2:
19: Set $\delta_n = \theta_n * (\delta_{average})$ where $\theta_n = (1 - V) * rand(1) + V$, $n = 1, 2, \ldots, \#\{Tr\}$, $0.1 \leq V \leq 0.9$;
20: Generate teacher signal $(x_n, y_n + (\delta_n|y_n|))$ and $(x_n, y_n - (\delta_n|y_n|))$;
21: Train a learner model by using DNNE with proper parameter settings (including the base number, scope, regularizing factor and architecture of RVFL networks);
22: Obtain PIs according to the trained model's outputs;
23: Calculate PICP, NMPIW and CWC on Te, respectively.

According to the results reported in [9], a range of 4–12 ensemble units is recommended (the number of base models is denoted by M). Note that both the scope setting for the random weights and biases and the number of hidden neurons (L) have an essential impact on the DNNE performance. In addition to these parameters, the penalty coefficient (λ) could also have an important impact on the performance of the DNNE technique.

The generation of the teacher signals for the upper-bound and lower-bound plays a key role to make good quality PIs in the proposed algorithm. These teacher signals are generated by modified outputs through adding or subtracting a product of a scalar δ and the absolute value of the output for each sample. A cross-validation scheme is applied in Phase 1 to estimate a proper value of the δ. To do this, the Trn set is used to generate the base models, and the Cr set is used to identify the most suitable value of δ that achieves the condition of (PICP > confidence level) for each partition. For each validation set, the algorithm starts with an initial value of δ and keeps reducing the value of δ by a step value until the value of PICP over the Cr reaches the pre-defined confidence level. In the end of going through the 10 cross-validation procedure, we obtained 10 different values of the $\delta_k, k = 1, 2, ..., 10$. In Phase 2, the teacher signals for the up-bound and low-bound of the target are generated by setting $y^+ = y_n + \delta_n|y_n|$ and $y^- = y_n - \delta_n|y_n|$, where $\delta_n = \theta_n * (\delta_{average})$ and $\theta_n = (1 - V) * rand(1) + V$, $0.1 \leq V \leq 0.9$ for each sample. It should be pointed out that the constructed PIs must guarantee the order relationship between the upper-bound and the lower-bound. Unfortunately, in our proposed algorithm, there are no constraints to force the upper bound to be greater than the lower bound for the constructed PIs. Therefore, as observed from the experiments, reverse cases could occur with very rare runs for some datasets.

4 Performance Evaluation

4.1 Datasets

Nine datasets are used to examine the performance of the proposed RAPI algorithm in this study. The effectiveness of this approach is verified and compared with the performance of the other existing methods (Delta, Bayesian, Bootstrap and LUBE) for PI generation using neural networks. Table 1 gives some statistics of the used datasets, and detailed descriptions on these datasets can be found in [7].

Table 1. Statistical description of the nine datasets [7].

Dataset	Target	Attributes	Samples
1	5-D function constant noise	5	300
2	5-D function nonconstant noise	5	500
3	Concrete compressive strength	8	1030
4	Plasma beta-carotene	12	315
5	Dry bulb temperature	3	867
6	Moisture content of raw materials	3	867
7	Steam pressure	5	200
8	Main steam temperature	5	200
9	Reheat steam	5	200

4.2   Experimental Setup and Parameter Settings

The quality of the constructed PIs using the RAPI algorithm depends on some parameters described in the previous section. For each dataset, the experimental and parameter settings are given as follows:

1. The dataset is split randomly into two sets: 70% for Tr and 30% for Te.
2. Tr set is split randomly into two sets: 70% for Trn and 30% for Cr.
3. The randomly initialized input weights and bias of the RVFL networks are taken from the uniform distribution over a symmetric interval $[-\alpha, \alpha]$. The range of input weight and bias is set for each dataset, as shown in Table 2.
4. The level of confidence is set to 90%, so μ is set to 0.9 and η is set to 50 in CWC.
5. The ten-fold cross-validation scheme is applied. Trn is used for training, Cr is used for cross-validation and to identify the most suitable value of δ for each subset. This procedure is performed for 10 times so that each subset has a suitable value of δ which meets the condition of PICP > confidence level.
6. An initial value of δ is set to 0.9 with decrement step ($\Delta = 0.1$).
7. In Phase 2, V was searched in the range [0.1–0.9] with step 0.1. Each dataset has one value from this range and is used to generate variable δ for each sample of the targets. This value is chosen by applying the whole range for each dataset and then choosing the value that generates best results.
8. Three parameters need to be set for the DNNE model (number of basis functions in RVFL networks L (hidden neurons), ensemble size M and penalty coefficient λ). The M of DNNE is changed between 4–12 for each dataset. Then, M with better results on the testing dataset is selected. L in RVFL networks and λ ranges is needed. The training and validation Root Mean Square Error (RMSE) is calculated in terms of λ and for different L. According to the equation $RMSE_{tr+cr} = (RMSE_{tr} + RMSE_{cr})/2$, the average values are calculated by applying this Eq. 100 times for each dataset. The ranges of L and λ are specified by choosing regions with minimum $RMSE_{tr+cr}$.

Table 2. Parameter settings in the experiments.

Dataset	L range	λ range	$\delta_{average}$	V	M	Input weights and biases
1	20–30	0.1–0.5	0.8	0.7	8	$[-1,1]$
2	60–70	0.1–0.5	0.2	0.1	5	$[-1,1]$
3	40–70	0.1–0.5	0.6	0.1	7	$[-50,50]$
4	10–40	0.1–0.5	0.8	0.9	8	$[-1,1]$
5	120–130	0.5–0.9	0.8	0.7	4	$[-100,100]$
6	80–100	0.5–0.9	0.7	0.6	4	$[-100,100]$
7	10–20	0.1–0.5	0.5	0.6	5	$[-1,1]$
8	5–20	0.1–0.5	0.7	0.5	5	$[-1,1]$
9	5–20	0.1–0.5	0.6	0.2	8	$[-1,1]$

After training the neural network ensemble with random weights with the generated teacher signals, the output weights values are adjusted and the model is set. The process of training and obtaining the output value is repeated 10 times. Each time, the range of L and λ is chosen randomly from the ranges that are specified in Table 2. The average values of the outputs for the upper and lower bounds are calculated. PICP, NMPIW and CWC are calculated to assess the PIs over the testing dataset.

4.3   Results and Discussion

Table 3 shows the results for the median of PICP and NMPIW for the test samples of the nine datasets. The results of the first four techniques (Delta, Bayesian, Bootstrap and LUBE) in Table 3 are obtained from [7]. The PIs obtained using the RAPI algorithm are compared with those constructed using other methods. The confidence level is set to 90%, so it is expected that PICPs will be equal or more than this confidence level.

The minimum PICP value for RAPI is 90.76%. This is slightly higher than the confidence level. As shown in Table 3, the mean of PICP for the Delta, Bayesian and Bootstrap methods for the nine datasets is less than the confidence level. The mean PICP for LUBE and RAPI is 92.05% and 93.45%, respectively. This implies that the suggested algorithm is superior to the other techniques in this regard. The median PICP values for all datasets for Delta, Bayesian, Bootstrap, LUBE and RAPI are 88.08%, 92.67%, 91.67%, 91.26% and 93.20%, respectively. Moreover, the standard deviation of PICP for RAPI is 2.4% which is lower than the Delta, Bayesian and Bootstrap methods, at 3.8%, 10.4% and 12.2%, respectively and is comparable to the value of the standard deviation of PICP for LUBE.

According to the obtained results, PIs constructed by RAPI for eight out of nine datasets are better than those constructed by the Delta technique with a coverage rate larger than the confidence level and a narrow width. Moreover, the PIs constructed by RAPI for all the datasets are better than those constructed using the Bayesian technique where the PICP value is larger than the confidence

Table 3. PI evaluation indices for the test samples of the nine datasets.

Approach name	Delta		Bayesian		Bootstrap		LUBE		RAPI	
Case study	PICP (%)	NMPIW (%)	PICP (%)	NMPIW (%)	PICP (%)	NMPIW (%)	PICP (%)	NMPIW (%)	PICP (%)	NMPIW (%)
1	85.56	79.37	100.00	120.12	85.56	85.47	90.00	72.22	93.33	71.36
2	86.67	11.29	92.67	16.54	98.00	24.53	92.00	23.09	98.00	15.58
3	89.00	25.42	88.67	22.17	91.26	27.52	91.26	35.57	93.20	34.27
4	90.39	36.12	97.31	36.20	93.85	37.51	91.15	33.91	92.63	24.71
5	88.08	54.73	100.00	87.72	58.46	32.09	94.62	69.39	90.76	68.03
6	95.79	47.04	94.74	60.60	85.26	33.13	89.47	41.80	91.53	49.50
7	91.67	43.45	76.67	13.08	96.67	50.79	93.33	45.05	93.33	41.07
8	83.33	15.85	75.00	8.52	91.67	32.62	96.67	24.71	91.67	42.14
9	85.00	25.58	76.67	15.99	96.67	50.79	90.00	31.52	96.67	41.65

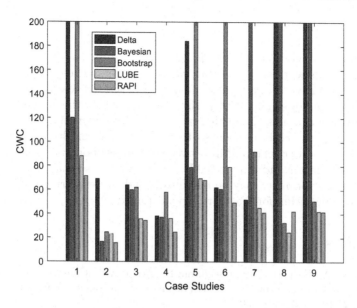

Fig. 1. The median CWC for the nine datasets generated by Delta, Bayesian, Bootstrap, LUBE and RAPI.

level with a narrow width. The behaviour of the Bayesian method fluctuates greatly for different datasets, so the standard deviation is very large compared with that of RAPI. In addition, for seven out of nine datasets, the PIs constructed using RAPI are better than those constructed using the Bootstrap technique with the PICP value being larger than the confidence level and a narrow width. The Bootstrap method obtained the highest PICP standard deviation compared with the other methods. As a consequence, this could lead to the construction of low quality PIs in some cases. The PICP values obtained by the Bayesian technique greatly fluctuate for the different datasets, for example, 58.46% for dataset 5 and 98.00% for dataset 2. As shown in Table 3, for seven out of nine datasets, the PIs constructed using RAPI are superior to those constructed using LUBE with the PICP value larger than the confidence level and a narrow width.

Figure 1 depicts the medians of CWC for the Delta, Bayesian, Bootstrap, LUBE and RAPI methods. The values of the medians of CWC for Delta, Bayesian, Bootstrap and LUBE in Fig. 1 are obtained from [7]. In some cases and because of the large results obtained for CWC, the axis for CWC is defined up to 200 for finer visualization [7], for example, CWC for the Delta and Bootstrap techniques in the first dataset. CWC for the RAPI algorithm is always less than 100, like the LUBE technique, which means that the PIs generated by these methods are reliable and sufficiently narrow. In 4 cases for Delta and Bayesian, and 3 cases for Bootstrap, the resulting CWCs for the generated PIs are larger than 100. As shown in Fig. 1, the values of the CWC obtained from our RAPI are the smallest for most of the datasets compared with the other four methods.

5   Conclusions

In this paper, a novel randomised algorithm is proposed to construct PIs using a neural network ensemble with random weights. The quality of the constructed PIs largely depends on the quality of generated teacher signals of the upper-bound and lower-bound. Experimental results over these nine benchmark datasets are quite promising. Further research in this direction include (i) adding a constraint in the cost function so that the outputs of the resulting leaner model will keep order relationship between the upper-bound and lower-bound; (ii) developing faster and more reliable learner models so that the uncertainty of a system's performance may be minimized; and (iii) exploring the implementation of the algorithm for real-world applications.

References

1. Khosravi, A., Nahavandi, S., Creighton, D., Atiya, A.F.: Comprehensive review of neural network-based prediction intervals and new advances. IEEE Trans. Neural Netw. **22**(9), 1341–1356 (2011)
2. Zhao, J.H., Dong, Z.Y., Xu, Z., Wong, K.P.: A statistical approach for interval forecasting of the electricity price. IEEE Trans. Power Syst. **23**(2), 267–276 (2008)
3. Khosravi, A., Nahavandi, S., Creighton, D.: Construction of optimal prediction intervals for load forecasting problems. IEEE Trans. Power Syst. **25**(3), 1496–1503 (2010)
4. Khosravi, A., Nahavandi, S., Creighton, D.: A prediction interval-based approach to determine optimal structures of neural network metamodels. Expert Syst. Appl. **37**(3), 2377–2387 (2010)
5. Van Hinsbergen, C.I., Van Lint, J., Van Zuylen, H.: Bayesian committee of neural networks to predict travel times with confidence intervals. Transp. Res. Part C: Emerg. Technol. **17**(5), 498–509 (2009)
6. Benoit, D.F., Van den Poel, D.: Benefits of quantile regression for the analysis of customer lifetime value in a contractual setting: an application in financial services. Expert Syst. Appl. **36**(7), 10475–10484 (2009)
7. Khosravi, A., Nahavandi, S., Creighton, D., Atiya, A.F.: Lower upper bound estimation method for construction of neural network-based prediction intervals. IEEE Trans. Neural Netw. **22**(3), 337–346 (2011)
8. Igelnik, B., Pao, Y.H.: Stochastic choice of basis functions in adaptive function approximation and the functional-link net. IEEE Trans. Neural Netw. **6**(6), 1320–1329 (1995)
9. Alhamdoosh, M., Wang, D.: Fast decorrelated neural network ensembles with random weights. Inf. Sci. **264**, 104–117 (2014)
10. Ming, L., Wang, D.: Insights into randomized algorithms for neural networks: practical issues and common pitfalls. Inf. Sci. **382**, 170–178 (2017)

Selection Mechanism in Artificial Bee Colony Algorithm: A Comparative Study on Numerical Benchmark Problems

Xinyu Zhou[1]([✉]), Hui Wang[2], Mingwen Wang[1], and Jianyi Wan[1]

[1] School of Computer and Information Engineering, Jiangxi Normal University,
Nanchang 330022, China
xyzhou@whu.edu.cn
[2] School of Information Engineering, Nanchang Institute of Technology,
Nanchang 330099, China

Abstract. Artificial bee colony (ABC) is a very effective and efficient swarm-based intelligence optimization algorithm, which has attracted a lot of attention in the community of evolutionary algorithms. Until now, many different variants of ABC have been proposed, and most of them are concentrated on improvement of the solution search equation. However, few works have been focused on the selection mechanism in the onlooker bee phase which is an important component of ABC. In this paper, hence, we present a comparative study on the selection mechanism to investigate its effect on the performance of ABC. Six different selection mechanisms are included in the comparison, and 21 well-known benchmark problems are used in the experiments. Results show that the fitness rank-based mechanisms perform better.

Keywords: Artificial bee colony · Selection mechanism · Fitness value · Selection pressure

1 Introduction

In the past few years, artificial bee colony (ABC) algorithm has attracted a lot of attention for its good optimization performance and simple algorithm structure. As a swarm-based intelligence optimization algorithm, ABC mimics the collective foraging behavior of a honeybee swarm, which was proposed by Karaboga in 2005 [1,2]. Several comparative studies have shown that the performance of ABC is competitive to that of some popular evolutionary algorithms [3–5], such as particle swarm optimization and differential evolution. The real-world applications of successfully employing ABC include image processing [6], filter design [7], artificial neural networks [8], flow shop scheduling [9], and vehicle routing problem [10].

A typical algorithm structure of ABC mainly consists of three phases: employed bee phase, onlooker bee phase, and scout bee phase. In the classic ABC, both employed bee and onlooker bee use the same solution search equation

© Springer International Publishing AG 2017
D. Liu et al. (Eds.): ICONIP 2017, Part IV, LNCS 10637, pp. 61–69, 2017.
https://doi.org/10.1007/978-3-319-70093-9_7

to search for new positions of food sources, which implies that the performance of ABC mainly depends on the solution search equation. Therefore, a large number of improved ABC variants focused on modification of the solution search equation. However, as parts of ABC algorithm structure, some components play no negligible role during the optimization process of ABC, such as the scout bee and the selection mechanism in onlooker bee phase. Unfortunately, there are few works on these two components, especially for the selection mechanism.

In this paper, we focus on the study of the selection mechanism in onlooker bee phase. In onlooker bee phase, the selection mechanism is used to determine which food sources can be selected for further exploitation produced by onlooker bees. Generally speaking, the better the fitness value a food source has, the bigger the selection probability it has. The idea behind this is that a good food source should has more chances to be searched by onlooker bees. So, in a good selection mechanism, the selection probability of a food source should be proportional to its fitness value, which is desired to keep moderate selection pressure. In the classic ABC, although the selection mechanism has been shown its effectiveness, it may run the risk of destroying selection pressure. Hence, some other selection mechanism have been proposed. In [11], Yu et al. proposed a rank-based selection scheme, Cui et al. presented a novel probability model to calculate the selection probability in [12], and Bao and Zeng proposed a tournament selection mechanism in [13].

To investigate the effect of the selection mechanism on the performance of ABC, we present a comparative study of six different selection mechanisms for performance comparison. Experiments are conducted on 21 well-known benchmark problems, and the experimental results are analyzed to guide how to design a suitable selection mechanism for further improvement of ABC. The rest of this paper is organized as follows. Section 2 describes a brief introduction to the classic ABC. Section 3 gives the detailed description on the selection mechanism. Section 4 presents the experimental results and discussions. Finally, the work is concluded in Sect. 5.

2   Classic ABC Algorithm

In the classic ABC, the colony of artificial bees consists of three different kinds of bees: employed bee, onlooker bee and scout bee. Accordingly, the search process of the classic ABC is divided into three phases: employed bee phase, onlooker bee phase, and scout bee phase. At first, similar to other evolutionary algorithms, ABC also starts with an initial population of SN randomly generated food sources. Each food source $X_i = (x_{i,1}, x_{i,2}, \cdots, x_{i,D})$ corresponds to a candidate solution to the optimization problem, and D denotes the problem dimension size. After initialization, the search process enters into the iteration loop of the three phases which are described as follows.

(1) Employed bee phase
In this phase, employed bee is responsible for exploring food sources near the hive, it generates a new food source $V_i = (v_{i,1}, v_{i,2}, \cdots, v_{i,D})$ in the neighborhood

of its parent position $X_i = (x_{i,1}, x_{i,2}, \cdots, x_{i,D})$ by using the following solution search equation.

$$v_{i,j} = x_{i,j} + \phi_{i,j} \cdot (x_{i,j} - x_{k,j}) \tag{1}$$

where $k \in \{1, 2, \cdots, SN\}$ and $j \in \{1, 2, \cdots, D\}$ are randomly chosen indexes. Moreover, k has to be different from i, and $\phi_{i,j}$ is a random number in the range $[-1, 1]$. If the new food source V_i is better than its parent X_i, then X_i is replaced with V_i. After all the employ bees have completed their searches, they would share the nectar information and position information of the food sources with the onlooker bees on the dance area. Note that an employed bee corresponds to a food source, thus the number of employed bees is the same as that of food sources.

(2) Onlooker bee phase

After the employed bee phase is finished, the onlooker bee phase starts. In this phase, an onlooker bee receives the information about the food sources from the employed bees, and then it would select a food source for further exploitation. If a food source has good fitness value, its selection probability is high. For a minimization optimization problem, the fitness value of a food source can be calculated by using the following formula in the classic ABC.

$$fit_i = \begin{cases} \frac{1}{1+f_i} & f_i \geq 0 \\ 1 + |f_i| & f_i < 0 \end{cases} \tag{2}$$

where fit_i denotes the fitness value, f_i is the objective function value. Then the selection probability p_i of food source X_i can be calculated by the following formula.

$$p_i = \frac{fit_i}{\sum_{j=1}^{SN} fit_j} \tag{3}$$

As seen, the probability p_i is proportional to the fitness value. The better a food source is, the higher chance to be selected. Based on the selection probability, the roulette selection is used to determine which food source can be selected. Once the onlooker bee completes the selection, it produces a modification on the chosen food source by using Eq. (1). As in the case of the employed bees, the greedy selection method is also employed to retain a better one from the old food source and the modified food source.

(3) Scout bee phase

If a food source cannot be further improved for at least *limit* times, it is considered to be exhausted, where *limit* is a control parameter. It is worth to note that, except for some common control parameters shared by other evolutionary algorithms as well, such as the population size, *limit* is the single specific parameter in ABC. Under this case, the food source needs to be abandoned, and a new food source is generated to replace it as follows.

$$x_{i,j} = a_j + rand_j \cdot (b_j - a_j) \tag{4}$$

where $[a_j, b_j]$ is the boundary constraint for the jth variable, and $rand_j \in [0, 1]$ is a random number.

3   Selection Mechanism in Onlooker Bee Phase

From the description of onlooker bee phase in the Sect. 2, it can be seen that the selection mechanism aims to pick out good food sources for onlooker bees to conduct exploitation. Hence, it is important that how to define "good". Generally speaking, fitness value is used a metric in the context of evolutionary algorithms. In other words, there is no doubt that how to calculate fitness value plays an important role. In the classic ABC, Eq. (2) is defined to achieve the goal of calculating fitness value. As seen, the selection probability of a food source is proportional to its fitness value. The better the fitness value a food source has, the bigger the selection probability it has. However, some studies have pointed out that although this mechanism has been shown its effectiveness in many cases, it may cause some problems in terms of selection pressure. First, in initial stage of the search process, some super-fit food sources have much better fitness values than other normal food sources. Accordingly, their selection probabilities are also much bigger. Once these super-fit food sources stagnate, the rest of food sources would converge to their positions by their huge attraction, and the real optimum would be skipped [13]. Second, with the progression of optimization, the differences between the fitness values of food sources are reduced in late stage [5,11]. As a result, it's difficult to pick out better food sources, which would cause the problem of losing population diversity.

To address the above issues, some other selection mechanisms have been proposed recently. In [5], Cui *et al.* proposed a new probability model to calculate the selection probability which is listed in Eq. (5). In there, $r(x_i)$ is the rank of the ith food source in ascending order among all food sources according to the objective function value. For example, the rank of the current best food source and worst food source is 1 and SN, respectively.

$$p_i = \frac{0.8}{(e^{r(x_i)/SN})\sqrt{r(x_i)}} \tag{5}$$

In [11], Yu *et al.* presented a rank-based selection mechanism. In their method, the probability of being selected depends on the fitness rank of a food source rather than its fitness value. The employed formula is listed as follows, in which the best food source has the best rank value SN.

$$p_i = \frac{r_i}{\sum_{i=1}^{SN} r_i} \tag{6}$$

In [13], Bao and Zeng compared and analyzed the selection mechanism. Three different mechanisms were introduced, they are rank selection, disruptive selection, and tournament selection. First, in the rank selection, each food source is sorted according to the objective value from the best to the worst. The formula is listed in Eq. (7), in which the best food source has the index $i = 1$ and the worst one $i = SN$. By doing this, some bad food sources may have more chance to be selected.

$$p_i = \frac{1}{SN} + a(t) \cdot \frac{SN + 1 - 2i}{SN \cdot (SN + 1)} \tag{7}$$

where $a(t) = 0.2 + \frac{3t}{4 \cdot N}$, $t \in \{1, 2, \ldots, N\}$ denotes the current generation index.

Second, the disruptive selection also attempts to provide more chances for bad food sources to survive in the selection mechanism. In this mechanism, the employed fitness function is listed in Eq. (8), in which $\bar{f}$ is the average objective function value of all food sources at the current generation. As seen, compared with the classic ABC, the disruptive selection not only tends to select good food sources but also tries to choose some bad food sources.

$$p_i = \frac{fit_i}{\sum_{i=1}^{SN} fit_i} \tag{8}$$

where $fit_i = |f_i - \bar{f}|$.

Third, similar to the classic ABC, the tournament selection favors good food sources. It randomly selects two food sources to compare their fitness values, the food source with better fitness value is assigned one score, and the variable a_i is defined to record the total scores. After any two of the food sources completing comparison, then the selection probability can be calculated by using the Eq. (9).

$$p_i = \frac{a_i}{\sum_{i=1}^{SN} a_i} \tag{9}$$

4   Experiments and Discussions

In the experiments, we compare the performance of the above six different selection mechanisms including the classic ABC. To have a fair comparison, these selection mechanisms are embedded into the classic ABC, which implies that they have the same framework except for the selection mechanism in onlooker bee phase. For convenience, the following algorithm names are used to denote the corresponding selection mechanisms.

- ABC (the classic ABC)
- ABC2 (the new probability model of Cui *et al.* [5])
- ABC3 (the rank-based mechanism of Yu *et al.* [11])
- ABC4 (the rank selection of Bao and Zeng [13])
- ABC5 (the disruptive selection of Bao and Zeng [13])
- ABC6 (the tournament selection of Bao and Zeng [13])

4.1   Benchmark Problems and Parameter Settings

The comparison are conducted on 21 well-known scalable benchmark problems, and the tested dimension size D is set to 30. Among these benchmark problems, the first 11 functions are unimodal type, while the remaining functions are multimodal type. All of these functions have the same global optimum zero. Due to the limit of paper space, only the function names are listed in Table 1, the detailed definitions can be referred to [12]. For the parameter settings, all of these six algorithms share the same settings for a fair comparison. To be specific, the number of food sources SN is set to 50, and the parameter *limit* is set to 200. For the stopping criteria, the maximum number of function evaluations *MaxFEs* is set to $5000 \cdot D$. Each algorithm is run 30 times per function, and the average results are recorded.

Table 1. The comparative results of ABC, ABC2, ABC3, ABC4, ABC5, and ABC6.

Function	ABC	ABC2	ABC3	ABC4	ABC5	ABC6
Sphere	7.09E−18	1.01E−19†	2.08E−21†	3.75E−22†	6.27E−21†	2.90E−21†
Schwefel 2.22	7.51E−11	8.34E−11$^\approx$	1.47E−11†	5.26E−12†	1.69E−11†	1.70E−11†
Schwefel 1.2	5.82E+03	4.03E+03†	5.97E+03$^\approx$	6.17E+03$^\approx$	5.00E+03†	5.74E+03$^\approx$
Schwefel 2.21	1.01E+01	8.25E+00†	1.09E+01$^\approx$	1.06E+01$^\approx$	9.35E+00$^\approx$	1.01E+01$^\approx$
Rosenbrock	3.47E−01	2.71E−01†	1.45E−01†	2.39E−01†	3.42E−01†	4.60E−01‡
Step	0.00E+00	0.00E+00$^\approx$	0.00E+00$^\approx$	0.00E+00$^\approx$	0.00E+00$^\approx$	0.00E+00$^\approx$
Quartic	9.74E−02	1.07E−01†	1.07E−01†	1.05E−01†	1.07E−01†	1.07E−01†
Elliptic	1.12E−09	1.60E−15†	3.73E−17†	6.64E−18†	2.29E−17†	4.15E−17†
SumSquare	1.60E−19	1.06E−20†	3.03E−22†	5.93E−23†	9.50E−22†	4.41E−22†
SumPower	7.60E−16	4.77E−10‡	9.76E−14‡	5.15E−15‡	2.94E−16†	1.35E−13‡
Exponential	6.53E−07	5.74E−07†	5.47E−07†	4.78E−07†	6.19E−07†	7.40E−07‡
Schwefel 2.26	3.82E−04	3.82E−04$^\approx$	3.82E−04$^\approx$	3.82E−04$^\approx$	3.82E−04$^\approx$	3.82E−04$^\approx$
Rastrigin	1.08E−14	8.17E−15$^\approx$	1.15E−13‡	1.85E−13‡	1.02E−13‡	4.70E−14‡
Ackley	3.27E−10	8.82E−10‡	2.39E−10†	1.17E−10†	8.99E−10†	2.43E−10†
Griewank	7.85E−14	2.47E−04$^\approx$	1.67E−15†	1.85E−15†	2.47E−04‡	3.00E−06$^\approx$
Penalized_1	1.08E−19	2.77E−21†	1.69E−22†	2.82E−22†	4.14E−22†	3.41E−22†
Penalized_2	9.70E−18	7.28E−20†	5.83E−21†	7.58E−22†	8.84E−21†	8.52E−21†
NCRastrigin	0.00E+00	0.00E+00$^\approx$	0.00E+00$^\approx$	0.00E+00$^\approx$	0.00E+00$^\approx$	0.00E+00$^\approx$
Alpine	2.53E−06	3.61E−07†	6.92E−07†	7.28E−07†	1.91E−06†	5.77E−07†
Levy	2.72E−12	2.41E−12$^\approx$	4.44E−13†	1.87E−13†	1.27E−13†	8.52E−13†
Bohachevsky	5.85E−14	7.59E−17†	0.00E+00†	0.00E+00†	1.85E−18†	0.00E+00†
w/l/t	–	12/2/7	14/2/5	14/2/5	13/3/5	11/4/6

4.2   Comparison Results and Discussions

The average results of these six algorithms are presented in the Table 1. In addition, the Wilcoxon's rank-sum test is employed to compare the significant differences between the classic ABC and other algorithms at 5% significance level. The signs "$\dagger$", "$\ddagger$", and "$\approx$" indicate the ABC variant is better than, worse than, and similar to the classic ABC, respectively. The last row "w/l/t" of Table 1 summarizes the comparison results.

As shown in Table 1, all of the five ABC variants achieve better results than the classic ABC on most of the benchmark functions. To be specific, both ABC3 and ABC4 are significantly better than the classic ABC on 14 benchmark functions, while they only lose on the functions SumPower and Rastrigin. For the ABC5, it wins ABC on 13 functions, but they ties on five functions. Similar to the ABC3 and ABC4, ABC2 also loses two functions, but it obtains better results than the classic ABC on 12 functions. For the ABC6, it defeats the classic ABC on 11 functions, but it also loses four functions. Overall, all of these five ABC variants are better than or at least comparable to the classic ABC on most cases.

Table 2. Average rank achieved by the Friedman test.

Algorithm	Average rank
ABC	4.43
ABC2	3.93
ABC3	2.90
ABC4	2.55
ABC5	3.55
ABC6	3.64

But among these ABC variants, there still exists some differences in terms of the accuracy of the results. To achieve a better description, the Friedman test is used to get the average rank of all these ABC algorithms, which is shown in Table 2. From Table 2, it can be seen that the ABC4 obtains the best average rank out of the six algorithms, and the ABC3 is the second best. It's not difficult to realize that both ABC3 and ABC4 are rank-based mechanisms. As seen, in the ABC3 (Eq. (6)), the selection probability is constant during the entire search process, while in the ABC4 (Eq. (7)), with increasing number of

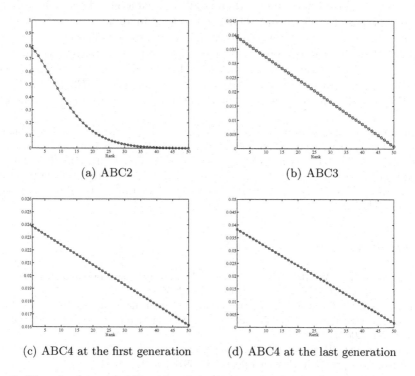

(a) ABC2

(b) ABC3

(c) ABC4 at the first generation

(d) ABC4 at the last generation

Fig. 1. The selection probability of each food source for the ABC2, ABC3, and ABC4.

generations, the search probability gradually increases slightly. For the remaining ABC algorithms, some of them are depend on the fitness values, while the others are also based on fitness rank. To be specific, the selection probability in the classic ABC is depend on the fitness value, and it may change at different generations. In the ABC2, the selection probability is kept the same during the search process, and the maximum value is less than 0.8. As for the ABC5, the selection probability is related to the fitness value and constantly changes. Although the selection probability in the ABC6 is based on the fitness rank, it also keeps changing at different generations. The Fig. 1 depicts the selection probability of each food source for the ABC2, ABC3, and ABC4. As seen, when the generation reaches the maximum value, the selection probability of ABC4 is very similar to that of ABC3.

5   Conclusions

In this paper, we present a comparative study on the selection mechanism in the onlooker bee phase of ABC. Six different selection mechanisms are investigated, and they are embedded into the classic ABC to get the same algorithm structure for a fair comparison. Experiments are conducted on 21 well-known benchmark problems. From the results, we can conclude that the fitness rank-based mechanisms perform better than the fitness value-based ones. Because the fitness rank-based mechanism is beneficial to keep moderate selection pressure, neither too large nor too small. However, it is worth to note that even the fitness rank-based mechanism has been shown its superiority, how to dynamically control the selection pressure is also important for designing selection mechanism, and this can be proved by the comparison results between ABC3 and ABC4. In fact, this would be a valuable research direction in the future for researchers.

Acknowledgments. This work was supported by the National Natural Science Foundation of China (Nos. 61603163, 61562042, and 61462045), the Science and Technology Foundation of Jiangxi Province (Nos. 20151BAB217007 and 20151BAB207027), the Science and Technology Plan Projects of Jiangxi Provincial Education Department (No. GJJ150318), and the Foundation of State Key Laboratory of Software Engineering (No. SKLSE2014-10-04).

References

1. Karaboga, D.: An idea based on honey bee swarm for numerical optimization. Technical report TR06, Erciyes University (2005)
2. Karaboga, D., Basturk, B.: A powerful and efficient algorithm for numerical function optimization: artificial bee colony (ABC) algorithm. J. Glob. Optim. **39**(3), 459–471 (2007)
3. Karaboga, D., Basturk, B.: On the performance of artificial bee colony (ABC) algorithm. Appl. Soft Comput. **8**(1), 687–697 (2008)
4. Wang, H., Wu, Z., Rahnamayan, S., Sun, H., Liu, Y., Pan, J.S.: Multi-strategy ensemble artificial bee colony algorithm. Inf. Sci. **279**, 587–603 (2014)

5. Cui, L., Li, G., Lin, Q., Du, Z., Gao, W., Chen, J., Lu, N.: A novel artificial bee colony algorithm with depth-first search framework and elite-guided search equation. Inf. Sci. **367**, 1012–1044 (2016)
6. Cuevas, E., Sención-Echauri, F., Zaldivar, D., Pérez-Cisneros, M.: Multi-circle detection on images using artificial bee colony (ABC) optimization. Soft. Comput. **16**(2), 281–296 (2012)
7. Bose, D., Biswas, S., Vasilakos, A.V., Laha, S.: Optimal filter design using an improved artificial bee colony algorithm. Inf. Sci. **281**, 443–461 (2014)
8. Yeh, W.C., Hsieh, T.J.: Artificial bee colony algorithm-neural networks for S-system models of biochemical networks approximation. Neural Comput. Appl. **21**(2), 365–375 (2012)
9. Pan, Q.K., Wang, L., Li, J.Q., Duan, J.H.: A novel discrete artificial bee colony algorithm for the hybrid flowshop scheduling problem with makespan minimisation. Omega **45**, 42–56 (2014)
10. Szeto, W., Wu, Y., Ho, S.C.: An artificial bee colony algorithm for the capacitated vehicle routing problem. Eur. J. Oper. Res. **215**(1), 126–135 (2011)
11. Yu, W.J., Zhan, Z.H., Zhang, J.: Artificial bee colony algorithm with an adaptive greedy position update strategy. Soft Comput., 1–15 (online) (2016)
12. Cui, L., Zhang, K., Li, G., Fu, X., Wen, Z., Lu, N., Lu, J.: Modified Gbest-guided artificial bee colony algorithm with new probability model. Soft Computing pp. 1–27 (online) (2017)
13. Bao, L., Zeng, J.C.: Comparison and analysis of the selection mechanism in the artificial bee colony algorithm. In: The Ninth International Conference on Hybrid Intelligent Systems. vol. 1, pp. 411–416. IEEE (2009)

Adaptive Fireworks Algorithm Based on Two-Master Sub-population and New Selection Strategy

Xiguang Li, Shoufei Han[(✉)], Liang Zhao, and Changqing Gong

School of Computer, Shenyang Aerospace University, Shenyang 110136, China
hanshoufei@gmail.com

Abstract. Adaptive Fireworks Algorithm (AFWA) is an effective algorithm for solving optimization problems. However, AFWA is easy to fall into local optimal solutions prematurely and it also provides a slow convergence rate. In order to improve these problems, the purpose of this paper is to apply two-master sub-population (TMS) and new selection strategy to AFWA with the goal of further boosting performance and achieving global optimization. Our simulation compares the proposed algorithm (TMSFWA) with the FWA-Based algorithms and other swarm intelligence algorithms. The results show that the proposed algorithm achieves better overall performance on the standard test functions.

Keywords: Adaptive fireworks algorithm · Two-master Sub-population · Selection strategy · Swarm intelligence algorithm · Standard test functions

1 Introduction

Fireworks Algorithm (FWA) [1] is a new group of intelligent algorithms developed in recent years based on the natural phenomenon of simulating fireworks sparking, and can solve some optimization problems effectively. Compared with other intelligent algorithms such as particle swarm optimization and genetic algorithm, the FWA algorithm adopts a new type of explosive search mechanism, which is explosive. In addition, to calculate the explosion amplitude and the number of explosive sparks through the interaction mechanism between fireworks.

However, many researchers quickly find that traditional FWA has some disadvantages in solving optimization problems, the main disadvantages include slow convergence speed and low accuracy, thus, many improved algorithms have been proposed. So far, research on the FWA has concentrated on improving the operators. One of the most important improvements of the FWA, the enhanced fireworks algorithm (EFWA) [2], the operators of the conventional FWA were thoroughly analyzed and revised. Based on the EFWA, an adaptive fireworks algorithm (AFWA) [3] was proposed, which was the first attempt to control the explosion amplitude without preset parameter by detecting the results of the search process. In [4], a dynamic search fireworks algorithm (dynFWA) was proposed in which divided the fireworks into core firework and non-core fireworks according to the fitness value and adaptive adjustment of explosion amplitude for the core firework. In addition, since the FWA was proposed,

© Springer International Publishing AG 2017
D. Liu et al. (Eds.): ICONIP 2017, Part IV, LNCS 10637, pp. 70–79, 2017.
https://doi.org/10.1007/978-3-319-70093-9_8

it has been applied to many areas [5], including digital filters design [6], nonnegative matrix factorization [7], spam detection [8], etc.

Aforementioned AFWA variants can improve the performance of FWA to some extent. However, inhibition of premature convergence and solution accuracy improvement is still a challenge issue for further research on AFWA.

In order to improve the above problems, in this paper, the searching range of AFWA is expanded by searching the mutual cooperation between the two groups of master sub-populations, to accelerate the convergence rate and improve the search ability of the algorithm. In addition, a new selection strategy is proposed to keep the diversity of the population. Based on this, an improved fireworks optimization algorithm (TMSFWA) is proposed to improve the convergence speed and precision.

The paper is organized as follows. In Sect. 2, the adaptive fireworks algorithm is introduced. The TMSFWA algorithm is presented in Sect. 3. The simulation experiments and results analysis are given in details in Sect. 4. Finally, the conclusion summarizes in final part.

2 Adaptive Fireworks Algorithm

The TMSFWA is based on the AFWA because its ideal is very simple and it works stably. In this section, we will briefly introduce the framework and the operators of the AFWA for further discussion.

In AFWA, there are two important components: the explosion operator (the sparks generated by the explosion) and the selection strategy.

2.1 Explosion Operator

Each firework explodes and generates a certain number of explosion sparks within a certain range (explosion amplitude). The numbers of explosion sparks (Eq. (1)) calculated according to the qualities of the fireworks.

For each firework X_i, its explosion sparks' number is calculated as follows:

$$S_i = m \times \frac{y_{\max} - f(X_i) + \varepsilon}{\sum_{i=1}^{N} (y_{\max} - f(X_i)) + \varepsilon} \tag{1}$$

where $y_{max} = max(f(X_i))$, m is a constant to control the number of explosion sparks, and ε is the machine epsilon to avoid S_i is equal to 0.

In AFWA, the calculation of the amplitude of the normal fireworks and the optimal firework (the value of the objective function is smallest) are different. The normal fireworks' explosion amplitudes are calculated just as in the previous versions of FWA:

$$A_i = A \times \frac{f(X_i) - y_{\min} + \varepsilon}{\sum_{i=1}^{N} (f(X_i) - y_{\min}) + \varepsilon} \tag{2}$$

where $y_{min} = minf(X_i))$, A is a constant to control the explosion amplitude, and ε is the machine epsilon to avoid A_i is equal to 0.

But for the optimal firework, its explosion amplitude is adjusted according to the search results in the last generation:

$$A_i(t+1) = \begin{cases} UB - LB & t = 0 \ or \ f(s_i) < f(x) \\ 0.5 \times (\lambda \times ||s_i - s^*||_\infty + A_i(t)) & otherwise \end{cases} \quad (3)$$

where UB and LB stand for the upper bound and lower bound of the search space respectively, $s_1...s_n$ denote all sparks generated in generation t, s^* denotes the best spark and x stands for fireworks in generation t, and the parameter λ is suggested to be fixed value of 1.3 empirically.

2.2 Selection Strategy

In AFWA, it applies a selection method, which is referred to as elitism-random selection method. In this selection process, the optima of the set will be selected first. Then, the other individuals are selected randomly.

3 The Proposed Algorithm (TMSFWA)

The proposed algorithm (TMSFWA) is a simple and easy to implement AFWA based on two-master sub-populations and new selection strategy.

3.1 Two-Master Sub-population

The realization of the two-master sub-populations' idea is: in a random initialization of a group of fireworks, it is divided into two independent sub-populations, one is the master sub-population, and the other is the assistant sub-population. The definitions are as follows.

Definition 1. Master sub-population is the optimal firework in the current fireworks population.

Definition 2. Assistant sub-population is the fireworks except the optimal in the current fireworks population.

For the assistant sub-populations are iteratively searched by the AFWA, but displacement operation of the master sub-population is calculated by two methods. Displacement operation of a master sub-population is calculated as the AFWA (Eq. (4)). As we known, the position of the master sub-population is the best information in the population, therefore, the other master sub-population is added into the AFWA, and its displacement operation is calculated by Eq. (5).

$$\Delta x_i^k = x_i^k + rand(0, A_i) \quad (4)$$

$$\Delta x_i^k = x_i^k - rand(0, A_i) \quad (5)$$

where $rand(0, A_i)$ represents a uniform random number within the amplitude A_i.

At the end of each iteration, the fitness values corresponding to the optimal positions of the two master sub-populations are compared, and the optimal fireworks are retained. The two master sub-populations complement each other and co-evolve to fully extend the search range and mine the useful information in the search domain to reduce the risk of the AFWA falling into the local optimal.

The Algorithm 1 is proposed to generate the explosion sparks with two-master sub-population.

Algorithm 1: Generating explosion sparks with two-master sub-population

---

Initialize the location of the explosion sparks: $X_j = X_i$
Calculate the number of explosion sparks S_i
Calculate the explosion amplitude A_i
Set $z = rand(1,d)$
For $k = 1:d$ do
 If $k \in z$ then
  If X_j^k is the optimal firework
  $X_j^k = X_j^k + rand(0, A_i)$
  $X_j^k = X_j^k - rand(0, A_i)$
  Else
   $X_j^k = X_j^k + rand(0, A_i)$
  End if
  If X_j^k out of bounds
   $X_j^k = X_{min}^k + |X_j^k| \% (X_{max}^k - X_{min}^k)$
  End if
 End if
End for

3.2 New Selection Strategy

In AFWA, it applies a selection method, which is referred to as elitism-random selection method. In this selection process, the optima of the set will be selected firstly. Then, the other individuals are selected randomly. Obviously, this method cannot ensure the diversity of the population. Based on this, this paper proposes a new selection strategy: elitism-tournament selection strategy.

The same as AFWA, elitism-tournament selection also requires that the current best location is always kept for the next iterations. And then, two individuals were randomly selected from the remaining individuals in the population. Each time the individual with the best fitness is placed in the next generation group, the next generation was obtained by repeating N−1 times [9].

From the above, we know that the elitism-tournament selection not only maintaining the competitive advantage, but also considering the diversity of the population. This method can maintain the diversity of the population, reflect the better global searching ability.

4 Experiments

4.1 Experiment Settings

Similar to AFWA, the number of fireworks in TMSFWA is set to 5, and the number of mutation sparks is also set to 5, the maximum number of sparks is set to 200 each generation.

In the experiment, the function of each algorithm is repeated 51 times, and the final results after the 300000 function evaluations are presented. In order to verify the performance of the algorithm proposed in this paper, we use the CEC2013 test set [10], including 28 different types of test functions. All experimental test functions dimensions are set to 30, d = 30.

Finally, we use Matlab R2014a software on a PC with a 3.2 GHz CPU (Intel Core i5-3470), and 4 GB RAM, and Windows 7 (64 bit).

4.2 Simulation Results and Analysis

Comparison of TMSFWA with FWA-Based algorithms. To assess the performance of TMSFWA, TMSFWA is compared with EFWA, dynFWA and AFWA, and EFWA parameters set in accordance with [2], AFWA parameters set in accordance with [3], dynFWA parameters set in accordance with [4].

For each test problems, each algorithm runs 51 times, all experimental test functions dimensions are set as 30, and their mean errors and total number of rank 1 are reported in Table 1. The best results among the comparisons are shown in bold. It can be seen that the proposed TMSFWA clearly outperforms among EFWA, AFWA and dynFWA on the test functions.

To clear show the advantages of TMSFWA, the convergence curves of mean objective function value which have great difference in evolution speed are plotted in Fig. 1. Evidently, TMSFWA has better solution accuracy, convergence rate and robustness than all the competitors on majority of cases.

Comparison of TMSFWA with other swarm intelligence algorithms. In order to measure the relative performance of the TMSFWA, a comparison among the TMSFWA and other swarm intelligence algorithms is conducted on the CEC 2013 single objective benchmark suite. The algorithms compared here are described as follows.

- Artificial bee colony (ABC) [11]: A powerful swarm intelligence algorithm. The results were reported in [12].
- Standard particle swarm optimization (SPSO2011) [13]: The most recent standard version of the famous swarm intelligence algorithm PSO. The results were reported in [5].
- Differential evolution (DE) [14]: One of the best evolutionary algorithms for optimization. The results were reported in [15].

Table 1. Mean errors and total number of rank 1 achieved by EFWA, AFWA, dynFWA and TMSFWA.

Functions	EFWA	AFWA	dynFWA	TMSFWA
	Mean error	Mean error	Mean error	Mean error
f1	7.82E−02	**00E+00**	**00E+00**	**00E+00**
f2	5.43E+05	8.93E+05	7.87E+05	**2.33E+05**
f3	1.26E+08	1.26E+08	1.57E+08	**6.86E+07**
f4	1.09E+00	1.15E+01	1.28E+01	**00E+00**
f5	7.9E−02	6.04E−04	5.42E−04	**00E+00**
f6	3.49E+01	2.99E+01	3.15E+01	**1.2E+01**
f7	1.33E+02	9.19E+01	1.03E+02	**7.7E+01**
f8	2.10E+01	2.09E+01	2.09E+01	**2.09E+01**
f9	3.19E+01	2.48E+01	2.56E+01	**1.99E+01**
f10	8.29E−01	4.73E−02	4.20E−02	**3.00E−02**
f11	4.22E+02	1.05E+02	1.07E+02	**8.45E+01**
f12	6.33E+02	1.52E+02	1.56E+02	**1.22E+02**
f13	4.51E+02	2.36E+02	2.44E+02	**1.95E+02**
f14	4.16E+03	2.97E+03	2.95E+03	**2.53E+03**
f15	4.13E+03	3.81E+03	3.9E+03	**3.77E+03**
f16	5.92E−01	4.97E−01	4.77E−01	**2.9E−01**
f17	3.10E+02	1.45E+02	1.48E+02	**1.19E+02**
f18	1.75E+02	1.75E+02	1.89E+02	**1.68E+02**
f19	1.23E+01	6.92E+00	6.87E+00	**5.68E+00**
f20	1.46E+01	1.30E+01	1.30E+01	**1.25E+01**
f21	3.24E+02	3.16E+02	2.92E+02	**2.97E+2**
f22	5.75E+03	3.45E+03	3.41E+03	**2.82E+03**
f23	5.74E+03	4.70E+03	4.85E+03	**4.6E+03**
f24	3.37E+02	2.70E+02	2.72E+02	**2.54E+02**
f25	3.56E+02	2.99E+02	2.97E+02	**2.84E+02**
f26	3.21E+02	2.73E+02	2.62E+02	**2.35E+02**
f27	1.28E+03	9.72E+02	9.92E+02	**8.6E+02**
f28	4.34E+02	4.37E+02	3.40E+02	**3.1E+02**
Total number of rank 1				
0	1	1	28	

- Covariance matrix adaptation evolution strategy (CMA-ES) [16]: A developed evolutionary algorithm. The results are based on the code from (https://www.lri.fr/ ∼hansen/purecmaes.m) using default settings.

The above four algorithms are using the default settings. The comparison results of among ABC, DE, CMA-ES, SPSO2011, and TMSFWA are presented in Table 2, where 'Mean error' is the mean error of best fitness value. The best results among

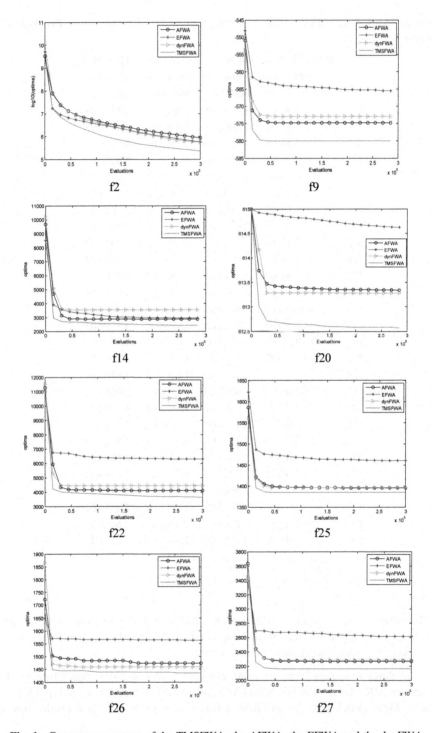

Fig. 1. Convergence curves of the TMSFWA, the AFWA, the EFWA and the dynFWA.

the comparisons are shown in bold. ABC beats other algorithms on 12 functions (some differences are not significant), which is the most, but performs poorly on other functions. CMA-ES performs extremely well on unimodal functions, but suffers from premature convergence on some complex functions. From the Table 3, the TMSFWA ranked the top three (25/28) more than the other four algorithms, and in terms of average ranking, the TMSFWA performs the best among these 5 algorithms on this benchmark suite due to its stability. DE and ABC take the second place and the third place respectively. The performances of CMA-ES and the SPSO2011 are comparable.

Table 2. Mean errors achieved by ABC, DE, CMA-ES, SPSO2011 and TMSFWA.

	ABC	DE	CMA-ES	SPSO2011	TMSFWA
f1	0.00E+00	1.89E−03	**0.00E+00**	**0.00E+00**	**0.00E+00**
f2	6.20E+06	5.52E+04	**0.00E+00**	3.38E+05	2.33E+05
f3	5.74E+08	2.16E+06	**1.41E+01**	2.88E+08	6.86E+07
f4	8.75E+04	1.32E−01	**0.00E+00**	3.86E+04	0.00E+00
f5	0.00E+00	2.48E−03	**0.00E+00**	5.42E−04	0.00E+00
f6	1.46E+01	7.82E+00	**7.82E−02**	3.79E+01	1.2E+01
f7	1.25E+02	4.89E+01	**1.91E+01**	8.79E+01	7.7E+01
f8	2.09E+01	**2.09E+01**	2.14E+01	2.09E+01	2.09E+01
f9	3.01E+01	**1.59E+01**	4.81E+01	2.88E+01	1.99E+01
f10	2.27E−01	3.42E−02	**1.78E−02**	3.40E−01	3.00E−02
f11	**00E+00**	7.88E+01	4.00E+02	1.05E+02	8.45E+01
f12	3.19E+02	**8.14E+01**	9.42E+02	1.04E+02	1.22E+02
f13	3.29E+02	**1.61E+02**	1.08E+03	1.94E+02	1.95E+02
f14	**3.58E−01**	2.38E+03	4.94E+03	3.99E+03	2.53E+03
f15	3.88E+03	5.19E+03	5.02E+03	3.81E+03	**3.77E+03**
f16	1.07E+00	1.97E+00	**5.42E−02**	1.31E+00	2.9E−01
f17	**3.04E+01**	9.29E+01	7.44E+02	1.16E+02	1.19E+02
f18	3.04E+02	2.34E+02	5.17E+02	**1.21E+02**	1.68E+02
f19	**2.62E−01**	4.51E+00	3.54E+00	9.51E+00	5.68E+00
f20	1.44E+01	1.43E+01	1.49E+01	1.35E+01	**1.25E+01**
f21	**1.65E+02**	3.20E+02	3.44E+02	3.09E+02	2.97E+02
f22	**2.41E+01**	1.72E+03	7.97E+03	4.30E+03	2.82E+03
f23	4.95E+03	5.28E+03	6.95E+03	4.83E+03	**4.6E+03**
f24	2.90E+02	**2.47E+02**	6.62E+02	2.67E+02	2.54E+02
f25	3.06E+02	2.89E+02	4.41E+02	2.99E+02	**2.84E+02**
f26	**2.01E+02**	2.52E+02	3.29E+02	2.86E+02	2.35E+02
f27	**4.16E+02**	7.64E+02	5.39E+02	1.00E+03	8.6E+02
f28	2.58E+02	4.02E+02	4.78E+03	4.01E+02	3.1E+02

Table 3. Total number of rank and average rankings.

	ABC	DE	CMA-ES	SPSO2011	TMSFWA
Total number of rank 1	12	5	9	3	8
Total number of rank 2	0	11	3	5	9
Total number of rank 3	3	7	0	8	8
Total number of rank 4	10	3	1	8	3
Total number of rank 5	3	2	15	4	0
Total number of rank	76	70	94	88	62
Average ranking	2.71	2.5	3.36	3.14	2.21

5 Conclusions

TMSFWA was developed by applying two-master sub-population and a new selection strategy to AFWA. We apply the CEC2013 standard functions to examine and compare the proposed algorithm TMSFWA with ABC, DE, SPSO2011, CMA-ES, AFWA, EFWA and dynFWA. The results clearly indicate that TMSFWA can perform significantly better than other seven algorithms in terms of solution accuracy. Overall, the research demonstrates that TMSFWA performed the best for solution accuracy.

References

1. Tan, Y., Zhu, Y.: Fireworks algorithm for optimization. In: Tan, Y., Shi, Y., Tan, K.C. (eds.) ICSI 2010. LNCS, vol. 6145, pp. 355–364. Springer, Heidelberg (2010)
2. Zheng, S., Janecek, A., Tan, Y.: Enhanced fireworks algorithm. In: Proceedings of 2013 IEEE Congress on Evolutionary Computation, Cancun, Mexico, pp. 2069–2077 (2013)
3. Zheng, S., Li, J., Tan, Y.: Adaptive fireworks algorithm. In: Proceedings of 2014 IEEE Congress on Evolutionary Computation, Beijing, China, pp. 3214–3221 (2014)
4. Zheng, S., Tan, Y.: Dynamic search in fireworks algorithm. In: Proceedings of 2014 IEEE Congress on Evolutionary Computation, Beijing, China, pp. 3222–3229 (2014)
5. Tan, Y.: Fireworks Algorithm Introduction, 1st edn. Science press, Beijing (2015)
6. Gao, H.Y., Diao, M.: Cultural firework algorithm and its application for digital filters design. Int. J. Model. Ident. Control **14**(4), 324–331 (2011)
7. Janecek, A., Tan, Y.: Using population based algorithms for initializing nonnegative matrix factorization. In: Tan, Y., Shi, Y., Chai, Y., Wang, G. (eds.) ICSI 2011. LNCS, vol. 6729, pp. 307–316. Springer, Heidelberg (2011). doi:10.1007/978-3-642-21524-7_37
8. Wen, R., Mi, G.Y., Tan, Y.: Parameter optimization of local-concentration model for spam detection by using fireworks algorithm. In: Proceedings of 4th International Conference on Swarm Intelligence, Harbin, China, pp. 439–450 (2013)
9. Chen, T.: On the Computational Complexity of Evolutionary Algorithms. University of Science and Technology of China, Anhui, China (2010, in Chinese)
10. Liang, J., Qu, B., Suganthan, P., et al.: Problem Definitions and Evaluation Criteria for the CEC 2013 Special Session on Real-Parameter Optimization (2013)
11. Karaboga, D., Basturk, B.: A powerful and efficient algorithm for numerical function optimization: artificial bee colony (ABC) algorithm. J. Glob. Optim. **39**(3), 459–471 (2007)

12. M, El-Abd.: Testing a particle swarm optimization and artificial bee colony hybrid algorithm on the CEC13 benchmarks. In: Proceedings of 2013 IEEE Congress on Evolutionary Computation, Cancun, Mexico, pp. 2215–2220 (2013)

13. Zambrano-Bigiarini, M., Clerc, M., Rojas, R.: Standard particle swarm optimization 2011 at CEC2013: a baseline for future PSO improvements. In: Proceedings of 2013 IEEE Congress on Evolutionary Computation, Cancun, Mexico, pp. 2337–2344 (2013)

14. Storn, R., Price, K.: Differential evolution - a simple and efficient heuristic for global optimization over continuous spaces. J. Glob. Optim. **11**(4), 341–359 (1997)

15. Padhye, N., Mittal, P., Deb, K.: Differential evolution: performances and analyses. In: Proceedings of 2013 IEEE Congress on Evolutionary Computation, Cancun, Mexico, pp. 1960–1967 (2013)

16. Hansen, N., Ostermeier, A.: Adapting arbitrary normal mutation distributions in evolution strategies: the covariance matrix adaptation. In: Proceedings of 1996 IEEE International Conference on Evolutionary Computation, Nagoya, Japan, pp. 312–317 (1996)

A Novel Osmosis-Inspired Algorithm for Multiobjective Optimization

Corina Rotar[1]($\boxtimes$), Laszlo Barna Iantovics[2], and Sabri Arik[3]

[1] "1 Decembrie 1918" University of Alba Iulia, Alba Iulia, Romania
crotar@uab.ro
[2] Petru Maior University, Targu Mures, Romania
[3] Istanbul University, Istanbul, Turkey

Abstract. Many real-life difficult problems imply more than one optimization criterion and often require multiobjective optimization techniques. Among these techniques, nature-inspired algorithms, for instance, evolutionary algorithms, mimic various natural process and systems and succeed to perform appropriately for hard optimization problems. Besides, in chemistry, osmosis is the natural process of balancing the concentration of two solutions. This process takes place at the molecular level. Osmosis's practical applications are multiple and target medicine, food safety, and engineering. However, osmosis process is not yet recognized as a rich source of inspiration for designing computational tools. At first glance, this well-known chemical process seems appropriate as a metaphor in nature-inspired computation as it can underlie the development of a search and optimization procedure. In this paper, we develop a novel algorithm called OSMIA (Osmosis inspired Algorithm) for multiobjective optimization problems. The proposed algorithm is inspired by the well-known physio-chemical osmosis process. For validation purposes, we have realized a case study in that we compared our proposed algorithm with the state-of-art algorithm NSGAII using some well known test problems. The conclusions of the case study emphasize the strengths of the proposed novel OSMIA algorithm.

Keywords: Nature-inspired algorithm · Multiobjective · Osmosis process

1 Introduction

Nature-inspired computing represents a significant research area of the Artificial Intelligence. It includes evolutionary algorithms [1], neural networks [2], cellular automata [3], emergent systems [4–6], artificial immune systems [7], membrane computing [8] and many others. Simply, nature-inspired computing is a growing field, developed by mostly imitating biological models, for the development of the computational models and techniques. Among these, a prolific research chapter includes nature-inspired algorithms for solving the multiobjective optimization problems (MOPs). Given the MOPs complexity and due to the proven potential of the nature-inspired algorithm for various complex problems, the developing and improving the nature-inspired techniques for MOPs represents a challenging task.

© Springer International Publishing AG 2017
D. Liu et al. (Eds.): ICONIP 2017, Part IV, LNCS 10637, pp. 80–88, 2017.
https://doi.org/10.1007/978-3-319-70093-9_9

The research presented in this paper focuses on two complementary directions involved and their functionality, and secondly, the adapting of the considered natural models into powerful computing models. The palette of the natural paradigms, which underlies the development of the nature-inspired metaheuristics, is diverse and encompasses the functioning of the brain, Darwinian evolution, self-replication, collective behavior, the vertebrate immune system, cell membranes, morphogenesis, and so on. There are, moreover, a lot of other natural phenomena which could lead to the development of computational methods. Of these, we identified a physicochemical process, the osmotic process that stands out by its strength and simplicity. Nevertheless, the physicochemical processes underlying the designing of the calculation methods are reduced in number compared with the patterns inspired by the biological systems and processes. Therefore, we identify as a new research domain to the identification of the manner in which the osmosis paradigm could generate new computational models. In this paper, we present such a novel algorithm that we called OSMIA.

The upcoming part of the paper is organized as follows: in the Sect. 2 we describe the natural paradigm and propose a novel osmosis-inspired algorithm for multiobjective optimization. Section 2.3 presents the experimental results. In Sect. 3 we discuss the main results and suggest further research directions.

2 Design of a New Metaheuristic Inspired by Nature

2.1 Natural Paradigm: Osmosis Process

Osmosis is the natural process of balancing the concentration of two solutions, a process that takes place at the molecular level. The process of osmosis is described as the diffusion of a solvent (usually water) through a semi-permeable membrane from a solution with low concentration of solute (high water potential, or, Hypotonic) in a solution with higher concentration solute (low potential of water, or, Hypertonic) to a certain concentration level/gradient of the solution. It is a physical process in which a solvent moves without receiving power through a semipermeable membrane (permeable to solvent but not the solution) separating the two different solutions. This effect can be measured by increased pressure of the hypertonic solution, compared to the hypotonic solution.

For instance, if two solutions of different concentration are separated by a membrane, which is permeable only to the smaller solvent molecules but not to the larger solute molecules, then the solvent will tend to diffuse across the membrane from the less concentrated to the more concentrated solution. As in Fig. 1, having two containers that communicate through a semipermeable membrane, one with less saline water and the other one with a saline solution of a higher concentration, the molecules of the pure water will migrate into the saline solution to reduce the concentration level until it reaches a balance. This phenomenon is caused by the normal diffusion of the molecules, and not by an external force.

Osmotic pressure is the pressure that must be applied to a solution to prevent the solvent migration, in the natural sense of diffusion, through a semi-permeable membrane. Considering Δh – the difference in height of the solution, ρ – the density of the

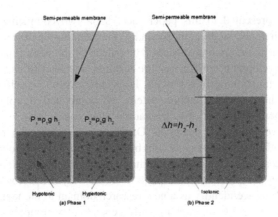

Fig. 1. Description of the osmosis process: before osmotic equilibrium (a); after osmotic equilibrium is attained (b)

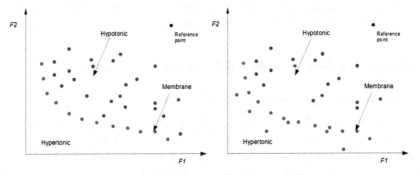

Fig. 2. First phase: molecules occupy the hypertonic region (*left*). During osmosis: the molecules diffuse through the membrane (*right*)

solution and g – the gravitational acceleration, the osmotic equilibrium is achieved when the osmotic pressure reaches the hydrostatic pressure, as follows:

$$P_{osmotic} = \rho \cdot g \cdot \Delta h \qquad (1)$$

2.2   OSMIA a Novel Osmosis-Inspired Algorithm

We observed that the natural process of osmosis can be considered as a model that underlies the development of a search and optimization procedure in multiobjective optimization problems. The strategy inspired by osmosis involves the management of three populations of molecules. These populations correspond to the water molecules from the hypotonic and hypertonic solutions.

Let us consider:

(a) *Hypotonic* = the set of molecules in the hypotonic environment
(b) *Membrane* = the set of molecules which form the membrane
(c) *Hypertonic* = the set of molecules in the hypertonic environment

Let us consider n – the dimension of the search space and m – the number of objectives. A molecule structure is given by the following formula.

$$mol = \{location_{Search}, location_{Obj}, mass, type\} \tag{2}$$

where: $location_{Search} = (x_1, x_2, \ldots, x_n)$ represents the location of the molecule in the search space; $location_{Obj} = (f_1, f_2, \ldots, f_m)$ represents the corresponding location in the objective space and type – represents an indicator of the actual state of the molecule (type = 0 if $mol \in Hypotonic$, type = 1 if $mol \in Membrane$, type = 2 if $mol \in Hypertonic$).

For a minimization problem, the mass of the molecule is computed as follows, signifying that those solutions, which have lower objective values, gain higher mass.

$$mass = 1 \left/ \left(1 + \sum_{i=1}^{m} f_i\right) \right. \tag{3}$$

In the first phase, a set of molecules is randomly generated. The size of the initial set is given *dim*. Each molecule corresponds to a possible solution in the search space. Some of these molecules form a virtual membrane. Establishing semi-permeable membrane is made after the evaluation of the solutions. The molecules are divided into two sets: Pareto non-dominated and dominated solutions. The molecules that correspond to non-dominant solutions will form the "semi-permeable membrane".

The hypotonic environment contains those molecules, which correspond to the dominated solutions. The molecules which correspond to the Pareto non-dominated solutions delimit the semi-permeable membrane. Those molecules, which diffuse through the membrane, according to the Pareto domination criterion, are classified as members of the hypertonic set. The osmotic process continues through the movement of molecules of the hypotonic environment. Hypotonic molecules diffuse toward different positions in the search space. If after the movement of a molecule, the new location of the molecule dominates the current position, this new position is retained further and the molecule is marked accordingly; otherwise, if the new position corresponds to a weaker solution, the old position is restored. A new position of a molecule is better if the corresponding candidate solution dominates the solution which corresponds to the molecule in the original location. If the new position dominates the previous one and the previous position corresponds to a non-dominated solution in current population, the molecule will be marked as a new member of the virtual membrane.

Osmosis Procedure/Cycle
 While (not osmotic equilibrium) Do
  For (each molecule in the Hypotonic solution) Do
  New:=Move the molecule
  if (New molecule *is located in the hypertonic solution*) **then**
   Hypertonic=Hypertonic U {New}
 EndWhile

The proposed algorithm inspired by osmosis repeats the osmotic procedure, which takes as long as the equilibrium between the hypertonic, respectively, hypotonic environments, is not reached. The osmosis is considered complete when the hydrostatic pressure balances the osmotic pressure, and therefore, the molecules will no longer flow from the hypotonic to the hypertonic fluid. Following figures describe the osmotic cycle for a bi-objective minimization problem (Fig. 3).

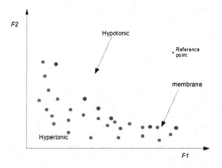

Fig. 3. Finale stage: osmosis process is done; the molecules occupy the hypotonic region

Let us consider: ρ_1, ρ_2, h_1, h_2 the pressures and heights of the two environments, at a certain time during the osmosis process, the equilibrium is reached when $P_1 = P_2$:

$$\rho_1 \cdot g \cdot h_1 = \rho_2 \cdot g \cdot h_2$$

Therefore, while $P_1 > P_2$, the molecules from the hypotonic environment will diffuse into the hypertonic environment.

The proposed Osmosis inspired Algorithm for Multicriteria Optimization (*OSMIA*) works as follows: while a termination condition is not true, the osmosis procedure runs and the virtual membrane is updated. The membrane consists of those molecules that correspond to the non-dominated solutions, and it is a dynamic structure, as long as the set of molecules varies along the osmosis process.

Figure 2 depicts a configuration of molecules in the objective space before the osmosis procedure starts. The reference point is computed as follows:

$$reference = (r_1, r_2, \ldots, r_m), \text{ where } r_i = \max\{f_i^j, j = 1 \ldots dim\}$$

In addition, for the hypertonic and hypotonic sets of molecules, the centroids are computed, as the arithmetic mean position of all the points in the corresponding set:

$$centroid(Hypertonic) = (H_1, H_2, \ldots, H_m)$$
$$centroid(hypotonic) = (h_1, h_2, \ldots, h_m)$$

Relative to this reference point, the pressures and heights of the two environments (hypotonic and hypertonic) can be computed. The *"heights"* of two environments are given by the Euclidian distances between the centroids of the Hypotonic and, respectively Hypertonic set and the reference point:

$$h_1 = \text{distance}(centroid(Hypotonic), reference)$$
$$h_2 = \text{distance}(centroid(Hypertonic), reference)$$

The "densities" of the two environments are computed by the following formulas:

$$\rho_1 = \frac{average\_mass_{hypotonic}}{Volume_{hypotonic}}$$
$$\rho_2 = \frac{average\_mass_{hypertonic}}{Volume_{hypertonic}}$$

The *average_mass* represents the arithmetic mean of the molecules' masses from the specific environment. The *Volume* is estimated by Ritter's algorithm [9] which finds the bounding sphere that contains all of a given molecules from the specific environment.

Algorithm OSMIA is:
Generate the set of molecules: **Hypotonic, Membrane, Hypertonic**
While (ending_condition*)
Compute: **reference**, ρ_1, ρ_2, h_1, h_2 //densities and heights

While($\rho_1 \cdot g \cdot h_1 > \rho_2 \cdot g \cdot h_2$)//**Osmosis cycle**
  For each *mol* from **Hypertonic U Hypertonic U Membrane**
    New=Move(mol)
    If *New dominates mol* and mol∈Membrane then
      Membrane=Membrane U {New}
    If *New dominates mol* and mol ∉Membrane then
      Hypertonic=Hypertonic U {New}
  Endfor
EndWhile
Update **Membrane** = set of non-dominated molecules
EndWhile
End
ending_condition may refer to attaining the maximum number of cycles

The *Move* procedure mimics the Brownian movement of the molecules within the given space. Therefore, the procedure varies the location (considered in the search space) of the current molecule.

For an intuitive approach, each molecule from the hypertonic and hypotonic sets shifts its position, with a given probability p, in each of the ith dimension of the search space. The probability of altering a specific coordinate is set to the value $p = 1/n$, where n – represents the given dimensionality of the search space. The alteration of the coordinates is influenced by one randomly selected element from the membrane (non-dominated solutions).

> **FunctionMove**(*mol1*): *mol2*
>   Input: *mol1*, Output: *mol2*
>   mol2=mol1
>   Randomly select *Membrane*(ind)
>   **for** each i from $\{1,...n\}$
>   **if** *rand*<1/n **then**
>   **if** mol2.x(i) <*Membrane*(ind).x(i) **then**
>   mol2.x(i)=Membrane(ind).x(i)+ *rand**(Max-Membrane(ind).x(i))
>   **if** mol2.x(i) > Membrane(ind).x(i) **then**
>   mol2.x(i) = *Membrane*(ind).x(i)-*rand**(Membrane(ind).x(i)-Min)
>   **endfor**
> **End**

2.3   Experimental Results

In order to illustrate the performance of the proposed OSMIA algorithm, we used several well-known test problems ZDT1, ZDT2, ZDT3 [10] used in the most of the researches and a state-of-art algorithm for multiobjective optimization: NSGAII [11]. NSGAII algorithm have many significant recent applications. For example we mention the multi-production and multi-echelon closed-loop pharmaceutical supply chain considering quality concepts [12], finding patterns in protein sequences [13]. For performance assessment, we compute the hypervolume metric (HV) [14]. The hypervolume metric corresponds to the size of the objective space, which contains the solutions that are Pareto-dominated by at least one of the members of the set. The higher the hyper-volume value is, the better outcomes the algorithm provides. Among the performance metrics, the hypervolume is popular as it captures both the convergence to the true Pareto front and the distribution over the objective space.

For an objective comparison with the popular algorithm NSGAII [11], we set the following parameters for the OSMIA: the size of molecule library is set to 50, and the maximum number of fitness evaluation is set to 10000. The most appropriate parameter values we have established experimentally. NSGAII's parameters (that are considered in different studies) are 100 individuals and 100 iterations per run. These settings assure the same maximum number of function evaluations for both algorithms. The algorithms run for 10 times and the hypervolume values are computed. The results are described in Table 1.

Table 1. Hypervolume: OSMIA versus NSGA2.

Test problem	HV	Mean value	Maximum	Standard dev
ZDT1	*OSMIA*	**0.847666**	0.85257	3.99E−03
	NSGA2	0.775539	0.804457	2.01E−02
ZDT2	*OSMIA*	**0.768684**	0.783413	2.31E−02
	NSGA2	0.579927	0.601901	2.41E−02
ZDT3	*OSMIA*	**0.634674**	0.650786	2.85E−02
	NSGA2	0.514411	0.631122	7.66E−02

The results presented in Table 1, show that for each test problem we considered, the proposed algorithm performs better than NSGAII.

3   Conclusions

In this paper, we explored a novel metaphor in Natural Computing, i.e. the natural process of osmosis, and we proposed a new metaheuristic for multiobjective optimization. The osmosis-inspired algorithm, OSMIA, is a population-based algorithm for optimization, which mimics the process of molecules' diffusion through a semi-permeable membrane. The convergence toward the problem's solutions is guided by the virtual membrane which collects, at each cycle, the set of Pareto non-dominated solutions. The molecules from the hypotonic environment diffuse through the virtual membrane, into the hypertonic environment by using a procedure that alters the original location and simulates Brownian movement. Each diffusion cycle is considered done when the osmotic equilibrium is attained. In natural paradigm, the osmotic equilibrium has attained the concentration is the same on both sides of a semi-permeable membrane. In artificial model, the osmotic equilibrium is considered achieved when the hydrostatic pressure balances the osmotic pressure.

OSMIA was compared with state of art algorithm for multiobjective optimization, NSGAII and the results showed that our proposal performs better in all test scenarios. As further research, we propose to investigate the recognized natural metaphor for different problems and to investigate OSMIA's performance for more difficult optimization problems.

An advantage of the proposed algorithm is given by the minimum number of user-defined parameters. Excluding the number of objectives, the number of variables, and the size of the initial set of molecules (*dim*), no other parameter is needed. Therefore, the Osmosis-inspired Algorithm can be considered a parameter free technique that may solve numerous optimization problems, which involve multiple criteria. Among these, we will investigate the problem of determination of the types of degradation that may affect heritage buildings due to multiple factors. The factors that may affect the heritage buildings include physical, chemical and biological actions. Also, we will consider OSMIA algorithm for a real-world problem such as identifying the optimal strategy to manage the waste, which results from interventions on buildings.

Acknowledgements. The authors gratefully acknowledge the financial support provided by the Romanian National Authority for Scientific Research, CNCS – UEFISCDI, under the Bridge Grant PN-III-P2-2.1-BG-2016-0302.

References

1. Eiben, A.E., Smith, J.E.: Introduction to Evolutionary Computing, vol. 53. Springer, Heidelberg (2003)
2. Rojas, R.: Neural Networks: a Systematic Introduction. Springer-Verlag New York, Inc., New York (1996)
3. Wolfram, S.: Universality and complexity in cellular automata. Phys. D: Nonlinear Phenom. **10**(1), 1–35 (1984)
4. Karaboga, D., Celal, O.: A novel clustering approach: artificial bee colony (ABC) algorithm. Appl. Soft Comput. **11**(1), 652–657 (2011)
5. Dorigo, M., Birattari, M., Blum, C., Clerc, M., Stützle, T., Winfield, A.F.T. (eds.): ANTS 2008. LNCS, vol. 5217. Springer, Heidelberg (2008). doi:10.1007/978-3-540-87527-7
6. Karaboga, D.: An idea based on honey bee swarm for numerical optimization, vol. 200, Technical report-tr06, Computer Engineering Department, Erciyes University (2005)
7. De Castro, L.N., Timmis, J.: Artificial immune systems: a new computational intelligence approach. Springer Science & Business Media (2002)
8. Păun, G.: Computing with membranes. J. Comput. Syst. Sci. **61**(1), 108–143 (2000)
9. Ritter, J.: An efficient bounding sphere. In: Glassner, A. (ed.) Graphics Gems. Academic Press, Boston, MA (1990)
10. Zitzler, E., Deb, K., Thiele, L.: Comparison of multiobjective evolutionary algorithms: empirical results. Evol. Comput. **8**(2), 173–195 (2000)
11. Deb, K., Pratap, A., Agarwal, S., Meyarivan, T.: A fast and elitist multi-objective genetic algorithm: NSGA-II. IEEE Trans. Evol. Comput. **6**(2), 182–197 (2002)
12. Moslemi, S., Zavvar Sabegh, M.H., Mirzazadeh, A., Ozturkoglu, Y., Maass, E.: Int J Syst Assur Eng Manage (2017). doi:10.1007/s13198-017-0650-4
13. González-Álvarez, D.L., Vega-Rodríguez, M.A., Rubio-Largo, Á.: A hybrid MPI/OpenMP parallel implementation of NSGA-II for finding patterns in protein sequences. Supercomput. **73**(6), 2285–2312 (2017). doi:10.1007/s11227-016-1916-3
14. Zitzler, E., Thiele, L., Laumanns, M., Fonseca, C.M., da Fonseca, V.G.: Performance assessment of multiobjective optimizers: an analysis and review. IEEE Trans. Evol. Comput. **7**(2), 117–132 (2003)

A Memetic Algorithm for Community Detection in Bipartite Networks

Xiaodong Wang and Jing Liu[(✉)]

Key Laboratory of Intelligent Perception and Image Understanding of Ministry
of Education, Xidian University, Xi'an 710071, China
neouma@163.com

Abstract. Community detection is a basic tool to analyze complex networks. However, there are many community detection methods for unipartite networks while just a few methods for bipartite networks (BNs). In this paper, we propose a memetic algorithm (MACD-BNs) to identify communities in BNs. We use MACD-BNs to optimize two extended measures, namely Baber modularity (Q_B) and modularity density (Q_D), on real-life and synthetic networks respectively so as to compare their performance. We conclude that Q_D are more effective than Q_B when the size of communities is heterogeneous while Q_B is more suitable to detect communities with similar size. Besides, we also make a comparison between MACD-BNs and other community detection method and the results show the effectiveness of MACD-BNs.

Keywords: Community detection · Measures · Comparison · Bipartite networks

1 Introduction

Bipartite networks (BNs) are a common type of complex networks in real-life world [1, 2]. A bipartite network consists of two types of nodes and there are links only between different types of nodes and no links among the same type of nodes [3].

An important property of complex networks is the community structure [4–7] and existing community detection methods in BNs can be generally classified into two categories [8]. One is to project bipartite networks into unipartite networks [9] and then obtain partitions by applying some classical community detection algorithms for unipartite networks, such as *BGLL* [10], label propagation methods (LP) [11]. The other is to handle BNs directly. Community detection can be modeled as an optimization problem. One can obtain partitions by maximizing or minimizing a predefined measure, such as Baber modularity (Q_B) [12], modularity density (Q_D) [13]. Through utilizing the essential property of BNs, Baber developed a recursive method, BRIM, to maximize Q_B [12]. In order to identify communities in large BNs, Liu *et al.* introduced an algorithm named as LP&BRIM which is based on two existing algorithms, LP and BRIM [14].

According to different views to communities in BNs, we simply classify those measures into two categories. Some researchers believe that a community should contain two types of nodes. There are more links in intra-communities and fewer links in inter-communities. Q_B and Q_D are such examples. Others think that a community should only consist of one type of nodes and those nodes are collected to form a community due to similar link patterns (such as having common neighbors) [8]. Most

© Springer International Publishing AG 2017
D. Liu et al. (Eds.): ICONIP 2017, Part IV, LNCS 10637, pp. 89–99, 2017.
https://doi.org/10.1007/978-3-319-70093-9_10

modified modularity, including modularity proposed by Guimerà et al. [3], modularity proposed by Murata [15], modularity of Suzuki and Wakita [16] are such examples.

However, there are fewer algorithms identifying communities by optimizing above measures in BNs and few efforts have been done to make a thoroughly comparison of these measures.

What should be noted is that optimizing those measures is NP-hard problem and we then employ Memetic Algorithms (MAs). MAs proposed by Moscato et al. are hybrid algorithms with the combination of Evolution Algorithms and local search methods and can deliver high-quality solutions in early time of the process [17].

In order to solve the two mentioned problems, we first propose a community detection method based on MAs, named as MACD-BNs. Then, we use MACD-BNs to optimize Q_B and Q_D separately so as to compare their performance. We conclude that when the size of communities is heterogeneous, optimizing Q_D can lead to better partitions; when the size of communities is equal and close, Q_B is more suitable. Second, we validate MACD-BNs's performance on real-life and synthetic networks by comparing it with other method.

The rest of paper is organized as follows. Section 2 gives detailed descriptions of two measures and MACD-BNs. Experiments and conclusion are given in Sects. 3 and 4, respectively.

2 MACD-BNs

In this section, two measures, Q_B and Q_D, are first introduced. Then, we give detailed descriptions about MACD-BNs.

2.1 Two Measures

Given an un-weighted network G with N nodes and M edges, we use X to denote the set of X-type nodes, Y to denote the set of Y-type nodes, and then $V = X \cup Y$ is the set of all nodes. A is the corresponding adjacency matrix and A_{ij} equals 1 if there is a link between nodes i and j, and 0 otherwise. For convenience, we renumber all nodes so that nodes 1, 2, ..., p are X-type nodes and nodes $p + 1, p + 2, ..., p + q$ are Y-type nodes, where p and q denote the number of X-type and Y-type nodes, respectively. Given a partition $\{V_1, V_2, ..., V_r\}$ of G (r is the number of communities), it must satisfy the following conditions:

$$V_i = V_i^X \cup V_i^Y, 1 \leq i \leq r, V_i^X \neq \emptyset \text{ and } V_i^Y \neq \emptyset \tag{1}$$

$$V_1^X \cup V_2^X ... \cup V_r^X = X \tag{2}$$

$$V_1^Y \cup V_2^Y ... \cup V_r^Y = Y \tag{3}$$

where V_i^X (V_i^Y) is the set of X-type (Y-type) nodes in community V_i.

If we define $L(V_i^X, V_i^Y) = \sum_{i \in V_i^X} \sum_{j \in V_i^Y} A_{ij}$, then Q_B can be written as

$$Q_B = \frac{1}{M} \sum_{i=1}^{r} L(V_i^X, V_i^Y) - \frac{d(V_i^X) \times d(V_i^Y)}{M} \tag{4}$$

where M is the number of edges of G, and $d(V_i^X)$ $(d(V_i^Y))$ is the total degree of X-type (Y-type) nodes in community V_i.

Q_D is defined as

$$Q_D = \sum_{i=1}^{r} \frac{L(V_i^X, V_i^Y) - L(\overline{V_i^X}, V_i^Y) - L(V_i^X, \overline{V_i^Y})}{|V_i^X| \times |V_i^Y|} \qquad (5)$$

where $|V_i^X|(|V_i^Y|)$ is the number of X-type (Y-type) nodes in community V_i and $L(\overline{V_i^X}, V_i^Y) = \sum_{i \in \overline{V_i^X}} \sum_{j \in V_i^Y} A_{ij}$, where $\overline{V_i^X} = V - V_i^X$.

In the following experiments, we maximize Q_B and Q_D independently so as to obtain partitions in BNs.

2.2   Representation and Initialization

We use $\{g^1, g^2, ..., g^N\}$ to denote an individual g, where the i-th gene g^i also denotes a node of G. Therefore, if node i is X-type node, g^i takes values from $\{p + 1, p + 2, ..., p + q\}$; otherwise, g^i takes values from $\{1, 2, ..., p\}$. In the decoding step, nodes i and g^i are assigned into the same community. This is the locus-based representation and we can decode an individual into a partition in linear time [18].

An objective is decomposable only if it is equivalent to the sum of some sub-objectives. From Eqs. (1) and (2), we can easily find that both Q_B and Q_D are decomposable. For Q_B, the modularity of a community V_i is defined as

$$L_B(V_i) = \frac{L(V_i^X, V_i^Y)}{M} - \frac{d(V_i^X) \times d(V_i^Y)}{M^2} \qquad (6)$$

For Q_D, the modularity density of a community V_i is defined as

$$L_D(V_i) = \frac{L(V_i^X, V_i^Y) - L(\overline{V_i^X}, V_i^Y) - L(V_i^X, \overline{V_i^Y})}{|V_i^X| \times |V_i^Y|} \qquad (7)$$

Inspired by the above observations and some local community detection methods [4, 19], we introduce Algorithm 1 to initialize a population. The basic idea is to find a local community iteratively by maximizing the sub-objective until all nodes have their own communities. The process starts from randomly selecting an unassigned node a as the seed node and then moving an unassigned node b into a's community which can lead to the maximum and positive gain of sub-objective at each step. Note that we choose node b from node a's neighbors. We repeat this process until no node can be assigned into the community. A special case may occur that all of a's neighbors have their own community and no neighbors can be assigned into a's community. For this case, we move node a to the community which most of its neighbors belong to.

What should be noted is that if node a is moved into community V_i, the gain of each sub-objective is easily calculated if we cache some intermediate variables. For Q_B, if node a is an X-type node, the gain of $L_B(V_i)$ is

$$\Delta L_B(V_i \cup a) = \frac{d_a^{in}}{M} - \frac{d_a \times d(V_i^Y)}{M^2} \tag{8}$$

otherwise,

$$\Delta L_B(V_i \cup a) = \frac{d_a^{in}}{M} - \frac{d_a \times d(V_i^X)}{M^2} \tag{9}$$

where d_a is the degree of node a and d_a^{in} is the number of neighbors of node a which are also in community V_i.

For Q_D, if node a is an *X-type* node, the gain of $L_D(V_i)$ is

$$\Delta L_D(V_i \cup a) = \frac{3 \times d_a^{in} - d_a}{(|V_i^X| + 1) \times |V_i^Y|} - \frac{L_D(V_i)}{|V_i^X| + 1} \tag{10}$$

otherwise,

$$\Delta L_D(V_i \cup a) = \frac{3 \times d_a^{in} - d_a}{|V_i^X| \times (|V_i^Y| + 1)} - \frac{L_D(V_i)}{|V_i^Y| + 1} \tag{11}$$

Algorithm 1: Initialization

Input:
  G: A bipartite network;
Output:
  A population P;

Step1: Randomly select an unassigned node a and assign it into a new and empty community C; Set $L(C)$ to $-\infty$ ($L(C)$ refers to $L_B(V_i)$ or $L_D(V_i)$);
Step2: Pick node b from a's unassigned neighbors and move it into community C, which leads to the maximum positive gain of $L(C)$; If there is no node meeting this condition, go to **Step 4**; otherwise, go to **Step 3**;
Step3: Update $L(C)$ and C, and let a:=b;
Step4: Repeat **Steps 2~3** until the size of community C is not increased at all;
Step5: Repeat **Steps 1~4** until all nodes are assigned.

2.3   Crossover and Mutation

As for the crossover operation, we use a hybrid strategy. If $u(0,1) < 0.5$ ($u(0,1)$ is employed to generate a value between 0 and 1), we employ the uniform crossover operator [17]; otherwise, we use the two-point crossover operator [18].

A well-designed mutation operator can help the algorithm skip the local optimum. Given an individual g, we perform mutation operation on some randomly selected nodes. For each selected node a, we randomly choose one of its neighbor b and then conduct the following operation

$$g^a = \begin{cases} g^{g^a} & if(u(0,1) < 0.5) \\ b & otherwise \end{cases} \tag{12}$$

2.4   Local Search Strategy

The local search strategy plays an important role in MAs and a well-designed local search operator can accelerate the algorithm. The basic idea is to move a node to one of its neighbors' community which can lead to the largest positive gain of the objective at each step; otherwise, to keep a node in its original community. This operation is applied iteratively on every node until no gain of objective can be obtained. Detailed procedures are presented in Algorithm 2. The idea from *BGLL* is based on the following observations.

For Q_B, if we move node a from community V_1 to V_2 and a is a *X-type* node, the gain is

$$\Delta Q_X = \frac{1}{M}(d_a^{in}(V_2) - d_a^{in}(V_1)) + \frac{1}{M^2} \times (d(V_1^Y) - d(V_2^Y)) \times d_a \qquad (13)$$

otherwise, if node a is *Y-type* and the gain is

$$\Delta Q_Y = \frac{1}{M}(d_a^{in}(V_2) - d_a^{in}(V_1)) + \frac{1}{M^2} \times (d(V_1^X) - d(V_2^X)) \times d_a \qquad (14)$$

where $d_a^{in}(V_2)$ $(d_a^{in}(V_1))$ is the number of neighbors of node a which are also in community V_2 (V_1).

For Q_D, if we move node a from community V_1 to V_2 and a is a *X-type* node, the gain is

$$\Delta D_X = \frac{L_D(V_1)}{|V_1^X| - 1} - \frac{L_D(V_2)}{|V_2^X| + 1} + \frac{3 \times d_a^{in}(V_2) - d_a}{(|V_2^X| + 1) \times |V_2^Y|} - \frac{3 \times d_a^{in}(V_1) - d_a}{(|V_1^X| - 1) \times |V_1^Y|} \qquad (15)$$

otherwise, if a is *Y-type*, the gain is

$$\Delta D_Y = \frac{L_D(V_1)}{|V_1^Y| - 1} - \frac{L_D(V_2)}{|V_2^Y| + 1} + \frac{3 \times d_a^{in}(V_2) - d_a}{(|V_2^Y| + 1) \times |V_2^X|} - \frac{3 \times d_a^{in}(V_1) - d_a}{(|V_1^Y| - 1) \times |V_1^X|} \qquad (16)$$

where $L_D(V_1)$ $(L_D(V_2))$ is defined in Eq. (7).

From the above equations, we can find that if we cache some intermediate variables, the gain of two objectives can easily be calculated. What should be noted is that the number of communities should be kept unchanged in the process, because $|V_1^X|$ $(|V_2^X|, |V_1^Y|$ and $|V_2^Y|)$ should be greater than 1. If not, some intermediate variables should be calculated again, which increases the computational complexity.

Algorithm 2: The local search strategy

Input:
 A partition g of the bipartite network;
Output:
 An improved partition;

Step1: Generate a random integer sequence from 1 to n and store the result in the array **Order**;
Step2: For each node a in the **Order**, we move it from community V_i to V_j ($i!=j$ and node a is in community V_i, V_j is the community which one of a's neighbor nodes belong to) only if this operation can lead to the maximum and positive gain of the objective and the number of *X-type* or *Y-type* nodes in community V_i is greater than 1;
Step3: Repeat **Steps1~2** until there is no positive gain of the objective can be obtained.

2.5   Framework of MACD-BNs

Algorithm 3 summarizes the details of MACD-BNs. After getting a population in the procedure of initialization, the algorithm enters into the main loop. We update the population by the operator of selection, crossover, mutation, local search, and then repeat the procedure until the number of iterations reaches the upper limit.

Note that if the best individual is not changed in the next two iterations, we perform the mutation operation on the best individual Q_{best} before executing the local strategy. If the output partition y of local search strategy is better than Q_{best}, we replace Q_{best} with y; otherwise, we kept Q_{best} unchanged.

Algorithm 3: MACD-BNs

Input:
    G: A bipartite network;
    Popsize: The size of population;
    T: The number of iterations;
    P_c: Crossover rate;
    P_m: Mutation rate;
    γ: The proportion of reserved elites;
Output:
    A partition of G with the maximum objective;

$P \leftarrow$ Initialization (*Popsize*);
Evaluate the objective of P and sort them in descending order;
for $i = 1$ **to** T **do**
    Find the best individual and term it as P_{best};
    $Q \leftarrow \varnothing$;
    $Q \leftarrow$ ReserveElite(P, γ); /*Reserve the best *popsize*$*\gamma$ individuals */
    $P_{parent} \leftarrow$ Select(P); /*Randomly select (*popsize* - *popsize*$*\gamma$) individuals by employing the tournament selection*/
    for $j = \gamma \times Popsize +1$ **to** $Popsize$ **do**
        Randomly select two individuals g_1 and g_2 from P_{parent};
        if (g_1 is identical to g_2) **then**
            $y \leftarrow$ Mutation(g_1, P_m);
        else
            if($u(0,1)<P_c$) **then** /* $u(0,1)$ randomly generates a value between 0 and 1*/
                $y \leftarrow$ Crossover(g_1, g_2);
            else
                $y \leftarrow$ Mutation(g_1, P_m);
            end if;
        end if;
        $Q \leftarrow Q \cup y$;
    end for;
    Evaluate each individual in Q and sort them in descending order according to their objective values;
    Find the best individual and term it as Q_{best};
    if (Q_{best} is identical to P_{best}) **then**
        $y \leftarrow$ Mutation(Q_{best}, P_m);
        $y \leftarrow$ LocalSearch(y);
        if(y has a higher objective value than Q_{best}) **then**
            $Q_{best} \leftarrow y$;
        end if;
    else
        $Q_{best} \leftarrow$ LocalSearch(Q_{best});
    end if;
    $P \leftarrow Q$;
end for;

---

3 Experiments

In this section, MACD-BNs is applied to some real-life and synthetic networks and the corresponding parameters setting are presented in Table 1. The normalized mutual information (*NMI*) [20] can measure the similarity between true partitions and detected ones. Usually, a higher value of *NMI* corresponds to a better partition. In the following experiments, we employ *NMI* to evaluate the quality of obtained partitions.

Table 1. The parameter setting.

Parameters	Meaning	Values
Popsize	The size of population	100
T	The number of iterations	80
P_c	Crossover probability	0.7
P_m	Mutation probability	0.3
γ	The proportion of reserved elites	0.1

3.1 Experiments on a Real-Life Network

The Southern Women network [8–10] is a benchmark to test the effectiveness of community detection methods in BNs because it has the real partition on its *X-type* nodes. Each algorithm is executed 100 times independently and for each obtained partition, we calculate the corresponding values of Q_B, Q_D and NMI_X respectively. Then, for each method, we calculate the average and standard deviation values of each criterion over 100 times and present the results in Table 2.

Table 2. Results obtained by different algorithms.

Methods	Q_B	Q_D	NMI_X	Best values[a]	Best NMI[b]
MACD-BNs(Q_B)	0.3454(0.0001)	−2.6098(1.2578)	0.4886(0.0441)	0.3455	0.4513
MACD-BNs(Q_D)	0.3184(0)	0.6759(0)	1(0)	0.6759	1
LP&BRIM	0.3199(0.0022)	0.5374(0.6906)	0.6405(0.0765)	0.3305	0.7432

[a]*Best Value* denotes the maximum value of each objective obtained by each algorithm (MACD-BNs(Q_B) and LP&BRIM acquire partitions by maximizing Q_B while MACD-BNs(Q_D) acquires partitions by maximizing Q_D);
[b]*Best NMI* denotes the maximum *NMI* value obtained by each algorithm.

Table 2 shows that MACD-BNs is much more stable than LP&BRIM with respect to the standard deviation of NMI_X. The first stage of LP&BRIM is to perform the LP algorithm on a bipartite network and the fact that LP is unstable [14] leads to the disturbance of LP&BRIM. Besides, we can find that MACD-BNs (Q_D) acquires the true partition ($NMI_X = 1$), which shows that optimizing Q_D can get better results than optimizing Q_B on the Southern Women network. Both LP&BRIM and MACD-BNs (Q_B) acquire partitions by maximizing Q_B, but MACD-BNs(Q_B) can achieve higher values of Q_B than LP&BRIM does, which reflects the effectiveness of MACD-BNs.

3.2   Experiments on Synthetic Networks

Two types of synthetic networks, easy and difficult cases, are generated by block model methods [8] to test MACD-BNs' performance. Note that the generated networks may contain some isolated nodes (a node having no neighbors is called the isolated node) and we remove those isolated nodes when computing the value of *NMI*.

For convenience, we introduce two new symbols: (K_X, K_Y) and λ. We use (K_X, K_Y) to denote a community K where K_X (K_Y) denotes the number of *X-type* (*Y-type*) nodes in community K and λ represents the mixing parameter. In the following experiments, we respectively set N, p, q and r to 2000, 1000, 1000 and 4 for the two types of networks (N, p, q and r respectively represent the size of network, the number of *X-type* nodes, the number of *Y-type* nodes and the number of communities in the network). Besides, λ varies from 0 to 1 at the interval of 0.05 and the higher the value of λ, the clear the community structure is in BNs.

The difference between two types of networks is the size of each community. For the easy cases, all nodes are divided into four communities with equal size and each community contains 250 *X-type* and *Y-type* nodes. For the difficult cases, all nodes are partitioned into four communities with different sizes, namely [(50, 300), (300, 150), (250, 350), (400, 200)]. All algorithms run 30 times on each network and each objective. We calculate the average *NMI* of each obtained partition and plot the results in Figs. 1 and 2.

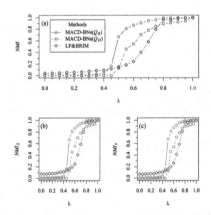

Fig. 1. The comparison of three algorithms with respect to *NMI* in easy cases where the size of communities is equal. (a), (b) and (c) denote the value of *NMI* between the true partitions and detected ones on all nodes, *X-type* nodes and *Y-type* nodes respectively.

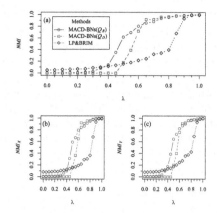

Fig. 2. The comparison of three algorithms with respect to *NMI* in difficult cases where the size of communities is heterogeneous. (a), (b) and (c) denote *NMI* values between the true partitions and detected ones on all nodes, *X-type* nodes and *Y-type* nodes respectively.

As can be seen from Fig. 1, the difference of three algorithms becomes clear when $0.5 < \lambda$ on the easy cases. When $0.5 \leq \lambda \leq 0.8$, both MACD-BNs($Q_B$) and MACD-BNs($Q_D$) obtain better results than LP&BRIM; when $0.8 < \lambda$, the results obtained by LP&BRIM and MACD-BNs(Q_B) are similar and close and both of the two algorithms achieve a little higher value of *NMI* than MACD-BNs(Q_D). Besides, in general, the performance of MACD-BNs(Q_D) is better than MACD-BNs(Q_B), which indicates Q_B is more suitable to identify communities with equal size than Q_D in BNs.

Figure 2 shows that MACD-BNs(Q_B) has better outputs than MACD-BNs(Q_D). When $0.5 \leq \lambda < 0.65$ and $0.65 \leq \lambda \leq 0.9$, MACD-BNs($Q_D$) has better performance. Roughly speaking, optimizing Q_D can lead to better results when the size of communities is different. Besides, LP&BRIM obtains close results with two other algorithms only when $0.9 \leq \lambda$. Therefore, both MACD-BNs(Q_B) and MACD-BNs(Q_D) have better performance compared with LP&BRIM in difficult cases.

In summary, Q_D is suitable to identify communities with heterogeneous size while Q_B is suitable to detect communities with equal or close size. On the other hand, MACD-BNs has higher quality outputs than LP&BRIM.

4   Conclusions

In this article, we propose a memetic algorithm, MACD-BNs, for identifying communities in BNs. We employ MACD-BNs to maximize two measures separately so as to compare their performance. Experiments on real-life and synthetic networks show that Q_B is suitable to detect communities with similar size while Q_D is more effective when the size of communities is heterogeneous. Moreover, we also compare MACD-BNs with other community detections methods in BNs so as to show the effectiveness of MACD-BNs.

However, MAs are criticized mainly for the speed and MACD-BNs is not scalable when the size of networks is too large. Besides, there are fewer researches about practical applications of community detection than those about community discovery methods. Therefore, our further works will focus on both aspects.

Acknowledgement. This work is partially supported by the Outstanding Young Scholar Program of National Natural Science Foundation of China (NSFC) under Grant 61522311, the Overseas, Hong Kong & Macao Scholars Collaborated Research Program of NSFC under Grant 61528205, and the Key Program of Fundamental Research Project of Natural Science of Shaanxi Province, China under Grant 2017JZ017.

References

1. Newman, M.E.: Networks: An Introduction. Oxford University Press, Oxford (2011)
2. Zha, H., He, X., Ding, C., Simon, H., Gu, M.: Bipartite graph partitioning and data clustering. Office of Scientific and Technical Information Technical Reports, pp. 25–32 (2001)
3. Guimerà, R., Sales-Pardo, M., Amaral, L.A.N.: Module identification in bipartite and directed networks. Phys. Rev. E **76**(3), 036102 (2007)
4. Fortunato, S.: Community detection in graphs. Phys. Rep. **486**(3), 75–174 (2009)
5. Newman, M.E., Girvan, M.: Finding and evaluating community structure in networks. Phys. Rev. E **69**(2), 026113 (2004)
6. Wang, X., Liu, J.: A layer reduction based community detection algorithm on multiplex networks. Phys. A **471**, 244–252 (2017)
7. Jiang, Z., Liu, J., Wang, S.: Traveling salesman problems with PageRank Distance on complex networks reveal community structure. Phys. A **463**, 293–302 (2016)
8. Larremore, D.B., Clauset, A., Jacobs, A.Z.: Efficiently inferring community structure in bipartite networks. Phys. Rev. E **90**(1), 012805 (2014)
9. Zhou, T., Ren, J., Medo, M., Zhang, Y.C.: Bipartite network projection and personal recommendation. Phys. Rev. E **76**(4), 046115 (2007)
10. Blondel, V.D., Guillaume, J.L., Lambiotte, R., Etienne, L.: Fast unfolding of communities in large networks. J. Stat. Mech: Theory Exp. **2008**(10), 10008 (2008)
11. Raghavan, U.N., Albert, R., Kumara, S.: Near linear time algorithm to detect community structures in large-scale networks. Phys. Rev. E **76**(3), 036106 (2007)
12. Barber, M.J.: Modularity and community detection in bipartite networks. Phys. Rev. E **76**(6), 066102 (2007)
13. Xu, Y., Chen, L., Li, B.: Density-based modularity for evaluating community structure in bipartite networks. Inf. Sci. **317**, 278–294 (2015)
14. Liu, X., Murata, T.: Community detection in large-scale bipartite networks, In: IEEE/WIC/ACM International Joint Conferences on Web Intelligence and Intelligent Agent Technologies, WI-IAT 2009, vol. 5, pp. 50–57 (2009)
15. Murata, T.: Modularities for bipartite networks. In: Proceedings of the 20th ACM conference on Hypertext and hypermedia, Italy, pp. 245–250 (2009)
16. Suzuki, K., Wakita, K.: Extracting multi-facet community structure from bipartite networks. In: International Conference on Computational Science and Engineering, CSE 2009, vol. 4, pp. 312–319 (2009)
17. Gong, M., Fu, B., Jiao, L.: Memetic algorithm for community detection in networks. Phys. Rev. E **84**(5), 056101 (2011)

18. Pizzuti, C.: A multiobjective genetic algorithm to find communities in complex networks. IEEE Trans. Evol. Comput. **16**(3), 418–430 (2012)
19. Clauset, A.: Finding local community structure in networks. Phys. Rev. E **72**(2), 026132 (2005)
20. Danon, L., Diaz-Guilera, A., Duch, J., Arenas, A.: Comparing community structure identification. J. Stat. Mech: Theory Exp. **2005**(09), 09008 (2005)

Complex-Valued Feedforward Neural Networks Learning Without Backpropagation

Wei Guo[1], He Huang[1(✉)], and Tingwen Huang[2]

[1] School of Electronics and Information Engineering, Soochow University,
Suzhou 215006, People's Republic of China
847697387@qq.com, cshhuang@gmail.com
[2] Texas A&M University at Qatar, Doha 5825, Qatar

Abstract. This paper presents an efficient learning algorithm for complex-valued feedforward neural networks with application to classification problems. It simplifies complex-valued neural networks learning by using the forward-only computation rather than traditional forward and backward computations. By incorporating the forward-only computation, the complex-valued Levenberg-Marquardt algorithm becomes more efficient. Comparison results of computation cost show that the proposed forward-only complex-valued learning algorithm can be faster than the traditional implementation of the Levenberg-Marquardt algorithm.

Keywords: Complex-valued neural networks · Forward-only computation · Levenberg-Marquardt algorithm · Computation cost · Classification

1 Introduction

Artificial neural networks (ANNs) have been proven as very efficient models in many applications, such as system identification and control [1], pattern recognition [2] and digital communications [3, 4]. Almost all supervised learning algorithms for ANNs originate from numerical optimization theory. The very efficient second-order Levenberg-Marquardt (LM) algorithm is one of the most widely used approaches. It combines the excellent convergence property of a second-order Newton method [5–9] near a solution with the consistently decreasing property of the first-order gradient descent when the state is far away from the solution. The Jacobian matrix is a key element in the LM algorithm, which is defined as the derivative of the output error vector with respect to the adjustable parameters [10]. With the increase of number of neurons, the Jacobian matrix becomes very large such that the speed advantage of the LM algorithm over the first-order error backpropagation (EBP) [11] algorithm is less evident.

In recent years, complex-valued signals are becoming ubiquitous in many areas. For example, some signals are naturally expressed in complex-valued form in the signal processing domain. Consequently, complex-valued neural networks appear as a natural choice for tackling problems such as channel equalization and time series prediction [12]. The complex-valued LM (CLM) algorithm is one of the most known and used methods to train complex-valued neural networks [14–18]. By incorporating

© Springer International Publishing AG 2017
D. Liu et al. (Eds.): ICONIP 2017, Part IV, LNCS 10637, pp. 100–107, 2017.
https://doi.org/10.1007/978-3-319-70093-9_11

the forward-only computation, the method proposed in [13] simplified real-valued neural networks training by using the forward-only computation instead of traditional forward and backward computation. As a result, it is much easier to implement with less computation cost.

Inspired by [13], in this paper, our attention focuses on presenting a learning algorithm for complex-valued feedforward neural networks (CVFNN). To improve the computational efficiency of the CLM algorithm, the proposed algorithm incorporates the method used in [13] into the CLM algorithm. During the implementation process, the calculation of Jacobian matrix only uses the forward computation, rather than the traditional combination of forward and backward computations. An advantage of the proposed algorithm is that the whole learning process requires only forward computation, which would greatly reduce the computational cost. The theoretical analysis and experimental results demonstrate that the proposed algorithm are faster than the traditional CLM algorithm [21].

2 Traditional Complex-Valued Backpropagation Algorithm

2.1 Description of Complex-Valued Feedforward Neural Networks

The CVFNN considered in this study consists of three layers an input layer, a hidden layer and an output layer. For a complex-valued input z, the hidden layer of the CVFNN is calculated by

$$h = f(W^{lh}z + a) \tag{1}$$

where $z = [z_1, z_2, \ldots, z_L]^T$ is an L dimensional complex-valued input vector, $a = [a_1, a_2, \ldots, a_M]^T$ is an M dimensional complex-valued hidden bias vector, $W^{lh} = [w_{i,j}^{lh}]$ is the complex-valued weight matrix from the input hidden layers with $i = 1, 2, \ldots, L$; $j = 1, 2 \ldots, M$, and $h = [h_1, h_2, \ldots, h_M]^T$ is an M dimensional complex-valued output vector of the hidden layer.

The output of the CVFNN is calculated by

$$y = f(W^{ho}h + b) \tag{2}$$

where $W^{ho} = [w_{i,j}^{ho}], i = 1, 2, \ldots, M; j = 1, 2 \ldots, N$ is the complex-valued output weight matrix from the hidden layer to the output layer, $b = [b_1, b_2, \ldots, b_N]^T$ is an N dimensional complex-valued output bias vector, $y = [y_1, y_2, \cdots, y_N]^T$ is an N dimensional complex-valued output vector.

For simplicity, it is denoted that the net input of the hidden neuron m by

$$net_m^{lh} = \sum_{l=1}^{L} w_{l,m}^{lh} z_l + a_m \tag{3}$$

where z_l is the l th input signal, $w_{l,m}^{Ih}$ is the complex-valued input weight from the l th input signal to the hidden neuron m, and a_m is the bias of the hidden neuron m.

The net input node of the output neuron n is calculated by

$$net_n^{ho} = \sum_{m=1}^{M} w_{m,n}^{ho} h_m + b_n \tag{4}$$

where h_m is the output of the hidden neuron m, $w_{m,n}^{ho}$ is the complex-valued output weight from the hidden neuron m to the output neuron n, and b_n is the bias of the output neuron n.

The objective function used in this study is the sum square error (SSE) E. For all patterns and outputs, it is calculated by

$$E = \frac{1}{2} \sum_{p=1}^{P} \sum_{n=1}^{N} e_{n,p}^* e_{n,p} = \frac{1}{2} e^H e \tag{5}$$

where $(.)^*$ is the conjugate of $(.)$, H represents Hermitian transpose operation, and $e_{n\,p}$ is the error of the output neuron n for the training pattern p which is defined as

$$e_{n,p} = d_{n,p} - y_{n,p} \tag{6}$$

where $d_{n,p}$ and $y_{n,p}$ are respectively the desired output and actual output of the output neuron n for the training pattern p. The training target of the CVFNN is to minimize the objective function E by adjusting all the involved parameters.

2.2   Complex-Valued Forward-Only Computation

Wirtinger formally developed the R-derivative and the conjugate R-derivative [20]. In the traditional CLM algorithm, the elements of Jacobian matrix are calculated in [20].

The method proposed in this paper aims to improve the efficiency of the computation of Jacobian matrix by removing the backpropagation process.

The notations of $\delta_{k,j}$ and $\delta_{k,j}^*$ can be introduced as the signal gain between neurons j and k, which are respectively given by

$$\delta_{k,j} = \frac{\partial F_{k,j}}{\partial net_j}, \qquad \delta_{k,j}^* = \frac{\partial F_{k,j}}{\partial net_j^*} \tag{7}$$

where $F_{k,j}$ is the output signal of the neuron k, net_j is the input signal of the neuron j.

For specificity, the notations of $\delta_{k,j}^h$ and $\delta_{k,j}^{h*}$ are introduced as the signal gain between the hidden neurons j and k, which are respectively given by

$$\delta_{k,j}^h = \frac{\partial h_k}{\partial net_j^{Ih}}, \qquad \delta_{k,j}^{h*} = \frac{\partial h_k}{\partial net_j^{Ih*}} \tag{8}$$

The notations of $\delta^o_{k,j}$ and $\delta^{o*}_{k,j}$ are adopted as the signal gain between the output neurons j and k, which are respectively given by

$$\delta^o_{k,j} = \frac{\partial y_k}{\partial net^{ho}_j}, \qquad \delta^{o*}_{k,j} = \frac{\partial y_k}{\partial net^{ho*}_j} \tag{9}$$

Furthermore, the notations of $\delta^{ho}_{k,j}$ and $\delta^{ho*}_{k,j}$ are adopted as the signal gain between the hidden neuron j and the output neuron k, which are respectively calculated by

$$\delta^{ho}_{k,j} = \frac{\partial y_k}{\partial net^{Ih}_j}, \qquad \delta^{ho*}_{k,j} = \frac{\partial y_k}{\partial net^{Ih*}_j} \tag{10}$$

Then, (7–14) can be respectively expressed by

$$\frac{\partial e_{n,p}}{\partial e_n} = -\frac{\partial y_{n,p}}{\partial net^{ho}_n} = -\delta^o_{n,n} \tag{11}$$

$$\frac{\partial e_{n,p}}{\partial b^*_n} = -\frac{\partial y_{n,p}}{\partial net^{ho*}_n} = -\delta^{o*}_{n,n} \tag{12}$$

$$\frac{\partial e_{n,p}}{\partial w^{ho}_{n,m}} = -\frac{\partial y_{n,p}}{\partial net^{ho}_n} h_m = -\delta^o_{n,n} h_m \tag{13}$$

$$\frac{\partial e_{n,p}}{\partial w^{ho*}_{n,m}} = -\frac{\partial y_{n,p}}{\partial net^{ho*}_n} h^*_m = -\delta^{o*}_{n,n} h^*_m \tag{14}$$

$$\frac{\partial e_{n,p}}{\partial a_m} = -\left(\delta^o_{nn} w^{ho}_{m,n} \delta^h_{mm} + \delta^{o*}_{nn} w^{ho*}_{m,n} \left(\delta^{h*}_{mm} \right)^* \right) = -\delta^{ho}_{mn} \tag{15}$$

$$\frac{\partial e_{n,p}}{\partial a^*_m} = -\left(\delta^o_{nn} w^{ho}_{m,n} \delta^{h*}_{mm} + \delta^{o*}_{nn} w^{ho*}_{m,n} \left(\delta^{h}_{mm} \right)^* \right) = -\delta^{ho*}_{mn} \tag{16}$$

$$\frac{\partial e_{n,p}}{\partial w^{Ih}_{l,m}} = -\left(\delta^o_{nn} w^{ho}_{m,n} \delta^h_{mm} z_l + \delta^{o*}_{nn} w^{ho*}_{m,n} \left(\delta^{h*}_{mm} \right)^* z_l \right) = -\delta^{ho}_{mn} z_l \tag{17}$$

$$\frac{\partial e_{n,p}}{\partial w^{Ih*}_{l,m}} = -\left(\delta^o_{nn} w^{ho}_{m,n} \delta^{h*}_{mm} z^*_l + \delta^{o*}_{nn} w^{ho*}_{m,n} \left(\delta^{h}_{mm} \right)^* z^*_l \right) = -\delta^{ho*}_{mn} z^*_l \tag{18}$$

From the above equations, it can be observed that all computation can be completed by the complex-valued forward-only computing process. During the implementation process, the calculation of Jacobian matrix only uses the forward computation, rather than the traditional combination of forward and backward computations. The signal gains between neurons improve the efficiency of Jacobian matrix computation.

3 Comparisons Between Traditional and the Proposed Algorithms

3.1 Theoretical Analysis

The traditional CLM algorithm needs the forward and backward computations. While, the proposed complex-valued forward-only computation removes the backward part by introducing an additional calculation in the forward computation. Here, the computational costs of the traditional and forward-only CLM algorithms are compared. From the theoretical analysis in Table 1, it can be seen that the proposed algorithm is faster than the traditional CLM algorithm.

Table 1. Analysis of computational cost between the traditional and proposed CLM Algorithms

	Traditional CLM algorithm	
	Forward part	Backward part
+/−	$3l \times m + 7m + 3m \times n + 7n$	$6m \times n \times (8m \times l + 8m + 1) + 4m + 4n$
×/÷	$4l \times m + 3m + 4m \times n + 3n$	$8m \times n \times (8m \times l + 8m + 1) + 4m + 4n$
Exp*	$2m + 2n$	0
	Forward-only CLM algorithm	
	Forward Part	Backward Part
+/−	$9l \times m + 19m + 9m \times n + 11n$	0
×/÷	$12l \times m + 15m + 12m \times n + 7n$	0
Exp*	$2m + 2n$	0
	Subtraction forward-only from traditional	
+/−	$8l \times m \times 6m \times n + 8m \times 6m \times n - 6l \times m - 8m$	
×/÷	$8l \times m \times 8m \times n + 8m \times 8m \times n - 8l \times m - 8m$	
Exp*	0	

*Exponential operation

3.2 Introduction of Data Sets

This section shows the performance of the proposed forward-only CLM algorithm compared with traditional CLM algorithm. Real-world classification problems from the UCI database, summarized in Table 2, are used to verify that the proposed algorithm is more efficient. As in [19], the data sets are firstly transferred from the real domain to the complex domain. The corresponding complex-valued attribute is obtained by using Euler formula

$$z = e^{i\phi} = \cos \phi + i \sin \phi \tag{19}$$

where a linear transformation $\phi = \pi \frac{x-a}{b-a}$ changes $x \in [a, b]$ to $\phi \in [0, \pi]$.

Table 2. Specification of classification benchmark problems

Datasets	Attributes	Training data	Testing data
Ionosphere	34	281	70
Biodegradation	41	844	211
Breast cancer	10	547	136
Liver disorders	6	278	69
Heart	13	216	54
PIMA	8	615	153
Sonar	60	167	41
Spambase	57	3681	920

3.3   Experimental Results and Analysis

The experiment results are shown in Tables 3 and 4. From the error rate comparison of forward-only complex-valued and real-valued LM algorithms in Table 3, the forward-only CLM algorithm reduces the error rates of the seven datasets. Table 3 shows that the forward-only CLM algorithm achieves more stable results and better variance reduction. Table 4 shows the comparison of the required computation time of

Table 3. Comparison of the error rate of forward-only real-valued and complex-valued computation

Dataset	Number of hidden nodes	Real-valued		Complex-valued	
		Error rate	Std	Error rate	Std
Ionosphere	30	0.1082	7.2	0.0826	3.4
Biodegradation	30	0.1289	3.4	0.1251	3.0
Breast cancer	30	0.0483	1.8	0.0424	1.0
Liver disorders	30	0.2898	7.8	0.2724	4.0
Heart	30	0.2407	6.2	0.2222	7.4
PIMA	30	0.2942	6.3	0.2747	4.6
Sonar	30	0.1826	4.6	0.1730	4.5
Spambase	30	0.0941	2.2	0.1019	0.8

Table 4. Comparison of the required computation time of traditional and forward-only CLM algorithms

Datasets	Number of hidden nodes	Computation time(s)	
		Traditional	Forward-only
Gamma	30	2960.58	2827.49
Spambase	30	12624.8	12294.3
Biodegradation	30	2356.25	2212.56
Biodegradation	35	3084.16	2907.29
Biodegradation	40	3842.27	3742.80

traditional and forward-only CLM algorithms. From Table 4, the forward-only CLM algorithm reduces the computation time and has a considerably improved efficiency. Theoretically, as the dataset become larger and the number of hidden neurons increases, the computational efficiency of the CLM algorithm becomes more pronounced.

4 Conclusion

An efficient computation of the CLM algorithm has been proposed in this paper for training CVFNNs. The theoretical analysis and experimental results show that the proposed forward-only complex-valued computation is more efficient and faster than traditional forward and backward computations. As the dataset is large and the number of hidden neurons increases, the proposed computation of the CLM algorithm can be much simpler and more efficient than traditional CLM algorithm.

Acknowledgements. This work was jointly supported by the National Natural Science Foundation of China under Grant nos. 61273122 and 61005047, and the Qing Lan Project of Jiangsu Province. This publication was made possible by NPRP grant: NPRP 8-274-2-107 from the Qatar National Research Fund (a member of Qatar Foundation). The statements made herein are solely the responsibility of the author[s].

References

1. Narendra, K.S., Parthasarathy, K.: Identification and control of dynamical systems using neural networks. IEEE Trans. Neural Netw. **1**, 4–27 (1990)
2. Kamimura, R., Konstantinov, K., Stephanopoulos, G.: Knowledge-based systems, artificial neural networks and pattern recognition: applications to biotechnological processes. Curr. Opin. Biotechnol. **7**, 231–234 (1996)
3. Ibnkahla, M.: Neural network predistortion technique for digital satellite communications. In: IEEE International Conference on Acoustics, pp. 3506–3509 (2000)
4. Birgmeier, M.: A digital communication channel equalizer using a Kalman-trained neural network. In: IEEE World Congress on IEEE International Conference on Neural Networks, pp. 3921–3925 (1995)
5. Ampazis, N., Perantonis, S.J.: Two highly efficient second-order algorithms for training feedforward networks. IEEE Trans. Neural Netw. **13**, 1064–1074 (2002)
6. Kim, C.T., Lee, J.J.: Training two-layered feedforward networks with variable projection method. IEEE Trans. Neural Netw. **19**, 371–375 (2008)
7. Toledo, A., Pinzolas, M., Ibarrola, J.J., Lera, G.: Improvement of the neighborhood based Levenberg-Marquardt algorithm by local adaptation of the learning coefficient. IEEE Trans. Neural Netw. **16**, 988–992 (2005)
8. Wu, J.M.: Multilayer Potts perceptrons with Levenberg-Marquardt learning. IEEE Trans. Neural Netw. **19**, 2032–2043 (2008)
9. Wilamowski, B.M.: Neural network architectures and learning algorithms: how not to be frustrated with neural networks. IEEE Ind. Electron. Mag. **3**, 56–63 (2009)
10. Hagan, M.T., Menhaj, M.B.: Training feedforward networks with the Marquardt algorithm. IEEE Trans. Neural Netw. **5**, 989–993 (1994)

11. Werbos, P.J.: Back-propagation: past and future. In: Proceedings of the IEEE International Conference on Neural Networks, pp. 343–353 (1988)
12. Hirose, A.: Complex-Valued Neural Networks: Advances and Applications. Wiley, Hoboken (2013)
13. Wilamowski, B.M., Yu, H.: Neural network learning without backpropagation. IEEE Trans. Neural Netw. **21**, 1793–1802 (2010)
14. Deng, J., Sundararajan, N., Saratchandran, P.: Communication channel equalization using complex-valued minimal radial basis function neural networks. IEEE Trans. Neural Netw. **13**, 687–696 (2014)
15. Valle, M.E.: Complex-valued recurrent correlation neural networks. IEEE Trans. Neural Netw. Learn. Syst. **25**, 1600–1612 (2014)
16. Wang, H., Duan, S., Huang, T., Wang, L., Li, C.: Exponential stability of complex-valued memristive recurrent neural networks. IEEE Trans. Neural Netw. Learn. Syst. **28**, 1–6 (2016)
17. Li, H., Adali, T.: Complex-valued adaptive signal processing using nonlinear functions. EURASIP J. Adv. Sig. Process. **2008**, 1–9 (2008)
18. Huang, T., Li, C., Yu, W.: Synchronization of delayed chaotic systems with parameter mismatches by using intermittent linear state feedback. Nonlinearity **22**, 569–584 (2009)
19. Amin, M.F.: Complex-valued neural networks: learning algorithms and applications (2012)
20. Van, A., Bos, D.: Complex gradient and Hessian. IEE Proc. Vis. Image Sig. Process. **141**, 380–383 (1994)
21. Huang, T., Li, C., Duan, S.: Robust exponential stability of uncertain delayed neural networks with stochastic perturbation and impulse effects. IEEE Trans. Neural Netw. Learn. Syst. **23**, 866–875 (2012)

Distributed Recurrent Neural Network Learning via Metropolis-Weights Consensus

Najla Slama$^{(\boxtimes)}$, Walid Elloumi, and Adel M. Alimi

REGIM-Lab: REsearch Groups in Intelligent Machines, National Engineering
School of Sfax (ENIS), University of Sfax, BP 1173, Sfax 3038, Tunisia
{Najla.slama.tn,walid.elloumi,adel.alimi}@ieee.org

Abstract. When data are shared among arbitrarily connected machines, the training process became an interesting challenge where each node is initialized with a specific scalar value, so it present a problem of computing their average taking into account interconnectivity between agents in order to ensure that the objective process converges as the centralized counterpart, the decentralized average consensus (DAC) is the most popular strategy due to its low-complexity. In this paper a random topology is choosing to validate a network of agents with a given probability of interconnectivity between every pair of neighbors nodes, the global regularized least-square problem requires an optimization procedure to solve it with decentralized fashion then, the question is what is the optimal output weight vector that we have to choose for the test task, here the DAC intervenes to encourage all agents having the same vectors or we will be on the case of local training, so we must choose appropriately the DAC strategy in order that all agents converge to the same state. The contribution key is to apply the Metropolis-Weights as a strategy of average consensus to compute the mean of the updates of nodes at each step with several tests, this protocol demonstrate convergence of the consensus algorithm for network without packet losses. Experimental results on prediction and identification tasks show a favorable performance in terms of accuracy and efficiency.

Keywords: Distributed learning · Metropolis-Weights · Recurrent neural network · Alternating direction method of multipliers

1 Introduction

A great deal of attention have drawn the computational intelligence technologies such as neural networks [23, 24], support vector machines (SVM) [10], fuzzy logic [5, 8] and evolutionary computation [9, 13, 25] and covers a large applications like pattern recognition [6, 7, 12, 14] and series prediction [1, 23]. When the data are not available on a centralized location and distributed between network agents, like computers in a peer-to-peer (p2p) networks, sensors, robotic swarms [2], virtually a centralized training can't deal with this kind of problems, when we have problem of getting the entire dataset for several cases, mentioned the following ones, first it can introduce one local failure, second the training data may not be shared among the nodes on account of

© Springer International Publishing AG 2017
D. Liu et al. (Eds.): ICONIP 2017, Part IV, LNCS 10637, pp. 108–119, 2017.
https://doi.org/10.1007/978-3-319-70093-9_12

its size to treated by an individual machine this is the case of big data concept. Recently, the researchers are directed towards the development of distributed learning algorithms, including approaches for training Multi Layer Perceptron [4]. SVM take a big interest using the ADMM presented in [10, 11]. In the big data analytics two recent works in [15, 16] based on the ADMM optimization technics.

On the context of distributed recurrent neural network, less attention is dedicated, presenting a set of recent advances such that [17, 19]. The context relying the decentralized training algorithms for recurrent neural networks present a gap, as a continuation to deal with this crack, we aim to develop an enhanced distributed learning algorithm for a specific case of recurrent neural networks with new consensus strategy. Examining the literature, DAC is the most adapted procedure of computing averaging that enforces the local measurements in the network topology to be equal and finally come up with one master learner model. DAC is general processes that are similar to a packed set of linear predictors [3] and have several particular strategies have to be applied in the right context; we choose Metropolis-Weights (MW) to be applied in our algorithm for the learning for echo sate networks with distributed and random mode. The MW is easy to compute, and guarantee asymptotic average consensus of the sequence of the time-varying communication graphs [3]. Recently Echo State Networks have waked a lot of interest. ESN has been successfully applied in time-series prediction, speech recognition and in dynamic pattern classification [21]. Several algorithms are proposed to train ESNs with centralized fashion, but a little and even one or two treat the data distributed case, [1] is the first and may be only approach proposed in this context.

The schema of the paper is composed of five sections; first, the ESN neural network is presented. In the second section the Metropolis weights-based distributed algorithms are used for ESNs. Then, experimental setup and numerical results on three realistic datasets are presented. Finally, conclusions and future works are drawn.

2 Echo State Networks ESNs

ESN networks are classical RNNs that have an unusual and simple learning algorithm called Reservoir Computing (RC) that is fixed in advance and built form standard non-linear neurons, the second basic part is the readout that is constructed like a linear layer where the weights are trained using ridge regression. To assure stability, the reservoir must satisfy the echo state property where the impact of an input on the state of the reservoir has to vanish after a period of time [20]. The connectivity in the reservoir is initialized randomly from the beginning of the training process, that make ESNs successful because it transform the training a simple linear regression counter to other RNNs. Referring to Georgopoulos [21], to realize reliable achievements on term of low test error, it is necessary to choose appropriately the reservoir size (large), the scarcity of connections and the random connectivity.

Training ESN is equal to solve the optimization problem that is given by:

$$\min \| WH - d \|_2^2 \tag{1}$$

The following regularized least-square problem is then solved to get the optimal objective weight vector:

$$W^* = \arg\min \frac{1}{2} \| HW - d \| + \frac{\lambda}{2} \| W \|_2^2 \tag{2}$$

W is the output weight matrix, H is the matrix of internal states; D is the output matrix and λ a positive scalar regularization.

3 Metropolis-Weights Based Distributed Learning

3.1 Metropolis-Weights Consensus Strategy

The consensus is the problem of finding linear iterations that achieve the average of some initial values given at the nodes. Introducing a set of L interconnected processes; each node has its initial measurement, noted by a scalar value $\beta_i(0)$. The consensus is used to compute asymptotically the average value at each node, requiring only local communication between them [22].

In such iteration n, the local update is given by:

$$\beta_i(n) = \sum_{j \in N_i} C_{ij} \beta_j(n-1) \tag{3}$$

where C is L * L connectivity matrix of the network and $C_{ij} \neq 0$ when the node i is connected to the node j. In this work, the connectivity of the undirected topology is given a priori. DAC is defined by a set of linear updating equations in the form:

$$\beta_i(n+1) = C_{ii}\beta_i(n) + \sum_{j \in N_i} C_{ij}(n)\beta_j(n) \tag{4}$$

where N_i is the set of the neighbors of node i. the choice of the connectivity matrix influences directly on the convergence to the global average.

$$\hat{\beta} = \frac{1}{L} \sum_{i=1}^{L} \beta_i(0) \tag{5}$$

All asymptotic values of $\beta_i(t)$ must be equal. For the MW strategy just the degree of connection must be known by an agent in order to be able to fix the weights on the adjacent edges. The Metropolis weight is defined as follows:

$$\mathbf{C_{ij}} = \begin{cases} 1/(\max\{d_i,d_j\} + 1 & \text{if i is connected to j} \\ 1 - \sum_{k \in N_i(t)} C_{ik}(t) & \text{if } i = j \\ 0 & \text{otherwise} \end{cases} \tag{6}$$

where d_i and d_j are the degrees defining the number of nodes connected to each one of the node i and j respectively. The process is iterated, so a stopped number of iterations is given or by a defined threshold δ [1].

The main contributions of this paper can be summarized as follows:

(1) We give a new hybrid method based ADMM and Metropolis-Weights strategy and the test through different applications show its performance.
(2) We vary the number of agents and for each simulation, we compare the proposed algorithm with the Max degree strategy based Consensus.
(3) We compare for each application and for each topology the proposed algorithm with the centralized one in order to demonstrate its performance.

Algorithm1. Average Consensus algorithm
```
For i=1 to n (number of nodes) do

For t=1 to T (number of iterations) do
```

$$x_i = \sum_{j=1}^{n} C_{ij} x_j$$

```
End for

End for
```

3.2   Distributed Learning for ESNs Based on Metropolis-Weights

In this section, we introduce our strategy to learn from distributed dataset, we consider a random network of nodes by two main parameters, the number of nodes L and the probability P of interconnectivity between neighbors, where such that the kth node, $k = 1...L$, has access to its own training set given by S_k.

This centralized repetitive processes repeated at every agent can be seen as a decentralized one process by solving directly the global optimization problem according to:

$$W^* = \arg \min_w \frac{1}{2} (\sum_{k=1}^{L} \|H_k W - d_k\|_2^2 + \frac{\lambda}{2} \|W\|_2^2) \tag{7}$$

The Metropolis-weights is the principal contribution that judged efficient and adaptable to our goals in the distributed scenario [22], and this work will approve this theoretical interpretation.

Alternating Direction Method of Multipliers ADMM
Many strategies are used in order to solve optimization problems such as [18]. Because
of the simplicity and the power of the alternating direction method of multipliers
(ADMM) what makes it adapted to distributed convex optimization, in particular to
problems of machine learning. It is a decomposition of the global problem on sub
problems, it will be solved by coordinate the solutions of local sub problems to find a
solution to a large global problem.

ADMM determine an optimization problem of the form:

$$\text{Minimise } f(x) + g(z)$$
$$\text{Subject to } Ax + Bz + c = 0 \tag{8}$$

The Lagrangian of the problem may be formulated as following:

$$L_\rho(x, z, y) = f(x) + g(z) + y^T(Ax + Bz + c) + \frac{\rho}{2}\|Ax + Bz + c\|_2^2$$

y is the vector of Lagrange multipliers, ρ is a scalar.

$$x^{k+1} = \arg\min_x L_\rho(x, z^k, y^k) \tag{9}$$

$$z^{k+1} = \arg\min_z L_\rho(x^{k+1}, z, y^k) \tag{10}$$

$$y^{k+1} = y^k + \rho(Ax^{k+1} + Bz^{k+1} - c) \tag{11}$$

The stop criteria can be achieved by a maximum number of iterations or by
absolute tolerance and relative tolerance. The original problem will be reformulate by
offering local variables w_k at each agent, and pushing them to be equal at the
Metropolis strategy step. The new reformulation is given:

$$W^* = \arg\min_w \frac{1}{2}\left(\sum_{k=1}^{L} \|H_k W - d_k\|_2^2 + \frac{\lambda}{2}\|w\|_2^2\right)$$
$$\text{Subject to } W_k = z, k = 1 \cdots L \tag{12}$$

The estimated measurments are given with the next equation:

$$\hat{w} = \frac{1}{L}\sum_{k=1}^{L} W_k[n+1] \tag{13}$$

$$\hat{y} = \frac{1}{L}\sum_{K=1}^{L} y_k[n] \tag{14}$$

The pseudo-code Metropolis-weights based ESNs (MW-ESN) is given below:

Algorithm2. MW-ESN algorithm

```
Inputs: Wk[0] (local), yk[0] (local), Number of iterations
T(global), Size of the network L (global), number of
neighbors for node k, Nk
For n=0: T do
Wk [n+1]= Wk[n]
For k=1:L do
```

$$W_k[n+1] = (1 - \sum_{j \in N_k} 1/(\max\{d_j, d_k\} + 1)) * W_k[n] + \frac{1}{\max\{d_j, d_k\} + 1} * \sum_{j \in N_k} W_j[n]$$

$$Err_{cons} = Err_{cons}^k \quad + \quad \left\| W_k[n+1] - W_k[n] \right\|_2^2$$

$$y_k[n+1] = (1 - \sum_{j \in N_k} 1/(\max\{d_j, d_k\} + 1)) * y_k[n] + \frac{1}{\max\{d_j, d_k\} + 1} * \sum_{j \in N_k} y_j[n]$$

$$Err_{cons} = Err_{cons}^k + \left\| y_k[n+1] - y_k[n] \right\|_2^2$$

```
Outputs:
```

$$\hat{w} = \frac{1}{L} \sum_{k=1}^{L} W_k[n+1]$$

$$\hat{y} = \frac{1}{L} \sum_{k=1}^{L} y_k[n]$$

d_j, d_k the number of nodes connected to j and k respectively, $\max\{d_j, d_k\}$ return the maximum degree in the set of connected nodes to the node k.

The proposed architecture

The process of the Fig. 1 is described as follow:

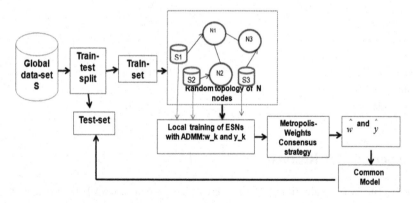

Fig. 1. The overall procedure of the proposed distributed learning for Recurrent Neural Network

Step 1: Generate a random topology of network composed by a number of agents L that are interconnected with a specified probability P.

Step 2: Divide the global data set according to the number of the agents.

Step 3: All agents share the same ESN architecture, each ESN will be trained with its local data set with distributed fashion according to ADMM as an optimization solver.

Step 4: The computation of global averages starting from local vectors. This operation was implemented with Metropolis Weights as a matrix of interconnectivity in order to have a common model (the same optimal output weight matrix).

Step 5: Starting from a common model, Test the proposed algorithm MW-ESN with standard artificial benchmarks applications.

4 Experimental Setup

For every kind of dataset we generate sequences of size 2000 elements that start from different initial conditions.

Neural Network parameters:

Parameter	Reservoir size	Spectral radius	Input scaling	Reservoir connectivity	Lambda
Value	300	0.9	0.5	0.25	10e−4

Metropolis-Weights-ADMM parameters:
For all of the time series, the designed ESN model consists of two inputs, one output and 300 internal neurons. The optimal parameters given by the next table.

Parameter	Value
ADMM iterations	300
MW iterations	300
MW termination Threshold	10e−6
Regularization parameter	0.02
Absolute tolerance	10e−6
Relative tolerance	10e−6

5 Experimental Results

The first criteria of evaluation is the test error. The Mean Square Error given by:

$$MSE = \frac{1}{n} \sum_{i=1}^{n} (\hat{Y}_i - Y_i)^2) \tag{15}$$

Desired output $Y_1, \ldots, Y_k$ and n is the number of samples of test. The second criteria of evaluation is the training time in order to measure the complexity.

a. Mackey-Glass (MGK) Data set

The dynamic predictive system is given by the next equation [1] (Table 2):

$$x[n] = \beta x[n] + \frac{\alpha x[n - \tau]}{1 + x^{10}[n - \tau]} \tag{16}$$

The prediction is for 10-step, given by: $\alpha = 0.2, \beta = -0.1, \gamma = 10 \, and \, \tau = 17$

$$d[n] = x[n + 10] \tag{17}$$

Table 1. MSE and St.dev for MGK

Training algorithm/ number agents	MSE	St-dev
Centralize	0.009	12.00E−05
MW-ESN(5)	0.009	18.00E−05
MW-ESN(10)	0.010	10.00E−05
MW-ESN(15)	0.010	9.926E−05
MW-ESN(20)	0.010	9.926E−05
MW-ESN(25)	0.009	13.00E−05
MW-ESN(30)	0.007	9.077E−05

Table 2. Tt and St.dev for MGK

Training algorithm/ number agents	Training time (Tt).(s)	St-dev
Centralize	28.37	0.8843
MW-ESN(5)	7.51	0.0078
MW-ESN(10)	6.31	0.0671
MW-ESN(15)	6.91	0.0408
MW-ESN(20)	5.61	0.0833
MW-ESN(25)	3.97	0.0987
MW-ESN(30)	4.76	0.1300

b. NARMA identification task

Is a very popular type of benchmark distinguished by a high rate of chaos. The dynamic Eq. (18) which generates this type depends on several parameters which makes NARMA models difficult to form the output y[n] is computed as follow (Table 4):

$$y[n] = 0.1 + 0.3y[n - 1] + 0.05y[n - 1] * \prod_{i=1}^{10} y[n - i] + 1.5x[n]x[n - 9] \tag{18}$$

Table 3. MSE and St.dev for NARMA

Training algorithm/ number agents	MSE	St-dev
Centralize	5.601e−05	2.443e−06
MW-ESN(5)	5.6e−05	2.443e−06
MW-ESN(10)	5.617e−05	2.403e−06
MW-ESN(15)	5.623e−05	2.395e−06
MW-ESN(20)	5.654e−05	2.343e−06
MW-ESN(25)	5.681e−05	2.509e−06
MW-ESN(30)	5.745e−05	2.101e−06

Table 4. Tt and St.dev for NARMA

Training algorithm/ number agents	Training time (Tt).(s)	St-dev
Centralize	25.52	0.7983
MW-ESN(5)	10.63	0.1498
MW-ESN(10)	4.887	0.1807
MW-ESN(15)	5.727	0.02855
MW-ESN(20)	6.2	0.1944
MW-ESN(25)	7.662	0.2624
MW-ESN(30)	5.048	0.1106

Table 5. MSE and St.dev for Lorenz

Training algorithm/ number agents	MSE	St-dev
Centralize	0.1374	0.001582
MW-ESN(5)	0.1373	0.001619
MW-ESN(10)	0.1377	0.00164
MW-ESN(15)	0.1373	0.001336
MW-ESN(20)	0.1375	0.001816
MW-ESN(25)	0.1376	0.001698
MW-ESN(30)	0.1377	0.001218

Table 6. Tt and St.dev for Lorenz

Training algorithm/ number agents	Training time (Tt).(s)	St-dev
Centralize	26.57	0.07897
MW-ESN(5)	11.69	0.03137
MW-ESN(10)	15.43	0.001582
MW-ESN(15)	14.68	0.06511
MW-ESN(20)	12.3	0.01396
MW-ESN(25)	9.306	0.04276
MW-ESN(30)	10.86	0.02559

c. Lorenz Attractor

Another chaotic time-series prediction task and a 3-dimensional time-series, defined in continuous time by the following set of differential equations:

$$\frac{dx}{dy} = \alpha(y - x) \quad \frac{dy}{dt} = -y - xy + rx \quad \frac{dz}{dt} = xy - bz \tag{19}$$

where $\alpha = 10$, $r = 28$ and $b = 8/3$. Our task is the prediction of $x(t)$ according to three inputs of x for precedent time steps. The Eq. (19) will be optimized using an ODE45 solver (Table 6).

Tables 1, 3 and 5 summarize the average errors for the three benchmark tasks, in fact we modify the number of nodes at each step and compute the DAC averages both for local training and DAC base training; as it clear that the local ESN values on the MSE term are less efficient than the centralized case, this is going to the small number of training set that have to be divided with the number of nodes, so the efficiency decreases with the augmentation of the number of nodes as it mentioned in Table 1. In the other side, the proposed approach can attend the performance in most situations, but with a number of 30 nodes, the MW-ESN is worse efficient than the centralized algorithm do to the large size of the topology for the Mackey-Glass data set. For both NARMA 10 and Lorenz attractor in the other tables, the proposed decentralized algorithm can excellently follow the performance of the centralized counterpart in all step of agents, a small gap in performance is present when examining big networks. In fact, the performance of MW-ESN is 1% worse than Centralize ESN 25 and 30 nodes in the datasets MKG and NARMA. In the literature, this gap can be treated by augmenting the number of ADMM iterations.

In the rest of tables, the Training time (Tt) requested by the centralize and decentralize algorithms was averaged throughout the agents. In all settings, the time asked in MW-ESN is significantly lower compared to the one in of the counterpart (Fig. 2).

As it clear in Fig. 3, compared to the Max-degree by computing the mean error, the MW is more efficient on term of MSE for the Mackey-Glass prediction task.

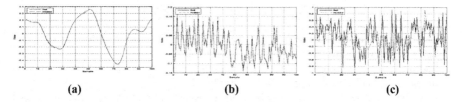

Fig. 2. Network outputs (Vs targets for (a) MGK, (b) NARMA10 and (c) Lorenz attractor)

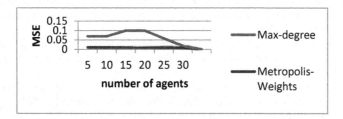

Fig. 3. Difference between the proposed strategy (MW-ESN) and the Max-degree DAC strategy on term of MSE (for Mackey-Glass OF DELAY = 17)

6   Conclusion

In this paper we start by the state of the art concerning distributed learning algorithms, in fact, we examine the recurrent type, then we present an approach of learning for Echo State Networks in decentralized way, a random network is built with a given probability of interconnectivity between each pair of nodes, and a modified number of agents at every simulation in order to generalize our work, the local ESNs are trained simultaneously with the ADMM optimization solver, followed by Metropolis-weight to encourage it to converge to a common model that was tested with three benchmarks data set. The proposed approach achieves the performances compared to the centralized algorithm on term of accuracy and. As future work this approach can be extended with modifying the optimization strategy with Particle Swarm Optimization in order to optimize the random ESN parameters.

Acknowledgements. The research leading to these results has received funding from the Ministry of Higher Education and Scientific Research of Tunisia under the grant agreement number LR11ES48.

References

1. Scardapane, S., Wang, D., Panella, M.: A decentralized training algorithm for echo state networks in distributed big data applications. Neural Netw. **78**, 65–74 (2016)
2. Goudarzi, A., Lakin, M.R., Stefanovic, D.: Reservoir computing approach to robust computation using unreliable nanoscale networks. In: Ibarra, O.H., Kari, L., Kopecki, S. (eds.) UCNC 2014. LNCS, vol. 8553, pp. 164–176. Springer, Cham (2014). doi:10.1007/978-3-319-08123-6_14

3. Fischione, C.: Distributed Estimation, Lecture 9 Principles of Wireless Sensor Networks (2009)
4. Zinkevich, M., Langford, J., Smola, A.J.: Slow learners are fast. In: Advances in Neural Information Processing Systems, pp. 2331–2339 (2009)
5. Baccour, L., Alimi, M.A., John, R.I.: Similarity measures for intuitionistic fuzzy sets: state of the art. J. Intell. Fuzzy Syst. 24(1), 37–49 (2013)
6. Ben Moussa, S., Zahour, A., Benabdelhafid, A., Alimi, A.M.: New features using fractal multi-dimensions for generalized Arabic font recognition. Pattern Recogn. Lett. 31(5), 361–371 (2010)
7. Bezine, H., Alimi, M.A., Derbel, N: Handwriting trajectory movements controlled by a bêta-elliptic model. In: Proceedings of the International Conference on Document Analysis and Recognition, ICDAR, p. 1228 (2003)
8. Elloumi, W., Baklouti, N., Abraham, A., Alimi, M.A.: The multi-objective hybridization of particle swarm optimization and fuzzy ant colony optimization. J. Intell. Fuzzy Syst. (JIFS) 27(1), 515–525 (2014)
9. Elloumi, W., El Abed, H., Abraham, A., Alimi, M.A.: A comparative study of the improvement of performance using a PSO modified by ACO applied to TSP. J. Appl. Soft Comput. (JASoC) 25, 234–241 (2014)
10. Lu, Y., Roychowdhury, V., Vandenberghe, L.: Distributed parallel support vector machines in strongly connected networks. IEEE Trans. Neural Netw. 19(7), 1167–1178 (2008)
11. Flouri, K., Beferull-Lozano, B., Tsakalides, P: Training a SVM-based classifier in distributed sensor networks. In: 14th European Signal Processing Conference, pp. 1 5 (2006)
12. Slimane, F., Kanoun, S., Hennebert, J., Alimi, A.M., Ingold, R.: A study on font-family and font-size recognition applied to Arabic word images at ultra-low resolution. Pattern Recogn. Lett. 34(2), 209–218 (2013)
13. Alimi, M.A.: Evolutionary computation for the recognition of on-line cursive handwriting. IETE J. Res. 48(5), 385–396 (2002). SPEC
14. Boubaker, H., Kherallah, M., Alimi, M.A.: New algorithm of straight or curved baseline detection for short arabic handwritten writing. In: Proceedings of the International Conference on Document Analysis and Recognition, ICDAR, p. 778 (2009)
15. Scardapane, S., Wang, D., Panella, M., Uncini, A.: Distributed learning for random vector functional-link networks. Inf. Sci. 301, 271–284 (2015)
16. Scardapane, S., Panella, M., Comminiello, D., Uncini, A.: Learning from Distributed Data Sources using Random Vector Functional-Link Networks. Procedia Comput. Sci. 53, 468–477 (2015)
17. Obst, O.: Distributed fault detection in sensor networks using a recurrent neural network. Neural Process. Lett. 40(3), 261–273 (2014)
18. Elloumi, W., Alimi, M.A.: A more efficient MOPSO for optimization. In: ACS/IEEE International Conference on Computer System and Applications (AICCSA) (2010)
19. Guijarro-Berdiñas, B., Martínez-Rego, D., Fernández-Lorenzo, S.: Privacy-preserving distributed learning based on genetic algorithms and artificial neural networks. In: Omatu, S., Rocha, M.P., Bravo, J., Fernández, F., Corchado, E., Bustillo, A., Corchado, J.M. (eds.) IWANN 2009. LNCS, vol. 5518, pp. 195–202. Springer, Heidelberg (2009). doi:10.1007/978-3-642-02481-8_27
20. Jaeger, H.: Tutorial on training recurrent neural networks, covering BPPT, RTRL, EKF and the "echo state network" approach, vol. 5. GMD-Forschungszentrum Informationstechnik (2002)
21. Georgopoulos, L., Hasler, M.: Distributed machine learning in networks by consensus. Neurocomputing 124, 2–12 (2014)

22. Xiao, L., Boyd, S., Lall, S.: Distributed average consensus with time-varying metropolis weights. Automatica (2006)
23. Dhahri, H., Alimi, M.A.: The modified differential evolution and the RBF (MDE-RBF) neural network for time series prediction. In: IEEE International Conference on Neural Networks - Conference Proceedings, p. 2938 (2006)
24. Bouaziz, S., Dhahri, H., Alimi, M.A., Abraham, A.: A hybrid learning algorithm for evolving Flexible Beta Basis Function Neural Tree Model. Neurocomputing **117**, 107–117 (2013)
25. Elbaati, A., Boubaker, H., Kherallah, M., Alimi, M.A., Ennaji, A., Abed, H.E.: Arabic handwriting recognition using restored stroke chronology. In: Proceedings of the International Conference on Document Analysis and Recognition, ICDAR, p. 411 (2009)

Bayesian Curve Fitting Based on RBF Neural Networks

Michael Li$^{(\boxtimes)}$ and Santoso Wibowo

Centre for Intelligent Systems and School of Engineering and Technology,
Central Queensland University, Rockhampton, QLD 4701, Australia
m.li@cqu.edu.au

Abstract. In this article, we introduce a novel method for solving curve fitting problems. Instead of using polynomials, we extend the base model of radial basis functions (RBF) neural network by adding an extra linear neuron and incorporating the Bayesian learning. The unknown function represented by datasets is approximated by a set of Gaussian basis functions with a linear term. The additional linear term offsets the localized behavior induced by basis functions, while the Bayesian approach effectively reduces overfitting. The presented approach is initially utilized to assess two numerical examples, then further on the method is applied to fit a number of experimental datasets of heavy ion stopping powers (MeV energetic carbon ions in various elemental materials). Due to the linear correction, the proposed method significantly improves accuracy of fitting and outperforms the conventional numerical-based algorithms. Through the theoretical results, the numerical examples and the application of fitting stopping powers data, we demonstrate the suitability of the proposed method.

Keywords: Bayesian learning · RBF · Curve fitting

1 Introduction

The concept of curve fitting was born centuries ago but its applications continue to rise when applied to many modern scientific domains. The basic idea of curve fitting involves producing a curve that best fits a set of data. From the curve, we can calculate the relevant analytical properties and make trend analysis by extrapolation.

Curve fitting is a process of constructing a curve with a regression function where ideally the curve produced should follow the shape of data points with a specified measure such as a squared Euclidean distance between the data and the model function. A curve fitting job typically implicates two correlative tasks: (i) To determine a suitable regression function with minimal parameters so that the function is able to describe the system represented reliably, and (ii) To resolve the problem of parameter values. To perform the first task, a strong grasp of the fundamental theory pertaining to numerical analysis is crucial. Much research has been conducted in curve fitting and its applications, and this can be demonstrated by the candidate regression functions which are readily available in some practical handbooks and scientific computing software. For instance, Arlinghaus [1] has compiled an encyclical manual containing a myriad of

© Springer International Publishing AG 2017
D. Liu et al. (Eds.): ICONIP 2017, Part IV, LNCS 10637, pp. 120–130, 2017.
https://doi.org/10.1007/978-3-319-70093-9_13

parametrized functions with real life applications ranging from epidemiology to astronomy. Several popular statistical and computing software packages have included a built-in feature that enables users to construct a curve with a provided set of data. However, these functions or their combinations may not always be able to best describe a phenomenon or a system modeling the real world, due to the complicated nature of problem. As for the second task, the Levenberg-Marquart algorithm is commonly used for determining the fitting coefficients. The determination of the optimal values of coefficients could be difficult and time consuming, because it often requires a lengthy iterative process starting from an initial guess.

To address the above challenges, some intelligent methods such as neural network techniques or machine learning algorithms have been considered to improve the fitting accuracy, where nonlinearity is approximated via a set of basis functions and the fitting coefficients are determined by a learning algorithm. Particularly a class of statistical regression model incorporating Bayesian probabilistic reasoning has been developed in recent years [2, 3]. The Bayesian approach is well-known for preventing overfitting through restricting flexibility in the statistical data analysis. It resolves the problems of curve fitting or regression analysis with two primary elements: (i) A full probabilistic description of the computational model; and (ii) Use of Bayes' theorem. The former consistently deals with uncertainties for data model and its parameters in terms of probability distributions, while the latter is used to make an information inference related to a learning process from data to model. Essentially Bayes' theorem takes the assumed prior knowledge about the model parameter, updates prior knowledge, and eventually gives the posterior probability distribution on the data.

Curve fitting based on Bayesian inference and linear regression models has been studied by several authors [4, 5]. The majority of existing research in Bayesian curve fitting highlight using piecewise polynomials or splines. Although these methods were competitive in terms of accuracy in performing challenging fitting tasks, they didn't produce a simple and universal empirical fitting formula for further application. Mathematically, curve fitting is a type of classical problem within function approximation framework, where the function with the unknown analytical form is approximated either by some simple functions or a set of basis functions. Much research has been conducted in the areas of function approximation and the regression model based on radial basis function (RBF) neural network is considered to be a preferred option due to its universal approximation capacity [6]. However, in practical applications, this technique requires a sufficiently large number of basis functions in a network configuration to ensure its best approximation property. Because the main objective of curve fitting is to find a relatively simple empirical formula with the minimum number of coefficients, the RBF neural network is not flawless. To effectively deal with this issue, an additional linear neuron in the RBF network architecture is introduced. The added on linear term has a correctional function through an adaptive training process.

In this paper, we propose a new method that incorporates Bayesian probabilistic inference in the RBF regression model for curve fitting. In particular, an additional linear term in RBF model has been introduced for a better approximation. Our approach will be utilized to investigate numerical examples then afterward, it is applied to fit stopping power curves versus the energy for MeV carbon energetic projectiles. Stopping power data has paramount significance in applications of two areas of major research - ion beam analysis technique and radiation therapy. This new approach is a universal method, due to the fact that it holds the weighted linear superposition from a series of Gaussian basis function at different centers with a linear correction.

The organization of this paper is as follows. In Sect. 2, the proposed method is described in details. Next, the benchmark numerical examples are tested, and computer simulation results for stopping power data are discussed in Sect. 3. Finally, Sect. 4 concludes the paper.

2 Bayesian Approach for Curve Fitting

2.1 Bayesian Probabilistic Linear Regression

In Bayesian data analysis, a key concept is uncertainty. Statistically each value of the observed quantities inevitably falls in a small uncertain range that arises from measurement errors or noises. Similarly, values of parameters of a statistical computational model may also be in uncertainty, due to the finite size of data set to derive them. Probabilistic modelling is the best way for dealing with uncertainties which are quantified through distributions. Consider a general regression problem where the input variable is a vector $\mathbf{x}$, the target variable is a scalar denoted by t, and an N-points sample data set $\{\mathbf{x}_i, t_i\}_{i=1}^{N}$ is given. The regression problem that fits the sample data set to an underlying function can be defined as follows:

$$t_i = y(\mathbf{x}_i; \mathbf{w}) + \varepsilon \quad i = 1, \ldots, N \tag{1}$$

where ε denotes the random error, and $\mathbf{w}$ denotes a vector of all adjustable parameters in the model. Under the linear regression model, the model function $y(\mathbf{x};\mathbf{w})$ is a linearly-weighted sum of M fixed basis functions $\varphi_j(\mathbf{x})$,

$$y(\mathbf{x}; \mathbf{w}) = \sum_{j=1}^{M} w_j \phi_j(\mathbf{x}) = \mathbf{w}^T \phi(\mathbf{x}) \tag{2}$$

The random error in data can be assumed to be a zero-mean Gaussian noise with variance β^{-1}:

$$\varepsilon \sim \mathcal{N}(0, \beta^{-1}) \tag{3}$$

Statistically it is a reasonable hypothesis that noises in data are Gaussian, since the underlying mechanisms generating physical data often include many stochastic processes while the summation of many random processes tends to have the normal distribution. Hence from Eqs. (1) and (3), the target variable t is a random variable and its conditional probability upon $\mathbf{x}$ and $\mathbf{w}$ satisfies a normal distribution with a mean equal to $y(\mathbf{x};\mathbf{w})$ and variance β^{-1}. It can be expressed as

$$p(t|\mathbf{x}, \mathbf{w}) = \mathcal{N}(t|y(\mathbf{x}, \mathbf{w}), \beta^{-1}) \tag{4}$$

As each data point is drawn independently and identically and its probability obeys the distribution of Eq. (4), the likelihood function of the entire dataset $\{\mathbf{x}_i, t_i\}_{i=1}^N$, is the product of the probability of each point occurrence and it is given by

$$p(\mathbf{t}|\mathbf{x}, \mathbf{w}, \beta) = \prod_{i=1}^{N} \mathcal{N}(t_i|y(\mathbf{x}_i, \mathbf{w}), \beta) \tag{5}$$

It is possible to make a point estimate on the model parameter $\mathbf{w}$ by using the maximum likelihood (ML) estimate in Eq. (5). However the ML method often leads to overfitting data. To control the model complexity, a prior distribution over $\mathbf{w}$ is introduced. For simplicity, we add an isotopic Gaussian distribution of the form

$$p(\mathbf{w}|\alpha) = \mathcal{N}(\mathbf{w}|0, \alpha^{-1}\mathbf{I}) \tag{6}$$

where $\mathbf{I}$ is the unit matrix, and α is termed as the hyperparameter of model.

The goal of curve fitting is to predict the corresponding value t^* of the target variable for a new test point $\mathbf{x}^*$, given the existing sample set $\{\mathbf{x}, \mathbf{t}\}$. Therefore it is necessary to evaluate the probability distribution of the predictive t^* i.e. $p(t^*|\mathbf{x}^*, \mathbf{x}, \mathbf{t})$. In a fully Bayesian treatment of the probabilistic model, in order to make a rigorous prediction for a new data point, it requires us to integrate the posterior probability distribution with respect to both the model parameter and hyperparameters. However, the triple integration for a complete marginalization is analytically intractable. As an approximate scheme, the practical Bayesian treatment assumes that the hyperparametrs α and β are known in advance. With this assumption, the expression of predictive distribution $p(t^*|\mathbf{x}^*, \mathbf{x}, \mathbf{t})$ can be derived through marginalizing over $\mathbf{w}$ [3],

$$p(t^*|\mathbf{x}^*, \mathbf{x}, \mathbf{t}) = \int p(t^*|\mathbf{x}^*, \mathbf{w})p(\mathbf{w}|\mathbf{x}, \mathbf{t}, \alpha, \beta)d\mathbf{w} \tag{7}$$

By using the Bayesian theorem, the posterior distribution $p(\mathbf{w}|\mathbf{x}, \mathbf{t}, \alpha, \beta)$ can be written as

$$p(\mathbf{w}|\mathbf{x}, \mathbf{t}, \alpha, \beta) \propto p(\mathbf{t}|\mathbf{x}, \mathbf{w}, \beta)p(\mathbf{w}|\alpha) \tag{8}$$

Substituting (4), (5), (6), and (8) into (7), we obtain

$$p(t^*|\mathbf{x}^*, \mathbf{x}, \mathbf{t}) = const \cdot \prod_{i=1}^{N} \int e^{-\frac{\beta}{2}(t^*-y^*)^2} \cdot e^{-\frac{\beta}{2}(t_i-y_i)} \cdot e^{-\frac{\alpha}{2}\mathbf{w}^T\mathbf{w}} d\mathbf{w}$$

$$= const \cdot \int e^{-\frac{1}{2}[\beta(t^*-y^*)^2 + \beta\sum_{i=1}^{N}(t_i-y_i)^2 + \alpha\mathbf{w}^T\mathbf{w}]} d\mathbf{w} \tag{9}$$

Through a series of algebraic and calculus manipulations, the predictive distribution $p(t^*|\mathbf{x}^*, \mathbf{x}, \mathbf{t})$ can be simplified as a normal distribution

$$p(t^*|\mathbf{x}^*, \mathbf{x}, \mathbf{t}) = \mathcal{N}(t^*|m(\mathbf{x}^*), s^2(\mathbf{x}^*)) \tag{10}$$

where m and s^2 are the mean and variance of the predictive distribution of t^*, they are given by

$$m(\mathbf{x}^*) = \beta\phi(\mathbf{x}^*)^T \mathbf{S} \sum_{i=1}^{N} \phi(\mathbf{x}_i)t_i \tag{11}$$

$$s^2(\mathbf{x}^*) = \beta^{-1} + \phi(\mathbf{x}^*)^T \mathbf{S}\phi(\mathbf{x}^*) \tag{12}$$

Here the matrix $\mathbf{S}$ is given by

$$\mathbf{S}^{-1} = \alpha\mathbf{I} + \beta \sum_{i=1}^{N} \phi(\mathbf{x}_i)\phi(\mathbf{x}_i)^T \tag{13}$$

From the above inference, the posterior probability distribution of the predictive value t^* has been derived. In the statistical sense, mean characterizes the central location in a set of points, where the highest probability event occurs for a normal distribution. Hence the mean value $m(\mathbf{x}^*)$ obtained from the predictive distribution Eq. (11) is the best approximation to the predictive t^*, which represents the predictive value of the target variable at the new test point $\mathbf{x}^*$.

2.2   Radial Basis Functions with Additional Linear Term

In the real-world problems of curve fitting, there may be few situations in which a single simple function is able to fit a complicated shape curve. Instead, the linear combination of a set of basis function could significantly raise the chance of flexibility and adaptive capacity. However, selecting suitable basis functions is a challenging task. A number of functions such as polynomial, sigmoid and radial basis function have been chosen as basis functions. Polynomial is a global function of the input variable so that changes caused by one point of input space will also affect all other points in the entire region, while majority of RBF is a localized function. The Gaussian function is typically selected as the basis function in many applications, due to its outstanding smoothness and infinite differentiability.

The Gaussian function is a type of localized functions. It decays quickly from the locations close to the centre. In general, its influence recedes with respect to the Mahalanobis distance from the centre, possibly suggesting that distant data points from centres with a large Mahalanobis distance will not succeed in activating that basis function. In particular, if the centres are sorted in ascending order, basis functions at the margins of lower boundary of first centre and upper boundary of last centre will not be able to efficiently deal with the behavior of the function to be approximated because other Gaussians have very small influence in the margin regions. In resolving these problems, we propose to add an extra linear term to the above regression model that efficiently helps to replicate the universal behavior of the functions.

With adding the extra linear term, the model regression function becomes

$$y(\mathbf{x}; \mathbf{w}) = \sum_{i=1}^{M} w_i \varphi_i(||\mathbf{x} - \mathbf{c}_i||) + k\mathbf{x} + h \tag{14}$$

where c_i is the centre parameter governing the location of the basis function in the input space, k is the linear coefficient, and h is the constant term.

The expression of (14) can be re-written in a matrix form as below

$$y(\mathbf{x}; \mathbf{w}) = \sum_{i=1}^{M} w_i \varphi_i(||\mathbf{x} - \mathbf{c}_i||) + k\mathbf{x} + h = \mathbf{\Phi}^T(r)\mathbf{w} \tag{15}$$

where

$$\mathbf{\Phi}^T(r) = [\varphi_1(r), \varphi_2(r), \ldots\ldots\ldots\varphi_N(r), \mathbf{x}, 1]$$
$$r = ||\mathbf{x} - \mathbf{c}_i||$$
$$\varphi_i(r) = \varphi(||\mathbf{x} - \mathbf{c}_i||)$$
$$\mathbf{w} = [w_1, w_2, \ldots\ldots.w_M, k, h]^T$$

With the above expressions, the predictive mean (11)–(13) can be re-written as follows,

$$\mathbf{m} = \beta \mathbf{S} \mathbf{\Phi}^T \mathbf{t} \tag{16}$$

$$s^2 = \beta^{-1} + \phi(\mathbf{x}^*)^T \mathbf{S} \phi(\mathbf{x}^*) \tag{17}$$

$$\mathbf{S}^{-1} = \alpha \mathbf{I} + \beta \mathbf{\Phi}^T \mathbf{\Phi} \tag{18}$$

where

$$\mathbf{\Phi} = \begin{bmatrix} \varphi_1(r_1) & \varphi_2(r_1) & \cdots & \varphi_N(r_1) & x_1 & 1 \\ \varphi_1(r_2) & \varphi_2(r_2) & \cdots & \varphi_N(r_2) & x_2 & 1 \\ \vdots & & & & & \\ \varphi_1(r_M) & \varphi_2(r_M) & \cdots & \varphi_N(r_M) & x_M & 1 \end{bmatrix} \tag{19}$$

$$\varphi_i(r_j) = \exp(-\frac{||\mathbf{x}_j - \mathbf{c}_i||^2}{2\sigma^2}) \tag{20}$$

It is interesting to note that the regression model with the Eq. (14) is equivalent to a RBF neural network with hybridizing the Gaussian neurons and an extra linear neuron. In this sense, the fixed centres can be selected by a training algorithm, for which we use k-means clustering method. In addition to the parameters of the basis functions, two hyperparameters α and β are also as inputs to the established Bayesian model. We adopt the grid search [7] for setting hyperparameters.

3   Experimental Results and Discussion

3.1   Numerical Experiments: Synthetic Data

In this section, two numerical examples are presented for the purpose of the test. The first test example is based on a widely studied problem in the machine learning paradigm, called the 'SinC' function problem. This 'SinC' function is the zero-order spherical Bessel function $J_0(x)$. A set of 1000 data points of $J_0(x)$ is sampled in the interval $[-10, 10]$, where x_is are uniformly distributed and the corresponding y_is are added in a zero-mean Gaussian noise with the standard deviation 0.2. Using a 7-basis functions Bayesian model with hyperparameters $\alpha = 10^{-4}$, $\beta = 25$, a regression has been performed for the above sample dataset. The regression result along with the original data are shown in Fig. 1, where the red solid line is the exact $J_0(x)$ function, the black dashed line denotes the regression curve that connects the means computed from the predictive distribution, and the light green shaded region crosses one standard deviation each side of the mean. The experimental results show that the Bayesian model presents a smooth fitting. The predictive curve appears a good fitting to the data points with noise. Individual segments deviate from the exact function with a relatively large error, which reflects the possible enlargement of perturbation from some data points due to large random noises.

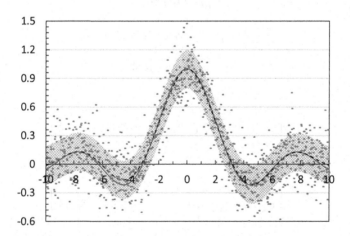

Fig. 1. Bayesian regression in the SinC synthetic data set with 1000 points. The red solid line denotes the exact SinC function, while the black dashed line represents the mean from Bayesian model (Color figure online)

The second example trialed is a typical nonlinear curve fitting problem in which Lorentzian shape line [8] spectral data at peak positions 18, 23 and 35, and with half width of 6 is added with the Gaussian random noises (zero mean and the standard deviation 1) and sampled in the interval [0, 60]. This dataset could potentially be difficult to fit using a conventional numerical technique due to its noise and multiple peaks. Considering a 7-basis function Bayesian model and assigning $\alpha = 2 \times 10^{-4}$ and $\beta = 25$, the dataset is fitted well and a high agreement between the fitted curve and the real spectrum is shown as in Fig. 2.

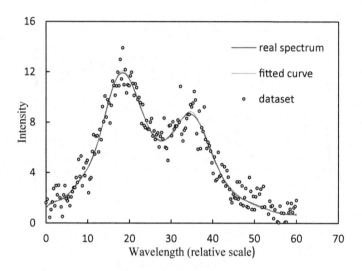

Fig. 2. Fitted curve for Lorentzian spectral data

3.2 Stopping Power Data Fitting: Real World Applications

This section demonstrates the effectiveness of the proposed method to make the empirical fitting of stopping power curves for MeV carbon projectiles in the target materials including C, Al, Si, Ti, Ni, Cu, Ag and Au. There are numerous data sets of heavy ion stopping power available. Carbon ion beam is the most extensively utilized incident beam because its stopping power characteristics have particular interests in ion beam analysis and radiation therapy. For instance, a beam of MeV energetic carbon ions could be applied with a large dose to deeply seated tumours to kill more malignant cells due to its superior depth dose distributions. The data to be fitted are primarily from the atomic and nuclear data compilations published by International Atomic Energy Agency (IAEA) (https://www-nds.iaea.org/stopping).

In the problems of curve fitting, the ultimate aim is to find a relative simple empirical formula so that a further application such as an interpolation can be carried out. Thus, it is desirable to have minimal amount of basis functions. Based on our empirical tests, we have found that a set of 4 basis functions is considered to be minimal for the current problem. The centre parameters of basis functions are set to {0.01495, 0.02832, 0.09345, 0.6563} according to k-means clustering, and σ is assigned to 1. The values of hyperparameter α, and β are set as 4×10^{-4} and 35, which is the outcome from a grid search. After these settings, it is straightforward to compute the predictive means according to Eqs. (16)–(18).

As the fitting results, the graphs of fitted stopping power curves along with the original measured points for C in C, Al, Si, Ti, Ni, Cu, Ag, and Au have obtained. Constrained to the page number limitation, only one representative figure is illustrated in Fig. 3. It can be observed from this figure that the fitting curve produced by the proposed Bayesian method fits the data points exceptionally well and. In addition, the fitting curve also reveals the typical features of data point distribution such as peaks.

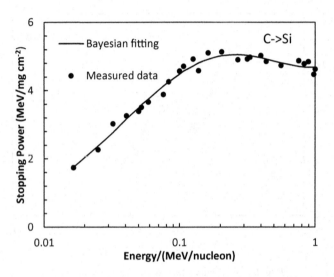

Fig. 3. The fitting curve from the proposed Bayesian method for stopping power of carbon projectile in Si target

From the fitting results, an empirical formula for carbon ion stopping power can be easily deduced. As the mean of the predictive distribution defines the fitted value of a new observation, by contrasting Eqs. (11) and (2), we obtained a simple empirical formula of carbon ion stopping power P upon incident energy E which is given by

$$P_{emp}(E) = \sum_{i=1}^{4} \lambda_i e^{-(E-E_i)^2/2} + kE + h \tag{21}$$

where $\lambda = \beta S \sum_{j=1}^{N} \phi(x_j)t_j$ and $E_i (i = 1, ..., 4)$ are constants. This is a simple 6-parameter empirical formula, which is convenient for interpolation at arbitrary energy positions. The norm $\|\lambda\|_2$ for the existing sample data ranges between 0.41 and 3.55, which are small values. This indicates that the performance of the developed method tends to be stable [7].

To demonstrate the effectiveness of our method, we compare our results with those from the conventional fitting based on numerical techniques. Paul's empirical fitting method [9] is selected for the comparison purposes. Paul's method was based on a two-stage 7 parameters fitting scheme with a Weibull-type function. Figure 4 compares the fittings from our Bayesian approach and Paul's empirical method. It can be seen that our Bayesian method prediction achieves a better fitting in the entire energy regions while Pauls' fitting slightly underestimates the stopping power around the peak.

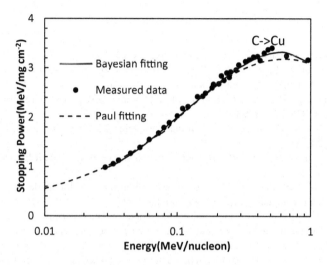

Fig. 4. Comparison of Bayesian fitting with MSE 1.92×10^{-3} and Paul's empirical prediction (with MSE 3.77×10^{-3}) of stopping power for C in Cu target

4 Conclusions

We have developed a theoretical framework based on Bayesian probabilistic model for nonlinear curve fitting. With the introduction of an extra linear term, the established linear regression model enhances the performance of fitting, where a universal approximation is constructed by overlapping a series of Gaussian functions with a linear term. Conceptually, a better approximation has achieved largely through the hybrid of Gaussian basis functions and a linear function. Relative to the ordinary linear regression model, the proposed method effectively refines the basic regression model with a dual correction – a linear term contribution, and Bayesian posterior information feedback which controls the possible overfitting.

The new model has been verified using numerical examples and carbon projectiles stopping power data fitting. Based on our simulation results, the proposed approach is well suited for fitting various heavy ions stopping power curves in batches. The proposed model has two exceptional advantages over the conventional numerical based technique. First, it is a non-specific method, minimalizing the number of parameters, ensuring a simple and fast computation for an automated data fitting job where a large quantity data is required to be processed. Second, this method is found to performing well with noisy data due to its ability to efficiently averaging noise during the training process. In addition, a simple empirical stopping power formula with 6-coefficients has been deduced for interpolation. One limitation of the proposed method is the setting of hyperparameters. The Bayesian approach helps to prevent overfitting but it gives rise to a new problem that requires a careful tuning for hyperparameters, which also is a tough task [10].

References

1. Arlinghaus, S.L.: Practical Handbook of Curve Fitting. CRC Press, Boca Raton (1994)
2. Gelman, A., et al.: Bayesian Data Analysis, 3rd edn. CRC Press, New York (2014)
3. Bishop, C.M.: Pattern Recognition and Machine Learning. Springer, Heidelberg (2006)
4. Denison, D.G.T., Mallick, B.K., Smith, A.F.M.: Automatic Bayesian curve fitting. J. R. Stat. Soc. **B60**(2), 333–350 (1998)
5. Chen, C., Yu, K.: Automatic Bayesian quantile regression curve fitting. Stat. Comput. **29**, 271–281 (2009)
6. Poggio, T., Girosi, F.: Networks for approximation and learning. Proc. IEEE **78**, 1481–1497 (1990)
7. Tang, X., Han, M.: Partial Lanczos extreme learning machine for single-output regression problem. Neurocomputing **72**(13), 3066–3076 (2009)
8. Chen, L., et al.: Effect of signal-to-noise and number of data points upon precision measure ment of peak amplitude, position and width in fourier transform spectrometry. Chemometr. Intell. Lab. Syst. **1**, 51–58 (1986)
9. Paul, H., Schinner, A.: An empirical approach to the stopping power of solids and gases for ion Li to Ar. Nucl. Instrum. Methods Phys. Res. B **179**, 299–315 (2001)
10. Bergstra, J., Benggio, Y.: Random search for hyperparameter optimization. J. Mach. Learn. Res. **13**, 281–305 (2012)

An Improved Conjugate Gradient Neural Networks Based on a Generalized Armijo Search Method

Bingjie Zhang[1], Tao Gao[1], Long Li[2], Zhanquan Sun[3], and Jian Wang[1(✉)]

[1] College of Science, China University of Petroleum, Qingdao 266580, China
bingjie_zhang_1993@163.com, GaoTao_1989@126.com, wangjiannl@upc.edu.cn
[2] College of Mathematics and Statistics, Hengyang Normal University,
Hengyang 421008, China
long_li1982@163.com
[3] Shandong Provincial Key Laboratory of Computer Networks,
Shandong Computer Science Center (National Supercomputer Center in Jinan),
Jinan 250014, Shandong, China
sunzhq@sdas.org

Abstract. In this paper, by constructing a generalized Armijo search method, a novel conjugate gradient (CG) model has been proposed to training a common three-layer backpropagation (BP) neural network. Compared with the classical gradient descent method, this algorithm efficiently accelerates the convergence speed due to the existence of the additional conjugate direction. Essentially, the optimal learning rate of each epoch is determined by the given inexact line search strategy. The presented model does not significantly increase the computational cost in dealing with real applications. Two benchmark simulations have been performed to illustrate the promising advantages of the proposed algorithm.

Keywords: Conjugate gradient method · Backpropagation · Generalized Armijo search · Neural networks

1 Introduction

Recently, the research of BP neural networks (BPNNs) has made great progress in pattern recognition, intelligent robot, automatic control, prediction, biology, medicine, economics and other fields [1–5]. As a supervised learning technique, the gradient descent method is widely applied during training neural networks [6,7]. In terms of gradient descent method, an adaptive optimal control method has been put forward for solving the Hamilton-Jacobi-Bellman equation [8]. For a model-free optimal control problem [9], the gradient descent scheme has also been employed to develop an adaptive optimal controller.

We notice that the updating direction based on gradient descent method depends on the negative gradient of the objective function [10]. There have

© Springer International Publishing AG 2017
D. Liu et al. (Eds.): ICONIP 2017, Part IV, LNCS 10637, pp. 131–139, 2017.
https://doi.org/10.1007/978-3-319-70093-9_14

been extensive studies to improve its convergent behaviors [11–13]. However, this method shows oscillatory behavior even with these changes in the training process while meeting steep valleys, which leads to poor efficiency. The essential reason of this poor efficiency is that this method has only first-order convergence.

To improve the convergent behavior, considerable reports have discussed the CG method and Newton method for BP algorithm [7,14]. Compared with the gradient descent method, Newton method is more effective, however, it needs to calculate the Hessian matrix and its inverse. As a compromise algorithm between these two methods, the CG method has the quadratic convergence and is still easy to compute without the information of second derivatives [15,16].

Due to this special advantage, the CG method has been an interesting research topic. A specific CG method is described in [17] to solve some linear systems whose coefficient matrices are positive definite. For solving massive nonlinear optimization problems, the nonlinear CG method is first proposed in [18]. Due to the diverse selections of the descent directions, there have been different CG methods, which are presented in [19–21]. There are three popular conjugate gradient methods which are presented as Fletcher-Reeves (FR) [18], Polak-Ribière-Polyak (PRP) [22,23] and Hestenes-Stiefel (HS) [17].

For purpose of solving the blind source separation problem, a PRP conjugate gradient method based on cyclic mode has been described in [24]. Its learning rate is obtained by exact line search method. In [25], the learning rate of batch mode conjugate gradient method (BCG) is a positive constant in the training process. Unfortunately, the proposed algorithm with fixed learning rate is prone to leading to poor performance. As an improvement, the existing exact line search method contributes to obtain the optimal learning step for each iteration [26]. However, it is more time-consuming due to the strong dependence on many functions and its gradients. Especially when the iteration point is far away form the optimum point, this search method is not effective and reasonable. In order to make up for these defects, the inexact line search methods are presented, including Wolf search method, Armijo search method and other methods. To solve unconstrained optimization problems, a new three terms CG method with generalized Armijo step size rule is proposed in [21]. To our best knowledge, the conjugate gradient network models with generalized Armijo search method is little referred to. As an inexact line search method, the generalized Armijo search method has many advantages. It not only can guarantee that the objective function has acceptable decrease, but also can make the eventual formed iterative sequence convergent.

Motivated by the line search strategy [21], we construct a novel conjugate gradient BP network model by virtue of a generalized Armijo search method. Instead of setting a fixed learning rate, the training produce automatically determine the optimal step-size under a low computational burden. During training, the sufficient descent direction of the objective function can be obtained by employing a given specific scheme to find the suitable conjugate coefficient. Based on these two learning advantages, the simulations demonstrate the competitive performance of the proposed algorithm.

To simplify the presentation, we introduce the following abbreviations with different BP algorithms: BPG for the gradient descent method, BPCG for the CG method and BPCGGA for the novel CG method with generalized Armijo search.

The other sections of the paper are following: In Sect. 2, we describe the network structure and the updating methods based on BPCGGA. Two supporting numerical experiments are shown in Sect. 3. Section 4 summarizes the paper.

2   Network Structure and BPCGGA Algorithm

Let us begin with a discussion of a BP neural network of three layers with p input neurons, n hidden neurons and one output neurons. The training sample set is given as $\left\{\mathbf{x}^j, O^j\right\}_{j=1}^{J} \subset \mathbb{R}^p \times \mathbb{R}$, where $\mathbf{x}^j$ and O^j are the input and the corresponding ideal output for the j-th sample, respectively. Denote the weight matrix connecting the input and hidden layers by $\mathbf{V} = (v_{i,j})_{n \times p}$, and write $\mathbf{v}_i = (v_{i1}, v_{i2}, \cdots, v_{ip})^T$ for $i = 1, 2, \cdots, n$. Let $\mathbf{u} = (u_1, u_2, \cdots, u_n)^T \in \mathbb{R}^n$ be the weight vector connecting the hidden and output layers. For simplicity, write $\mathbf{w} = \left(\mathbf{u}^T, \mathbf{v}_1^T, \cdots, \mathbf{v}_n^T\right)^T \in \mathbb{R}^{n(p+1)}$. Denote the given activation functions for the hidden and output layers by g, $f : \mathbb{R} \to \mathbb{R}$, respectively. For convenience, the vector valued function is defined as

$$G(\mathbf{z}) = (g(z_1), g(z_2), \cdots, g(z_n))^T, \quad \forall \, \mathbf{z} \in \mathbb{R}^n. \tag{1}$$

For any given input $\mathbf{x} \in \mathbb{R}^p$, the output of the hidden neurons is $G(\mathbf{Vx})$, and the final actual output is denoted by $y = f(\mathbf{u} \cdot G(\mathbf{Vx}))$. For any fixed weight $\mathbf{w}$, the error of the neural networks is

$$E(\mathbf{w}) = \frac{1}{2} \sum_{j=1}^{J} (O^j - f(\mathbf{u} \cdot G(\mathbf{Vx}^j)))^2. \tag{2}$$

The gradients of the error function with respect to $\mathbf{u}$ and $\mathbf{v}_i$ are, respectively, given by

$$E_{\mathbf{u}}(\mathbf{w}) = -\sum_{j=1}^{J} (O^j - y^j) f'(\mathbf{u} \cdot G(\mathbf{Vx}^j)) G(\mathbf{Vx}^j), \tag{3}$$

$$E_{\mathbf{v}_i}(\mathbf{w}) = -\sum_{j=1}^{J} (O^j - y^j) f'(\mathbf{u} \cdot G(\mathbf{Vx}^j)) u_i g'(\mathbf{v}_i \cdot \mathbf{x}^j) \mathbf{x}^j, \tag{4}$$

where $y^j = f(\mathbf{u} \cdot G(\mathbf{Vx}^j))$, $i = 1, 2, \cdots, n$; $j = 1, 2, \cdots, J$.
    Write

$$E_{\mathbf{V}}(\mathbf{w}) = \left(E_{\mathbf{v}_1}(\mathbf{w})^T, E_{\mathbf{v}_2}(\mathbf{w})^T, \cdots, E_{\mathbf{v}_n}(\mathbf{w})^T\right)^T, \tag{5}$$

$$E_{\mathbf{w}}(\mathbf{w}) = \left(E_{\mathbf{u}}(\mathbf{w})^T, E_{\mathbf{V}}(\mathbf{w})^T\right)^T. \tag{6}$$

For brevity, we give the following notations:

$$E_{\mathbf{w}}^m = E_{\mathbf{w}}(\mathbf{w}^m), \quad m \in \mathbb{N}, \tag{7}$$

where m is the index of the training cycle.

The description of the BPG and BPCG algorithms can be found in many papers [6, 7, 16, 23, 25]. Therefore, we just describe the BPCGGA algorithm in this section.

Before the training process, we give an arbitrary initial weight $\mathbf{w}^0$. The BPCGGA algorithm is given iteratively by the sequential expressions

$$\mathbf{w}^{m+1} = \mathbf{w}^m + \eta^m \mathbf{d}^m, \quad m \in \mathbb{N}, \tag{8}$$

$$\mathbf{d}^m = \begin{cases} -E_{\mathbf{w}}^m, & m = 0, \\ -E_{\mathbf{w}}^m + \beta^m \mathbf{d}^{m-1}, & m \geq 1, \end{cases} \tag{9}$$

where the conjugate direction coefficient β^m $(m \geq 1)$ satisfies the following expressions to guarantee the sufficient descent direction of error function

$$\begin{cases} (E_{\mathbf{w}}^m)^T E_{\mathbf{w}}^m > |\beta^m (E_{\mathbf{w}}^m)^T \mathbf{d}^{m-1}|, \\ |(E_{\mathbf{w}}^m)^T \mathbf{d}^m| \geq (1 + \delta^m)|\beta^m| \, \|E_{\mathbf{w}}^m\| \, \|\mathbf{d}^{m-1}\|. \end{cases} \tag{10}$$

Actually it gives a range of β^m:

$$\beta^m \in [-\underline{\beta}^m(\delta^m), \overline{\beta}^m(\delta^m)], \tag{11}$$

$$\overline{\beta}^m(\delta^m) = \frac{1}{1 + \delta^m + \cos\theta^m} \frac{\|E_{\mathbf{w}}^m\|}{\|\mathbf{d}^{m-1}\|}, \tag{12}$$

$$\underline{\beta}^m(\delta^m) = \frac{1}{1 + \delta^m - \cos\theta^m} \frac{\|E_{\mathbf{w}}^m\|}{\|\mathbf{d}^{m-1}\|}, \tag{13}$$

where θ^m is the angle between $E_{\mathbf{w}}^m$ and $\mathbf{d}^{m-1}, \delta^m \geq \delta > 0$ (δ^m is the positive constant for the m-th training process and δ is a small positive constant).

Let $\mu_1, \mu_2 \in (0, 1)$ $(\mu_1 \leq \mu_2), \gamma_1$ and γ_2 are positive constants. With the generalized Armijo search, the learning rate η^m in form (8) satisfies

$$E(\mathbf{w}^m + \eta^m \mathbf{d}^m) \leq E(\mathbf{w}^m) + \mu_1 \eta^m (E_{\mathbf{w}}^m)^T \mathbf{d}^m, \tag{14}$$

and

$$\eta^m \geq \gamma_1 \quad \text{or} \quad \eta^m \geq \gamma_2 \eta_*^m > 0, \tag{15}$$

where η_*^m satisfies

$$E(\mathbf{w}^m + \eta_*^m \mathbf{d}^m) > E(\mathbf{w}^m) + \mu_2 \eta_*^m (E_{\mathbf{w}}^m)^T \mathbf{d}^m. \tag{16}$$

The proposed algorithm, BPCGGA, is distinguished from the other two common algorithms, BPG and BPCG, on both the learning rate and descent direction. The details are listed as follows (Table 1).

Table 1. Comparison of three methods.

Methods	Descent direction	Learning rate
BPG	$\mathbf{d}^m = -E_{\mathbf{w}}^m, \quad m \in \mathbb{R}.$	Constant value [6]
BPCG	$\mathbf{d}^m = \begin{cases} -E_{\mathbf{w}}^m, & m = 0, \\ -E_{\mathbf{w}}^m + \alpha^m \mathbf{d}^{m-1}, & m \geq 1, \end{cases}$ where α^m satisfies the FR, PRP or HS formulas	Constant value [25]
BPCGGA	$\mathbf{d}^m = \begin{cases} -E_{\mathbf{w}}^m, & m = 0, \\ -E_{\mathbf{w}}^m + \beta^m \mathbf{d}^{m-1}, & m \geq 1, \end{cases}$ where β^m satisfies (11)	Generalized Armijo search

We note that a constant learning rate is one common choice for BPG and BPCG. Although it requires no additional computational burden to find the optimal learning rate, it is prone to poor convergent performance. As an improvement, some exact line search methods, such as Fibonacci search method, bisection method, golden section line search and so on, have been proposed in solving complex optimization problems [27,28]. These methods can automatically reach an optimal learning rate in each training epoch, however, they are more time-consuming in many real applications. Particularly, they are rarely used in training BP neural network models. As a trade-off, a generalized Armijo search method is presented in this paper to look for the suitable learning rate under a considerable less computational cost.

The BP algorithms based on the above exact line search methods waste enormous amount of computation time, which is meaningless for BPNNs, thus, in the next section, we just compare two common algorithms, BPG and BPCG (take PRP formula as an example), with the proposed algorithm, BPCGGA.

3 Simulations

In this section, we research the performance behavior of BPG, BPCG and BPCGGA for two data sets, including 4-bit parity problem and the MNIST handwritten digital data set. For all nodes of the following network structures, we use the logistic activation function $g(t) = 1/(1+\exp(-2t)), (t \in \mathbb{R})$. To clarify the convergence property, only one training result is given for each example.

3.1 Example 1: Parity Problem

The 4-bit parity problem is considered to ensure the convergence property of BPCGGA in this example. The performance behavior has been compared clearly for BPG, BPCG and BPCGGA. The inputs and the objective outputs of the training samples is described in Table 2.

The three neural networks based on BPG, BPCG and BPCGGA have the identical structure with 5 input nodes, 50 hidden nodes and one output node. We

Table 2. 4-bit parity problem.

Input				Output	Input				Output
1	1	1	1	0	−1	1	−1	−1	1
−1	1	1	1	1	1	1	−1	−1	0
1	1	1	−1	1	1	−1	1	1	1
−1	1	1	−1	0	−1	−1	1	1	0
−1	−1	−1	1	1	1	−1	−1	1	0
1	−1	1	−1	0	−1	−1	1	−1	1
−1	−1	−1	−1	0	1	1	−1	1	1
1	−1	−1	−1	1	−1	1	−1	1	0

stochastically choose the initial weights in the interval $[−0.5, 0, 5]$ with uniform distribution. The learning rates of BPG and BPCG are set to be the same value, 0.37, and the positive constant δ of BPCGGA is with 0.05. The termination criteria of the training procedure are as follows: the maximum iterations reach 5,000, or the error is less than 0.001.

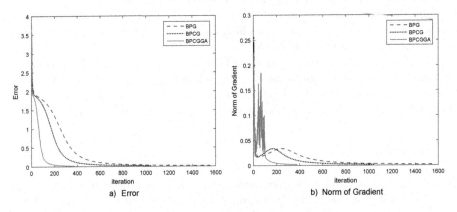

a) Error   b) Norm of Gradient

Fig. 1. Performance comparisons for different learning schemes 1: (a) Error, (b) Norm of Gradient.

From Fig. 1, it can be observed that the error function decreases monotonically and that all the norm of the gradient of error function finally tends to zero, which verify the monotonicity and the convergent behavior of the proposed algorithm, BPCGGA. Moreover, we can see that BPCGGA uses less number of iterations than BPG and BPCG with the same termination error, just as this novel CG method with generalized Armijo search has fast convergent rates due to the existence of the additional conjugate direction and the optimal learning rate of each epoch that is determined by the given inexact line search strategy.

3.2   Example 2: The MNIST Handwritten Digital Data Set

The MNIST Handwritten Digital Data Set is widely used as a classification problem, which has a training set of 60,000 examples and a test set of 10,000 examples. It is a subset of a larger set available from NIST. The digits have been size-normalized and centered in a fixed-size image. Each normalized image with 28×28 pixels are represented as a 784×1 vector in this experiment.

For fairly comparing the different performance of BPG, BPCG and BPCGGA, we construct three identical network architectures, $784 - 50 - 10$ (input, hidden and output nodes), and set the same training parameters. Let initial weight values be randomly selected in $[-0.5, 0.5]$. The learning rates of BPG and BPCG are both with 0.5, and the parameter δ of BPCGGA is fixed as 0.05. The stop criteria for these three networks are set to be: the maximum iterations, 5,000, or the error is less than 0.001.

The very similar performance (cf. Fig. 1) on the errors and the gradient norms of error function can be observed in Figs. 2 and 3, which clearly illustrates the effectiveness of the presented algorithm in this paper. And they also demonstrate

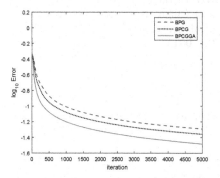

Fig. 2. Error curves for different algorithms.

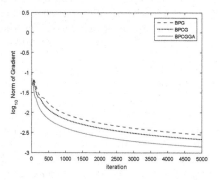

Fig. 3. The norm sequences of gradients based on different training algorithms.

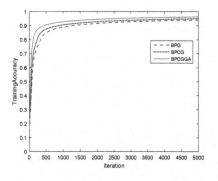

Fig. 4. Comparisons on training performance of different learning schemes.

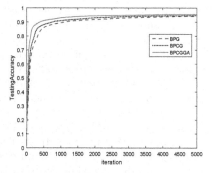

Fig. 5. Comparisons on generalization of different learning schemes.

the faster convergence behavior of BPCGGA than other two algorithms, BPG and BPCG. Importantly, it is easy to see that BPCGGA performs best among these algorithms on both training and testing accuracies for the MNIST dataset (Figs. 4 and 5).

4   Conclusions

In this paper, we propose the BPCGGA algorithm for three-layer BPNNs. The simulation results show that the error function is monotonically decreasing and that both the error function and the gradient norm of weights tend to zero. Additionally, this algorithm performs much better than the BPG and BPCG algorithms, which can be concluded that the convergence rates are mainly influenced by the learning rate and the updating direction.

Acknowledgments. This work was supported in part by the National Natural Science Foundation of China (No. 61305075, 11401185), the China Postdoctoral Science Foundation (No. 2012M520624), the Natural Science Foundation of Shandong Province (Nos. ZR2013FQ004, ZR2013DM015, ZR2015AL014), the Specialized Research Fund for the Doctoral Program of Higher Education of China (No. 20130133120014) and the Fundamental Research Funds for the Central Universities (Nos. 13CX05016A, 14CX05042A, 15CX05053A, 15CX08011A, 15CX02064A).

References

1. Xia, Y., Hu, R.: Fuzzy neural network based energy efficiencies control in the heating energy supply system responding to the changes of user demands. J. Netw. Intell. **2**, 186–194 (2017)
2. Erkaymaz, O., Şenyer, İ., Uzun, R.: Detection of knee abnormality from surface EMG signals by artificial neural networks. In: 2017 25th Signal Processing and Communications Applications Conference (SIU) (2017)
3. Li, X.D., Rakkiyappan, R.: Impulsive controller design for exponential synchronization of chaotic neural networks with mixed delays. Commun. Nonlinear Sci. Numer. Simul. **18**, 1515–1523 (2013)
4. Li, X.D., Rakkiyappan, R., Velmurugan, G.: Dissipativity analysis of memristor-based complex-valued neural networks with time-varying delays. Inf. Sci. **294**, 645–665 (2015)
5. Rumelhart, D.E., Hinton, G.E., Williams, R.J.: Learning representations by back-propagating errors. Nature **323**, 533–536 (1986)
6. Park, D.C., Elsharkawi, M.A., Marks, R.J., Atlas, L.E., Damborg, M.J.: Electric-load forecasting using an artificial neural networks. IEEE Trans. Power Syst. **6**, 442–449 (1991)
7. Saini, L.M., Soni, M.K.: Artificial neural network-based peak load forecasting using conjugate gradient methods. IEEE Trans. Power Syst. **17**, 907–912 (2002)
8. Luo, B., Wu, H.N., Li, H.X.: Adaptive optimal control of highly dissipative nonlinear spatially distributed processes with neuro-dynamic programming. IEEE Trans. Neural Netw. Learn. Syst. **26**, 684–696 (2015)

9. Luo, B., Liu, D.R., Wu, H.N., Wang, D., Lewis, F.L.: Policy gradient adaptive dynamic programming for data-based optimal control. IEEE Trans. Cybern. **99**, 1–14 (2016)
10. Zhang, H.S., Tang, Y.L.: Online gradient method with smoothing ℓ_0 regularization for feedforward neural networks. Neurocomputing **224**, 1–8 (2017)
11. Lu, C.N., Wu, H.T., Vemuri, S.: Neural network based short-term load forecasting. IEEE Trans. Power Syst. **8**, 336–342 (1993)
12. Papalexopoulos, A.D., Hao, S.Y., Peng, T.M.: An implementation of a neural-network-based load forecasting-model for the EMS. IEEE Trans. Power Syst. **9**, 1956–1962 (1994)
13. Zhang, H.S., Mandic, D.P.: Is a complex-valued stepsize advantageous in complex-valued gradient learning algorithms? IEEE Trans. Neural Netw. Learn. Syst. **27**, 2730–2735 (2016)
14. Goodband, J.H., Haas, O.C.L., Mills, J.A.: A comparison of neural network approaches for on-line prediction in IGRT. Med. Phys. **35**, 1113–1122 (2008)
15. Hagan, M.T., Demuth, H.B., Beale, M.H.: Neural Network Design. PWS Publisher, Boston (1996)
16. Nocedal, J., Wright, S.J.: Numerical Optimization. Springer, New York (2006). doi:10.1007/978-0-387-40065-5
17. Hestenes, M.R., Stiefel, E.L.: Method of conjugate gradients for solving linear systems. National Bureau of Standards, Washington (1952)
18. Fletcher, R., Reeves, C.M.: Function minimization by conjugate gradients. Comput. J. **7**, 149–154 (1964)
19. Dai, Y.H., Yuan, Y.X.: Nonlinear Conjugate Gradient Methods. Shanghai Scientific, Technical Publishers, Shanghai (2000)
20. Gonzalez, A., Dorronsoro, J.R.: Natural conjugate gradient training of multilayer perceptrons. Neurocomputing **71**, 2499–2506 (2008)
21. Sun, Q.Y., Liu, X.H.: Global convergence results of a new three terms conjugate gradient method with generalized Armijo step size rule. Mathematica Numerica Sinica **26**, 25–36 (2004)
22. Polak, E., Ribiere, G.: Note sur la convergence de directions conjugates. Revue Francaise d'Informatique et de Recherche Operationnelle **16**, 94–112 (1969)
23. Polyak, B.T.: The conjugate gradient method in extremal problems. USSR Comput. Math. Math. Phys. **9**, 94–112 (1969)
24. Shen, X.Z., Shi, X.Z., Meng, G.: Online algorithm of blind source separation based on conjugate gradient method. Circuits Syst. Sig. Process. **25**, 381–388 (2006)
25. Wang, J., Wu, W., Zurada, J.M.: Deterministic convergence of conjugate gradient method for feedforward neural networks. Neurocomputing **74**, 2368–2376 (2011)
26. Magoulas, G.D., Vrahatis, M.N., Androulakis, G.S.: Effective backpropagation training with variable stepsize. Neural Netw. **10**, 69–82 (1997)
27. Orozco-Henao, C., Bretas, A.S., Chouhy-Leborgne, R., Herrera-Orozco, A.R., Marín-Quintero, J.: Active distribution network fault location methodology: a minimum fault reactance and Fibonacci search approach. Electr. Power Energy Syst. **84**, 232–241 (2017)
28. Vieira, D.A.G., Lisboa, A.C.: Line search methods with guaranteed asymptotical convergence to an improving local optimum of multimodal functions. Eur. J. Oper. Res. **235**, 38–46 (2014)

Removing Bias from Diverse Data Clusters for Ensemble Classification

Sam Fletcher[(✉)] and Brijesh Verma

Centre for Intelligent Systems, School of Engineering and Technology,
Central Queensland University, Brisbane, QLD 4000, Australia
{s.fletcher,b.verma}@cqu.edu.au

Abstract. Diversity plays an important role in successful ensemble classification. One way to diversify the base-classifiers in an ensemble classifier is to diversify the data they are trained on. Sampling techniques such as bagging have been used for this task in the past, however we argue that since they maintain the global distribution, they do not engender diversity. We instead make a principled argument for the use of k-Means clustering to create diversity. When creating multiple clusterings with multiple k values, there is a risk of different clusterings discovering the same clusters, which would then train the same base-classifiers. This would bias the ensemble voting process. We propose a new approach that uses the Jaccard Index to detect and remove similar clusters before training the base-classifiers, reducing classification error by removing repeated votes. We demonstrate the effectiveness of our proposed approach by comparing it to three state-of-the-art ensemble algorithms on eight UCI datasets.

Keywords: Ensemble · Classification · Clustering · Diversity · Bias · Voting

1 Introduction

By combining the predictions of multiple diverse classifiers, an ensemble of classifiers can perform better than any one classifier can. We propose a novel ensemble classifier that can more accurately classify unseen data than the current state-of-the-art. The proposed algorithm f takes some $n \times m$ data X as input, where the rows $x \in X$ represent independently and identically distributed samples from some universe $\mathcal{X}$, each with a label y that represents the classification category that x belongs to from some output space Y. $f(X)$ is trained using X, learning the underlying patterns in the data that determine what category $y \in Y$ any particular datum x will belong to. The aim of ensemble classification is then to use $f(X)$ to predict the label y of unseen data $z \in Z$, which comes from the same universe $\mathcal{X}$ as X [9]. The columns of X represent features A that describe the properties of each row (i.e., record) x, and it is these features that a classification algorithm uses to learn how to classify unseen data [9]. Examples of

© Springer International Publishing AG 2017
D. Liu et al. (Eds.): ICONIP 2017, Part IV, LNCS 10637, pp. 140–149, 2017.
https://doi.org/10.1007/978-3-319-70093-9_15

classification algorithms include support vector machines [28] and decision trees [21]. By training multiple classifiers, and then combining the predictions made by each classifier into a final overall prediction, an ensemble of classifiers can outperform any individual classifier [8,25].

1.1 Contributions

We propose a generalized ensemble classification algorithm that uses data diversity and base-classifier diversity to build a decision model with low classification error. Our novel contributions are:

- We maximize data diversity by creating subsets of data with large differences in distribution, using k-Means clustering for a large range of k values, $k = 1, \ldots, K$. We ensure that each cluster has sufficient data for training the base-classifiers by bounding K such that if there are k clusters, there are at least an average of k^2 records in each cluster.
- During the incremental clustering process, we compare the new clusters to all previous clusters, and remove (i.e., prune) the new cluster if it is sufficiently similar to a previous cluster. This prevents the prediction voting process from being biased by repetitious votes.
- When classifying a new record z, rather than inputting it into all base-classifiers, we only input the record into the classifiers built using the K clusters that are closest to the new record. This is done regardless of which clusterings the K clusters came from.

After presenting related work in Sect. 2, we introduce the proposed approach in Sect. 3, exploring each component in separate subsections. We then empirically test the proposed algorithm in Sect. 4, before concluding the paper in Sect. 5.

2   Related Work

The notion of "diversity" when building ensemble classifiers has been researched extensively in the past [1,6,14,24]. Despite being difficult to define precisely [6], the overall concept is straight-forward: if all the base-classifiers in an ensemble make the same predictions, they are also making the same mistakes, and if they are all making the same mistakes, there is no advantage in having more than one of them. By diversifying the predictions that the base-classifiers make, the ensemble can perform better than the sum of its parts. There are several types of diversity that an ensemble algorithm can achieve.

Data diversity is achieved by sampling subsets of data from an original dataset, in a way that causes the predictions made by classifiers trained on the subsets to differ from one another. This can be achieved by selecting either records or features (or both) from the original dataset. Duplicating records or features across the new sets of data is viable (such as bagging [2] or random feature subspaces [29]), as is creating mutually-exclusive sets (such as clustering

[22, 30, 32]). The manipulation of data has successfully diversified the data if the end result is that a diversity of predictions is outputted [14].

Classifier diversity (or "structural diversity" [24] or "heterogeneous ensembling" [18]) has a similar goal of diversifying the predictions that the base-classifiers output. Classifier diversity is achieved by using different classifier algorithms that learn from the data in different ways. By using a variety of classifiers, each with their own advantages and disadvantages, the outputs of the classifiers are diverse [4].

As an example, the Random Forest algorithm [3] uses bagging [2] and random feature subspaces [10, 29] to achieve data diversity. It only builds an ensemble of decision trees though, and thus does not target classifier diversity. In this paper, we use both types of diversity in our proposed approach. These are explored below in Sect. 3.

3   Proposed Approach

We first provide an overview of the proposed approach in Sect. 3.1. We then investigate each novel component of the approach one-by-one in the proceeding subsections.

3.1   Overview

The approach can be summarized in the following steps:

- **Step 1:** Calculate the largest number of clusters K we can partition the training data X into without reducing the average number of records in each cluster below the square of the number of clusters.
- **Step 2a:** For $k = 1, \ldots, K$ partition the data X using k-Means clustering.
- **Step 2b:** For each new cluster created, compare its similarity (in terms of the records it contains) to all previously created clusters, using the Jaccard Index [11]. Remove a new cluster if it is very similar to a previous cluster.
- **Step 3a:** For each remaining cluster, check if all the records in the cluster have the same class label y (i.e., the cluster is homogeneous). If so, skip 3b, and future records that are filtered to this cluster will be predicted to have the same label that all the training records had. In effect, the cluster will output v votes for the label y, rather than training v base-classifiers from the homogeneous data.
- **Step 3b:** For each cluster not addressed by 3a, build v base-classifiers using the data in the cluster. Examples of base-classifiers include a decision tree [21], a support vector machine [28], a naive Bayes model [20], a discriminant analysis model [19], a k-nearest neighbors model [31] and a randomly under-sampled boosted model (RUSBoost) [27]. Finding the optimal set of base-classifiers is part of future work.
- **Step 4:** Predict the label of new records z by filtering them into the K closest clusters, using the base-classifiers built from those clusters to each vote on a label, and then using the majority vote as the final prediction.

The specifics of each of these steps are explored below in the following subsections.

3.2   Achieving Data Diversity

To produce a diverse set of base-classifiers, we diversify the training data. Previous research supports using bagging to achieve this [5,16,26], however we argue that the benefits of bagging are in reducing the variance of the models [2], not in promoting diversity. Because bagging maintains the distribution of the underlying data with increasing detail as the sample size increases, it does not provide a diverse range of distributions to the base-classifiers. This problem can be avoided by clustering the data instead, finding regions of data with many attributes in common, and few attributes in common with other regions of data. We achieve this using k-Means clustering [12].

We *could* find a single optimal value for k, and limit our ensemble of base-classifiers to a single clustering. Instead though, we propose using a range of values for k, and building a much larger ensemble. By using values of k ranging from 1 to some upper bound K, we increase diversity further by finding clusters of different sizes (and different distributions) in the training data.

3.3   Choosing K

We need an appropriate K value that gives us a large set of clusters to build many diverse classifiers from, but also provides enough data in each cluster to meaningfully train the base-classifiers with. We balance these two goals with the following heuristic:

1. Let n_k equal the average number of records in each cluster created from k-Means clustering. For a number of clusters k, $n_k = n/k$, where n is the total number of records in the dataset.
2. We limit the maximum size of k such that $n_k \geq k^2$. That is, each cluster has an average number of records equal to at least the square of the number of clusters.
3. Thus we have $n/k \geq k^2$. Re-arranging this formula gives us: $n \geq k^3$.
4. The maximum number of clusters K is therefore:

$$K = \lfloor \sqrt[3]{n} \rfloor .$$

5. We define the minimum k value for k-Means clustering at $k = 1$.
6. Our proposed ensemble classifier therefore executes k-Means clustering K times, for $k = 1, \ldots, K$.

This balances the number of clusters with the size of each cluster. It is based on a similar concept used to bound k-Means clustering when using it on its own [12,23].

3.4   Pruning Repeated Clusters

In Sects. 3.2 and 3.3, we described how the proposed algorithm creates an increasing number of clusters, from 1 to K, using k-Means clustering. This results in a total of $\sum_{i=1}^{K} i$ clusters, or in other words, $K(K+1)/2$ clusters. This creates a

Table 1. Average difference in classification error compared to when $\theta = 0.9$, across the eight datasets presented in Table 2.

θ	0.5	0.6	0.7	0.8	0.9	1.0	No pruning
Error difference compared to 0.9	+0.056	+0.032	+0.023	+0.016	0.000	+0.007	+0.011

large set of clusters from which to then train base-classifiers from, which will in turn be used to vote on predicted class labels.

However there is a risk in using *all* of these clusters to train classifiers. If two clusters, made during different clusterings (i.e., when $k = i$, and then when $k = j$ such that $i \neq j$), contain all the same records, then the classifiers built from those two clusters will be very similar, maybe even identical (since there is zero data diversity). Not only does this waste computation time, but it also means that when voting on the final predicted label, the votes from these classifiers are doubling up (i.e. getting two votes), biasing the ensemble towards their output.

We remove this bias using the following process: as we grow k towards K, each cluster we create is compared to all previous clusters to check if it is sufficiently diverse. For each new cluster c created in clustering k, we compare the records in c (X_c) to the records of each cluster u in the set of accepted clusters U using the Jaccard Index [11]:

$$ J(c, u) = \frac{X_c \cap X_u}{X_c \cup X_u}; \forall u \in U. $$

Computationally, we can calculate $J(c, u)$ using the indexes of the records in X, rather than comparing the contents of each $x \in X$. If there are no records in common, $J(c, u) = 0$; and $J(c, u) = 1$ if both clusters contain precisely the same set of records. If, for some u, $J(c, u) > \theta$, c is not added to U. Here, θ represents a threshold of similarity, which we empirically demonstrate is ideally placed at $\theta = 0.9$ in Table 1. Table 1 presents the average difference in classification error, across the eight datasets presented in Table 2, when $\theta = 0.5, 0.6, 0.7, 0.8, 1.0$ (and when there is no pruning) compared to when $\theta = 0.9$.

Table 2. Details of the eight datasets we use in our experiments, taken from the UCI Machine Learning Repository [15].

Dataset	Records	Features	Labels
Sonar	208	60	2
Heart	270	13	2
Bupa	345	6	2
Ionosphere	351	34	2
WBC	683	9	2
PimaDiabetes	768	8	2
Vehicle	846	18	4
Segmentation	2310	19	7

3.5   Achieving Classifier Diversity

We then build a collection of base-classifiers from the data in each non-pruned cluster. For our experiments in this paper, we use the following six classifiers: a decision tree, a support vector machine, a naive Bayes model, a discriminant analysis model, a k-nearest neighbors model and a randomly under-sampled boosted model. This collection of classifiers is independent of the proposed ensemble framework, and future work will involve investigating the optimal amount and types of classifiers to use.

This diverse collection of classifiers enables the ensemble to discover correlations and patterns in the data that would not be discovered if we limited ourselves to a single classifier, such as what Random Forest does [3]. By discovering different patterns with different classifiers, we diversify the errors made by each base-classifier, which in turn reduces the final classification error (as discussed in Sects. 1 and 2). We can see in Table 5 (presented later in Sect. 4) that we are able to outperform Random Forest on almost all datasets.

3.6   Classifying New Records

Once the ensemble has been built and trained (Steps 1–3 in Sect. 3.1), our model is ready to predict the label of unseen records. When inputting an unseen record z into our ensemble classifier, we propose finding the K clusters in U whose centroids are closest to z, and using the base-classifiers made from those K clusters to predict the label of z.

This approach means that clusters made from different clusterings (when k had different values) are not treated differently from clusters built from the same clustering; if the centroids of two clusters from the same clustering are closer to z than the centroids of two clusters from different clusterings, the closer clusters are used. We also do not want to use the base-classifiers made from every cluster to classify z. Many of the base-classifiers were trained on data that had very different distributions to the distributions that z follows, and did not contain any records that resemble z. To use those classifiers to predict z therefore makes little sense.

4   Experiments and Results

Here we present experiments that cover both individual components of the proposed algorithm, and the overall performance of the algorithm compared to the current state-of-the-art. All experiments are performed using stratified five-fold cross-validation, repeated ten times and aggregated. We perform our experiments on eight datasets from the UCI Machine Learning Repository [15]. The details of the datasets are presented in Table 2. We use Matlab's implementation of k-Means and the base-classifiers for our experiments [17]. We use the default settings in all cases, except for the following:

– The maximum number of iterations for k-Means is increased from 100 to 500, to ensure that centroid convergence occurs.

- Regularization is turned off for discriminant analysis models, to prevent the software throwing an error if a feature with zero variance is inputted.
- Kernel smoothing density estimation is used when building naive Bayes models, instead of using Gaussian distributions, to avoid the software throwing an error if a feature with zero variance is inputted.

4.1 Assessing Cluster Size and Pruning

The first step in our proposed algorithm is to define K. We argue that $\sqrt[3]{n}$ is an appropriate value of K, and this is supported empirically by the results seen in Table 3. In Table 3, we compare $K = \sqrt[3]{n}$ to one smaller value ($\sqrt[4]{n}$) and one larger value ($\sqrt{n}$) of K. Classification error is lowest when $K = \sqrt[3]{n}$ for six of the eight datasets, and very close to lowest for the remaining two (within one standard deviation).

As part of Step 2, we propose removing repeated clusters using the Jaccard Index. Based on the empirical results of Table 1, we recommend setting the similarity threshold to $\theta = 0.9$. This removal of repeated clusters represents a large saving in computation time, preventing v (in our experiments, $v = 6$)

Table 3. The classification error for three different values of K.

Dataset	$K = \sqrt[4]{n}$	$K = \sqrt[3]{n}$	$K = \sqrt{n}$
Sonar	0.1712	**0.1295**	0.1481
Heart	**0.1793**	**0.1778**	0.2252
Bupa	0.2817	**0.2609**	0.2916
Ionosphere	0.0814	**0.0797**	0.0866
WBC	0.0562	**0.0322**	**0.0301**
PimaDiabetes	0.2393	**0.2306**	0.2521
Vehicle	**0.2373**	**0.2352**	0.2532
Segmentation	0.0515	0.0325	**0.0275**

Table 4. The change in classification error with and without cluster pruning.

Dataset	With pruning	Without pruning
Sonar	**0.1295**	0.1501
Heart	**0.1778**	0.1881
Bupa	**0.2609**	0.2817
Ionosphere	**0.0797**	0.0832
WBC	**0.0322**	**0.0340**
PimaDiabetes	**0.2306**	0.2432
Vehicle	**0.2352**	0.2454
Segmentation	0.0325	**0.0269**

redundant classifiers from being trained per removed cluster. It also represents the removal of a high number of repeated votes. The reduction in classification error because of this removal of biased votes can be seen in Table 4. For seven of the eight datasets, the classification error after pruning repeated clusters is lower than or equal to the error without this pruning. On average across all datasets, the average reduction in error is 1.1 percentage points.

4.2   Comparison with Other Ensemble Algorithms

Table 5 presents the classification error our proposed algorithm achieves on eight datasets, compared to the classification error of three other algorithms. One of these algorithms, Random Forest, is included as a benchmark ensemble algorithm due to its reputation as a consistently high-performing algorithm [7]. The other two algorithms represent the current state-of-the-art in ensemble classification, with the results presented being the results the authors reported in their respective papers: Kuncheva and Rodriguez [13]; and Zhang and Suganthan [33]. In both cases, we present the results for the highest performing version of their proposed algorithms; the naive Bayes (NB) version of Kuncheva and Rodriguez's [13], and the version of Zhang and Suganthan's that uses oblique rotation forests with axis-parallel splits (MPRRoF-P) [33].

Out of the eight datasets, the approach proposed in this paper has the lowest classification error in five cases. It has the second-lowest in two cases, and the third-lowest for one dataset (Segmentation). Interestingly, as we saw in Table 4, the Segmentation dataset is also the only dataset for which our proposed cluster pruning does not perform well. This explains the sub-par performance compared to the state-of-the-art for this dataset.

Table 5. The classification error results for four ensemble algorithms, including our proposed approach.

Dataset	Proposed approach	Random forest	Kuncheva 2014 (NB)	Zhang 2015 (MPRRoF-P)
Sonar	**0.1295**	0.1460	0.238	0.1923
Heart	**0.1778**	0.1810	0.195	**0.1763**
Bupa	**0.2609**	0.2727	0.328	N/A
Ionosphere	0.0797	0.0703	0.083	**0.0530**
WBC	**0.0322**	0.0390	0.040	**0.0333**
PimaDiabetes	**0.2306**	0.2396	0.245	0.2474
Vehicle	0.2352	0.2435	0.275	**0.2219**
Segmentation	0.0325	**0.0200**	0.036	**0.0196**

5   Conclusion

Diversity is crucial for building a high-performing ensemble classifier. By performing K clusterings of different sizes, and using these clusters as training data for a diverse set of base-classifiers, the error of the ensemble classifier is reduced. Not only that, but by first pruning repeated clusters, biased votes can be removed from the majority voting process, further reducing classification error. On average, pruning repeated clusters reduces classification error by 1.1%. Looking forward, we plan to investigate what factors affect classifier diversity, and how classifier diversity impacts the performance of ensemble classification.

Acknowledgments. This research was supported by the Australian Research Council's Discovery Project funding scheme (Project Number DP160102639).

References

1. Asafuddoula, M., Verma, B., Zhang, M.: An incremental ensemble classifier leaning by means of a rule-based accuracy and diversity comparison. In: International Joint Conference on Neural Networks, p. 8. IEEE, Anchorage (2017)
2. Breiman, L.: Bagging predictors. Mach. Learn. **24**(2), 123–140 (1996)
3. Breiman, L.: Random forests. Mach. Learn. **45**(1), 5–32 (2001)
4. Britto, A.S., Sabourin, R., Oliveira, L.E.S.: Dynamic selection of classifiers - a comprehensive review. Pattern Recogn. **47**(11), 3665–3680 (2014)
5. Chang, K.H., Parker, D.S.: Complementary prioritized ensemble selection. In: International Joint Conference on Neural Networks, pp. 863–872 (2016)
6. Didaci, L., Fumera, G., Roli, F.: Diversity in classifier ensembles: fertile concept or dead end? In: Zhou, Z.-H., Roli, F., Kittler, J. (eds.) MCS 2013. LNCS, vol. 7872, pp. 37–48. Springer, Heidelberg (2013). doi:10.1007/978-3-642-38067-9_4
7. Fernández-Delgado, M., Cernadas, E., Barro, S., Amorim, D., Amorim Fernández-Delgado, D.: Do we need hundreds of classifiers to solve real world classification problems? J. Mach. Learn. Res. **15**(1), 3133–3181 (2014)
8. Gopika, D., Azhagusundari, B.: An analysis on ensemble methods in classification tasks. Int. J. Adv. Res. Comput. Commun. Eng. **3**(7), 7423–7427 (2014)
9. Han, J., Kamber, M., Pei, J.: Data Mining: Concepts and Techniques. Morgan Kaufmann Publishers, San Diego (2006)
10. Ho, T.K.: The random subspace method for constructing decision forests. IEEE Trans. Pattern Anal. Mach. Intell. **20**(8), 832–844 (1998)
11. Jaccard, P.: Etude comparative de la distribution florale dans une portion des Alpes et du Jura. Bull. Soc. Vaudoise Sci. Nat. **37**, 547–579 (1901)
12. Jain, A.K.: Data clustering: 50 years beyond K-means. Pattern Recogn. Lett. **31**(8), 651–666 (2010)
13. Kuncheva, L.I., Rodríguez, J.J.: A weighted voting framework for classifiers ensembles. Knowl. Inf. Syst. **38**(2), 259–275 (2014)
14. Kuncheva, L.I., Whitaker, C.J.: Measures of diversity in classifier ensembles and their relationship with the ensemble accuracy. Mach. Learn. **51**(2), 181–207 (2003)
15. Lichman, M.: UCI Machine Learning Repository (2013). http://archive.ics.uci.edu/ml/
16. Mao, S., Jiao, L., Xiong, L., Gou, S., Chen, B., Yeung, S.K.: Weighted classifier ensemble based on quadratic form. Pattern Recogn. **48**(5), 1688–1706 (2015)

17. MathWorks: MATLAB and Statistics and Machine Learning Toolbox
18. Mendes-Moreira, J., Soares, C., Jorge, A.M., Sousa, J.F.D.: Ensemble approaches for regression. ACM Comput. Surv. **45**(1), 1–40 (2012)
19. Mika, S., Ratsch, G., Weston, J., Schölkopf, B., Muller, K.R.: Fisher discriminant analysis with kernels. In: IEEE Signal Processing Society Workshop, pp. 41–48. IEEE (1999)
20. Ng, A.Y., Jordan., M.I.: On discriminative vs. generative classifiers: a comparison of logistic regression and naive Bayes. In: Advances in Neural Information Processing Systems, pp. 841–848. NIPS (2002)
21. Quinlan, J.R.: C4.5: Programs for Machine Learning, 1st edn. Morgan Kaufmann, Burlington (1993)
22. Rahman, A., Verma, B.: A novel layered clustering based approach for generating ensemble of classifiers. IEEE Trans. Neural Netw. **22**(5), 781–792 (2011)
23. Rahman, M.A., Islam, M.Z.: A hybrid clustering technique combining a novel genetic algorithm with K-Means. Knowl.-Based Syst. **71**(1), 345–365 (2014)
24. Ren, Y., Zhang, L., Suganthan, P.N.: Ensemble classification and regression - recent developments, applications and future directions. IEEE Comput. Intell. Mag. **11**(1), 41–53 (2016)
25. Rokach, L.: Ensemble-based classifiers. Artif. Intell. Rev. **33**(1), 1–39 (2010)
26. Santucci, E., Didaci, L., Fumera, G., Roli, F.: A parameter randomization approach for constructing classifier ensembles. Pattern Recogn. **69**(1), 1–13 (2017)
27. Seiffert, C., Khoshgoftaar, T.M., Hulse, J.V., Napolitano, A.: RUSBoost: a hybrid approach to alleviating class imbalance. IEEE Trans. Syst. Man Cybern. Part A Syst. Hum. **40**(1), 185–197 (2010)
28. Suykens, J.A.K., Vandewalle, J.: Least squares support vector machine classifiers. Neural Process. Lett. **9**(3), 293–300 (1999)
29. Tan, C., Li, M., Qin, X.: Random subspace regression ensemble for near-infrared spectroscopic calibration of tobacco samples. Anal. Sci. **24**(5), 647–653 (2008)
30. Verma, B., Rahman, A.: Cluster oriented ensemble classifier: impact of multi-cluster characterisation on ensemble classifier learning. IEEE Trans. Knowl. Data Eng. **24**(4), 605–618 (2012)
31. Weinberger, K., Blitzer, J., Saul, L.: Distance metric learning for large margin nearest neighbor classification. In: Advances in Neural Information Processing Systems, pp. 1473–1480 (2006)
32. Yang, Y., Jiang, J.: Hybrid sampling-based clustering ensemble with global and local constitutions. IEEE Trans. Neural Netw. Learn. Syst. **27**(5), 952–965 (2016)
33. Zhang, L., Suganthan, P.N.: Oblique decision tree ensemble via multisurface proximal support vector machine. IEEE Trans. Cybern. **45**(10), 2165–2176 (2015)

An Efficient Algorithm for Complex-Valued Neural Networks Through Training Input Weights

Qin Liu, Zhaoyang Sang, Hua Chen, Jian Wang, and Huaqing Zhang[✉]

College of Science, China University of Petroleum, Qingdao 266580, China
lqin_1994@163.com, {sangzhy,chenhua,wangjiann1,zhhq}@upc.edu.cn

Abstract. Complex-valued neural network is a type of neural networks, which is extended from real number domain to complex number domain. Fully complex extreme learning machine (CELM) is an efficient algorithm, which owes faster convergence than the common complex back-propagation (CBP) neural networks. However, it needs more hidden neurons to reach competitive performance. Recently, an efficient learning algorithm is proposed for the single-hidden layer feed-forward neural network which is called the upper-layer-solution-aware algorithm (USA). Motivated by USA, an efficient algorithm for complex-valued neural networks through training input weights (GGICNN) has been proposed to train the split complex-valued neural networks in this paper. Compared with CELM and CBP, an illustrated experiment has been done in detail which observes the better generalization ability and more compact architecture for the proposed algorithm.

Keywords: Complex-valued · Neural networks · Extreme learning machine · Complex backpropagation · Gradient

1 Introduction

Complex-valued neural network (CVNN) is one of typical network models which deals with problems in complex space. It is an extensive version of conventional real-valued neural networks [1]. CVNN has been widely used in the fields of solving signal processing, classification problems and channel equalization applications [2,3]. A series of complex-valued neural network models have been proposed in past decades which mainly contain the complex backpropagation [6–10], fully complex extreme learning machine [16,17], and fractional-order complex-valued neural networks [5]. We note that there are two main categories for CVNN models: the split CVNN and the fully CVNN, which are valid for both backpropagation and recursive training models.

For the fully CVNN [19], a typical characteristic is that the activation functions are fully complex-valued. According to the Liouvilles theorem, there exists an inevitable conflict between the boundedness and the differentiability of the common activation function, such as, Sigmoid and Gaussian functions. On the

© Springer International Publishing AG 2017
D. Liu et al. (Eds.): ICONIP 2017, Part IV, LNCS 10637, pp. 150–159, 2017.
https://doi.org/10.1007/978-3-319-70093-9_16

contrary, the split CVNN employs a pair of activation functions to active the real and imaginary parts of the neurons' input, respectively. This training scheme may essentially avoid the above singularity conflict. For backpropagation scheme, the split and fully complex-valued neural networks has gradually been an attractive research topic [8–12]. Accordingly, there are two types of complex BP algorithm: one is the the split complex BP algorithm (SCBPA) [8,9] for split CVNN, and another is the fully complex BP algorithm (FCBPA) [10–12] for fully CVNN. Similar to the common real BP network model, SCBPA employs the gradient descent method to iteratively update the total weights. The input weights (connecting input and hidden layers) and the output weights (connecting hidden and output layers) are randomly initialized in complex domain.

Extreme learning machine (ELM) was first proposed in [14], which mainly focused on the efficient training for single hidden layer neural networks (SHLNNs). Compare with the inherent drawbacks of BP networks, slow convergence and local minimum, it was a breakthrough in finding the optimal solutions of SHLNNs. During training, the input weights were randomly assigned, and the output weights were immediately then expressed as a least square solution by using the Moore-Penrose generalized inverse formula [4,13]. This non-iteration updating scheme resulted in an extremely fast training speed. As a neural extension, the complex counterpart of the classical ELM, complex extreme learning machine (CELM), was first presented in [16], it extended the searching space from real domain to complex domain. More importantly, CELM can deals with the complex problems, which are widely existed in practical applications. However, wo note that it requires more hidden neurons to achieve the competitive performance than those in applying the common complex BP neural networks (CBPNN). This then leads to more testing time and increases the computational burden in dealing with complex problems.

An upper-layer-solution-aware algorithm (USA) was referred in [15], which can be treated as an intermediate of BP and ELM models. It efficiently overcomes the slow convergence speed of BP neural network and the high requirements of hidden neurons of ELM. It implies the gradient descent method to adjust the input weights and evaluates the output weights by Moore-Penrose generalized inverse. Motivated by this training scheme, we propose an efficient algorithm for complex-valued neural networks through training input weights (GGICNN) to train single hidden layer complex-valued neural networks. During training process, the input weights are iteratively updated by the gradient descent method, and the output weights are always considered to be a nonlinear mapping of the input weights, which is simultaneously expressed by Moore-Penrose generalized inverse. For given complex problem, numerical simulations demonstrate the better performance of the proposed algorithm than its counterparts, SCBPA and CELM.

The rest of the paper is organized as follows. Section 2 presents a simple review of the SCBPA and CELM algorithms. Section 3 describes a novel algorithm to train single hidden layer complex-valued neural networks. These different algorithms have been detailedly compared on a specific complex problem in Sect. 4. Finally, some conclusions are given in Sect. 5.

2   Related Works

We consider a single hidden layer complex-valued neural network with L input neurons, M hidden neurons and one output neuron. Given a series of complex-valued training samples $(\mathbf{z}_q, d_q), q = 1, \cdots, Q$, where $\mathbf{z}_q$ is the q-th input and d_q is the corresponding ideal output. The input matrix $\mathbf{z} = (\mathbf{z}_1, \mathbf{z}_2, \cdots, \mathbf{z}_Q) \in \mathbb{C}^{L \times Q}$, each vector can be denoted by $\mathbf{z}_q = \mathbf{z}^{q,R} + i\mathbf{z}^{q,I} \in \mathbb{C}^L$, where $\mathbf{z}^{q,R} = [z_1{}^{q,R}, z_2{}^{q,R}, \cdots, z_L{}^{q,R}]^T$, $\mathbf{z}^{q,I} = [z_1{}^{q,I}, z_2{}^{q,I}, \cdots, z_L{}^{q,I}]^T$. The target vectors $\mathbf{D} = \mathbf{D}^R + i\mathbf{D}^I = [d_1, d_2, \cdots, d_Q]^T \in \mathbb{C}^Q$, where $d^q = d^{q,R} + id^{q,I} \in \mathbb{C}$. We write $\mathbf{w}_m = \mathbf{w}_m^R + i\mathbf{w}_m^I = [w_{m1}, w_{m2}, \cdots, w_{mL}]^T \in \mathbb{C}^L, m = 1, \cdots, M$, as the complex input weight vector connecting the input neurons and the mth hidden neuron. $\mathbf{V} = \mathbf{V}^R + i\mathbf{V}^I = [v_1, v_2, \cdots, v_M]^T \in \mathbb{C}^M$ is the complex output weight vector between the hidden neurons and the output neuron, where $v_m = v_m^R + iv_m^I \in \mathbb{C}, m = 1, \cdots, M$. Specially, we assume $g(x)$ and $f(x)$ $(x \in \mathbb{R})$ are the activation functions of hidden and output layers, separately. The output of a standard single hidden layer complex-valued neural network can be evaluated as follows:

$$o^q = \sum_{m=1}^{M} f(v_m \cdot g(\mathbf{w}_m \cdot \mathbf{z}_q)). \tag{1}$$

2.1   Split-Complex Back-Propagation Algorithm

SCBPA [8,9], which is an effective learning algorithm for training split CVNN, is an extension of the BPA in complex domain. For simplicity, we write all the weight vectors in an abbreviated form

$$\mathbf{W} = \left((\mathbf{w}_1)^T, (\mathbf{w}_2)^T, \ldots, (\mathbf{w}_M)^T, \mathbf{V}^T\right)^T \in \mathbb{C}^{M(L+1)}. \tag{2}$$

For the qth sample, $\mathbf{z}_q$, the input of the mth hidden neuron is as follows:

$$u_m^q = u_m^{q,R} + iu_m^{q,I} = \sum_{l=1}^{L}\left(w_{ml}^R z_l^{q,R} - w_{ml}^I z_l^{q,I}\right) + i\sum_{l=1}^{L}\left(w_{ml}^I z_l^{q,R} + w_{ml}^R z_l^{q,I}\right)$$

$$= \begin{pmatrix}\mathbf{w}_m^R \\ -\mathbf{w}_m^I\end{pmatrix} \cdot \begin{pmatrix}\mathbf{z}^{q,R} \\ \mathbf{z}^{q,I}\end{pmatrix} + i\begin{pmatrix}\mathbf{w}_m^I \\ \mathbf{w}_m^R\end{pmatrix} \cdot \begin{pmatrix}\mathbf{z}^{q,R} \\ \mathbf{z}^{q,I}\end{pmatrix}, \tag{3}$$

where $m = 1, 2, \cdots, M$, "$\cdot$" denotes the inner product of two vectors, i is the imaginary unit. For SCBPA, we consider the following popular real-imaginary-type activation strategy. The output of the mth hidden neuron is

$$h_m^q = h_m^{q,R} + ih_m^{q,I} = g\left(u_m^{q,R}\right) + ig\left(u_m^{q,I}\right), \tag{4}$$

where g is a real function. Similarly, the input of the output neuron is stated as:

$$s^q = s^{q,R} + is^{q,I} = \sum_{m=1}^{M}\left(v_m^R h_m^{q,R} - v_m^I h_m^{q,I}\right) + i\sum_{m=1}^{M}\left(v_m^I h_m^{q,R} + v_m^R h_m^{q,I}\right)$$

$$= \begin{pmatrix}\mathbf{V}^R \\ -\mathbf{V}^I\end{pmatrix} \cdot \begin{pmatrix}\mathbf{h}^{q,R} \\ \mathbf{h}^{q,I}\end{pmatrix} + i\begin{pmatrix}\mathbf{V}^I \\ \mathbf{V}^R\end{pmatrix} \cdot \begin{pmatrix}\mathbf{h}^{q,R} \\ \mathbf{h}^{q,I}\end{pmatrix}, \tag{5}$$

and the final output of the network is

$$o^q = o^{q,R} + io^{q,I} = f\left(s^{q,R}\right) + if\left(s^{q,I}\right), \tag{6}$$

where $\mathbf{h}^{q,R} = \left(h_1^{q,R}, h_2^{q,R}, \ldots, h_M^{q,R}\right)^T$ and $\mathbf{h}^{q,I} = \left(h_1^{q,I}, h_2^{q,I}, \ldots, h_M^{q,I}\right)^T$ are the output vectors of hidden layer. The mean squared error can be represented as follows:

$$E(\mathbf{W}) = \frac{1}{2}\sum_{q=1}^{Q}(o^q - d^q)(o^q - d^q)^*$$

$$= \frac{1}{2}\sum_{q=1}^{Q}\left[(o^{q,R} - d^{q,R})^2 + (o^{q,I} - d^{q,I})^2\right], \tag{7}$$

where $(\cdot)^*$ stands for the complex conjugate.

Given an initial weight $\mathbf{W}^0$, the weight sequence is iteratively updated based on gradient descent method,

$$\mathbf{W}^{n+1} = \mathbf{W}^n + \Delta\mathbf{W}^n, n = 0, 1, \cdots, \tag{8}$$

where $\Delta\mathbf{W}^n = \left((\Delta\mathbf{w}_1^n)^T, \cdots, (\Delta\mathbf{w}_M^n)^T, (\Delta\mathbf{V}^n)^T\right)^T$, and

$$\Delta\mathbf{w}_m^n = -\eta\left(\frac{\partial E\left(\mathbf{W}^n\right)}{\partial\mathbf{w}_m^R} + i\frac{\partial E\left(\mathbf{W}^n\right)}{\partial\mathbf{w}_m^I}\right), \tag{9}$$

$$\Delta\mathbf{V}^n = -\eta\left(\frac{\partial E\left(\mathbf{W}^n\right)}{\partial\mathbf{V}^R} + i\frac{\partial E\left(\mathbf{W}^n\right)}{\partial\mathbf{V}^I}\right), \tag{10}$$

where $n \in N$; $m = 1, 2, \cdots, M$; $\eta > 0$ is the learning rate.

2.2 Fully Complex Extreme Learning Machine

The CELM [16], which is a popular learning technique for training SHLNNs, is an extension of the ELM in complex domain. The complex input weight matrix connecting the input layer to the hidden layer is randomly assigned in a given interval, which is

$$\mathbf{w} = \left((\mathbf{w}_1)^T, (\mathbf{w}_2)^T, \ldots, (\mathbf{w}_M)^T\right)^T \in \mathbb{C}^{M \times L}, \tag{11}$$

where $\mathbf{w} = \mathbf{w}^R + i\mathbf{w}^I$. The complex input matrix of hidden layer is

$$\mathbf{U} = \mathbf{U}^R + i\mathbf{U}^I = \mathbf{w}\mathbf{z} = \begin{pmatrix} \mathbf{w}_1 \cdot \mathbf{z}_1 & \mathbf{w}_1 \cdot \mathbf{z}_2 & \cdots & \mathbf{w}_1 \cdot \mathbf{z}_Q \\ \mathbf{w}_2 \cdot \mathbf{z}_1 & \mathbf{w}_2 \cdot \mathbf{z}_2 & \cdots & \mathbf{w}_2 \cdot \mathbf{z}_Q \\ \vdots & \vdots & \ddots & \vdots \\ \mathbf{w}_M \cdot \mathbf{z}_1 & \mathbf{w}_M \cdot \mathbf{z}_2 & \cdots & \mathbf{w}_M \cdot \mathbf{z}_Q \end{pmatrix}_{M \times Q}. \tag{12}$$

Different from SCBPA, CELM is used to train the fully CVNN. The activation functions of the fully CVNN are fully complex-valued. The complex output matrix of hidden layer is

$$
\mathbf{H} = \begin{pmatrix} g\left(\mathbf{w}_1 \cdot \mathbf{z}_1\right) & g\left(\mathbf{w}_1 \cdot \mathbf{z}_2\right) & \cdots & g\left(\mathbf{w}_1 \cdot \mathbf{z}_Q\right) \\ g\left(\mathbf{w}_2 \cdot \mathbf{z}_1\right) & g\left(\mathbf{w}_2 \cdot \mathbf{z}_2\right) & \cdots & g\left(\mathbf{w}_2 \cdot \mathbf{z}_Q\right) \\ \vdots & \vdots & \ddots & \vdots \\ g\left(\mathbf{w}_M \cdot \mathbf{z}_1\right) & g\left(\mathbf{w}_M \cdot \mathbf{z}_2\right) & \cdots & g\left(\mathbf{w}_M \cdot \mathbf{z}_Q\right) \end{pmatrix}_{M \times Q} , \tag{13}
$$

where "·" denotes the inner product of two vectors. g is the complex activation function, which has a variety of options, such as circular functions, hyperbolic functions and so on [18], while linear function $f(x) = x$ is the activation function of output layer.

In practice, the number of training samples Q is usually much more than the number of the hidden neurons M, thus, the least-squares solution $\mathbf{V}$ of the linear system $\mathbf{H}^T \mathbf{V} = \mathbf{D}$ can be computed by using the method of singular value decomposition, that is

$$
\mathbf{V} = \left(\mathbf{H}^T\right)^\dagger \mathbf{D}, \tag{14}
$$

where complex matrix $\left(\mathbf{H}^T\right)^\dagger$ is the Moore-penrose generalized inverse of complex matrix $\mathbf{H}^T$ [4,13],

$$
\left(\mathbf{H}^T\right)^\dagger = \left(\mathbf{H}^* \mathbf{H}^T\right)^{-1} \mathbf{H}^*. \tag{15}
$$

For this algorithm, there is no iteration during the whole process, thus, it has the advantages of faster learning speed and avoiding falling into local minimum. However, CELM needs more hidden neurons and training time in order to achieve better performance.

3   The Proposed GGICNN Algorithm

In the training process, GGICNN updates the input weights $\mathbf{w}$ by the gradient descent method, and the output weights $\mathbf{V}$ are always solved by the least-square method. The essential idea of this algorithm is that $\mathbf{V}$ is considered as a function of $\mathbf{w}$. Same to (11), an arbitrary initial value $\mathbf{w}^0$ is randomly given. Denote the vector valued function $\mathbf{G}(\mathbf{x}) = (g(x_1), g(x_2), \cdots, g(x_M))^T \in \mathbf{R}^M$, where $\mathbf{x} = (x_1, x_2, \cdots, x_M)^T \in \mathbf{R}^M$.

According to complex input matrix (12), we consider the real-imaginary-type activation function. The complex output matrix of hidden layer is

$$
\mathbf{H} = \mathbf{H}^R + i\mathbf{H}^I = \mathbf{G}\left(\mathbf{U}^R\right) + i\mathbf{G}\left(\mathbf{U}^I\right) = \begin{pmatrix} h_1^1 & h_1^2 & \cdots & h_1^Q \\ h_2^1 & h_2^2 & \cdots & h_2^Q \\ \vdots & \vdots & \ddots & \vdots \\ h_M^1 & h_M^2 & \cdots & h_M^Q \end{pmatrix}_{M \times Q} . \tag{16}
$$

The activation function of output layer is $f(x) = x$. Thus, the actual output of the network is

$$\mathbf{O} = \mathbf{H}^T \mathbf{V} = (o_1, o_2, \cdots, o_Q) \in \mathbb{C}^Q. \tag{17}$$

Similar to the ELM model, we employ the least-square method on the linear system $\mathbf{O} = \mathbf{H}^T \mathbf{V}$ to get the least-squares solution $\mathbf{V}$,

$$\mathbf{V} = \left(\mathbf{H}^T\right)^\dagger \mathbf{D} = (\mathbf{H}^* \mathbf{H}^T)^{-1} \mathbf{H}^* \mathbf{D}, \tag{18}$$

where $\left(\mathbf{H}^T\right)^\dagger$ is the Moore-penrose generalized inverse.

By (16), (17) and (18), we can immediately observe that the output weight matrix $\mathbf{V}$ is a function of the output weight matrix $\mathbf{w}$.

The aim of the network training is to find the optimal weights $\mathbf{w}$ that can minimize the error function,

$$E(\mathbf{w}) = \frac{1}{2} \sum_{q=1}^{Q} (o^q - d^q)(o^q - d^q)^* \tag{19}$$

$$= \frac{1}{2} \sum_{q=1}^{Q} [(o^{q,R} - d^{q,R})^2 + (o^{q,I} - d^{q,I})^2].$$

The gradients of the error function $E(\mathbf{w})$ with respect to $\mathbf{w}^R$ and $\mathbf{w}^I$ are, respectively,

$$\frac{\partial E(\mathbf{w})}{\partial \mathbf{w}_m^R} = \sum_{q=1}^{Q} \left[(o^{q,R} - d^{q,R})((\frac{\partial v_m^R}{\partial h_m^{q,R}} \cdot h_m^{q,R} + v_m^R - \frac{\partial v_m^I}{\partial h_m^{q,R}} \cdot h_m^{q,I})g'(u_m^{q,R})\mathbf{z}^{q,R} \right.$$

$$+ (\frac{\partial v_m^R}{\partial h_m^{q,I}} \cdot h_m^{q,R} - (\frac{\partial v_m^I}{\partial h_m^{q,I}} \cdot h_m^{q,I} + v_m^I))g'(u_m^{q,I})\mathbf{z}^{q,I})$$

$$+ (o^{q,I} - d^{q,I})((\frac{\partial v_m^I}{\partial h_m^{q,R}} \cdot h_m^{q,R} + v_m^I + \frac{\partial v_m^R}{\partial h_m^{q,R}} \cdot h_m^{q,I})g'(u_m^{q,R})\mathbf{z}^{q,R}$$

$$\left. + (\frac{\partial v_m^I}{\partial h_m^{q,I}} \cdot h_m^{q,R} + v_m^R + \frac{\partial v_m^R}{\partial h_m^{q,I}} \cdot h_m^{q,I})g'(u_m^{q,I})\mathbf{z}^{q,I}) \right], \tag{20}$$

$$\frac{\partial E(\mathbf{w})}{\partial \mathbf{w}_m^I} = \sum_{q=1}^{Q} \left[(o^{q,R} - d^{q,R})(-(\frac{\partial v_m^R}{\partial h_m^{q,R}} \cdot h_m^{q,R} + v_m^R - \frac{\partial v_m^I}{\partial h_m^{q,R}} \cdot h_m^{q,I})g'(u_m^{q,R})\mathbf{z}^{q,I} \right.$$

$$+ (\frac{\partial v_m^R}{\partial h_m^{q,I}} \cdot h_m^{q,R} - (\frac{\partial v_m^I}{\partial h_m^{q,I}} \cdot h_m^{q,I} + v_m^I))g'(u_m^{q,I})\mathbf{z}^{q,R})$$

$$+ (o^{q,I} - d^{q,I})(-(\frac{\partial v_m^I}{\partial h_m^{q,R}} \cdot h_m^{q,R} + v_m^I + \frac{\partial v_m^R}{\partial h_m^{q,R}} \cdot h_m^{q,I})g'(u_m^{q,R})\mathbf{z}^{q,I}$$

$$\left. + (\frac{\partial v_m^I}{\partial h_m^{q,I}} \cdot h_m^{q,R} + v_m^R + \frac{\partial v_m^R}{\partial h_m^{q,I}} \cdot h_m^{q,I})g'(u_m^{q,R})\mathbf{z}^{q,R}) \right], \tag{21}$$

where $1 \leq m \leq M$.

Starting from any initial weight $\mathbf{w}^0$, GGICNN updates the weight matrix $\mathbf{w}$ by

$$\mathbf{w}^{n+1} = \mathbf{w}^n + \Delta\mathbf{w}^n, n = 0, 1, \cdots, \tag{22}$$

where $\Delta\mathbf{w}^n = \left((\Delta\mathbf{w}_1^n)^T, \cdots, (\Delta\mathbf{w}_M^n)^T \right)^T$, with

$$\Delta\mathbf{w}_m^{n,R} = \mathbf{w}_m^{n+1,R} - \mathbf{w}_m^{n,R} = -\eta \frac{\partial E(\mathbf{w}^n)}{\partial \mathbf{w}_m^R}, \tag{23}$$

$$\Delta\mathbf{w}_m^{n,I} = \mathbf{w}_m^{n+1,I} - \mathbf{w}_m^{n,I} = -\eta \frac{\partial E(\mathbf{w}^n)}{\partial \mathbf{w}_m^I}, \tag{24}$$

where $n \in N$; $m = 1, 2, \cdots, M$; $\eta > 0$ is the learning rate.

4   Numerical Experiment

In order to evaluate the performance of GGICNN, two numerical simulations have been done on the complex nonminimum-phase channel model. The results have been compared with those of SCBPA and CELM.

An equalization model is used to evaluate the GGICNN equalizer performance. It is a complex nonminimum-phase channel of order 3 with nonlinear distortion for 4-QAM signaling which was introduced in [2]. The input of the equalizer is given by

$$x_n = y_n + 0.1y_n^2 + 0.05y_n^3 + v_n, v_n \sim \mathcal{N}(0, 0.01) \tag{25}$$

$$y_n = (0.34 - i0.27)s_n + (0.87 + i0.43)s_{n-1} + (0.34 - i0.21)s_{n-2} \tag{26}$$

The target output of the equalizer is the sequence $\{s_n\}(n \in [1, 1000])$, where $s_0 = -0.7 - 0.7i$, $s_{-1} = -0.7 + 0.7i$, the real and imaginary parts of the sequence are randomly selected from the set $\{\pm 0.7\}$.

Simulation 1

The three complex-valued neural networks based on the SCBPA, CELM and GGICNN have the same structure with 3 input neurons, 4 hidden neurons and one output neuron. The training dataset was generated by the following case: the input matrix

$$\mathbf{z} = \begin{pmatrix} x_3 & x_6 & \cdots & x_{1000} \\ x_2 & x_5 & \cdots & x_{999} \\ x_1 & x_4 & \cdots & x_{998} \end{pmatrix}_{3 \times 998},$$

the target vector

$$\mathbf{D} = [s_2, s_3, \cdots, s_{999}]^T.$$

The activation functions of these algorithms are chosen as $tanh(\cdot)$. The learning rate $\eta = 0.1$. We choose the initial weights (both the real part and imaginary part) in a interval $[-1, +1]$. The maximum iteration is set to be 100, for SCBPA

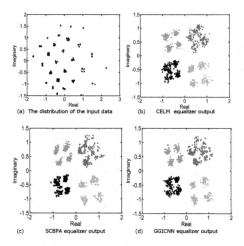

Fig. 1. Input data and the output of different equalizers.

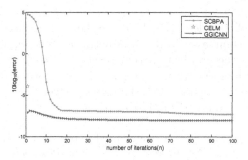

Fig. 2. The mean square errors of different learning algorithms.

and GGICNN. From Fig. 1, we observe that CELM, SCBPA and the proposed GGICNN algorithm can divide the output into four parts clearly. This then shows the competitive classification ability on the complex-valued problem. In addition, we have also compare the training errors for these different models. Figure 2 shows that the mean square errors of the SCBPA and GGICNN are monotonically decreasing. It apparently indicates that the proposed GGICNN performs much better than SCBPA and CELM. For the GGICNN algorithm, since the weight updating formula is too complex, we calculate the gradient by the finite difference method. Thus, it can be seen that the error of GGICNN grows a bit in Fig. 2 at the beginning but reduces in the following iterations.

Simulation 2

In this simulation, we assume the hidden neurons are 4 for SCBPA, GGICNN and 10 for CELM, respectively. The training parameters are identical to those in Simulation 1. Each network configuration was run 5 times to calculate the mean values of errors both on training and testing sets. Table 1 shows the square errors and running time comparison for the three algorithms. It can be seen that GGICNN trains faster than the SCBPA under the same network structure, and requires fewer hidden neurons than CELM algorithm. Importantly, we observe that GGICNN performs best among these algorithms.

Table 1. Mean square error comparison for different algorithms

Algorithms	Hidden neurons	Number of iterations	Training error	Testing error	Training time (s)	Testing time (s)
SCBPA	4	100	0.168	0.199	9.870	19.951
CELM	10	1	0.205	0.145	0.001	0.002
GGICNN	4	100	0.150	0.117	0.023	0.052

5   Conclusions

In this paper, an efficient learning algorithm, GGICNN, has been proposed for single hidden layer complex-valued neural networks. The input weights are iteratively updated by the gradient descent method, and the output weights are expressed by the Moore-Penrose generalized inverse formula. Numerical simulations are employed to demonstrate the better convergence and generalization abilities of the proposed algorithm in this paper.

Acknowledgements. This work was supported in part by the National Natural Science Foundation of China (No. 61305075), the China Postdoctoral Science Foundation (No. 2012M520624), the Natural Science Foundation of Shandong Province (No. ZR2013FQ004, ZR2013DM015, ZR2015AL014), the Specialized Research Fund for the Doctoral Program of Higher Education of China (No. 20130133120014) and the Fundamental Research Funds for the Central Universities (Nos. 14CX05042A, 15CX05053A, 15CX02079A, 15CX08011A, 15CX02064A).

References

1. Hirose, A.: Complex-Valued Neural Networks. World Scientific, Singapore (2003)
2. Cha, I., Kassam, S.A.: Channel equalization using adaptive complex radial basis function networks. IEEE J. Sel. Areas Commun. **13**, 122–131 (1995)
3. Aizenberg, I.: Complex-Valued Neural Networks with Multi-valued Neurons. Springer, Berlin (2011). doi:10.1007/978-3-642-20353-4. Finance, A.: Multivariate nonlinear analysis and prediction of Shanghai stock market. Discret. Dyn. Nat. Soc. 47–58 (2008)
4. Serre, D.: Matrices: Theory and Applications. Springer, New York (2002). doi:10.1007/978-1-4419-7683-3
5. Rakkiyappan, R., Velmurugan, G., Cao, J.: Stability analysis of fractional-order complex-valued neural networks with time delays. Chaos Solitons Fractals Interdisc. J. Nonlinear Sci. Nonequilibrium Complex Phenom. **78**, 297–316 (2015)

6. Leung, H., Haykin, S.: The complex backpropagation algorithm. IEEE Trans. Signal Process. **39**, 2101–2104 (1991)
7. Nitta, T.: An extension of the back-propagation algorithm to complex numbers. Neural Netw. Off. J. Int. Neural Netw. Soc. **10**, 1391–1415 (1997)
8. Zhang, H., Zhang, C., Wu, W.: Convergence of batch split-complex backpropagation algorithm for complex-valued neural networks. Discret. Dyn. Nat. Soc. **2009**, 332–337 (2009)
9. Zhang, H., Xu, D., Zhang, Y.: Boundedness and convergence of split-complex backpropagation algorithm with momentum and penalty. Neural Process. Lett. **39**, 297–307 (2014)
10. Zhang, H., Liu, X., Xu, D., Zhang, Y.: Convergence analysis of fully complex backpropagation algorithm based on Wirtinger calculus. Cogn. Neurodyn. **46**, 5789–5796 (2014)
11. Zhang, H., Mandic, D.P.: Is a complex-valued stepsize advantageous in complex-valued gradient learning algorithms? IEEE Trans. Neural Netw. Learn. Syst. **27**, 1–6 (2015)
12. Xu, D., Dong, J., Zhang, H.: Deterministic convergence of Wirtinger-gradient methods for complex-valued neural networks. Neural Process. Lett. 1–12 (2016)
13. Rao, C.R., Mitra, S.K.: Generalized Inverse of Matrices and its Applications. Wiley, New York (1971)
14. Huang, G.B., Zhu, Q.Y., Siew, C.K.: Extreme learning machine: theory and applications. Neurocomputing **70**, 489–501 (2006)
15. Yu, D., Deng, L.: Efficient and effective algorithms for training single-hidden-layer neural networks. Pattern Recogn. Lett. **33**, 554–558 (2012)
16. Li, M.B., Huang, G.B., Saratchandran, P., Sundararajan, N.: Fully complex extreme learning machine. Neurocomputing **68**, 306–314 (2005)
17. Shukla, S., Yadav, R.N.: Regularized weighted circular complex-valued extreme learning machine for imbalanced learning. IEEE Access 3048–3057 (2016)
18. Kim, T., Adal, T.: Approximation by fully complex multilayer perceptrons. Neural Comput. **15**, 1641–1666 (2003)
19. Suresh, S., Savitha, R., Sundararajan, N.: Supervised Learning with Complex-Valued Neural Networks. Studies in Computational Intelligence. Springer, Heidelberg (2013). doi:10.1007/978-3-642-29491-4

Feature Selection Using Smooth Gradient $L_{1/2}$ Regularization

Hongmin Gao[1], Yichen Yang[1], Bingyin Zhang[1], Long Li[2], Huaqing Zhang[1], and Shujun Wu[1(✉)]

[1] College of Science, China University of Petroleum, Qingdao 266580, China
hongmin_gao_1997@163.com, yichen_yang_1997@163.com,
bingyin_zhang_1994@163.com, zhhq@upc.edu.cn, wushujun1981@163.com
[2] College of Mathematics and Statistics, Hengyang Normal University,
Hengyang 421008, China
long_li1982@163.com

Abstract. In terms of $L_{1/2}$ regularization, a novel feature selection method for a neural framework model has been developed in this paper. Due to the non-convex, non-smooth and non-Lipschitz characteristics of $L_{1/2}$ regularizer, it is difficult to directly employ the gradient descent method in training multilayer perceptron neural networks. A smoothing technique has been considered to approximate the original $L_{1/2}$ regularizer. The proposed method is a two-stage updating approach. First, a multilayer network model with smoothing $L_{1/2}$ regularizer is trained to eliminate the unimportant features. Second, the compact model without regularization has been simulated until there is no improvements for the performance. The experiments demonstrate that the presented algorithm significantly reduces the redundant features while keeps a considerable model accuracy.

Keywords: $L_{1/2}$ regularizer · Feature selection · Neural networks · Gradient descent · Non-smooth

1 Introduction

The size of the data set is usually measured by the number of features (properties) and samples. Large number of features (high dimensionality) easily leads to "the curse of the dimension". It is already an empirical "axiom" that the dimension of the feature space should not be too high [1], especially in the field of machine learning. In general, an essential task of many learning machines is to estimate unknown parameters which are strongly related to the dimension of the feature space. If the dimensionality of the feature space is too high, it probably requires considerable training samples to estimate these parameters with a competitive accuracy. However, it is difficult to obtain enough training samples in practical applications. This then decreases the predicting accuracy of the given model, which in turn affects the generalization performance of learning machines.

© Springer International Publishing AG 2017
D. Liu et al. (Eds.): ICONIP 2017, Part IV, LNCS 10637, pp. 160–170, 2017.
https://doi.org/10.1007/978-3-319-70093-9_17

Feature selection is one of the commonly used data dimensionality reduction methods [2]. The purpose is to select the smallest subset of features according to selection criteria, then the compact learning machines with reduced features can achieve the similar or even better generalization ability on given tasks such as classification and regression problems. Essentially, it is an optimization problem for searching the optimal or suboptimal subset in the feature space. After feature selection, a simplified data set usually produces more accurate model and are easier to be interpreted [3–5]. It can be divided into two categories for feature selection strategies: feature sorting and feature subset search. In the traditional feature selection algorithm, the representative algorithms based on feature sorting are Laplacian Score (LS) [6] and Fisher Score (FS) [7]. These two typical methods are separately with unsupervised and supervised learning models. For feature selection, it is an attractive strategy by introducing regularization items to induce more discriminant features [8]. One advantage is that these methods largely preserve the spatial distribution of similar samples in the data set.

In general, there are three common regularization forms which penalize the unknown parameters with L_0, L_1 and L_2 norms [9–11]. The L_0 regularization method is one of the earliest strategy in coping with high dimensionality reduction. Unfortunately, it has to face NP hard optimization problem since it is constrained by the number of unknown parameters. The Lasso (L_1 norm) regularizer which was first presented in [12] which provides an efficient feature selection method by solving a quadratic programming problem. It has become one of very popular feature selection methods in dealing with various data sets, even it builds less sparse solutions than the L_0 regularizer. For L_2 regularizer, although it possesses the smoothing properties, it does not have the ability to achieve the sparse solutions. Recently, a novel regularizer, $L_{1/2}$ norm, has attracted more researchers for data analysis. It considers the $L_{1/2}$ norm of the unknown parameters to reach sparse solutions. More simulations demonstrate the effectiveness of this method, which is more sparse than the L_1 regularizer and easily to be fast solved by employing an iterative half threshlding algorithm [13–15]. Here, we note that all of these regularizers are mainly focused on dealing with linear problems.

Neural network [16–18] has become a hot research topic in the development of artificial intelligence, which has been widely used in the field of feature selection due to its excellent nonlinear mapping and approximation abilities for any continuous function [18,19]. Many feature selection methods based on neural networks have been surveyed [20]. In addition, the gradient-based algorithm with penalty term has been an efficient method to improve the generalization performance and decrease the redundant features as well [21].

One latest research on feature selection is emerged by employing Lasso penalty term for multilayer perceptron neural networks [22]. It provides one promising design on network model with Lasso regularizer. Simulations display the better prediction accuracies than other feature selection methods along with a more compact constructed model.

Motivated by the above descriptions, we propose a new feature selection model for feed-forward neural networks with $L_{1/2}$ regularizer. Due to the introduction of the $L_{1/2}$ regularizer, it leads to a non-smooth and non-Lipschitz optimization problem. As a remedy, a smoothing technique has been employed in this paper to approximate the $L_{1/2}$ regularizer, which guarantee the continuous differential property of the established model.

The rest of this paper is organized as following: In Sect. 2, we introduce the feature selection method based on Lasso penalty for multilayer perceptron neural networks. In Sect. 3, we propose two algorithms which employ $L_{1/2}$ regularizer and smoothing $L_{1/2}$ regularizer to replace the Lasso penalty. The supporting numerical experiments are shown in Sect. 4. Some useful comments are concluded in Sect. 5.

2 Related Works

Multilayer perceptron is the typical model of feedforward neural networks. Figure 1 shows the structure of a three-layer feedforward neural network which consists of p inputs, n hidden and q output nodes, respectively. We suppose that $\{\mathbf{x}^j, \mathbf{y}^j\}_{j=0}^{J-1} \subset \mathbb{R}^p \times \mathbb{R}^q$ is a given set of training samples, where $\mathbf{x}^j$ and y^j are the input and the corresponding ideal output for the jth sample, respectively. Suppose $\mathbf{V} = (v_{r_1}, v_{r_2}, \cdots, v_{r_n})^T$ is the weight matrix between the input and hidden layers. Let $\mathbf{v}_{r_k} = (v_{k1}, v_{k2}, \cdots, v_{kp})$ for $k = 1, 2, \cdots, n$ is the weight vector between input nodes with the kth hidden node. Suppose $\mathbf{U} = (u_1, u_2, \cdots, u_q)^T$ is the weight matrix between the hidden and output layers. Let $\mathbf{u}_t = (u_{t1}, u_{t2}, \cdots, u_{tn})$ for $t = 1, 2, \cdots, q$. Then for simplicity, we combine the weight matrix $\mathbf{U}$ and $\mathbf{V}$ as follows $\mathbf{w} = (\mathbf{u}_1^T, \mathbf{u}_2^T, \cdots, \mathbf{u}_q^T, \mathbf{v}_{r_1}^T, \mathbf{v}_{r_2}^T, \cdots, \mathbf{v}_{r_n}^T)^T \in \mathbb{R}^{n(p+q)}$.

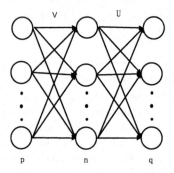

Fig. 1. Architecture of a classical single hidden layer neural network.

Let g and f be the activation functions of hidden and output layers, respectively. For any finite dimensional vector $\mathbf{z}$, we define the following vector-valued function $G(\mathbf{z}) = (g(\mathbf{z}_1), g(\mathbf{z}_2), \cdots, g(\mathbf{z}_n))$, $F(\mathbf{z}) = (f(\mathbf{z}_1), f(\mathbf{z}_2), \cdots, f(\mathbf{z}_q))$. For input sample $\mathbf{x} \in \mathbb{R}^p$, the actual output of network is with $\mathbf{O} = F(\mathbf{U}(G(\mathbf{Vx})))$. Thus, for given training samples $\{\mathbf{x}^j, \mathbf{y}^j\}_{j=0}^{J-1} \subset \mathbb{R}^p \times \mathbb{R}^q$, the empirical error without penalty can be evaluated as follows

$$\tilde{E}(\mathbf{w}) = \frac{1}{2}\sum_{j=0}^{J-1} \| F(\mathbf{U}(G(\mathbf{Vx}^j))) - \mathbf{y}^j \|^2 = \frac{1}{2}\sum_{j=0}^{J-1} \| \mathbf{O}^j - \mathbf{y}^j \|^2 \qquad (1)$$

An effective model on feature selection has been proposed in [22], which integrates the Lasso penalty into the multilayer perceptron network. The objective function is reformulated as follows

$$E(\mathbf{w}) = \tilde{E}(\mathbf{w}) + \lambda\Psi(\mathbf{v}), \qquad (2)$$

where $\Psi(\mathbf{v}) = \sum_{i=1}^{p}\sum_{j=1}^{n} |v_{ij}|$. According to the gradient descent method, the weights are iteratively updated

$$\mathbf{u}_t^{m+1} = \mathbf{u}_t^m - \eta\tilde{E}_{\mathbf{u}_t}(\mathbf{w}^m), \qquad (3)$$

$$\mathbf{v}_{r_k}^{m+1} = \mathbf{v}_{r_k}^m - \eta\tilde{E}_{\mathbf{v}_{r_k}}(\mathbf{w}^m) - \lambda\eta\frac{\partial\Psi(\mathbf{v})}{\partial\mathbf{v}_{r_k}}, \qquad (4)$$

3   Algorithms

Let us begin with a discussion of a three-layer feedforward neural network consisting of p inputs nodes, n hidden nodes and q output nodes, shown in Fig. 2. The training sample set $\{\mathbf{x}^j, \mathbf{y}^j\}_{j=0}^{J-1} \subset \mathbb{R}^p \times \mathbb{R}^q$, the weight matrix between the hidden layer and output layer $\mathbf{U} = (u_1, u_2, \cdots, u_q)^T$ and the activation functions for the hidden and output layers are identical to those in Sect. 2.

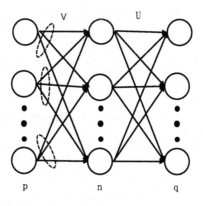

Fig. 2. The architecture of network

Suppose $\mathbf{V} = (v_{c_1}, v_{c_2}, \cdots, v_{c_p})$ is the weight matrix between the input layer and hidden layer. Let $\mathbf{v}_{c_i} = (v_{1i}, v_{2i}, \cdots, v_{ni})^T$ for $i = 1, 2, \cdots, p$ is the weight vector between the ith input node with the hidden nodes. For simplicity, we write $\mathbf{w} = (\mathbf{u}_1^T, \mathbf{u}_2^T, \cdots, \mathbf{u}_q^T, \mathbf{v}_{c_1}, \mathbf{v}_{c_2}, \cdots, \mathbf{v}_{c_p})^T \in \mathbb{R}^{n(p+q)}$. For convenience, we introduce the Kronecker product, which denoted by $\otimes$, is an operation on two matrices of arbitrary size resulting in a block matrix. We choose the ith element of the total samples with $\mathbf{x}_i = (\mathbf{x}_{i1}, \mathbf{x}_{i2}, \cdots, \mathbf{x}_{iJ})$, $i = 1, 2, \cdots, p$. The empirical error without penalty is as follows

$$\tilde{E}(\mathbf{w}) = \frac{1}{2} \sum_{j=0}^{J-1} \| F(\mathbf{U}(G(\sum_{i=1}^{p} \mathbf{v_i} \otimes \mathbf{x_i}))) - \mathbf{y}^j \|^2 . \tag{5}$$

3.1 $L_{1/2}$ Regularization for Input Layer

We use the $L_{1/2}$ regularizer to express the norm of each input weight vector. Then, the penalty term is given by

$$\Phi(\mathbf{v}_{c_i}) = \sum_{i=1}^{p} \|\mathbf{v}_{c_i}\|_{\frac{1}{2}} \tag{6}$$

The empirical error of neutral networks and the penalty term constitute the error function as follows

$$E(\mathbf{w}) = \tilde{E}(\mathbf{w}) + \lambda\Phi(\mathbf{v}_{c_i}) \tag{7}$$

where λ is the coefficient of $L_{1/2}$ regularization.

The weight sequences are updated by

$$\mathbf{v}_{c_i}^{m+1} = \mathbf{v}_{c_i}^m - \eta\tilde{E}_{\mathbf{v}_{c_i}}(\mathbf{w}^m) - \lambda\eta \bigtriangledown \Phi(\mathbf{v}_{c_i}^m) \tag{8}$$

$$\mathbf{u}_t^{m+1} = \mathbf{u}_t^m - \eta\tilde{E}_{\mathbf{u}_t}(\mathbf{w}^m) \tag{9}$$

$$\mathbf{w}^{m+1} = \mathbf{w}^m - \eta E_{\mathbf{w}}(\mathbf{w}^m) \tag{10}$$

where $m \in \mathbb{N}$; $i = 1, 2, \cdots, p$; $t = 1, 2, \cdots, q$, the learning rate $\eta \geq 0$ is a constant.

3.2 Smoothing $L_{1/2}$ Regularization for Input Layer

Since the error function is non-differentiable at the origin, we use smoothing approximation to replace the error function when the weight vector approaches the origin. For any finite dimensional vector $\mathbf{z}$, the smoothing function is constructed as follows

$$h(\mathbf{z}) = \begin{cases} \| \mathbf{z} \|, & \| \mathbf{z} \| \geq \alpha \\ \frac{\|\mathbf{z}\|^2}{2\alpha} + \frac{\alpha}{2}, & \| \mathbf{z} \| < \alpha \end{cases} \tag{11}$$

where α is a small fixed positive constant.

The total error function can be replaced as follows:

$$E\left(\mathbf{w}\right) = \tilde{E}\left(\mathbf{w}\right) + \lambda \sum_{i=1}^{p} h(\mathbf{v}_{c_i}) \tag{12}$$

Combined with formula (7), the weight vectors updating is giving by

$$\mathbf{v}_{c_i}^{m+1} = \mathbf{v}_{c_i}^{m} - \eta \tilde{E}_{\mathbf{v}_{c_i}}^{m} - \lambda \eta \bigtriangledown h\left(\mathbf{v}_{c_i}^{m}\right) \tag{13}$$

where $m \in \mathbb{N}; i = 1, 2, \cdots, p$. We note that the updating formula of $\{\mathbf{u}_i^m\}$ is identical to that in (9), $m \in \mathbb{N}$.

4   Simulations

In this section, 6 data sets are used for experiments. A brief description of used data sets is given in Table 1. Iris, Glass, Ecoli, Balance and Fertility are low dimension data sets, while Sonar is higher dimension data set.

Table 1. Description of all data sets used

Data set	Data set size	Dimensionality	Classes
Iris	150	4	3
Glass	214	10	7
Ecoli	336	7	8
Balance	625	4	3
Fertility	100	9	2
Sonar	208	60	2

We use cross-validation to verify the effectiveness of our method. In numerical experiments, 4 parameters are needed to be chosen, namely the regularization coefficient λ, the number of hidden layer nodes, the threshold of eliminating a feature E_0 and the learning rate η. We can choose different number of the features by changing these parameters. The λ and the η are determined by cross-validation. In this paper, the λ ranges from 0 to 2, it depends on the number of selected features. The parameter E_0 has been set to different values for different data sets. If the norm of the connecting weight vectors of the input node is smaller than the E_0, the corresponding feature is eliminated.

4.1   Comparisons with and Without Smoothing Technique Based on $L_{1/2}$ Penalty

Figures 3 and 4 show the results of the method without and with the smoothing term on Iris and Glass data sets. We can see that the obvious oscillations have occurred during the feature selection without smoothing term. However, when smoothing term is used, the oscillations are significantly reduced. In this paper, we then employ the method with the smoothing penalty term.

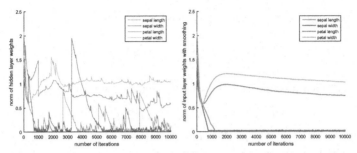

(a) $L_{1/2}$ method without smooth-(b) $L_{1/2}$ method with smoothing
ing term   term

Fig. 3. $L_{1/2}$ for Iris data set with $\lambda = 0.6$

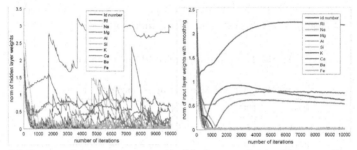

(a) $L_{1/2}$ method without smooth-(b) $L_{1/2}$ method with smoothing
ing term   term

Fig. 4. $L_{1/2}$ for Iris data set with $\lambda = 0.5$

4.2 Comparison with Weight Decay and Lasso

In this paper, we compare the proposed method with the methods using Weight
Decay and Lasso regularization. We test on Iris set with Weight Decay method,
Lasso method and smoothing $L_{1/2}$ method. The results are shown in Table 2,
where FS shows the frequency rate of the each selected feature of Iris data set.
TrAcc represents the training accuracy, while TeAcc represents the testing accu-
racy. AvgSN shows the average number of selected features. Table 2 shows that
less features are selected by the method using $L_{1/2}$ with the same parameters.
The method using $L_{1/2}$ is better than others.

4.3 Data Set with Low and Higher Dimension

We have summarized the training accuracy, the testing accuracy and the average
number with different penalty coefficients based on different data sets in Table 3.
When the penalty coefficient is zero, there is no penalty in feature selection.
As the penalties increase, the average number of selected features decreases.
However, the predictive performance is not reduced obviously.

Table 2. Feature selection on Iris data with WD, Lasso and $L_{1/2}$ method of different regularization coefficient λ

Penalty (λ)	Reg	FS (%)				TrAcc	TeAcc	AvgSN
		1	2	3	4			
0.2	WD	100.00	100.00	100.00	100.00	99.53	95.22	4.00
	Lasso	46.00	100.00	100.00	100.00	99.50	95.36	3.43
	L0.5	16.00	88.00	100.00	100.00	99.27	95.18	3.00
0.4	WD	100.00	100.00	100.00	100.00	99.67	95.21	4.00
	Lasso	26.00	100.00	100.00	100.00	99.48	95.38	3.26
	L0.5	1.00	50.00	87.00	99.00	98.65	95.38	2.37
0.6	WD	100.00	100.00	100.00	100.00	99.63	95.31	4.00
	Lasso	15.00	98.00	100.00	100.00	99.37	95.32	3.13
	L0.5	1.00	7.00	70.00	89.00	97.52	95.01	1.67

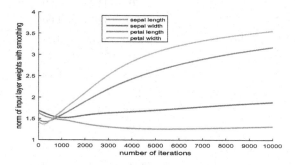

Fig. 5. Feature selection on Iris data with $L_{1/2}$ of $\lambda = 0$

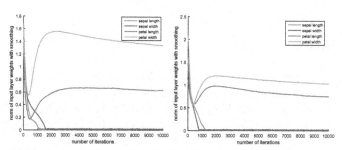

(a) Feature selection on Iris data with $L_{1/2}$ of $\lambda = 0.4$

(b) Feature selection on Iris data with $L_{1/2}$ of $\lambda = 0.6$

Fig. 6. Feature selection on Iris data with $L_{1/2}$ of different λ

The result curves based on different penalty coefficients with $\lambda = 0$, 0.4, and 0.6 are shown in the Figs. 5, 6(a) and 6(b). We can see norm of the first and the second features is decreased with the increase of λ. Therefore, with the increase of λ, the average number of features is decreased.

Table 3. Feature selection on different data set with $L_{1/2}$ of different λ

Dataset	Penalty (λ)	TrAcc	TeAcc	AveSN
Iris	0	99.67	95.52	3.98
	0.2	99.27	95.18	3.00
	0.4	98.65	95.38	2.37
	0.6	97.52	95.01	1.67
Glass	0	68.55	66.77	9.95
	0.5	67.56	69.27	5.00
	1	68.33	66.67	3.17
Balance	0	98.67	93.33	4.00
	1	99.02	93.47	4.00
	2	99.15	92.77	4.00
Fertility	0	99.50	80.50	8.94
	1	97.92	79.67	5.47
	2	91.58	85.22	1.10
Ecoli	0	62.41	61.39	6.96
	1	62.88	60.95	4.67
	2	63.26	61.94	2.33
Sonar	0	98.72	72.70	59.86
	1	99.43	75.24	13.80
	2	98.93	74.60	6.33

5   Conclusions

In this paper, we propose a novel feature selection method that takes $L_{1/2}$ regularization of each neural network input node as a penalty term in the form of a group based on the Lasso. However, the penalty is non-differentiable at the origin, in order to solve this problem, we smooth the penalty items, and give supporting numerical experiments to prove our advantages.

In numerical experiments, we select 6 data sets to verify feature selection by classification problem. And the results show that with the penalty coefficient increases continuously, the average number of features selected reduced significantly, while the decrease of testing accuracy is not obviously. Less features are selected with the higher testing accuracy achieved simultaneously.

After comparing with Weight Decay method and Lasso method, we can see the strong sparseness of smoothing $L_{1/2}$ regularization. That is, the proposed method can choose less features with the same regularization coefficient λ than Weight Decay and Lasso methods, while the accuracies are similar.

Acknowledgements. This work was supported in part by the National Natural Science Foundation of China (Nos. 61305075, 11401185), the China Postdoctoral Science Foundation (No. 2012M520624), the Natural Science Foundation of Shandong Province (Nos. ZR2013FQ004, ZR2013DM015, ZR2015AL014), the Specialized Research Fund for the Doctoral Program of Higher Education of China (No. 20130133120014), the Fundamental Research Funds for the Central Universities (Nos. 14CX05042A, 15CX05053A, 15CX08011A, 15CX02064A) and the University-level Undergraduate Training Program for Innovation and Entrepreneurship (No. 20161349).

References

1. Jain, A.K., Duin, R.P.W., Mao, J.: Statistical pattern recognition: a review. IEEE Trans. Pattern Anal. Mach. Intell. **22**, 4–37 (2000)
2. Sun, Y., Todorovic, S., Goodison, S.: Local-learning-based feature selection for high-dimensional data analysis. IEEE Trans. Pattern Anal. Mach. Intell. **32**, 1610–1626 (2010)
3. Liu, H., Yu, L.: Toward Integrating Feature Selection Algorithms for Classification and Clustering. IEEE Educational Activities Department (2005)
4. Bu, H.L., Xia, J., Han, J.B.: A summary of feature selection algorithms and prospect. J. Chaohu Coll. (2008)
5. Chakraborty, R., Pal, N.R.: Feature selection using a neural framework with controlled redundancy. IEEE Trans. Neural Netw. Learn. Syst. **26**, 35–50 (2014)
6. He, X., Cai, D., Niyogi, P.: Laplacian score for feature selection. In: International Conference on Neural Information Processing Systems, pp. 507–514. MIT Press (2005)
7. Gu, Q., Li, Z., Han, J.: Generalized Fisher Score for Feature Selection, pp. 266–273 (2012)
8. Yao, Y.: Statistical Applications of Linear Programming for Feature Selection via Regularization Methods (2008)
9. Burden, F., Winkler, D.: Bayesian regularization of neural networks. Methods Mol. Biol. **458**, 25 (2008)
10. Ticknor, J.L.: A Bayesian regularized artificial neural network for stock market forecasting. Expert Syst. Appl. **40**, 5501–5506 (2013)
11. Han, M., Li, D.: An norm 1 regularization term ELM algorithm based on surrogate function and Bayesian framework. Acta Autom. Sinica **37**, 1344–1350 (2011)
12. Tibshirani, R.J.: Regression shrinkage and selection via the LASSO. J. Roy. Stat. Soc. Ser. B **58**, 267–288 (1996)
13. Xu, Z.B., Guo, H.L., Wang, Y., Zhang, H.: Representative of L1/2 regularization among Lq $(0 \leq q \leq 1)$ regularizations: an experimental study based on phase diagram. Acta Autom. Sinica **38**, 1225–1228 (2012)
14. Liu, C., Liang, Y., Luan, X.Z., Leung, K.S., Chan, T.M., Xu, Z.B., Zhang, H.: A improve direct path seeking algorithm for L1/2 regularization, with application to biological feature selection. In: International Conference on Biomedical Engineering and Biotechnology, pp. 8–11. IEEE (2012)

15. Wu, W., Yang, J.: L1/2 regularization methods for weights sparsication of neural networks (in Chinese). Sci. Sinica **45**, 1487–1504 (2015)
16. Wu, W., Shao, H., Li, Z.: Convergence of batch BP algorithm with penalty for FNN training. In: King, I., Wang, J., Chan, L.-W., Wang, D.L. (eds.) ICONIP 2006. LNCS, vol. 4232, pp. 562–569. Springer, Heidelberg (2006). doi:10.1007/11893028_63
17. Haykin, S.: Neural Networks: A Comprehensive Foundation. Prentice-Hall, Upper Saddle River (1994)
18. Hornik, K., Stinchcombe, M., White, H.: Multilayer feedforward networks are universal approximators. Neural Netw. **2**, 359–366 (1989)
19. Kim, Y., Kim, J.: Gradient LASSO for feature selection. In: International Conference on Machine Learning, p. 60. ACM (2004)
20. Challita, N., Khalil, M., Beauseroy, P.: New feature selection method based on neural network and machine learning. In: Multidisciplinary Conference on Engineering Technology, pp. 81–85. IEEE (2016)
21. Setiono, R.: A penalty-function approach for pruning feedforward neural networks. Neural Comput. **9**, 185 (1997)
22. Sun, K., Huang, S.H., Wong, D.S., Jang, S.S.: Design and application of a variable selection method for multilayer perceptron neural network with LASSO. IEEE Trans. Neural Netw. Learn. Syst. **28**, 1386–1396 (2017)

Top-k Merit Weighting PBIL for Optimal Coalition Structure Generation of Smart Grids

Sean Hsin-Shyuan Lee[1]([✉]), Jeremiah D. Deng[1], Lizhi Peng[2],
Martin K. Purvis[1], and Maryam Purvis[1]

[1] Department of Information Science, University of Otago, Dunedin, New Zealand
sean.hslee@postgrad.otago.ac.nz,
{jeremiah.deng,martin.purvis,maryam.purvis}@otago.ac.nz
[2] Shandong Provincial Key Laboratory of Network Based Intelligent Computing,
University of Jinan, Jinan, China
penglizhi.jn@gmail.com

Abstract. The cooperation of agents in smart grids to form coalitions could bring benefit both for agent itself and the distribution power system. To tackle the problem as a game of partition form function poses significant computing challenges due to the huge search space for the optimization problem. In this paper, we propose a stochastic optimization approach using Population Based Incremental Learning (PBIL) algorithm with top-k Merit Weighting and a customized strategy for choosing the initial probability to solve the problem. Empirical results show that the proposed algorithm gives competitive performance compared with a few stochastic optimization algorithms.

Keywords: Coalition Structure Generation · Smart grids · Optimization

1 Introduction

Since the implementation of smart grids is becoming more wide-spread and the penetration of renewable energy keeps increasing, there is a growing research interest on decentralized electricity systems recently [6,14]. Among those, one of key concerns is how to organize and operate the distributed agents of small scale demand and supply, such as households with solar panels or wind turbines, form decentralized power systems [5].

While coordinating a multi-agent system, particularly when a grand coalition is infeasible or less efficient, a major concern is how to divide agents into disjoint coalitions so that the system can enhance its performance or obtain a maximized total profit from the consequent coalitions. A Coalition Structure Generation (CSG) [9], *aka* "Coalitional game in Partition Form", which has received much awareness in the literature, is a primary paradigm to deal with the optimal problem. However, the main challenge for applying CSG is the computational complexity in searching a optimal solution by exiting approaches, e.g. Dynamic

© Springer International Publishing AG 2017
D. Liu et al. (Eds.): ICONIP 2017, Part IV, LNCS 10637, pp. 171–181, 2017.
https://doi.org/10.1007/978-3-319-70093-9_18

Programming (DP), which restricts its application within small scale. Although there are many studies have been devoted to improve the efficiency of DP, such as Sandholm et al. [12] present a partial search algorithm which guarantees the solution to be within a bound from optimum. While a global optimum still being unreachable for large scale CSG then the stochastic optimization (SO) algorithms for CSG may provide a competitive solution efficiently. For instance, Mohamed et al. [8] prove that, in the context of smart grids, the Particle Swarm Optimization is a promising SO technique due to its ability to reach the global optimum with relative simplicity and computational proficiency contrasted with the DP techniques.

To our knowledge this work is one of the few attempts [9] in employing SOs to solve the CSG problem. We propose to use a Population Based Incremental Learning (PBIL) [1] algorithm and its variants for CSG. Specifically, a new PBIL-based algorithm that applies particle weighting according to particles' merit is proposed, and achieves competitive performance compared with a few SO counterparts.

Furthermore, the applications of Multi-agent Systems and Game Theory in smart grids have been widely investigated in literature [11]. For example, Saad et al. [10] utilize a dynamic programming approach [15] to explore an optimal coalition formation of minimizing the transmission loss. Based on Saad's study, Chakraborty et al. seek to prove that their hierarchical coalition formation mechanism is a scalable and optimal coalition formation [3]. However, the algorithms used in the aforementioned studies still enclosed in the category of DP.

The rest of the paper is organized as follows. In Sect. 2 we present the model to form a coalition game of agents in smart grids to be solved in partition form. We then proposes our SO-based solutions, especially the new Top-k Merit Weighting PBIL (PBIL-MW) algorithm, to search for the optimal formation. Experiment results are then presented in Sect. 4, with the algorithm's performance compared with a few other SO algorithms in terms of the approximation to the global optimum, and the convergence speed. Finally, we conclude the paper and point to possible further research.

2   Problem Formation for Smart Grids

In this study we assume that every agent who is willing to participate in the coalition should generate sufficient renewable energy to meet its own requirement in the long term. However, due to the intermittency of renewable energy, every agent may face shortage frequently. Therefore, binding an agreement to share the energy with others is a favorable, plausible and economic way in comparison with enlarge the facility (such as wind turbine or solar panel), install larger backup storage or trade with power company. As the energy consumption and generation are both dynamic, we need to employ a fast optimization technique to derive the optimal coalition formation continuously.

2.1   Coalition Criterion

For every period of time (e.g. per hour), any agent Ag_i with surplus can share its energy with deficit ones within the team. The goal of that team is to optimize the maximum benefit by forming coalition. For the stability of smart grid, we require that every feasible sub-coalition must have a power surplus.

For example, Ag_1 and Ag_2 each has 1.2 and 0.7 kWh surplus respectively, but Ag_3 has a deficit of 0.9 kWh. With a net joint surplus, Ag_1 and Ag_3 can form a feasible coalition $\{Ag_1, Ag_3\}$. On the contrary, a coalition such as $\{Ag_2, Ag_3\}$ is infeasible. As a result, a grand coalition cannot always be a viable solution and a model of CSG [13] should be shaped and needs to be solved. Furthermore, since our experiment is focused on coalition within a local area, such as a city, therefore the transmission cost and power loss are ignored.

2.2   Coalition Evaluation

Let $\mathcal{S}$ denote a set of n cooperative agents, a coalition C_k is a subset of $\mathcal{S}$. A coalition structure CS is a collection of coalitions, where $CS = \{C_1, C_2, \cdots, C_m\}$ such that $\cup_k C_k = \mathcal{S}$, $C_k \neq \emptyset$ and $C_k \cap C'_k = \emptyset$, if $k \neq k'$. A coalition structure CS is said to be globally optimal for a characteristic function game G, if CS gives the maximal overall characteristic value [4]:

$$v(CS) = \sum_{C_k \in CS} v(C_k). \tag{1}$$

In our study, $v(C_k)$ is the fitness function of C_k given by

$$v(C_k) = \begin{cases} 0 & \text{if } |C_k| = 1, \\ Q_d \times P_r & \text{if } |C_k| > 1 \text{ and } Q(C_k) \geq 0, \\ -9999 & \text{otherwise,} \end{cases} \tag{2}$$

where $|C_k|$ denotes the size of coalition C_k, P_r represents the price difference for trading with power utilities, Q_d is the total power need for deficit agents in the C_k, and $Q(C_k)$ is the net surplus within the C_k accordingly. Furthermore, for giving penalty to an infeasible coalition, we let $v(C_k) = -9999$.

Our aim is to find the CS and its optimal profit. For example, a set of 3 agents, there are five possible CS:
$\{Ag_1, Ag_2, Ag_3\}$, $\{\{Ag_1, Ag_2\}\{Ag_3\}\}$, $\{\{Ag_1, Ag_3\}, \{Ag_2\}\}$,
$\{\{Ag_1\}, \{Ag_2, Ag_3\}\}$, and $\{\{Ag_1\}, \{Ag_2\}, \{Ag_3\}\}$.

The total number of CS for a given set S is known as a Bell's number $(Bell(n))$, which is proven to satisfy $(n/4)^{n/2} \leq Bell(n) < n^n$. For example, $Bell(3) = 5$, $Bell(4) = 15$, $Bell(6) = 203$, $Bell(10) = 115,975$ and $Bell(12) = 4.213597 \times 10^6$. The Bell's number implies that an approach of exhausted search will take super-polynomial time with respect to S.

Certainly, an exhaustive search for exploring the optimal coalition structure can solve this problem, but will suffer from time and memory complexity when the number of agents increases, and is almost infeasible for large number of agents. Hence, as in [9], other alternative methods are vital for obtaining the optimal, or near-optimal coalition structure in practical.

3   Algorithmic Solutions

3.1   Population-Based Incremental Learning

In order to demonstrate the potential ability of our proposed algorithm, the procedure of standard PBIL is described as follows.

Let $\mathbf{P}$ denote the probability vector with real values in the range $[0, 1]$ for each component P_j. $\mathbf{P}$'s length, denoted by l, is equal to that of genotype. In PBIL, every component P_j represents the threshold of gaining a value 1 in the j-th component of genotype. $\mathbf{P} = (P_1, P_2, \cdots, P_l)$.

Initially, the probability vector, denoted by $\mathbf{P}^{(0)}$, is simply set with $P_j = 0.5$. For each iteration, a n rows by l columns matrix R with entries of random numbers inside $[0, 1]$ is drawn, where n is the size of population's samples. Let $\mathbf{G}$ denote the population matrix, $\mathbf{G}_i$ represent its i-th genotype, corresponding to a CS. Since $\mathbf{P}$ and R are given, every entry G_{ij} in $\mathbf{G}_i$ is assigned as 1 if $R_{ij} < P_j$, otherwise $G_{ij} = 0$. Let f_i denote the fitness value of $\mathbf{G}_i$, which is evaluated by Eq. (2) with G_{ij} given, such that $f_i = -\sum_{C_k \in CS_{(i)}} v(C_k)$.

Thereafter, $P_j^{(t+1)}$ is then assessed upon the minimum and maximum fitnesses with a pair of parameters named learning rate γ and negative learning rate ϵ, respectively. In terms of fitness values obtained from current step, there are two updating options for assigning the next $\mathbf{P}^{(t+1)}$ vector given by

$$P_j^{(t+1)} = \begin{cases} (1 - \gamma)P_j^{(t)} + \gamma \; \acute{G}_j & \text{if } \acute{G}_j = \ddot{G}_j, \\ (1 - \gamma)P_j^{(t)} + \epsilon(\acute{G}_j - P_j^{(t)}) + \gamma \; \acute{G}_j & \text{if } \acute{G}_j \neq \ddot{G}_j, \end{cases} \qquad (3)$$

where $\acute{\mathbf{G}}$ and $\ddot{\mathbf{G}}$ represent the vector $\mathbf{G}_i$ when $i = \arg\min_j f_j$ or $\arg\max_j f_j$, respectively.

After successive $\mathbf{P}^{(t+1)}$-updating iterations, the procedure can be stopped by a given repetitions reached or a criteria of convergence met. A pseudocode of PBIL is shown in Algorithm 1. Further to the standard PBIL, a variation of PBIL [2], the Top-k PBIL (T-PBIL), includes the genotype of the second highest

---

Algorithm 1. PBIL

1: Initialize probability vector $\mathbf{P}^{(0)}$
2: **repeat**
3:     Generate a population $\mathbf{G}_n$ from $\mathbf{P}^{(t)}$
4:     Evaluate the fitness f_i of each member $\mathbf{G}_i$ for all $i \in n$
5:     Find $\acute{\mathbf{G}}$ and $\ddot{\mathbf{G}}$
6:     Update $P_j^{(t+1)}$ according to Eq. (3)
7: **until** termination condition has been met

---

fitness rather than subtracts the worst one is considered. Thus, the method for updating the new $P(t + 1)$ can be given by

$$P_j^{(t+1)} = (1 - \gamma)P_j^{(t)} + \frac{\gamma}{k} \sum_{m=1}^{k} G_m, \tag{4}$$

Note G_m's are sorted. A pseudocode of T-PBIL is shown in Algorithm 2.

Algorithm 2. T-PBIL

1: Initialize probability vector $\mathbf{P}^{(0)}$
2: **repeat**
3:    Generate a population $\mathbf{G}_n$ from $\mathbf{P}^{(t)}$
4:    Evaluate and rank the fitness f_i of each member $\mathbf{G}_i$
5:    Sort all G_m's according to their fitness
6:    Update $\mathbf{P}^{(t+1)}$ according to Eq. (4)
7: **until** termination condition has been met

3.2 Merit Weighting PBIL (PBIL-MW)

As we know from the aforementioned algorithm of PBIL, the probability vector $\mathbf{P}^{(t+1)}$ is updated depending on whether each pairs of elements G_{ij} drawn from $\acute{\mathbf{G}}$ and $\ddot{\mathbf{G}}$ are equal or not. Hence, only $\min(f_n)$ and $\max(f_n)$ are utilized to choose the two alternative learning rules for updating $P_i^{(t+1)}$. Here we propose a scheme that every fitness values f_i of $\mathbf{G}_i$ may also give contribution towards updating of $\mathbf{P}^{(t+1)}$ to increase the diversity of new genotypes.

The proposed adaptive algorithm incorporates the aforementioned concept using a weighted average, and works as follow. The initial steps are the same as original PBIL until every fitness individual f_i has been computed in first iteration. Then f_i is ranked and the weights are given by

$$w_i = \frac{f_i'}{\sum_{i=1}^{k} f_i'}, \qquad 2 \le k \le n \tag{5}$$

where $f_i' = \max_j(f_j) - f_i$, and k is the number of chosen particles with the highest fitness values. Now that every w_i has been obtained, the probability vector $\mathbf{P}^{(t+1)}$ is given by

$$\mathbf{P}^{(t+1)} = (1 - \gamma)\,\mathbf{P}^{(t)} + \gamma \sum_{i=1}^{k}(w_i \mathbf{G}_i). \tag{6}$$

Note that, since every fitness f_i is considered and its weight w_i is given accordingly, there is only one learning rate γ that needs to be chosen in our approach. The pseudocode of PBIL-MW is shown in Algorithm 3. Furthermore, it should be clarified that the aim of our study is to search for the maximal profit, differing from general optimization problems (i.e. to obtain the minimum), hence the fitness function is inversed in our study.

Algorithm 3. PBIL-MW

1: Initialize probability vector $\mathbf{P}^{(0)}$
2: **repeat**
3:     Generate a population $\mathbf{G}_n$ from $\mathbf{P}^{(t)}$
4:     Evaluate and rank the fitness f_i of each member $\mathbf{G}_i$
5:     Obtain w_i from Eq. (5)
6:     Update $\mathbf{P}^{(t+1)}$ according to Eq. (6)
7: **until** termination condition has been met

3.3   Initialization Treatment for Coalition Formation

To make the PBIL algorithms work for the coalition formation problem, there is a key treatment for giving the initial probability of the coalition formation which is described below.

Assume a graph of n nodes. We consider all possible connections, full mesh: $N = n(n-1)/2$, which is the length of gene probability vector in PBIL. Suppose for each of the N-connections the probability for its being "1" (connected) is p. $\pi^{(0)}$ is the probability of having all singletons:

The probability of having m connections, is

$$\pi^m = C_N^m (1-p)^m p^{N-m} \tag{7}$$

which gives a binomial distribution.

It can be shown that for a conventional probability threshold set as 0.5, the probability of having $n-1$ connections in the network potentially forming a grand coalition is very large. To avoid this we set a threshold such that the mean value of the number of connections is $E = (n-1)/2$. Now that the binomial distribution's mean is given by $E = Np$, we have

$$p = \frac{E}{N} = \frac{(n-1)/2}{n(n-1)/2} = \frac{1}{n}, \tag{8}$$

suggesting $1/n$ as the threshold for initializing the probability vector.

4   Experiment

4.1   Data

The data we used in this study are composed of two different sources. The first part are power consumption of smart-meter readings in New Zealand. The other part are power generated by commercialized facilities of wind turbines and solar panels which are coupled with meteorological data of New Zealand[1]. The power status of all members, i.e. agents of experiment data, are then given by subtracting consumption from power generated. Eventually we use the power

---

[1] Meteorological data obtained from "CliFlo: NIWA's National Climate Database on the Web".

Table 1. Power status for 12 agents in 4 case studies.

	A1	A2	A3	A4	A5	A6	A7	A8	A9	A10	A11	A12
Case I	−0.348	−1.558	−0.074	0.569	0.523	1.122	0.184	0.261	−0.450	−0.376	−0.222	−0.384
Case II	−2.020	−0.762	0.082	0.552	0.570	−0.309	−1.435	−0.889	−0.193	0.129	−0.808	2.074
Case III	−1.541	−1.084	−0.160	−0.196	0.596	−0.435	−1.611	−1.790	−0.167	0.337	1.791	3.131
Case IV	−3.414	−0.774	0.244	0.842	0.844	−1.881	−0.479	−0.901	−1.452	−0.324	0.121	0.287

data (in kWh) on hourly basis, and the price $P_r = 20$ (¢/kWh). Furthermore, we know that once the whole group has a power surplus at a given hour then the grand coalition will make a trivial solution. Thus, we consider only cases with power deficits. Four cases are shown in Table 1.

4.2 Results

In our experiments, cases with a group size of 12 actually takes over 200 s to complete the exhaustive search (ES), while the size growing to 13 demanding time increases up to 20 min more, which becomes infeasible for larger smart grids.

We assess the effect of using different k values for top-k weightings in PBIL-MW, and to compare with the other two PBIL algorithms. The result is shown in Fig. 1. The k value for each relevant algorithm is appended after the algorithm name, e.g., T-PBIL-2 stands for T-PBIL using top 2 for averaging, and PBIL-MW-2 for PBIL-MW with top 2 for weighted averaging. From the figure it can be seen that PBIL variants perform better than the original PBIL. Within about 120 iterations all PBIL variants manage to achieve the global optimum, although the convergence speeds differ. PBIL-MW seems to converge faster than T-PBIL.

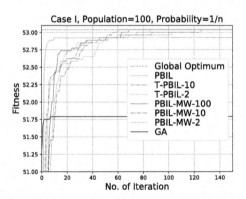

Fig. 1. Comparisons of optimums and converge speeds among different top-k T-PBIL and weights of PBIL-MW.

Hereafter, a four cases study in terms of PBIL-MW, PBIL and GA are carried through. For simplicity, we use all particles in merit weighting for PBIL-MW. Figure 2 is an example of optimized connections and profits found for samples in case I.

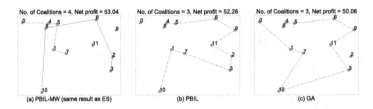

Fig. 2. Different connections and profits found in terms of PBIL-MW, PBIL and GA.

Since the ground-truths of four cases have been obtained by ES, the three algorithms were experimented and compared, using different initial probability threshold settings. In order to address the potentiality of gaining the optimal structure by heuristic algorithms, a size of 100 populations, a maximum of 200 iterations and 4 initial probabilities p, (=0.5, 0.1, $1/n \approx 0.083$, 0.05), are given to a 12-agents coalition game as the standard setup in experiment. Since we consider all possible connections, therefore, let *len* denote a full mesh of all agents' connections, ($len = 12 \times (12 - 1)/2 = 66$), which has been given as the length of gene probability vector for all three algorithms. Each of the three approaches is executed with 6 given initial probabilities and 20 runs for 4 cases.

The implementation of GA follows the standard operation procedure [7], the parameters used here are: mutation-probability $= 1/len$, crossover $=$ uniform. For PBIL, $\gamma = 0.1$, $\epsilon = 0.075$, mutation-probability $= 0.02$ and mutation-shift $= 0.05$ are used. For PBIL-MW, only one parameter $\gamma = 0.05$ is used.

Table 2 is a statistical summary which shows the values of mean, variance and p-value of results obtained by different methods. For giving the p-value of student's t-test in hypothesis testing, the significance level α is set to 0.05 and the result of PBIL-MW with $p = 1/n$ is chosen to be the principle sample while the test were performing.

By comparing the results obtained by the same methods in Table 2, we can clearly see that our suggestion $p = 1/n$ consistently leads to the best outcome, regardless the algorithm is PBIL, PBIL variants, or GA.

Moreover, by comparing the probabilities given in experiment, we can easily know that $p = 1/n$ is actually a better choice for initial the probability while searching for the optimal structure. Hence the result proves that our suggestion of initializing probability by $1/n$ is reasonable and effective strategy.

Furthermore, from Fig. 3 we know that although GA can find the better local optimum at the first few iterations, it is trapped there for the rest of iterations. The PBIL too, can get a better local optimum than PBIL-MW initially, but again, it is confined. Though the convergence speed is the least, PBIL-MW has a better diversity which help to find the global optimum of fitness.

Time consumed for experiments[2] of the three algorithms are similar and approximate to 6 s, which is faster than ES obviously.

---

[2] The code of the experiments is written and testing in Python 3.6 on Windows 7 PC with Intel core i5-4570 CPU and 16 GB RAM.

Table 2. Performance of PBIL-MW compared with PBIL and GA, p-values with statistical significance (against PBIL-MW $1/n$) are highlighted in bold.

Algorithm	PBIL-MW				PBIL				Genetic algorithm			
Init. prob.	0.5	0.1	$1/n$	0.05	0.5	0.1	$1/n$	0.05	0.5	0.1	$1/n$	0.05
Case I Ground-truth = 53.04												
Mean	39.44	53.04	53.04	53.04	36.54	52.93	52.96	52.89	3.07	52.52	52.51	52.26
Std dev.	5.61	0	0	0	19.49	0.27	0.23	0.30	102.02	0.77	1.19	0.57
p-value	**0.00**	-	-	-	**0.00**	**0.04**	0.08	**0.02**	**0.02**	**0.00**	**0.03**	**0.00**
Case II Ground-truth = 67.90												
Mean	2.75	67.90	67.90	67.71	54.29	67.30	67.04	67.11	−995.5	66.73	66.04	67.03
Std dev.	11.97	0	0	0.46	23.09	0.66	0.64	0.74	83.89	1.13	1.98	0.97
p-value	**0.00**	-	-	**0.04**	**0.01**	**0.00**	**0.00**	**0.00**	**0.00**	**0.00**	**0.00**	**0.00**
Case III Ground-truth = 114.80												
Mean	109.2	114.8	114.8	114.8	109.2	114.7	114.7	114.4	100.9	114.0	113.4	114.4
Std dev.	1.26	0	0	0	2.77	0.16	0.03	1.6	33.50	1.86	2.19	1.29
p-value	**0.00**	-	-	-	**0.00**	0.17	0.17	0.17	**0.04**	**0.04**	**0.01**	0.09
Case IV Ground-truth = 45.10												
Mean	0	45.10	45.10	45.02	20.86	44.50	44.75	44.43	−3784.85	43.40	42.87	42.79
Std dev.	0	0	0	0.25	21.02	0.70	0.73	1.25	673.04	2.64	3.11	2.58
p-value	**0.00**	-	-	0.09	**0.00**	**0.00**	**0.03**	**0.02**	**0.00**	**0.01**	**0.00**	**0.00**

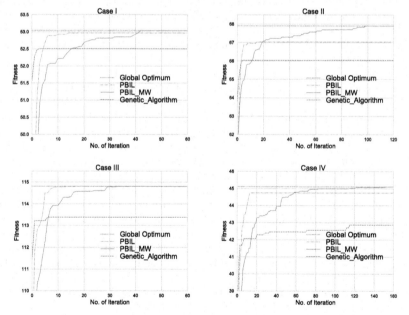

Fig. 3. Optima and converge speeds of the algorithms for all 4 cases.

5   Conclusions

In this paper, we propose a novel heuristic PBIL-MW algorithm and a strategy for choosing the initial probability for solving a coalition game of agents in partition form. The recommended approach, based on stochastic optimization, distributes participated agents to collaborate and act upon coalitions, for the purpose of obtaining the maximum profit hourly, and therefore could bring considerable payoff for members within group in the long term. Furthermore, by restraining the sub-coalition to form in the condition of power-surplus, we can secure the power distribution network from fluctuated demand and supply. Meanwhile, the scheme of our approach could enable a decentralized power system and improve the penetration of renewable energy. We have shown the potential of PBIL-MW in solving the problem of CSG, such that a large number of agents in a group to form viable coalitions will still be feasible with an optimization solution found within limited time, and which was a primary concerned in the implementation of CSG. Our result shows that PBIL-MW has a potential to outperform the PBIL and GA when searching for the optimal solution. For future work, we will further improve the PBIL-based algorithm and compare it with more heuristic and optimization methods using larger-scale smart grids.

References

1. Baluja, S.: Population-based incremental learning. A method for integrating genetic search based function optimization and competitive learning. Technical report, Carnegie-Mellon Univ, Dept Of Computer Science, Pittsburgh, PA (1994)
2. Baluja, S., Caruana, R.: Removing the genetics from the standard genetic algorithm. In: Machine Learning: Proceedings of the Twelfth International Conference, pp. 38–46 (1995)
3. Chakraborty, S., Nakamura, S., Okabe, T.: Scalable and optimal coalition formation of microgrids in a distribution system. In: 2014 IEEE PES Innovative Smart Grid Technologies Conference Europe (ISGT-Europe), pp. 1–6. IEEE (2014)
4. Chalkiadakis, G., Elkind, E., Wooldridge, M.: Computational aspects of cooperative game theory. Synth. Lect. Artif. Intell. Mach. Learn. **5**(6), 1–168 (2011)
5. Fadlullah, Z.M., Nozaki, Y., Takeuchi, A., Kato, N.: A survey of game theoretic approaches in smart grid. In: 2011 International Conference on Wireless Communications and Signal Processing (WCSP), pp. 1–4. IEEE (2011)
6. Hiremath, R., Shikha, S., Ravindranath, N.: Decentralized energy planning; modeling and application–a review. Renew. Sustain. Energy Rev. **11**(5), 729–752 (2007)
7. Marsland, S.: Machine Learning: An Algorithmic Perspective. CRC Press, Boca Raton (2015)
8. Mohamed, M.A., Eltamaly, A.M., Alolah, A.I.: PSO-based smart grid application for sizing and optimization of hybrid renewable energy systems. PLoS ONE **11**(8), e0159702 (2016)
9. Rahwan, T., Michalak, T.P., Wooldridge, M., Jennings, N.R.: Coalition structure generation: a survey. Artif. Intell. **229**, 139–174 (2015)
10. Saad, W., Han, Z., Poor, H.V.: Coalitional game theory for cooperative micro-grid distribution networks. In: 2011 IEEE International Conference on Communications Workshops (ICC), pp. 1–5. IEEE (2011)
11. Saad, W., Han, Z., Poor, H.V., Basar, T.: Game-theoretic methods for the smart grid: an overview of microgrid systems, demand-side management, and smart grid communications. IEEE Signal Process. Mag. **29**(5), 86–105 (2012)

12. Sandholm, T., Larson, K., Andersson, M., Shehory, O., Tohmé, F.: Coalition structure generation with worst case guarantees. Artif. Intell. **111**(1–2), 209–238 (1999)
13. Shoham, Y., Leyton-Brown, K.: Multiagent Systems: Algorithmic, Game-Theoretic, and Logical Foundations. Cambridge University Press, Cambridge (2008)
14. Wolsink, M.: The research agenda on social acceptance of distributed generation in smart grids: renewable as common pool resources. Renew. Sustain. Energy Rev. **16**(1), 822–835 (2012)
15. Yeh, D.Y.: A dynamic programming approach to the complete set partitioning problem. BIT Numer. Math. **26**(4), 467–474 (1986)

Towards a Brain-Inspired Developmental Neural Network by Adaptive Synaptic Pruning

Feifei Zhao[1,2], Tielin Zhang[1], Yi Zeng[1,2,3(✉)], and Bo Xu[1,2,3]

[1] Institute of Automation, Chinese Academy of Sciences, Beijing, China
{zhaofeifei2014,tielin.zhang,yi.zeng}@ia.ac.cn
[2] University of Chinese Academy of Sciences, Beijing, China
[3] Center for Excellence in Brain Science and Intelligence Technology,
Chinese Academy of Sciences, Shanghai, China

Abstract. It is widely accepted that appropriate network topology should be empirically predefined before training a specific neural network learning task. However, in most cases, these carefully designed networks are easily falling into two kinds of dilemmas: (1) When the data is not enough to train the network well, it will get an underfitting result. (2) When networks have learned too much patterns, they are likely to lead to an overfitting result and have a poor performance on processing new data or transferring to other tasks. Inspired by the synaptic pruning characteristics of the human brain, we propose a brain-inspired developmental neural network (BDNN) algorithm by adaptive synaptic pruning (BDNN-sp) which could get rid of the overfitting and underfitting. The BDNN-sp algorithm adaptively modulates network topology by pruning useless neurons dynamically. In addition, the evolutional optimization method makes the network stop on an appropriate network topology with the best consideration of accuracy and adaptability. Experimental results indicate that the proposed algorithm could automatically find the optimal network topology and the network complexity could adaptively increase along with the increase of task complexity. Compared to the traditional topology-predefined networks, trained BDNN-sp has the similar accuracy but better transfer learning abilities.

Keywords: Brain-inspired developmental neural network · Brain-inspired pruning rules · Structural plasticity · Network adaptability · Synaptic pruning

1 Introduction

Recently, neural network models, such as deep neural networks (DNN), have been widely used in many different domains including computer vision (CV) and natural language processing (NLP)[1,2], etc. Inspired by the hierarchical

---

Feifei Zhao and Tielin Zhang contributed equally to this work and should be considered as co-first authors, and the corresponding author is Yi Zeng.

© Springer International Publishing AG 2017
D. Liu et al. (Eds.): ICONIP 2017, Part IV, LNCS 10637, pp. 182–191, 2017.
https://doi.org/10.1007/978-3-319-70093-9_19

information processing mechanism in the brain, DNN has achieved big improvements on the tasks of image classification [3], face recognition [4] and video prediction [5]. However, there are still many opportunities to improve the models. Firstly, it is hard to measure the complexity of data before processing by the model. There is no criterion to evaluate the complexity of the data, for example, the data with simple pattern types but large number of samples, or complex patterns and small number of samples. Secondly, the huge number of hyperparameters on DNN leads the model time-consuming and easier to overfit. More importantly, the topologies of DNN are usually empirically set for given tasks which may not be suitable for other task complexities [6]. Thirdly, the training time of DNN mostly depends on the personal settings of specific variables such as iteration time, patch size or learning rate, and these variables are usually set by experiences or training tricks. The network usually could not be predefined at the best network state.

For the traditional DNNs, they only depend on the integration of lost function minimization and weights regulations [7] to avoid overfitting. As shown in Fig. 1(A), the precision error differences between training sample and test sample could be used to find the best fitting point, and the weights regulations will make the differences smaller. However, these methods are still with predefined network topology and even need more tricks to train a network well.

Some existing pruning neural network methods make the dynamical optimization possible, such as evolutionary artificial neural network (EANN). These kinds of methods usually calculate a fitness value to evaluate the performance of the network, and the fitness values usually include the descriptions of classification accuracy (e.g. the reciprocal of error, the mean square error [8,9] or the cross entropy error [10]) and network scales (e.g., the number of neurons or connections [11,12]). These methods focus on finding an appropriate model complexity which has the best balance between the network size and the test accuracy. However, the test accuracy is actually acceptable among a wide range of network size, as the solid black line shown in Fig. 1(B). The maximal accuracy could not reflect the comprehensive performance of network such as adaptability.

From information processing perspective, a complex neural network is considered to be a major part of the brain. In human brain, the process of eliminating synapses is important during the development period. Research has evidenced that the development period contains synaptic overgrowth in infant brain, then surplus synapses are gradually eliminated throughout childhood and adolescence [13]. Synapses that are rarely used are more likely to be eliminated during the pruning process [14]. The process of overgrowth first and then elimination is actually an efficient way for brain to achieve optimal development [13,15].

Inspired by the pruning mechanism in human brain, we propose a version of brain-inspired developmental neural network algorithm to optimize network topology structure by iteratively pruning useless neurons until the fitness function reaches the peak (BDNN-sp for short). The most appropriate network size is the red dashed line in Fig. 1(B). The network could automatically find the optimal topology and the network complexity could adaptively increase along with the increase of the task complexity, as shown in Fig. 1(C).

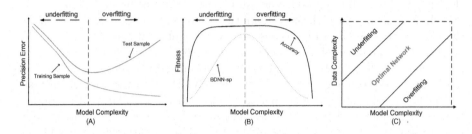

Fig. 1. Underfitting and overfitting in traditional DNN **(A)** and BDNN-sp **(B)**. The optimal network **(C)** has the very model complexity which is complex enough for the data complexity and will not cause overfitting or underfitting. (Color figure online)

This paper is organized as follows. Section 2 introduces the detailed methodology of the proposed BDNN-sp algorithm. Section 3 verifies the adaptability and transfer learning ability of the BDNN-sp algorithm on various complexities of tasks. A conclusion is provided in Sect. 4.

2   Methodology of the BDNN-sp Algorithm

The BDNN-sp algorithm is a member and a branch of a series of Brain-inspired Developmental Neural Networks (BDNN). It mainly contains two parts, the first part is the brain inspired neuron pruning method on dynamically modulating network topology, and the second part is the main developmental learning process of the BDNN-sp network.

2.1   Brain-Inspired Neural Network Pruning Rules

When learning a new task, a proportion of synapses are strengthened with increase of the volume of dendritic spines, and a proportion of synapses are weakened. Critically, these enlarged dendritic spines play an important role in performing this task, and they persist for a long time despite the subsequent learning of other tasks [16]. Meanwhile, the useless dendritic spines even neurons are gradually eliminated during the learning process, as shown in Fig. 2(A).

There are lots of synapses eliminated during the development process of the brain. The reasons can be summarized as the following hypotheses, such as saving spaces for storing new knowledge, or reducing the number of variables for better satisfying the outside environment, or increasing the response speed to some specific stimulus. The pruning process of the brain plays a main role during the procedure of brain and cognitive function development [13,17,18]. In this paper, inspired by the pruning mechanism of the human brain, in which the more weaker synapses even neurons are easier to be eliminated, we optimize the topology of artificial neural network (ANN) by pruning unimportant neurons for simplification. The basic process of pruning ANN is shown in Fig. 2(B).

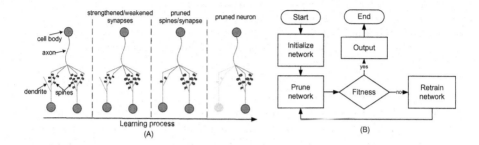

Fig. 2. The pruning process in human brain **(A)** and in BDNN-sp **(B)**.

We firstly build a three-layer ANN as initial network and ensure that the network is with enough variables (or enough complexity). Then, we shrink the complexity of the network by pruning the least important neurons. The reason is that unimportant neurons have little effect on the output. Then, we iteratively retrain the remaining network and prune unimportant neurons until the fitness function reaches the peak.

On the neuron level, every time we eliminate a proportion of the most unimportant neurons through pruning all the input and output connections of these neurons. We sum the total input and output weights of each neuron to evaluate its importance. Since presynaptic neuron may have both excitatory and inhibitory effects on postsynaptic neuron, then the importance degree of each neuron is calculated as shown in Eq. (1), where N_{in} is the number of connections sent to neuron j, and N_{out} is the number of connections sent from neuron j. w_{ij} is the weight from neuron i to j, and w_{jk} is the weight from neuron j to k.

$$I_j = \sum_{i}^{N_{in}} |w_{ij}| + \sum_{k}^{N_{out}} |w_{jk}| \tag{1}$$

After pruning the proportion of unimportant neurons, we retrain the remaining network through back propagation (BP) method to update remaining weights. Then, we evaluate the adaptability of network by fitness function which is introduced in Sect. 2.2. This process will be iteratively executed until the network reaches the best fitness. The framework of BDNN-sp algorithm is shown in Algorithm 1.

2.2   Fitness Function in BDNN-sp

This subsection introduces the fitness function which is the criteria of the adaptability for the network. This fitness function contains some evaluative conditions of network stated as the following:

Algorithm 1. The framework of the BDNN-sp algorithm.

Input: Initial neural network with enough complexity;
Output: Pruned neural network;
 1: Calculate the importance degree I_j for each neuron j;
 2: Prune a proportion of the most unimportant neurons;
 3: Retrain the remaining neural network by BP;
 4: Calculate the fitness function F;
 5: Repeat Step 1 to Step 4 until fitness function F reaches the peak;

---

(1) **Classification Performance**

We measures the classification performance of network by test accuracy. It is calculated by the ratio of accurate number to the total number of test samples, as shown in Eq. (2), where N_c represents the number of test samples that are correctly classified, and N_s is the total number of test samples. Classification accuracy is the benchmark to evaluate the classification performance, as well as the adaptability of the network.

$$A = \frac{N_c}{N_s} = 1 - error \tag{2}$$

(2) **Network Stability**

Information entropy is one of the most widely used measurement theory of information [19]. Generally, entropy represents uncertainty or disorder. If the outcome of an event has big probability, the entropy is small because it gives little new information. If the outcome of an event is unpredictable (small probability), the entropy will be large because it may contain some new information [20]. This paper proposes an entropy-like measurement to calculate the stability of the neural network.

Firstly, we define the difference between un-retrained weights and retrained weights as probabilities distribution, as shown in Eq. (3), where w_p represents the weights after pruning the most unimportant neurons in the network, and w_r represents the weights after retraining this pruned network. p represents the variation of the remaining weights. If the pruned neurons play little role in the network, then the p will be small and entropy will be large. After pruning to a certain extent, the remaining network is greatly influenced by the pruned neurons, then p will be fluctuant greatly and the entropy will be small. The equation of the entropy is shown in Eq. (4), where N_c is the number of remaining connections.

$$p = |w_r - w_p| \tag{3}$$

$$H(p) = -\sum_i^{N_c} p_i \log_2 p_i \tag{4}$$

Then we normalize the $H(p)$ as stability function, as shown in Eq. (5). S is effective for representing the stability of the network because it can reflect the change of overfitting and underfitting. At the beginning, the initial network complexity is so excess that it is easier to overfit. With the pruning of unimportant neurons, the network becomes smaller and the S will gradually increase until the network reaches the best stability. After that, if we go on pruning the network, it will be too small to learn the training data well. Then underfitting occurred, the S will start to decrease.

$$S = \frac{H(p)}{\max_i \left(|p_i \log_2 p_i|\right) * N_c} \tag{5}$$

(3) **Mean of Weights**

The mean of weights is calculated by Eq. (6), where n is the number of neurons in input layer, and m is the number of neurons in hidden layer, w_{ij} is the connection weight from neuron i to neuron j, w_{jk} is the connection weight from neuron j to neuron k. Mean of weights could reflect the change of connection weights. The more information it needs to represent, the larger weight value it will be with. Hence the mean of weights increases along with the increase of pruning ratio and task's complexity. Mean of weights is useful to balance the distribution of weights. When the network is complex enough, the mean of weights will be small, which leads the network easier to overfit.

$$E = \frac{\sum_{i=1}^{n} |w_{ij}| + \sum_{j=1}^{m} |w_{jk}|}{n + m} \tag{6}$$

Fitness function is a trade-off among the conditions introduced above, as shown in Eq. (7), where α, β and γ are the learning rates. Fitness function plays the most important role in measuring the adaptability of network and will help to prune the neural network to appropriate network size. Supervised by this fitness function, the network will be minimized as much as possible, and the network after pruning will make the best use of the remaining connections for the best adaptability.

$$F = \alpha A + \beta S + \gamma E \tag{7}$$

3   Experimental Results

The performance of the proposed BDNN-sp algorithm, especially the adaptability, will be verified and tested on different complexities of the Mixed National Institute of Science and Technology classification datasets (MNIST). To verify the adaptability of our BDNN-sp algorithm, we test it on different complexities of tasks. For example, different number of training samples or different groups of classes are tested. Besides, we also compare the transfer learning ability of pruned network with un-pruned network on another different complexities of tasks. The activation function of neuron is sigmoid function, and the learning rate η is equal to 1. Here, we set $\alpha = \beta = \gamma = \frac{1}{3}$.

The MNIST dataset contains 10 classes of hand written digits from 0 to 9. The total number of images contains 60,000 training samples and 10,000 test samples.

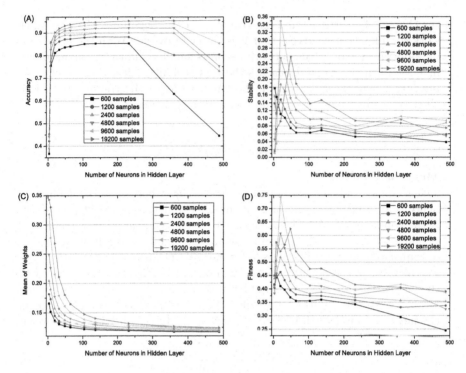

Fig. 3. The accuracy **(A)**, the stability **(B)**, the means of weights **(C)** and the fitness **(D)** of BDNN-sp on different number of MNIST training samples. The x axis shows the remaining number of neurons after pruning.

Each image is represented by a $28 * 28$ length of vector. As a result, the initial ANN is with 784 neurons in its input layer, and the number of neurons in output layer is equal to the class number. We give 1,000 neurons in the hidden layer at the beginning of the ANN training.

Here we test 600, 1200, 2400, 4800, 9600, 19200 training samples with 10 classes. Then we calculate the accuracy A, the stability S, the mean of weights E and the fitness F as shown in Fig. 3.

As Fig. 3(A) shows, the classification accuracy on different number of training samples are changing with the iterative hidden neuron pruning process. Accuracy represents the test accuracy with the 10,000 test samples for testing. From the Fig. 3(A), when the network is too small (i.e. 0 to 50 neurons are left) or too big (400 to 500 neurons are left), underfitting or overfitting occurred and the accuracy is low. The range of acceptable network size with proper accuracy is different from different number of training samples. As shown in Fig. 3(B), the performance of stability has clear peak which is corresponding to the network size with the best stability, and underfitting and overfitting are corresponding to the left and right side of the peak respectively. Figure 3(C) shows the change of means of network weights. It falls sharply at first and then keeps steady with the pruning of the network, and with more number of training samples, the higher mean of weights will be. The overfitting occurred when the network is too large

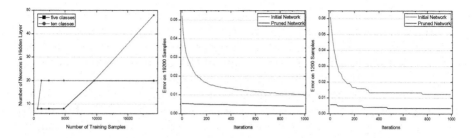

Fig. 4. (A): The final network size on different complexities of tasks. **(B)**: The change of error on 19,200 samples during iteration. **(C)**: The change of error on 1,200 samples during iteration. (Color figure online)

and the mean of weights is small. Figure 3(D) shows the change of fitness F, it has clear peak and the most appropriate network size has the best fitness. The more training samples are involved, the larger size of the network will be required.

Then the most appropriate network size on different number of training samples will be obtained. Figure 4(A) depicts the number of neurons in hidden layers after neural network pruning. With the increase of training samples, the number of neurons in hidden layer increases. The red line in Fig. 4(A) represents the training samples in 10 classes, and the black line represents the training samples in 5 classes. From the result, we could see that the network sizes of 5 classes are smaller than the ones in 10 classes which is consistent with the prediction. The result indicates that the pruned network complexity will increase along with the increase of task's complexity. In all, the model could automatically modulate network size for different complexities of tasks through the BDNN-sp algorithm.

Besides, we also test the transfer learning ability on another different complexities of tasks. We first train the initial network and pruned the network on 9,600 training samples. Then, we take out the pruned network and un-pruned network which both have been trained to the same accuracy 88.72%. Finally, we test pruned network and un-pruned network on untrained 1,200 and 19,200 samples. Figure 4(B) depicts the change of error on 19,200 samples, and Fig. 4(C) depicts the change of error on 1,200 samples. The black line in Fig. 4(B) and Fig. 4(C) represents the error of pruned network, and the red line represents the error of initial un-pruned network. Obviously, pruned network performs better than un-pruned network on both simple and complex tasks. Thus we can conclude that the pruned network has better transfer learning ability than un-pruned network.

4   Conclusion

This paper proposes a neuron pruning method to adaptively modulate the topology structure of neural networks. Pruning process iteratively eliminates unimportant neurons and retrains the remaining network until its fitness reaches the peak. In order to determine the most appropriate network topology, we propose a new BDNN-sp algorithm with fitness function which integrates classification performance, stability and mean of weights. The experimental results show that the BDNN-sp model could reflect the network states of overfitting and underfitting. In addition, to verify the adaptability of BDNN-sp, we test it on different complexities of MNIST classification tasks. The experimental results show that the network complexity increases along with the increase of the task complexity. By comparing with the initial network, the size of network could be greatly reduced while the accuracy is with little reduction. On the transfer learning tasks, the BDNN-sp model performs better than un-pruned network.

Acknowledgment. This study was funded by the Strategic Priority Research Program of the Chinese Academy of Sciences (XDB02060007), and Beijing Municipal Commission of Science and Technology (Z161100000216124).

References

1. Hinton, G.E., Osindero, S., Teh, Y.-W.: A fast learning algorithm for deep belief nets. Neural Comput. **18**(7), 1527–1554 (2006)
2. Lecun, Y., Bengio, Y., Hinton, G.: Deep learning. Nature **521**(7553), 436–444 (2015)
3. He, K., Zhang, X., Ren, S., Sun, J.: Delving deep into rectifiers: surpassing human-level performance on imagenet classification. In: Proceedings of the IEEE International Conference on Computer Vision, pp. 1026–1034 (2015)
4. Lawrence, S., Giles, C.L., Tsoi, A.C., Back, A.D.: Face recognition: a convolutional neural network approach. IEEE Trans. Neural Netw. **8**(1), 98–113 (1997)
5. Deng, L., Hinton, G., Kingsbury, B.: New types of deep neural network learning for speech recognition and related applications: an overview. In: Proceedings of the IEEE International Conference on Acoustics, Speech and Signal Processing, pp. 8599–8603 (2013)
6. Zhou, Z.-H., Feng, J.: Deep forest: towards an alternative to deep neural networks (2017)
7. Molchanov, P., Tyree, S., Karras, T., Aila, T., Kautz, J.: Pruning convolutional neural networks for resource efficient transfer learning. In: Proceedings of the International Conference on Learning Representations (ICLR) (2017)
8. Angeline, P.J., Saunders, G.M., Pollack, J.B.: An evolutionary algorithm that constructs recurrent neural networks. IEEE Trans. Neural Netw. **5**(1), 54–65 (1994)
9. Yao, X., Liu, Y.: Evolving artificial neural networks through evolutionary programming. In: Proceedings of the 5th Annual Conference on Evolutionary Programming, pp. 257–266 (1996)
10. Park, J.C., Abusalah, S.T.: Maximum entropy: a special case of minimum cross-entropy applied to nonlinear estimation by an artificial neural network. Complex Syst. **11**, 289–308 (1997)
11. Vonk, E., Jain, L.C., Johnson, R.: Using genetic algorithms with grammar encoding to generate neural networks. In: Proceedings of IEEE International Conference on Neural Networks **4**, 1928–1931 (1995)
12. Ioan, I., Rotar, C., Incze, A.: The optimization of feed forward neural networks structure using genetic algorithms. In: Proceedings of the International Conference on Theory and Applications of Mathematics and Informatics (ICTAMI) **8**, 223–234 (2004)

13. Chechik, G., Meilijson, I., Ruppin, E.: Synaptic pruning in development: a computational account. Neural Comput. **10**(7), 1759–1777 (1998)
14. Johnston, M.V., Ishida, A., Ishida, W.N., Matsushita, H.B., Nishimura, A., Tsuji, M.: Plasticity and injury in the developing brain. Brain Dev. **31**(1), 1–10 (2009)
15. Pascual-Leone, A., Amedi, A., Fregni, F., Merabet, L.B.: The plastic human brain cortex. Annu. Rev. Neurosci. **28**(28), 377–401 (2005)
16. Hayashi-Takagi, A., Yagishita, S., Nakamura, M., Shirai, F., Wu, Y., Loshbaugh, A.L., Kuhlman, B., Hahn, K.M., Kasai, H.: Labelling and optical erasure of synaptic memory traces in the motor cortex. Nature **525**(7569), 333–338 (2015)
17. Chechik, G., Meilijson, I., Ruppin, E.: Neuronal regulation: a biologically plausible mechanism for efficient synaptic pruning in development. Neurocomputing **26–27**(98), 633–639 (1999)
18. Chechik, G., Meilijson, I., Ruppin, E.: Synaptic pruning in development: a novel account in neural terms. In: Bower, J.M. (ed.) Computational Neuroscience, pp. 149–154. Springer, Boston (1998). doi:10.1007/978-1-4615-4831-7_25
19. Shannon, C.E.: A mathematical theory of communication. Bell Syst. Tech. J. **27**(3), 379–423 (1948)
20. Shannon, C.E.: Prediction and entropy of printed English. Bell Syst. Tech. J. **30**, 50–64 (1951)

Using Word Mover's Distance with Spatial Constraints for Measuring Similarity Between Mongolian Word Images

Hongxi Wei[✉], Hui Zhang, Guanglai Gao, and Xiangdong Su

School of Computer Science, Inner Mongolia University, Hohhot 010021, China
cswhx@imu.edu.cn

Abstract. In the framework of bag-of-visual-words, visual words are independent each other, which results in discarding spatial relations and lacking semantic information of visual words. To capture semantic information of visual words, a deep learning procedure similar to word embedding technique is used for mapping visual words to embedding vectors in a semantic space. And then, word mover's distance (WMD) is utilized to measure similarity between two word images, which calculates the minimum traveling distance from the visual embeddings of one word image to another one. Moreover, word images are partitioned into several sub-regions with equal sizes along rows and columns in advance. After that, WMDs can be computed from the corresponding sub-regions of the two word images, separately. Thus, the similarity between the two word images is the sum of these WMDs. Experimental results show that the proposed method outperforms various baseline and state-of-the-art methods, including spatial pyramid matching, latent Dirichlet allocation, average visual word embeddings and the original word mover's distance.

Keywords: Visual word embeddings · Word mover's distance · Spatial information · Keyword spotting · Query-by-example

1 Introduction

How to access the content from a large number of scanned historical document images is still a challenging task. Because of aging, the historical document images are often degradation and poor quality. Therefore, robust optical character recognition (OCR) tools are not available yet. When OCR is infeasible, keyword spotting technology is an alternative approach. In the keyword spotting technology, all scanned historical document images are generally segmented into individual word images to form a word image collection. As for a given query keyword, relevant word images can be detected in the collection of word images by image matching [1].

In the traditional keyword spotting, profile-based features were widely used to represent word images [2] and dynamic time warping (DTW) algorithm was utilized to accomplish image matching [3]. Though the DTW algorithm works well, it is so time-consuming that cannot be suited for real-time image matching for a large collection of word images. Hence, this study focuses on how to represent word images so as to realize real-time image matching.

© Springer International Publishing AG 2017
D. Liu et al. (Eds.): ICONIP 2017, Part IV, LNCS 10637, pp. 192–201, 2017.
https://doi.org/10.1007/978-3-319-70093-9_20

In recent years, Bag-of-Visual-Words (BoVW) has been attracted much more attention and shown advantages in keyword spotting on historical documents [4, 5]. In the BoVW framework, word images are represented as visual histograms with a fixed-length. In this way, cosine similarity (or Euclidean distance) between word images can be calculated on their histograms. At the retrieval stage, when a query keyword is provided, the corresponding cosine similarities can be calculated for a collection of word images. By this way, a ranking list of word images can be formed in descending order of the cosine similarities. Thus, the BoVW-based representation approach is competent for the task of keyword spotting on a large number of word images. However, local descriptors (i.e. visual words) within one word image are independent each other in the BoVW-based representation, which results in not only discarding spatial relations between visual words but also lacking semantic information of visual words.

In this paper, an approach has been proposed to capture semantic information of visual words. To be specific, a deep learning procedure similar to word embedding is used for mapping visual words to embedding vectors in a semantic space. Consequently, the semantic relatedness between visual words can be measured by calculating Euclidean distance or cosine similarity on their embedding vectors. In order to distinguish from the original word embeddings proposed by Mikolov et al. [6], the embedding vectors in this study are called *visual word embeddings*. Kusner et al. [7] recently proposed word mover's distance (WMD), a distance function between two documents, which calculates the minimum traveling distance from the word embeddings of one document to another one. In our study, the WMD is used for measuring the similarity between two word images. Through this way, the semantic information of visual words is integrated into image matching.

Additionally, all word images are partitioned into a certain quantity of sub-regions with equal sizes along rows and columns in advance. In the image matching phase, only the corresponding sub-regions of the two word images are matched each other. Thus, WMDs can be computed from the corresponding sub-regions of the two word images, separately. Finally, the similarity of the two word images is the sum of these WMDs. By this means, such the spatial relations can be added to the procedure of image matching. Hence, the above-mentioned drawbacks of the BoVW-based representation can be overcome using the proposed WMD with spatial constraints.

The rest of the paper is organized as follows. The related work is presented in Sect. 2. The proposed method is described detailedly in Sect. 3. Experimental results are shown in Sect. 4. Section 5 provides the conclusions and future work.

2   Related Work

In the keyword spotting technology, several manners for providing query keywords have been proposed in the literature, which can be divided into query-by-example (QBE) and query-by-string (QBS) approaches [8]. In the QBS approach, query keywords are provided by textual strings [9, 10]. But, the QBS approach needs to learn a model to map from textual strings to images on a certain number of annotated word images. When there is no such annotated word images, the QBE approach can be competent. The QBE approach [11, 12] requires that an example image of a query keyword is provided for being retrieved. In this study, we concentrate on the QBE based approach for realizing keyword spotting on historical Mongolian document images.

In our previous work, visual language model (VLM) [13] was proposed for representing the corresponding word images segmented from a collection of historical Mongolian documents. Therein, each word image was represented as a probability distribution of visual words and query likelihood model was used to calculate similarity between two word images. Although the VLM (e.g. bigram visual language model) can provide the spatial orders between the neighboring visual words, there is still lacking semantic information of visual words. Therefore, a latent Dirichlet allocation (LDA) based word image representation approach presented in another our previous work [14]. In the LDA-based representation, topics were treated as probability distributions over visual words. Each word image was viewed as a probabilistic mixture over these topics. Thus, the LDA-based representation can provide the semantic information of visual words. However, the semantic information in the LDA-based representation is latent, which cannot be used for measuring the semantic relatedness between visual words directly. Consequently, the semantic information needs to be obtained in more obvious form.

In the last few years, word embedding techniques have been shown significant improvements in various natural language processing (NLP) tasks, such as word analogy [6], information retrieval [15], and so forth. Word2vec [6] and GloVe [16] are examples of successful implementations of word embeddings that respectively use neural networks and matrix factorization to learn embedding vectors. Because GloVe incorporates co-occurrence statistics of words that frequently appear together within the documents. Pennington et al. [16] have proved that GloVe outperforms Word2vec on word analogy, word similarity and named entity recognition tasks. Therefore, GloVe is utilized to generate embedding vectors for visual words in this paper. In this manner, visual words will be mapped as vectors in a semantic space. The generated embedding vectors are named *visual word embeddings*. As far as we know, this is the first time to learn and generate embedding vectors on visual words.

After that, a common approach for representing a word image is to take a centroid of its visual word embeddings. And then, an inner product or cosine between the centroids can be calculated for measuring similarity [17]. However, taking a simple centroid is not a good approximation for representing a word image. A more reasonable approach is to calculate similarity between visual words from one word image to another one. Consistent with this, Kusner et al. [7] proposed a word mover's distance that can calculate similarity between two documents on their embedding vectors. This study is partly motivated by the word mover's distance to measure similarity between two word images.

To integrate the spatial relations of visual words into image matching, spatial pyramid matching (SPM) has been proposed by Lazebnik et al. [18]. The SPM method partitions an image into several sub-regions with equal sizes and computes visual histograms in each sub-region. Our study is also inspired by the SPM method. All word images are partitioned into a number of sub-regions with equal sizes along rows and columns. At the stage of image matching, only the corresponding sub-regions of the two word images are matched each other. In this paper, similarity between two word images is measured by using the WMD with spatial constraints.

3   Proposed Method

In our study, the handling objects are word images. Hence, each scanned image in a collection of historical Mongolian documents should be segmented into individual word images in advance. And the QBE approach is used in the retrieval phase. The details of the proposed method are presented in the following subsections.

3.1   Obtaining Visual Words

Given a collection of word images, SIFT descriptors are extracted from each word image. And then, *k-means clustering algorithm* is applied on these SIFT descriptors so as to generate a certain number of clusters. Thus, the center of each cluster is taken as a visual word. By this way, a codebook can be formed. Figure 1 shows the procedure for constructing a codebook.

After that, each SIFT descriptor will be assigned the label of the closet center (i.e. visual word) according to the codebook. Thus, the corresponding visual words can be obtained from the collection of word images.

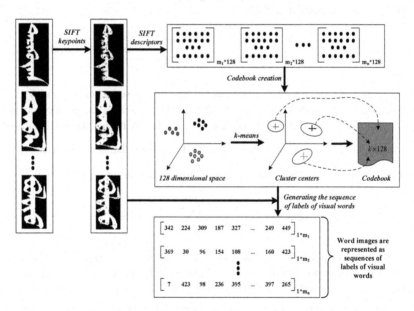

Fig. 1. The procedure for constructing a codebook.

3.2   Generating Visual Word Embeddings

After obtaining visual words, one word image can be represented as a sequence of labels of visual words along the writing direction (see Fig. 1). On a collection of word images, a training corpus of visual words can be collected by concatenating the corresponding sequences of labels of visual words one after another. And then, a GloVe tool (http://nlp.stanford.edu/projects/glove/) is utilized to generate embedding vectors of visual words on the training corpus.

In our study, the parameters of GloVe are set to as follows. The size of embedding vector and context window are set to 200 and 15, separately. And the number of iterations is set to 15. After generating visual word embeddings, the semantic relatedness between visual words can be measured by calculating Euclidean distance (or cosine similarity) on their embedding vectors.

3.3   Word Mover's Distance

Our work is based on the original word mover's distance (WMD) between text documents proposed by Kusner et al. in [7]. The average time complexity of the original WMD is $O(n^3 \log n)$, where n denotes the number of vocabularies on a collection of documents. For documents with many unique words, solving the WMD optimal transport problem may become prohibitive. So, Kusner et al. also introduced relaxed and much faster WMD versions.

In our case, the first relaxation is to sum the distances of the visual word embeddings $\vec{w}$ in a query keyword q to the closest visual word embeddings $\vec{w}'$ of a word image d. Thus, the WMD from the query keyword q to the word image d (denoted by $RWMD_{Q2D}$) can be defined as follows.

$$RWMD_{Q2D}(q \rightarrow d) = \sum_{w \in q} \frac{count(w)}{\sum_{t \in q} count(t)} \cdot \min_{w' \in d} distance\left(\vec{w}, \vec{w}'\right) \qquad (1)$$

where w and w' are the visual words occurred in q and d, separately. $\vec{w}$ and $\vec{w}'$ are the corresponding visual embedding vectors of w and w'. $count(w)$ means the occurrence frequency of the visual word w in q. And $\sum_{t \in q} count(t)$ means the total number of visual words in q. $distance\left(\vec{w}, \vec{w}'\right)$ denotes the Euclidean distance between two visual embedding vectors $\vec{w}$ and $\vec{w}'$, and its formulation is as follows.

$$distance\left(\vec{w}, \vec{w}'\right) = \sqrt{\sum_{i=1}^{K} (w_i - w_i')^2} \qquad (2)$$

where w_i and w_i' denote the ith elements in the visual embedding vectors $\vec{w}$ and $\vec{w}'$. K indicates the dimension of the visual word embeddings. In our study, K equals to 200.

Similarly, the second relaxed form is to sum the distances of the visual word embeddings $\vec{w}'$ of d to the closest visual word embeddings $\vec{w}$ of q. The corresponding WMD from the word image d to the query keyword q (denoted by $RWMD_{D2Q}$) can be defined as the following formula.

$$RWMD_{D2Q}(d \rightarrow q) = \sum_{w' \in d} \frac{count(w')}{\sum_{t' \in d} count(t')} \cdot \min_{w \in q} distance\left(\vec{w}', \vec{w}\right) \qquad (3)$$

In (1) and (3), the time complexity for getting the optimal solution is $O(n^2)$, which is faster than the original WMD. Kusner et al. found the maximum of $RWMD_{Q2D}$ and

RWMD$_{D2Q}$ to be the best relaxation of the original WMD. Therefore, the final WMD between two word images q and d can be calculated by the following equation.

$$\text{WMD}(q, d) = \max\{\text{RWMD}_{Q2D}(q \to d), \text{RWMD}_{D2Q}(d \to q)\} \quad (4)$$

In this way, when a query keyword and a collection of word images are given, a ranking list of word images can be formed according to (4).

3.4 Integrating Spatial Information

In order to integrate spatial information into word images representation, all word images are partitioned into a quantity of sub-regions along rows and columns. All sub-regions within one word image have the equal sizes. Figure 2 depicts an example for partitioning a word image into three sub-regions along rows.

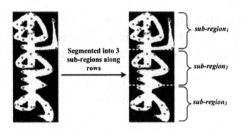

Fig. 2. An example for partitioning a word image.

At the image matching phase, the corresponding sub-regions between two word images are matched, respectively. Thus, the RWMD$_{Q2D}$ and RWMD$_{D2Q}$ between two word images can be rewritten as follows.

$$\text{RWMD}_{Q2D}(q \to d) = \sum_{j=1}^{N} RWMD_{Q2D}(q_j \to d_j) \quad (5)$$

$$\text{RWMD}_{D2Q}(d \to q) = \sum_{j=1}^{N} RWMD_{D2Q}(d_j \to q_j) \quad (6)$$

where q_j and d_j denote the jth sub-regions of the two word images, severally. N indicates the number of sub-regions. Particularly wish to point out, the difference between the adopted spatial information in this study and the SPM is regardless of partition levels.

4 Experimental Results

4.1 Dataset and Baselines

To evaluate the performance, a collection of Mongolian historical documents has been collected, which consists of **100** scanned Mongolian Kanjur images with **24,827** words.

Each page has been annotated manually to form the ground truth data. Twenty meaningful words are selected and taken as query keywords. The dataset and the query keywords are the same as in [13, 14]. Evaluation metric is *mean average precision* (MAP).

For constructing a codebook, SIFT descriptors are extracted from the **24,827** word images and the total number of the SIFT descriptors is **2,283,512**. After that, the k-means clustering algorithm has been performed on those descriptors. Therein, we vary the number of clusters from **500** to **10,000** with **500** as an interval. In the following subsections, the appropriate number of clusters will be determined.

In this section, *spatial pyramid matching, average visual word embeddings, visual language model* and *latent Dirichlet allocation* are taken as baselines for comparison. The details of the baseline methods are as follows.

Spatial pyramid matching (SPM): The standard SPM method [18] is utilized to accomplish the aim of image matching between a given query keyword and each word image in a collection.

Average visual word embeddings (AVWE): After generating visual embedding vectors, a word image (denoted by W) can be represented as a centroid (denoted by W_{cent}) of the embedding vectors of its visual words using the following equation [17]:

$$W_{cent} = \frac{1}{|W|} \sum_{j=1}^{|W|} v_j \tag{7}$$

where v_j is the embedding vector of the corresponding visual word and $|W|$ is the number of visual words within the word image W. Under the circumstance, Euclidean distance can be calculated and used for measuring similarity between word images.

Visual language model (VLM): Each word image can be represented as probability distribution of visual words. Query likelihood model is utilized to rank word images. In our previous work [13], the best performance of VLM can attain to **31.75%**.

Latent Dirichlet allocation (LDA): A LDA-based topic model is adopted to obtain the semantic relations between visual words. In our previous work [14], the best performance is **43.78%**. At present, the LDA-based representation method is the state-of-the-art for keyword spotting on the same dataset.

4.2   Performance of SPM and AVWE

For Mongolian language, its writing direction is from top to bottom. So, horizontal partitions are more important than vertical partitions. In the SPM method, we have tested nine types of one-level partitions and five types of two-level partitions. Their MAPs are shown in Figs. 3 and 4, respectively.

From Figs. 3 and 4, we can see that one-level partitions and two-level partitions both obtain the best performance when the number of clusters is **500**. In various one-level partitions, the best performance is **37.71%** and the manner of the partition is **9 * 2**. Correspondingly, the best performance of the two-level partitions is **38.38%** when the

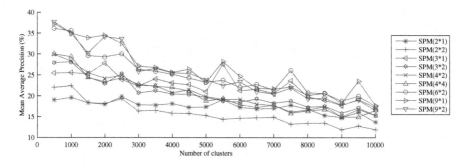

Fig. 3. The performance of SPM with one-level partitions.

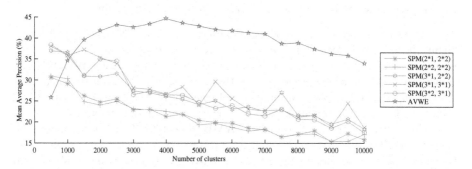

Fig. 4. The performance of SPM with two-level partitions.

first level and the second level are **3 * 2** and **3 * 1**, severally. Indeed, the horizontal partitions are more crucial than the vertical partitions for Mongolian word images.

Additionally, the performance of AVWE has been tested. In Fig. 4, the best performance of AVWE is **44.61%** when the number of clusters is **4,000**. Therefore, the AVWE is superior to the SPM. The MAP is improved from 38.38% to 44.61%. It indicates that the semantic information of visual words is more important than the spatial information in our case. Meanwhile, the AVWE is superior to the LDA-based method as well. So, the proposed visual embeddings can capture much more semantic information than the LDA-based representation method.

4.3 Performance of the Relaxed WMD Without Spatial Constraints

For comparison, we have also tested the performance of the relaxed WMD without spatial constraints. According to (1), (3) and (4), word images can be ranked for a given query keyword. In Fig. 5, the best performance is **38.16%** when the number of clusters is **7500**. Although the relaxed WMD is superior to the VLM (31.75%) and the one-level partition based SPM (37.71%), it is inferior to the other baseline methods including the tow-level partition based SPM (38.38%), the LDA-based representation method (43.78%) and the AVWE (44.61%).

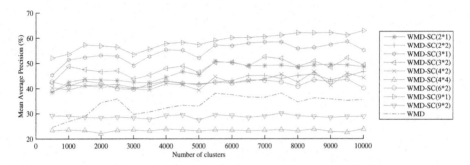

Fig. 5. The performance of the relaxed WMD and the proposed WMD-SC.

4.4 Performance of the Proposed WMD with Spatial Constraints

Here, we tested the performance of the proposed relaxation WMD with spatial constraints (denoted by **WMD-SC**). There are nine partition styles for word images, which are the same as in Fig. 3. The corresponding results of the proposed **WMD-SC** are shown in Fig. 5. As for the proposed **WMD-SC**, the various partition styles are consistently superior to the WMD without spatial constraints except for 4 * 4 and 9 * 2. Therefore, the spatial information plays an important part in our study.

In Fig. 5, the best performance of the **WMD-SC** is **62.88%** when the number of clusters is **10,000** and word images are partitioned into **9** sub-regions along rows only. Therefore, the performance of the proposed **WMD-SC** is increased by **44%** (from 43.78% to 62.88%) against to the state-of-the-art method (i.e. LDA-based representation method) on the same dataset.

5 Conclusion and Future Work

In this paper, a novel method has been proposed for measuring similarity between Mongolian word images. On the one hand, the spatial information is obtained by partitioning word images into sub-regions along rows and columns. Only the corresponding sub-regions between two word images are matched. On the other hand, embedding vectors of visual words are generated by utilizing a deep learning tool. After that, the relaxed word mover's distance is used for calculating similarity on the corresponding sub-regions between two word images. Therefore, the proposed method can combine the spatial information of visual words with the semantic relatedness so as to attend the aim of measuring similarity between word images. And the performance of the proposed method outperforms various baseline methods and the state-of-the-art method.

In our future work, the corresponding semantic relatedness between the visual embeddings will be utilized to attain the aim of query expansion. The proposed method will be validated on the other datasets of historical documents.

Acknowledgement. This paper is supported by the National Natural Science Foundation of China under Grant 61463038.

References

1. Rath, T.M., Manmatha, R.: Word spotting for historical manuscripts. Int. J. Doc. Anal. Recogn. **9**(2), 139–152 (2007)
2. Rath, T.M., Manmatha, R.: Features for word spotting in historical manuscripts. In: Proceedings of ICDAR 2003, pp. 218–222. IEEE Press, New York (2003)
3. Rath, T.M., Manmatha, R.: Word image matching using dynamic time warping. In: Proceedings of CVPR 2003, pp. 521–527. IEEE Press, New York (2003)
4. Shekhar, R., Jawahar, C.V.: Word image retrieval using bag of visual words. In: Proceedings of DAS 2012, pp. 297–301. IEEE Press, New York (2012)
5. Aldavert, D., Rusinol, M., Toledo, R., Llados, J.: A study of bag-of-visual-words representations for handwritten keyword spotting. Int. J. Doc. Anal. Recogn. **18**(3), 223–234 (2015)
6. Mikolov, T., Sutskever, I., Chen, K., Coorado, G.S., Dean, J.: Distributed representations of words and phrases and their compositionality. In: Proceedings of NIPS 2013, pp. 3111–3119. MIT Press, Massachusetts (2013)
7. Kusner, M.J., Sun, Y., Kolkin, N.I., Weinberger, K.Q.: From word embeddings to document distances. Proc. Mach. Learn. Res. **37**, 957–966 (2015)
8. Fornes, A., Frinken, V., Fischer, A., Almazan, J., Jackson, G., Bunke, H.: A keyword spotting approach using blurred shape model-based descriptors. In: Proceedings of HIP 2011, pp. 83–89. ACM Press, New York (2011)
9. Aldavert, D., Rusinol, M., Toledo, R., Llados, J.: Integrating visual and textual cues for query-by-string word spotting. In: Proceedings of ICDAR 2013, pp. 511–515. IEEE Press, New York (2013)
10. Rothacker, L., Fink, G.A.: Segmentation-free query-by-string word spotting with bag-of-features HMMs. In: Proceedings of ICDAR 2015, pp. 661–665. IEEE Press, New York (2015)
11. Wei, H.X., Gao, G.L., Su, X.D.: A multiple instances approach to improving keyword spotting on historical Mongolian document images. In: Proceedings of ICDAR 2015, pp. 121–125. IEEE Press, New York (2015)
12. Wei, H.X., Zhang, H., Gao, G.L.: Representing word image using visual word embeddings and RNN for keyword spotting on historical document images. In: Proceedings of ICME 2017, pp. 1374–1379. IEEE Press, New York (2017)
13. Wei, H.X., Gao, G.L.: Visual language model for keyword spotting on historical Mongolian document images. In: Proceedings of CCDC 2017, pp. 1765–1770. IEEE Press, New York (2017)
14. Wei, H., Gao, G., Su, X.: LDA-based word image representation for keyword spotting on historical Mongolian documents. In: Hirose, A., Ozawa, S., Doya, K., Ikeda, K., Lee, M., Liu, D. (eds.) ICONIP 2016. LNCS, vol. 9950, pp. 432–441. Springer, Cham (2016). doi:10.1007/978-3-319-46681-1_52
15. Zamani, H., Croft, W.B.: Embeddings-based query language models. In: Proceedings of ICTIR 2016, pp. 147–156. ACM Press, New York (2016)
16. Pennington, J., Socher, R., Manning, C.D.: GloVe: global vectors for word representation. In: Proceedings of EMNLP 2014, pp. 1532–1543. ACL Press, Stroudsburg (2014)
17. Nalisnick, E., Mitra, B., Craswell, N., Caruana, R.: Improving document ranking with dual word embeddings. In: Proceedings of WWW 2016, pp. 83–84. ACM Press, New York (2016)
18. Lazebnik, S., Schmid, C., Ponce, J.: Beyond bags of features: spatial pyramid matching for recognizing natural scene categories. In: Proceedings of CVPR 2006, pp. 2169–2178. IEEE Press, New York (2006)

A Multimodal Vigilance Monitoring System Based on Fuzzy Logic Architecture

Ahmed Snoun[✉], Ines Teyeb, Olfa Jemai, and Mourad Zaied

National Engineering School of Gabes (ENIG),
RTIM: Research Team in Intelligent Machines, University of Gabes,
Gabès, Tunisia
ahmedsnoun3@gmail.com,
{ines.teyeb.tn,olfa.jemai,mourad.zaied}@ieee.org

Abstract. This paper deals with the problem of vigilance level monitoring. A novel method of hypovigilance detection is presented in this work. It is based on the analysis of eyes' blinking and head posture. The fusion task of both systems is achieved by the fuzzy logic technique which allows us to obtain five vigilance levels. This paper contains two key contributions. The first is the amelioration of our previous works in the classification field employing fast wavelet network classifier (FWT) by using another classification system based on a deep learning architecture. It provides more accurate results than the wavelet network classifier. The second resides in the conception of a driver alertness control system able to detect five vigilance levels which is different from previous works of the literature characterized by two, three or four levels. Experiments, using different datasets, prove the good performance of our new approach.

Keywords: Head pose · Fuzzy · Sleep · Eyes' blinking · Deep learning

1 Introduction

The hypovigilance state during driving is the origin of several road accidents. According to a study done by the National Institute of Sleep and Vigilance, 34% of accidents are due to the lack of sleep. In fact, this state is mainly shown by a decrease in vigilance level with the appearance of different behavioral signals such as a heaviness in tasks' accomplishments, reflexes' reduction, yawning, eyelids' heaviness, and the difficulty of keeping the head in frontal position compared to the field of view.

Consequently, the use of an assistive system which controls a driver's alertness position and alerts him in situation of vigilance decrease, can be efficient to avoid accidents and ensure the driver's safety.

Many works were proposed in the literature to detect this dangerous state. These different research activities can be divided into two main categories which are vehicle-oriented approaches and driver-oriented ones.

For the first category, drowsiness may be detected via different measurements such as pressure on the acceleration pedal, analysis of movement of the steering

© Springer International Publishing AG 2017
D. Liu et al. (Eds.): ICONIP 2017, Part IV, LNCS 10637, pp. 202–211, 2017.
https://doi.org/10.1007/978-3-319-70093-9_21

wheel and the angle value of car movement compared to the lane position, etc. When the previously cited indices reach a specific value, it means a significantly low probability that the person is sleepy. This approach does not seem very efficient because it may depend on the way of driving and the shape and characteristics of the road. For the second approach, the vigilance control may be released by analyzing physiological and behavioral signs. For physiological signs studying, invasive techniques are exploited. These techniques are used in most cases, electroencephalographic data for cerebral activity measurement which was used by Picot for hypovigilance detection. He developed a muti-variable system based on cerebral activity control and eyes' blinking analysis. It allowed the classification of driver's vigilance into three levels [1].

However, behavioral signs, are analyzed by a non-invasive technique, which uses, purely visual indicators relating to the vigilance decline. These signs like yawning, blinking, gaze direction or head pose, can be captured from an embedded camera in the car. As an example, we cite the research activities of Akrout [2] who developed a system based on the analysis of eyes' blinking, head pose and yawning detection using a video camera for facial signs capturing. These different inputs are fused by an expert system based on binary rules (0, 1) and provided four possible vigilance levels. On the practical level, the non-invasive techniques are easier to be exploited under real conditions as they have less constraints than the other techniques during deployment.

The structure of our paper is as follows: We begin by presenting the general process of our system as well as the explanation of its key steps. The second part is devoted to evaluating the performance of our hypovigilance detection system and we conclude with a conclusion and perspectives.

2   Proposed Approach for Hypovigilance Detection System

The general process of our proposed system for vigilance control is illustrated in Fig. 1.

After segmentation of the captured video into frames, the interest areas (head and eyes) are detected and tracked using Viola and Jones algorithm [3] thanks to its fiability in real time applications. Our system is based on a multi-variable approach. In fact, it is composed of two sub systems which are, the eyes' blinking analysis and the head posture estimation. The first parameter of vigilance monitoring is the eyes' closure duration which presents a significant sign for heavy eyelids detection, where the driver has the desire to close his eyes for a moment because he feels drowsy. If the eyes' closure duration exceeds a predefined time T, that means that the driver is in an hypovigilance state [4–6].

The second parameter is the head movement angle. If the angle exceeds a specific value, it means that the driver is in an hypovigilant state. Also, the frequency of head movement is an efficient sign for fatigue detection [7–9].

Our eyes classification system is based on a deep learning architecture and especially the Transfer Learning classifier to recognize if eyes are closed or

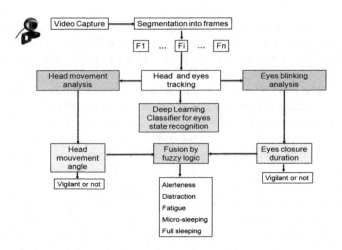

Fig. 1. Architecture of our proposed system

opened. Its principle will be detailed later in next sections. This technique seems to be more efficient than the wavelet network classifier which was used in our previous works [4–6], more details about the wavelet networks can be founded in [10,11]. For head movement angle computing, we used the Pythagorian Theorem which was applied in the triangle located between both eyes bounding rectangles. More details about this part are found in [7–9]. If we consider each sub system separately, we may obtain a general decision (vigilant or not) without mentioning exactly the vigilance level. Also we may have a false decision. However, when fusing both parameters, we obtain a more precise decision with different vigilance levels such as distraction, fatigue, micro and full sleep.

2.1 Transfer Learning Architecture for Eyes Classification

The transfer learning belongs to the deep learning approaches which refers to a set of automatic learning methods that are based on the artificial Neural Network. This new type of learning is used to model the data with a high level of abstraction. Indeed, this technique provides a significant and rapid progress in fields of signal analysis, object recognition, and computer vision. Figure 2 illustrates the general architecture of the Transfer learning. This architecture includes a set of independent processing layers which are:

- A convolution layer
- A correlation layer
- A pooling layer

Each input image is transformed into a feature vector thanks to the convolutive part of the pre-trained network. This vector is used subsequently to train a new classifier. As a pre-trained model, we have used the Alexnet model [12]. It is a Convolutional Neural Netowrk that is trained on large benchmarks. These

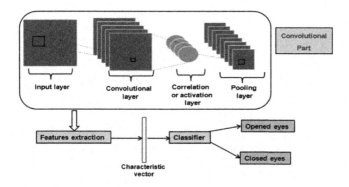

Fig. 2. Architecture of the transfer learning

types of network are adapted to other classification problems with big or small databases. Despite its simplicity, this method has many advantages:

- The image is transformed into a small vector, which contains generally relevant characteristics
- The user has the freedom to use the final classifier of his choice

In addition, several types of classifiers can be used to train the features vector which is generated by the convolutive part of the pre-training model, such as the Softmax and the Linear SVM (Support Vector Machine) which was used in our work.

2.2 Fuzzy Logic for Fusion of both Vigilance Monitoring Systems

As we have already said in the introduction section, our objective is the conception of a multi-variable system for hypovigilance detection based on head posture estimation and eyes' blinking analysis. To achieve this task, we used the fuzzy logic method as a fusion technique. Its principle is mentioned in Fig. 3.

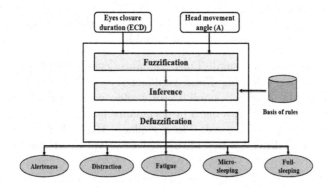

Fig. 3. Principle of our proposed fusion task

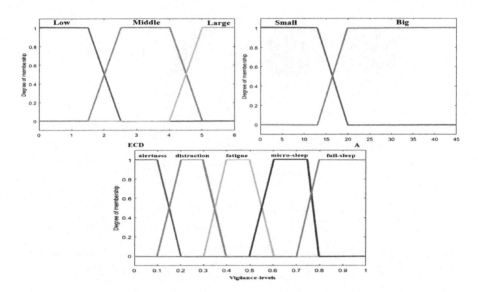

Fig. 4. Membership functions

Fuzzification. This module is the operation of transforming a crisp set to a fuzzy set. It defines suitable membership functions allowing to translate crisp input or measured values into linguistic concepts. For most cases, the trapezoidal and triangular membership functions are used. We have used the trapezoidal shape [13] in this paper as it is shown in Fig. 4.

Inference and rule basis. Once the inputs are fuzzified, the corresponding inputs fuzzy sets are passed to the inference engine. Using linguistic rules retrieved from the rule base, this module processes the current inputs. Several techniques have been used for the inference. The most common used are the following:Max-Min and Max-product fuzzy inference methods. In this paper, we used the Max-Min fuzzy inference method because it is the most widely used rule of inference and is simple to implement. In our case, the inference will merge both input variables, the eyes closure duration and the head movement angle, to find the alertness level of the output variable.

- **If** ECD = low **And** A = small **Then** vigilance level = **Alertness**
- **If** ECD = low **And** A = big **Then** vigilance level = **Distraction**
- **If** ECD = middle **And** A = small **Then** vigilance level = **Fatigue**
- **If** ECD = middle **And** A = big **Then** vigilance level = **Micro-sleep**
- **If** ECD = large **And** A = small **Then** vigilance level = **Full-sleep**
- **If** ECD = large **And** A = big **Then** vigilance level = **Full-sleep**

Defuzzification. At the output of the fuzzy inference the result is always a fuzzy set. In order to be used in the real world, the fuzzy output needs to be transformed to the crisp domain by the defuzzifier. This needs the use of a suitable membership functions. In this paper we have used the center of gravity method. The threshold value of angle (A) is fixed experimentally to 16. Concerning the duration of eye closure, Sarbjit [14] considered that the person was in a state of full sleep if he keeps his eyes closed for 5 to 6 s. But, if this duration is between 2 and 3 s, the person is in a micro-sleep state. However, Horng [15] affirmed that the driver is drowsy if he closes his eyes for 5 successive frames. According to the research activities of Sharabaty [16], the maximum duration of eyes' normal blinking is 0.5 s. If the closing time exceeds this value, then we are talking about a state of prolonged closure. According to these studies, we have categorized the duration of eye closure into three sub-intervals:

- low = [0, 2s]; middle = [2s, 4s]; large = [5s, 6s]
- small (A ≤ 16) ; big (A > 16)

3   Experimental Results

3.1   Influence of Camera Position on Eyes' Tracking Performance

The position of the camera may influence the quality of eyes' detection and tracking. We check the good eyes tracking rate using the Viola and Jones algorithm in Yaw DD [17] datatsets and our appropriate basis. When the camera is placed in parallel position to the eyes' axis of the driver, the correct tracking rate is more efficient than the case of dash or mirror position (Table 1).

Table 1. Different camera positions in Yaw DD datasets and our appropriate basis

Dataset	Mirror position	Dash position	Parallel position (our basis)
CTR	51%	83%	89%

3.2   Performance of the Eyes' Classification System

To test the efficiency of our proposed classification system based on deep learning architecture, we used the Closed Eyes in the Wild benchmark (CEW) [18], ZJU basis [18] and our appropriate basis. We compared two techniques of classification; the first method is based on fast wavelet network classifier (FWNC). It is used in our previous works [5–7]. The second method is our new classifier based on transfer learning architecture based on Alexnet Model (TLA). Classification results are mentioned in Table 2. Results show that the classification system based on deep learning architecture performs better than our classic version of fast wavelet network. Also, to prove the efficiency of our classifier, we compared it to other methods. The results are cited in Table 3.

Table 2. Results of classification rate on CWE, ZJU and our appropriate dataset

Dataset	CEW	ZJU	Our basis
FWNC	74.45%	61.75%	90.91%
TLA(our approach)	94.01%	82.93%	100%

Table 3. Results of classification rates on CEW and ZJU datasets

Dataset	CEW	ZJU
Nearest Neighbor (NN)	74.31%	84.74%
SVM	82.85%	89.62%
Adaboost	87.09%	92.06%
FWNC	74.45%	61.75%
TLA (our approach)	94.01%	82.93%

3.3 Evaluation of Our Hypovigilance Detection System

For performance evaluation of our proposed hypovigilance detection system, we built our appropriate dataset composed of 45 videos (9 recorded video for each vigilance level), under different light conditions. The recorded participants are students, workers and volunteers. They are of different sexes and aged between 20 and 60 years old. All the videos are available in MP4 format with a resolution of 640×480 and frequency of 30 frames/s. Besides, we used the MiraclHB datasets [2]. Figure 5 shows a video series of both datasets. Our built dataset contains five vigilance levels which are alertness state, distraction, fatigue, micro-sleeping and full sleeping. However in MiraclHB, we found two main levels which are the alertness and the full sleeping state. Results of correct hypovigilance detection rate on both cited datasets are summarized in Fig. 6; Our hypovigilance detection system, which is based on the fusion of eyes' closure duration and head movement angle, provides good results equal to 91.11% as global correct detection rate. The detection of full-sleeping state is usually correct (100%) because, whatever the angle value of head movement (big or small), this state exists as long as the eyes' closure duration is large. So it does not depend on the angle value of head motion.

Examples of our private datasets   Examples of MiraclHB datasets

Fig. 5. Series of our private basis and MiraclHB datasets

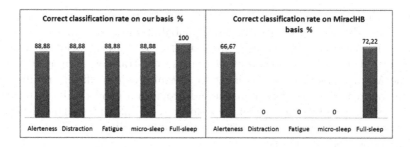

Fig. 6. Results of vigilance levels classification rate on MiraclHB datasets and our appropriate basis

Table 4. Comparison of our results with those of the literature

Authors	Results	Data	Method	VL
Our approach (2017)	91.1%	Video	**Fuzzy logic**	**5**
Akrout and Mahdi (2015)[2]	87.75%	Video	Expert system	4
Celine et al. (2015)[19]	98.4%	Video, audio and mechanic	Bayesian network	2
Liang and Lee (2014)[20]	90%	Video and driving performance data	Hybrid Bayesian network	2
Picot et al. (2012)[1]	81.7%	Video and ECG	Fuzzy logic	3

However, the distraction state detection depends on the angle value of head movement and not on the eyes' closure duration. In fact, the detection rate decrease of this state refers to the quality of eyes' detection and tracking by Viola and Jones algorithm. In other words, false eyes' detection may influence the head movement angle computing which is based on Pythagorian theorem applied in the triangle created between both bounding eyes' rectangles.

For MiraclHB results, we have 72,22% as a correct detection rate of full sleepiness state, 66,67% for alertness state and 0% for the other levels. These different rates may be explained by the fact that the MiraclHB is oriented to detect specific levels and even the data inputs of our system and those of MiraclHB are different. Besides, the false eyes' classification of our algorithm in some cases may influence our system decision in MiraclHB basis.

Table 4 shows a comparison between our hypovigilance detection system and four other approaches. This comparison argues in favor of our system point of view number of levels of vigilance detected. We mean by VL, vigilance level.

4   Conclusion

This paper is a contribution from our part in the field of vigilance monitoring by proposing a multimodal system based on eyes' blinking analysis and head posture estimation. Our system allows us to classify the driver's vigilance state into five different levels with efficient results. This classification is different from the work carried out in the literature allowing the detection of four, three or two levels. This advantage presents the major contribution of our paper. The second contribution is manifested in the approach of eyes' state recognition which is based on a deep learning architecture by using the transfer learning technique and the Alexnet model. This classification system provides efficient results. Our system was tested on various datasets and provides a good results in terms of eyes' classification rate and correct detection rate of different vigilance levels. In our future work, we aim at extending our system by adding other inputs like the head movement duration.

Acknowledgments. The authors would like to acknowledge the financial support of this work by grants from General Direction of Scientific Research (DGRST), Tunisia, under the ARUB program.

References

1. Picot, A., Charbonnier, S., Caplier, A.: Using retina modelling to characterize blinking: comparison between EOG and video analysis. Mach. Vis. Appl. **23**, 1195–1208 (2012)
2. Akrout, B., Mahdi, W.: Spatio-temporal features for the automatic control of driver drowsiness state and lack of concentration. Mach. Vis. Appl. (MVA) **26**, 1–13 (2015). Springer
3. Viola, P., Jones, M.: Robust real-time object detection. Int. J. Comput. Vis. **57**, 137–154 (2001). http://dx.doi.org/10.1023/B:VISI.0000013087.49260.f
4. Teyeb, I., Jemai, O., Bouchrika, T., Ben Amar, C.: Detecting driver drowsiness using eyes recognition system based on wavelet network. In: 5th International Conference on Web and Information Technologies (ICWIT13) Proceedings, May 09–12, pp. 245–254, Hammamet, Tunisia (2013)
5. Jemai, O., Teyeb, I., Bouchrika, T., Ben Amar, C.: A novel approach for drowsy driver detection using eyes recognition system based on wavelet network. Int. J. Recent Contrib. Eng. (IJES) Sci. IT **1**, 46–52 (2013)
6. Teyeb, I., Jemai, O., Mourad, Z., Ben Amar, C.: A novel approach for drowsy driver detection using head posture estimation and eyes recognition system based on wavelet network. In: The Fifth International Conference on Information, Intelligence, Systems and Applications (IISA 2014) Proceedings, pp. 379-384, 07–09 July, Chania, Greece (2014). doi:10.1109/IISA.2014.6878809
7. Teyeb, I., Jemai, O., Zaied, M., Ben Amar, C.: A drowsy driver detection system based on a new method of head posture estimation. In: Corchado, E., Lozano, J.A., Quintián, H., Yin, H. (eds.) IDEAL 2014. LNCS, vol. 8669, pp. 362–369. Springer, Cham (2014). doi:10.1007/978-3-319-10840-7_44
8. Teyeb, I., Jemai, O., Mourad, Z., Ben Amar, C.: A multi level system design for vigilance measurement based on head posture estimation and eyes blinking. In: Eighth International Conference on Machine Vision (ICMV 2015), Proceedings of SPIE, vol. 9875, 98751p, 8 December (2015). doi:10.1117/12.2229616
9. Teyeb, I., Jemai, O., Mourad, Z., Ben Amar, C.: Vigilance measurement system through analysis of visual and emotional drivers signs using wavelet networks. In: Proceedings of the 15th International Conference on Intelligent Systems Design and Applications (ISDA 2015), pp. 140–147, 14–16 December, Marrakech, Maroc (2015)

10. Guedri, B., Zaied, M., Ben Amar, C.: Indexing and images retrieval by content. In: International Conference on High Performance Computing and Simulation (HPCS), 4–8 July, Istanbul, Turkey, pp. 369–37 (2011)
11. Ejbali, R., Zaied, M., Ben Amar, C.: Multi-input multi-output beta wavelet network: modeling of acoustic units for speech recognition. Int. J. Adv. Comput. Sci. Appl. (IJACSA) Sci. Inf. Organ. (SAI) **3**(4), 38–44 (2012)
12. Krizhevsky, A., Sutskever, I., Hinton, G.E.: Imagenet classification with deep convolutional neural networks. In: Advances in Neural Information Processing Systems 25 (NIPS 2012), pp. 1097–1105 (2012)
13. Jemai, O., Ejbeli, R., Zaied, M., Ben Amar, C.: A speech recognition system based on hybrid wavelet network including a fuzzy decision support system. In: International Conference on Machine Vision (ICMV 2014), Proceedings of SPIE, 19–21 November, Milan, vol. 9445, pp. 944503-1–7, doi:10.1117/12.2180554 (2015)
14. Sarbjit, S., Nikolaos, P.: Monitoring driver fatigue using facial analysis techniques. In: IEEE Conference on Intelligent Transportation Systems, Proceedings (ITSC), pp. 314–318, Tokyo, Japan (1999)
15. Horng, W., Chen, C., Chang, Y.: Driver fatigue detection based on eye tracking and dynamic template matching. In: IEEE International Conference on Networking, Sensing and Control, pp. 7–12, Taipei, Taiwan (2004)
16. Sharabaty, H., Jammes, B., Esteve, D.: EEG analysis using HHT : one step toward automatic drowsiness scoring. In: 2nd International Conference on Advanced Information Networking and Applications - Workshops (AINA Workshops 2008), pp. 826–831, Okinawa (2008)
17. Abtahi, S., Omidyeganeh, M., Shirmohammadi, S., Hariri, B.: A yawning detection dataset. In: Proceedings of ACM Multimedia Systems, pp. 24–28, 19–21 March, Singapore (2014)
18. Song, F., Tan, X., Liu, X., Chen, S.: Eyes closeness detection from still images with multi-scale histograms of principal oriented gradients. Pattern Recogn. **47**(9), 2825–2838 (2014)
19. Celine, C., Abdullah, R., Mohamed, K., Fakhri, K.: A multi-modal driver fatigue and distration assessment system. Int. J. Intell. Transp. Syst. Res. **14**, 1–22 (2015)
20. Liang, Y., Lee, J.D.: A hybrid Bayesian network approach to detect driver cognitive distraction transport. Res. Part C: Emerg. Technol. **38**, 146–155 (2015)

Shape-Based Image Retrieval Based on Improved Genetic Programming

Ruochen Liu[(⊠)], Guan Xia, and Jianxia Li

Laboratory of Intelligent Perception and Image Understanding of Ministry
of Education, Xidian University, Xi'an 710071, China
ruochenliu@xidian.edu.cn

Abstract. Two-stage genetic programming algorithm based on a novel coding strategy (NTGP) is proposed in this paper, in which the generation of individual tree is not random but according to a special rule. This rule assigns each function operator a weight and the assignments of these weights based on the frequencies of function operators in good individuals. The greater weight of a function is, the more possibly it will be selected. By using the new coding strategy, the image feature database can be rebuilt. For two-stage genetic programming algorithm, in the first stage, the feature weight vector is obtained, GP is used to construct new features for the next step. While in the second stage, GP is used to induce an image matching function based on the features provided by the first stage. Based on these models, one can retrieve target images from the image database with much better performance. Three benchmark problems are used to validate performance of the proposed algorithm. Experimental results demonstrate that the proposed algorithm can obtain better performance.

Keywords: Two-stage genetic programming · Image retrieval · Special rule for generation of individual tree

1 Introduction

Image retrieval is an important field in computer vision. Most traditional methods for image retrieval utilize adding meta-data to the images so that retrieval can be performed on annotation words. In lecture [1], a genetic programming (GP) method was used in image retrieval to obtain a better feature matching strategy.

In traditional algorithms, matching strategy in shape-based image retrieval evaluates the similarity according to the distance between two images. Each item in image features is treated equally. Image retrieval process compares the distances between a target feature vector and other feature vectors. We can pick up some smallest ones according to these distances, then their corresponding images are the result of image retrieval. In this process, some similar images may have very similar image features. The image retrieval system can hardly deal with these images.

Through the analysis above, we proposed a weight strategy. In this strategy, each dimension of a feature vector is not treated equally. Each dimension has a special weight. When a target image is retrieved, new feature vectors is obtained by using each dimension of the old feature vectors times the corresponding weight, and then, matching strategy operates on the new feature vectors.

D. Liu et al. (Eds.): ICONIP 2017, Part IV, LNCS 10637, pp. 212–220, 2017.
https://doi.org/10.1007/978-3-319-70093-9_22

Considering the method of GP-based image retrieval. We designed the new algorithm as two stages. In the first stage, GP algorithm is used to evaluate a weight vector. In the second stage, GP algorithm is used to evaluate matching strategy.

In the remainder of this paper, Sect. 2 develops a new coding strategy with two-stage GP structure. Section 3 discusses the experimental results, Sect. 4 makes some conclusions and gives directions for future research.

2 Proposed a Novel Two-Stage GP Strategies

2.1 The Process of the Proposed Algorithm

The structure of our new algorithm is shown as follows. The first stage of the algorithm is described in 1 to 5, and the second stage is in 6 to 11.

1: Extract features of each image and construct the feature vectors of these images.
2: Set parameters for the first stage. Initialize a population for the first stage.
3: Calculate the fitness of each individual, and then pick up better individuals.
4: Based on the chosen population, conduct mutation and crossover operators, and then compute their fitness again.
5: Check the best individual, and then judge whether it meets our demand; if it does not meet our demand, repeat 4; otherwise, conduct 6.
6: Decode the best individual, and reconstruct features vectors of images in the database based on the result decoded from first stage.
7: Set the parameters for the second stage. Initialize a new population.
8: Calculate the fitness of each individual, and then pick up better individuals.
9: Conduct the crossover and mutation operators; then calculate fitness again.
10: If meet the ending condition, decode the best individual; otherwise, repeat 9.
11: Construct the final matching model.

2.2 The First GP Stage is Performed for Evaluating Feature Weights

Traditionally, the comparison between two features of two images can be described as follows.

$$f = \sum_{k=1}^{N} abs\left(I_i^k - I_j^k\right) \tag{1}$$

where I_i^k represents the kth feature value of the ith image.

In this equation, each dimension of a feature vector is equally important. While in most cases, for example, in the feature vector using the FD (Fourier Shape Descriptor) feature extracting method, the first dimension is much more important than others. Because of this, we introduced a new method to compare two features of two images.

$$f = \sum_{k=1}^{N} w_k * abs\left(I_i^k - I_j^k\right) \tag{2}$$

where w_k represents the kth weight in the decoding vector.

Weight w_k will influence the matching value. The different dimension of a feature vector plays different roles in matching process. In traditional GP, two feature vectors of two images are used to process objects without considering the details. In this new method, the special value of weight vector must be known. We need to know the special numbers of weighted values, it also equals to the number of image features. The GP tree is composed of one operator "abs" and a series of random numbers.

2.3   Special Initialization Method

Traditionally, individuals are randomly generated. Every node of each parse tree is randomly selected from a function set and a terminal set. In this paper, we use another method which assigns every function operator a weight. As the function operators usually appear with unequal frequency, the weights are assigned based on the frequencies of function operators in good individuals. The greater weight of a function operator is, the more possibly it will be selected. The weight of a function operator represents the selection probability of this function operator. The strategy is introduced as follows.

Algorithm 1: Update weights of each function operator

Input: *function set*, the presetting function operators set; *tree set*, the operators appearing in the individual parse tree; *fitness* ; *threshold* ; V, the frequencies of function operators in the selected tree set; W, original weights; N, the number of tree set; M, the number of function operators; w, a parameter that controls the proportion of W to V in the new W is set as 0.5 in this paper.

Output: New weights.

```
 1:   V := zeros(1, M)
 2:   if fitness > threshold then
 3:       V := count(function set, tree set)
 4:       for i = 1 to M do
 5:           W(i) := W(i) * w * (1 − fitness) + (1 − w) * fitness * V(i)
 6:       end for
 7:   end if
 1:   function count(function set, tree set)
 2:       n := zeros(1, m)
 3:       v := zeros(1, m)
 4:       for i = 1 to N do
 5:           j := find(function set, tree set)    // j is the index of the ith operator
 6:           n(j) := n(j) + 1
 7:       end for
 8:       for i = 1 to M do
 9:           v(i) := n(i) / N
10:       end for
11:       return v
12:   end function
```

The function operator selection procedure will be executed as Algorithm 2 which presents how a function operator is selected based on the weights. In this paper, we used Hierarchic function sets to obtain a better initial population.

Algorithm 2: The method for selecting function operator

Input: FS, the predetermined function set; W, weight vector when the individual is going to be created; M, size of FS.

Output: index of selected function operator in FS

1:   $SUM := sum(W)$
2:   $Wheel := zeros(1, M)$
3:   for $i = 1$ to M do
4:      $W(i) := W(i) / SUM$
5:      if $i = 1$ then
6:         $Wheel(i) := W(i)$
7:      else
8:         $Wheel(i) := Wheel(i-1) + W(i)$
9:      end if
10: end for
11: $RAND := rand()$ // $RAND$ distributes in $(0,1)$
12: for $i = 1$ to M do
13:    if $RAND < Wheel(i)$ then
14:       $Index := i$
15:       break
16:    end if
17: end for
18: return $Index$ // index of selected function operator in FS

2.4 Special Genetic Operators

For the crossover operator, firstly, generate a cross node randomly according to the number of the nodes in the individual tree. Secondly, swap the right parts of the cross node in two selected parents. For the mutation operator, choose the mutation node randomly, then change the value of respond node by adding 0.5.

2.5 Fitness Function

For the binary classification problems, it can be described as follows [2, 3].

$$fitness = \frac{1}{N} \sum_{i=1}^{N} \frac{precision_i + recall_i}{2} \qquad (3)$$

$$recall = p(R|S) = \frac{P(S \cap R)}{P(S)} = \frac{s}{s+v} \qquad (4)$$

$$precision = p(S|R) = \frac{P(S \cap R)}{P(R)} = \frac{s}{s+u} \tag{5}$$

where S represents a set of all images related to the target image, and R represents a set of images that have been retrieved; s is the number of images related to the target image in once retrieval process; u is the number of images not related to the target image in once retrieval process; v is the number of images which are related to the target image but have not been retrieved in once retrieval process.

3 Experiments

3.1 Database Construction

The image dataset we choose in paper are as follows.

Mpeg (part) image dataset composes of six kinds of objects. And each kind of object contains twenty different images. It is chosen from part B of MPEG-7image dataset.

Bicego image dataset is composed of seven kinds of objects. And each of them contains ten different images.

Plane image dataset is composed of seven kinds of planes. Each kind of plane has different rotate angle or size in details. Since all the shapes of these images are similar with each other, image retrieval becomes very difficult.

This paper uses four kinds of image feature extracting methods which are FD (Fourier Shape Descriptor) [4], CSM (the Contour Sequence Moment) [5], CHEN [6] and Combo [7] respectively.

3.2 Parameter Settings

Table 1 shows the details of parameter settings for new coding two-stage GP.

Table 1. Setting for new coding two-stage GP (NTGP)

Items	First stage	Second stage
Maximum numbers of generations	50	20
Population size	50	20
Function set	'abs'	'+', '−', '*', '/', 'diff', 'div', 'cos'
Terminal set	Rand, 1/n	'x1', 'x2', rand
Crossover probability	0.65	0.65
Mutation probability	0.30	0.30
Elitism probability	0.05	0.05
Minimum depth of initial individual	2	3
Maximum depth of initial individual	2	4
Maximum depth of evolution	2	7
Tree type	Full	Ramped half-half
Selection style	Tournament	Tournament
Tournament size	4	4
Random number interval	[−N, N]	[−N, N]

For the comparison, we choose standard GP (GP, for short), a two-stage GP (2SGP, for short) [8] to confirm the advantages of our strategies.

The differences between two-stage GP(2SGP) and NTGP are showed in Table 2, other setting of parameters is same as Table 1. The differences between standard GP and NTGP are the maximum numbers of generations and population size, for standard GP, the two parameters both are set to 70, other setting of parameters is same as the second stage of NTGP.

Table 2. Setting for two-stage GP (2SGP)

Items	First stage	Second stage
Function set	'+', '−', '*', '/', 'diff', 'div', 'cos'	'+', '−', '*', '/', 'diff', 'div', 'cos'
Terminal set	'x1', 'x2', rand	'x1', 'x2', rand
Minimum depth of initial individual	3	3
Maximum depth of initial individual	4	4
Maximum depth of evolution	7	7
Tree type	Ramped half-half	Ramped half-half

3.3   Experimental Results and Analysis

We independently run each algorithm 20 times on each image dataset. The number in Table 3 represents the average precision of the algorithm under the corresponding strategy and image dataset.

From Table 3, we can know that new coding two-stage GP obtains better results in 10 groups of experiments. Especially for the MPEG dataset, our strategy has an explicit improvement. It is also concluded that NTSGP is better than our approach in Bicego dataset using Chen and FD methods for extracting feature.

For more detailed results of our experiments, we conduct each experiment on variant retrieval numbers. According to the function curves of *precision* and *recall* on retrieval number, we can see, when retrieval number varies from one to maximum value of dataset, *recall* will increase gradually and *precision* in reverse. When the area surrounded by the curve and the axes is bigger, the performance of this algorithm is better (Figs. 1, 2 and 3).

In the 12 pictures, we can find that this new algorithm (NTGP) in 9 groups of experiments obtains better results because the blue curve is higher than other curves. Especially in MPEG dataset using CSM (Fig. 1), Bicego dataset using Combo (Fig. 2), and plane datasets using CSM, Combo and FD (Fig. 3) obtain a considerable improvement. However, in MPEG using Combo and CHEN (Fig. 1), and in Bicego datasets using FD and CHEN (Fig. 2), these curves of this new algorithm equal or approximately equal to those curves of other algorithms. It is also concluded that green curves representing TGP algorithm is approximately same to red curves representing GP algorithm in 11 groups of experiments. Hence, TGP algorithm does not make retrieve performance improve.

Table 3. The average results and standard deviation

Algorithm	MPEG (part)	Bicego	Plane
CSM	0.7075	0.8511	0.5979
GP + CSM	0.6804(0.349)	0.8629(0.010)	0.7128(0.366)
NYSGP + CSM	0.7482(0.015)	0.8778(0.004)	0.6199(0.004)
NTGP + CSM	**0.8685**(0.014)	**0.8956**(0.014)	**0.7805**(0.024)
Chen	0.6954	0.5550	0.2737
GP + Chen	0.7134(0.364)	0.6039(0.000)	0.2836(0.146)
NTSGP + Chen	0.7099(0.005)	**0.6411**(0.005)	0.2818(0.002)
NTGP + Chen	**0.7702**(0.020)	0.6048(0.030)	**0.3115**(0.014)
Combo	0.8483	0.7775	0.5683
GP + Combo	0.6222(0.319)	0.7729(0.000)	0.4589(0.010)
NTSGP + Combo	0.8172(0.005)	0.8222(0.002)	0.4606(0.009)
NTGP + Combo	**0.8713**(0.017)	**0.8646(0.039)**	**0.7571**(0.072)
FD	0.9187	0.8286	0.7114
GP + FD	0.9202(0.472)	0.8465(0.000)	0.7250(0.373)
NTSGP + FD	0.9237(0.003)	**0.8636(0.092)**	0.7617(0.021)
NTGP + FD	**0.9345**(0.010)	0.8395(0.014)	**0.8168**(0.017)

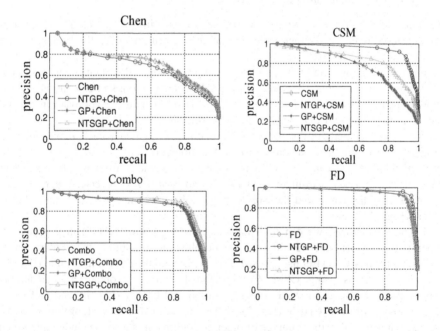

Fig. 1. MPEG (part) (Color figure online)

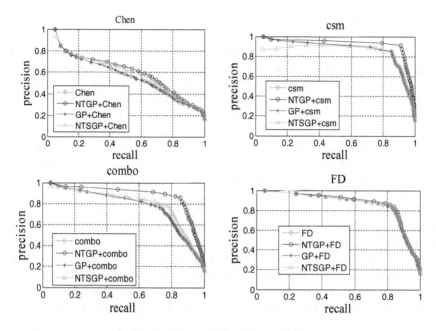

Fig. 2. Bicego (Color figure online)

Comparing the new algorithm (NTGP) with standard GP algorithms, we can find that the first stage in NTGP algorithm plays a very important role in improving performance. And comparing with 2SGP, we can find Special initialization method also plays a very important role in NTGP algorithm.

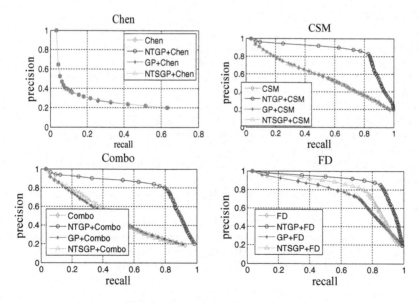

Fig. 3. Plane (Color figure online)

4   Conclusion

In this paper, a two-stage genetic programming by using a new coding strategy for image retrieval was proposed. In the first stage, the weight vector is obtained; GP is used to construct new features for the next step, which is a very good supplement for the second stage. While in the second stage, GP is used to induce an image matching function based on the features provided by the first GP. Then some experiments are conducted to verify its performance. Based on these novel image feature abstracting methods, we did several experiments over three image datasets. Results show that new coding two-stage GP can induce relatively better solution. Through these experiments, our algorithm results in better solutions in most datasets. Furthermore, in some datasets, the improvement is quite obvious.

Although this algorithm proposed have some advantages compared with standard GP, two-stage GP, there are still some problems inherently. For example, the fitness of individuals for our new algorithm varies very little and even no in an evolution procedure. Some other methods for our algorithm should be also taken into account. In this paper, just traditional GP was used in the second stage. And in the first stage, only general structure changes were discussed. However, we can improve these two aspects by introducing some new methods by which much better matching functions can be built.

Acknowledgments. This work was supported by the National Natural Science Foundation of China (No. 61373111); the Fundamental Research Funds for the Central Universities (Nos. K50511020014, K5051302084); and the Provincial Natural Science Foundation of Shaanxi of China (No. 2014JM8321).

References

1. Torres, R.S., Falcao, A.X.: A genetic programming framework for content-based image retrieval. Pattern Recogn. **42**(2), 283–292 (2009)
2. Abbasi, S., Mokhtarian, F.: Curvature scale space image in shape similarity retrieval. Multimedia Syst. **7**(6), 467–476 (1999)
3. Bai, X., Yang, X.: Learning context-sensitive shape similarity by graph transduction. IEEE Trans. Pattern Anal. Mach. Intell. **32**(5), 861–874 (2010)
4. Zhang, D.S., Lu, G.J.: Shape-based image retrieval using generic Fourier descriptor. Sig. Process. Image Commun. **17**(10), 825–848 (2002)
5. Gupta, L., Srinath, M.D.: Contour sequence moments for the classification of closed planar shape. Pattern Recogn. **20**(3), 267–272 (1987)
6. Chen, C.: Improved moment invariants for shape discrimination. Pattern Recogn. **26**(5), 683–686 (1993)
7. Sun, J.D., Zhang, Z.S.: Shape retrieval based on combination moment invariants. In: Proceedings of Information Technology and Environmental System Sciences (2008)
8. Huang, J.J., Tzeng, G.H., Ong, C.S.: Two-stage genetic programming (2SGP) for the credit scoring model. Appl. Math. Comput. **174**(2), 1039–1053 (2006)

An AI-Based Hybrid Forecasting Model for Wind Speed Forecasting

Haiyan Lu[1(✉)], Jiani Heng[2], and Chen Wang[3]

[1] University of Technology Sydney, 15 Broadway, Sydney,
NSW 2007, Australia
Haiyan.lu@uts.edu.au
[2] Dongbei University of Finance and Economics, Dalian, China
hengjnl3@lzu.edu.cn
[3] Lanzhou University, Lanzhou, China
chenwangl5@lzu.edu.cn

Abstract. Forecasting of wind speed plays an important role in wind power prediction for management of wind energy. Due to intermittent nature of wind, accurately forecasting of wind speed has been a long standing research challenge. Artificial neural networks (ANNs) is one of promising approaches to predict wind speed. However, since the results of ANN-based models are strongly dependent on the initial weights and thresholds values which are usually randomly generated, the stability of forecasting results is not always satisfactory. This paper presents a new hybrid model for short term forecasting of wind speed with high accuracy and strong stability by optimizing the parameters in a generalized regression neural network (GRNN) using a multi-objective firefly algorithm (MOFA). To evaluate the effectiveness of this hybrid algorithm, we apply it for short-term forecasting of wind speed from four wind power stations in Penglai, China, along with four typical ANN-based models, which are back propagation neural network (BPNN), radical basis function neural network (RBFNN), wavelet neural network (WNN) and GRNN. The comparison results clearly show that this hybrid model can significantly reduce the impact of randomness of initialization on the forecasting results and achieve good accuracy and stability.

Keywords: Wind speed forecasting · Artificial neural network · Firefly algorithm

1 Introduction

As one of the most promising renewable energy, wind energy has attracted great attention from researchers [1]. Almost every government across the world has introduced incentives to support wind energy development [2]. Wind is more competitive than other renewable energy sources. In 2015, the global installed capacity of wind power exceeds 60,000 MW, and in China alone, the total new installed capacity of wind power was 30,500 MW. The total global capacity reached 432,419 MW by the end of 2015 [2].

© Springer International Publishing AG 2017
D. Liu et al. (Eds.): ICONIP 2017, Part IV, LNCS 10637, pp. 221–230, 2017.
https://doi.org/10.1007/978-3-319-70093-9_23

Due to intermittent nature of wind, accurate forecasting of wind power is highly desirable for maximizing the usage of wind source and maintaining the security of power grids [3]. However, wind speed is generally regarded as one of the hardest weather parameters to predict because of its nonstationary and nonlinear fluctuations [4]. Great effort has been devoted to studies on wind speed forecasting in last decade. The reported forecasting methods fall into four categories [5]: (a) physical models; (b) statistical models; (c) spatial correlation models; and (d) artificial intelligence based models. With the rapid development of AI technology, AI-based forecasting methods become very promising, such as ANNs [6], fuzzy logic methods [7] and support vector machines (SVMs) [8].

Since the results of ANN-based models are strongly dependent on the initial weights and thresholds values, which are usually randomly generated, the stability of forecasting results is not always satisfactory. This has substantial impact on the effectiveness of existing ANN-based forecasting models (ANNs). One promising approach to overcome this limitation is to build hybrid models by controlling the initial weight and threshold values in ANNs through optimization algorithms to mitigate the impact of randomness of initialization and threshold values. Most existing hybrid models of this kind only focus on improvement of accuracy of ANNs and use single-objective optimization algorithms. However, the quality of a forecasting model is not only reflected by its accuracy, but also determined by its stability.

The multi-objective firefly algorithm (MOFA) [9] has found many applications recently as an effective search algorithm [10]. It has shown its superiority over some optimization algorithms [11], such as vector evaluated genetic algorithm (VEGA), multi-objective artificial bee colony (ABC) algorithm and strength Pareto evolutionary algorithm (SPEA).

This paper presents a new hybrid model for short term forecasting of wind speed with high accuracy and strong stability based on a generalized regression neural network (GRNN) and a multi-objective firefly algorithm (MOFA). With this hybrid algorithm, the impact of randomness of initialization is relaxed by using optimized initial weights and thresholds rather than random ones.

The remainder of the paper is organized as follows: Sect. 2 describes the construction of the hybrid model based on GRNN and MOFA. Section 3 presents the implementation of this hybrid model. Section 2 presents the evaluation of this hybrid model by comparing its performance with the other four competing models. The comparison results are described and discussed. Lastly, Sect. 4 concludes the paper.

2   Construction of MOFA-GRNN

In order to simultaneously achieve higher accuracy and strong stability, a generalized regression neural network GRNN is chosen as the base model from accuracy perspective [11] and a multi-objective flower firefly algorithm, MOFA [12], is adopted to optimize the initial weight and threshold of the GRNN, to achieve a greater stability. Figure 1 illustrates the idea of MOFA-GRNN with a flowchart of MOFA and the structure of GRNN.

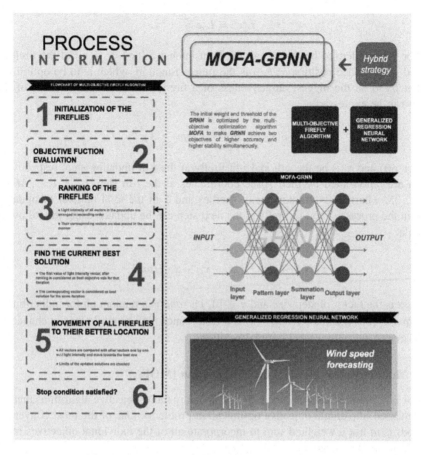

Fig. 1. Flowchart of MOFA and structure of GRNN

2.1 Flower Firefly Algorithm FA

FA was developed as a metaheuristic inspired by the flashing behavior of fireflies [12]. Let $X_{i,k}$ and $X_{j,k}$ be the k-th component of i-th and j-th firefly in D-dimension space, respectively, $k = 1,...,D$, where D is the dimension of each firefly vector, i.e. the number of components in each firefly vector. Then the distance between i-th and j-th fireflies can be expressed as

$$r_{ij} = \left\| X_i - X_j \right\| = \sqrt{\sum\nolimits_{k=1}^{d} \left(X_{i,k} - X_{j,k} \right)} \tag{1}$$

Let I_0 be the maximum fluorescent brightness of a firefly; γ is the light intensity absorption coefficient, usually in theory $\gamma \in [0, +\infty]$, in practical problems $\gamma \in [0.01, 100]$. The fluorescent brightness of the firefly can be expressed as

$$L(r) = L_0 e^{-\gamma r^2} \tag{2}$$

For the problem to be optimized, the light intensity $I(x_i) \propto f(x_i)$ of the firefly is located at the space position γ; $f(x_i)$ is the fitness function, which is to be optimized.

Set β_0 to be the biggest attractiveness of a firefly, then the attractiveness of this firefly can be expressed as

$$\beta(r) = \beta_0 e^{-\gamma r^2} \quad (m \geq 1) \tag{3}$$

Set $x_i(t)$ and $x_j(t)$ be the i-th and j-th firefly at the t-th updating step, respectively, which are represented by their space coordinates at the t-th step, $i, j = 1,\ldots, M$ and $t = 1,\ldots,S$, where M is the number of fireflies and S in the total number of updating steps in the process. The i-th firefly at the next step can be updated using the following formula:

$$x_i = x_i + \beta_0 e^{-\gamma r^2}(x_i + x_j) + \alpha * \left(rand - \frac{1}{2}\right) \tag{4}$$

where α is the factor of step size and $\alpha \in [0, 1]$; $rand$ is a random uniform distribution and $rand \in [0, 1]$, and $a * \left(rand - \frac{1}{2}\right)$ is a disturbance term to minimize the chance for the search to fall into a local optimal position.

2.2 Multi-objective Flower Firefly Algorithm (MOFA)

MOFA is a modified version of FA. There are many multi-objective algorithms that can be converted into single-objective optimization problems. One of the most convenient methods is to use a weighted sum to incorporate all of the individual objectives into a composite single objective:

$$f = \sum_{i=1}^{m} w_i f_i, \quad \sum_{i=1}^{m} w_i = 1, w_i > 0 \tag{5}$$

where m denotes the total number of objectives, and the w_i $(i = 1, \ldots, m)$ are non-negative weights. For the purpose of accurately acquiring the probability function (PF) such that the solutions are uniformly distributed along the front, the weights w_i must obey a uniform distribution; they can be obtained as random numbers from a uniform distribution or from a low-discrepancy sequence.

2.3 Standard GRNN

The generalized regression neural network (GRNN) was proposed by Specht in 1991. Its theoretical basis is nonlinear (kernel) regression analysis. The regression analysis of the non-independent variable with respect to the independent variable X is in fact the calculation of the maximum probability value of Y. Set the random variable x and y is the joint probability density function $f(x, y)$, which is a conditional mean,

$$\hat{Y} = E[y|X] = \frac{\int_{-\infty}^{+\infty} yf(X,y)dy}{\int_{-\infty}^{+\infty} f(X,y)dy} \tag{6}$$

where the unknown probability distribution function $f(x, y)$ can be obtained by the x and y estimation of sample observations.

The generalized regression neural network (GRNN) comprises of four layers as shown in Fig. 1: the input layer, pattern layer, summation layer, and output layer.

The number of neurons in the input layer is equal to the dimension of the input vector in the learning sample. Each neuron is a simple distribution unit, which directly transfers the input variables to the pattern layer.

The number of neurons in the pattern layer is equal to the number of learning samples and each neuron corresponds to different samples. The transfer function of each neuron shows as follow:

$$p_i = exp\left[-\frac{(X - X_i)^T (X - X_i)}{2\sigma^2}\right] \quad (i = 1, 2, \ldots, n) \tag{7}$$

The output of i-th neuron is the square of Euclidean metric between the input variable X and the corresponding sample X_i. Among them: X is the network input variables; X_i is the i-th neuron corresponding to the learning sample.

The summation layer uses two types of neurons to calculate the weight of inter-connection. One type uses the following formula:

$$\sum_{i=1}^{n} exp\left[-\frac{(X - X_i)^T (X - X_i)}{2\sigma^2}\right] \tag{8}$$

which is an arithmetic sum for all neurons of the pattern layer. The connection weights of each neuron in the pattern layer are 1 and the transfer function can be described as follow:

$$S_S = \sum_{i=1}^{n} p_t \quad t = 1, 2, \ldots n \tag{9}$$

The other type uses the following formula:

$$\sum_{i=1}^{n} Y_i exp\left[-\frac{(X - X_i)^T (X - X_i)}{2\sigma^2}\right] \tag{10}$$

which is a weighted sum for all neurons of the pattern layer.

The connection weight of the j-th element y_{ij} in i-th output sample Y_i is the sum of the i-th neurons of summation layer and the j-th neurons of pattern layer. The transfer function can be described using the following formula:

$$S_{wt} = \sum\nolimits_{i=1}^{n} p_t w_t \quad t = 1, 2, \ldots n \tag{11}$$

The number of neurons in the output layer is equal to the dimension of output vectors of the learning sample k, each neuron will output division and layer, the elements corresponding to the estimation results of output neurons.

$$\hat{Y} = S_S / S_{wo}, \quad o = 1, 2, \ldots, k \tag{12}$$

2.4   Optimisation of GRNN by MOFA

To improve the forecasting accuracy and increased the stability of GRNN, the weight and threshold in a GRNN are optimized by a multi-objective firefly algorithm.

Fitness function

The bias-variance framework is used to determine the fitness functions of the optimization algorithm to obtain increased accuracy and stability. The performance bias can indicate the accuracy of the forecasting model. A smaller absolute value of **Bias**$(\hat{Y})$ indicates a more accurate forecasting performance of the forecasting model, and a formula for **Bias**$(\hat{Y})$ that considers the average difference between the observed and predicted values over all observed and predicted data can be defined as follows:

$$\mathbf{Bias}(\hat{Y}) = y - E(\hat{Y}) \tag{13}$$

$$E(\hat{Y}) = \frac{1}{N} \sum\nolimits_{n=1}^{N} \hat{y}_n \tag{14}$$

$$y = \frac{1}{N} \sum\nolimits_{n=1}^{N} \hat{y}_n \tag{15}$$

where $E(\hat{Y})$ represents the expectation of the predicted value over all predicted data, y denotes the expectation of the observed values, y_n denotes the observed value of the n-th datum, $\hat{y}_n$ denotes the predicted value of the n-th datum, and N is the total number of data used for the performance evaluation and comparison.

The performance variance can indicate the forecasting stability of the model. A smaller $\mathbf{Var}(\hat{Y})$ indicates a more stable forecasting performance. The performance variance is formulated as follows:

$$\mathbf{Var}(\hat{Y}) = E(\hat{Y} - E(\hat{Y})^2) \tag{16}$$

Thus, in this hybrid model, the fitness function for accuracy and stability can be defined as follows:

$$min \begin{cases} f_1(x) = \mathbf{Bias}(\hat{y}) \\ f_2(x) = \mathbf{Var}(\hat{y}) \end{cases} \tag{17}$$

3 Experiments

Each ANN experiment was repeated 50 times, and the average value was then taken to avoid the effects of uncertainty and to ensure that the final results were reliable and independent of the initial random weight values of the ANNs and the optimization algorithm. The simulations of the algorithms applied to all data sets were performed in the MATLAB R2012a environment running on Windows 8 with a 64-bit 2.50 GHz Intel Core i7 4870HQ CPU and 16 GB of RAM. The experimental parameters are shown in Table 1.

Table 1. Experimental parameter values

Model	Experimental parameter	Default value
BPNN	Neuron number in the input layer	12
	Neuron number in the hidden layer	1–8
	Neuron number in the output layer	1
	Learning velocity	0.1
	Maximum number of trainings	1000
	Training requirements precision	0.00004
RBFNN	Interval of the neural network	0.01
	Number of samples	400
WNN	Iteration time	100
	Learning rate	0.1
	Training requirement accuracy	0.00004
	Maximum generation	1000
	Population size	50
	P_a	0.25
	Convergence tolerance	10^{-5}
MOFA-GRNN	Maximum generation	10000
	Population size	100
	A	0.1
	β_0	1.0/3.5
	Γ	0.001
	Convergence tolerance	10^{-5}
GRNN	Neuron number of the input layer	4
	Neuron number of the hidden layer	9
	Neuron number of the output layer	1
	Radial basis function expansion	0.1 to 2.0
	Maximum number of training	1000
	Training requirement precision	0.00002

From Table 2 the forecasting results of the BP neural network, RBF neural network, Wavelet neural network, Generalized neural network, and MOFA-GRNN, it can be seen that the proposed model demonstrated good accuracy in the forecasting of wind

Table 2. Forecasting results

Sites	Metric	One step forecast				
		BPNN	RBFNN	WNN	GRNN	Hybrid model
Site 1	AE	0.0152	0.0018	−0.0125	−0.0064	0.0010
	MAE	0.1883	0.2230	0.2609	0.1851	0.1146
	MSE	0.0371	0.0518	0.0717	0.0356	0.0135
	MAPE	2.696%	3.189%	3.738%	2.650%	1.639%
Site 2	AE	−0.0174	0.0081	0.0112	−0.0012	−0.0042
	MAE	0.2040	0.2162	0.2779	0.2265	0.1158
	MSE	0.0445	0.0495	0.0825	0.0542	0.0141
	MAPE	2.819%	2.994%	3.833%	3.142%	1.602%
Site 3	AE	−0.0199	0.0087	−0.0061	−0.0115	0.0112
	MAE	0.2066	0.2164	0.2490	0.2453	0.1137
	MSE	0.0459	0.0497	0.0667	0.0631	0.0136
	MAPE	2.808%	2.921%	3.380%	3.323%	1.536%
Sites	Metric	Two step forecast				
		BPNN	RBFNN	WNN	GRNN	Hybrid model
Site 1	AE	0.0646	0.0241	0.0761	0.0001	0.0105
	MAE	0.3361	0.3512	0.5511	0.3414	0.2352
	MSE	0.1166	0.1275	0.3130	0.1218	0.0572
	MAPE	4.812%	5.032%	7.888%	4.879%	3.362%
Site 2	AE	0.0145	0.0330	−0.0196	0.0032	−0.0027
	MAE	0.3499	0.3975	0.5689	0.3285	0.2395
	MSE	0.1284	0.1680	0.3389	0.1154	0.0602
	MAPE	4.832%	5.481%	7.855%	4.534%	3.307%
Site 3	AE	0.0870	0.0668	0.0365	0.0210	0.0219
	MAE	0.3665	0.3934	0.5503	0.2981	0.2356
	MSE	0.1402	0.1628	0.3174	0.0935	0.0580
	MAPE	4.968%	5.311%	7.443%	4.042%	3.189%
Sites	Metric	Three step forecast				
		BPNN	RBFNN	WNN	GRNN	Hybrid model
Site 1	AE	0.0092	0.0324	0.0052	0.0323	0.0002
	MAE	0.4501	0.5613	0.6371	0.4250	0.3829
	MSE	0.2108	0.3248	0.4187	0.1865	0.1506
	MAPE	6.449%	8.021%	9.102%	6.074%	5.478%
Site 2	AE	0.0087	0.0298	−0.0075	−0.0391	0.0342
	MAE	0.4772	0.5872	0.6477	0.4368	0.3950
	MSE	0.2375	0.3601	0.4383	0.2000	0.1627
	MAPE	6.596%	8.109%	8.948%	6.033%	5.453%
Site 3	AE	−0.0742	0.0080	0.0130	−0.0240	−0.0585
	MAE	0.4724	0.5966	0.6843	0.4434	0.4054
	MSE	0.2351	0.3720	0.4878	0.2057	0.1712
	MAPE	6.388%	8.066%	9.257%	6.003%	5.484%

speed, where AE is average error, MAE is mean absolute error, MSE is mean squared error. However, the hybrid neural network in wind speed forecasting is significantly better than the other single neural networks with a decrease of 1.84% to 4.55% compared to the other neural networks. In the prediction process, the MAE of hybrid model is lower than that of the other three networks, the reduction of MSE is approximately 0.3, the reduction of MAE is approximately 0.33, and the operation time of the BP neural network is far less than that of the other three networks. The prediction error of the BP neural network can be significantly smaller than those of the other three networks.

The wind speed forecasting results of one-step forecasting, two-step forecasting, and three-step forecasting were given. Figure 2 shows the wind speed forecasting for one step forecasting, of the proposed neural network speed prediction results from the assessment; The mean absolute percentage errors (MAPE) were 1.59%, 3.29% and 5.47%, respectively. Thus, the hybrid neural network is more accurate in forecasting the three step forecasting wind speed, and the effect is relatively poor when the time is longer than three step forecasting. With the forecasting time step gets bigger, the precision is gradually reduced, and the prediction of the wind speed obviously deviates.

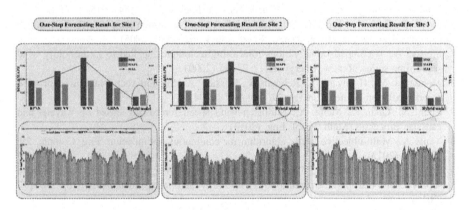

Fig. 2. Forecasting results of one-step forecasting

4 Conclusion

This paper aims to propose a better algorithm to forecast wind speed accurately, which is a long standing challenge due to the fact that wind speed is affected by various environmental factors, so wind speed data presents high fluctuations, autocorrelation and stochastic volatility. The new contribution of this paper is to consider both accuracy and stability of an algorithm.

A new hybrid model for short term forecasting of wind speed is proposed by optimizing the parameters in a generalized regression neural network (GRNN) using a multi-objective firefly algorithm (MOFA) to achieve high accuracy and strong stability. To evaluate its effectiveness, this hybrid algorithm was used for short-term forecasting of wind speed from four wind power stations in Penglai, China, along with four typical

ANN-based models, which are back propagation neural network (BPNN), radical basis function neural network (RBFNN), wavelet neural network (WNN) and GRNN. The comparison results clearly show that this hybrid model can significantly reduce the impact of randomness of initialization on the forecasting results and achieve good accuracy and stability.

References

1. Alboyaci, B., Dursun, B.: Electricity restructuring in Turkey and the share of wind energy production. Renew. Energy **33**(11), 2499–2505 (2008)
2. Cordeiro, M., Valente, A., Leitão, S.: Wind energy potential of the region of Trásos-Montes and Alto Douro Portugal. Renew. Energy **19**(1–2), 185–191 (2000)
3. Liu, H., Tian, H., Chen, C., Li, Y.: A hybrid statistical method to predict wind speed and wind power. Renew. Energy **35**(8), 1857–1861 (2010)
4. Abdel-Aal, R., Elhadidy, M., Shaahid, S.: Modeling and forecasting the mean hourly wind speed time series using GMDH-based abductive networks. Renew. Energy **34**(7), 1686–1699 (2009)
5. Lei, M., Shiyan, L., Chuanwen, J., Hongling, L., Yan, Z.: A review on the forecasting of wind speed and generated power. Renew. Sustain. Energy Rev. **13**(4), 915–920 (2009)
6. Flores, P., Tapia, A., Tapia, G.: Application of a control algorithm for wind speed prediction and active power generation. Renew. Energy **30**(4), 523–536 (2005)
7. Sfetsos, A.: A comparison of various forecasting techniques applied to mean hourly wind speed time series. Renew. Energy **21**(1), 23–35 (2000)
8. Mohandes, M., Halawani, T., Rehman, S., Hussain, A.: Support vector machines for wind speed prediction. Renew. Energy **29**, 939–947 (2004)
9. Chen, N., Qian, Z., Meng, X.: Multistep wind speed forecasting based on wavelet and gaussian processes. Math. Probl. Eng. (2013)
10. Yang, X.: Multi-objective firefly algorithm for continuous optimization. Eng. Comput. **29**, 175–184 (2013)
11. Shu, T., Gao, X., Chen, S., Wang, S., Lai, K., Gan, L.: Weighing efficiency-robustness in supply chain disruption by multi-objective firefly algorithm. Sustainability **8**, 250–277 (2016)

Parameter Identification for a Class of Nonlinear Systems Based on ESN

Xianshuang Yao, Zhanshan Wang[✉], and Huaguang Zhang

College of Information Science and Engineering, Northeastern University,
Shenyang 110819, China
zhanshan_wang@163.com

Abstract. In this paper, a new identification method based on echo state network (ESN) is proposed to identify the parameters of a class of discrete-time nonlinear systems. Through analyzing the characteristics of output signals, the identification method can determine the maximal delay time of the given nonlinear system. To obtain the better prediction and identification accuracy of this method, an online learning algorithm is proposed to train the output weights of ESN. Simulation examples show the effectiveness of the proposed identification method.

Keywords: Echo state network · Parameter identification · Discrete-time nonlinear systems

1   Introduction

In many applications, it is important to obtain the models of real systems. These models can be effectively applied in system analysis, prediction or simulation, design of controllers, supervision, fault detection, and so on [1,2]. System identification is usually used to construct a mathematical relationship model. The model can estimate the real outputs with a certain expected accuracy subject to the actual input signals. In recent years, many identification methods based on the neural networks (NN) have been widely used to identify nonlinear systems [3–7], for example, fuzzy stochastic NN (FSNN) [3], radial basis function NN (RBFNN) [4], recurrent wavelet NN (RWNN) [5], fuzzy wavelet NN (FWNN) [6], RNN [7], etc. By analyzing the characteristics of NN, the NN has the following advantages: parallel computation, distribution of data processors, universal approximation ability and generalization abilities.

Echo state networks (ESN) [8] is an improved model of recurrent neural network (RNN) [9,10]. ESN uses an interconnected recurrent grid of processing neurons called dynamical reservoir to replace the hidden layer of RNN. The advantage of ESN over RNN is that only the output weights need to be trained, while the reservoir weights and input weights usually are given randomly. Compared with RNN, ESN not only can provide a simple and distinctive learning method, but also obtain the higher accuracy learning result [8]. Thus, ESN has

© Springer International Publishing AG 2017
D. Liu et al. (Eds.): ICONIP 2017, Part IV, LNCS 10637, pp. 231–238, 2017.
https://doi.org/10.1007/978-3-319-70093-9_24

been attracted much attention in purely input-driven applications, for example, time-series prediction and classification [11,12], dynamic pattern recognition [13], system modeling or identification [14], filtering or control [15], big data application [16], etc. Therefore, combining with the advantages of ESN, the ESN can overcome the drawbacks of NN. For example, due to the structure features of the reservoir, ESN has the deep learning ability to utilize fully the input information for the limited available data. In addition, compared with the ANN, the ESN can reduce the computational burden because of training less parameters of network. Therefore, a new identification method based on ESN is proposed to identify the parameters of a class of discrete-time nonlinear systems.

In order to obtain better prediction and identification accuracy, the output weights of ESN should be trained. For calculating the output weights, the matrix pseudo-inverse method [11,12] and gradient-based learning method [17] are usually used. However, the matrix pseudo-inverse method relies on the collected state matrix and collected output matrix. The gradient-based learning algorithm is usually used to adjust online the output weights, but it suffers from slow convergence in the applications and the danger of trapping at local minima [18]. Therefore, in order to solve above mentioned problems, in this paper, an online learning output weights algorithm is given to train the output weights of ESN.

The remaining part of this paper is organized as follows. In Sect. 2, the brief description of ESNs are given. In Sect. 3, the proposed identification method based on ESN is introduced to identify the parameters of discrete-time nonlinear systems. Simulation example of the proposed identification method is performed in Sect. 4. Finally, the conclusion is given in Sect. 5.

2 ESN

ESN is a particular recurrent neural networks. The reservoir is used to replace the hidden layer of RNN. The structure of ESN is illustrated in Fig. 1. It comprises an input layer, a reservoir and an output layer. The number of input neurons, reservoir neurons and output neurons is K, N and L, respectively. Let $u = u(n)$ denote the external input vector, $x = x(n)$ denote the reservoir state, and

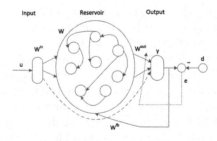

Fig. 1. Structure of ESN

$y = y(n)$ denote the output vector. n is discrete time step. W^{in}, W, W^{fb} and W^{out} denote the input weights, reservoir weights, feedback weights and output weights, respectively, and their sizes are successively $N \times K$, $N \times N$, $N \times L$ and $L \times (K + N)$. The discrete-time model of the standard ESN is given as follows [19]:

$$x(n + 1) = f(W^{in}u(n + 1) + Wx(n) + W^{fb}y(n)) \qquad (1)$$
$$y(n) = g(W^{out}[x(n); u(n)]) \qquad (2)$$

where f denotes reservoir state activation function (usually select $tanh$), and g denotes the output activation function (usually the identity).

Considering the leaky integrator units of reservoir, leaky integrator ESN (Leaky-ESN) is proposed to improve the performance of the standard ESN. Therefore, the reservoir state update equation of Leaky-ESN is modified as follows [11]:

$$x(n + 1) = (1 - a)x(n) + af\left(W^{in}u(n + 1) + Wx(n) + W^{fb}y(n)\right) \qquad (3)$$

where $a \in (0, 1]$ is a constant and denotes the leaking rate.

3   Parameter Identification Method Based on ESN

3.1   The Model of Identification Method

According to the performance and structure of ESN, a new identification method based on ESN is proposed to identify a class of discrete-time nonlinear systems. Its structure is illustrated in Fig. 2.

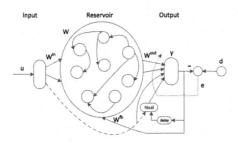

Fig. 2. Structure of ESN

For a class of discrete-time nonlinear system, its expression is shown as follows:

$$y(n) = f_r\left(u(n), u(n - 1), \cdots, u(n - \tau_u), y(n - 1), \cdots, y(n - \tau_y)\right) \qquad (4)$$

where f_r denotes the nonlinear function, τ_u and τ_y denote the time delay of $u(n)$ and $y(n)$, respectively.

In Fig. 2, the estimated output of the identified discrete-time nonlinear system is given as follows:

$$y(n) = W^{out}[x(n); u(n); \varphi(n)] \tag{5}$$

where $\varphi(n) \in u(n)y(n-1), u^2(n), y^2(n-1)$.

Before determining the time delay of a given nonlinear system, we describe the autocorrelation of the output signal. The definition of the autocorrelation coefficient for output signal is shown as follows [12]:

$$\eta_k = \frac{\sum_{t=1}^{n-k}(x_t - \bar{x})(x_{t+k} - \bar{x})}{\sum_{t=1}^{n}(x_t - \bar{x})^2} \tag{6}$$

where k denotes the interval of output signal ($k = 0, 1, \cdots, \bar{K}$), $\bar{K}$ is the maximum interval of output signal, $\bar{x}$ denotes the mean value of input signal.

For a given output signal, we use Eq. (6) to calculate the autocorrelation coefficients $\eta_k(k = 0, 1, 2, \cdots, \bar{K})$, which $\eta_{k1}(k1 > 0, k1 \in k)$ is the effective value. We gradually increase the value of $\bar{K}$ to obtain the period of the autocorrelation coefficients. After determining the period of output signal, we give a reasonable threshold ξ, to meet a certain performance requirement. Comparing the autocorrelation coefficients η_k with threshold ξ, if $\eta_i > \xi, i \in k$, the corresponding i denotes the time delay τ.

3.2   Identifying the Parameters of Nonlinear System

In this subsection, we specifically illustrate how to identify the parameters of discrete-time nonlinear system. According to the characteristics of ESN, the $W^{out}(n)$ can be broken down as follows:

$$W^{out}(n) = [W_x^{out}(n) \ W_u^{out}(n)] \tag{7}$$

According to Eq. (5), the output can be modified as follows:

$$y(n) = W_x^{out}x(n) + W_u^{out}u(n) + W_\varphi^{out}\varphi(n) \tag{8}$$

For the $W_x^{out}x(n)$ in Eq. (8), we have

$$
\begin{aligned}
&W_x^{out}x(n) \\
=&W_x^{out}f\left(W^{in}u(n) + Wx(n-1) + W^{fb}y(n-1)\right) \\
=&W_x^{out}\left(W^{in}u(n) + Wx(n-1) + W^{fb}y(n-1) + \varepsilon_0\right) \\
=&W_x^{out}W^{in}u(n) + W_x^{out}W^{fb}y(n-1) + W_x^{out}Wx(n-1) + W_x^{out}\varepsilon_0 \tag{9}
\end{aligned}
$$

For the $W_x^{out} W x(n-1)$ in Eq. (9), we have

$$
\begin{aligned}
& W_x^{out} W x(n-1) \\
=& W_x^{out} W f\Big(W^{in} u(n-1) + W x(n-2) + W^{fb} y(n-2)\Big) \\
=& W_x^{out} W \Big(W^{in} u(n-1) + W x(n-2) + W^{fb} y(n-2) + \varepsilon_1\Big) \\
=& W_x^{out} W W^{in} u(n-1) + W_x^{out} W W^{fb} y(n-2) + \\
& W_x^{out} W^2 x(n-2) + W_x^{out} W \varepsilon_1
\end{aligned}
\tag{10}
$$

For the $W_x^{out} W^\tau x(n-\tau)$, we have

$$
\begin{aligned}
& W_x^{out} W^\tau x(n-\tau) \\
=& W_x^{out} W^\tau f\Big(W^{in} u(n-\tau) + W x(n-\tau-1) + W^{fb} y(n-\tau-1)\Big) \\
=& W_x^{out} W^\tau \Big(W^{in} u(n-\tau) + W x(n-\tau-1) + W^{fb} y(n-\tau-1) + \varepsilon_\tau\Big) \\
=& W_x^{out} W^\tau W^{in} u(n-\tau) + W_x^{out} W^\tau W^{fb} y(n-\tau-1) + \\
& W_x^{out} W^{\tau+1} x(n-\tau-1) + W_x^{out} W^\tau \varepsilon_\tau
\end{aligned}
\tag{11}
$$

Inserting Eqs. (9)–(11) into Eq. (8), we can obtain

$$
\begin{aligned}
y(n) =& (W_u^{out} + W_x^{out} W^{in}) u(n) + \cdots + W_x^{out} W^\tau W^{in} u(n-\tau) \\
& + W_x^{out} W^{fb} y(n-1) + \cdots + W_x^{out} W^\tau W^{fb} y(n-\tau-1) + W_\varphi^{out} \varphi(n) \\
& + W_x^{out} W^{\tau+1} x(n-\tau-1) + W_x^{out} \varepsilon_0 + W_x^{out} W \varepsilon_1 + \cdots + W_x^{out} W^\tau \varepsilon_\tau
\end{aligned}
\tag{12}
$$

where $\tau = \max(\tau_u, \tau_y)$.

Simplifying Eq. (12), we have

$$
\begin{aligned}
y(n) =& a_0 u(n) + \cdots + a_\tau u(n-\tau) + b_1 y(n-1) + \cdots \\
& + b_\tau y(n-\tau) + c_0 \varphi(n) + \Sigma.
\end{aligned}
\tag{13}
$$

The parameters in Eq. (13) are given as follows:

$$
a_0 = W_u^{out} + W_x^{out} W^{in}
$$

$$
\vdots
$$

$$
a_\tau = W_x^{out} W^\tau W^{in}
$$

$$
b_1 = W_x^{out} W^{fb}
$$

$$
\vdots
$$

$$
b_\tau = W_x^{out} W^\tau W^{fb}
$$

$$
c_0 = W_\varphi^{out}
$$

$$
\Sigma = W_x^{out} W^{\tau+1} x(n-\tau-1) + W_x^{out} \varepsilon_0 + \cdots + W_x^{out} W^\tau \varepsilon_\tau
\tag{14}
$$

3.3 Training the Output Weights of ESN

To obtain the better identification parameters, the output weights of ESN should be trained. The objective of training ESN is to find the appropriate output weights, such that the error

$$e(n) = y(n) - d(n) \tag{15}$$

or mean-square error (MSE)

$$E(n) = \frac{1}{2}\|y(n) - d(n)\|^2 \tag{16}$$

is minimized and the output $y(n)$ can match the teacher output $d(n)$ as much as possible. Here, $E(n)$ denotes the mean-square error, $\|\cdot\|$ denotes the Euclidean distance (or norm).

In order to overcome the drawbacks of the matrix pseudo-inverse method and gradient-based learning method, a new update rule for training the output weights of ESN is given as follows:

$$W^{out}(n+1) = W^{out}(n) + \lambda_W e(n)[x(n); u(n)]^T \tag{17}$$

where λ_W denotes the learning rate.

4 Simulation Examples

In this section, ESN with the learning law (17) is used for identification of the discrete-time nonlinear system. A discrete-time nonlinear system is select to verify the effectiveness of this proposed identification method. In this paper, the discrete-time nonlinear system is given as follows:

$$y(n) = 0.3u(n) + 0.05y(n-1) + 0.1y^2(n-1) \tag{18}$$

According to the definition (6), the autocorrelation coefficient of output signal is determined, and the maximal delay time of system is also determined, i.e. $\tau = 1$. Through the proposed parameter identification method, the change curve of parameters in the training process are shown in Figs. 3, 4 and 5. The change curve of identification accuracy is shown in Fig. 6.

From Figs. 3, 4 and 5, we can see that, the obtained parameters are gradually close to the actual parameters in the iteration process. When the identification error is 1.23e−04, the parameters are obtained as follows:

$$a_0 = 0.2989, b_1 = 0.052, c_0 = 0.1007$$

Therefore, the simulation results show that the parameter identification method is effective.

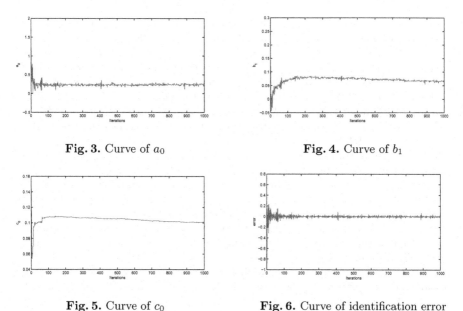

Fig. 3. Curve of a_0 **Fig. 4.** Curve of b_1

Fig. 5. Curve of c_0 **Fig. 6.** Curve of identification error

5   Conclusion

In this paper, a new parameter identification method based on ESN is proposed to identify a class of discrete-time nonlinear systems. Through analyzing the characteristics of output signals, the identification method can determine the maximal delay time of the given nonlinear system. An online learning algorithm is developed to train the output weights of ESN. The simulation results show that this proposed parameter identification method based on ESN can significantly improve parameter identification accuracy.

Acknowledgments. This work was supported by the National Natural Science Foundation of China (Grant Nos. 61473070, 61433004, 61627809), the Fundamental Research Funds for the Central Universities (Grant No. N160406002), and SAPI Fundamental Research Funds (Grant No. 2013ZCX01).

References

1. Nelles, O.: Nonlinear System Identification: From Classical Approaches to Neural Networks and Fuzzy Models. Springer, Berlin (2001). doi:10.1007/978-3-662-04323-3
2. Ahmadi, G., Teshnehlab, M.: Designing and implementation of stable sinusoidal rough-neural identifier. IEEE Trans. Neural Netw. Learn. Syst. **28**, 1774–1786 (2017)
3. Jiang, X.M., Mahadevan, S., Yuan, Y.: Fuzzy stochastic neural network model for structural system identification. Mech. Syst. Sig. Process. **82**, 394–411 (2017)

4. Ko, C.N.: Identification of non-linear systems using radial basis function neural networks with time-varying learning algorithm. IET Sig. Process. **6**, 91–98 (2012)
5. Zhao, H.Q., Gao, S.B., He, Z.Y., Zeng, X.P., Jin, W.D., Li, T.R.: Identification of nonlinear dynamic system using a novel recurrent wavelet neural network based on the pipelined architecture. IEEE Trans. Ind. Electron. **61**, 4171–4182 (2014)
6. Yilmaz, S., Oysal, Y.: Fuzzy wavelet neural network models for prediction and identification of dynamical systems. IEEE Trans. Neural Netw. **21**, 1599–1609 (2010)
7. Ahn, C.K.: L_2-L_∞ nonlinear system identification via recurrent neural networks. Nonlinear Dyn. **62**, 543–552 (2010)
8. Jaeger, H., Haas, H.: Harnessing nonlinearity: predicting chaotic systems and saving energy in wireless telecommunication. Science **304**, 78–80 (2004)
9. Jaeger, H.: A tutorial on training recurrent neural networks, covering BPTT, RURL, EKF and the 'Echo State Network' approach. Technical report, German National Research Center for Information Technology (2002)
10. Zhang, H.G., Wang, Z.S., Liu, D.R.: A comprehensive review of stability analysis of continuous-time recurrent neural networks. IEEE Trans. Neural Netw. Learn. Syst. **25**, 1229–1262 (2014)
11. Jaeger, H., Lukoševičius, M., Popovici, D., Siewert, U.: Optimization and applications of echo state networks with leaky-integrator neurons. Neural Netw. **20**, 335–352 (2007)
12. Lun, S.X., Yao, X.S., Hu, H.F.: A new echo state network with variable memory length. Inf. Sci. **370–371**, 103–119 (2016)
13. Skowronski, M.D., Harris, J.G.: Automatic speech recognition using a predictive echo state network classifier. Neural Netw. **20**, 414–423 (2007)
14. Lun, S.X., Wang, S., Guo, T.T., Du, C.J.: An I-V model based on time warp invariant echo state network for photovoltaic array with shaded solar cells. Sol. Energy **105**, 529–541 (2014)
15. Han, S.I., Lee, J.M.: Fuzzy echo state neural networks and funnel dynamic surface control for prescribed performance of a nonlinear dynamic system. IEEE Trans. Ind. Electron. **61**, 1099–1112 (2014)
16. Scardapane, S., Wang, D.H., Panella, M.: A decentralized training algorithm for Echo State Networks in distributed big data applications. Neural Netw. **78**, 65–74 (2016)
17. Koprinkova-Hristova, P., Oubbati, M., Palm, G.: Heuristic dynamic programming using echo state network as online trainable adaptive critic. Int. J. Adapt. Contr. Sig. Process. **27**, 902–914 (2013)
18. Man, Z., Wu, H.R., Liu, S., Yu, X.: A new adaptive backpropagation algorithm based on Lyapunov stability theory for neural networks. IEEE Trans. Neural Netw. **17**, 1580–1591 (2006)
19. Jaeger, H.: The 'echo state' approach to analysing and training recurrent neural networks-with an erratum note. Technical report, German National Research Center for Information Technology (2010)

Personalized Web Search Based on Ontological User Profile in Transportation Domain

Omar ElShaweesh[1(✉)], Farookh Khadeer Hussain[1], Haiyan Lu[1],
Malak Al-Hassan[2], and Sadegh Kharazmi[3]

[1] University of Technology, Sydney, Australia
omar.g.elshaweesh@student.uts.edu.au
[2] The University of Jordan, Amman, Jordan
[3] Redbubble, Melbourne, Australia

Abstract. Current conventional search engines deliver similar results to all users for the same query. Because of the variety of user interests and preferences, personalized search engines, based on semantics, hold the promise of providing more efficient information that better reflects users' needs. The main feature of building a personalized web search is to represent user interests in terms of user profiles. This paper proposes a personalized search approach using an ontology-based user profile. The aim of this approach is to build user profiles based on user browsing behavior and semantic knowledge of specific domain ontology to enhance the quality of the search results. The proposed approach utilizes a re-ranked algorithm to sort the results returned by the search engine to provide a search result that best relates to the user query. This algorithm evaluates the similarity between a user query, the retrieved search results and the ontological concepts. This similarity is computed by taking into account a user's explicit browsing behavior, semantic knowledge of concepts, and synonyms of term-based vectors extracted from the WordNet API. A set of experiments using a case study from a transport service domain validates the effectiveness of the proposed approach and demonstrates promising results.

Keywords: Fuzzy · Personalization · Ontology · Web search · User profile

1 Introduction

With the rapid growth of information on the World Wide Web, search engines are the primary means of finding relevant answers to users' questions. However, the information provided to all users for a specific query is similar. As users' interests and preferences differ, it is important to help users to find relevant information through personalized systems [1]. Personalization can be achieved by modeling each individual by managing, building and representing user preferences in the form of user profiles [2].

The main challenge of personalized web search is to model user needs and interests with the aim of enhancing search quality. Numerous approaches to modeling user needs have been developed in the web search domain by collecting user information explicitly through direct interaction between systems and users, or implicitly based on browsing documents or past queries [3–7].

© Springer International Publishing AG 2017
D. Liu et al. (Eds.): ICONIP 2017, Part IV, LNCS 10637, pp. 239–248, 2017.
https://doi.org/10.1007/978-3-319-70093-9_25

Modeling a user's interests and preferences presents a number of issues: (i) there is a limited understanding of the user's intent [8]; and (ii) the content of the web is described in natural language which often lacks clarity due to the different meanings of words. To overcome these limitations, a number of research studies have attempted to incorporate ontology within the information retrieval process [9]. Employing ontology to personalized web search systems in this manner is performed in two different ways. One is by building an ontological user profile, which represents user preferences, and using it for the re-ranking process. The other is to reformulate user queries in such a way that the probability of retrieving relevant documents is significantly increased. Ontology-based personalized search approaches have proven effectiveness and efficiency in representing users' interests and generating more personalized results than traditional keyword-based approaches [10]. However, building more effective ontology-based user profiles is still an open research topic.

The main objective of the proposed web search personalization approach in this paper is to create an ontology-based user profile that implicitly uses a new weighting method. This method attempts to assign interest scores to concepts in the targeted domain ontology using a semantic fuzzy classification technique and considering the browsing behavior of users while scoring the concepts.

The effectiveness of the proposed personalized search approach has been validated in a case study in a transport service domain. It achieves highly effective results in terms of the quality of search results generated.

The rest of this paper is organized as follows. Related work is presented in Sect. 2. Section 3 describes the proposed approach. Section 4 discusses the experimental results and evaluations. Lastly, the paper is concluded in Sect. 5.

2   Related Work

Most personalization approaches are based on modeling a user's interests, which is known as a user profile. The user-related data is either collected explicitly by asking direct questions of the user and obtaining feedback about his/her interests, or implicitly by monitoring the user's browsing activities. In personalized search systems, a user profile is useful in three different ways [11]. Firstly, it is used as part of the retrieval process. Secondly, it is used to re-rank the retrieved results from the search engine. Lastly, it is used in the query expansion process.

One of the most important representations of users' interests in a personalized retrieval system is the use of ontology, which is a promising solution for solving word ambiguity and the cold start problem [2, 12, 13].

A number of personalized web search engine use ontology for the query expansion process. A new generation of ontologies based on fuzzy theory with two degrees of uncertainty was proposed in [9]. The concepts and relations between concepts were assigned by an expert in the related domain, and the proposed ontology was used for the query expansion process. The results show that the fuzzy ontology approach has increased the precision result compared to the crisp ontology approach. An online information retrieval system named SIRO was proposed in [14]. The authors combined domain ontology with service ontology to assist user query reformulation. In this system, the relation between every concept in the domain ontology and the tasks in the service ontology is established. When a user makes a query, the query is expanded by the addition of new terms in the service ontology which are related to the matching concepts. A comparison of the proposed system with other systems showed a considerable improvement in the results. An individual fuzzy ontology method for integration into the query reformulating process was proposed by [4]. Three main components were introduced: automatic individual fuzzy ontology building, query reformulation based on fuzzy ontology, and document classification. The individual fuzzy ontology is built automatically by assigning an initial membership value to the ontology concepts and relations; the values are then updated continuously based on the user's previous queries. The user's negative terms are considered in the query expansion process by removing negative preferences from the original query terms. This method achieved improvement in the quality of the search results.

Other research uses ontology as the main component for representing the user's interests by assigning an interest value to each concept in the ontology. This profile is then used to re-rank the results that are relevant to the user's query. For example, [7] created a Dynamic Category Interest Tree (DCIT) in which a user interest score is added implicitly to each category in the tree through the browsing history of the user and collaborative filtering. To match the user's document of interest with the corresponding category, a fuzzy classification approach is used. The final results obtained by Google for the user's query will be re-ranked based on the weighted tree. The authors in [12] proposed a semantic framework for personalized search using semantic web technique. Three issues were considered: firstly, how to discover user interests; secondly, how to model the user profile; and lastly, how to expand the user query. The results of the proposed approach show an improvement in search accuracy and efficiency compared to existing approaches like SPWS and SOnP.

3   The Proposed Approach

To provide personalized search results that reflect users' interests, the initial results delivered by a search engine are re-ordered based on the degree to which the retrieved results match the user's ontological profile. The proposed approach performs three main tasks. Firstly, it receives the user's query and expands it using a WordNet API. Secondly, the expanded query is passed to the keyword-based search engine and the related results are retrieved. Lastly, the retrieved results are re-ranked based on the ontological user profile, and the personalized results are presented to the user. The proposed approach effectively enhances the overall quality of the search result by offering users the most suitable result with the least search effort.

The proposed approach has two main features, namely the use of ontology and the ontological user profile.

The use of ontology is considered to be the knowledge base of the proposed approach. It offers a structured and unambiguous representation of knowledge in a well-defined format using ontology modeling languages such as Ontology Web Language (OWL). Semantic knowledge embedded in ontology can be exploited to support the semantic analysis of heterogeneous content and to return useful information for better decision making [10, 15]. The success of semantically enhanced approaches in web search and the variety of user needs inspired the idea of developing our personalized web search approach using ontology.

An ontological user profile is created based on existing reference domain ontology and a user's browsing history, where the user's interests are closely associated with relevant concepts in reference ontology. The ontological user profile is beneficial for the proposed approach in two ways: (i) concepts in user profiles can have interest scores that indicate the importance of concepts to users, and (ii) the search results can be re-ranked using an ontological user profile.

In this paper, the transport service domain ontology previously developed by [16] is considered as the reference ontology. The transport service ontology is based on metadata extracted from transport service websites. The major mission of transport service metadata is to extract meaningful information regarding the transport service from downloaded web pages. An additional description of the reference ontology is presented in Sect. 4.1.

The proposed approach comprises two main components: the building ontology-based user profile and the re-ranking algorithm. These components are explained in detail in the sections that follow.

3.1 Building an Ontology-Based User Profile

In this component, a user profile is built as a structured data record which maintains the related data of users, including preferences and interests. Users' preferences and interests are represented according to their search history (i.e. web log). The search history of users is mainly used to extract the URLs of web pages that have been visited at the given website, and the URLs are analyzed to fetch the metadata of those web pages, i.e. keywords and descriptions of web pages based on the *meta* tag in the HTML.

User profiles maintain two types of data: (i) instances, i.e. items, which are interesting web pages for an active user (e.g. transport service items in the considered domain), and (ii) concepts, in an ontology by which the service items are categorized. Each concept is annotated with a weight reflecting a user's degree of interest.

It is important to emphasize that one of the main contributions of this paper is that a user's interest score values of concepts are computed semantically based on a semantic fuzzy-based classification approach. This approach considers the overlapping nature of concepts, thus it calculates the relevance degree between each web page visited and every concept, which in turn is used to find the user's interest scores for concepts.

Computing a user's interest scores of concepts consists of four main steps, explained below.

Step 1: Building user-weighted vectors for visited pages and concepts.

A user-weighted vector must be built for each user in terms of visited pages and concepts. Pages and concepts are ultimately represented as a vector of terms. To represent each visited web page in this way, the description of its metadata is pre-processed to remove stop and stemmed words using the Porter stemmer [17]. The web page is then represented as a vector of terms which includes both keywords and distinguished terms extracted from the description of the page itself.

In contrast, each concept in the reference ontology is also represented as a vector of distinguished terms. These terms include all the terms extracted from the metadata of keywords and descriptions of all the web pages that are classified under a specific concept.

Intuitively, synonyms of terms can add meaningful knowledge that may help to determine more accurate content. Therefore, each vector, either for the web page or for the concept, is expanded using the WordNet API, where the synonyms for each term are added to the corresponding vector. Formally, the vector for a specific webpage p_j is defined as: $\overrightarrow{p_j} = \left(w_{t_1,j}, w_{t_2,j}, \ldots, w_{t_n,j}, w_{s_1,j}, w_{s_2,j}, \ldots, w_{s_m,j}\right)$, where $w_{t_1,j}$ represents the weight of the first term in the vector space of a webpage p_j, and $w_{s_l,j}$ represents the synonym of a specific term in the vector space of a webpage p_j.

Formally, the vector for a specific concept c_i is defined as follows:
$\overrightarrow{c_i} = \left(w_{t_1,i}, w_{t_2,i}, \ldots, w_{t_n,i}, w_{s_1,i}, w_{s_2,i}, \ldots, w_{s_m,i}\right)$, where $w_{t_1,i}$ represents the weight of the first term in the vector space of a concept c_i, and $w_{s_i,i}$ represents the synonym of a specific term in the vector space of a concept c_i.

Step 2:   Find weight values for all terms in the defined vectors of both the web page and the concept.

In this step, a weight value for each term in the defined vectors of the web page and the concept is calculated. Formally, the weight values of a specific term and its synonym are computed using Eqs. 1 and 2 as follows:

$$w(t) = occ(t) + occ(t)\sqrt{1/occ(syn_t)} \tag{1}$$

$$w(syn_t) = occ(syn_t) + occ(t)\sqrt{1/occ(syn_t)} \tag{2}$$

where $w(t)$ is the weight of term t, $occ(t)$ is the number of occurrences of term t, $occ(syn_t)$ is the number of words with the same meaning as term t (i.e. synonyms), $w(syn_t)$ is the weight of the synonym of term t.

Step 3:   Find the similarity between each visited URL and concepts, then store the result in a similarity matrix as output.

In this step, the similarity degree between each visited web page and concepts for a target user is computed using cosine similarity. The defined weighted vectors of a specific web page and concept, as previously presented in step 1, are considered to find their relevance (i.e. similarity) degree. The computed similarity results are stored as output in a similarity matrix named page-concept, i.e. $SIM[P \times C]$ matrix. The similarity $sim(p_j, c_i)$ between p_j and c_i is computed using the following equation:

$$sim(p_j, c_i) = \cos\left(\overrightarrow{p_j}, \overrightarrow{c_i}\right) = \frac{\overrightarrow{p_j} \cdot \overrightarrow{c_i}}{\left\|\overrightarrow{p_j}\right\|_2 \times \left\|\overrightarrow{c_i}\right\|_2} = \frac{\sum_{t_k \in T_{p_j, c_i}} w_{p_j, t_k} \times w_{c_i, t_k}}{\sqrt{\sum_{t_k \in T_{p_j, c_i}} w_{p_j, t_k}^2} \sqrt{\sum_{t_k \in T_{p_j, c_i}} w_{c_i, t_k}^2}} \tag{3}$$

where $p_j.c_i$ denotes the dot-product between the two vectors $\vec{p_j}$ and $\vec{c_i}$, T_{p_j,c_i} is the set of terms available in both vectors $\vec{p_j}$ and $\vec{c_i}$, w_{p_j,t_k} and w_{c_i,t_k} is the weight value of term t_k in the vector of p_j and the vector of concept c_i, respectively.

The similarity between a page and concept is calculated by computing the cosine of the angle between two vectors. The cosine similarity results are between 0 and 1, therefore a high matching value between a web page and a concept indicates high similarity, and vice versa.

Step 4:   Find a user's interest score for a specific concept.

In this step, the user's interest score of a concept c_i is computed based on the following equation:

$$IS(c_i) = \sum\nolimits_{\forall p \in P, c=c_i} sim(p_j, c_i) \Big/ \sum\nolimits_{\forall c \in C} \sum\nolimits_{\forall p \in P} sim(p_j, c_i) \qquad (4)$$

where P is the set of all visited URLs in a user profile, C is the set of all concepts in the reference ontology, $\sum_{\forall p \in P, c=c_i} sim(p_j, c_i)$ is the sum of all the relevance degrees of web pages for a concept c_i, and $\sum_{\forall c \in C} \sum_{\forall p \in P} sim(p_j, c_i)$ is the sum of all relevance degrees of web pages for all concepts.

The output of this step is a user-concept interest score matrix (i.e. $IS[U \times C]$), in which each entry denotes the user's interest score for a concept c_i.

Not all search history URLs have the same degree of interest for users, however. Users may browse web pages for a few seconds or enter a wrong page, so these pages cannot be considered as pages of interest compared to web pages that the user prints or saves. For a more personalized process, a mechanism is introduced to determine the level of user interest in the URLs in the search history. This mechanism collects a user's behavior while browsing web pages and allocates a weight status value to every URL. The status value is then considered to calculate the user's interest score of concepts available in the reference domain ontology.

To identify the weight status of the web pages visited, three different status values are defined, namely browse, search result and favorites. The browse status occurs when a user browses a web page. The search result status occurs when a user browses a retrieved URL based on a query. The favorite's status represents the occasions on which a user saves or prints a web page. A weight status of 1, 0.75 and 0.5 is assigned for the favorites, search result and browse status, respectively.

The page's weight status has been investigated in this paper, as can be seen later in the experiments section. To demonstrate this, each entry in the similarity matrix $SIM[P, C]$, as described in step 3, is weighted by multiplying the similarity of p_j and c_i by the corresponding weight status value of p_j. The user's interest scores of concepts are then computed using Eq. (4), based on the obtained weighted similarity. The experiments show that using the page's weight status improves the quality of the results. More detail is presented in the experimental evaluation section.

Algorithm 1: Re-ranking of the retrieved search results

Input: ontological user profile for a target user (u), set of retrieved search results for a query (Q)
Output: re-ranked search results, i.e. personalized results for the target user (u)
Process:
  Set $C = \{c_1, c_2, ..., c_m\}$
  Set $URL = \{URL_1, URL_2, ..., URL_n\}$
For each $URL_j \in URL$
$c =$ find the highest matching concept to URL_j
   compute $sim(Q, c)$
   compute $sim(Q, URL_j)$
     $IS(URL_j) = sim(Q, URL_j) \times sim(Q, c) \times IS(c)$
End For
Sort *URLs* based on interest score *IS*

3.2   Search Personalization – Re-ranking Algorithm

Personalized search results are provided by re-ordering the results returned by the search engine for a given query based on the ontological user profile. Algorithm 1 presents the algorithmic steps which describe the re-ranking process of results retrieval. As can be seen in Algorithm 1, the input should be the ontological user profile and the retrieved search results for a specific query. Each retrieved result (i.e. URL) is first represented in a weighted vector and expanded as described in step 1.1, and the weight values of the terms computed according to Eqs. (1) and (2). Secondly, the highest similarity concept is obtained for each retrieved URL, based on the similarity matrix $[P \times C]$, as presented in step 2. Lastly, the user's interest score for each URL in the retrieved search result is computed using the following equation:

$$IS(URL_j) = sim(Q, URL_j) \times sim(Q, c) \times IS(c) \tag{5}$$

where $sim(Q, URL_j)$ is the similarity between the query Q and a retrieved search result URL_j, $sim(Q, c)$ is the similarity between the query Q and the concept c with the highest match with URL_j, and $IS(c)$ is the interest score of the concept c which has the highest match with URL_j. The similarity in Eq. (5) is calculated using cosine similarity, as presented in Eq. (3).

4   Experimental Evaluation

The goal of the experimental evaluation is to measure the performance of the proposed approach. In this regard, the proposed approach is compared with the Google ranking scheme in terms of the improvement in precision of different *top-n* results.

4.1   Data Set

This paper utilizes a transport service domain ontology based on the work conducted by [16]. The considered ontology was created by conducting a survey of websites of transport service companies.

The concepts relating to the transport service ontology are separated into two groups: abstract concepts and actual concepts. The abstract concepts refer to the abstract domain and the sub-domain of service concepts, whereas the actual concepts relate to real services (e.g. transport web pages); that is, the actual service concepts can link to the metadata of a service description entity (i.e. SDE). All service concepts have two fundamental properties: service name and service description.

4.2   Collecting Experimental Data

We evaluated the proposed approach against the most popular search engine, Google, and invited ten users to participate in our experiments. We asked users to search and browse the web using a given set of terms related to the transportation domain. Users were required to write down the web pages they visited that were of interest. To capture user behavior on web pages and help us build the user profile, we used Chromium browser automation software, which records all users' browsing behavior.

Next, users were asked to insert six different queries from the previously selected set into the Google search engine, and every user chose the results of interest from the first 30 results returned by Google for each query result. We called this collection of pages *the interesting pages data set*. Users also used our model to retrieve the data set of interest for the same queries. The top 5, 10, 15 and 20 links were then compared.

To evaluate the performance of the proposed approach, we calculate the precision for the top 5, 10, 15, and 20 links as follows.

$$Precision = \frac{number\ of\ interesting\ pages\ retrieved}{total\ number\ of\ pages\ retrieved} \tag{6}$$

4.3   Experimental Results

Average Precision of Top-n Results
In this experiment, we attempted to discover the improvement in precision obtained by the proposed approach and the Google search results at different *top-n*. In general, the results show that the personalized system based on the proposed approach achieved high precision at all *top-n*, and particularly at top 5 and 10.

Figure 1 shows the average precision for the proposed approach without considering user behavior (PWB), the proposed approach with user behavior considered (PB), and Google search engine. As can be seen from Fig. 1, the proposed approach with PB shows an improvement in precision of 13%, 12.5%, 10.5% and 5% for the top 5, 10, 15 and 20 links, respectively over Google search engine. The proposed PWB approach demonstrates improvements in precision of 9.5%, 10%, 8.5% and 4.5% for the top 5, 10, 15 and 20 links respectively, over Google search engine. However, user behavior has demonstrates a positive impact on the search result.

The Impact of User Behavior on Precision
In this experiment, the improvement in precision is found when user behavior is involved in the construction of user profiles and used to rank the retrieved results.

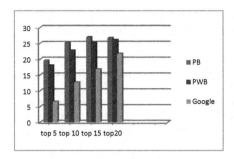
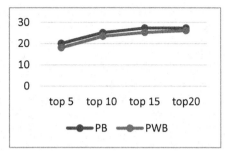

Fig. 1. Average precision for top-n results

Fig. 2. The effect of user behavior on precision value

Figure 2 shows the average precision for personalized search results with and without user behavior. The results show that the proposed personalized approach based on user behavior achieves slightly better precision at all *top-n*. There is an improvement in precision of 1.5%, 2.5%, 2% and 0.05% for the top 5, 10, 15 and 20 links respectively.

5   Conclusion

In this paper, we propose an approach for personalized search results by re-ranking retrieved search results based on ontological user profile. We use a semantic fuzzy classification to assign an interest weight to each concept to build ontological user profiles. The experiments show that the proposed approach achieves higher performance than Google search engine. In addition, considering user behavior results in an improvement in precision.

In future, more participants could be invited to conduct large-scale experiments, and user behavior analysis techniques employed to obtain even more accurate personalized results. Lastly, the feasibility of the proposed approach shows its capacity to be applied to other real world system applications.

References

1. Sieg, A., Mobasher, B., Burke, R.D.: Learning ontology-based user profiles: a semantic approach to personalized web search. IEEE Intell. Inform. Bull. **8**, 7–18 (2007)
2. Gauch, S., Speretta, M., Chandramouli, A., Micarelli, A.: User profiles for personalized information access. In: Brusilovsky, P., Kobsa, A., Nejdl, W. (eds.) The Adaptive Web. LNCS, vol. 4321, pp. 54–89. Springer, Heidelberg (2007). doi:10.1007/978-3-540-72079-9_2
3. Akhlaghian, F., Arzanian, B., Moradi, P.: A personalized search engine using ontology-based fuzzy concept networks. In: The International Conference on Data Storage and Data Engineering, pp. 137–141. IEEE, Bangalore (2010)
4. Baazaoui-Zghal, H., Ghezala, H.B.: A fuzzy-ontology-driven method for a personalized query reformulation. In: The IEEE International Conference on Fuzzy Systems, pp. 1640–1647. IEEE Press, Beijing (2014)

5. Daoud, M., Tamine-Lechani, L., Boughanem, M.: Using a concept-based user context for search personalization. In: Proceedings of the 2008 International Conference of Data Mining and Knowledge Engineering, London (2008)

6. Ferreira-Satler, M., Romero, F.P., Menendez-Dominguez, V.H., Zapata, A., Prieto, M.E.: Fuzzy ontologies-based user profiles applied to enhance e-learning activities. Soft. Comput. **16**, 1129–1141 (2012)

7. Nanda, A., Omanwar, R., Deshpande, B.: Implicitly learning a user interest profile for personalization of web search using collaborative filtering. In: IEEE/WIC/ACM International Joint Conferences on Web Intelligence (WI) and Intelligent Agent Technologies (IAT), vol. 2, pp. 54–62 (2014)

8. Jiang, X., Tan, A.H.: Learning and inferencing in user ontology for personalized Semantic Web search. Inf. Sci. **179**, 2794–2808 (2009)

9. Hourali, M., Montazer, G.A.: An intelligent information retrieval approach based on two degrees of uncertainty fuzzy ontology. Adv. Fuzzy Syst. **2011**, 7 (2011)

10. Al-Hassan, M., Lu, H., Lu, J.: A semantic enhanced hybrid recommendation approach: a case study of e-Government tourism service recommendation system. Decis. Support Syst. **72**, 97–109 (2015)

11. Micarelli, A., Gasparetti, F., Sciarrone, F., Gauch, S.: Personalized search on the world wide web. In: Brusilovsky, P., Kobsa, A., Nejdl, W. (eds.) The Adaptive Web. LNCS, vol. 4321, pp. 195–230. Springer, Heidelberg (2007). doi:10.1007/978-3-540-72079-9_6

12. Duong, T.H., Uddin, M.N., Nguyen, C.D.: Personalized semantic search using ODP: a study case in academic domain. In: Murgante, B., Misra, S., Carlini, M., Torre, Carmelo M., Nguyen, H.-Q., Taniar, D., Apduhan, Bernady O., Gervasi, O. (eds.) ICCSA 2013. LNCS, vol. 7975, pp. 607–619. Springer, Heidelberg (2013). doi:10.1007/978-3-642-39640-3_44

13. Calegari, S., Pasi, G.: Ontology-based information behaviour to improve web search. Future Internet **2**, 533–558 (2010)

14. Baazaoui, H., Aufaure, M.A., Soussi, R., Laboratoy, R.G., de la Manouba, E.C.U.: Towards an on-line semantic information retrieval system based on fuzzy ontologies. J. Digital Inf. Manag. **6**, 375 (2008)

15. Sánchez, D., Batet, M., Isern, D., Valls, A.: Ontology-based semantic similarity: a new feature-based approach. Expert Syst. Appl. **39**, 7718–7728 (2012)

16. Dong, H., Hussain, F.K., Chang, E.: A service search engine for the industrial digital ecosystems. IEEE Trans. Ind. Electron. **58**, 2183–2196 (2011)

17. Porter, M.F.: An algorithm for suffix stripping. Program **14**, 130–137 (1980)

Adaptive Dynamic Programming for Human Postural Balance Control

Eric Mauro[✉], Tao Bian, and Zhong-Ping Jiang

Department of Electrical and Computer Engineering, Tandon School of Engineering,
New York University, 5 Metrotech Center, Brooklyn, NY 11201, USA
{eric.mauro,tbian,zjiang}@nyu.edu
http://engineering.nyu.edu/tandon

Abstract. This paper provides a basis for studying human postural balance control about upright stance using adaptive dynamic programming (ADP) theory. Previous models of human sensorimotor control rely on *a priori* knowledge of system dynamics. Here, we provide an alternative framework based on the ADP theory. The main advantage of this new framework is that the system model is no longer required, and an adaptive optimal controller is obtained directly from input and state data. We apply this theory to simulate human balance behavior, and the obtained results are consistent with the experiment data presented in the past literature.

Keywords: Optimal control · Motor learning · Adaptive dynamic programming

1 Introduction

Human postural balance control is a commonly studied problem that is always under investigation due to the complexities of the human central nervous system (CNS). There are many motivations within medical and robotics research to study human balance, such as understanding the degeneration of the human control system caused by aging and neuromuscular disorders such as Parkinsons disease [1,2]. Postural instability leads to additional health risks, especially in the associated risk of falling [3,4]. Research in human balance models can lead to better medical information and diagnoses, and it could also be used to develop better dynamic models for robotics and prostheses [5].

While there are many problems related to human balance and movement, maintaining upright stance is one particular noteworthy problem. The upright position is an inherently unstable equilibrium position and requires constant control, and one hypothesis is that the CNS uses feedback with minimal muscle forces to automatically correct posture [6]. While the equations that define the motion are complex and nonlinear, it is relatively simple to analyze using linearization about the vertical stance as an operating point. Many researchers have studied human balance using principles of optimal control [7–10]. However,

© Springer International Publishing AG 2017
D. Liu et al. (Eds.): ICONIP 2017, Part IV, LNCS 10637, pp. 249–257, 2017.
https://doi.org/10.1007/978-3-319-70093-9_26

most existing optimal control design methods rely on precise knowledge of system dynamics. These methods are not entirely suited for the problem of posture control because human dynamics are complex and vary from person to person.

ADP is an iterative learning approach for optimal control in a system with unknown dynamics [11–17]. It combines the concepts of reinforcement learning and optimal control to find an adaptive optimal controller using only online state and input data. In the past decade, ADP has been developed for continuous-time deterministic systems [12,15] and was extended to stochastic systems for sensorimotor control [18,19]. Since the human body and CNS are personalized systems that are difficult to model, ADP is well suited for studying the human postural balance control.

The paper is organized as follows. In Sect. 2, we derive the nonlinear dynamic model and linearize it about the upright equilibrium position. In Sect. 3, we give a brief overview on ADP theory. In Sect. 4, we apply the ADP approach to design an optimal control policy for a simulated human balance model. Finally, concluding remarks and future work are contained in Sect. 5.

2   Dynamic Model

When viewed from the side with only forward and backward motion, the human body acts like a multi-segmented pendulum with actuators at each joint. However, there are some challenges to modeling this kind of system. First of all, the upright equilibrium position is unstable and the CNS always needs to apply control. For the human body, this is a much easier task compared to an under-actuated system. Second, the system is made up of nonlinear equations which make system modeling and control more complicated. Lastly, each additional segment adds more complexity to the pendulum dynamics as there are more degrees of freedom. The model for the human system is made up of three segments characterized by movement about the ankle, knee, and hip. The following covers the derivation of the nonlinear and linearized models.

The system dynamics can be derived from Lagrangian mechanics. Indeed, to find the system equations, the Euler-Lagrange equation must be set to equal to the external forces [20]. For the human body, the joint torques act as the control input u. Namely,

$$\frac{\mathrm{d}}{\mathrm{d}t}\frac{\partial L}{\partial \dot{\theta}} - \frac{\partial L}{\partial \theta} = u \tag{1}$$

where

$$L = \sum_{i=1}^{3}(\frac{1}{2}m_i v_i^2 + \frac{1}{2}I_i\dot{\theta}^2 - m_i g y_i)$$

is the Lagrangian function, and m_i, v_i, I_i and y_i correspond to the mass, velocity, moment of inertia, and vertical position with $i = 1, 2, 3$ representing the ankle, knee, and hip respectively.

Equation (1) can be rewritten in matrix notation as in [20]:

$$M(\theta)\ddot{\theta} + N(\theta)\dot{\theta}^2 + G(\theta) = u \tag{2}$$

where M is a symmetric inertia matrix, N is an anti-symmetric matrix, and G is a gravitation vector.

While an inverted pendulum can be modeled with joint angles with respect to the vertical, the human balance control system is more accurately modeled using segment angles. To be specific, the state variable in (2) can be converted from joint angles referenced to the vertical to segment angles (lower leg, thigh, and torso) referenced to each other using a transformation matrix F such that

$$\theta = F\phi = \begin{bmatrix} 1 & 0 & 0 \\ 1 & 1 & 0 \\ 1 & 1 & 1 \end{bmatrix} \phi. \tag{3}$$

Then, the nonlinear model (2) in terms of segment angle becomes

$$M(\phi)F\ddot{\phi} + N(\phi)F\dot{\phi}^2 + N_s(\dot{\phi}) + G(\phi) = u \tag{4}$$

with an additional matrix N_s from the $\dot{\theta}^2$ term in (2). The model (4) can be easily linearized around the upright equilibrium position ($\phi = 0$, $\dot{\phi} = 0$) using the segment angles as the state of the linearized model, $x \triangleq [\phi_1, \phi_2, \phi_3, \dot{\phi}_1, \dot{\phi}_2, \dot{\phi}_3]^T$ where ϕ_1, ϕ_2, and ϕ_3 represent the angles of the lower leg, thigh, and torso respectively. The linearization of (4) becomes

$$\dot{x} = Ax + Bu \tag{5}$$

where

$$A = \begin{bmatrix} 0 & I_3 \\ F^{-1}\hat{M}^{-1}\hat{G}F & 0 \end{bmatrix}, \quad B = \begin{bmatrix} 0 \\ F^{-1}\hat{M}^{-1} \end{bmatrix}.$$

Additionally, $\hat{M}$ is a constant symmetric inertia matrix and $\hat{G}$ is a constant diagonal gravitation matrix with constants k_i and h_i such that

$$\hat{M} = \begin{bmatrix} I_1 + l_1^2 h_1 & l_1 l_2 k_2 & l_1 l_3 k_3 \\ l_1 l_2 k_2 & I_2 + l_2^2 h_2 & l_2 l_3 k_3 \\ l_1 l_3 k_3 & l_2 l_3 k_3 & I_3 + l_3^2 h_3 \end{bmatrix}, \quad \hat{G} = \begin{bmatrix} l_1 k_1 g & 0 & 0 \\ 0 & l_2 k_2 g & 0 \\ 0 & 0 & l_3 k_3 g \end{bmatrix}.$$

3 Adaptive Dynamic Programming

While the system matrices A and B in (5) can be measured, they vary from person to person. In practice, these system matrices are unknown, and a control policy must be computed without exact knowledge of the system dynamics. First, we review an online, off-policy ADP algorithm which will be used to study the human balance control. The design philosophy is based on the principles of policy iteration [21] and reinforcement learning [22]. The ADP algorithm presented here is found in more detail in [12,13].

3.1   Problem Formulation

The objective is to find a control input u for system (5) that minimizes the cost function

$$J = \int_0^\infty (x^T Q x + u^T R u) dt \tag{6}$$

with $Q = Q^T \geq 0$, $R = R^T > 0$, and $(A, Q^{1/2})$ observable.

When A and B are accurately known, the minimum cost is $x^T(0) P^* x(0)$ where $P^* \in \mathbb{R}^{n \times n}$ is a unique, symmetric, positive-definite solution to the algebraic Riccati equation (ARE)

$$A^T P + P A + Q - P B R^{-1} B^T P = 0. \tag{7}$$

The optimal controller is

$$u = -K^* x \tag{8}$$

where the feedback matrix K^* in (8) can then be determined by

$$K^* = R^{-1} B^T P^*. \tag{9}$$

3.2   Policy Iteration

P^* appears nonlinearly in (7) and thus is difficult to solve directly, especially for high-order systems. To tackle this issue, the author of [21] has proposed a successive approximation algorithm, also known as policy iteration, to approximate P^*:

1. Choose K_0 such that $A - BK_0$ is Hurwitz. Let $k = 0$.
2. Solve P_k from

$$(A - BK_k)^T P_k + P_k(A - BK_k) + Q + K_k^T R K_k = 0. \tag{10}$$

3. Improve control policy by

$$K_{k+1} = R^{-1} B^T P_k. \tag{11}$$

4. Let $k \leftarrow k + 1$, go to Step 2.

It has also been shown in [21] that $\{P_k\}$ and $\{K_k\}$ solved iteratively from (10) and (11) hold the following properties:

(1) $A - BK_k$ is Hurwitz for all k,
(2) $P^* \leq P_{k+1} \leq P_k \leq \cdots \leq P_0$, and
(3) $\lim_{k \to \infty} K_k = K^*$, $\lim_{k \to \infty} P_k = P^*$.

While this policy iteration is useful for solving P^*, it still relies on precise system knowledge. Therefore, it is not well suited for modeling human balance control where the system varies from person to person.

3.3   ADP for Continuous-Time Linear Systems

While the policy iteration relies on full knowledge of the A and B matrices, when A and B are unknown, equivalent iterations can be achieved using online state and input measurements [12]. Define the following notation:

$$P \in \mathbb{R}^{n \times n} \rightarrow \hat{P} \in \mathbb{R}^{\frac{1}{2}n(n+1)}, \ x \in \mathbb{R}^n \rightarrow \bar{x} \in \mathbb{R}^{\frac{1}{2}n(n+1)}$$

where

$$\hat{P} = [p_{11}, \ 2p_{12}, \ \ldots, \ 2p_{1n}, \ p_{22}, \ p_{23}, \ \ldots, \ 2p_{n-1,n}, \ p_{nn}]^T,$$
$$\bar{x} = [x_1^1, \ x_1 x_2, \ \ldots, \ x_1 x_n, \ x_2^2, \ x_2 x_3, \ \ldots, \ x_{n-1} x_n, \ x_n^2]^T$$

Also define the matrices (for a positive integer l) $\delta_{xx} \in \mathbb{R}^{l \times \frac{1}{2}n(n+1)}$, $I_{xx} \in \mathbb{R}^{l \times n^2}$, $I_{xu} \in \mathbb{R}^{l \times mn}$, $\Theta_k \in \mathbb{R}^{\times[\frac{1}{2}n(n+1)+mn]}$, $\Xi_k \in \mathbb{R}^l$ such that

$$\delta_{xx} = \left[\bar{x}(t_1) - \bar{x}(t_0), \bar{x}(t_2) - \bar{x}(t_1), \ldots, \bar{x}(t_l) - \bar{x}(t_{l-1})\right]^T,$$
$$I_{xx} = \left[\int_{t_0}^{t_1} x \otimes x d\tau, \int_{t_1}^{t_2} x \otimes x d\tau, \ldots, \int_{t_{l-1}}^{t_l} x \otimes x d\tau\right]^T,$$
$$I_{xu} = \left[\int_{t_0}^{t_1} x \otimes u d\tau, \int_{t_1}^{t_2} x \otimes u d\tau, \ldots, \int_{t_{l-1}}^{t_l} x \otimes u d\tau\right]^T,$$
$$\Theta_k = [\delta_{xx}, \ -2I_{xx}(I_n \otimes K_k^T R) - 2I_{xu}(I_n \otimes R)],$$
$$\Xi_k = -I_{xx} vec(Q + K_k^T R K_k)$$

where $0 \leq t_0 < t_1 < \ldots < t_l$ and the vec operator denotes the vectorization of an $m \times n$ matrix into an $mn \times 1$ column vector by concatenating the column vectors of the matrix.

If Θ_k has full column rank, then P_k and K_{k+1} can be updated from

$$\begin{bmatrix} \hat{P}_k \\ vec(K_{k+1}) \end{bmatrix} = (\Theta_k^T \Theta_k)^{-1} \Theta_k^T \Xi_k \tag{12}$$

The off-policy policy-iteration-based ADP algorithm is given below [12]:

0. Choose K_0 as a stabilizing feedback matrix. Set $k = 0$.
1. Apply $u = -K_k x + e$ as the input, where e is the exploration noise. Compute δ_{xx}, I_{xx}, and I_{xu} until Θ_k has full column rank.
2. Solve P_k and K_{k+1} from (12).
3. Let $k \leftarrow k + 1$ and repeat Steps 1 and 2 until $||P_k - P_{k-1}|| \leq \epsilon$ for $k \geq 1$ where the constant $\epsilon > 0$ is a predefined small threshold.
4. Use $u = -K_k x$ as the approximated optimal control policy.

In the following section, we will study the human balance control behavior using the above ADP algorithm.

4   Results and Discussion

In this section, we conduce a series of simulation studies to investigate how the CNS learns and controls the balance of the human body. The simulation was separated into two stages: learning and implementation. In the learning stage, several trials were performed on the linearized model (5) to obtain state data x with the following exploration noise on the input

$$e = 10 \sum_{i=1}^{3} sin(\omega_i t) \tag{13}$$

where ω_i for $i = 1, 2, 3$ were equally spaced frequencies selected in the range [0, 200]. After each trial, the feedback matrix was updated for the next trial. After several trials, the final learned feedback matrix was implemented on the input of the nonlinear model (4).

4.1   Selection of Parameters

Human model parameters from [20] were used for the simulation. The Q and R matrices are weighting matrices that limit the variations in the state and input respectively, and they can be chosen to model the human CNS. For this case, R was set to the identity matrix, and Q was selected through observation of the 2-D simulation. Q could however be selected to penalize movements from the center of mass or upright stance [9].

Selection of an initial stabilizing K_0 is difficult for an inverted pendulum as the dynamics are inherently unstable. However, it is assumed that the CNS in healthy adults has some estimate of the control policy. Such initial estimation can be obtained in several ways. For example, recent work in value iteration-based ADP has provided a framework for modeling sensorimotor learning in an initially unstable environment [11,18]. For simplicity, in our simulation, K_0 was preselected such that $A - BK_0$ was stable.

Figure 1 shows the trajectories for ten trials. One can see that after only a few learning trials with noise (13), the iterated solution P_k and feedback matrix K_k converge to their optimal LQR values of P^* and K^*. It is well-known that the policy iteration algorithm converges quadratically [21]. This models the learning process of the human CNS. After each trial, the control policy updates and gets closer to optimal.

Figure 2 shows the comparison of both the linear and nonlinear trajectories implemented with feedback matrix

$$K_{10} = \begin{bmatrix} 738.4050 & 81.0269 & 100.6732 & 273.6253 & 138.1745 & 53.1615 \\ 460.9864 & 486.1430 & -187.4425 & 178.7882 & 109.9136 & 32.7221 \\ 272.5167 & 379.4538 & 527.5426 & 116.6796 & 93.3560 & 69.3561 \end{bmatrix}$$

which is reasonably close to the optimal feedback matrix

$$K^* = \begin{bmatrix} 738.4050 & 81.0270 & 100.6732 & 273.6253 & 138.1745 & 53.1615 \\ 460.9864 & 486.1430 & -187.4424 & 178.7882 & 109.9135 & 32.7221 \\ 272.5167 & 379.4538 & 527.5426 & 116.6796 & 93.3560 & 69.3561 \end{bmatrix}$$

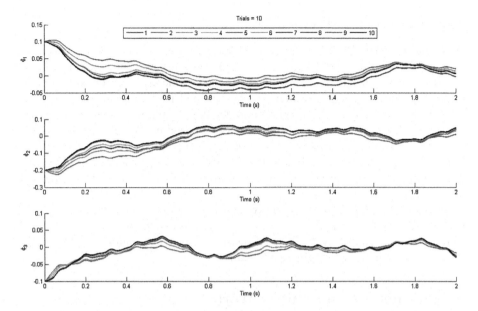

Fig. 1. Trajectories of segment angles with added noise (13) for ten trials.

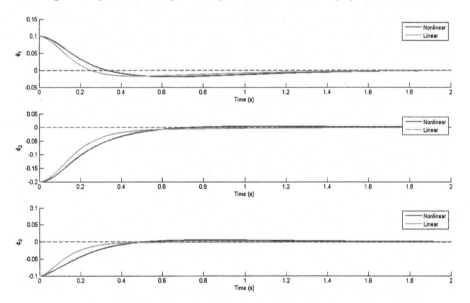

Fig. 2. Comparison of linear and nonlinear trajectories with input $u = -K_{10}x$.

from the LQR solution. For an initial starting position close to the upright equilibrium position, both models' trajectories are very similar. For this localized region, it is possible to study human balance with these methods.

4.2   Comparison to Previous Methods

While other methods [7–10,20] use similar techniques to model human balance using optimal control and state estimation, they are typically not flexible in terms of personalized dynamics. These models typically counter some uncertainty by combining state estimation with optimal feedback control [8,20], but even these models make assumptions about unknown parameters and dynamics. Even though an optimal control policy can be formed using statistically average human parameters, each person responds to perturbations from upright stance in slightly different ways. Using ADP, the concepts and biological implications of previous methods can be extended to more personalized cases where physical parameters are not easily known.

Furthermore, the ADP method provides a new model of the human learning process. Not only can it be used in the same manner as previous methods to study dynamics and possibly diagnose medical disorders, it can be used to study how humans adapt to changing environment.

5   Conclusions and Future Work

This work presents only a basis for studying human balance with ADP. While it has shown that balance can be studied in a small localized region about upright stance, the methods presented here are simplified to the deterministic case for simulation. Certain issues with ADP, such as robustness, have already been addressed by previous work [18,19] and should be applied to this work. Further work must also be done in experimentation. The Sensory Organization Test (SOT) is one experimental procedure that has been used to study balance in varying sensory environments [8,9,23] and can also be used here. The simulation is useful for presenting ADP as a feasible method of study. However, it is best studied using real data. Compared to previous methods, ADP for human balance is an interesting computational learning mechanism that could be used to study the CNS and medical disorders such as Parkinson's disease.

References

1. Horak, F., Nutt, J., Nashner, L.: Postural inflexibility in parkinsonian subjects. J. Neurol. Sci. **111**(1), 46–58 (1992)
2. Vaugoyeau, M., Hakam, H., Azulay, J.P.: Proprioceptive impairment and postural orientation control in Parkinson's disease. Hum. Mov. Sci. **30**(2), 405–414 (2011)
3. Adkin, A.L., Frank, J.S., Jog, M.S.: Fear of falling and postural control in Parkinson's disease. Mov. Disord. **18**(5), 496–502 (2003)
4. Alexander, N.B.: Postural control in older adults. J. Am. Geriatr. Soc. **42**(1), 93–108 (1994)
5. Laschi, C., Johansson, R.S.: Bio-inspired sensory-motor coordination. Auton. Rob. **25**(1), 1–2 (2008)
6. Nashner, L.M., McCollum, G.: The organization of human postural movements: a formal basis and experimental synthesis. Behav. Brain Sci. **8**(01), 135 (1985)

7. Golliday, C., Hemami, H.: Postural stability of the two-degree-of-freedom biped by general linear feedback. IEEE Trans. Autom. Control **21**(1), 74–79 (1976)
8. Kuo, A.D.: An optimal control model of human balance: can it provide theoretical insight to neural control of movement? In: Proceedings of the 1997 American Control Conference, vol. 5, pp. 2856–2860 (1997)
9. Kuo, A.D.: An optimal state estimation model of sensory integration in human postural balance. J. Neural Eng. **2**(3), S235–S249 (2005)
10. Winter, D.A.: Human balance and posture control during standing and walking. Gait Posture **3**(4), 193–214 (1995)
11. Bian, T., Jiang, Z.P.: Value iteration and adaptive dynamic programming for data-driven adaptive optimal control design. Automatica **71**, 348–360 (2016)
12. Jiang, Y., Jiang, Z.P.: Computational adaptive optimal control for continuous-time linear systems with completely unknown dynamics. Automatica **48**(10), 2699–2704 (2012)
13. Jiang, Y., Jiang, Z.P.: Robust Adaptive Dynamic Programming. Wiley IEEE Press, Hoboken (2017)
14. Jiang, Z.P., Jiang, Y.: Robust adaptive dynamic programming for linear and nonlinear systems: an overview. Eur. J. Control **19**(5), 417–425 (2013)
15. Lewis, F.L., Vrabie, D., Vamvoudakis, K.G.: Reinforcement learning and feedback control: using natural decision methods to design optimal adaptive controllers. IEEE Control Syst. **32**(6), 76–105 (2012)
16. Vamvoudakis, K.G.: Non-zero sum nash q-learning for unknown deterministic continuous-time linear systems. Automatica **61**, 274–281 (2015)
17. Vrabie, D., Vamvoudakis, K.G., Lewis, F.L.: Optimal adaptive control & differential games by reinforcement learning principles. Institution of Electrical Engineers (2013)
18. Bian, T., Jiang, Z.P.: Model-free robust optimal feedback mechanisms of biological motor control. In: 2016 12th World Congress on Intelligent Control and Automation (WCICA), June 2016
19. Jiang, Y., Jiang, Z.P.: Adaptive dynamic programming as a theory of sensorimotor control. Biol. Cybern. **108**(4), 459–473 (2014)
20. van der Kooij, H., Jacobs, R., Koopman, B., Grootenboer, H.: A multisensory integration model of human stance control. Biol. Cybern. **80**(5), 299–308 (1999)
21. Kleinman, D.: On an iterative technique for Riccati equation computations. IEEE Trans. Autom. Control **13**(1), 114–115 (1968)
22. Sutton, R.S., Barto, A.G.: Introduction to Reinforcement Learning, 1st edn. MIT Press, Cambridge (1998)
23. Peterka, R.J., Black, F.O.: Age-related changes in human posture control: sensory organization tests (1989)

Dynamic Multi Objective Particle Swarm Optimization Based on a New Environment Change Detection Strategy

Ahlem Aboud[(⊠)], Raja Fdhila, and Adel M. Alimi

REGIM-Lab.: REsearch Groups on Intelligent Machines, National Engineering School of Sfax (ENIS), University of Sfax, BP 1173, 3038 Sfax, Tunisia
{aboud.ahlem.tn,raja.fdhila,adel.alimi}@ieee.org

Abstract. The dynamic of real-world optimization problems raises new challenges to the traditional particle swarm optimization (PSO). Responding to these challenges, the dynamic optimization has received considerable attention over the past decade. This paper introduces a new dynamic multi-objective optimization based particle swarm optimization (Dynamic-MOPSO). The main idea of this paper is to solve such dynamic problem based on a new environment change detection strategy using the advantage of the particle swarm optimization. In this way, our approach has been developed not just to obtain the optimal solution, but also to have a capability to detect the environment changes. Thereby, Dynamic-MOPSO ensures the balance between the exploration and the exploitation in dynamic research space. Our approach is tested through the most popularized dynamic benchmark's functions to evaluate its performance as a good method.

Keywords: Dynamic optimization · Dynamic multi-objective problems · Particle swarms optimization · Dynamic environment · Time varying parameters

1 Introduction

In previous years optimization problems are limited below static research space related to various applications [1–7] and many other works are based on meta-heuristics such as the swarm intelligence approach, including particle swarm optimization (PSO), ant colony optimization (ACO) and bee-inspired methods [8–16].The domain of optimization presents a big challenge to resolve single or multi-objective problems in order to maximize or minimize the fitness function. The field related to evolutionary multi objective optimization has a several amounts of research interest in many real-world applications, to resolve relatively two objectives or more that has in conflict with one another. Due to multi-objectivity, the goal of solving Multi-objective Problems (MOPs) isn't always discovering one optimal solution but a set of solutions. Although dynamic and multi-objective optimization have separately received an immense interest. In the literature, dynamism tasks are related to the objective function, constraints and the parameters of a predefined problem that change over the time. Dynamic multi-objective problems pose big challenges associated with the evolutionary computation approaches [17–20].

© Springer International Publishing AG 2017
D. Liu et al. (Eds.): ICONIP 2017, Part IV, LNCS 10637, pp. 258–268, 2017.
https://doi.org/10.1007/978-3-319-70093-9_27

Hence, the PSO methods are proven as a good technique to solve a single and multi-objective problem in a static environment. Adapting MOPSO to resolve such as the problem is not yet treated; this is why our paper is presented.

The remaining work is delineated as follows: Sect. 2 presents an overview of the elementary concepts of the dynamic optimization domain. Section 3 describes the trends of dynamic optimization methods followed by our suggested approach called the Dynamic-MOPSO in Sect. 4. A comparative study is provided to be the topic by Sect. 5 followed by discussion part. The paper outline with a conclusion and some suggested ideas for future work in Sect. 6.

2 Overview of Dynamic Multi Objective Optimization Problem

The process of dynamic multi-objective optimization problems (DMOOP) [21, 22] is totally different from the static MOOP [12, 13]. In most cases, we notice that the new definition of optimality needs is to adjust a set of optimal solutions at each instance of time. A dynamic optimization problem can be defined as a dynamic problem f_t such as presented by the mathematical presentation (see Eq. 1: example of the minimization problem), which needs an optimization approach D, at a given optimization period $\left[t^{begin}, t^{end}\right]$, f_t is termed a dynamic optimization problem in the research period $\left[t^{begin}, t^{end}\right]$ if during this period the underlying objective landscape that D helps to represent f_t changes and D has to reply to this change by providing new optimal solutions.

$$\text{Min } F(x, k(t)), \; x = (x_1 \ldots x_n), \; k(t) = (k_1(t), \ldots, k_{nm}(t))$$
$$\text{Where} : g_i(x, t) \leq 0, \; i = 1 \ldots n_g \text{ and } h_j(x, t) \leq 0, \; j = n_g + 1 \ldots n_h \qquad (1)$$
$$x \in [x_{min}, \; x_{max}]$$

The main goal of the dynamic optimization approach can be explained as the problem of locating a vector of decision variables $x*(i, t)$, that is presented to be a Pareto optimal solution based on the Pareto dominance relation between solutions (see Eq. 2), that satisfies an absolute set called the Pareto optimal set of solutions at instance, t, denoted as POS (t)* (see Eq. 3) and improve a function vector whose dynamic values represent the best solutions that change over the time.

$$f(j, t) \prec f(i, t) * \backslash f(j, t) \in F^M \qquad (2)$$

$$POS(t)* = \left\{ x_i^* | \nexists f(j, t) \prec f(\chi_i^*, t)*, f(\chi_j, t) \in F^M \right\} \qquad (3)$$

The generated Pareto Optimal Front at time t, denoted as POF(t)$*$ is the set of the best solutions with respect to the objective space at time step t, so when solving DMOOPs the purpose is to detect the change of the best optimal front at every time instance such as defined in Eq. 4:

$$POF(t)^* = \{f(t) = f_1(\chi^*, t), f_2(\chi^*, t), \ldots, f_{n_m}(\chi^*, t)\}, \forall \chi^* \in POF^*(t) \qquad (4)$$

3   Trends of Dynamic Optimization Approaches

In previous works, Farina *et al.* [18] suggested four types to categorize the DMOOPs which outlined in table (see Table 1). Many works are done, including various types and one of the active research areas in the last few years is Evolutionary Dynamic Optimization (EDO) domain [23, 24]. Many researchers have been highlighted the intention in evolutionary computation (EC).

Table 1. Categories of DMOOPs.

	Pareto optimal Set (POS)	
	No change	Change
Pareto optimal front (POF)		
No change	Type IV	Type I
Change	Type III	Type II

Many real-world problems are time-dependent parameters that involve optimization in a dynamic environment [25, 26]. To deal with the various changes in the environment, many EDO methods take into consideration as a reactive strategy. In this context, we are facing two conditions: either the algorithm has to define a methodology to find the change in the environment, or made it well known before the optimization process. To detect changes in the environment, we have typically followed one of the particular consecutive approaches: the first one is detecting modification changes by re-evaluated detectors or, the second, detecting changes based on a set of the state behaviors defined by the algorithm itself.

Many methods are related to the first strategy that introduces the change detection as a process of re-evaluating frequent existing solutions. In order to follow the previous context, function values and their feasibility must be a part of the swarm. Some existing optimization methods manage separately the detectors in the search population to ensure flexibility to maintain a high convergence during the run time and to ensure the exploitation in the research area. As a result, many diversity-based approaches is implemented such as the Dynamic Non-dominated Sorting Genetic Algorithm II (D-NSGA-II) [27], the Dynamic Constrained NSGA-II (DC-NSGA-II) [28] the individual Diversity Multi-objective Optimization EA (IDMOEA) and others [29]. One of re-evaluating detectors advantage is ensuring robustness in the time-varying environment. To maximize the performance of the algorithm, an important number of detectors are used to entail additional function evaluations. As a consequence, the used methodology requires to become informed of the most optimal number of used detectors.

To detect changes based on the behaviors of the algorithm, researchers must define a monitoring method to calculate the average of the best optimal solution founded over the time. The benefits of this method, are there are no detectors and does not require any additional function evaluations. Because no detector is used, but there may be no support that assures changes are detected and the algorithm response unnecessarily when no change occurs [30]. Many others approach is treated to predict change parameters in the environment [31].

4 The Proposed Approach of the Dynamic-MOPSO Based on a New Environment Change Detection Strategy

The Dynamic Multi-objective Particle Swarm Optimization denoted by the Dynamic-MOPSO is developed based on the advantage of the fashionable particle swarm optimization technique that was in 1995 developed by Kennedy and Eberhart [12], when every particle in the population represents a candidate solution and characterized by specific parameters to be optimized in the quest research process by way of updating their position (see Eq. 5) and velocity (see Eq. 6) at each generation. PSO is an ideal evolutionary computation approach which can be able to resolve single and multi-objective problems for static search spaces.

$$X(k + 1) = X(k) + V(k + 1) \tag{5}$$

$$V(k + 1) = w * v(k) + c_1 * \text{rand}() * (p_{id}(k) - X(k)) + c_2 * \text{rand}() * (p_{gd}(k) - X(k)) \tag{6}$$

As far as, the use of the standard MOPSO as an optimization method for dynamic problem has many negative consequences for the constraints, number of variables or their domain and the objective function of the defined problem and it can cause the problem of stagnation in local optima and many solutions can disappear over the time and cause the problem of lack of diversity and convergence after each change. As consequences, the MOPSO cannot be carried through to dynamic environment without any modifications to keep swarm diversity. In order to resolve a dynamic problems, a specific approach will be able to identify when a change in the dynamic research space has taken place after which react to such change to track the most beneficial set of solutions and to adapt in the new modified environment and this why our approach is developed. Our motivation is presented through the architecture in figure (see Fig. 1) that is investigated in type I of DMOOPs, when the problem has a change in the optimal decision variables $x_i(t)^*$ when the optimal objective function does not change.

Two major problems should be resolved to handle the changed parameters: first, the way to discover that a change has occurred, second the way to respond or react appropriately to the change. Our proposed approach started with the process of MOPSO, considering the example of Dynamic Multi-objective Problem (DMOP) such as presented in the above mathematical presentation (see Eq. 1). Our proposed technique is developed to resolve a dynamic multi objective problem with dynamic parameters. In this paper, we specifically consider the environmental change that may have an effect in the parameters of the problem after each interval of time the problem which is defined as follows: $t = \frac{1}{n_t} \frac{\tau}{\tau_t}$; where n_t, τ_t and τ represent the severity, the frequency of change, and the iteration counter, respectively.

The Dynamic-MOPSO presents two main steps which are the dynamic detection and the reaction strategy that will be detailed step by step, then after the evaluation step of the fitness function F (i) of each particle p in the swarm S, our proposed approach presents an environment change detection strategy that can be able to detect change based on the process to re-evaluate the set optimal solutions $POF(t)^*$ that evaluates each candidate solutions x after each interval of time τ_t.

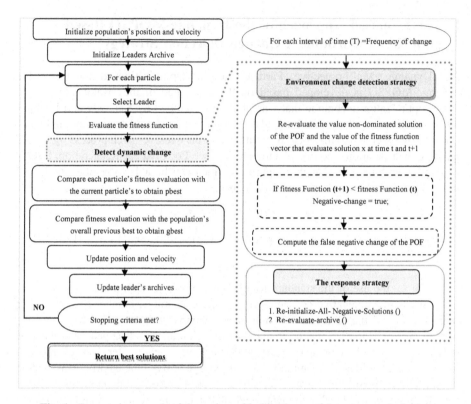

Fig. 1. Proposed approach of dynamic multi objective particle swarm optimization

Environmental change detection strategy: is the first step that aims to detect the change in the research space and we preserve the set of non-dominated solutions denoted by $POF(t)^*$ as a detector of the observed change that is caused by the influence of the parameters change. As a result, the detection of dynamic parameters after each interval of time presented by the iteration counter, in our case the interval of time is defined to $(t = 10)$ this parameter presents the speed of change which is severe when the value is small, moderate if frequency and severity are equal, and slight environmental changes when the value is very high. In our case, the frequency of change is defined as the value of severity to ensure a moderate environmental change.

The response change strategy: are the second and the important step that is elaborated to maintain convergence and diversity in the research area. After change detection steps, a tested process is implemented to verify the number of false negative changes of optimal solutions, after each generation cycle, we need to compute the number of individuals in the population which has a negative change in the value of the fitness function $F(i)$. In our algorithm the reactive strategy is defined by the re-initialization of all the solutions (particles) that presents a negative change in the non-dominated solution, to ensure that this dominated solution cannot lead particles to trap in local optima. Another primordial step is to re-evaluate the archive in order to update the best optimal solution at each time instance until the end of the optimization process.

5   Experiments and Results

Our experimental studies based on the benchmark functions of the FDA, DIMP and dMOP suites functions, which are classified into the type I of dynamic multi-objective optimization problem [18].

5.1   Parameters Setting

The parameterization used for testing algorithms details in Table 2:

Table 2. Parameters setting

Parameters	
Common parameters of dynamic-MOPSO and OMOPSO	Swarm size = 200 Archive size = 100 Independent runs = 30 Mutation probability = 1.0/number of problems' variable Acceleration Coefficients (c1, c2) = Rand (1.5, 2.0) Inertia weight (w) = Rand(0.1, 0.5) Max iteration = 200
Parameters of NSGAII	Swarm size = 100 Max iteration = 25000 Crossover Probability = 0.9 Mutation Distribution Index = 20 Crossover Distribution Index = 20

5.2   Performance Metric

Each dynamic optimization process needs to be evaluated adopting a quality indicator to maintain diversity using the spread (Δ) as a performance metrics, the generational distance (GD) to measure the convergence and the hyper-volume (HV) to measure both of them.

- **The GD:** used to measure the convergence of the approximated best solutions towards the true POF. The GD is defined in the Eq. 7:

$$GD = \frac{\sqrt{\sum_{i=1}^{n_{POF*}} d_i^2}}{n_{POF*}} \tag{7}$$

- **The Spread** (Δ): The metric Δ measures the diversity between consecutive solutions inside the Pareto front PF. Mathematically, Δ is presented in Eq. 8:

$$\Delta = \sum_{i=1}^{|POF|} \frac{\left|dist_i - \overline{dist}\right|}{|POF|} \tag{8}$$

- **The HV or S-metric:** computes the scale of the location that is dominated by a set of non-dominated solutions, based on a reference vector. Mathematically, HV is defined in Eq. 9:

$$HV = \bigcup_i vol_i | i \in POF \tag{9}$$

5.3   Results and Discussion

The present section is yielding to analyze the results of the experiments and to evaluate the effect of environment change parameters defined by means of the severity (nt) and the frequency (τt) of change was both set to 10. We started out by producing the qualitative results, such as presented in figure (see Fig. 2) which gives the shape of the Pareto optimal fronts and shows that adapting MOPSO to dynamic environment without take in consideration of the parameters change cause a problem that all solutions converge to the local optima. So, the dynamic-MOPSO cover this problem, the figure shows that our method are more effectively and correctly to converge to the true POF and keep diversity in dynamic research space.

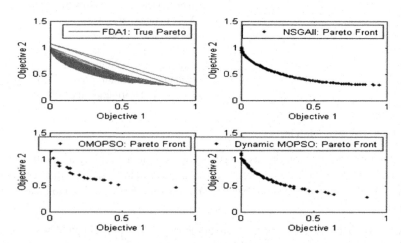

Fig. 2. The generated Pareto fronts of NSGAII, OMOPSO and dynamic-MOPSO algorithms for FDA1 function.

To more explain the performance measure of our approach, the quantitative results are generated applying quality indicator such as the GD, the Spread and the HV respectively. From the results shows in Table 3, we can be concluded that our approach which called the Dynamic-MOPSO obtains the leading results that are highlighted in bold face for the tested FDA1 and dMOP3 problems compared with the standard OMOPSO and the NSGAII algorithms.

Table 3. Quantitative results of tested approach OMOPSO, NSGAII and dynamic-MOPSO

DMOOPs	Quality indicators	Omopso	Nsgaii	Dynamic-mopso
FDA1	GD	2.68e	2.29e	**1.32e**
	Δ	7.21e	7.27e	**3.94e**
	HV	5.57e	5.81e	**7.74e**
DIMP2	GD	4.14e	**1.19e**	3.13e
	Δ	1.76e	1.60e	**1.19e**
	HV	**5.36e**	1.51e	3.55e
dMOP3	GD	4.69e	2.95e	**1.24e**
	Δ	7.68e	8.83e	**5.92e**
	HV	3.57e	1.87e	**5.24e**

The following figures (see Figs. 3 and 4) present the performance of HV over FDA1 and dMOP3 functions respectively. These two figures show that our approach can achieve a good trade-off between convergence and diversity for solving dynamic multi-objective problems.

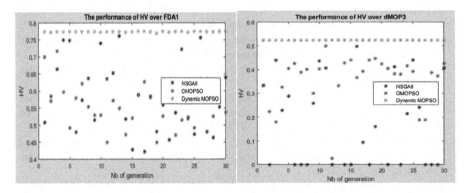

Fig. 3. The performance of HV over FDA1. **Fig. 4.** The performance of HV over dMOP3.

At the conclusion of the experimental studies, the OMOPSO and NSGAII algorithms are an important method which might be used in previous works, however, cannot be applied to dynamic environments without any changes for that reason the Dynamic-MOPSO system is presented as a new approach to deal with this problem and the previous results proved that our method can keep exploration and exploitation during the optimization process in a dynamic environment.

6 Conclusions and Future Research Directions

As a conclusion, the Dynamic-MOPSO approach is implemented to solve the problem categorized into the first type of dynamic multi-objective problems. So, the traditional OMOPSO is a simple and easy algorithm, but cannot have the ability to resolve DMOOP without any modification to keep swarm exploration and exploitation in a time varying environment. The no-adaptation of MOPSO for dynamic change can cause lack of convergence and diversity in the research space. But the distinction is that the new Dynamic-MOPSO approach overcomes this limitation and achieve the goal of high convergence precision and diversity by combining the simplicity of MOPSO and their efficiency to optimize dynamic problems. As a future work, we have to enhance the proposed approach that will be used as an efficient global search technique to address feature selection tasks to optimize the online learning process.

Acknowledgements. The research leading to these results has received funding from the Ministry of Higher Education and Scientific Research of Tunisia under the grant agreement number LR11ES48.

References

1. Ben Moussa, S., Zahour, A., Benabdelhafid, A., Alimi, M.A.: New features using fractal multi-dimensions for generalized Arabic font recognition. Pattern Recogn. Lett. **31**(5), 361–371 (2010)
2. Bezine, H., Alimi, M.A., Derbel, N.: Handwriting trajectory movements controlled by a bêta-elliptic model. In: 7th IEEE International Conference on Document Analysis and Recognition, pp. 1228–1232. IEEE, Edinburgh, UK (2003)
3. Alimi, M.A.: Evolutionary computation for the recognition of on-line cursive handwriting. IETE J. Res. **48**(5), 385–396 (2002)
4. Boubaker, H., Kherallah, M., Alimi, M.A.: New algorithm of straight or curved baseline detection for short Arabic handwritten writing. In: 10th International Conference on Document Analysis and Recognition, pp. 778–782. IEEE, Barcelona, Spain (2009)
5. Slimane, F., Kanoun, S., Hennebert, J., Alimi, M.A., Ingold, R.: A study on font-family and font-size recognition applied to Arabic word images at ultra-low resolution. Pattern Recogn. Lett. **34**(2), 209–218 (2013)
6. Elbaati, A., Boubaker, H., Kherallah, M., Alimi, M.A., Ennaji, A., Abed, H.E.: Arabic handwriting recognition using restored stroke chronology. In: 10th International Conference on Document Analysis and Recognition, pp. 411–415. IEEE, Barcelona, Spain (2009)
7. Baccour, L., Alimi, M.A., John, R.I.: Similarity measures for intuitionistic fuzzy sets: state of the art. J. Intell. Fuzzy Syst. **24**(1), 37–49 (2013)
8. Fdhila, R., Hamdani, T.M., Alimi, M.A.: Distributed MOPSO with a new population subdivision technique for the feature selection. In: The 5th International Symposium Computational Intelligence and Intelligent Informatics, pp. 81–86. IEEE, Floriana, Malta (2011)
9. Fdhila, R., Hamdani, T.M., Alimi, M.A.: A multi objective particles swarm optimization algorithm for solving the routing pico-satellites problem. In: Systems, Man, and Cybernetics, pp. 1402–1407. IEEE, Seoul, South Korea (2012)

10. Fdhila, R., Walha, C., Hamdani, T.M., Alimi, M.A.: Hierarchical design for distributed MOPSO using sub-swarms based on a population pareto fronts analysis for the grasp planning problem. In: The 13th International Conference on Hybrid Intelligent Systems, pp. 203–208. IEEE, Gammarth, Tunisia (2013)
11. Chouikhi, N., Fdhila, R., Ammar, B., Rokbani, N., Alimi, M.A.: Single-and multi-objective particle swarm optimization of reservoir structure in echo state network. In: The International Joint Conference on Neural Networks, pp. 440–447. IEEE, Vancouver, BC, Canada (2016)
12. Eberhart, R., Kennedy, J.: Particle swarm optimization. In: Proceedings of the 1995 IEEE International Conference on Neural Networks, pp. 1942–1948. IEEE Service Center, Piscataway, New Jersey (1995)
13. Fdhila, R., Hamdani. T., Alimi. M.A.: A new distributed approach for MOPSO based on population Pareto fronts analysis and Dynamic. In: Systems Man and Cybernetics (SMC), pp. 947–954. IEEE, Istanbul (2010)
14. Fdhila, R., Hamdani, T.M., Alimi, M.A.: A new hierarchical approach for MOPSO based on dynamic subdivision of the population using Pareto fronts. In: IEEE International Conference on Systems, Man, and Cybernetics, pp. 947–954. IEEE, Istanbul, Turkey (2010)
15. Fdhila, R., Hamdani, T.M., Alimi, M.A.: Population-based distribution of MOPSO with continuous flying pareto fronts particles. J. Inf. Process. Syst. (2016, accepted paper)
16. Fdhila, R., Ouarda, W., Alimi, M.A., Abraham, A.: A new scheme for face recognition system using a new 2-level parallelized hierarchical multi objective particle swarm optimization algorithm. J. Inf. Assur. Secur. 11(6), 385–394 (2016)
17. Helbig, M., Engelbrecht, A.P.: Dynamic multi-objective optimization using PSO. In: Alba, E., Nakib, A., Siarry, P. (eds.) Metaheuristics for Dynamic Optimization. SCI, vol. 433, pp. 147–188. Springer, Heidelberg (2013). doi:10.1007/978-3-642-30665-5_8
18. Farina, M., Deb, K., Amato, P.: Dynamic multiobjective optimization problems: test cases, approximations, and applications. In: Transactions on Evolutionary Computation, pp. 425–442. IEEE, USA (2004)
19. Fdhila, R., Hamdani, T.M., Alimi, M.A.: Optimization algorithms, benchmarks and performance measures: from static to dynamic environment. In: The 15th International Conference on Intelligent Systems Design and Applications, pp. 597–603. IEEE, Marrakech, Morocco (2015)
20. Aboud, A., Fdhila, R., Alimi, M.A.: MOPSO for dynamic feature selection problem based big data fusion. In: the IEEE International Conference on Systems, Man, and Cybernetics, pp. 003918–003923. IEEE, Budapest, Hungary (2016)
21. Fdhila, R., Elloumi, W., Hamdani, T.M.: Distributed MOPSO with dynamic Pareto front driven population analysis for TSP problem. In: the 6th International Conference Soft Computing and Pattern Recognition, pp. 294–299. IEEE, Tunis, Tunisia (2014)
22. Hu, X., Eberhart, R.: Tracking dynamic systems with PSO: where's the cheese? In: Proceedings of the workshop on particle swarm optimization. Purdue School of Engineering and Technology. IEEE, Indianapolis (2001)
23. Du, W., Li, B.: Multi-strategy ensemble particle swarm optimization for dynamic optimization. In: Information Sciences, pp. 3096–3109. Elsevier, Huangshan Road, Hefei, Anhui, China (2008)
24. Branke, J., Kaussler, T., Smidt, C., Schmeck, H.: A multi-population approach to dynamic optimization problems. In: Parmee, I.C. (ed.) Evolutionary Design and Manufacture. Springer, London (2000). doi:10.1007/978-1-4471-0519-0_24
25. Dhahri, H., Alimi, M.A.: The modified differential evolution and the RBF (MDE-RBF) neural network for time series prediction. In: IEEE International Conference on Neural Networks - Conference Proceedings, pp. 2938–2943. IEEE, Vancouver, BC, Canada (2006)

26. Bouaziz, S., Dhahri, H., Alimi, M.A., Abraham, A.: A hybrid learning algorithm for evolving flexible beta basis function neural tree model. Neurocomputing **117**, 107–117 (2013)

27. Deb, K., Rao, N.U.B., Karthik, S.: Dynamic multi-objective optimization and decision-making using modified NSGA-II: a case study on hydro-thermal power scheduling. In: Obayashi, S., Deb, K., Poloni, C., Hiroyasu, T., Murata, T. (eds.) EMO 2007. LNCS, vol. 4403, pp. 803–817. Springer, Heidelberg (2007). doi:10.1007/978-3-540-70928-2_60

28. Chen, H., Li, M., Chen, X.: Using diversity as an additional-objective in dynamic multiobjective optimization algorithms. In: Second International Symposium on Electronic Commerce and Security, pp. 484–487. IEEE, Nanchang City, China (2009)

29. Hatzakis, I., Wallace, D.: Dynamic multi-objective optimization with evolutionary algorithms: a forward-looking approach. In: Proceedings of the Genetic and Evolutionary Computation Conference, pp. 1201–1208. ACM, Seattle, Washington, USA (2006)

30. Hu, X., Eberhart, R.: Adaptive particle swarm optimisation: detection and response to dynamic systems. In: IEEE Congress on Evolutionary Computation, pp. 1666–1670. IEEE, Honolulu, HI, USA, USA (2002)

31. Zhou, A., Jin, Y., Zhang, Q.: A population prediction strategy for evolutionary dynamic multiobjective optimization. Trans. Cybern. **44**(1), 40–53 (2014)

Multi Objective Particle Swarm Optimization Based Cooperative Agents with Automated Negotiation

Najwa Kouka$^{(\boxtimes)}$, Raja Fdhila, and Adel M. Alimi

REGIM-Laboratory: REsearch Groups in Intelligent Machines,
National Engineering School of Sfax (ENIS), University of Sfax,
BP 1173, 3038 Sfax, Tunisia
{najwa.kouka.tn, raja.fdhila, adel.alimi}@ieee.org

Abstract. This paper investigates a new hybridization of multi-objective particle swarm optimization (MOPSO) and cooperative agents (MOPSO-CA) to handle the problem of stagnation encounters in MOPSO, which leads solutions to trap in local optima. The proposed approach involves a new distribution strategy based on the idea of having a set of a sub-population, each of which is processed by one agent. The number of the sub-population and agents are adjusted dynamically through the Pareto ranking. This method allocates a dynamic number of sub-population as required to improve diversity in the search space. Additionally, agents are used for better management for the exploitation within a sub-population, and for exploration among sub-populations. Furthermore, we investigate the automated negotiation within agents in order to share the best knowledge. To validate our approach, several benchmarks are performed. The results show that the introduced variant ensures the trade-off between the exploitation and exploration with respect to the comparative algorithms.

Keywords: Multi objective optimization problems · Particle swarm optimization · Multi agent system · Distributed architecture · Automated negotiation

1 Introduction

Optimization problems have received appreciable attention over the past decade and presented as an active research field that is encountered in various fields of technology such as image processing [1], path planning [2] and handwriting recognition [3–8].

The optimization can be incorporated into other intelligent tools of soft computing such as the neural network [9, 10] and the fuzzy system [11] to produce better and faster result. In fact, Swarm intelligence (SI) is considered as an adaptable concept for the optimization problem. One of the most dominant algorithms in SI is a particle swarm optimization (PSO). Although PSO has been widely used for solving many well-known numerical test problems, but it suffers from the premature convergence. Thereby, several strategies have been developed responding to this limitation, such as the distributed evolutionary (DE) [12, 13]. Hence, with the DE, a parallel optimization process can be formed which offers the ability to resolve a high-dimensional problem. Frequency, the

© Springer International Publishing AG 2017
D. Liu et al. (Eds.): ICONIP 2017, Part IV, LNCS 10637, pp. 269–278, 2017.
https://doi.org/10.1007/978-3-319-70093-9_28

DE presented at the population level, in which the population is distributed within the search space. The DE increase the diversity of solutions, thereby solve the premature convergence. Due to the several issues that address the distributed sub-populations, such as the communications protocol, it becomes required to endow a novel system with the capability to communicate, cooperate and reach agreements within the different sub-populations. These trends have led to the incorporation of Multi-Agent System (MAS) [14] as a distributed model. MAS allows building a distributed PSO with greater ease and reliability. In this work, we investigate a new distributed MOPSO based cooperative agents (MOPSO-CA) to optimize MOP. The MAS is advantageously used to elaborate a new variant of distributed MOPSO. Additionally, applying the automated negotiation [15], in order to share the best knowledge among sub-populations. In this way, the good information obtained by each sub-population is exchanged among the sub-populations; thereby the diversity of the population is increased simultaneously.

The organization of the remainder of this paper is as follows: Then, we introduce the main concepts of our approach. We present the related work in Sect. 3. Section 4 details the purpose of our approach. The experimental result is discussed in Sect. 5. Finally, the conclusion and future work are then summarized in Sect. 6.

2 Theoretical Foundation

In this section, we briefly present the main concept that will be employed throughout this article. First, we need to define a multi-objective optimization problem (MOP) and its basic concept, then the MOPSO.

2.1 Introduction to Multi Objective Optimization Problem

A MOP has a number of objective functions, which are to be minimized or maximized simultaneously. Those objectives are often immeasurable and conflicting with each other. MOP typically contains a set of constraints, that any feasible solution must satisfy, including the set of the optimal solution [16]. Subsequently, MOP can be written mathematically as follow:

$$\left\{ \begin{array}{ll} min/\max(f_i) & i = 1, 2, \ldots, k \\ g_j(x) \geq 0 & j = 1, 2, \ldots, J \\ h_p(x) \geq 0 & p = 1, 2, \ldots, H \\ x_i^l \leq x_i \leq x_i^u & i = 1, 2, \ldots, n \end{array} \right\} \tag{1}$$

In order to define the concept of optimization, we introduce a few useful terminologies:

Dominance relationship. A solution x dominates solution y, if x is no worse than y in all objectives, and x is strictly better than y in at least one objective.

Pareto optimal. Is a non-dominated solution, which are equally good when compared to other solutions, means there exists no other feasible solution, which would decrease some criterion without causing a simultaneous increase in at least one other criterion.

Pareto front. The plot of the objective functions whose non-dominated vectors are in the Pareto optimal set is called the Pareto.

Convergence. The Pareto-front, which is as close as to the true Pareto-front, is considered best. Ideally, the true Pareto-front should contain the best-known Pareto-front.

Diversity. Pareto-front should provide solutions, which are uniformly distributed and diversified across the Pareto-front.

2.2 Introduction to Multi Objective Particle Swarm Optimization

PSO is a population-based search algorithm introduced by Kennedy and Eberhart [17]. In PSO, each particle in the population is a solution to the problem. Instead, the particles are "flown" through hyper dimensional search space to search out a new optimal solution through two Eqs. (2) and (3).

$$\vec{v_i}(t) = W\,\vec{v_i}(t-1) + C_1 r_1\left(\vec{x}_{pbest_i} - \vec{x_i}(t)\right) + C_2 r_2\left(\vec{x}_{leader} - \vec{x_i}\right) \tag{2}$$

$$\vec{x_i}(t) = \vec{x_i}(t-1) + \vec{v_i}(t) \tag{3}$$

where leader presents the global best solution in the population, the pbest position is the best personal solution of a given particle. Additionally, r_1 and r_2 are random values, W is the inertia weight, c_1 is the cognitive learning factor and c_2 is the social learning factor. In order to adopt the PSO for the optimization of MOP (MOPSO), a few modifications must be made under the original PSO. First, the aim is to discover a set of the optimal solution, not even one. Second, an external archive is kept, where all non-dominated solutions found at each iteration are saved in.

3   Related Works

MOPSO is one of the dominant techniques to find the promised solutions much faster than the other algorithms. Instead, it suffers from the premature convergence. This problem tends to converge to local optima, such that problem led the MOPSO to fail to find the Pareto-optimal solutions. In order to solve this problem, it is obvious that the original algorithm has to be modified. One of the interesting methods that have the ability to overcome the problem of premature convergence is the distributed evolutionary (DE). The granularity of the DE may be at the population level. Since the distributed population based on the idea of dividing the entire population into sub-populations, each of which is processed by one processor. In fact, we summarize a few outputs.

A new version of MOPSO [18] adopts the Pareto ranking to dynamic subdivide the population. There are a few variants of hierarchical architecture are proposed in [19–21] that is proposed by Fdhila. Its main idea is to have a 2-levels that adopts a bidirectional dynamic exchange of particles between MOPSOs. Indeed, these variants improve its efficacy in many real applications such as the feature selection [22], the routing Pico-satellites problem [23], the grasp planning problem [24], the TSP problem [2], the Face Recognition [25]. The organization and communication between sub-population play an important role in the DE. In fact, there are varieties of methods that attempt to address this deficiency. One of the most notable methods is the incorporation of MAS as a model of DE. Indeed, the MAS is adopted to model, manage and coordinate the process of optimization among different sub-population. Different methods are improved in [26–29] these methods applied the MAS as a model of DE, in which the MAS achieves the purpose of communication, organization and cooperation.

4   Description of the Proposed Approach

4.1   Motivation

The adaptation of MOPSO with DE makes evident the notion of using MAS could be the straightforward way to recover MOPSO in order to overcome the premature convergence. The hybridization between the MAS and the MOPSO algorithm could balance between the local exploitation and the global exploration. Indeed, we propose to use not one, but several sub-populations (each with a dynamic size). Each sub-population overfly within a specific region of the search space. In addition, it has its own set of particle and particle guides kept in the local archive. It is known that the use of disconnected sub-population led the algorithm unable to converge to the true Pareto front. This issue makes the using of a good strategy of communication is necessary. In this context, the automated negotiation (AN) is used. The AN used to ensure the changing information, considering that we are instead a decision conflict between agents, which are the optimal solution to be selected. In fact, the AN accurate the selection of global leader, since the solution is largely depends on the guiding points. In this way, the AN guarantees the set of Pareto optimal solution since each agent tends to exploit the sub-search space, while ensuring that an exploration is reached within the search space by sharing best knowledge (among sub-populations).

4.2   Main Process

The main process of our algorithm is illustrated in Fig. 1. At the first stage, the population creates, initializes and updates its own particles. Additionally, the leaders set is generated and saved in the external archive. Once the initialization is completed, the Pareto ranking divides the population, as a result, a dynamic number of fronts (F0, F1, ..., Fn) are generated. Each of these fronts plays the role of the sub-population. These sub-populations distributed among agents. Then, for a maximum number of iterations, each agent performs the execution of MOPSO in its own sub-population, including the selection of gbest (global leader) (see Fig. 2) by the AN, the update of position and velocity and, finally, the local archive is updated too. However, as AN process we adopt a multi-lateral negotiation since we have k cooperative agents as negotiators. In this model, we assume that our domain has one issue (defining the best solution in order to share among sub-populations).

The general process of the negotiation process is as follows: once all agents attending the evolving process, a new negotiation session has begun, first, each agent sends a call for proposal CFP to other agents. Next, the agent responds to the CFP by

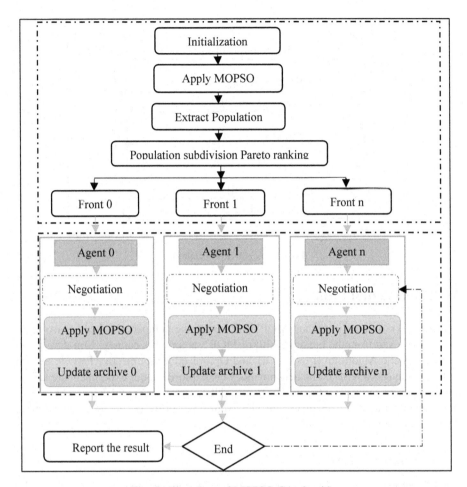

Fig. 1. Flowchart of MOPSO-CA algorithm

making an offer (best local solution: based on the dominance operator). In the turn, the agent evaluates the incoming offer using the fitness function. Consequently, the accepted proposal is the offer which accepted by all agents. In result, the accepted proposal becomes the best global solution that used during the update position. Therefore, each particle of sub-population adjusts its trajectory according to its own experience (p_{best}), the experience of its neighbors (l_{best}), and the experience of best global solution among sub-populations (g_{best}). So the new equation for velocity is presented in Eq. (4).

$$\vec{v_i}(t) = w\vec{v_i}(t-1) + c_1 r_1 \left(\vec{x}_{pbest_i} - \vec{x}_i(t)\right) + c_2 r_2 \left(\vec{x}_{lbest} - \vec{x}_i\right) + c_3 r_3 \left(\vec{x}_{gbest} - \vec{x}_i\right) \quad (4)$$

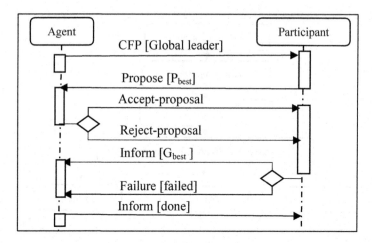

Fig. 2. Negotiation protocol

5   Experimental Studies

In order to know how important MOPSO-CA was, we compared it against two algorithms that taken from the literature [30] of MOP's algorithms namely: Non-dominated Sorting Genetic Algorithm II (NSGAII), and Optimized Multiobjective Particle Swarm Optimization (OMOPSO). The benchmarks were explored in our experiment are DTLZ family (DTLZ5 and DTLZ6) and UF family (UF1, UF2, UF3 and UF10), which have sufficient complexity to evaluate the algorithm's performance, in terms of solution diversity and convergence rate.

5.1   Performance Metric

Several performance evaluations are available to compare the performance of the presented approach. In the present context, we choose the following three metrics [31]: Spread (SP), Inverted Generational Distance (IGD) and Hypervolume (HV) which used to evaluate the diversity, the convergence and the both (convergence and diversity) respectively.

5.2   Experimental Setting

To evaluate the performance of the comparative algorithms, 30 runs of each algorithm for each test function are performed; a population with 200 individuals is fixed, and the archive size is set to 100. Further, the parameters of different algorithms detailed as the following, for MOPSO-CA and MOPSO, an acceleration coefficients c1, c2 = Rand (1.5, 2.0), and inertia weight w = Rand (0.1, 0.5). For NSGAII the max evaluations = 25000 and crossover probability = 0.9.

5.3   Experimental Results

In this section, we analyze the results obtained by the algorithms. Derivatives figures (Figs. 3, 4 and 5) show the graphical results generated by the comparative algorithms. (see Figs. 3 and 4) show the Pareto front produced by the comparative algorithms for DTLZ5 and UF10 respectively; clearly, we can conclude that NSGAII, OMOPSO and MOPSO-CA cover the entire true Pareto front of DTLZ5 on one hand. On the other hand, we can see that only the MOPSO-CA may cover the true Pareto front of UF10.

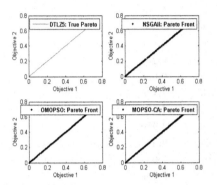

Fig. 3. Pareto fronts obtained by the algorithms for DTLZ5

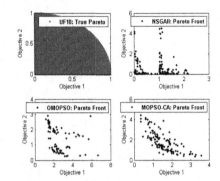

Fig. 4. Pareto fronts obtained by the algorithms for UF10

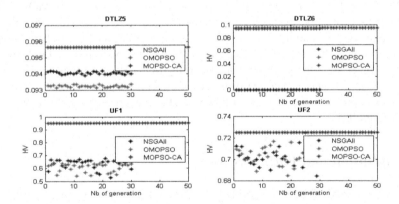

Fig. 5. Performance (HV) over DTLZ5, DTLZ6, UF1 and UF2 problem

For more precise the accuracy of the solution, statistical values are provided in Table 1. From these values, it can be seen, that the average performance of MOPSO-CA is better than NSGAII and OMOPSO with respect to the HV metric.

Regarding the SP metric, we can conclude that MOPSO-CA has the better spread of solutions for UF1, UF3, UF10, DTLZ5 and DTLZ6. On the other hand, the OMOPSO has the best SP value for UF2. Regarding the IGD metric, we can conclude that

Table 1. Performance metrics (mean value) for the different test functions

	Metric	NSGAII	OMOPSO	MOPSO-CA
UF1	SP	1:37	9:77e	**1:32e**
	IGD	1:85e	1:99e	**0:00e**
	HV	6:37e	6:05e	**9:53e**
UF2	SP	7:77e	**5:18e**	6:40e
	IGD	1:28e	**1:11e**	3:79e
	HV	7:02e	7:04e	**7:25e**
UF3	SP	9:82e	6:62e	**6:56e**
	IGD	1:09e	6:43e	**0:00e**
	HV	2:43e	3:85e	**4:31e**
UF10	SP	8:63e	6:79e	**6:20e**
	IGD	7:57e	2:18e	**0:00e**
	HV	6:14e	1:64e	**7:14e**
DTLZ5	SP	4:57e	1:82e	**1:68e**
	IGD	5:78e	8:15e	**3:95e**
	HV	9:40e	9:32e	**9:57e**
DTLZ6	SP	7:98e	1:28e	**1:03e**
	IGD	2:69e	7:41e	**2:13e**
	HV	0:00e	9:35e	**9:48e**

MOPSO-CA is relatively better than other algorithms for UF1, UF3, UF10, DTLZ5 and DTLZ6, since MOPSO-CA have the minimum IGD values. On the other hand, OMOPSO was the best for UF2. Hence, to have a deep dissection of the MOPSO-CA, the HV values for UF1 and UF2, DTLZ5 and DTLZ6 were plotted (see Fig. 5).

Graphically, it is intuitive that MOPSO-CA can achieve the best tradeoff between convergence and diversity (the higher value, the better performance for HV) with respect to other algorithms. Meanwhile, according to the HV values, we can conclude that MOPSO-CA gets better performance of different test function. Clearly, our MOPSO-CA produces the best trade-off between the convergence and diversity, within the tested problems.

6   Conclusion and Future Work

In this paper, the MOPSO-CA algorithm proposed to solve MOP. In this algorithm, the sub-populations, Pareto ranking, MAS and automated negotiation are used. MAS improve the performance of distributed MOPSO but strategies for communication between agents are very important. Thus, the efficiency of synchronous knowledge (most successful solution) exchange strategies has been achieved by using the automated negotiation. Through experiments, it can be concluded that MOPSO-CA can outstanding performances in terms of convergence and diversity qualities. As a future work, we will explore more the feature of MAS to increase the intelligence level of particles. In addition, our proposed approach can be incorporated in many real-world problems.

Acknowledgement. The research leading to these results has received funding from the Ministry of Higher Education and Scientific Research of Tunisia under the grant agreement number LR11ES48.

References

1. Ghamisi, P., Couceiro, M.S., Martins, F.M.L., Benediktsson, J.A.: Multilevel image segmentation based on fractional-order darwinian particle swarm optimization. IEEE Trans. Geosci. Remote Sens. **52**(5), 2382–2394 (2014)
2. Fdhila, R., Elloumi, W., Hamdani, T.M.: Distributed MOPSO with dynamic Pareto front driven population analysis for TSP problem. In: The 6th International Conference Soft Computing and Pattern Recognition, pp. 294–299. IEEE, Tunis (2014)
3. Ben Moussa, S., Zahour, A., Benabdelhafid, A., Alimi, M.A.: New features using fractal multi-dimensions for generalized Arabic font recognition. Pattern Recogn. Lett. **31**(5), 361–371 (2010)
4. Bezine, H., Alimi, M.A., Derbel, N.: Handwriting trajectory movements controlled by a Bêta-elliptic model. In: 7th International Conference on Document Analysis and Recognition, pp. 1228–1232. IEEE, Edinburgh (2003)
5. Alimi, M.A.: Evolutionary computation for the recognition of on-line cursive handwriting. IETE J. Res. **48**(5), 385–396 (2002)
6. Boubaker, H., Kherallah, M., Alimi, M.A.: New algorithm of straight or curved baseline detection for short arabic handwritten writing. In: 10th International Conference on Document Analysis and Recognition, pp. 778–782. IEEE, Barcelona (2009)
7. Slimane, F., Kanoun, S., Hennebert, J., Alimi, M.A., Ingold, R.: A study on font-family and font-size recognition applied to Arabic word images at ultra-low resolution. Pattern Recogn. Lett. **34**(2), 209–218 (2013)
8. Elbaati, A., Boubaker, H., Kherallah, M., Alimi, M.A., Ennaji, A., Abed, H.E.: Arabic handwriting recognition using restored stroke chronology. In: 10th International Conference on Document Analysis and Recognition, pp. 411–415. IEEE, Barcelona (2009)
9. Dhahri, H., Alimi, M.A.: The modified differential evolution and the RBF (MDE-RBF) neural network for time series prediction. In: IEEE International Conference on Neural Networks - Conference Proceedings, pp. 2938–2943. IEEE, Vancouver (2006)
10. Bouaziz, S., Dhahri, H., Alimi, M.A., Abraham, A.: A hybrid learning algorithm for evolving flexible Beta basis function neural tree model. Neurocomputing **117**, 107–117 (2013)
11. Baccour, L., Alimi, M.A., John, R.I.: Similarity measures for intuitionistic fuzzy sets: state of the art. J. Intell. Fuzzy Syst. **24**(1), 37–49 (2013)
12. Bahareh, N., Mohd, Z., Ahmad, N., Mohammad, N.R., Salwani, A.: A survey: particle swarm optimization based algorithms to solve premature convergence problem. J. Comput. Sci. **10**(9), 1758–1765 (2014)
13. Gong, Y.-J., et al.: Distributed evolutionary algorithms and their models: a survey of the state-of-the-art. Appl. Soft Comput. **34**, 286–300 (2015)
14. Wooldridge, M.: An Introduction to Multiagent System, 2nd edn. Wiley, Chichester (2009)
15. Jennings, N.R., Faratin, P., Lomuscio, A.R., Parsons, S., Sierra, C., Wooldridge, M.: Automated negotiation: prospects, methods and challenges. Int. J. Group Decis. Negot. **10**(2), 199–215 (2001)

16. Deb, K., Deb, K.: Multi-objective optimization. In: Burke, E., Kendall, G. (eds.) Search Methodologies Introductory Tutorials in Optimization and Decision Support Techniques, pp. 403–449. Springer, Boston (2014). doi:10.1007/978-1-4614-6940-7_15

17. Kennedy, J., Eberhart, R.: Particle swarm optimization. In: IEEE International Conference on Neural Networks, pp. 1942–1948. IEEE, Perth (1995)

18. Fdhila, R., Hamdani, T., Alimi, M.A.: A new distributed approach for MOPSO based on population Pareto fronts analysis and dynamic. In: Systems Man and Cybernetics, pp. 947–954. IEEE, Istanbul (2010)

19. Fdhila, R., Hamdani, T.M., Alimi, M.A.: A new hierarchical approach for MOPSO based on dynamic subdivision of the population using Pareto fronts. In: IEEE International Conference on Systems, Man, and Cybernetics, pp. 947–954. IEEE, Istanbul (2010)

20. Fdhila, R., Hamdani, T.M., Alimi, M.A.: Optimization algorithms, benchmarks and performance measures: from static to dynamic environment. In: The 15th International Conference on Intelligent Systems Design and Applications, pp. 597–603. IEEE, Marrakech (2015)

21. Fdhila, R., Hamdani, T.M., Alimi, M.A.: Population-based distribution of MOPSO with continuous flying Pareto fronts particles. J. Inf. Process. Syst. (2016)

22. Fdhila, R., Hamdani, T.M., Alimi, M.A.: Distributed MOPSO with a new population subdivision technique for the feature selection. In: The 5th International Symposium Computational Intelligence and Intelligent Informatics, pp. 81–86. IEEE, Floriana (2011)

23. Fdhila, R., Hamdani, T.M., Alimi, M.A.: A multi objective particles swarm optimization algorithm for solving the routing pico-satellites problem. In: Systems, Man, and Cybernetics, pp. 1402–1407. IEEE, Seoul (2012)

24. Fdhila, R., Walha, C., Hamdani, T.M., Alimi, M.A.: Hierarchical design for distributed MOPSO using sub-swarms based on a population Pareto fronts analysis for the grasp planning problem. In: The 13th International Conference on Hybrid Intelligent Systems, pp. 203–208. IEEE, Gammarth (2013)

25. Fdhila, R., Ouarda, W., Alimi, M.A., Abraham, A.: A new scheme for face recognition system using a new 2-level parallelized hierarchical multi objective particle swarm optimization algorithm. J. Inf. Assur. Secur. **11**(6), 385–394 (2016)

26. Kouka, N., Fdhila, R., Alimi, M.A.: A new architecture based distributed agents using PSO for multi objective optimization. In: 13th International Conference on Applied Computing (2016)

27. Ilie, S., Bădică, C.: Multi-agent approach to distributed ant colony optimization. Sci. Comput. Program. **78**(6), 762–774 (2013)

28. Takano, R., Yamazaki, D., Ichikawa,Y., Hattori, K., Takadama, K.: Multiagent-based ABC algorithm for autonomous rescue agent cooperation. In: IEEE International Conference on Systems, Man, and Cybernetics, pp. 585–590. IEEE, San Diego (2014)

29. Yingchun, C., Wei, W.: MAS-based distributed particle swarm optimization. In: 8th International Conference on Wireless Communications, Networking and Mobile Computing, pp. 1–4. IEEE, Shanghai (2012)

30. Godinez, A.C., Espinosa, L.E.M., Montes, E.M.: An experimental comparison of multiobjective algorithms: NSGA-II and OMOPSO. In: Conference Electronics, Robotics and Automotive Mechanics, pp. 28–33. IEEE, Morelos (2010)

31. Zhang, Q., Zhou, A., Zhao, S., Suganthan, P.N., Tiwari, S.: Multiobjective optimization test instances for the CEC 2009 special session and competition. Technical report, CES-487 (2009)

Emergency Materials Scheduling in Disaster Relief Based on a Memetic Algorithm

Yongwei Qin and Jing Liu[(✉)]

Key Laboratory of Intelligent Perception and Image Understanding of Ministry of Education, Xidian University, Xi'an 710071, China
qinyongwei8888@163.com, neouma@163.com

Abstract. In the case of large-scale natural disaster, it is very important to determine emergency materials scheduling quickly and efficiently from multiple emergency logistics centers as supply points supplying to multiple disasters affected points. In order to improve the efficiency of supplying organization and reduce casualties and economic losses, a mathematical model is first constructed in this paper to minimize the total emergency cost including emergency response system cost and the loss caused by untimely rescue. Then, a memetic algorithm (MA) using natural coding is proposed to solve the problem, and the experimental results show that the performance of this algorithm is much better than that of genetic algorithm (GA). At the same time, the proposed model and algorithm provide robust support for decision makers when quick responses are necessary for disaster relief activities.

Keywords: Natural disaster · Emergency materials scheduling · Memetic algorithm

1 Introduction

In recent years, several devastating natural disasters have occurred in different countries and areas, especially in China, such as earthquakes, floods, and hurricanes. The emergency events have a severe impact on the people's life and possessions [1]. For example, the great Wenchuan Earthquake has resulted in 69, 225 deaths and 379, 640 injuries [2]. In order to reduce the losses of lives and assets, the science of disaster and crisis management is very important [3, 4].

The relevant researchers proposed the decision-making models with the aim of reducing the impacts of disasters and providing a quick and efficient response to emergency materials scheduling [8, 15]. Liu *et al.* [12] proposed a combined optimization model and a fast solving algorithm for a kind of material scheduling. Hong *et al.* [13] built a variety of materials scheduling model of single supply to multi demand. Wu and Yang [14] studied one kind of material scheduling from single demand to minimize supply points number and emergency start time. Yu and Zhang [5] proposed the corresponding scheduling methods of two stages, and a polynomial algorithm was proposed to minimize the weighted makespan. Researchers considered a single machine scheduling and the objective is to minimize the makespan [6, 7]. In [9], researchers proposed the relevant mathematical model, and the proposed model

© Springer International Publishing AG 2017
D. Liu et al. (Eds.): ICONIP 2017, Part IV, LNCS 10637, pp. 279–287, 2017.
https://doi.org/10.1007/978-3-319-70093-9_29

improved the ability of emergency materials operation. *Liu and Xiong* [10] discussed the dispatching of emergency materials with multiple rescue-single object questions to minimize the time spending on emergency logistics scheduling.

In a word, most researchers commonly focus on the completion time, the number of supply activities, and emergency response system cost. Moreover, the existing researches pay attention to single supply point or single demand point or one kind of material. Considering realistic affected scope, the number of supply points and types of required materials, we proposed a kind of emergency materials scheduling model of multi-supply points to multi-demand points. The objective of this model is to minimize the total emergency cost including emergency response system cost and the loss caused by untimely rescue.

In this paper, we proposed a memetic algorithm [11] based on the natural coding for emergency materials scheduling in disaster relief to optimize the proposed model. The new local search operator and the repair operator are designed to obtain high-quality solution more efficiently. The experiments show that, with appropriate set of genetic parameters, MA can find optimal scheduling schemes with less number of iterations compared with the traditional GA, and thus made emergency materials scheduling be more efficiently.

The rest of this paper is organized as follows. Section 2 describes the problem and model of emergency materials scheduling. In Sect. 3, the details of the proposed algorithm are introduced. Experimental results and discussions are given in Sect. 4. Finally, conclusions and considerations for future work are given in Sect. 5.

2 Scheduling Problem and Model

Based on the analysis of emergency materials scheduling system, we make the following definitions and descriptions about the emergency scheduling problem and model:

Suppose that there are m emergency materials supply points S_1, S_2,..., S_m, and n demand points D_1, D_2, ..., D_n. The m supply points are responsible for supplying emergency materials to the n demand points. The unit emergency materials response time for supply point S_i to demand point D_j is t_{ij} ($1 \leq i \leq m$ and $1 \leq j \leq n$). The unit emergency materials objective time for the demand point D_j is t_j ($1 \leq j \leq n$). The amount of emergency materials required by D_j is DQ_j ($1 \leq j \leq n$). x_{ij} ($1 \leq i \leq m$ and $1 \leq j \leq n$) indicates the emergency materials contributed by S_i for D_j. c_{ij} indicates the unit cost of emergency materials without delaying from supply point S_i to demand point D_j. w_{ij} indicates unit cost of unit emergency materials with delaying unit time from supply point S_i to demand point D_j. s_i indicates the real emergency materials supply total quantities of supply point S_i. SQ_i shows the supply ability of the supply point S_i.

To simplify model, the problem assumptions are as follows:

(1) Quantity of emergency materials scheduling is integer, and independent of each other.

(2) The time t_{ij} is constant, without considering transport capacity of road and uncertainties.
(3) In the process of transportation, emergency materials are not damaged or lost.
(4) Each demand point can be served by several supply points.
(5) The total amount of emergency materials is greater than or equal to that of incidents' required emergency materials.

Based on optimization objective and assumption, following model is established.

$$\min \sum_{j=1}^{n} \sum_{i=1}^{m} c_{ij}.x_{ij} + \sum_{j=1}^{n} \sum_{i=1}^{m} w_{ij}.x_{ij}.(t_{ij} - t_j) \tag{1}$$

$$s.t. \quad \sum s_i = \sum DQ_j (i = 1, 2, \ldots, m, j = 1, 2, \ldots, n) \tag{2}$$

$$\sum_{i=1}^{m} x_{ij} = DQ_j (j = 1, 2, \ldots, n) \tag{3}$$

$$\sum_{j=1}^{n} x_{ij} = s_i (i = 1, 2, \ldots, m) \tag{4}$$

$$s_i \leq SQ_i (i = 1, 2, \ldots, m) \tag{5}$$

$$x_{ij} \geq 0 \ (i = 1, 2, \ldots, m, j = 1, 2, \ldots, n) \tag{6}$$

In the above proposed model, the objective is to minimize the total emergency cost. Objective function (1) consists of two parts. The first part $\sum_{j=1}^{n} \sum_{i=1}^{m} c_{ij}.x_{ij}$ indicates the emergency response system cost without delaying. The second part represents the cost caused by untimely rescue. Constraint (2) shows that the total supply quantities of emergency supply points are equal to the total demand quantities of emergency demand points. Constraint (3) states that the scheduling quantities of supply points are equal to the demand quantities of every demand point. Constraint (4) states the scheduling quantities from every supply point to all the demand points are equal to the actual supply quantities. Constraint (5) shows the actual supply quantities do not surpass the supply ability. Constraints (6) states that the transport quantities are not-negative integer.

In this paper, in order to obtain the results more conveniently, we let the value of fitness function be equal to the reciprocal value of objection function multiplied by 100,000. The objective function $f(x_{ij})$ is given in (1). Equation (7) is the fitness function.

$$Fitness (x_{ij}) = \frac{100000.0}{f(x_{ij})} \tag{7}$$

3 Description of MA

In this section, we give a detailed description of the proposed memetic algorithm for optimizing the proposed model.

3.1 Parameter Coding

In the disaster relief of emergency materials scheduling, we adopts the natural number coding, and a chromosome represents a kind of scheme of transport emergency materials. A chromosome consists of several genes, and the value of every gene represents the quantity of transport emergency materials from supply points to demand points. The chromosome is given as follows:

$$(x_{11}, \ x_{12}, \ \ldots, \ x_{1n}, \ x_{21}, \ x_{22}, \ \ldots, \ x_{2n}, \ x_{m1}, \ x_{m2}, \ \ldots, \ x_{mn}).$$

where x_{ij} is an integer greater than or equal to 0, indicating the quantity of supply point S_i to demand point D_j.

3.2 Fitness Function

Fitness is used to evaluate the individual's adaptation degree to the environment, and the greater adaptation degree of the individual is, the greater chance of being selected and passing down to next generation is. On the contrary, the lower adaptation degree of the individual will have little chance to pass down to next generation. The fitness function refers to the function of evaluating the individual's adaptation degree. The fitness function $Fitness\,(x_{ij})$ is shown as Eq. (7).

3.3 Initial Population

In the beginning, on the basis of constraints (2)–(6), each individual is initialized according to each supply point's emergency materials s_i' and every demand point's request DQ_j. That is, for each element x_{ij} in an individual, x_{ij} is randomly initialized to be an integer over the range $[0, \ max\{SQ_1, SQ_2, \ldots, SQ_3\}]$ and each produced individual should meet constraint (2)–(6). The population is a set of chromosomes and each chromosome is evaluated by calculating the fitness function value of the individual according to (7).

3.4 Selection Operator

Selection is very important for keeping the diversity of the population, and it decides which individual can be passed to the next generation. Widely used methods of selection are Roulette, Stochastic and Tournament. In this paper, we adopt the Roulette. The probability of selection for each chromosome is based on a fitness value relative to the total fitness value of the population. The selection module ensures that more number of highly fit chromosomes is selected. The selection process is repeated as many times as the population size.

3.5   Crossover Operator

The crossover operator is a very important component in a population-based algorithm. It is used to generate one or more new offspring individual to discover new promising search areas. Traditional methods of crossover are single-point crossover, two-point crossover, and uniform crossover. In this paper, we adopt single-point crossover. Crossover randomly chooses a point and exchanges the subsequence before and after that point between two chromosomes to create two offspring, and the crossover probability means how many couples will be picked for mating.

3.6   Mutation Operator

Mutation means randomly change the value of gene at any position. Its aim is to prevent all solution to fall into local optima and to preserve diversity in the search. Widely used methods of mutation are basic position, reverse, and uniform probability. In this paper, we adopt the basic position mutation. During the process, a random number "q" is generated for each individual. If mutation probability ($P_m < q$), then the particular individual will undergo the mutation process.

3.7   Repair Operator

Except the selection operator, crossover operator, and mutation operator, we design a repair operator which is to repair the exceeding or insufficient emergency materials scheduling. In order to make each individual of the population satisfy the constraints (2)–(6), the repair operator is used after the crossover operator and mutation operator. If a new individual after mutation does not satisfy the constraints (2)–(6), the repair operator is performed on the individual, so that the individual after repair satisfies the constraints, which make emergency materials scheduling more effectively and efficiently.

3.8   Local Search Operator

First, we sort all individuals in the population decreasingly according to the fitness of each individual, and then we select 10% individuals with higher fitness in the population. Second, for each selected chromosome, we change the sign of an arbitrary single gene. If the change produces a chromosome with better fitness and the chromosome satisfies the constraints, we keep this change and add the individual to the population. Otherwise, we reject this change and change the sign of another arbitrary single gene, repeating the above process until the stop criteria of local search are met.

3.9   Implementation of MA

Hence, based on the above description, we summarize the process of MA in Algorithm 1, where SQ_i is the supply ability of supply point S_i. DQ_j indicates the demand quantity of the demand point D_j, P_c is crossover rate, P_m is mutation rate, N_p represents population scale.

Algorithm 1: MA
Input:
Data: SQ_i, t_{ij}, t_j, w_{ij}, DQ_j;
Parameters: N_p, P_o, P_m;
Output:
The emergency materials scheduling sequences of superior individual and cost;
$t := 0$;
Initialize population P(t) randomly;
Evaluate fitness(P(t));
while (*stopping criteria are not satisfied*)**do**
$P'(t) := Selection operator((P(t))$;
Crossover operator(P'(t));
Mutation operator(P'(t));
Repair operator (P'(t));
Local search operator(P'(t));
Evaluate fitness(P'(t));
$P(t+1) := (P'(t))$;
$t := t+1$;
end while.

4   Experimental Result

Assume that a disaster burst in certain place, and there are three supply points S_1, S_2, S_3, and three demand points D_1, D_2, D_3. The supply ability of S_1, S_2 and S_3 is respectively SQ_1, SQ_2 and SQ_3. The demand quantity of D_1, D_2 and D_3 is respectively DQ_1, DQ_2, and DQ_3. The unit emergency material objective time of demand point t_1, t_2 and t_3 is respectively 3 h, 4 h and 5 h. The unit emergency material cost c_{ij} (10 thousand yuan/1000 kg) from supply point S_i to demand point D_j without delaying is given in Table 1. The average cost w_{ij} of transporting unit emergency materials with delaying unit time from supply point S_i to demand point D_j is given in Table 2. The average time t_{ij} of transporting unit emergency material from supply point S_i to demand point D_j is given in Table 3.

Table 1. The unit emergency material cost c_{ij} from supply point S_i to demand point D_j without delaying

S_i	D_j		
	D_1	D_2	D_3
S_1	3	4	5
S_2	2	3	4
S_3	5	3	5

Table 2. The average cost w_{ij} of transporting unit emergency materials with delaying unit time from supply point S_i to demand point D_j

S_i	D_j		
	D_1	D_2	D_3
S_1	1	2	3
S_2	5	7	8
S_3	5	6	3

Table 3. The average time t_{ij} of transporting unit emergency material from supply point S_i to demand point D_j

S_i	D_j		
	D_1	D_2	D_3
S_1	5	6	5
S_2	3	4	5
S_3	4	5	7

Table 4. The performance comparisons of MA with GA

Index	SQ_i	DQ_j	Optimal scheduling	Cost MA/GA
I	$SQ_1 = 250$	$DQ_1 = 100$	97 50 95	**2805/3164**
	$SQ_2 = 150$	$DQ_2 = 300$	1 147 0	
	$SQ_3 = 200$	$DQ_3 = 100$	2 103 5	
II	$SQ_1 = 160$	$DQ_1 = 150$	13 19 128	**2647/2819**
	$SQ_2 = 200$	$DQ_2 = 200$	118 67 15	
	$SQ_3 = 240$	$DQ_3 = 150$	19 114 7	
III	$SQ_1 = 155$	$DQ_1 = 255$	11 6 137	**3937/4341**
	$SQ_2 = 250$	$DQ_2 = 150$	156 21 73	
	$SQ_3 = 325$	$DQ_3 = 255$	88 123 45	
IV	$SQ_1 = 350$	$DQ_1 = 100$	95 27 198	**3619/3873**
	$SQ_2 = 120$	$DQ_2 = 315$	3 117 0	
	$SQ_3 = 245$	$DQ_3 = 200$	2 171 2	
V	$SQ_1 = 150$	$DQ_1 = 350$	8 26 116	**4736/5043**
	$SQ_2 = 425$	$DQ_2 = 280$	318 7 98	
	$SQ_3 = 400$	$DQ_3 = 250$	24 247 36	
VI	$SQ_1 = 250$	$DQ_1 = 255$	77 17 156	**6587/7082**
	$SQ_2 = 350$	$DQ_2 = 485$	168 46 135	
	$SQ_3 = 555$	$DQ_3 = 325$	10 422 34	
VII	$SQ_1 = 289$	$DQ_1 = 236$	111 15 136	**3005/3690**
	$SQ_2 = 365$	$DQ_2 = 325$	23 250 10	
	$SQ_3 = 295$	$DQ_3 = 150$	23 60 4	
VIII	$SQ_1 = 365$	$DQ_1 = 245$	165 5 185	**4185/4643**
	$SQ_2 = 275$	$DQ_2 = 375$	77 192 0	
	$SQ_3 = 258$	$DQ_3 = 188$	3 178 3	

On the basis of the above information, we conducted the experiments with GA and MA, respectively. Compared with GA, MA adopts local search operator besides the same selection operator, crossover operator, mutation operator and repair operator as GA. The relative parameter settings of GA and MA are listed as follows: $N_p = 20$, $P_c = 0.9$, $P_m = 0.2$, iterations number = 1000. All algorithms are implemented using Microsoft Visual Studio 2010. The performance comparisons of MA with GA based on the above information are shown in Table 4.

It can be seen from Table 4 that the optimal scheduling scheme obtained by MA is much better than that of GA. and the optimal value of cost obtained by MA is also much better than that of GA. Based on the above analysis, it can be concluded that MA is more efficient than GA.

5 Conclusion

In this paper, we discuss the problem of emergency materials scheduling, and establish an optimization model, aiming to minimize the total emergency cost including emergency response system cost and the loss caused by untimely rescue. Then, a memetic algorithm is designed to solve the model. The experimental results show that our algorithm has a better performance than GA. In the future, in order to make emergency material scheduling more reasonable and efficient, more efficient models need to be established, and the model can consider the priority of different kinds of emergency materials, and the effect of roads on emergency scheduling.

Acknowledgements. This work is partially supported by the Outstanding Young Scholar Program of National Natural Science Foundation of China (NSFC) under Grant 61522311, the overseas, Hong Kong & Macao Scholars Collaborated Research Program of NSFC under Grant 61528205, and the Key Program of Fundamental Research Project of Natural Science of Shanxi Province, China under Grant 2017JZ017.

References

1. Ademola, A., Adebukola, D., Adeola, C.S., Cajetan, A., Christiana, U.: Effect of natural disaster on social and economic well-being: a study in Nigeria. Int. J. Risk Reduct. **17**, 1–12 (2016)
2. Wang, Z.F.: A preliminary report on the great WenChuan Earthquake. Earthq. Eng. EngVib. **7**(2), 225–234 (2008)
3. Pearce, L.: Disaster nanagement and community planning, and public participation: how to achieve sustainable hazard mitigation. Nat. Hazards **28**(2), 211–228 (2003)
4. Silovs, M., Malahova, J., Jemeljanovs, V., Ketners, K.: Wind-related disasters management and prevention improvement strategy. Procedia-Soc. Behav. Sci. **213**, 515–520 (2015)
5. Yu, X.Y., Zhang, Y.L.: The emergency scheduling engineering in single resource center. Syst. Eng. Procedia **5**, 107–112 (2012)
6. Zhao, C.L.: Single machine scheduling with general job-dependent aging effect and maintenance activities to minimize makespan. Appl. Math. Model. **34**(3), 837–841 (2010)

7. Yang, S.J., Yang, D.L.: Minimizing the makespan on single-machine scheduling with aging effect and variable maintenance activities. Omega **38**(6), 528–533 (2010)
8. Ivgin, M.: The decision-making models for relief asset management and interaction with disaster mitigation. Int. J. Disaster Risk Reduct. **5**, 107–116 (2013)
9. Pradhananga, R., Mutlu, F., Pokharel, S., Holguin-Veras, J.: An integrated resource allocation and distribution model for pre-disaster planning. Comput. Ind. Eng. **91**, 229–238 (2016)
10. Liu, H.Z., Xiong, J.Q.: Research on the city emergency logistics scheduling decision based on cloud theory-based genetic algorithm. Adv. comput. Sci. Environ. Ecoinformatics and Educ. **217**, 181–185 (2011)
11. Deulkar, K., Narvekar, M.: An improved memetic algorithm for web search. Procedia Comput. Sci. **45**, 52–59 (2015)
12. Liu, C.L., He, J.M., Shi, J.J.: The study on optimal model for a kind of emergency materials dispatch problem. Chin. J. Manage. Sci. **6**(3), 29–36 (2001)
13. Hong, H.L., Liu, N., Zhang, G.C., Yu, H.H.: A model for distribution of multiple emergency commodities to multiple affected areas based on loss of victims of calamity. J. Syst. Manage. Sci. **6**(3), 29–36 (2001)
14. Wu, S.H., Yang, J.J.: The study of material dispatch problem for the one-time used-up emergency response system. Logistics Technol. **7**, 47–49 (2009)
15. Zhou, Y.W., Liu, J., Gan, X.H.: A multi-objective evolutionary algorithm for multi-period dynamic emergency resource scheduling problems. Transp. Part E. **99**, 77–95 (2017)

Robot Path Planning Based on A Hybrid Approach

Zhou Jiang and Zhigang Zeng[✉]

School of Automation, Huazhong University of Science and Technology,
Wuhan 430074, China
{kebbcy,zgzeng}@hust.edu.cn

Abstract. In this paper, an optimal method based on combination of improved genetic algorithm (IGA) and improved artificial potential field (IAPF) for path planning of mobile robot is proposed. This method consists of two steps. Firstly, free space model of mobile robot is established by using grid-based method and IGA is employed to find a global optimal collision-free path which is usually the shortest through known static environment. Secondly, according to the path obtained by IGA, IAPF is utilized to generate a real-time path to avoid dynamic obstacles. This ensures that robot can avoid obstacles as well as move along the optimal path. Simulation experiments are carried out to verify the superiority of the proposed algorithm.

Keywords: Path planning · Genetic algorithm · Artificial potential field

1 Introduction

Recently, how to solve the robot path planning problem has aroused increasing interest of many researchers. Robot needs to automatically generate an optimal collision-free path which is from the start location to the goal location with respect to some given criteria. In this work, we focus the research on optimum path planning problem on real-time obstacle-avoidance and shortest distance criteria.

Various optimization methods have been proposed to solve this conundrum, which are classified as two major categories: traditional methods and intelligent methods. Notable conventional path planning methods include artificial potential field method [1] and visibility graph method [2]. Prominent intelligent path planning methods include particle swarm optimization [3], ant colony algorithm [4] and genetic algorithm [4–9]. Each method has its own merits and shortcomings. We can't find a way to solve this problem completely. So, researchers have been continuously exploring to improve existing methods or combine different methods to learn the advantages of the those algorithms.

Recently, because of the strong power of algorithms (GA) to find global optimum for optimization problems, GA have been widely used as an alternative method to generate the optimal path. Hu and Yang. [10] proposed a knowledge-based GA. A parallel elite GA for global path planning is proposed in [11].

© Springer International Publishing AG 2017
D. Liu et al. (Eds.): ICONIP 2017, Part IV, LNCS 10637, pp. 288–295, 2017.
https://doi.org/10.1007/978-3-319-70093-9_30

However, there are some problems associated with these methods. Firstly, the initial population is not chosen very well, which can make GA behave poorly. Secondly, there are not sufficient heuristic knowledge based genetic operators. Thirdly, GA has a poor real-time property to respond to emergency situations that may be encountered. In this paper, to circumvent these drawbacks of conventional GA, we present a hybrid method based on the combination of IGA and IAPF.

In this paper, a robot path planner based on IGA and IAPF is proposed, which can ensure that robot can find the optimal path and avoid obstacles in real time by taking the advantages of the hybrid algorithm.The remainder of this paper is organized as follows: In Sect. 2, an IGA is proposed to find a global optimal collision-free path. Section 3 elucidates the procedure of how to combine IGA and IAPF to get a real-time obstacle-avoidance path. Section 4 shows the results of some simulations to demonstrate the performance and merit of the proposed methods. The conclusions are given in Sect. 5.

2   Proposed Improved Genetic Algorithm

This section presents an effective IGA which can solve the global path planning problem.

2.1   Environment Modeling and Chromosome Encoding

In IGA, a grid-based representation is used for the work space of robot motion, as shown in Fig. 1. The shadow grids represent obstacle areas and the blank grids represent free areas. In order to treat the robot as a point, the girds' boundaries equal to their real boundaries add to a safety distance that is defined with consideration of the size of the robot. Therefore, a path can be encoded as a sequence of coordinates which starts from the initial grid and ends at the goal grid with a series of intermediate grids. For example, if coordinates (1, 1) are the start coordinates and coordinates (20, 20) are the goal coordinates, then the feasible path shown in Fig. 1 can be encoded as a series of coordinates {(1, 1), (1, 14), (17, 14), (20, 20)}. The coordinates encoding (17, 14) represents the grid of 17th rows and 14th columns in grid map. The individual uses the real coordinates as genes, which can reduce calculation time because no time is costed for encoding and decoding of the genes in the population.

2.2   Initialization of the Population and Cost Function

The initial population is generated using the approach proposed in [12]. This method uses the greedy algorithm based on heuristic Euclidean distance which can help build initial paths fast and feasibly. The convergent speed of GA can be speeded up by reasonable initialization.

In the case of robot path planning, the optimal path should be shortest and collision-free. To take into account these criteria, a cost function is used and this

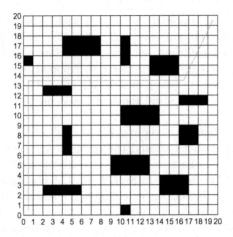

Fig. 1. Grid-based environment with obstacles

problem becomes a search for a path which has a lower cost. The cost function is defined as follows:

$$F_{cost} = \sum_{i=1}^{N} (d_i + \beta_i C) \qquad (1)$$

where N denotes the number of path segments in a individual, d_i is the length of the i_{th} path segment, C is a constant to penalize the infeasible solution, β_i represents the feasibility of the segment, and if the i_{th} path segment is feasible, β_i equals to 0, else it equals to 1.

2.3   Genetic Operators

Crossover. We developed a new smart operator based on one-point random crossover, it randomly select a node denoted as X excluding the starting and ending points from Parent 1, and the distance between node X and the each node which is in Parent 2 excluding the start and goal is compared, the node Y in Parent 2 which has the minimum distance with node X is selected as the crossover point of Parent 2, then cross operation is performed between X and Y. Through this operation, we can try to ensure that the population after crossing is more reasonable.

Mutation. Mutation is a significant operator to avoid the phenomenon of premature because it can increase the diversity of the population. In mutation operator, randomly choose one node and replace it with a new node which is not the obstacle node and not in the original path.

Improvement. This operation is designed to improve the quality of a path. One node P is chosen randomly, perform a local search in the neighboring grids of P, for each node, the cost function is calculated and P is replaced by the best node which has a minimum cost.

Deletion. Deletion is applied to decrease the cost of a path. Randomly choose one node Z in the path and remove it, then connect its two adjacent nodes, if the cost of the new path is lower, delete the node Z.

These operators are very significant during evolution. They ensure that the algorithm runs quickly in a desired direction.

2.4   Outline of the Improved Genetic Algorithm

A flow chart of the proposed IGA is shown in Fig. 2. Initial population are generated randomly through the greedy method. Then perform the genetic operators on them until some stop criterion is satisfied. The termination conditions are that algorithm reach the maximum generation or the optimal solution remains unchanged for certain generations. To speed up the convergence rate of IGA, reproduction strategy with elite policy is performed, 20% of the best individuals called elites are duplicated immediately to next generation. In order to increase the powerful ability to find global optimal solution, increasing the diversity of population by adding random initial population is very important. So in each generation, 20% of new random initial individuals are put in the population pool.

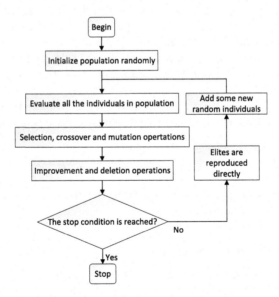

Fig. 2. Flow chart of proposed improved genetic algorithm

3   Collision Avoidance of Proposed Hybrid Algorithm

During the running of the robot, some unexpected dynamic obstacles may be encountered which may cause the planning path to become infeasible, so it is significant for robots to have the ability to avoid real-time obstacles. Because artificial potential field has elegant mathematical analysis and it is easy to calculate, this method is attractive in real-time obstacle avoidance. In this section, a new potential function is adopted and a hybrid algorithm is proposed.

3.1   Potential Function and Virtual Force

Attractive Potential Function and Virtual Force. In this paper, attractive potential function proposed in [1] is adopted, which is defined as a function of distance and velocity between the robot and the goal.

Repulsive Potential Function and Virtual Force. A new repulsive potential function is proposed to avoid moving obstacles. The new repulsive potential function is defined as follows:

$$U_{rep}(q,v) = \begin{cases} 0, \text{ if } (\rho_{obs} - R_{obs}) > \rho_0 \text{ or } v_{ro} \le 0 \\ \alpha_q(\frac{1}{\rho_{obs}-R_{obs}} - \frac{1}{\rho_0})(X - X_g)^2 + \alpha_v v_{ro}, \text{ else} \end{cases} \quad (2)$$

where ρ_{obs} is the distance from robot to the center of obstacle, R_{obs} is the radius of obstacle, ρ_0 is a constant which denotes the distance of influence that the obstacle has on the robot. $(X - X_g)^2$ represents the square of distance from robot to goal, by introducing this distance, the potential field can be guaranteed to be minimum at the target point, α_q and α_v are scale coefficients, v_{ro} denotes the relative velocity in the direction from robot to obstacle between robot and obstacle.

By computing the negative gradient of the repulsive potential function with regard to position and velocity, the virtual repulsive force is obtained as follows:

$$\begin{aligned} F_{rep}(q,v) &= -\bigtriangledown U_{rep}(q,v) \\ &= -\bigtriangledown_p U_{rep}(q,v) -- \bigtriangledown_v U_{rep}(q,v) \end{aligned} \quad (3)$$

The total virtual force obtained by adding the attractive force and repulsive force can guide the robot without collision.

3.2   Proposed Hybrid Algorithm

A flow chart of the proposed hybrid algorithm is shown in Fig. 3. First, IGA is performed to get a global optimum path such as $\{P_0, P_1,..., P_i,..., P_n\}$ which is usually shortest, P_0 is the beginning and P_n is the destination. Then follow this path for real-time obstacle avoidance by using IAPF. For example, P_1 is initially set as the local target point of IAPF, and robot moves toward the point P_1, if robot arrives P_1, then next point P_2 is set as the local target point, this operator is repeated until robot arrives the goal point.

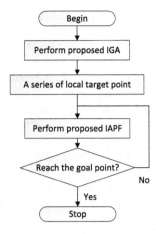

Fig. 3. Flow chart of proposed hybrid algorithm

4   Simulation Results

In this section, some simulation studies are presented to validate the superiority of the hybrid algorithm.

4.1   Simulation Experiments of Proposed IGA

A simulation is performed to compare IGA with the conventional genetic algorithm (CGA), the parameter setting for both GA is same as: population size is 50, probability for mutation is 0.1, and probability for rest genetic operators is 0.9. The number of max generations is 50. The choice of parameters is determined by performance and efficiency of the algorithm. The length of optimal path which is showd in Fig. 4 found by IGA is 27.4171, whereas the length of optimal path found by CGA is 34.3092, and sometimes CGA can not converge to feasible solution.

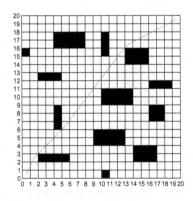

Fig. 4. The global optimum path obtained by IGA

4.2   Simulation Experiments of Proposed Hybrid Algorithm

A simulation is performed to verify that the robot can avoid real-time obstacle. The path obtained by hybrid algorithm is shown in Fig. 5. As shown in the left part of Fig. 5, the triangles denotes local target point obtained by IGA, the blue line segments represent global optimal path, and the red curve trajectory is the path obtained by hybrid algorithm when a dynamic obstacle is not encountered. The right part of Fig. 5 represents the real trajectory while a dynamic obstacle is encountered. A series of pink circles indicate the trajectory of the dynamic obstacle whose initial position is (8, 14) and direction of motion is downward. In Fig. 5, robot walks almost along the global optimal path and when a obstacle is encountered abruptly, robot can adjust its path to avoid obstacles in real time. As shown in Fig. 6, when robot moves into the vicinity of dynamic obstacle, robot turns in order to avoid obstacle.

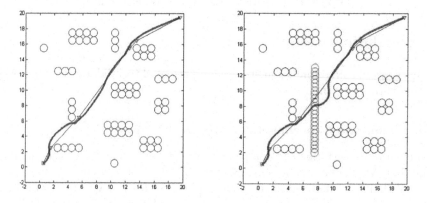

Fig. 5. The path obtained by hybrid algorithm (Color figure online)

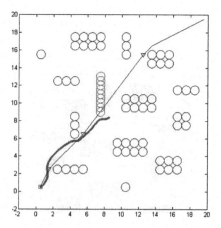

Fig. 6. The trajectories of robot and obstacles (Color figure online)

5   Conclusions

In this paper, a hybrid algorithm based on the combination of IGA and IAPF is proposed. This method can make full use of known environmental information to find a global optimal path. It can also cope with the dynamic obstacles encountered suddenly in the process of robot motion. So this ensures that the robot can avoid obstacles in real time as well as move along the optimal path. The results of simulation experiments demonstrate the effectiveness of path planning based on the hybrid algorithm.

Acknowledgments. The work was supported by the Natural Science Foundation of China under Grants 61673188 and 61761130081, the National Key Research and Development Program of China under Grant 2016YFB0800402, the Foundation for Innovative Research Groups of Hubei Province of China under Grant 2017CFA005.

References

1. Ge, S.S., Cui, Y.J.: Dynamic motion planning for mobile robots using potential field method. Auton. Robots **13**, 207–222 (2000)
2. Alexopoulos, C., Griffin, P.M.: Path planning for a mobile robot. IEEE Trans. Syst. Man. Cybern. **22**, 318–322 (1992)
3. Zhang, Y., Gong, D.W., Zhang, J.H.: Robot path planning in uncertain environment using multi-objective particle swarm optimization. Neurocomputing **103**, 172–185 (2013)
4. Byerly, A., Uskov, A.: A new parameter adaptation method for genetic algorithms and ant colony optimization algorithms. In: IEEE International Conference on Electro Information Technology, Grand Forks, pp. 668–763 (2016)
5. Zeng, X.P., Li, Y.M., Q, J.: A dynamic chain-like agent genetic algorithm for global numerical optimization and feature selection. Neurocomputing **74**, 1214–1228 (2013)
6. Mathias, H.D., Ragusa, V.R.: An empirical study of crossover and mass extinction in a genetic algorithm for pathfinding in a continuous environment. In: IEEE Congress on Evolutionary Computation, Vancouver, pp. 4111–4118 (2016)
7. Bao, Y.Q., Wu, H.Y., Chen, Y.: The multi-robot task planning based on improved GA with elite set strategy. In: IEEE International Conference on Robotics and Biomimetics, QingDao, pp. 1367–1371 (2016)
8. Lu, N.N., Gong, Y.L., Pan, J.: Path planning of mobile robot with path rule mining based on GA. In: Chinese Control and Decision Conference, YinChuan, pp. 1600–1604 (2016)
9. Shehata, H.H., Schlattmann, J.: Non-dominated sorting genetic algorithm for smooth path planning in unknown environments. In: IEEE International Conference on Autonomous Robot Systems and Competitions, Espinho, pp. 14–21 (2014)
10. Hu, Y., Yang, S.X.: A knowledge based genetic algorithm for path planning of a mobile robot. In: IEEE International Conference on Robotics and Automation, New Orleans, pp. 4350–4355 (2004)
11. Tsai, C.C., Huang, H.C., Chan, C.K.: Parallel elite genetic algorithm and its application to global path planning for autonomous robot navigation. IEEE Trans. Ind. Electron. **58**, 4813–4821 (2011)
12. Alajlan, M., Koubâa, A., Châari, I., Bennaceur, H., Ammar, A.: Global path planning for mobile robots in large-scale grid environments using genetic algorithms. In: International Conference on Individual and Collective Behaviors in Robotics, Sousse, pp. 1–8 (2013)

A Portable System of Visual Fatigue Evaluation for Stereoscopic Display

Yue Bai[1,2], Jun-Dong Cho[2], Ghulam Hussain[2], and Song-Yun Xie[1(✉)]

[1] School of Electronic and Information,
Northwestern Polytechnical University, Xi'an, China
syxie@nwpu.edu.cn
[2] Department of Electrical and Computer Engineering,
Sungkyunkwan University, Suwon, Korea

Abstract. Stereoscopic display is contributing to realistic three dimensional (3D) effect which has been widely prevalent and successfully commercialized. However, visual fatigue is still an unsolved issue for these applications and has negative effects on viewers. In this paper, we proposed a method based on analysis of the production theory of 3D display and measurement of biological signals to evaluate visual fatigue. Given that two types of methods have a complementary relationship, we designed a Fuzzy Fusion of Visual Fatigue (FFVF) model using vergence-accommodation conflict (VAC) and electroencephalogram (EEG) signal. By utilizing the fuzzy theory as a fusion method to multiple features, our proposed FFVF model shows a high Pearson correlation value of 0.9676 with questionnaire results while maintaining high stability. This kind of portable human-friendly 3D viewing evaluation technology can be widely deployed.

Keywords: EEG · Fuzzy theory · Visual fatigue · Vergence-accommodation conflict (VAC)

1 Introduction

Due to the conflict between the image theory of stereoscopic display and the human vision system, the discrepancy between vergence and accommodation also called vergence-accommodation conflict (VAC) is inevitable. Viewer feels visual discomfort and pain due to functional overload of brain which is an important cause driven by VAC [1]. Along with strong depth disparity and other external factors, visual fatigue will always exit. Subjective evaluation is an effective way to detect and determine the extent of different symptoms for visual fatigue [2]. However, the sensitivity of subjects to visual fatigue is variable and the results may be biased. Objective methods are utilized to measure a precisely level of the fatigue, which can be roughly divided into two methods: (1) analyzing 3D videos and (2) measuring biological signals.

Firstly, analyzing characteristics of current 3D display and the external environment to estimate visual fatigue [3–6]. Analyzing the inherent characteristics

© Springer International Publishing AG 2017
D. Liu et al. (Eds.): ICONIP 2017, Part IV, LNCS 10637, pp. 296–306, 2017.
https://doi.org/10.1007/978-3-319-70093-9_31

of 3D techniques show stable and objective results of visual fatigue measurement. However, the amount of visual fatigue could be changed depending on the viewer and the viewers status. Secondly, biological signals are used for the reliable assessment of products by providing objective measurements, even for slight changing external stimuli. So it is suitable for measuring factors of the visual fatigue level of individual viewers. Some previous researches focused on blinking rate, Electroencephalogram (EEG), cognitive functions (Event-related potential: ERP), Punctum Maximum Accommodation (PMA), electrocardiogram (ECG) and photoplethysmogram (PPG) etc. [7–9]. Zou et al. assessed three EEG activities, θ, α and β during a monotonous and repetitive random dot stereogram (RDS) based task in a conventional stereoscopic 3D display. Results of EEG data showed stable of θ activity and a significant increase of α activity, and a significant decrease of β activity over time [10]. But the results have a little conflict with CX Chen's results in his experiments [11], which show the energy in α and β frequency bands significantly decreased. It can be concluded that biological signals have unstable results and strongly affected by task type.

In our research, we chose gravity frequency of power spectrum (GF), power spectral entropy (PSE) and ratio algorithms $(\alpha + \theta)/\beta$ (R) of EEG as biological features, which showed remarkable correlation with fatigue status after compared between time domain, frequency domain and non-linear dynamic analysis. We use the portable EEG acquisition equipment, focused on EEG measurement at the posterior and frontal sites, the former is supposed to be related to visual processing, and the latter is to cognitive functions (such as attention and executive functions). Using three electrodes at the posterior and frontal (P8, F3, F7 with p-value < 0.05) to detect visual fatigue is enough and exact by correlation test and t-test. The portable device overcome limitations of poor portability, operational complexity and high cost.

Based on the above advantages and disadvantages, it is suitable to combine two kinds of objective features (physical feature of 3D display: VAC, biological features of EEG: GF, PSE and R) with one subjective feature (Subjective evaluation: questionnaire score (QS) parameters) to design a Fuzzy Fusion of Visual Fatigue (FFVF) model. The multi-modalities FFVF by using fuzzy theory as fusion method is a novel way to measure visual fatigue more accurately and easily, the simple and convenient system could provide elicitation to commercialization and availability.

2   Methods and Experiment

2.1   Visual Fatigue Prediction

Parameters of the 3D display are measured before experiment, thus we defined analyzing 3D videos as visual fatigue prediction (VFP). Prior researches show effect factors of disparity magnitude such as display methods, viewing distance, display size, video resolution, both are closely related to the accuracy of visual system [12].

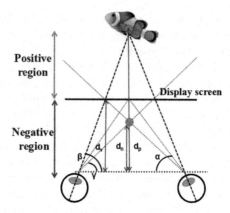

Fig. 1. Human vision system

As following equation, the values of $Disp_{cm}$ will change if the resolution and the size of showing device is different.

$$Disp_{cm} = Disp_{pixel} \times \frac{Display \quad width}{Resolution \quad width} \tag{1}$$

The amount of VAC was obtained by the disparity of viewpoint, as depicted in Fig. 1, α represents the viewpoint angle on the display screen. β and γ represent viewpoint angles in the positive and negative region separately. Then the VFP is defined as follows:

$$VAC = VFP = \begin{cases} \alpha - \gamma, & \text{if crossed disparity} \\ \beta - \alpha, & \text{if uncrossed disparity} \end{cases} \tag{2}$$

We use the disparity magnitude $Disp_{pixel}(x, y)$ to measure viewpoint angles α, β and γ. By using (1) we transform $Disp_{pixel}$ to $Disp_{cm}$, then there are following relationship:

$$\begin{cases} d_p : \frac{l}{2} = d_p - d_v : \frac{D_{cm}(x,y)}{2} \\ d_n : \frac{l}{2} = d_v - d_n : \frac{D_{cm}(x,y)}{2} \end{cases} \tag{3}$$

where d_v is the viewing distance, d_p is the positive perceived distance and d_n is negative perceived distance, and l is inter-ocular distance. Based on the trigonometric ratio, we substitute parameters in (2) as follows:

$$VAC = VFP = \begin{cases} \tan^{-1}\left(\frac{2d_v}{l}\right) - \tan^{-1}\left(\frac{2d_v}{l+D_{cm}(x,y)}\right), & \text{if crossed disparity} \\ \tan^{-1}\left(\frac{2d_v}{l-D_{cm}(x,y)}\right) - \tan^{-1}\left(\frac{2d_v}{l}\right), & \text{if uncrossed disparity} \end{cases} \tag{4}$$

2.2   Visual Fatigue Obtain

Subjects. 15 adults took part in the experiment: 7 females, 8 males; mean age 25.31 (SD $= 2.81$). All subjects were in good physical health and none of them were

Fig. 2. Experiment environment

reported of any visual disease. These subjects were notified have an excellent sleep
(6–8 h) before the experiment. For insure good physically and mentally condition,
wine, tea, coffee and drugs were prohibited. Before beginning experiment all of
them gave their informed written consent to take part in and were briefed on the
purpose of the experiment and experimental procedures. The test during 9:00–
11:00 AM in order to control the potentially physiological rhythms and this period
shown more awareness.

Apparatus. Stereoscopic images were shown in full HD resolution (1080p) on
a 67.57 cm LG D2743 (Width * High = 59 cm * 33 cm), an active display, subjects
wear matching shuttered glasses. Subjects sit in a soft chair and quite environment
with comfortable temperature, the distance between subjects and screen is 130–
150 cm. The setup and environment of the experiment are show in Fig. 2. EEG
signals are acquired using portable device Emotive EPOC on the micro-voltage
levels, this headset device has two reference nodes (CMS and DRL) and 14 other
electrodes.

Measures. In order to make the system has a wide range of adaptability, the
experiment include three trails which show the different types of 3D movies
randomly: documentary (Ocean Wonderland), action (Avengers) and animation
(Frozen). In each trail, subjects watching time is 20 min, before and after watching could have 2 min as relax. At the beginning of the experiment, subjects need
completed the questionnaire, then, operator help subject wear Emotive, and subjects close eyes and measured brain signals in 5 min if everything is ready, as well
as after finished watching the 3D movie. During the recorded, all subjects tried to
relax and avoid unnecessary movements. The proposed experiment process of data
acquisition and analysis are show in Fig. 3.

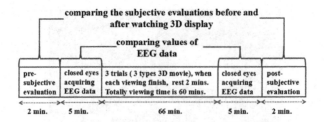

Fig. 3. The experiment procedure

Table 1. The fatigue symptoms of questionnaire

Eye fatigue	Body fatigue
Eyestrain	Nausea
Watery eyes	Dizziness
Bleary eyes	Headache
Difficult focusing	Vomiting
Double vision	Numbing

The subjective evaluation questionnaire consisted of five questions on eye fatigue and five questions on body fatigue, and participants answered on a five-point rating scale: 1: Very severe, 2: Severe, 3: Moderate, 4: Comfortable, 5: Very comfortable. Table 1 shows the fatigue symptoms in the questionnaire. Our proposed experiment composition related to participants and the questionnaire was designed following ITU recommendation [13].

2.3   Data Processing

The raw EEG data of each trial for each subject was processed with a 50 Hz notch filter and 0.5–30 Hz band filter (IIR digital filter). After that choose 1 min EEG data without obvious interference and discontinuity in the close eyes period to the following analysis [14]. Then the gravity frequency of power spectrum (GF) was calculated by:

$$GF = \sum_{w=w_1}^{w_2} (p(\hat{w}) \cdot w) / \sum_{w=w_1}^{w_2} p(\hat{w}) \tag{5}$$

where $p(\hat{w})$ is power spectrum, w_1, w_2 represent frequency upper and lower limits respectively, namely from 0.5–30 Hz. Then the power spectral entropy (PSE) is given by:

$$PSE = - \sum_i p_n(i) \cdot \log_2(p_n(i)) \tag{6}$$

where $p_n(i)$ represents the probability density distribution of power spectrum at w_i. $\sum_i p_i = 1$. Every segment data was used band-pass filter separately, then acquired four frequency wavebands. The ratio algorithm $(\alpha + \theta)/\beta$ (R) could be calculated by:

$$R = (p_\alpha + p_\theta)/ + p_\beta \tag{7}$$

where p_α, p_β and p_θ represent the power spectrum of α, β and θ respectively.

2.4 Fuzzy-Fusion Method

The fundamental purpose of fuzzy theory method is to define an uncertain state based on the relationship between the characteristic of datasets [15]. Since a defined uncertain state includes information on the relationship between input datasets, the results of defuzzification can be used as a weight value. The entire process of obtaining a weight value to each input factors using fuzzification and defuzzification has five steps as depicted in Fig. 4.

The first step, drawing a fuzzy rule table. The fuzzy rule table describes the characteristic of input data sets and defines the level of fuzzy output depending on the relationship between the input and output. In this study, we perform a quality measurement for five factors: GF, PSE, R, VFP and QS. The two features (F_1 and F_2) are extracted from the corresponding values of each factor, they are used as the inputs for the fuzzy system to produce the quality (weight) values. Each factor indicates the variation of EEG, VAC and Questionnaire score (QS) before and after watching 3D display. For the EEG feature we choose the difference value of the first three electrodes as F_1 and F_2 after using Pearson correlation analysis results. The membership function and rules is given by operators control action and knowledge, we adopted the triangular membership function according to processing speed and complexity of problem [16] and applied center of gravity (COG) as defuzzification methods. After acquired weight values of each modality, we normalized the weight values as follows:

$$W_i = \frac{w_i}{\sum_{k=1}^{n} w_{EEG_k} + w_{VFP} + w_{QS}} \tag{8}$$

where i is EEG_k, VFP and QS, k is GF, PSE and R. Finally, we proposed visual fatigue evaluation system obtain one final output by combining EEG, VFP and QS with a corresponding weight is given as follows:

$$FFVF = \sum_{k=1}^{n}(EEG_k \times W_k) + VFP \times W_{VFP} + QS \times W_{QS} \tag{9}$$

Fig. 4. Process flow of fuzzy theory

3   Results and Analysis

3.1   VFP Analysis

In the above mentioned formulation (1), by programming in Visual Studio 2013, we executed sum of squared differences (SSD) algorithm to calculate the disparity map in each frame of video segments. After equalizing acquired VFP factors, the linear relevant fitting methods are applied to verify our proposed VFP. As shown in Fig. 5, VFP shows a high fitting coefficient ($R^2 = 0.8964$, adjusted $R^2 = 0.8816$) in linear relevant fitting with QS.

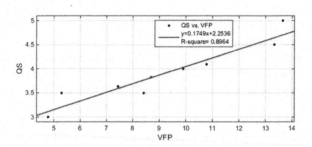

Fig. 5. Linear relevant fitting of Questionnaire Score (QS) and Visual Fatigue Prediction (VFP)

VFP produces more stable results than biological signals due to the results of a certain value are affected by only one parameter. On the other hand, the results of each biological signal composite all possibility of viewers condition. Thus, the proposed combine VFP factors in FFVF method will increase reliability and maintain stability.

3.2   EEG Signal Analysis

GF and PSE show significantly decreases in several brain regions after long time of watching 3D display, meanwhile R values show significantly increase. In order to select the best representative channel of each EEG feature, we calculated the Pearson correlation between each variation of factors and Questionnaire Score (QS) respectively. Table 2 shows the top three r-value electrodes of GF, PSE and R

Table 2. The Pearson correlation of GF, PSE, R

Factors	GF			PSE			R		
Channel	P8	F3	FC5	F3	P8	P7	F7	FC5	P8
r-value	0.7248	0.6991	0.5236	0.7324	0.7081	0.6028	0.8442	0.8081	0.7274
p-value	0.0022	0.0037	0.0451	0.0012	0.0031	0.0173	7.5×10^{-5}	0.0002	0.0021

respectively, strength attachment electrodes of the posterior and the frontal have higher variation than others, except temporal cortex (T7, T8) which is irrelevant with visual processing and prefrontal cortex (AF3, AF4) which is too close to the eyes and therefore easily influenced by eye movements.

3.3   Fusion Process and Results

We conduct a quality measurement for three features to detect a more accurate visual fatigue level. For EEG signals, if the tendency of difference value between before and after watching 3D display in the three nodes are consistently similar, that means we acquired brain signal is stable and representative, so the data is acceptable as consequence of F_1 and F_2. Instead, if the quality of the acquired EEG signal is poor or variation, which is frequently caused by EEG signal noise related to the movement of the head or facial muscle. For example, the differences among the values of the P8, F3 and FC5 nodes are considerable, which makes the influence of the F_1 and F_2 of the EEG signal significant. Thus, the two input of GF, PSE and R factors are negative proportion with output data.

For the VAC features, VFP is leveraged for penalizing the values over the threshold of a visual comfortable zone, the stronger stereoscopic effect the higher visual fatigue level, so the maximum value of VFP (F_1) is positive proportion with output data, whereas the minimum value of the VFP (F_2) is negative proportion.

For the subjective evaluation features, in the case of a higher users preference and greater number of users watching 3D movies, we can assume that the user is more accustomed to 3D content and he (or she) can perform a more accurate and objective QS. That is because body and mind will fight it with procrastination due to dislike. So the two input of QS factors is both positive proportion with output data. The detail information of two input features as indicated in Table 3.

Table 3. Two input features for calculate the weight values

Factors	Feature	Explanation of feature
EEG	F_1	Difference of GF between P8-F3
GF	F_2	Difference of GF between P8-FC5
EEG	F_1	Difference of PSE between F3-P8
PSE	F_2	Difference of PSE between F3-P7
EEG	F_1	Difference of R between F7-FC5
R	F_2	Difference of R between F7-F8
VAC	F_1	Max value of the VFP
VPF	F_2	Min value of the VFP
QS	F_1	User preference for watching 3D content
	F_2	Number of users watching 3D movies

After acquired all weight of each factor, utilizing function (9) then obtain the finally visual fatigue level of the FFVF model. The proposed method for measuring eye fatigue was implemented by MATLAB program (MATLAB R2014b). The final values of visual fatigue using proposed FFVF were calculated Pearson correlation with subjective evaluation, the results showed r = 0.9676 and p = 3.8×10^{-9}.

4   Discussion

Our experiment demonstrated that EEG power spectrum will change with the difference of alertness and vigilant state, the subjects' before state shows a more sober and vigilant state than after watching 3D display. The comparisons shown in Fig. 6(a–c) by paired-t-test.

Figure 6(a) indicate that GF parameters significantly decreased after watching 3D display, especially in FC5, O1, P8 channel. Similarity, as shown in Fig. 6(b), the PSE factors also decreased which represent the complexity of the disorder time sequence signals and the level of chaos of multi-frequency components. However, as shown in Fig. 6(c), the R factors have clearly increased which related to the mental alertness level, that was more reliable fatigue indicator since it showed a distinct indication of increasing fatigue as the scale between the slow wave and fast wave activities increased. Figure 6(d) is the comparison of subjects answer to their con-

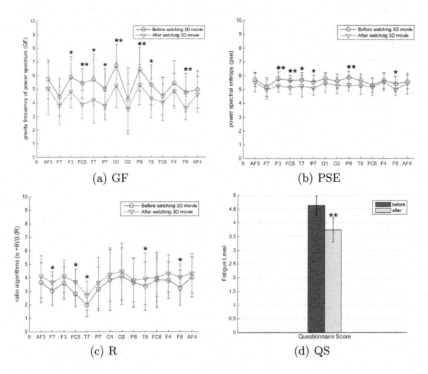

(a) GF

(b) PSE

(c) R

(d) QS

Fig. 6. Comparison of each channel of GF, PSE, R and subjective evaluation between before and after watching (*$p \leq 0.05$, **$p \leq 0.005$)

Table 4. Comparison of multiple linear regression and fuzzy fusion theory

	Pearson correlation	R^2	Adjusted R^2	p-value
Multiple linear regression	0.9429	0.8891	0.8806	p < 0.001
Proposed FFVF system	0.9676	0.9362	0.9313	p < 0.001

dition between before and after watching 3D movie, which indicates that after 1 h watching 3D movie can lead to fatigue on mental and physiology.

Figures 5 and 6 demonstrate that validity of type and number of proposed parameters necessity to build this system. Furthermore, in terms of performance, our proposed model verified $R^2 = 0.9362$, adjusted $R^2 = 0.9313$ ($p < 0.001$) showed better results than multiple linear regression analysis as Table 4.

5 Conclusion

In this paper, we proposed a novel visual fatigue evaluation system by using fuzzy theory to combine EEG signals and physical features of 3D display. EEG signals can reflect human physical condition more accurately than other biological signals, we chose three EEG features which change obviously during watching 3D display. Analysis of the production theory of 3D display and vergence-accommodation conflict (VAC) indicated a direct relationship to the fundamental cause of visual fatigue, then we acquired objective physical features by equalizing VAC parameters in our experiment video. Comparing with prior researches, our work has following advantages: (1) With analyzing the brain regions associated with visual fatigue, we verified 3 most relevant electrodes to visual fatigue, which are chosen as the system input is enough. Thus our work reduced the computational complexity and improved the accuracy. (2) Exploration of the source of watching fatigue, obtaining one final value for the variation of visual fatigue level by two types features fusion, meanwhile, considering the subjective opinion, the system also includes subjects evaluation. (3) Proposed weight values of each feature based on the fuzzy theory, the dynamic evaluation overcame external and internal factors change, adjusted to different subjects and different watching conditions.

The evaluation system give visual fatigue level with high accuracy and stability, but in the way of fusion method of fuzzy theory, we determined membership function and rules by expert experience and prior knowledge. For future work, we should consider set system parameters automatically by combine neurol control to fuzzy control. Moreover, considering the integration of a variety of patterns, further optimize the FFVF system performance for real-time visual fatigue monitoring.

Acknowledgments. This work was supported in part by National Natural Science Foundation of China (61273250), the Fundamental Research Funds for the Central Universities (No. 3102017jc11002).

References

1. Park, M., Mun, S.: Overview of measurement methods for factors affecting the human visual system in 3D displays. J. Disp. Technol., 1 (2015)
2. Shibata, T., Kim, J., Hoffman, D.M., Banks, M.S.: The zone of comfort: predicting visual discomfort with stereo displays. J. Vis. 11(8), 74–76 (2011)
3. Jeong, H.G., Ko, Y.H., Han, C., Oh, S.Y., Park, K.W., Kim, T.: The impact of 3D and 2D TV watching on neurophysiological responses and cognitive functioning in adults. Eur. J. Public Health 25(6), 1047–1052 (2015)
4. Wenzel, M.A., Kraft, R., Blankertz, B.: EEG-based usability assessment of 3D shutter glasses. J. Neural Eng. 13(1), 9 (2015)
5. Cho, J.D., Hussain, G., Park, J.H., et al.: Visual fatigue measurement model based on multi-area variance in a stereoscopy. In: Proceedings of the 2016 18th International Conference on Advanced Communication Technology (ICACT), pp. 1–2 (2016)
6. Kim, N., Park, J., Oh, S.: Visual clarity and comfort analysis for 3D stereoscopic imaging contents. Int. J. Comput. Sci. Netw. Secur. (IJCSNS) 11(2), 227–231 (2011)
7. Kim, N., Park, J., Oh, S.: Visual clarity and comfort analysis for 3D stereoscopic imaging contents. In: Park, D.S., Chao, H.C., Jeong, Y.S., Park, J. (eds.) Advances in Computer Science and Ubiquitous Computing. LNEE, vol. 373, pp. 227–231. Springer, Singapore (2011). doi:10.1007/978-981-10-0281-6_29
8. Frey, J., Appriou, A., Lotte, F., Hachet, M.: Classifying EEG signals during stereoscopic visualization to estimate visual comfort. Comput. Intell. Neurosci. 2016, 1–11 (2016)
9. Park, S., Won, M.J., Mun, S., Whang, M.: Does visual fatigue from 3D displays affect autonomic regulation and heart rhythm? Int. J. Psychophys. 92(1), 42–48 (2014)
10. Zou, B., Liu, Y., Guo, M., Wang, Y.: EEG-based assessment of stereoscopic 3D visual fatigue caused by vergence-accommodation conflict. J. Disp. Technol. 11(12), 1076–1083 (2015)
11. Chen, C.X., Li, K., Wu, Q.Y., et al.: EEG-based detection and evaluation of fatigue caused by watching 3DTV. Displays 34(2), 81–88 (2013)
12. Oh, C., Ham, B., Choi, S., Sohn, K.: Visual fatigue relaxation for stereoscopic video via nonlinear disparity remapping. IEEE Trans. Broadcast. 61(2), 142–153 (2015)
13. ITU-R: Methodology for the subjective assessment of the quality of television pictures. ITU-R. Technical report BT-500-13 (2012)
14. Chen, C.X., Wang, J., Li, K., et al.: Assessment visual fatigue of watching 3DTV using EEG power spectral parameters. Displays 35(5), 266–272 (2014)
15. Zedeh, L.A.: Knowledge representation in fuzzy logic. IEEE Trans. Knowl. Data Eng. 1(1), 89–100 (1989)
16. Feng, G.: A survey on analysis and design of model-based fuzzy control systems. IEEE Trans. Fuzzy Syst. 14(5), 676–697 (2006)

A Swarm Optimization-Based Kmedoids Clustering Technique for Extracting Melanoma Cancer Features

Amin Khatami[1]([✉]), Saeed Mirghasemi[2], Abbas Khosravi[1], Chee Peng Lim[1], Houshyar Asadi[1], and Saeid Nahavandi[1]

[1] Institute for Intelligent Systems Research and Innovation, Deakin University, Geelong, VIC 3216, Australia
{skhatami,abbas.khosravi,chee.lim,houshyar.asadi, saeid.nahavandi}@deakin.edu.au
[2] School of Engineering and Computer Science, Victoria University of Wellington, Wellington, New Zealand
saeed.mirghasemi@ecs.vuw.ac.nz

Abstract. Melanoma is a dangerous type of skin cancers. It is alarming to see the increase of this noxious disease in modern societies, however, it can be cured by surgical excision if it is detected early. In this paper, a swarm-based clustering technique for detecting melanoma is developed. Meaningful colour features from images are extracted, and a new objective function is introduced by applying an efficient and fast linear transformation to detect Melanoma. Specifically, the proposed technique consists of three main phases. The first phase is a pre-processing stage to organize data into proper attributes, while the subsequent two phases comprise iterative swarm optimisation procedures. The iterative swarm optimisation procedures involve a linear transformation to convert the existing colour components into a new colour space, formulation of the Kmedoids objective function, and error minimisation of the particle swarm optimisation (PSO) solutions. The Otsu threshold technique is utilised to provide binary images. The proposed technique is efficient and effective due to its linearity and simplicity.

Keywords: Melanoma skin cancer · Kmedoids clustering · PSO · Otsu threshold technique · Colour space

1 Introduction

Skin cancer is a fast-growing medical problem in recent years. New researches have revealed a 20% increase on average in the rate of diagnosed cases of skin cancer in modern societies; for instance, Australia has the highest incidence rate of 50–60 per 100,000 [16]. Exposure of skin to ultraviolet radiation (UV) of the sun is the major reason causing a cell to become cancerous. One of the most dangerous types of skin cancer is Melanoma. It develops in the cells producing

© Springer International Publishing AG 2017
D. Liu et al. (Eds.): ICONIP 2017, Part IV, LNCS 10637, pp. 307–316, 2017.
https://doi.org/10.1007/978-3-319-70093-9_32

melanin. According to Mayo clinic[1], the exact cause of this disease is not clear, however exposure to UV from sunlight raises the risk of developing Melanoma. Melanoma is a familiar word to most Australians; over 1,200 mortality per year. However, if Melanoma is diagnosed in the early stage, it can be successfully cured by surgical excision.

Recently, a number of comparative studies have been conducted to segment and cluster skin images for Melanoma detection [3,15]. Celebi et al. [3] introduced an unsupervised algorithm utilizing a modified version of the JSEG algorithm [5] to detect lesion border in skin images. Two independent steps, color quantization and spatial segmentation, were considered to separate the segmentation process. In [15], the Support Vector Machine (SVM) was used to classify features extracted from the lesion.

Colour-based features are mostly analytical parameters determined from different colour channels, such as averages and standard deviations of Red, Green, Blue (RGB) or Hue Saturation Value (HSV) colour channels [1,2,9–11,17]. A thresholding technique was proposed by Ganster et al. [7] for Melanoma detection by classifying the darker pixels as infected lesions. They applied a threshold to convert a gray-scale image into a binary one. The main difficulty of this technique is the choice of the threshold setting. Crisp and Tao [4] developed a clustering technique to detect Melanoma regions in an image. The number of clusters, i.e. k, was considered as an input factor, and the use of an inappropriate k could result in poor outputs.

Motivations: The aforementioned difficulties motivate us to develop an efficient and effective colour-based technique by utilizing the Kmedoids clustering technique and PSO to identify the important features describing a colour model to detect Melanoma. As demonstrated in the experimental part, by defining a new objective function, a significant reduction in the number of clusters can be achieved, resulting in an efficient thresholding stage. In addition, we use RGB as the original colour model because the colours associated with Melanoma skin cancer regions usually include shades of tan, brown, or black, and occasional patches of red, white, or blue [6].

Contribution: The main contribution of this study is formulation of a linear transformation that leads to an efficient detection technique. Another contribution is definition of an evolutionary-based clustering cost function using the linear transformation to distinguish between Melanoma and healthy pixels. Applying the Manhattan metric along with the linear conversion is also another novelty of this study, which has not been considered in previous researches in this field.

The rest of the paper is organized as follows. Section 2 describes the details of the proposed technique. Section 3 presents the experimental results, which is followed by concluding remarks and suggestions for future work in Sect. 4.

---

[1] http://www.mayoclinic.org/diseases-conditions/melanoma/basics/definition/con-20 026009.

2   The Proposed Technique

Defining a new colour system is a necessity for different approaches [12–14]. This study mainly concentrates on introducing a new colour space to detect Melanoma skin cancer pixels as the object of interest in a background. To achieve this goal, a pre-processing stage is introduced, which is followed by a PSO-based clustering technique to obtain the optimal weights denoted by W for describing the colour space. The achieved optimal W is used for a conversion equation to feed an objective function of the clustering technique to produce the proposed colour space. The following two transformation equations are introduced in this study to achieve the Melanoma skin cancer colour models. In the experimental section, the colour models obtained by these transformations are compared with another one using different accuracy measures.

$$y = W \times x \tag{1}$$

$$y = (W \times (x \times W)^T)^T \tag{2}$$

where W is a 3×3 conversion matrix, x represents a matrix of features extracted from the RGB colour model, and y is the converted image to the new colour model. These conversions transform a point in the original space, x, to the corresponding point in the new space, y.

Kmedoids is the clustering technique used in the study. It is a partitioning method considering medoids as the cluster centres [8]. Medoids are representative objects of the clusters that have minimal average dissimilarity to all objects in the cluster. Kmedoids uses an objective function as a measure to cluster the available data samples. The following equation is the objective function used in Kmedoids:

$$Cost(X, V) = \sum_{i=1}^{c} \sum_{j=1}^{n} d^2(x_j, v_i) \tag{3}$$

where $X = \{x_1, x_2, \ldots, x_n\}$ is the data samples, n is the number of data samples, c is the number of clusters. $V = \{v_1, v_2, \ldots, v_c\}$ denotes the cluster centres, v_i is the centre of the ith cluster, and $d^2(x_j, v_i)$ is a measure between x_j and v_i. The following measures, Euclidean (4) and Manhattan distances (5), are two metrics used in this objective function.

$$d(\mathbf{x}, \mathbf{v}) = \|\mathbf{x} - \mathbf{v}\|_2 = \sqrt{\sum_{i=1}^{n} (x_i - v_i)^2} \tag{4}$$

$$d(\mathbf{x}, \mathbf{v}) = \sum_{i=1}^{n} |x_i - v_i| \tag{5}$$

The following error function is the fitness function defined in this study, which is a measure of the conversion error.

$$Fitness\ Value = (e_{n1} + e_{n2}) \tag{6}$$

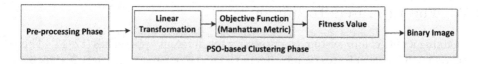

Fig. 1. The overview of the proposed method

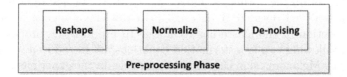

Fig. 2. The first stage of the proposed algorithm

where e_{n1} is the number of pixels which are not part of the first cluster and e_{n2} is the number of pixels which are not part of the second cluster.

Figure 1 shows the procedure for Melanoma detection our proposed technique. It contains three main phases. The pre-processing phase involves re-shaping, normalising, and de-noising, as shown in Fig. 2. As the images exist in different shapes, a re-shape function is first activated during pre-processing. Then, the normalisation function is called to normalise the image features between 0 and 1. Next, the de-noising function is used to remove hair-pixels of the object. This step is a necessity for most biomedical images containing hair pixels on skin images. In the second phase, we extract some pixels of the object (lesion region) and non-object (non-lesion region) manually from the pre-processed output. Each of them is in the shape of 25×25 pixels as a sub-image. Therefore, each feature sample contains a $25 \times 25 \times 3$ matrix. The third dimension (i.e., 3) represents the R, G, and B components. These matrices are considered as the feature components for clustering using PSO. Equation 1, captured by a 3×3 W matrix, is the linear transformation used in the objective function of Kmedoids, Eq. 8. In fact, Kmedoids and PSO parameters are initialized at the first stage of this phase. Each W matrix corresponds to a PSO particle. Each particle is a potential solution. Accordingly, all particles have nine dimensions, i.e., $x_i = (x_{i1}, x_{i2}, \ldots, x_{i9})$. PSO starts with a random W matrix and the features depicted in Fig. 4(d, e). The main aim of this PSO-based clustering phase is to produce optimal features for Eq. 1. To obtain the best features of this colour model, the Kmedoids objective function, Eq. 8, segments the images, and the fitness function of PSO, Eq. 6, calculates the clustering error. This iterative phase generally finds the best W matrix for Eq. 1 resulting in a suitable colour model for detecting Melanoma pixels. At the end of this phase, the achieved W matrix is generalised to all images in the database to obtain the corresponding converted images. In the last phase, the Otsu threshold technique is employed to produce binary images. Figure 3 depicts the procedure step-by-step. The following pseudo-code describes the details of the procedure in Fig. 3. The objective

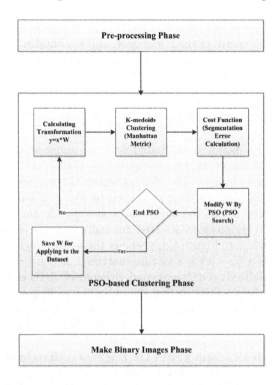

Fig. 3. The flowchart of the proposed method

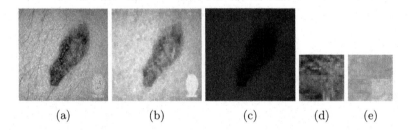

| (a) | (b) | (c) | (d) | (e) |

Fig. 4. (a) Original image. (b) Removing hair pixels (de-noising). (c) The output of the PSO-based clustering phase. (d) Object region. (e) Non-object region (Color figure online)

is to find the best transformation matrix aiming to an efficient colour space for Melanoma detection.

$$J(U,V) = \sum_{k=1}^{n} \sum_{i=1}^{c} \|y_k - v_i\| \tag{7}$$

$$J(U,V) = \sum_{k=1}^{n} \sum_{i=1}^{c} \|(x * W)_k - v_i\| \tag{8}$$

where c is the number of clusters. $V = \{v_1, v_2, \ldots, v_c\}$ is the cluster prototypes set in which v_i is the centre of the *ith* cluster, and U contains the data samples. Note that both linear metric and linear transformation are used to define the objective function, resulting in an efficient detection technique.

3 Experimental Results

The proposed colour model is evaluated using a public database introduced by Diepgen and Yihune et al. i.e., Dermatology Online Atlas DermIS[2]. This database consists of 64 images. The ground truths of the images are provided.

Firstly, the original images were re-shaped to 256 × 256 dimensions. Then, hair pixels were removed, and the object and non-object features were extracted. As such, two 25 × 25 regions were selected manually from the chosen images, and they were known as lesion. Figure 4(d, e) shows the feature regions. We put them along each other as a 50 × 25 × 3 feature matrix. The third dimension showed the three features R(Red), G(Green), and B(Blue) of RGB colour space. This feature matrix was fed to PSO to find the optimal weight matrix. The optimal w yields a significant decrease in the number of bins in histogram of the images. This leads to a proper automatically selected threshold, and result in a robust Otsu technique. It shows the fundamental shape of the distribution and also the numbers of bins with a uniform width. The height of each rectangle indicates the number of elements in the bin. With respect to distributions, it is clear that the proposed technique is effective for defining a colour space to detect Melanoma.

Figure 5 depicts the diffusion of the colour components of Fig. 4(a) in the RGB colour space and the diffusion of the colour components of Fig. 4(c) once

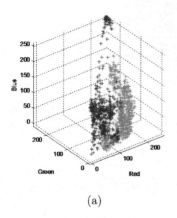
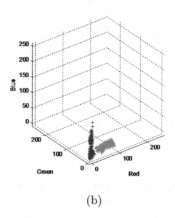

(a)   (b)

Fig. 5. Diffusion of colour components of Fig. 4(a) before and after conversion. (a) The original image (Fig. 4(a)). (b) The converted image (Fig. 4(c)). (Color figure online)

---

[2] Dermatology Information System published online at: http://www.dermis.net/doia/, 2012, Accessed: 08 Nov 2012.

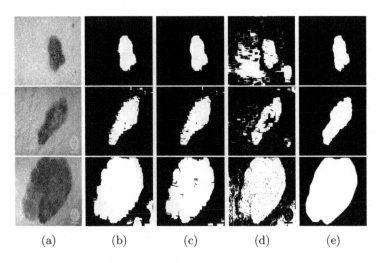

(a)   (b)   (c)   (d)   (e)

Fig. 6. (a) Original image. (b) Result with Eq. 2. (c) Result with Eq. 1. (d) Result with SVM. (f) Ground truth.

converted. It is obvious from these 3-D plot that the new colour space has two important attributes. Firstly, it tends to differentiate between the colour components of Melanoma and healthy pixels as much as possible. Secondly, the variance of colour distribution in the healthy and non-healthy pixels is reduced.

Table 1. The comparison in terms of precision, recall, and F-measure

	Precision%	Recall%	F-measure%
Manhattan metric + linear conversion	94	76	88
Euclidean metric + linear conversion	96	71	87
Euclidean metric + non-linear conversion	95	73	86
SVM	66	87	68

To convert the original images into the new colour model, we implemented two different transformations, linear and non-linear. As depicted in Fig. 6, and shown in Table 1, the accuracy rate of linear conversion (Eq. 1) is better than that of nonlinear conversion (Eq. 2) by 1% for both precision and F-measure.

We also examined the selected linear transformation with two different metrics, Eq. 4 as Euclidean and Eq. 5 as Manhattan distances fed into Eq. 8. As shown in Fig. 7 and Table 1, the Manhattan distance generally performed better than the Euclidean distance, i.e., by 2% for both precision and F-measure. A support vector machine (SVM) classifier with radial basis kernel was employed to compare its output with our experiment results. As shown in Fig. 6, our proposed technique outperformed the SVM by 28% in precision and 20% in F-measure. This improvement shows that defining the objective function based on

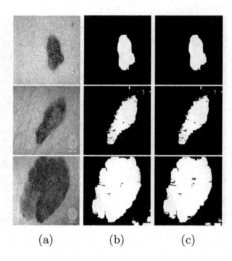

(a)   (b)   (c)

Fig. 7. (a) Original image. (b) Result with Minhattan distance (Eq. 5). (c) Result with Eq. 1.

the linear transformation and metric functions leads to proper diffusion of colour components, resulting in an efficient and effective colour space for Melanoma detection.

The following matrices are the optimal W setting calculated by the proposed technique using both linear transformation and Manhattan distance coupled with the objective function:

$$W = \begin{pmatrix} -0.147 & -0.361 & -0.774 \\ 1.674 & 2.942 & 4.212 \\ -1.169 & -1.917 & -2.671 \end{pmatrix} \tag{9}$$

W matrix was generalised to all images. The final segmentation of the lesion regions was obtained by using the Otsu threshold technique to produce the corresponding binary images.

4   Conclusions

ABCDE features are the main information utilized to detect Melanoma. Among them, colour-based features are useful analytical parameters, playing an important role for Melanoma detection. This study presents a new colour model for Melanoma skin cancer detection. We have proposed a method comprising Kmedoids and PSO for clustering Melanoma lesion in skin images. Noise such as hair on or around the lesion is always a problem in medical images. It should be removed in order to detect the object accurately. The experimental results show that applying a linear transformation and linear metric to define the new objective function for the Kmedoids method optimized by the fitness function defined in PSO leads to an efficient and effective technique for Melanoma detection. The comparison results after de-noising the images show that, on average, the accuracy rates of precision and F-measure are 96% and 86% using linear conversion of the objective function, as compared with 94% and 86% using nonlinear conversion, respectively. Comparing both Manhanttan and Euclidean distances with the objective function, the former performed better with a 2% higher F-measure rate. A significant improvement in terms of precision and F-measure can be seen in Table 1 for the proposed technique, as compared with the SVM classifier.

For further work, we will convert all images to the HSV colour space, and examine the performance of the proposed clustering technique. We will also study automatic feature selection and extraction methods, instead of manual feature selection. In addition to general features like area, border, shape, and colour, the high level features should also be integrated for automated Melanoma detection.

References

1. Babaie, M., Kalra, S., Sriram, A., Mitcheltree, C., Zhu, S., Khatami, A., Rahnamayan, S., Tizhoosh, H.R.: Classification and retrieval of digital pathology scans: a new dataset. arXiv preprint arXiv:1705.07522 (2017)
2. Cascinelli, N., Ferrario, M., Bufalino, R., Zurrida, S., Galimberti, V., Mascheroni, L., Bartoli, C., Clemente, C.: Results obtained by using a computerized image analysis system designed as an aid to diagnosis of cutaneous melanoma. Melanoma Res. **2**(3), 163–170 (1992)
3. Celebi, M.E., Aslandogan, Y.A., Bergstresser, P.R.: Unsupervised border detection of skin lesion images. In: International Conference on Information Technology: Coding and Computing, ITCC 2005, vol. 2, pp. 123–128. IEEE (2005)
4. Crisp, D.J., Tao, T.C.: Fast region merging algorithms for image segmentation. In: The 5th Asian Conference on Computer Vision (ACCV2002), Melbourne, Australia, pp. 23–25 (2002)
5. Deng, Y., Manjunath, B.: Unsupervised segmentation of color-texture regions in images and video. IEEE Trans. Pattern Anal. Mach. Intell. **23**(8), 800–810 (2001)
6. Faziloglu, Y., Stanley, R.J., Moss, R.H., Van Stoecker, W., McLean, R.P.: Colour histogram analysis for melanoma discrimination in clinical images. Skin Res. Technol. **9**(2), 147–156 (2003)
7. Ganster, H., Pinz, A., Röhrer, R., Wildling, E., Binder, M., Kittler, H.: Automated melanoma recognition. IEEE Trans. Med. Imaging **20**(3), 233–239 (2001)
8. Kaufman, L., Rousseeuw, P.J.: Partitioning Around Medoids (Program PAM). Finding Groups in Data: an Introduction to Cluster Analysis, pp. 68–125 (1990)
9. Khatami, A., Babaie, M., Khosravi, A., Tizhoosh, H., Salaken, S.M., Nahavandi, S.: A deep-structural medical image classification for a radon-based image retrieval. In: 2017 IEEE 30th Canadian Conference on Electrical and Computer Engineering (CCECE), pp. 1–4. IEEE (2017)
10. Khatami, A., Khosravi, A., Lim, C.P., Nahavandi, S.: A wavelet deep belief network-based classifier for medical images. In: Hirose, A., Ozawa, S., Doya, K., Ikeda, K., Lee, M., Liu, D. (eds.) ICONIP 2016. LNCS, vol. 9949, pp. 467–474. Springer, Cham (2016). doi:10.1007/978-3-319-46675-0_51
11. Khatami, A., Khosravi, A., Nguyen, T., Lim, C.P., Nahavandi, S.: Medical image analysis using wavelet transform and deep belief networks. Expert Syst. Appl. (2017)

12. Khatami, A., Mirghasemi, S., Khosravi, A., Lim, C.P., Nahavandi, S.: A new pso-based approach to fire flame detection using k-medoids clustering. Expert Syst. Appl. **68**, 69–80 (2017)
13. Khatami, A., Mirghasemi, S., Khosravi, A., Nahavandi, S.: An efficient hybrid algorithm for fire flame detection. In: 2015 International Joint Conference on Neural Networks (IJCNN), pp. 1–6. IEEE (2015)
14. Khatami, A., Mirghasemi, S., Khosravi, A., Nahavandi, S.: A new color space based on k-medoids clustering for fire detection. In: 2015 IEEE International Conference on Systems, Man, and Cybernetics (SMC), pp. 2755–2760. IEEE (2015)
15. Maglogiannis, I., Zafiropoulos, E., Kyranoudis, C.: Intelligent segmentation and classification of pigmented skin lesions in dermatological images. In: Antoniou, G., Potamias, G., Spyropoulos, C., Plexousakis, D. (eds.) SETN 2006. LNCS, vol. 3955, pp. 214–223. Springer, Heidelberg (2006). doi:10.1007/11752912_23
16. Siascope, T., Consensus, E., Expert, M., Solarscan, T., Ogorzaek, M., Nowak, L., Surwka, G., Alekseenko, A.: Jagiellonian University Faculty of Physics, Astronomy and Applied Computer Science Jagiellonian University Dermatology Clinic, Collegium Medicum Poland (2005)
17. Yang, J., Fu, Z., Tan, T., Hu, W.: Skin color detection using multiple cues. In: Proceedings of the 17th International Conference on Pattern Recognition, ICPR 2004, pp. 632–635. IEEE (2004)

A Deep Learning-Based Model for Tactile Understanding on Haptic Data Percutaneous Needle Treatment

Amin Khatami[1]([✉]), Yonghang Tai[1], Abbas Khosravi[1], Lei Wei[1],
Mohsen Moradi Dalvand[1], Min Zou[2], and Saeid Nahavandi[1]

[1] Institute for Intelligent Systems Research and Innovation, Deakin University,
Geelong, VIC 3216, Australia
{skhatami,yonghang,abbas.khosravi,Lei.Wei,mohsen.m.dalvand,
saeid.nahavandi}@deakin.edu.au
[2] Urology Surgery Department, Yunnan First People's Hospital, Kunming, China
576665318@qq.com

Abstract. Tactile understanding during surgery is essential in medical simulation. To improve a remote surgical operation one step further, in this paper, we develop a sequence classification technique, categorising different tissues, evaluating on biomechanics data. The importance of the proposed model is emphasised when problems such as a delay is occurring during simulation. Monitoring, predicting, and understanding the sense of tissue which is supposed to be involved in operation is vital during surgery. To achieve this, different deep structural techniques are investigated to find the effect of deep features for tactile and kinaesthetic understanding. The experimental results reveal that residual networks outperform others with respect to different terms. The results are accurate and fast which enables the technique to perform in real-time.

Keywords: Tactile understanding · Sequence classification, Residual networks, Time series

1 Introduction

Haptic feedback, which is a combination of kinaesthetic and tactile, refers to a touch sense feeling. This sense of touch, which is a basic element of haptic communication and is an important part of our daily interaction with the environment, enables one to feel a variety of sensations such as pain, temperature, smoothness, hardness, etc. Haptic communication recreates the touch sense by applying forces to users, generating better understanding of the environment, practically.

Haptic understanding can be applied to a wide variety of applications [2,3,14,16], however unlike vision, the understanding of the haptic processing is still rudimentary. Regarding the medical domain, the impact of applying deep learning techniques is widely seen in literature [1,7–10]. Following the trend,

© Springer International Publishing AG 2017
D. Liu et al. (Eds.): ICONIP 2017, Part IV, LNCS 10637, pp. 317–325, 2017.
https://doi.org/10.1007/978-3-319-70093-9_33

recently, Goa et al. [3] used a Long Short-Term Memory (LSTM) model to propose a surface classification scheme with different haptic adjectives for increasing the capability of tactile understanding. A deep Recurrent Neural Network, short RNN, model was also developed in [13] learning an appropriate representation of force feedback. A full Convolutional Network, short FCN, model was proposed in [16] to classify the haptic signals recognising the surface materials. A similar approach has reported in [15] as well.

In this paper, inspired by the usefulness of the deep structural techniques, a real-time tactile understanding model is proposed for percutaneous needle treatment based on biomechanics data. The aim of the model is to recognise the organ of a body based on the force feedback captures by sensors. There are several contributions as follows; (1) an accurate and fast real-time deep structural-based tactile understanding model is investigated for percutaneous needle treatment. The importance of this prediction is highlighted when, for instance, a delay occurs during operation. Dropping the speed of the network might also be considered as another issue. (2) Providing a classification-based dataset for deep learning techniques based on the data collected and published by Yunnan First Peoples Hospital. (3) Finally, we show the force feedback data is well generalised on a ResNet model, rather than FCN and LSTM.

This paper is organised as follows: Sect. 2 contains a brief explanation of the methodologies used to conduct the study, followed by demonstrating the proposed real-time model. Experimental results along with a discussion and comparison are conducted in Sect. 3, followed by a conclusion in Sect. 4.

2 Methodology

The impact of utilising three types of deep structural networks containing convolutional, recurrent, and residual networks, evaluated on a haptic dataset is investigated here.

2.1 Fully Convolutional Networks (FCNs)

A fully convolutional network, short FCN, is a well-known technique in the deep learning domain. In this paper, the FCN is utilised as a feature extractor, followed by a batch normalisation [6] layer and a global average pooling layer [11] with *softmax* classifier at the end. The convolution is calculated by (1);

$$y = \mathbf{v} \bigotimes \mathbf{kernel} \tag{1}$$

where $\mathbf{v}$ is the input, obtained from the previous layer, **kernel** is a filter, and $\bigotimes$ is the convolution operator. Batch normalisation layer is calculated by (2), where y is the feature maps.

$$s = BN(y) \tag{2}$$

note that *ReLU* is considered as the activation function.

2.2 Long Short-Term Memory (LSTM) Network

LSTM-based RNNs was originally proposed in [5]. These networks are able to learn sequences of observations which causes the model well-suited to time series applications. An LSTM model calculates a mapping from a sequence of $x = (x_1, x_2, ..., x_T)$ as input to a sequence of $h = (h_1, h_2, ..., h_T)$ as output, by following the below equations iteratively from $t = 1$ to T:

$$f_t = \sigma(W_f.[h_t - 1, x_t] + b_f) \tag{3}$$

$$i_t = \sigma(W_i.[h_t - 1, x_t] + b_i) \tag{4}$$

$$\tilde{C}_t = tanh(W_C.[h_t - 1, x_t] + b_C) \tag{5}$$

$$C_t = f_t \odot C_{t-1} + i_t \odot \tilde{C}_t \tag{6}$$

$$o_t = \sigma(W_o.[h_{t-1}, x_t] + b_o) \tag{7}$$

$$h_t = o_t \odot tanh(C_t) \tag{8}$$

where the W terms are weight matrices. The b is bias vector, and σ and $tanh$ are activation functions. Also, C, i, f, and o are cell activation, input, forget, and output gate vectors, respectively.

2.3 Residual Network

Residual network is another deep structural networks which, as seen in Fig. 1, adds shortcut connections between the blocks making the capability of flowing gradient directly through the bottom layers. The following equations are used to build each residual block;

$$\begin{aligned} h_1 &= block_1(x) \\ h_2 &= block(h_1) \\ h_3 &= block(h2) + input \\ \hat{h} &= ReLU(h_3) \end{aligned} \tag{9}$$

3 Experimental Results

The impact of utilising several deep structural networks, different in learning procedure, is investigated in this section. As follows, an explanation of how to prepare the data is followed by describing how to preprocess the data set suitable for a learning process. We define some scenarios, followed by statistical analyses to report the best well-adapted model and structure on our data. Finally, we conclude that the proposed technique is accurate and fast which is practicable for a real-time system.

Accessing Publicly Available Data: The authors have made contact with surgeons from Yunnan First Peoples Hospital and accessed the data from a

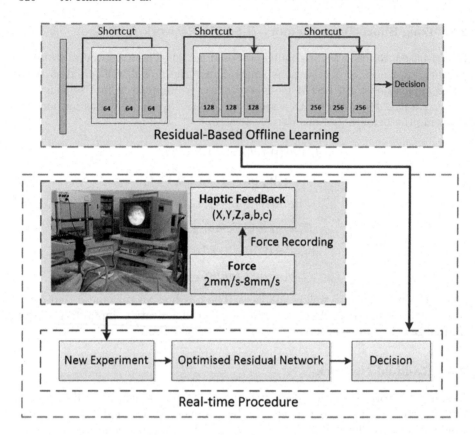

Fig. 1. The black diagram of our real-time tactile understanding simulator.

publicly available website[1]. According to the introduction of the dataset, all data and images are collected on a custom-built instrument and are based on animal organs purchased from a supermarket. Professional surgeons were invited to perform the procedures to ensure the credibility of the acquired data. The room setup was mimicking a surgical theatre but is not a real one. The authors have also requested additional details about the experiments from the surgeons, as explained in this paper.

Haptic Feedback Data: According to the introduction of the dataset, the platform mainly consists of the surgical instrument, force sensor acquisition system, stepper motor drive system, container for soft tissue and computer. Trocar is used as a surgical instrument, ATI Nano-17 was selected as the force sensor, it is installed on the handle of the trocar to collect the information of the insertion force in real-time, the precision is 0.01N. Stepper motor is mainly composed of slider, subdivision, controller, the maximum motion range is 200 mm, motion accuracy is 0.16 um, the velocity can be programmed to control, controllable

---

[1] (http://civ.ynnu.edu.cn/ChineseShow.aspx?ID=1)

range of 1–30 mm/s. There are two stepper motor drive systems, one was used to control the velocity of the insertion, another one was used to simulate the motion platform of human respiratory to achieve dynamic puncture. And the size of the container is 12 mm × 7 mm × 7 mm. Computer can read and display the insertion force collected by the sensor in real-time. It includes acquisition software and analysis software. Temperature is 18–21 centigrade; and the puncture needle is 18G trocar (COOK Corp.) needle for percutaneous/biopsy surgery.

3.1 Force Feedback Dataset

The raw haptic data contains a range of force feedback captured by the sensors ATI Nano 17 for vertical injecting on different kind of porcine organ. A visualisation of the samples is illustrated in Fig. 2, showing the distribution of the data. Because the prediction of the forward percutaneous is investigated, the negative values are removed. We subsample the data with 2000 length, and by following with sliding windows, a dataset with the dimension of $(3363, 2000)$ is created for both categories. To make the dataset smooth, the mean value is

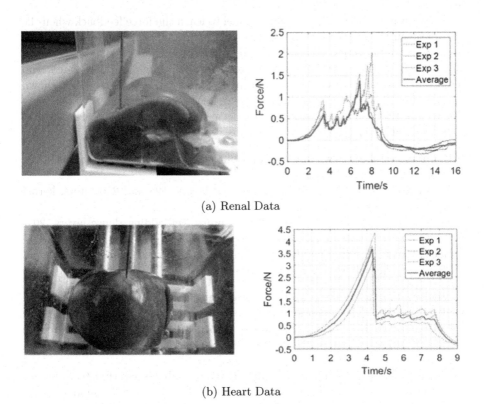

(a) Renal Data

(b) Heart Data

Fig. 2. A visualisation of data with different experiments.

subtracted. After normalisation, the dataset is split into train and test, using a five-fold cross validation. The training data samples are utilised for training the proposed model, and the test data is utilised for the performance network evaluation.

To measure the performance of the proposed approach, Area Under the Receiver Operating Characteristic Curve (AUC) [4] metric is used. AUC is a metric showing the sensitivity versus specificity (true positive rate versus false positive rate). Precision, recall, and F-measure are the other terms utilised for evaluation. Moreover, prediction accuracy which is the number of accurate predictions over the total number of data samples is also considered as a metric.

3.2   Results and Discussion

Understanding the amount of force, felt at fingertips, gives much information to the user, resulting in a better performance in operating systems. Finding this sense of touch during medical surgery representing the tactile understanding of organs of a body is vital because the organs are very close to each other, and the feeling of the sense prevents such problems as misunderstanding, which might occur.

At an offline procedure, we train a model to learn the force feedback when the sensors touch the surfaces of the organs. A proper analysis is performed to find an accurate model which is capable of a real-time procedure, as seen below. After tuning the model, at an online procedure, a new experiment is performed, and based on the predicted feedback, the type of the organ is predicted and shown on a monitor. The following strategies are defined to conduct the research. Note that to find the proper network for each scenario, different hyper-parameters of each case study were examined.

- **Scenario 1:** CNN with one convolutional layer, 128 filters, kernel size of 8, stride 1, 20% Drop-Out.
- **Scenario 2:** CNN with two convolutional layers, 128 and 256 filters, kernel sizes of 8, 5, stride 1, 20% Drop-Out.
- **Scenario 3:** LSTM with one layer, 10 units, embedding dense layers, 20% drop-Out.
- **Scenario 4:** Residual network with three convolutional blocks, 64, 128, and 256 filters, kernel sizes of 8, 5, and 3.

Figure 1 illustrates the residual network utilised in this study. Experimental results show that this network, defined as scenario 4, performs better than the rest of scenarios. Batch normalisation is applied to accelerate the convergence speed and improves generalisation. We also perform early stopping at epoch 4 to avoid overfitting as well.

The Fig. 3 and 95% confidence level of H-test (p-value is less than 0.05) proves the superiority of the proposed model, compared with the FCN models. Note that the H-test cannot reject the hypothesis against that of LSTM, however by considering other criteria, the superiority is achieved.

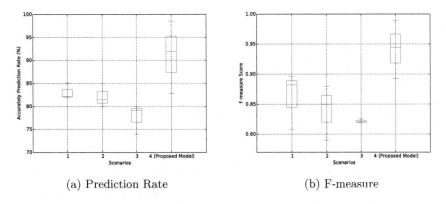

(a) Prediction Rate   (b) F-measure

Fig. 3. A comparison among the scenarios based on accurate prediction rate and F-measure score.

Table 1. A comparison among the four case studies, evaluated by precision, recall, and F-measure with a 5-fold cross validation procedure.

		Precision	Recall	F-measure
Scenarios	Strategies	Mean $\pm$ std	Mean $\pm$ std	Mean $\pm$ std
1	CNN1	0.796 ± 0.046	0.979 ± 0.028	0.861 ± 0.038
2	CNN2	0.771 ± 0.02	0.95 ± 0.017	0.85 ± 0.032
3	LSTM	0.786 ± 0.015	0.860 ± 0.012	0.821 ± 0.002
4	**Residual network**	$\mathbf{0.900 \pm 0.079}$	$\mathbf{0.993 \pm 0.009}$	$\mathbf{0.942 \pm 0.039}$

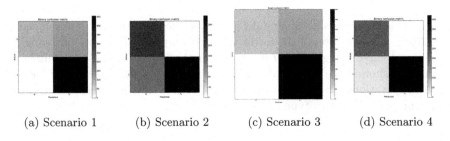

(a) Scenario 1   (b) Scenario 2   (c) Scenario 3   (d) Scenario 4

Fig. 4. The confusion matrices for fold 1 experiment.

In term of F-measure criterion, as illustrated in Fig. 3 and the statistical analysis of H-test, the advantage of using the residual network on the force data is highlighted. They show that the results of our proposed model significantly outperforms, compared with that of the other scenarios. Table 1 compares the experiments in terms of the precision, recall, and F-measure. Note that the mean and median of recall related to our proposed model is 0.993 and 1, respectively, and this should be taken into account because recall criterion is very important in the medical domain. The confusion matrix of the fold 1 for all the scenarios are depicted in Fig. 4. It is clear that both scenario 1 and the proposed model

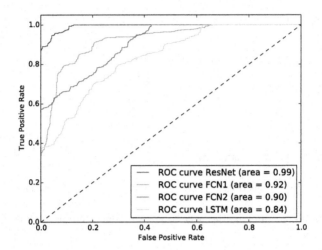

Fig. 5. ROC curve comparison among the scenarios for fold 1 experiment.

have high rate of recall, showing the effect of convolutional approach, evaluating on the dataset.

Figure 5 illustrates the ROC curve which is related to fold 1 experiment of all scenarios. Note that micro-averaging technique implemented in [12] in Python, is used to plot the ROC. Regarding computational costs.

4   Conclusion

A robust tactile understanding model for percutaneous therapy was developed in this study. Classification of the forces which are captured by haptic instruments were investigated using different deep structural methods. A real-time mimicked simulation showed the ability of implementing the proposed technique for real surgical operations. The experimental results showed that deep residual networks works properly with this type of data.

References

1. Babaie, M., Kalra, S., Sriram, A., Mitcheltree, C., Zhu, S., Khatami, A., Rahnamayan, S., Tizhoosh, H.R.: Classification and retrieval of digital pathology scans: a new dataset. arXiv preprint (2017). arXiv:1705.07522
2. Chu, V., McMahon, I., Riano, L., McDonald, C.G., He, Q., Perez-Tejada, J.M., Arrigo, M., Darrell, T., Kuchenbecker, K.J.: Robotic learning of haptic adjectives through physical interaction. Robot. Auton. Syst. **63**, 279–292 (2015)
3. Gao, Y., Hendricks, L.A., Kuchenbecker, K.J., Darrell, T.: Deep learning for tactile understanding from visual and haptic data. In: Robotics and Automation (ICRA), 2016 IEEE International Conference on, pp. 536–543. IEEE (2016)
4. Goodenough, D.J., Rossmann, K., Lusted, L.B.: Radiographic applications of receiver operating characteristic (ROC) curves 1. Radiology **110**(1), 89–95 (1974)

5. Hochreiter, S., Schmidhuber, J.: Long short-term memory. Neural Comput. **9**(8), 1735–1780 (1997)
6. Ioffe, S., Szegedy, C.: Batch normalization: accelerating deep network training by reducing internal covariate shift. In: International Conference on Machine Learning, pp. 448–456 (2015)
7. Khatami, A., Babaie, M., Khosravi, A., Tizhoosh, H., Salaken, S.M., Nahavandi, S.: A deep-structural medical image classification for a radon-based image retrieval. In: Electrical and Computer Engineering (CCECE), 2017 IEEE 30th Canadian Conference on, pp. 1–4. IEEE (2017)
8. Khatami, A., Khosravi, A., Lim, C.P., Nahavandi, S.: A wavelet deep belief network-based classifier for medical images. In: Hirose, A., Ozawa, S., Doya, K., Ikeda, K., Lee, M., Liu, D. (eds.) ICONIP 2016. LNCS, vol. 9949, pp. 467–474. Springer, Cham (2016). doi:10.1007/978-3-319-46675-0_51
9. Khatami, A., Khosravi, A., Nguyen, T., Lim, C.P., Nahavandi, S.: Medical image analysis using wavelet transform and deep belief networks. Expert Syst. Appl (2017)
10. Khatami, A., Mirghasemi, S., Khosravi, A., Lim, C.P., Nahavandi, S.: A new PSO-based approach to fire flame detection using k-medoids clustering. Expert Syst. Appl. **68**, 69–80 (2017)
11. Lin, M., Chen, Q., Yan, S.: Network in network. arXiv preprint (2013). arXiv:1312.4400
12. Pedregosa, F., Varoquaux, G., Gramfort, A., Michel, V., Thirion, B., Grisel, O., Blondel, M., Prettenhofer, P., Weiss, R., Dubourg, V., et al.: Scikit-learn: machine learning in python. J. Mach. Learn. Res. **12**(Oct), 2825–2830 (2011)
13. Sung, J., Salisbury, J.K., Saxena, A.: Learning to represent haptic feedback for partially-observable tasks. arXiv preprint (2017). arXiv:1705.06243
14. Vander Poorten, E.B., Demeester, E., Lammertse, P., Vander Poorten, E.V.: Haptic feedback for medical applications, a survey. In: Proceedings of the Actuator, pp. 519–525, June 2012
15. Wang, Z., Yan, W., Oates, T.: Time series classification from scratch with deep neural networks: a strong baseline. arXiv preprint (2016). arXiv:1611.06455
16. Zheng, H., Fang, L., Ji, M., Strese, M., Özer, Y., Steinbach, E.: Deep learning for surface material classification using haptic and visual information. IEEE Trans. Multimedia **18**(12), 2407–2416 (2016)

Measuring Word Semantic Similarity Based on Transferred Vectors

Changliang Li[1(✉)], Teng Ma[1], Yujun Zhou[2], Jian Cheng[1], and Bo Xu[1]

[1] Institute of Automation, Chinese Academy of Sciences,
Beijing, People's Republic of China
{changliang.li,mateng,Jian.cheng,xubo}@ia.ac.cn
[2] University of Chinese Academy of Sciences,
Beijing, People's Republic of China
zhouyujun2014@ia.ac.cn

Abstract. Semantic similarity between words has now become a popular research problem to tackle in natural language processing (NLP) field. Word embedding have been demonstrated progress in measuring word similarity recently. However, limited to the distributional hypothesis, basic embedding methods generally have drawbacks in nature. One of the limitations is that word embeddings are usually by predicting a target word in its local context, leading to only limited information being captured. In this paper, we propose a novel transferred vectors approach to compute word semantic similarity. Transferred vectors are obtained via a reasonable combination of the source word and its nearest neighbors on semantic level. We conduct experiments on popular both English and Chinese benchmarks for measuring word similarity. The experiment results demonstrate that our method outperforms previous state-of-the-art by a large margin.

Keywords: Semantic similarity · Word embeddings · Transferred vectors · Nearest neighbors

1 Introduction

The study of semantic similarity between words has been an important part of natural language processing and artificial intelligence field, such as word sense disambiguation [1], semantic text similarity [2], machine translation [3], information extraction [4] and opinion mining [5].

A number of semantic similarity methods have been proposed. For example, [6] proposed an approach for measuring semantic similarity between words using multiple information sources. These methods can be divided into two categories: knowledge-based methods and corpus-based methods [1]. More recently, word embeddings have been demonstrated outstanding performance in measuring English word similarity across several benchmark datasets [7].

However, basic embedding methods are based on the distributional hypothesis. A word representation is a mathematical object associated with each word, often a vector. An effective approach for word representation is to learn distributed representations. A distributed representation is dense, low dimensional and real-valued

D. Liu et al. (Eds.): ICONIP 2017, Part IV, LNCS 10637, pp. 326–335, 2017.
https://doi.org/10.1007/978-3-319-70093-9_34

vector. Distributed word representations are also called word embeddings. Each dimension of the embeddings represents a latent feature of the word, hopefully capturing useful syntactic and semantic properties [8]. They generally have drawbacks in nature. One of the limitations is that word embeddings are usually learned by predicting a target word in its local context, leading to only limited information being captured.

In this paper, we deal with this issue by introducing a novel word representation in vector space, called transferred vectors. We combine the word and its nearest words on semantic level to obtain the transferred vectors. We use the transferred vectors to measure word similarity, instead of only using word's embeddings. The main contribution of our work is to propose a reasonable combination way, rather than simply adding or using synonyms, to make the transferred vectors better capturing word semantic information and more robust via considering more information. Through learning, we can get an optimal combination parameter, which decides how much degree we depend on the word itself (source word) and how much degree we depend on its neighbours. The method is simple to implement but powerful.

In order to fully evaluate our method's effectiveness. We conduct experiments on both English and Chinese benchmarks, which are popularly used in related research works [9, 10]. Chinese benchmark is from NLPCC&ICCPOL-2016 Task 3 "measuring Chinese word similarity", which tries to evaluate the study on word similarity for Chinese language. English benchmark is Wordsim-353, which has been popularly used to evaluate measuring word similarity methods. The experimental results demonstrate that our model outperforms previous state-of-the-art approaches by a large margin, no matter on which language data.

The remainder of this paper is arranged as follows. In Sect. 2, we introduce some related work. Section 3 introduces our method based on transferred vectors. Section 4 presents the experiments on Chinese and English benchmarks, the experimental results are also discussed in this section. Some conclusions are summarized in Sect. 5.

2 Related Work

In this section, we describe some related works on measuring word similarity.

A variety of methods have been proposed to compute lexical semantic similarity. These methods can be divided into two categories: knowledge-based methods and corpus-based methods [1]. Usually, knowledge-based approaches rely mainly on artificial semantic resources to identify similarities between two words with structural semantic thesaurus. As the essential resources, a lot of thesaurus or lexical knowledge bases, including WorldNet [11], Tongyici Cilin [12, 13], HowNet [14, 15], etc. are all used to measure semantic distance between a pair of words. Although these measures are interpretable and effective, these semantic measures have serious drawbacks that lacks context information and only can compute the pairs when both members are existed in the lexicons.

Due to these limitations, Corpus-based methods are then proposed to utilize context information around words. The corpus-based word similarity study mostly uses a statistical description of the context, that is, the conclusion that the context of a word can provide sufficient information for the word definition. The meaning of words is modelled by using its distribution in contexts, and then the semantic similarity between two words will be calculated by comparing these distributions [16]. There are many statistical models to train words, but the curse of dimensionality can't be avoided in the face of large data sets. Therefore, machine learning algorithm is applied for distributed representation of words and semantic similarity can be measured with cosine similarity [17]. Distributed representation is firstly proposed by Hinton in 1986 [18]. Then different neural network models are presented to study the word embedding. As a famous open tool, Word2vec is a highly efficient tool proposed by Google in 2013 to characterize the word as a real value vector. It can simplify the processing of the text content into vector computing in K-dimensional vector space, spatial similarity can be used to represent similarity in text semantics [19, 20].

However, limited to the distributional hypothesis, basic embedding methods generally have drawbacks in nature, one of the limitations is that the word embedding is usually learned by predicting the target word with its context, which results in only local co-occurrence information being captured [21]. Thus, in order to overcome this limitation, several studies have focused on this field. [22] integrated the paragraph information into a Word2vec-based model, so that they can capture the paragraph level information. For polysemy, [23] proposed an approach of converting word embeddings into sense level and using knowledge from a large semantic network, which allowed them to take advantages of structured knowledge. [24] introduced the concept of feature embeddings induced by analyzing a large unannotated corpus and then learning embeddings for the manually crafted features. Unfortunately, more or less, all these methods have their own limitations. Measuring word similarity is still an open and challenging problem.

3   Transferred Vector Approach

In this section, we firstly introduce our Transferred Vector Approach. Then we describe the details of parameter learning.

3.1   Transferred Vector

The main idea of our method is to use a transferred vector to represent a word, which can reveal the word semantic better, not just relying on its own embeddings. To some extent, this method plays a role in reducing the risk of semantic deviation while considering more information.

There is a vivid example in life to explain the idea. If we want to know a person, besides just judging from his/her own behavior, we would like to see his/her friends to judge on a bigger picture. This would be helpful to get a more comprehensive understanding of a person. Back to our work, Fig. 1 gives the overall illustration of our method. It is split into three steps as follows:

1. Finding one word's nearest neighbors on semantic level;
2. Learning an optimal way to combine the word and its neighbors to represent the transferred vectors;
3. Measuring word similarity via computing the distance between transferred vectors in vector space.

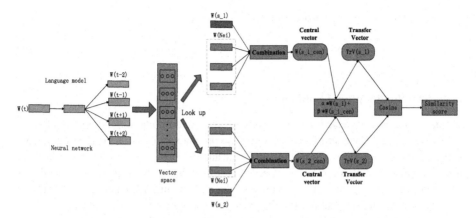

Fig. 1. The overall illustration of our method.

We describe our method in details as following.

We firstly describe word vector representations employed in our method. Each word is represented as a d dimensional vector. All the word vectors are stacked in a word embedding matrix $L \in R^{d \times |M|}$, $|M|$ is the size of the vocabulary.

Given a word pair (w_{s_1}, w_{s_2}), each word is represented as a word vector in L.

Firstly, we search K words through word embedding matrix, which are closely to the word w_{s_i} on semantic level. We use $Nei_k \in R^d$ to represent these words, where Nei is short for neighbor. The search process is implemented by calculating Cosine value between word embeddings:

$$Max_K \, sim(w_i, Nei_k) = max_K \, cos\,(V(w_i), V(Nei_k)) \tag{1}$$

where $k \in R^{|M|}$ and max_K means to select top-K words, which achieve highest similarity score.

Then, we seek the center node of the K neighbors. This neighbor center node is not a word in real sense. It is a virtual node that represents the overall semantic mean of these words.

$$w_{s\_i\_cen} = aveg(Nei_1, \ldots, Nei_k) \tag{2}$$

Finally, we obtain the transferred vector via combining the source word w_{s_i} and neighbor center node $w_{s\_i\_cen}$. Due to that both parts play role in capturing semantic information on different degree, we introduce a combination factor α, β to determine how much we depend on w_{s_i} and $w_{s\_i\_cen}$. The transferred vector is represented as formula:

$$TrV_{s\_i} = [\alpha, \beta] \begin{bmatrix} w_{s\_i} \\ w_{s\_i\_cen} \end{bmatrix} \tag{3}$$

As a result, we get the word similarity as shown:

$$sim\left(w_{s\_i}, w_{s\_j}\right) = cos\left(TrV_{s\_i}, TrV_{s\_j}\right) \tag{4}$$

3.2 Searching Optimal Solution

In this subsection, we describe how to find the optimal combination factor α, β. Given a word pair (w_1, w_2), we aim to obtain the same predicted score sim_p as the target score sim_g, which is the gold score of the word pair annotated by humans. In this paper, we use Spearman's rank correlation coefficient to evaluate the statistical dependence between automatic computing results and the golden human labelled data without making any assumptions about the frequency distribution of the variables.

Spearman's rank correlation coefficient is defined as follow.

$$r_R = 1 - \frac{6\sum_{i=1}^{n} d_i^2}{n(n^2 - 1)} \tag{5}$$

where n is the number of word pairs in training dataset. d_i is the difference between each rank of corresponding values of predicted R_{Xi} and gold rank R_{Yi}. R_{Xi} and R_{Yi} are defined as follows.

$$R_{Xi} = rank\left(sim_{p1}, sim_{p2}, \ldots, sim_{pN}\right) \tag{6}$$

$$R_{Yi} = rank\left(sim_{g1}, sim_{g2}, \ldots, sim_{gN}\right) \tag{7}$$

Our goal is to find an optimal parameter to maximize Spearman's rank correlation coefficient. We employ Nelder mead simplex algorithm [25] to search the optimal parameter to maximize the Spearman's rank correlation coefficient. n is the total number of word pairs annotated by human in training dataset.

4 Experiment

In this work, we collect Baidu Baike documents to train Chinese word embeddings, due to its good coverage of topics and word usages, and its clean organization of document by topic. Similarly, we use Wikipedia documents to train English word embeddings. We employed Word2vec tool to train word embeddings.

We have trained 3.77 million Chinese words embeddings and 2.16 million English words embeddings which cover all the words in the test dataset.

4.1 Chinese Word Similarity Task

NLPCC&ICCPOL-2016 Task 3 provides a benchmark for evaluating the performance of semantic similarity calculating approaches for word pairs. Sample dataset includes 40 word and test dataset includes 500 word pairs with similarity scores estimated by humans. And any candidate tested approach is required to give similarity scores for each test word pairs.

The task employ Spearman's rank correlation coefficient to evaluate the statistical dependence between automatic computing results and the golden human labeled data. The larger Spearman's rank correlation coefficient is, the better the result is.

Table 1 gives some examples in NLPCC&ICCPOL-2016 Task 3 dataset. Table 1 lists eight examples, which are selected from different perspectives, which contain not only words with high similarity, but also words with very low similarity. Through these examples, we can get a good understanding of the dataset.

Table 1. Some examples in NLPCC2106 dataset

w1	w2	Score
日 期 (date)	时 间 (time)	6.0
权 限 (authority)	权 利 (power)	4.0
节 日 (festival)	假 日 (holiday)	6.8
免 费 (free)	便 宜 (cheap)	5.2
理 论 (theory)	结 论 (conclusion)	3.1
作 用 (effect)	功 能 (function)	8.2
食 物 (food)	水 果 (fruit)	3.5
商 业 (commerce)	工 业 (industry)	4.8

We give top 2 methods in NLPCC&ICCPOL-2016 Task 3.

1. The first baseline adopt the lexicon-based method by using Tongyici Cilin, and their method is derived from the algorithm of Tian and Zhao [12], which takes advantage of the structure information and coding rules to estimate the similarity of a word pair. They propose two improvements:

   - For polysemantic words, they take the average value of similarity scores across all senses rather than the biggest similarity score.
   - For OOV words that are not recorded in Tongyici Cilin, they split the words into characters and extract those words that contain the character. Then they calculate the similarities of those words and take the average value.

2. The second baseline presents a combining strategy based on a variety of semantic resources using different similarity computing methods. Their work can be divided into three steps:

   - Compute the similarity score based on HowNet(sim1), then compute the cosine similarity between a pair of word that are pre-trained using the Word2vec tool (sim2). Last, the similarity score (S) is the average value: $S = (sim1 + sim2)/2$.
   - If both members of a pair inside the same frame according to the Chinese FrameNet [26], then the score: $S = (S + 10.0)/2$.
   - If both words are in DaCilin, Synonym dictionary (Tongyici Cilin) and Antonym dictionary, the score is further updated according to heuristic rules.

We use the sample data given as training dataset to seek the optimal combination factor. And as a result, we obtain $(\alpha, \beta) = (0.611, 0.389)$. Table 2 gives the result on Chinese benchmark. In Table 2, W means Word2vec, TrV means transferred vector method, T means translation method. We can see that our method outperforms previous best result by a large margin. Besides, our method is data-driven, without employing any external resources such as Hownet, Tongyici Cilin and so on. This shows that our method is very effective in measuring word similarity. Besides comparing with other works, we also compare to only using word embedding method. We can see that, our approach improves the result by 0.122. This is a strong evidence that indicates that our method's efficiency.

Table 2. Spearman's rank correlation between similarity scores assigned by various models or methods and by human annotators with NLPCC&ICCPOL dataset.

Method	Spearman correlation
Baseline1	0.436
Baseline2	0.518
W	0.461
TrV	**0.583**

Besides, we also validate the influence brought by k value. Figure 2 gives the illustration of K value. We can see that when k = 4, the result is best. When k gets higher, the performance doesn't get better. This is easy to understood that maybe some noise information is included when k get higher.

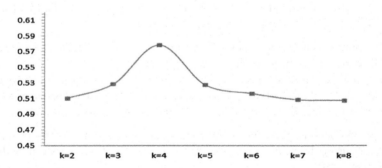

Fig. 2. The illustration of K value

4.2   English Word Similarity Task

We select wordsim-353 as benchmark, which a gold standard resource for evaluating distributional semantic models that improves on existing resources in several important ways. The WordSimilarity-353 test collection contains two sets of English word pairs along with human-assigned similarity judgments. The collection can be used to train and/or test computer algorithms implementing semantic similarity measures (i.e., algorithms that numerically estimate similarity of natural language words). The first set (set1) contains 153 word pairs along with their similarity scores assigned by 13 subjects. The second set (set2) contains 200 word pairs, with their similarity assessed by 16 subjects.

All the subjects in both sets possessed near-native command of English. Their instructions were to estimate the relatedness of words in pairs on a scale from 0 (totally unrelated words) to 10 (very much related or identical words). Each set provides the raw scores assigned by each subject, as well as the mean score for each word pair. For convenience, a combined set (combined) is provided that contains a list of all 353 words, along with their mean similarity scores. The combined set is merely a concatenation of the two smaller sets. In our experiment, we use all the 353 word pairs as test data. Besides, we use RG dataset as training data to learn the optimal combination factor. RG dataset [27] is similar to wordsim-353, and it includes 65 word pairs annotated by human. Via Nelder mead simplex algorithm, we obtain $(\alpha, \beta) = (0.617, 0.383)$ which is close to the value in measuring Chinese word similarity task. Table 3 gives the result on wordsim-353.

Table 3. Spearman's rank correlation between similarity scores assigned by various models or methods and by human annotators with wordsim-353 dataset.

Method	Spearman correlation
HSMN	0.628
HSMN + stem	0.628
HSMN + sim-RNN	0.645
HSMN + scm-RNN	0.652
W	0.616
TrV	**0.671**

Besides comparing to computing cosine value between word embeddings, we also report baseline result provided by [28] (referred as HSMN). HSMN is global context-aware neural language model. Furthermore, we reported three more improved versions based on HSMN, which are HSMN + stem, HSMN + simRNN, HSMN + scmMRNN [11].

We can see that our method improves word embeddings method by a large margin. Combined with the result on Chinese task, we can make a safe conclusion that our method reliable and robust in measuring word semantic similarity.

Besides, like the experiment on Chinese dataset, we also validate the influence brought by k value. Figure 3 gives the illustration of K value. We can see that when k = 2, the result is best. This is similar to the result in Chinese task.

The experiment results validate our idea that taking into account more semantic information and combine them in a reasonable way could better capture semantic information on a bigger picture.

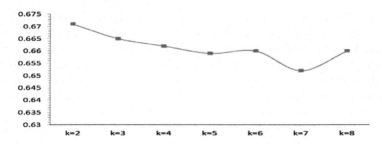

Fig. 3. The illustration of K value on wordsim-353

5 Conclusion

In this work, we introduce a novel transferred vector method for measuring word similarity. This method makes utilize of one word's closest neighbors on semantic level, and combine them in a reasonable way. Our model achieves great improvement on both Chinese and English word similarity benchmarks. Our model can be a new clue to other NLP tasks. A promising future work for us is to extend our model for measuring phrase and sentence similarity. And another interesting research point is the combination factor, which maybe achieve excellent performance when it is in a fixed range. We will validate it on a larger data and wider field.

References

1. Mihalcea, R., Corley, C., Strapparava, C.: Corpus-based and knowledge-based measures of text semantic similarity. In: AAAI, pp. 775–780 (2006)
2. Islam, A., Inkpen, D.: ACM Trans. Knowl. Discov. Data, **2**(2), Article 10 (2008)
3. Gao, J., Nie, J.-Y., Xun, E., Zhang, J., Zhou, M., Huang, C.: Improving query translation for cross-language information retrieval using statistical models. In: International ACM SIGIR Conference on Research & Development in Information Retrieval (2001)
4. Hassan, H., Hassan, A., Emam, O.: Unsupervised information extraction approach using graph mutual reinforcement. In: Conference on Empirical Methods in Natural Language Processing (2006)
5. Cambria, E., Schuller, B., Xia, Y., Havasi, C.: New avenues in opinion mining and sentiment analysis. Intell. Syst. IEEE **28**(2), 15–21 (2013)
6. Li, Y., McLean, D., Bandar, Z.A., O'shea, J.D., Crockett, K.: Sentence similarity based on semantic nets and corpus statistics. IEEE Trans. Knowl. Data Eng. **18**(8), 1138–1150 (2006)
7. Luong, M., Socher, R., Manning, C.: Better word representations with recursive neural networks for morphology. In: CoNLL-2013, pp. 104–113 (2013)
8. Turian, J., Ratinov, L., Bengio, Y.: Word representations: a simple and general method for semi-supervised learning. In: ACL 2010 Proceedings of the 48th Annual Meeting of the Association for Computational Linguistics, pp. 384–394 (2010)
9. Panchenko, A., Morozova, O., Naets, H.: A semantic similarity measure based on lexico-syntactic patterns. In: Conference on Natural Language Processing (KONVENS 2012), Vienna (Austria), pp. 174–178 (2012)
10. Wu, Y., Li, W.: Chinese word similarity measurement. NLPCC-ICCPOL 2016 shared task 3. In: Proceedings of NLPCC 2016 (2016)

11. Fellbaum, C.: WordNet. Wiley Online Library (1998)
12. Tian, J.L., Zhao, W.: Words similarity algorithm based on Tongyici Cilin in semantic web adaptive learning system. J. Jilin Univ. **28**(06), 602–608 (2010)
13. Mei, J.J., Zhu, Y.M., et al.: Tongyici Cilin. Shanghai Lexicon Publishing Company, Shanghai (1983)
14. Dong, Z., Dong, Q.: HowNet and the Computation of Meaning, pp. 85–95. World Scientific, Singapore (2006)
15. Liu, Q., Li, S.: Word similarity computing based on HowNet. Comput. Linguist. Chin. Lang. Process. **7**(2), 59–76 (2002)
16. Miller, G.A., Charles, W.G.: Contextual correlates of semantic similarity. Lang. Cogn. Process. **6**, 1–28 (1991)
17. Bengio, Y., Courville, A., Vincent, P.: Representation learning: a review and new perspectives. IEEE Trans. Pattern Anal. Mach. Intell. **35**, 1798–1828 (2013)
18. Hinton, G.E.: Learning distributed representations of concepts. In: Proceedings of the Eighth Annual Conference of the Cognitive Science Society, Amherst, MA, pp. 1–12 (1986)
19. Mikolov, T., Chen, K., Corrado, G., Dean, J.: Efficient estimation of word representations in vector space. arXiv preprint arXiv:1301.3781, pp. 1–12 (2013)
20. Mikolov, T., Sutskever, I., et al.: Distributed representations of words and phrases and their compositionality. In: Proceedings of NiPS, pp. 3111–3119 (2013)
21. Levy, Q., Goldberg, Y.: Neural word embedding as implicit matrix factorization. In: Advances in Neural Information Processing Systems, pp. 2177–2185 (2014)
22. Le, Q., Mikolov, T.: Distributed representations of sentences and documents. In: Proceedings of the 31st International Conference on Machine Learning (ICML 2014), ICML 2014, pp. 1188–1196 (2014)
23. Iacobacci, I., Pilehvar, M.T., Navigli, R.: SensEmbed: learning sense embeddings for word and relational similarity. In: Proceeding of ACL, pp. 95–105 (2015)
24. Chen, W., Zhang, Y., Zhang, M.: Feature embedding for dependency parsing. In: Proceedings of COLING 2014, The 25th International Conference on Computational Linguistics: Technical Papers, pp. 816–826 (2014)
25. Nelder, J.A., Mead, R.: A simplex method for function minimization. Comput. J. **7**, 308–313 (1965)
26. Liu, K.: Research on Chinese FrameNet construction and application technologies. J. Chin. Inf. Process. **6**, 47 (2011)
27. Rubenstein, H., Goodenough, J.B.: Contextual correlates of synonymy. Commun. ACM **8**(10), 627–633 (1965)
28. Huang, E.H., Socher, R., Manning, C.D., Ng, A.Y.: Improving word representations via global context and multiple word prototypes. In: Annual Meeting of the Association for Computational Linguistics (ACL) (2012)

Multi-population Based Search Strategy Ensemble Artificial Bee Colony Algorithm with a Novel Resource Allocation Mechanism

Liu Wu[1], Zhiwei Sun[2], Kai Zhang[1], Genghui Li[1], and Ping Wang[3(✉)]

[1] College of Computer Science and Software Engineering, Shenzhen University, Shenzhen 518060, Guangdong, China
[2] School of Computer Engineering, Shenzhen Polytechnic, Shenzhen 518060, Guangdong, China
[3] College of Information Engineering, Shenzhen University, Shenzhen 518060, Guangdong, China
wangping@szu.edu.cn

Abstract. Artificial bee colony algorithm (ABC) is a simple yet effective biologically-inspired optimization method for global numerical optimization problems. However, ABC often suffers from slow convergence due to its solution search equation performs well in exploration but badly in exploitation. Moreover, all food sources are assigned with almost equal computing resources so that good solutions are not being fully exploited. In order to address these issues, we propose a multi-population based search strategy ensemble ABC algorithm with a novel resource allocation mechanism (called MPABC_RA). Specifically, in employed bee phase, all food sources are divided into three subgroups according to their quality. Then each subgroup uses different search equations to find better solutions. By this way, better tradeoff between exploitation and exploration can be obtained. In addition, the superior solutions in onlooker bee phase are allocated with more resources to evolve. And onlooker bees fully exploit the area between the locations of the selected superior solutions and the current best solution by a novel search equation. We compare MPABC_RA with four state-of-the-art ABC variants on 22 benchmark functions, the experimental results show that MPABC_RA is significantly better than the compared algorithms on most test functions in terms of solution accuracy, convergence rate and robustness.

Keywords: Artificial bee colony algorithm · Multi-population · Search strategy ensemble · Resource allocation · Global numerical optimization

1 Introduction

Global optimization problems (GOPs) have been appeared in many engineering and scientific fields over the past decades. In general, GOPs can be characterized as discontinuity, multimodal, non-convexity and non-differentiability etc. As the complexity of GOPs increases, traditional optimization algorithms are difficult to meet their requirements. Therefore, some swarm-based intelligence algorithms have been proposed and have shown great potential to handle these complex optimization problems,

© Springer International Publishing AG 2017
D. Liu et al. (Eds.): ICONIP 2017, Part IV, LNCS 10637, pp. 336–345, 2017.
https://doi.org/10.1007/978-3-319-70093-9_35

such as Particle Swarm Optimization (PSO) [1], Ant Colony Optimization (ACO) [2], Artificial Bee Colony (ABC) algorithm [3] and firefly algorithm (FA) [4].

In this paper, we concentrate on ABC algorithm, which is inspired by the cooperative foraging behavior of the honey bee and firstly designed by Karaboga in [3]. Due to its simple structure, easy implementation and brilliant performance, ABC has attracted a great deal of attention and has been successfully applied to many practical applications [5–7]. However, ABC faces the challenge of slow convergence due to its solution search equation [8] and computing resource allocation mechanism. Thus, numerous ABC variants have been proposed in the past decade, which can be mainly classified into two categories as follows:

1. Invention of new solution search equations. Inspired by PSO, Zhu and Kwong proposed a gbest-guided ABC (GABC) [9] by adding the information of global best solution into the solution search equation to enhance the exploitation ability of ABC. In EABC [10], Gao et al. presented two new search equations respectively for employed bee phase and onlooker bee phase to balance the exploration and exploitation. Kiran and Findik [11] added the directional information to ABC and designed a new search strategy to select a search equation according to the previous directional information (dABC). Moreover, since different search equations have distinct advantages on different problems or at different stages on the same problems, some methods using multiple search equations were proposed to enhance the comprehensive performance of ABC, such as MEABC [12], ABCVSS [13] and MuABC [14].
2. Combination with auxiliary techniques. Gao et al. proposed a chaotic map, opposition-based learning method [15] and the orthogonal learning method [16] to enhance the performance of ABC. In RABC [17], the exploration phase is realized by original ABC and the exploitation phase is completed by the rotational direction method. Akay and Karaboga [18] proposed two control parameters, i.e., modification rate (MR) and scaling factor (SF), to control frequency and magnitude of perturbation.

This paper mainly focuses on the first category. A new ABC variant with multi-population based search strategy ensemble and a novel resource allocation mechanism is proposed, called MPABC_RA, which can significantly enhance the performance of ABC.

The rest of this paper is organized as follows. Section 2 briefly describes the original ABC and MPABC_RA are presented in Sect. 3. The experimental results and discussions are demonstrated in Sect. 4. Finally, Sect. 5 concludes this paper.

2   The Original ABC Algorithm

ABC is a novel swarm intelligence optimization algorithm by simulating the collective foraging behavior of honey bee [3]. In ABC, the location of a food source denotes a possible solution of the optimization problem, and the nectar amount of a food source can be regarded as the quality(fitness value) of the related solution. In the colony, bees can be categorized into three types: employed bees, onlooker bees and scout bees. Employed bees and onlooker bees account for half of the colony respectively. Scout bees are changed from employed bees, which abandon their food sources to randomly seek for new ones, and then turn to employed bees again. The ABC algorithm can be sequentially divided into four phases, i.e., initialization phase, employed bee phase, onlooker bee phase and scout bee phase. After the initialization phase, ABC enters a circulation of employed bee phase, onlooker bee phase and scout bee phase until the termination condition is met. The details of each phase are described as follows.

2.1   Initialization Phase

Let $X_i = (X_{i,1}, X_{i,2}, \cdots, X_{i,D})$ be the ith food source in the population, all the initial food sources are produced randomly according to Eq. (1)

$$X_{i,j} = X_j^{\min} + rand(0, 1) \cdot (X_j^{\max} - X_j^{\min}) \tag{1}$$

where $i \in \{1, 2, \cdots, SN\}, j \in \{1, 2, \cdots, D\}$; SN is the number of food sources, D is the dimension size of problem; $X_j^{\min}$ and $X_j^{\max}$ are the lower and upper bounds of the jth dimension respectively. The fitness value of each food source is measured by Eq. (2)

$$fit_i = \begin{cases} \frac{1}{1+f(X_i)} & if \ (f(X_i) \geq 0) \\ 1 + |f(X_i)| & otherwise \end{cases} \tag{2}$$

where fit_i denotes the fitness value of the ith food source X_i, $f(X_i)$ is the objective function value of the food source X_i for the optimization problem.

2.2   Employed Bee Phase

Each employed bee possesses a unique food source and search around it for a candidate food source. The search equation of employed bees is given as follows:

$$V_{i,j} = X_{i,j} + \phi_j(X_{i,j} - X_{k,j}) \tag{3}$$

where V_i is the ith candidate food source; X_k is randomly selected from the population, which is different from X_i; ϕ_j is a random number in $[-1, 1]$. If the candidate food source V_i is better than its parent X_i, then V_i replaces X_i. Otherwise, the X_i is kept to enter into the next generation.

2.3   Onlooker Bee Phase

After finishing the search tasks, employed bees share their food source information about quality and location with onlooker bees. An onlooker bee will choose a food source with probability P_i, which is calculated by Eq. (4). Obviously, the better the food source is, the bigger the selection probability is. And then it employs Eq. (3) to do further search in the neighborhood of the selected food source. Similarly, the greedy selection will be performed again as in the case of the employed bees phase.

$$P_i = fit_i / \sum_{i=1}^{SN} fit_i \tag{4}$$

2.4 Scout Bee Phase

If a food source with maximum *counter* value cannot be improved over a preset threshold *limit*, it will be discarded by its employed bee. And then this employed bee will become a scout bee to reproduce a new food source randomly by Eq. (1).

Noted that, in employed bee phase and onlooker bee phase, if the jth dimension of V_i violates the predefined boundary constraints, it will be reset according to Eq. (1).

3 The Proposed Algorithm (MPABC_RA)

In this section, we propose a multi-population based search strategy ensemble method and a novel resource allocation mechanism to improve the performance of ABC.

3.1 Multi-population Based Search Strategy Ensemble Approach

In original ABC, all employed bees use the same search equation to find better solutions in the vicinity of their own food source locations, which makes ABC do well in exploration but badly in exploitation. Actually, in terms of the quality of food sources, the search strategy of employed bees should be different. Like the learning method of students in a class, students with plain performance are used to learning from the well-performed students and the highest achiever; the well-performed students tend to take the highest achiever as an example; the highest achiever is a self-learner, nevertheless, other students may have a good impact or adverse effect to him. Therefore, we propose a multi-population based search strategy ensemble approach for employed bee phase as follows.

Firstly, we divide the population into three subpopulations in each generation according to the ascending order of the objective function value of all food sources, namely pop_1, pop_2 and pop_3. The current best food source is in the pop_1, while the pop_2 contains the top $T = r \cdot SN$ food sources except the best one, where $r \in (0, 1]$. And the rest of the food sources are in the pop_3. In this paper, the top T food sources are called superior food sources. Then, in order to improve the performance of ABC, each subpopulation is assigned a unique search strategy, which is listed as below.

$$V_{i,j} = \begin{cases} X_{i,j} + \phi_j \cdot \omega \cdot (X_{i,j} - X_{k,j}), & if\ X_i \in pop_1 \\ X_{i,j} + \phi_j \cdot (X_{best,j} - X_{i,j}), & if\ X_i \in pop_2 \\ X_{superior,j} + \phi_j \cdot \omega \cdot (X_{superior,j} - X_{k,j}), & otherwise \end{cases} \tag{5}$$

where X_{best} is the current best food source; $X_{superior}$ is randomly selected from the superior food sources; X_k is randomly chosen from the population, $k \neq i$ or $k \neq sup\ erior$; ω controls the range of the perturbation when a randomly selected food source X_k is used. ω is defined as follow:

$$\omega = 0.2 + 0.8 \cdot \left(\frac{MAXFES - FES}{MAXFES}\right) \tag{6}$$

where *FES* denotes the current number of function evaluations and *MAXFES* is the maximal number of function evaluations. Obviously, ω is decreases from 1 to 0.2

linearly with the number of function evaluations *FES*, which is conducive to narrow the search range of food sources gradually.

3.2   A Novel Resource Allocation Mechanism

In onlooker bee phase of ABC, food sources are selected to search according to the probability P as shown by Eq. (4), which can ensure that the food source with higher quality will be assigned more computing resources theoretically. Nevertheless, if there is no significant difference between the fitness values of all food sources, the computing resources will be allocated to all food sources almost evenly, which leads to the good food sources cannot be fully exploited and weakens the exploitation of ABC further. Moreover, as described in Eq. (5), the superior food sources have great guidance for the whole colony. Considering these facts, a novel computing resource allocation mechanism is put forward for onlooker bee phase, which exploits the superior food sources only. The details are given as follows.

As we all know, the worse one in the top T food sources should be allocated with more computing resources since they have more room to be improved than the better one. Therefore, a novel selecting method is proposed as Eq. (7).

$$P_s = fit_{T-s+1} / \sum_{s=1}^{T} fit_s \qquad (7)$$

where fit_s and fit_{T-s+1} denote the fitness value of sth and $(T - s + 1)$th food source in the top T food sources respectively. Obviously, the onlooker bee only search around the top T food sources and the selected probabilities of the superior food sources are in inverse proportion to their fitness values.

After selecting a superior food source, we proposed a new search strategy for onlooker bees, which exploits the location information between the selected food source and the current best one. The mathematically expression is given as follow.

$$V_{i,j} = \lambda \cdot X_{i,j} + (1 - \lambda) \cdot X_{best,j} + \phi_j \cdot \omega \cdot (X_{best,j} - X_{k,j}) \qquad (8)$$

where X_i represents the selected superior food source; ω is defined as Eq. (6). λ controls the base vector to make it close to X_i or X_{best}, which is set as Eq. (9).

$$\lambda = 0.6 \cdot \left(\frac{MAXFES - FES}{MAXFES}\right)^4 \qquad (9)$$

It is obvious that onlooker bees search towards the current best food source gradually, which fully exploits the area between the locations of the selected superior food source and the current best one. Besides, λ can help the superior food sources avoid being trapped in the local optimum in the early stage of evolution.

3.3   The Complete Proposed Algorithm

The multi-population based search strategy ensemble approach and a novel resource allocation mechanism are integrated with the framework of original ABC to form the

MPABC_RA algorithm. The pseudo-code of the complete MPABC_RA is demonstrated in Algorithm 1.

Compared with the original ABC, the operation added in MPABC_RA is the population sorting, whose computational complexity is $O(SN \cdot \log(SN))$. Since the complexity of the original ABC is $O(SN \cdot D)$ [19, 20], the computational complexity of MPABC_RA remains to be $O(SN \cdot D)$.

Algorithm 1. The procedure of the MPABC_RA algorithm

 1: **Initialization:** Generate SN solutions that contain D variables according to Eq.(1)
 2: **while** $FES < MAXFES$ **do**
 3:       Divide population into pop_1, pop_2 and pop_3 by the objective function value
 4:       **for** $i = 1$ to SN **do**   // employed bee phase
 5:          Generate a candidate solution V_i for X_i using Eq.(5)
 6:          **if** $f(V_i) \le f(X_i)$
 7:             Replace X_i by V_i
 8:             $counter(i) = 0$
 9:          **else**
10:             $counter(i) = counter(i) + 1$
11:          **end if**
12:          $FES = FES + 1$
13:       **end for** // end employed bee phase
14:       Calculate probability P for the top $T = r \cdot SN$ solutions using Eq.(7)
15:       **for** $i = 1$ to SN **do**   // onlooker bee phase
16:          Select a solution X_s from the superior solutions according to probability P
17:          Generate a new solution V_s using Eq.(8)
18:          **if** $f(V_s) \le f(X_s)$
19:             Replace X_s by V_s
20:             $counter(s) = 0$
21:          **else**
22:             $counter(s) = counter(s) + 1$
23:          **end if**
24:          $FES = FES + 1$
25:       **end for** // end onlooker bee phase
26:       Find out the solution X_{max}^G with the max $counter$ value   // scout bee phase
27:       **if** $counter(max) > limit$
28:          Replace X_{max}^G by a new solution generated according to Eq.(1)
29:          $FES = FES + 1$, $counter(max) = 0$
30:       **end if** //end scout bee phase
31: **End while**

---

4 Experiments and Results

In order to investigate the performance of MPABC_RA, a set of 22 benchmark functions with dimensions $D = 30$ are used (In our previous experiments, the overall results is similar when $D = 50$ and $D = 100$). The detailed description of these benchmark functions can be found in [21]. If the objective function value of the best solution obtained by an algorithm is less than the acceptable value in a run, this run is called a *successful run*.

In our experiment, three metrics are considered to evaluate the performance of MPABC_RA, which are described as follows.

(1) The mean and standard deviation (Mean/std): they are applied to evaluate the accuracy of the best objective function value for each algorithm. The smaller the value of Mean and std is, the higher quality/accuracy of the solution has.

(2) The average *FES* (AVEN): it is required to reach the acceptable value for the first time in a run and is adopted to evaluate the convergence rate. The smaller the value of AVEN is, the faster the convergence rate is. It is noted that AVEN will be only recorded for the successful runs.

(3) The successful rate (SR%): it is employed to evaluate the robustness of different algorithms. The greater the value of SR is, the better the robustness is. If SR is equal to 0, the AVEN will be denoted by "NA".

MPABC_RA is compared with four state-of-the-art ABC variants, i.e., GABC [9], EABC [10], ABCVSS [13], dABC [11]. To make a fair comparison, for all compared algorithms, *MAXFES* is used as the termination condition, which is set to 5000D. *SN* is set to 50, *limit* is set to $SN \cdot D$. The detailed parameters setting of the four compared algorithms are same as used in their original paper, and r in MPABC_RA is set to 0.1. Each algorithm will be run 25 times for all test functions. Furthermore, in order to show the significant differences between MPABC_RA and other algorithms, the Wilcoxon's rank sum test at 5% significance level is conducted. The experimental results are shown in Table 1 and the best results are marked with boldface.

Table 1 indicates that MPABC_RA is significantly better than all compared algorithms in terms of accuracy and convergence rate on unimodal functions $f_1 - f_6$, while the most of algorithms have the similar results on f_8 except EABC. On f_7, all the algorithms obtain the global optimal solution. Besides, MPABC_RA is superior to all competitors on f_9 and has the better convergence rate on f_{10}. As for multimodal functions $f_{11} - f_{22}$, with regard to solution accuracy and robustness, MPABC_RA is better than or at least comparable to competitors on all functions excluding that ABCVSS is superior to MPABC_RA on f_{14} and f_{15}. Moreover, MPABC_RA shows the best convergence rate on all multimodal functions except f_{20} in terms of AVEN. Overall, MPABC_RA outperforms GABC, EABC, ABCVSS and dABC on 14, 10, 9 and 18 cases out of 22 functions respectively. In contrast, MPABC_RA is only beaten by ABCVSS on 2 functions, while it is faster than all other competitors on almost all the test functions excluding f_7 and f_{20}. In addition, the convergence curves of mean

Table 1. Comparisons of MPABC_RA with other ABC variants on 22 test functions with 30D

Alg	GABC Mean(std) SR/AVEN	EABC Mean(std) SR/AVEN	ABCVSS Mean(std) SR/AVEN	dABC Mean(std) SR/AVEN	MPABC_RA Mean(std) SR/AVEN
f_1	4.76e-33(3.39e-33) - 100/49,778	2.48e-63(9.78e-63) - 100/27,726	9.69e-33(4.19e-32)- 100/50,954	4.22e-13(3.05e-13) - 100/97,950	**3.37e-102(6.96e-102)** 100/20,364
f_2	1.35e-26(1.03e-26) - 100/75,950	6.40e-59(2.00e-58) - 100/40,038	1.86e-23(9.26e-23)- 100/76,866	1.40e-07(3.26e-07) - 12/144,180	**5.23e-98(1.57e-97)** 100/26,817
f_3	3.23e-34(3.18e-34) - 100/45,310	7.37e-65(1.11e-64) - 100/25,658	4.36e-32(1.73e-31)- 100/48,374	2.15e-14(1.89e-14) - 100/88,030	**1.96e-103(4.34e-103)** 100/19,024
f_4	1.57e-53(5.00e-53) - 100/13,818	3.23e-32(1.30e-31) - 100/7,962	1.15e-37(5.76e-37)- 100/16,490	1.43e-25(2.79e-25) - 100/28,026	**4.85e-130(2.42e-129)** 100/7,169
f_5	5.15e-18(1.29e-18) - 100/77,086	1.07e-33(7.43e-34) - 100/42,278	2.60e-16(1.06e-15)- 100/74,638	4.52e-08(1.33e-08) - 0/NA	**6.70e-53(6.44e-53)** 100/31,439
f_6	2.45e-01(1.15e-01) - 100/109,150	2.48e+01(4.40e+00) - 0/NA	2.22e-01(9.25e-02) - 100/10,7774	1.18e+00(6.08e-01) - 100/30,900	**1.33e-03(1.59e-03)** 100/30,900
f_7	**0.00e+00(0.00e+00)=** 100/9,998	**0.00e+00(0.00e+00)=** 100/7,466	**0.00e+00(0.00e+00)=** 100/9,834	**0.00e+00(0.00e+00)=** 100/12,898	**0.00e+00(0.00e+00)** 100/7,591
f_8	**7.18e-66(1.51e-76)** - 100/150	8.23e-66(1.59e-66) - 100/150	**7.18e-66(2.15e-81)=** 100/150	**7.18e-66(8.84e-71)=** 100/150	**7.18e-66(2.15e-81)** 100/53
f_9	2.85e-02(5.64e-03) - 100/43,678	1.50e-02(4.17e-03) - 100/24,850	2.57e-02(8.01e-03) - 100/42,102	6.43e-02(1.33e-02) - 100/93,142	**1.17e-02(3.85e-03)** 100/17,073
f_{10}	5.58e-01(1.80e+00) = 68/77,426	3.84e+00(1.58e+01) = 48/70,008	5.66e-02(1.70e-01)= 96/84,575	1.33e-01(**1.62e-01**) - 60/107,940	3.77e-01(8.99e-01) 68/68,605
f_{11}	**0.00e+00(0.00e+00)=** 100/67,694	**0.00e+00(0.00e+00)=** 100/33,510	**0.00e+00(0.00e+00)=** 100/53,306	1.32e-12(2.92e-12) - 100/108,280	**0.00e+00(0.00e+00)** 100/29,247
f_{12}	**0.00e+00(0.00e+00)=** 100/76,878	**0.00e+00(0.00e+00)=** 100/36,794	**0.00e+00(0.00e+00)=** 100/60,718	6.62e-11(2.54e-10) - 100/121,970	**0.00e+00(0.00e+00)** 100/29,943
f_{13}	**0.00e+00(0.00e+00)=** 100/63,410	6.35e-09(3.15e-08) - 96/38,267	5.27e-14(2.60e-13) - 100/65,918	7.06e-04(2.46e-03) - 88/117,000	**0.00e+00(0.00e+00)** 100/30,431
f_{14}	6.91e-12(**7.43e-13**) - 100/64,526	5.89e-12(7.93e-13) = 100/38,214	**7.28e-13(1.39e-12)** + 100/51,174	8.66e-12(4.31e-12) - 100/88,378	6.40e-12(9.28e-13) 100/33,869
f_{15}	1.62e-14(3.09e-15) - 100/89,338	1.09e-14(4.55e-15) - 100/46,990	**6.08e-15(1.91e-15)** + 100/80,518	4.12e-07(1.65e-07) - 100/117,000	7.11e-15(**4.83e-30**) 100/35,480
f_{16}	**1.57e-32(5.59e-48)** - 100/45,126	**1.57e-32(5.59e-48)** - 100/24,354	2.57e-32(4.82e-32) - 100/47,798	4.68e-14(3.78e-14) - 100/90,358	**1.57e-32(2.79e-48)** 100/18,364
f_{17}	4.49e-33(2.80e-33) - 100/49,474	**1.50e-33(0.00e+00)** - 100/25,830	2.19e-33(2.42e-33) = 100/50,526	3.35e-13(5.00e-13) - 100/99,938	**1.50e-33(3.49e-49)** 100/20,311
f_{18}	4.19e-07(9.29e-07) - 36/136,210	1.67e-16(3.62e-16) - 100/42,558	1.08e-15(4.66e-15) - 100/83,474	7.43e-06(4.13e-06) - 0/NA	**2.49e-50(1.08e-49)** 100/41,561
f_{19}	1.42e-31(1.78e-32) - 100/49,914	**1.35e-31(2.23e-47)** = 100/26,850	1.59e-30(6.53e-30) = 100/51,590	1.01e-11(1.27e-11) - 100/106,840	**1.35e-31(2.23e-47)** 100/22,626
f_{20}	4.72e-02(4.46e-02) - 0/NA	**0.00e+00(0.00e+00)** = 100/50,314	2.84e-16(1.42e-15) - 100/96,362	2.07e-02(2.87e-02) - 0/NA	**0.00e+00(0.00e+00)** 100/54,716
f_{21}	**-78.332(0.00e+00)** = 100/16,158	-78.332(8.70e-15) - 100/8,710	**-78.332(4.35e-14)** = 100/12,510	**-78.332(0.00e+00)=** 100/32,030	-78.332(4.35e-14) 100/7,892
f_{22}	-29.998(1.24e-03) - 100/21,630	**-30.000(2.70e-06)** = 100/12,226	**-30.000(3.23e-06)** = 100/18,874	-29.999(6.04e-04)- 100/25,794	**-30.000(0.00e+00)** 100/5,462
+/=/-	0/8/14	0/12/10	2/11/9	0/4/18	

"+", "=", and "-" respectively denote that the performance of the corresponding algorithm is better than, similar to,and worse than that of MPABC_RA according to the Wilcoxon's rank test at a 5% significance level.

objective function value for some representative functions are presented in Fig. 1, which clearly illustrates that MPABC_RA has better solution accuracy, convergence rate and robustness than all the competitors on most test functions.

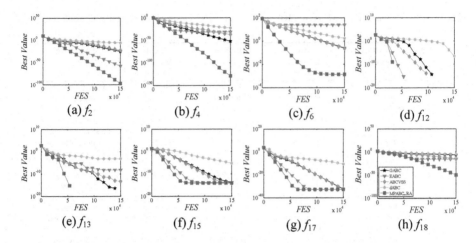

Fig. 1. Convergence curve of ABCs on some representative test functions

5 Conclusion and Future Works

In this paper, in order to improve the convergence rate and enhance the exploitation ability of ABC, we propose a multi-population based search strategy ensemble ABC algorithm with a novel resource allocation mechanism (called MPABC_RA). In employed bee phase, food sources are divided into three subpopulations by their quality, and each subpopulation uses different search strategy to exhibit its unique search ability. In onlooker bee phase, only the superior food sources can be selected by onlooker bees and the low-quality superior food sources will be allocated with more computing resources. Besides, onlooker bees employ a novel search equation to fully exploit the area between the locations of the selected superior solutions and the current best solution. The performance of MPABC_RA is validated by comparing with other outstanding ABC variants (i.e., GABC, EABC, ABCVSS and dABC) on 22 benchmark functions with $30D$.

In the future, the proposed multi-population based search strategy ensemble and resource allocation mechanism will be employed to other evolutionary algorithms. MPABC_RA can also be applied to solve some practical optimization problems.

Acknowledgments. This work is supported by the National Natural Science Foundation of China (Grant nos. 61402293, 61602316), Shenzhen Technology Plan (Grant nos. JCYJ20150324141711694), Seed Funding from Scientific and Technical Innovation Council of Shenzhen Government (Grant no. 827-000035).

References

1. Poli, R., Kennedy, J., Blackwell, T.: Particle swarm optimization. Swarm Intell. **1**, 33–57 (1995)

2. Dorigo, M., Maniezzo, V., Colorni, A.: Ant system: optimization by a colony of cooperating agents. IEEE Trans. Syst. **26**, 29–41 (1996)
3. Karaboga, D.: An idea based on honey bee swarm for numerical optimization. Technical report-TR06, Erciyes University, Engineering Faculty, Department of Computer Science (2005)
4. Yang, X.S.: Firefly algorithm, stochastic test functions and design optimization. Int. J. Bio-Insp. Comput. **2**(2), 78–84 (2010)
5. Szeto, W.Y., Wu, Y.Z., Ho, S.C.: An artificial bee colony algorithm for the capacitated vehicle routing problem. Eur. J. Oper. Res. **215**, 126–135 (2011)
6. Pan, Q.K., Tasgetiren, M.F., Suganthan, P.N., Chua, T.J.: A discrete artificial bee colony algorithm for the lot-streaming flow shop scheduling problem. Inf. Sci. **181**, 2455–2468 (2011)
7. Gao, W.F., Liu, S.Y., Jiang, F.: An improved artificial bee colony algorithm for directing orbits of chaotic systems. Appl. Math. Comput. **218**, 3868–3879 (2011)
8. Cui, L.Z., Li, G.H., Lin, Q.Z., Chen, J.Y., Lu, N., Zhang, G.J.: Artificial bee colony algorithm based on neighboring information learning. In: ICONIP, pp. 279–289 (2016)
9. Zhu, G., Kwong, S.: Gbest-guided artificial bee colony algorithm for numerical function optimization. Appl. Math. Comput. **217**(7), 3166–3173 (2010)
10. Gao, W.F., Liu, S.Y., Huang, L.L.: Enhancing artificial bee colony algorithm using more information-based search equations. Infrom. Sci. **270**(1), 112–133 (2014)
11. Kiran, M.S., Findik, O.: A directed artificial bee colony algorithm. Appl. Soft Comput. **26**, 454–462 (2015)
12. Wang, H., Wu, Z., Rahnamayan, S., Sun, H., Liu, Y., Pan, J.: Multi-strategy ensemble artificial bee colony algorithm. Inform. Sci. **279**, 587–603 (2014)
13. Kiran, M.S., Hakli, H., Gunduz, M., Uguz, H.: Artificial bee colony algorithm with variable search strategy for continuous optimization. Inform. Sci. **300**, 140–157 (2015)
14. Gao, W.F., Huang, L.L., Liu, S.Y., Chan, F.T.S., Dai, C.: Artificial bee colony algorithm with multiple search strategies. Appl. Math. Comput. **271**, 269–287 (2015)
15. Gao, W.F., Liu, S.Y.: A modified artificial bee colony algorithm. Comput. Oper. Res. **39**(3), 687–697 (2012)
16. Gao, W.F., Liu, S.Y., Huang, L.L.: A novel artificial bee colony algorithm based on modified search equation and orthogonal learning. IEEE Trans. Cybern. **43**(3), 1011–1024 (2013)
17. Kang, F., Li, J.J., Ma, Z.Y.: Rosenbrock artificial bee colony algorithm for accurate global optimization of numerical functions. Inform. Sci. **181**(16), 3508–3531 (2011)
18. Akay, B., Karaboga, D.: A modified artificial bee colony algorithm for real-parameter optimization. Inform. Sci. **192**(1), 120–142 (2012)
19. Cui, L.Z., Li, G.H., Zhu, Z.X., Lin, Q.Z., Wen, Z.K., Lu, N., Wong, K.C., Chen, J.Y.: A novel artificial bee colony algorithm with adaptive population size for numerical function optimization. Inf. Sci. **414**, 53–67 (2017)
20. Li, G.H., Cui, L.Z., Fu, X.H., Wen, Z.K., Lu, N., Lu, J.: Artificial bee colony algorithm with gene recombination for numerical function optimization. Appl. Soft Comput. **52**, 146–159 (2017)
21. Cui, L.Z., Zhang, K., Li, G.H., Fu, X.H., Wen, Z.K., Lu, N., Lu, J.: Modified Gbest-guided artificial bee colony algorithm with new probability model. Soft Comput. (2017). doi:10. 1007/s00500-017-2485-y

Grammatical Evolution Using Tree Representation Learning

Shunya Maruta[1], Yi Zuo[2(✉)], Masahiro Nagao[3], Hideyuki Sugiura[1],
and Eisuke Kita[1]

[1] Graduate School of Information Sciences, Nagoya University, Nagoya, Japan
maruta.shunya@c.mbox.nagoya-u.ac.jp,
sugiura.hideyuki@h.mbox.nagoya-u.ac.jp, kita@is.nagoya-u.ac.jp
[2] Institute of Innovation for Future Society, Nagoya University, Nagoya, Japan
zuo@nagoya-u.jp
[3] Graduate School of Environmental Science, Nagoya University, Nagoya, Japan
nagao@urban.env.nagoya-u.ac.jp

Abstract. Grammatical evolution (GE) is one of the evolutionary computations, which evolves genotype to map phenotype by using the Backus-Naur Form (BNF) syntax. GE has been widely employed to represent syntactic structure of a function or a program in order to satisfy the design objective. As the GE decoding process parses the genotype chromosome into array or list structures with left-order traversal, encoding process could change gene codons or orders after genetic operations. For improving this issue, this paper proposes a novel GE algorithm using tree representation learning (GETRL) and presents three contributions to the original GE, genetic algorithm (GA) and genetic programming (GP). Firstly, GETRL uses a tree-based structure to represent the functions and programs for practical problems. To be different from the traditional GA, GETRL adopts a genotype-to-phenotype encoding process, which transforms the genes structures for tree traversal. Secondly, a pointer allocation mechanism is introduced in this method, which allows the GETRL to pursue the genetic operations like typical GAs. To compare with the typical GP, however GETRL still generates a tree structure, our method adopts a phenotype-to-genotype decoding process, which allows the genetic operations be able to be apply into tree-based structure. Thirdly, due to each codon in GE has different expression meaning, genetic operations are quite different from GAs, in which all codons have the same meaning. In this study, we also suggest a multi-chromosome system and apply it into GETRL, which can prevent from overriding the codons for different objectives.

Keywords: Grammatical evolution · Tree representation · Multiple chromosomes · Pointer allocation · Genotype-phenotype map

1 Introduction

Genetic algorithm (GA) is one of the most popular algorithms in evolutionary computation [1, 4, 14], which has been considerably applied to optimization,

© Springer International Publishing AG 2017
D. Liu et al. (Eds.): ICONIP 2017, Part IV, LNCS 10637, pp. 346–355, 2017.
https://doi.org/10.1007/978-3-319-70093-9_36

adaptation and learning problems. However, some problems e.g., symbolic regression, syntactic problems and automatic generation program, are still hard to be solved for GAs to represent schema information. Therefore, genetic programming (GP) was proposed by Koza [5]. GP evolves a population of computer programs by using Lisp language to automatically solve problems without requiring the user to know or specify the form or structure of the solution in advance. Despite the advantages, GP also has several limitations. Recombination problem is the well-known one which limits its applicability and performance. First, sub-tree crossover in GP sometimes generated the invalid individual which was translated into incorrect function or program. Second, sub-tree crossover also had a tendency for parse trees to grow larger and larger, which would cause program size bloat.

Grammatical evolution (GE) is another tree-based evolutionary algorithm, which was presented by [2,8,10,12,13,15]. The main features of GE are to present an evolutionary process, map genotype to phenotype in a genetic algorithm approach, and translate rules using the Backus-Naur Form (BNF) syntax. Therefore, GE is very attractive in two folds: (1) The use of the translations rule can avoid the generation of the invalid phenotype. (2) The genotype-to-phenotype mapping can capture important schema information. In past few years, many works about the GE representations empirically measured the locality of GE, and identified that standard GE has low locality and compromises search effectiveness [3,9,11,16]. To enhance GE representations, several studies have sufficiently investigated grammar-guided GP and tree-based GE [6,9]. Murphy et al. [7] examined the behavior of tree-adjunct grammars to grammatical evolution. Whigham et al. [17] investigated the application of context-free grammar genetic programming. However, several limitations have revealed during the evolutionary process, such as, the genetic operators in GE overriding the original meaning of each codon in chromosome and violating the better partial structures of the phenotypes.

For these issues, this paper proposes a novel GE-based algorithm using tree representation learning (GETRL) and presents three contributions to the original GE and GP. First, GETRL uses a tree-based structure to represent the functions and programs for practical problems. To be different from the traditional GA, GETRL adopts a genotype-to-phenotype encoding process, which transforms the linear gene structures into tree traversal. In contrast to the original GE, which caused the overriding of codons when reading codons in array sequence with leftmost derivation, GETRL can represent translation process in advance to assign the codons as nodes of tree structure following the level-order traversal. Second, we introduce a pointer allocation mechanism to learn tree structures. During the translation process of genotype-to-phenotype mapping, a pointer is employed to map tree structure from the genotype chromosome. The pointer can link the nodes in virtually continuous address, which are distributed discretely in the linear genotype chromosome. When the sequence of chromosome is changed by genetic operators, the pointer remembers the address of the right position and finds the exact nodes in the parse tree. To compare with the typical

GP, the non-terminal nodes are labeled and linked to the corresponding codons in chromosome, which allows the GETRL to pursue the genetic operations by using the pointers. Third, we also introduce a multi-chromosomes system into GETRL, which can improve the strategy of genetic operators (e.g., crossover and mutation). Due to there are different kinds of nonterminals in BNF grammar definition, this paper uses multi-chromosomes each sub-chromosome is assigned to individual nonterminals to represent different objectives. Furthermore, the multi-chromosomes system can also prevent from overriding the codons for different objectives.

In Sect. 2, we describe the GETRL. In Sect. 3, we present the experiments, results and discussions. In Sect. 4, we conclude and summarize the paper.

2   Method

2.1   Tree Representation

The original GE used the binary strings to define individuals in genotype (Fig. 1). GE adopted a genotype-to-phenotype process, where individuals were denoted as integer-form using the grammar as shown in Table 1. From the leftmost unused integer number, the genotype was referred to g_i for $i = 0$ to n, and translation of genotype was started from symbol <expr> (see Fig. 2). We assumed that the leftmost untranslated symbol in the phenotype is α and the number of the substitution rules for α is s_α. When the remainder r_i was calculated from g_i mod s_α, and the symbol α was replaced with the $(r_i + 1)$-th symbol in the candidate rule list. Following this translation rules, We can obtain the phenotype "$1/x - x$" from the genotype "6214331513" according to pre-order walk traverse as shown in Fig. 2.

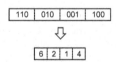

Fig. 1. Mapping from binary genotype to integer genotype.

Table 1. Translation rule in simple example

Rule	Candidate rule list	Rule no.
(A)	<expr> ::= <expr><op><expr> | <var>	(0) | (1)
(B)	<op> ::= + | - | * | /	(0) | (1) | (2) | (3)
(C)	<var> ::= 1 | x	(0) | (1)

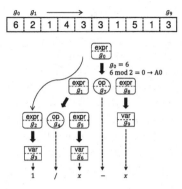

Fig. 2. Pre-order walk traverse of symbols from genotype.

2.2   Pointer Allocation Mechanism

Our proposal introduced a pointer allocation mechanism to map the tree structure from the chromosomes as shown in Fig. 3. The codons of <expr> are assigned into the tree structure following the linear-order walk traverse. The pointer p is firstly allocated to the header of chromosome e as follows:

$$p = e[0] \tag{1}$$

and

$$p \rightarrow left = e[1], \tag{2}$$

$$p \rightarrow right = e[2] \tag{3}$$

where $e[\ ]$ denotes the <expr> array. The p points the root of binary tree, and $p.left$ and $p.right$ points the left node and right node, respectively. As a tree is a self-referential data structure, the linear genotype structure can be allocated as a linked lists as follows:

$$p = p \rightarrow left, \tag{4}$$

$$p = p \rightarrow right. \tag{5}$$

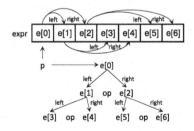

Fig. 3. Pointer allocation mechanism for tree structure.

2.3 Multiple Chromosome System

For different kinds of nonterminals, arrays $e[\ ], o[\ ]$ and $v[\ ]$ denote `<expr>` array, `<op>` array and `<var>` array, which were allocated to each sub-chromosome (Fig. 4). According to Table 1, our proposal separated the nonterminals into the recursive nonterminals and non-recursive nonterminals. `<expr>` is the recursive nonterminal, and `<op>` and `<var>` are the non-recursive nonterminals.

Due to recursive nonterminals were recursively replaced by themselves as shown in translation rule (A0), this type of rule is called recursive rule. When pointer p was allocated to any $e[i]$, allocation strategy for `<expr> ::= <expr><op><expr>` could be defined as follows. The `<expr>` in the root is represented as

$$p = e[i], \tag{6}$$

and the two `<expr>` in the both sides are represented as

$$p \to left = e[2 * i + 1], \tag{7}$$

$$p \to right = e[2 * i + 2]. \tag{8}$$

Here, `<op>` in rule (A0) is non-recursive nonterminal, and is replaced by non-recursive nonterminals alone i.e. rule (A1). Due to non-recursive nonterminals would be translated into phenotype by terminals, this type of rule is called non-recursive rule. When pointer p was allocated to any $e[i]$, allocation strategy for `<op>` in `<expr> ::= <expr><op><expr>` and `<expr> ::= <var>` could be defined as follows:

$$p \to op = o[i] \tag{9}$$

and

$$p \to var = v[i]. \tag{10}$$

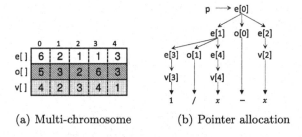

(a) Multi-chromosome   (b) Pointer allocation

Fig. 4. Pointer allocation mechanism in multi-chromosome.

3 Experiment

3.1 Problems Setting

In the experiments, we used three kinds of symbolic regression and Santa Fe Ant Trail problem to compare the performance and effectiveness of GETRL with that of original GE and GE using multi-chromosome (GEMC). Parameter setting for

Table 2. Parameter settings

Parameter	Values
Maximum generations	500
Population size	200
Selection	Tournament selection
Tournament size	5
Number of elites	5
Crossover	Uniform crossover
Crossover rate (CR)	0.0, 0.1, 0.2, 0.3, 0.4, 0.5
Mutation rate (MR)	0.03, 0.05, 0.1

GE, GEMC and GETRL was shown in Table 2 and the fitness was based on root mean square error. In order to obtain the most approximative target function, we examined respective methods for 50 runs, and selected the best parameter for illustrating the results and discussions.

Symbolic Regression. The grammar used for symbolic regression is given as follow:

$$N = \{\text{expr}, \text{op}, \text{val}, \text{num}, \text{char}\},$$
$$T = \{\text{+, -, *, /, ^, 1, x, y}\},$$
$$S = \{\text{expr}\},$$

and three kinds of symbolic regression are listed bellow.

Ex. 1:
$$f_1(x) = x + x^2 + x^3 + x^4.$$

Ex. 2:
$$f_2(x) = x^4 - 2x^3 + 3x^2 - 4x + 5.$$

Ex. 3:
$$f_3(x, y) = (x - y)^5.$$

Santa Fe Ant Trail Problem (Ex. 4). The grammar of Santa Fe Ant Trail problem [8] is given as follow:

$$N = \{\text{code}, \text{op}\},$$
$$T = \{\text{if, else, food\_ahead, turn\_left,}$$
$$\text{turn\_right, move}\},$$
$$S = \{\text{code}\},$$

and the fitness is calculated by Eq. (11),

$$f = F - F_{max} \tag{11}$$

where $F_{max} = 89$ denotes all the pieces of food, and F denotes the obtained pieces of food.

3.2 Results

Ex. 1. Ex.1 presented the standard quartic symbolic regression problem, which is widely used as the benchmark. However, either of these three methods can find the appropriate function to the target function, GETRL showed a better convergence than GE and GEMC around the 10th–50th generation (Fig. 5(a)).

Ex. 2. Ex.2 presented another quartic symbolic regression problem. Ex.2 used the same BNF syntax as Ex.1, but the expression is more complicated than Ex. 1. In Ex.2 (Fig. 5(b)), GETRL also showed a better convergence than GE and GEMC from the 10th generation. The convergence speed of GETRL to find the appropriate function to the target function around the 100th generation is also much faster than GE and GEMC, in which the appropriate function can be found to the target function around the 200th generation. Due to the proposed pointer allocation strategy, GETRL outperformed GE and GEMC in preventing the invalid individuals, especially at the beginning of search.

Ex. 3. Ex.3 presented a symbolic problem with two variables. However, GETRL showed a worse convergence than GE and GEMC at the beginning, GETRL became much superior than GE and GEMC from the 20th generation, and got the appropriate function around 480th generation (Fig. 5(c)). Accounting for its pointer allocation strategy, GETRAL allowed the offspring to remain the effective schemata and showed a dramatically performance than GE and GEMC.

Ex. 4. Figure 5(d) showed the comparison of GE, GEMC and GETRL in Santa Fe Ant Trail problem. GETRL outperformed GE around the 10th generation,

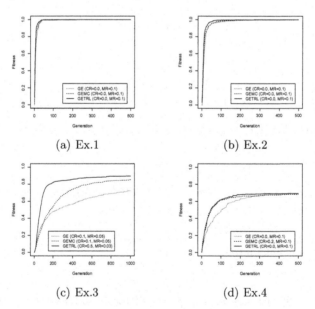

(a) Ex.1   (b) Ex.2

(c) Ex.3   (d) Ex.4

Fig. 5. Comparison of convergence.

and outperformed GEMC around the 100th generation. The pointer allocation strategy in GETRL contributed to this problem into two folds. One is the effectiveness of remaining the schemata from parents. The other one is the utility of preventing to read invalid genes, which can lead it to producing more efficient program than GE and GEMC.

3.3   Discussion

Symbolic Regression. Figure 6 showed the effectiveness of GETRL comparing with GE and GEMC. First, we subtracted average number of <expr> using in each generation from the previous generation. We estimated the difference of the average number between two generations, and the smaller change of this value indicated the more similarity of offspring inherited from the parents. The difference in each generation of Ex.1, Ex.2 and Ex.3 were represented in Figs. 6(a), (b) and (c), respectively. Second, we also investigated individuals to read the invalid genes resulting death. The number of dead individuals in each generation of Ex.1, Ex.2 and Ex.3 were represented in Fig. 6(d), (e) and (f). Due to the pointer allocation mechanism, GETRL was able to outperform GE and GEMC in inheriting better schemata from parents and preventing invalid individuals in offspring.

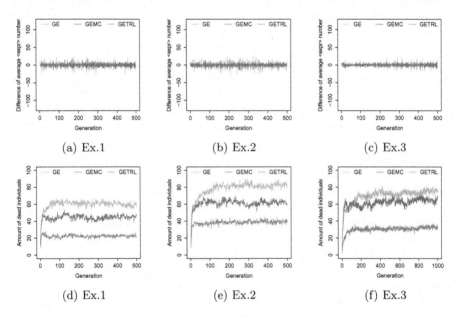

Fig. 6. Symbolic regression problem. (a), (b) and (c): difference of average <expr> number in Ex.1, Ex.2 and Ex.3; (d), (e) and (f): amount of dead individuals in Ex.1, Ex.2 and Ex.3.

Santa Fe Ant Trail Problem. As shown in Fig. 7, average amount of foods eaten was calculated for 50 runs. The maximum values and minimum values in these 50 runs were extracted, which denoted foods eaten by using the best program

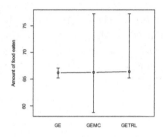

Fig. 7. Amount of foods eaten in Santa Fe Ant Trail problem.

and the worst program generated from GE, GEMC, and GETRL, respectively. As GETRL obtained best results in either of the maximum, minimum and average amount of foods eaten, the pointer allocation mechanism in GETRL could lead to the production of a more superior program than GE and GEMC.

4   Conclusion

This paper described an improved algorithm of grammatical evolution using tree representation learning (GETRL). GETRL presented a novel approach to the GE, which adopted pointer allocation mechanism and multiple chromosomes system. The pointer allocation mechanism was helpful for recording the position of individual genes in the chromosome and mapping the tree structure from genotype instead of list or array structure. The multiple chromosomes system was used for enhancing the effectiveness of GETRL, and better preserving the partial structures of solutions when genetic operators were applied. We compared the performance of GETRL with that of the original GE and GE using multi-chromosomes for the symbolic regression problems and Santa Fe Ant Trail problem. Due to GETRL perfectly resolved the poor search properties, which are resulted from the agnostic link between the linear representation in GE and the derivation tree of phenotype, our method showed faster convergence in finding the appropriate solutions.

For future work, there remained two limitations to be addressed. First, however the point allocation mechanism can prevent reading of invalid genes. As the tree structure is a fixed binary tree, it is easy to cause the pointer index out of range. Next, the recursive rules of BNF syntax design must contain two recursive nonterminals such as `<expr><op><exper>` or `<code><code>`, for fitting to binary tree structure. Therefore, we plan to propose a self-adaptive pointers and memory allocation strategy with two operations *malloc* and *free*. According to any grammar, the memory can be dynamically allocated by *malloc* to generate a dynamic tree structure. After genetic operations, the unused memory can be deallocated by *free* to prevent from out of range.

References

1. Bäck, T., Schwefel, H.P.: An overview of evolutionary algorithms for parameter optimization. Evol. Comput. **1**(1), 1–23 (1993)

2. Brabazon, A., O'Neill, M.: Biologically Inspired Algorithms for Financial Modelling. Natural Computing Series. Springer-Verlag New York Inc., Secaucus (2006). doi:10.1007/3-540-31307-9
3. Byrne, J., O'Neill, M., McDermott, J., Brabazon, A.: An analysis of the behaviour of mutation in grammatical evolution. In: Esparcia-Alcázar, A.I., Ekárt, A., Silva, S., Dignum, S., Uyar, A.Ş. (eds.) EuroGP 2010. LNCS, vol. 6021, pp. 14–25. Springer, Heidelberg (2010). doi:10.1007/978-3-642-12148-7_2
4. Holland, J.H.: Adaptation in Natural and Artificial Systems: An Introductory Analysis with Applications to Biology, Control and Artificial Intelligence. MIT Press, Cambridge (1992)
5. Koza, J.R.: Genetic Programming: On the Programming of Computers by Means of Natural Selection. MIT Press, Cambridge (1992)
6. Lourenço, N., Pereira, F.B., Costa, E.: Unveiling the properties of structured grammatical evolution. Genet. Program Evolvable Mach. **17**(3), 251–289 (2016)
7. Murphy, E., O'Neill, M., Galvan-Lopez, E., Brabazon, A.: Tree-adjunct grammatical evolution. In: IEEE Congress on Evolutionary Computation, pp. 1–8 (2010)
8. O'Neill, M., Ryan, C.: Grammatical evolution. IEEE Trans. Evol. Comput. **5**(4), 349–358 (2001)
9. O'Neill, M., Brabazon, A., Nicolau, M., Garraghy, S.M., Keenan, P.: πGrammatical evolution. In: Deb, K. (ed.) GECCO 2004. LNCS, vol. 3103, pp. 617–629. Springer, Heidelberg (2004). doi:10.1007/978-3-540-24855-2_70
10. O'neill, M., Ryan, C., Keijzer, M., Cattolico, M.: Crossover in grammatical evolution. Genet. Program Evolvable Mach. **4**(1), 67–93 (2003)
11. Ryan, C., Azad, A., Sheahan, A., O'Neill, M.: No coercion and no prohibition, a position independent encoding scheme for evolutionary algorithms – the Chorus system. In: Foster, J.A., Lutton, E., Miller, J., Ryan, C., Tettamanzi, A. (eds.) EuroGP 2002. LNCS, vol. 2278, pp. 131–141. Springer, Heidelberg (2002). doi:10.1007/3-540-45984-7_13
12. Ryan, C., Collins, J.J., Neill, M.O.: Grammatical evolution: evolving programs for an arbitrary language. In: Banzhaf, W., Poli, R., Schoenauer, M., Fogarty, T.C. (eds.) EuroGP 1998. LNCS, vol. 1391, pp. 83–96. Springer, Heidelberg (1998). doi:10.1007/BFb0055930
13. Ryan, C., O'Neill, M., Collins, J.: Grammatical evolution: solving trigonometric identities. In: Proceedings of Mendel 1998: 4th International Conference on Genetic Algorithms, Optimization Problems, Fuzzy Logic, Neural Networks and Rough Sets, pp. 111–119 (1998)
14. Schwefel, H.P.P.: Evolution and Optimum Seeking: The Sixth Generation. Wiley, New York (1993)
15. Shaker, N., Nicolau, M., Yannakakis, G.N., Togelius, J., O'Neill, M.: Evolving levels for super mario bros using grammatical evolution. In: 2012 IEEE Conference on Computational Intelligence and Games (CIG), pp. 304–311 (2012)
16. Thorhauer, A., Rothlauf, F.: On the locality of standard search operators in grammatical evolution. In: Bartz-Beielstein, T., Branke, J., Filipič, B., Smith, J. (eds.) PPSN 2014. LNCS, vol. 8672, pp. 465–475. Springer, Cham (2014). doi:10.1007/978-3-319-10762-2_46
17. Whigham, P.A., Dick, G., Maclaurin, J., Owen, C.A.: Examining the "best of both worlds" of grammatical evolution. In: Proceedings of the 2015 Annual Conference on Genetic and Evolutionary Computation, GECCO 2015, pp. 1111–1118. ACM, New York (2015)

Application of Grammatical Swarm to Symbolic Regression Problem

Eisuke Kita[1]([✉]), Risako Yamamoto[2], Hideyuki Sugiura[2], and Yi Zuo[3]

[1] Graduate School of Informatics, Nagoya University, Nagoya, Japan
kita@is.nagoya-u.ac.jp
[2] Graduate School of Information Sciences, Nagoya University, Nagoya, Japan
[3] Institute of Innovation for Future Society, Nagoya University, Nagoya, Japan
zuo@nagoya-u.jp

Abstract. Grammatical Swarm (GS), which is one of the evolutionary computations, is designed to find the function, the program or the program segment satisfying the design objective. Since the candidate solutions are defined as the bit-strings, the use of the translation rules translates the bit-strings into the function or the program. The swarm of particles is evolved according to Particle Swarm Optimization (PSO) in order to find the better solution. The aim of this study is to improve the convergence property of GS by changing the traditional PSO in GS with the other PSOs such as Particle Swarm Optimization with constriction factor, Union of Global and Local Particle Swarm Optimizations, Comprehensive Learning Particle Swarm Optimization, Particle Swarm Optimization with Second Global best Particle and Particle Swarm Optimization with Second Personal best Particle. The improved GS algorithms, therefore, are named as Grammatical Swarm with constriction factor (GS-cf), Union of Global and Local Grammatical Swarm (UGS), Comprehensive Learning Grammatical Swarm (CLGS), Grammatical Swarm with Second Global best Particle (SG-GS) and Grammatical Swarm with Second Personal best Particle (SG-GS), respectively. Symbolic regression problem is considered as the numerical example. The original GS is compared with the other algorithms. The effect of the model parameters for the convergence properties of the algorithms are discussed in the preliminary experiments. Then, except for CLGS and UGS, the convergence speeds of the other algorithms are faster than that of the original GS. Especially, the convergence properties of GS-cf and SP-GS are fastest among them.

Keywords: Grammatical Swarm · Particle Swarm Optimization · Symbolic Regression

1 Introduction

Meta-heuristic algorithms are well known techniques for solving the optimization problems defined with complicated objective functions and constraint conditions.

© Springer International Publishing AG 2017
D. Liu et al. (Eds.): ICONIP 2017, Part IV, LNCS 10637, pp. 356–365, 2017.
https://doi.org/10.1007/978-3-319-70093-9_37

Specially, Evolutionary Computation (EC) [1] and Particle Swarm Optimization (PSO) [2] are studied widely.

Genetic Algorithm (GA) [3], which is one of the popular evolutionary algorithms, are designed to find the optimal solution of the function. The search process starts from the definition of the set of the candidate solutions. The candidate solutions ("individuals") are defined by the set of the binary numbers (bit-string). The population is given as the set of the individuals. The use of genetic operations such as selection, crossover and mutation updates the population in order to find the optimal solution. Genetic Programming (GP) [4] is also well-known evolutionary algorithm. The aim of GP is to determine the function, the program or program segment satisfying the design objective. The candidate solutions are defined in the tree structures. The use of the genetic operations modifies and updates the edges and the leafs of the tree structures in order to find the function or the program satisfying the design objective.

Ryan et al. have presented Grammatical Evolution (GE) [5, 6]. GE is designed to find the function or the program satisfying the design objective, which is similar to GP. The candidate solutions are defined by the bit-strings, like GA. The bit-strings of the candidate solutions are translated into the function or the program according to the translation rules defined in Backus Naur Form (BNF). The bit-strings of the candidate solutions are updated by GA.

O'Neill et al. have presented Grammatical Swarm (GS) [7]. The candidate solutions are defined by the bit-strings and then, the bit-strings are updated by Particle Swarm Optimization (PSO) [2]. Several improved algorithms for PSO have been presented. The aim of this study is to improve the search performance of GS by changing the original PSO in the GS with the other PSO algorithms.

The remaining part of this paper is organized as follows. The background of this study is explained in Sect. 2. The improved algorithms for GS are described in Sect. 3. The numerical examples are shown in Sect. 4. The conclusions are summarized again in Sect. 5.

2   Background

2.1   Grammatical Evolution (GE)

Algorithm. GE algorithm is described as follows.

1. Individuals are generated randomly in order to define the initial population.
2. The genotype of each individual is translted into the phenotype according to the translation rules.
3. Individuals' fitness is evaluated.
4. Convergence criterion is confirmed. If the criterion is satisfied, the best individual's phenotype is output. Otherwise, the process goes to the next step.
5. The individuals in the population are updated by Genetic Algorithms.
6. The process goes to step 2.

Translation of Genotype to Phenotype. A translation rules are made of the tuple N, T, P and S, which are the set of all non-terminal symbols, the set of all terminal symbols, the set of the production rules which map N to N or P, and the initial start symbol, respectively.

Using the tuple N, T and S as

```
N = { <expr>, <op>, <var> }
T = { +, -, *, /, X, Y }
S = { <expr> }
```

the tuple P is shown in Table 1. The symbol "|" denotes the "or", which is the separations between the candidate rules. The production rule (A) denotes that `<expr>` is replaced with `<expr><op><expr>` or `<var>`. In other words, `<expr>` has two candidates `<expr><op><expr>` and `<var>`. `<op>` has four candidates +, -, * and /. `<var>` has two candidates X and Y.

Table 1. Example of translation rules in GE.

(A)	`<expr>` ::= { `<expr><op><expr>`	(A0)
		| `<var>` }	(A1)
(B)	`<op>` ::= { +	(B0)
	| -	(B1)
	| *	(B2)
	| / }	(B3)
(C)	`<var>` ::= { X	(C0)
	| Y }	(C1)

The candidate solutions are defined by the set of the binary or decimal numbers. The genotype x of the candidate solution is defined by the set of the decimal numbers.

According to the Table 1, the genotype

$$\{x_i\} = \{x_1, x_2, x_3, x_4, x_5, x_6\} = \{4, 1, 2, 3, 7, 1\} \tag{1}$$

is translated into the phenotype as follows.

1. The start symbol is $\alpha = S = $ `<expr>`.
2. The leftmost unused number is $x_1 = 4$. The symbol $\alpha = $ `<expr>` has two candidates; $x_\alpha = 2$. Since $x_l = x_1 \% x_\alpha = 4\%2 = 0$, the symbol $\alpha = $ `<expr>` is replaced with the 0-th candidate `<expr><op><expr>` (A0).
3. The leftmost unused number is $x_2 = 1$. The leftmost non-terminal symbol $\alpha = $ `<expr>` has two candidates; $x_\alpha = 2$. Since $x_l = x_2 \% x_\alpha = 1\%2 = 1$, the symbol $\alpha = $ `<expr>` is replaced with the 1-st candidate `<var>` (A1).
4. According to the similar translation, finally, the phenotype X*Y is generated.

2.2 Particle Swarm Optimization (PSO)

Update Algorithms. Each particle denotes the candidate solution. The position vector of the particle is the set of the design variables of the candidate solution. The velocity vector of the particle is the update values of the position vector. The position and the velocity vectors of the particle i at the time t are represented with the variables $\boldsymbol{x}_i(t)$ and $\boldsymbol{v}_i(t)$, respectively.

$$\boldsymbol{x}_i(t) = \{x_{i1}(t), x_{i2}(t), \cdots, x_{in}(t)\} \tag{2}$$
$$\boldsymbol{v}_i(t) = \{v_{i1}(t), v_{i2}(t), \cdots, v_{in}(t)\} \tag{3}$$

The variable n denotes the number of the design variables. The variable $\boldsymbol{v}_i(t+1)$ is calculated from the Eq. (4).

$$v_i(t+1) = wv_i(t) + c_1 r_1 (\boldsymbol{x}_i^{pbest}(t) - \boldsymbol{x}_i(t))$$
$$+ c_2 r_2 (\boldsymbol{x}^{gbest}(t) - \boldsymbol{x}_i(t)) \tag{4}$$

The variables $\boldsymbol{x}_i^{pbest}(t)$ and $\boldsymbol{x}^{gbest}(t)$ denote the best particle which has been found by the particle i ever and the best particle which has been found by all particles ever, respectively.

$$\boldsymbol{x}_i^{pbest}(t) = \begin{cases} \boldsymbol{x}_i(t) & if\ f(\boldsymbol{x}_i(t)) < f(\boldsymbol{x}_i^{pbest}(t-1)) \\ \boldsymbol{x}_i^{pbest}(t-1) & otherwise \end{cases} \tag{5}$$

$$\boldsymbol{x}^{gbest}(t) = \underset{\boldsymbol{x}_i^{pbest}(t)}{\operatorname{argmin}}\ f(\boldsymbol{x}_i^{pbest}(t)),\ 1 \le i \le N \tag{6}$$

The variables r_1 and r_2 denote the random number in $[0, 1]$ and the variables c_1 and c_2 mean the acceleration coefficients. The variable w is the inertia weight defined by Eq. (7) [12].

$$w = w_{max} - \frac{w_{max} - w_{min}}{t_{max}} t \tag{7}$$

The variable t_{max} denotes the maximum iteration of the simulation. The position vector $\boldsymbol{x}_i(t)$ is updated by the following equation.

$$\boldsymbol{x}_i(t+1) = \boldsymbol{x}_i(t) + \boldsymbol{v}_i(t+1) \tag{8}$$

Algorithm. The PSO algorithm is summarized as follows.

1. Particles are randomly generated in order to define the swarm.
2. Fitness of particles is calculated.
3. If the convergence criterion is satisfied, the results is output. Otherwise, the process goes to the next step.
4. The variables $\boldsymbol{x}_i^{pbest}(t)$ and $\boldsymbol{x}^{gbest}(t)$ are updated by Eqs. (5) and (6), respectively.
5. The velocity vector $\boldsymbol{v}_i$ and the position vector $\boldsymbol{x}_i$ of the particles are updated by Eqs. (4) and (8), respectively.
6. The process goes to step 2.

3 Grammatical Swarm (GS)

3.1 Algorithm

GS algorithm is as follows.

1. Particles are generated randomly in order to define the initial swarm.
2. Since the particles are defined in the set of the real-valued variables, the real values are rounded off in order to define the genotype of the integer numbers.
3. The genotype is translated into the phenotype according to the translation rules.
4. Fitness of individuals is evaluated.
5. Convergence criterion is confirmed. If the criterion is satisfied, the best individual's phenotype is output. Otherwise, the process goes to the next step.
6. The variables $x_i^{pbest}(t)$ and $x^{gbest}(t)$ are updated by Eqs. (5) and (6), respectively.
7. The velocity vector v_i and the position vector x_i of the particles are updated by Eqs. (4) and (8), respectively.
8. The process goes to step 3.

3.2 Improved Algorithms of Grammatical Swarm

Grammatical Swarm with Constriction Factor (GS-cf). PSO with constriction factor (PSO-cf) has been presented by Clerc and Kennedy [8] for improving the original PSO. The position vector of the particle i at the time t is updated by Eq. (8) and the velocity vector is by Eq. (9), respectively.

$$v_i(t+1) = K[v_i(t) + c_1 r_1(x_i^{pbest}(t) - x_i(t)) + c_2 r_2(x^{gbest}(t) - x_i(t))] \quad (9)$$

The variable $K \in [0,1]$ is called as the constriction factor which is defined as follows.

$$K = \frac{2}{|2 - \varphi - \sqrt{\varphi^2 - 4\varphi}|}, \varphi = C_1 + C_2 > 4 \quad (10)$$

Union of Global and Local Grammatical Swarm (UGS). Union of Global And Local PSOs (UPSO) has been presented by Parsopoulos and Vrahatis [9]. The position vector of the particle i at the time t is updated by Eq. (8) and the velocity vector is by Eq. (11), respectively.

$$v_i(t+1) = u G_i(t+1) + (1-u) L_i(t+1) \quad (11)$$

The variables $G_i(t)$ and $L_i(t)$ denote the global and local best particles at the time t, respectively. The variable u is the uniform factor between them. Besides, they are updated with the following equations

$$\left.\begin{aligned} G_i(t+1) &= K[G_i(t) + c_1 r_1(x_i^{pbest}(t) - x_i(t)) + c_2 r_2(x^{gbest}(t) - x_i(t))] \\ L_i(t+1) &= K[L_i(t) + c_1 r_1(x_i^{pbest}(t) - x_i(t)) + c_2 r_2(x^{gbest}(t) - x_i(t))] \end{aligned}\right\} \quad (12)$$

The variable $K \in [0,1]$ is the constriction factor defined in Eq. (10).

Comprehensive Learning Grammatical Swarm (CLGS). Comprehensive Learning PSO (CLPSO) has been presented by Liang et al. [10]. The position and the velocity vectors of the particles are updated with the following equations.

$$\boldsymbol{x}_i(t+1) = \boldsymbol{x}_i(t) + \boldsymbol{v}_i(t+1) \tag{13}$$

$$\boldsymbol{v}_i(t+1) = \{v_{i1}(t+1), v_{i2}(t+1), \cdots, v_{iD}(t+1)\} \tag{14}$$

The variable $v_{id}(t)$ denotes the d-th element of the velocity vector of the particle i at the time t. CLPSO updates the variable $v_{id}(t)$ by the following equation.

$$\boldsymbol{v}_{id}(t+1) = w\boldsymbol{v}_{id}(t) + c_3 r_{id}(\boldsymbol{x}_{Id}^p best(t) - \boldsymbol{x}_{id}(t)) \tag{15}$$

The variable $x_{Id}^p best(t)$ are the position vector of the particle selected from two particles as follows.

1. Two particles a1 and a2 are chosen randomly from the swarm.
2. If the fitness of the particle a1 is larger than that of the particle a2, $\boldsymbol{x}_a^{pbest}(t) = \boldsymbol{x}_{a1}^{pbest}$. Otherwise, $\boldsymbol{x}_a^{pbest}(t) = \boldsymbol{x}_{a2}^{pbest}$.
3. The processes from the step 4 to 6 are performed for $d = 1, 2, \cdots, n$.
4. The learning probability $P_c \in [0, 1]$ is specified.
5. The random number $P \in [0, 1]$ is generated.
6. If $P > P_c$, $x_{Id}^p best(t) = x_{id}^p best(t)$. Otherwise, $x_{Id}^p best(t) = x_{ad}^p best(t)$.

Grammatical Swarm with Second Global Best Particle (SG-GS). PSO with Second Global best Particle (SG-PSO) has been presented by Shin and Kita [11]. The velocity vector of the particle is updated with the position vectors of the global best particle *gbest*, the personal best particle *pbest* and the global second best particle *gsecbest* as follows.

$$\boldsymbol{v}_i(t+1) = w\boldsymbol{v}_i(t) + c_1 r_1(\boldsymbol{x}_i^{pbest}(t) - \boldsymbol{x}_i(t)) + c_2 r_2(\boldsymbol{x}^{gbest}(t) - \boldsymbol{x}_i(t))$$
$$+ c_4 r_3(\boldsymbol{x}^{gsecbest}(t) - \boldsymbol{x}_i(t)) \tag{16}$$

The variable c_3 is the third acceleration coefficient and the variable r_3 is the random number in $[0, 1]$.

The SG-PSO algorithm is summarized as follows.

1. Particles are generated randomly in order to define the initial swarm.
2. Real values in the genotype are rounded to the integer numbers.
3. The genotype is translated into the phenotype according to the translation rules.
4. Fitness of individuals is evaluated.
5. Convergence criterion is confirmed. If the criterion is satisfied, the best individual's phenotype is output. Otherwise, the process goes to the next step.
6. The variables $\boldsymbol{x}_i^{pbest}(t)$, $\boldsymbol{x}^{gbest}(t)$ and $\boldsymbol{x}^{gsecbest}(t)$ are updated.
7. The velocity vector $\boldsymbol{v}_i$ and the position vector $\boldsymbol{x}_i$ of the particles are updated by Eqs. (16) and (8), respectively.
8. The process goes to step 2.

Grammatical Swarm with Second Personal Best Particle (SG-GS).
PSO with Second Personal best Particle (SP-PSO) has presented by Shin and
Kita [11]. The velocity vector of the particle is updated with the position vectors
of the global best particle $gbest$, the personal best particle $pbest$ and the personal
second best particle $psecbest$ as follows.

$$v_i(t+1) = wv_i(t) + c_1r_1(x_i^{pbest}(t) - x_i(t)) + c_2r_2(x^{gbest}(t) - x_i(t))$$
$$+ c_5r_3(x_i^{psecbest}(t) - x_i(t)) \tag{17}$$

The variable c_4 is the third acceleration coefficient and the variable r_3 is the
random number in $[0, 1]$.

The SP-PSO algorithm is summarized as follows.

1. Particles are generated randomly in order to define the initial swarm.
2. The real values are rounded off to the integer numbers.
3. The genotype is translated into the phenotype according to the translation
   rules.
4. Fitness of individuals is evaluated.
5. Convergence criterion is confirmed. If the criterion is satisfied, the best individual's phenotype is output. Otherwise, the process goes to the next step.
5. The variables $x_i^{pbest}(t)$, $x^{gbest}(t)$ and $x^{psecbest}(t)$ are updated.
6. The velocity vector v_i and the position vector x_i of the particles are updated
   by Eqs. (17) and (8), respectively.
7. The process goes to step 2.

4   Numerical Example

The search performance of the algorithms is compared in the symbolic regression
problem. The aim of this problem is to find the function $\bar{f}$ which can approximate
the given set of data $\{(x_1, y_1), (x_2, y_2), \cdots, (x_n, y_n)\}$. When the actual function
is given as f, $y_i = f(x_i)$.

The actual function f is given as follows.

$$f(x) = x^4 - 2x^3 + 3x^2 - 4x + 5 \tag{18}$$

The set of data is defined by calculating $y_i = f(x_i)$ at $\{x_i\} = -10, -9.9, \cdots, 10$.
Error is estimated as follows.

$$E = \sqrt{\frac{1}{201} \sum_{i=1}^{201} (f_{(x_i)} - \bar{f}_{(x_i)})^2} \tag{19}$$

The set of all non-terminal symbols N, the set of all terminal symbols T, and
the initial start symbol S are given as follows.

```
N = { <expr>, <op>, <var>, <num>, <X> }
T = { +, -, *, /, 1, 2, 3, 4, 5, 6, 7, 8, 9, x }
S = { <expr> }
```

Table 2. Translation rules for symbolic regression problem.

```
<expr> ::= { <expr><expr><op>
           | <var>              }
   <op> ::= { +
            | -
            | *
            | / }
  <var> ::= { <num>
            | <X>    }
  <num> ::= { 1
            | 2
            | 3
            | 4
            | 5
            | 6
            | 7
            | 8
            | 9 }
    <X> ::= { x }
```

The set of the production rules P which map N to N or P is shown in Table 2. The parameters are shown in Table 3. The effect of the parameters C_1, C_2, C_3, C_4 and C_5 for the convergence properties of the algorithms are discussed in the preliminary experiments. The sets of the parameters for the fastest convergence property in each algorithm are summarized in Table 4.

Table 3. Parameters for algorithms in symbolic regression problem.

Maximum iteration	5000
Swarm size	100
Vector length	100
C_1 (Original, GS-cf, UGS, SG-GS, SP-GS)	$2.0, 2.1, \cdots, 2.9$
C_2 (Original, GS-cf, UGS, SG-GS, SP-GS)	$2.0, 2.1, \cdots, 2.9$
C_3 (CLGS)	$0.5, 1.0, \cdots, 5.0$
C_4 (SG-GS)	$0.5, 1.0, \cdots, 5.0$
C_5 (SP-GS)	$0.5, 1.0, \cdots, 5.0$

The convergence properties of all algorithms are compared in Fig. 1. The figure is plotted with the average fitness of the best particles as the vertical axis and the number of the iteration as the horizontal axis, respectively. Except for CLGS and UGS, the convergence speeds of the other algorithms are faster than that of the original GS. Especially, the convergence properties of GS-cf and SP-GS are fastest among them.

Table 4. Parameters for fastest convergence property in each algorithm.

Algorithm	C_1	C_2	C_3	C_4	C_5
Original	0.5	1.5	–	–	–
GS-cf	1.5	3.0	–	–	–
UGS	7.0	8.0	–	–	–
CLGS	–	–	0.5	–	–
SG-GS	0.5	1.5	–	0.1	–
SP-GS	0.5	1.5	–	–	0.2

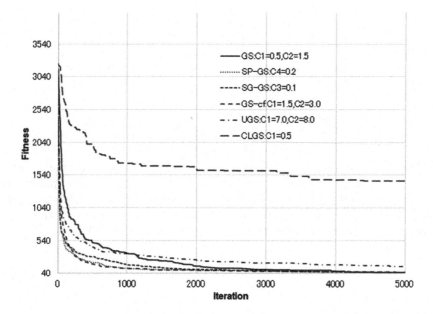

Fig. 1. Comparison of convergence histories of best particles

5   Conclusions

Grammatical Evolution (GE) is designed to find the function or the program satisfying the design objective. GE uses Genetic Algorithm (GA) for finding the optimal solution. Grammatical Swarm (GS) adopts the Particle Swarm Optimization (PSO), instead of GA.

In this paper, the original PSO in GS is updated with five improved PSO algorithms. The improved GS algorithms are named as Grammatical Swarm with constriction factor (GS-cf), Union of Global And Local Grammatical Swarm (UGS), Comprehensive Learning Grammatical Swarm (CLGS), Grammatical Swarm with Second Global best Particle (SG-GS) and Grammatical Swarm with Second Personal best Particle (SG-GS).

The algorithms are compared in the symbolic regression problem. Except for CLGS and UGS, the convergence speeds of the other algorithms are faster than that of the original GS. Especially, the convergence properties of GS-cf and SP-GS are fastest among them.

References

1. Schwefel, H.-P.: Evolution and Optimum Seeking. Wiley, Hoboken (1995)
2. Kennedy, J., Eberhart, R.C.: Particle swarm optimization. In: Proceedings of IEEE the International Conference on Neural Networks, pp. 1942–1948 (1995)
3. Hollan, J.H.: Adaptation in Natural and Artificial Systems. The University of Michigan Press, Ann Arbor (1975)
4. Koza, J.R.: Genetic Programming: On the Programming of Computers by Means of Natural Selection. MIT Press, Cambridge (1992)
5. Ryan, C., Collins, J.J., O'Neill, M.: Grammatical evolution: evolving programs for an arbitrary language. In: Banzhaf, W., Poli, R., Schoenauer, M., Fogarty, T.C. (eds.) EuroGP 1998. LNCS, vol. 1391, pp. 83–96. Springer, Heidelberg (1998). doi:10.1007/BFb0055930
6. Ryan, C., O'Neill, M.: Grammatical Evolution: Evolutionary Automatic Programming in an Arbitrary Language. Springer, Heidelberg (2003). doi:10.1007/978-1-4615-0447-4
7. O'Neill, M., Brabazon, A.: Grammatical swarm. In: Deb, K. (ed.) GECCO 2004. LNCS, vol. 3102, pp. 163–174. Springer, Heidelberg (2004). doi:10.1007/978-3-540-24854-5_15
8. Clerc, M., Kennedy, J.: The particle swarm-explosion, stability, and convergence in a multidimensional complex space. IEEE Trans. Evol. Comput. **6**, 58–73 (2002)
9. Parsopoulos, K.E., Vrahatis, M.N.: UPSO: a unified particle swarm optimization scheme. Lect. Ser. Comput. Comput. Sci. **1**, 868–873 (2004)
10. Liang, J.J., Qin, A.K., Suganthan, P.N., Baskar, S.: Comprehensive learning particle swarm optimizer for global optimization of multimodal functions. IEEE Trans. Evol. Comput. **10**, 281–295 (2006)
11. Shin, Y.B., Kita, E.: Search performance improvement of particle swarm optimization by second best particle information. Appl. Math. Comput. **246**, 346–354 (2014)
12. Eberhart, R.C., Shi, Y.: Evolving artificial neural networks. In: Proceedings of 1998 International Conference on Neural Networks and Brain, pp. 1423–1447 (1998)

Bi-MOCK: A Multi-objective Evolutionary Algorithm for Bi-clustering with Automatic Determination of the Number of Bi-clusters

Meriem Bousselmi[1]([✉]), Slim Bechikh[1,2], Chih-Cheng Hung[2,3],
and Lamjed Ben Said[1]

[1] SMART Lab, Computer Science Department, University of Tunis,
Tunis, Tunisia
meriem.bousselmi@gmx.com
[2] Kennesaw State University, Marietta, GA, USA
[3] Anyang Normal University, Anyang, China

Abstract. Bi-clustering is one of the main tasks in data mining with many possible applications in bioinformatics, pattern recognition, text mining, just to cite a few. It refers to simultaneously partitioning a data matrix based on both rows and columns. One of the main issues in bi-clustering is the difficulty to find the number of bi-clusters, which is usually pre-specified by the human user. During the last decade, a new algorithm, called MOCK, has appeared and shown its performance in data clustering where the number of clusters is determined automatically. Motivated by the interesting results of MOCK, we propose in this paper a new algorithm, called Bi-MOCK, which could be seen as an extension of MOCK for bi-clustering. Like MOCK, Bi-MOCK uses the concept of multi-objective optimization and is able to find automatically the number of bi-clusters thanks to a newly proposed variable string length encoding scheme. The performance of our proposed algorithm is assessed on a set of real gene expression datasets. The comparative experiments show the merits and the outperformance of Bi-MOCK with respect to some existing recent works.

Keywords: Bi-clustering · Number of bi-clusters · Multi-objective evolutionary algorithms · Variable-size encoding

1 Introduction

Bi-clustering, also called co-clustering, is a data mining task referring to partitioning a data matrix based on both rows and columns in a simultaneous way. Each row corresponds to one object while each column expresses one feature. A bi-cluster could be seen as a subset of objects sharing some similarities with respect to a subset of features [1]. Bi-clustering is a very interesting task in many applications such as text mining, Web mining, collaborative filtering, and bioinformatics [2]. For instance, in microarray data analysis, the goal of bi-clustering

© Springer International Publishing AG 2017
D. Liu et al. (Eds.): ICONIP 2017, Part IV, LNCS 10637, pp. 366–376, 2017.
https://doi.org/10.1007/978-3-319-70093-9_38

is to find a set of genes (objects) that have similar behaviors under a set of experimental conditions (features), which is very important for biologists. Several algorithms have been proposed to solve the bi-clustering problem. These algorithms can be classified into two main families [3]: (1) Systematic search approaches such as greedy algorithms, divide-and-conquer algorithms, and enumeration ones; and (2) Stochastic search algorithms such as neighborhood-based methods, evolutionary metaheuristics, and hybrid ones.

Although several representatives of all these approaches have shown promising results in solving the bi-clustering problem, almost all existing works share the same shortcoming that is the difficulty of determining *the number of bi-clusters*, which is usually pre-specified by the human user. Motivated by this observation, we propose in this paper a new bi-clustering algorithm, called Bi-MOCK, that has the ability to partition the data matrix while automatically finding the number of bi-clusters. Bi-MOCK is an extended version of MOCK [4], which is a multi-objective evolutionary algorithm for clustering with automated determination of the number of clusters. The rest of this paper is structured as follows. Section 2 gives the problem statement. Section 3 describes in detail the Bi-MOCK algorithm. Section 4 presents the comparative experimental results. Section 5 concludes this paper and offers some avenues for future research.

2 Related Works

2.1 Bi-clustering Problem Definition

To conveniently define the bi-clustering problem, we describe the bi-clustering in the context of microarray data analysis. In this context, we have a data matrix M where the ith row represents the ith gene, the jth column corresponds to the jth experimental condition, and the cell m_{ij} represents the expression level of the ith gene under the jth condition. The bi-clustering problem could be defined as the problem of finding a set of coherent bi-clusters, where a bi-cluster is a subset of genes showing similar behavior under a subset of conditions.

Formally, a bi-cluster is defined as follows. Let $I = \{1, \ldots, n\}$ be a set of indices of n genes, $J = \{1, \ldots, m\}$ be a set of m conditions, and M (I, J) be a data matrix associated with I and J, a bi-cluster $B(I', J')$ is a submatrix from M (I, J) such that $I' \subseteq I$ and $J' \subseteq J$. In this way, the bi-clustering problem could be stated as follows. Given a data matrix M (I, J), construct a group of bi-clusters B_{opt} associated with M (I, J) such that:

$$f(B_{opt}) = max_{B \in BC(M)} f(B), \qquad (1)$$

where f is an objective function measuring the quality (i.e., the degree of coherence) of a group of bi-clusters B and $BC(M)$ is the set of all the possible groups of bi-clusters associated with M (I, J). It is important to note that bi-clustering has been reported to be an NP-hard combinatorial problem [1]. Also, bi-clustering could be modeled as a multi-objective problem where several objectives are optimized simultaneously such as the *MSR* (Mean Squared Residue), the bi-cluster *Size*, and the Average *RV* (Average Row Variance) [5].

2.2 MOCK

MOCK is a Multi-Objective Evolutionary Algorithm (MOEA) that clusters data into k partitions where k is found automatically during the search process [4]. Indeed, MOCK could be seen as the adaptation of the MOEA *PESA*-II (Pareto Envelope-based Selection Algorithm-II) to the clustering problem as follows. First, solution encoding is performed by using the locus-based adjacency representation proposed in [6]. In this representation, each population individual is a vector of n genes and each gene g_i can take a value j in the range $\{1, \ldots, n\}$. In this way, a value of j assigned to the ith gene means that there is a link between data items i and j, i.e., i and j belong to the same cluster. This representation is adopted in MOCK because it does not need to specify the number of clusters in advance, since this number will be automatically deduced in the decoding step. Secondly, MOCK uses two conflicting and complementary objective functions that should be minimized: one based on compactness and the other based on connectedness. The interaction between the two objective functions allows finding a balance between compactness and connectedness, thereby calibrating the number of clusters k automatically. The interested reader could refer to [4,7] for detail about the method allowing the selection of the preferred solution from the Pareto front (i.e., the set of the best obtained clustering solutions).

3 Bi-MOCK

3.1 Overview and Solution Encoding

Bi-MOCK is an extension of MOCK for the bi-clustering problem. Bi-MOCK extends the solution representation of MOCK to the case of bi-clustering by adding a subset of columns (conditions) to each chromosome. In this way, each chromosome corresponds to a particular clustering problem that is based on a specific subset of conditions (features). Since the latter ones vary from one chromosome to another, Bi-MOCK outputs a set of partitions corresponding to different subsets of columns, which is conform to the definition of the concept of bi-cluster. For this reason we can say that Bi-MOCK outputs a set of bi-clusters where each chromosome corresponds to a set of bi-clusters having the same conditions.

Figure 1 illustrates the solution representation in Bi-MOCK. According to this figure, we have three bi-clusters that are based on three conditions, i.e., $C1$, $C5$, and $C8$. For example, genes 1, 2 and 4 build a bi-cluster because they are linked through their positions. As previously noted, a value of j assigned to the ith gene means that i and j belong to the same bi-cluster. It is worth noting that if we modify the subset of conditions, we obtain new bi-clusters related to the new settled conditions. It is also important to note that the minimum number of conditions is set to two according to the related literature [3].

The main working principle of Bi-MOCK is as follows. Bi-MOCK uses two populations: (1) an Internal Population (*IP*) and External Population (*EP*) with a varying but limited size. The *EP* plays the role of an archive storing a set of

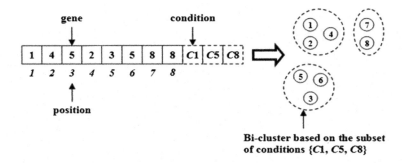

Fig. 1. Bi-MOCK solution encoding. The number of conditions randomly varies during the search process. A value of j assigned to the ith gene means that i and j belong to the same bi-cluster.

diversified non-dominated solutions, i.e., the best found solutions. The *IP* has the role of exploring solutions by means of crossover and mutation operators. The selection process is done by updating the *EP* with solutions from the *IP* following some rules. First, each solution s from *IP* dominating one or several solutions from *EP* is inserted into *EP* and all *EP* solutions dominated by s are removed from *EP*. Secondly, for the case where s is non-dominated with respect to all *EP* solutions, there are two possible scenarios. In fact, if the *EP* size does not exceed the limit, we add s to *EP*; otherwise s replaces the most crowded solution in *EP* in order to preserve the *EP* diversity and come up with a well-diversified and well-converged Pareto front.

3.2   Objective Functions

In Bi-MOCK, two conflicting objective functions are optimized. These objectives are among the most used ones in the bi-clustering field. The first objective is the Mean Squared Residue (*MSR*), which measures the coherence of a bi-cluster. In fact, the residue is an indicator of the degree of coherence of an element following shifting and scaling patterns, with respect to the remaining ones in the bi-cluster. The *smaller* the residue value is, the *better* the coherence is. The *MSR* measures the coherence of a bi-cluster using the variance between all data points in the bi-cluster summed with the means of rows and columns expression values.

Mathematically, the *MSR* of a bi-cluster $B(I', J')$ can be expressed as follows:

$$MSR(B(I', J')) = \frac{1}{|I'||J'|} \sum_{i \in I', j \in J'} (R(w_{ij}))^2, \qquad (2)$$

where $R(w_{ij})$ is the residue value of an entry w_{ij} of a bi-cluster $B(I', J')$, which is computed as follows:

$$R(w_{ij}) = w_{ij} - w_{iJ'} - w_{I'j} + w_{I'J'}. \qquad (3)$$

such that:

$w_{iJ'}$ is the mean of the ith row

$$w_{iJ'} = \frac{1}{|J'|} \sum_{j \in J'} w_{ij}, \tag{4}$$

$w_{I'j}$ is the mean of the jth column

$$w_{I'j} = \frac{1}{|I'|} \sum_{i \in I'} w_{ij}, \tag{5}$$

and $w_{I'J'}$ is the mean of all elements in the bicluster $B(I', J')$

$$w_{I'J'} = \frac{1}{|I'||J'|} \sum_{i \in I', j \in J'} w_{ij}. \tag{6}$$

The second objective function is the *Size* of the bi-cluster, which should be maximized. Analytically, the *Size* of a bi-cluster $B(I', J')$ is expressed as follows:

$$Size(B(I', J')) = \frac{|I'||J'|}{|I||J|}. \tag{7}$$

We recall that $|I'|$ and $|J'|$ are the dimensions of the bi-cluster while $|I|$ and $|J|$ are the dimensions of the whole data set. It is worth noting that higher the bi-cluster *Size* is, the higher *MSR* value becomes. Since biologists prefer large bi-clusters with low *MSR* values (i.e., with high coherence), the two objectives *MSR* and *Size* are conflicting, i.e., improving one of them causes the deterioration of the other. This allows calibrating the *Size* of the bi-cluster in an automated way.

It is important to note that both *MSR* and *Size* evaluate a single bi-cluster. Since each chromosome may encode more than one bi-cluster in Bi-MOCK, the evaluation of each chromosome is performed based on the mean values of *MSR* and *Size*. It is also worth noting that once the stopping criterion is met, Bi-MOCK outputs a set of chromosomes each encoding several bi-clusters. Bi-clusters having *MSR* values less than or equal to a user-specified threshold δ (δ-biclusters) are accepted and the others are rejected.

3.3 Variation Operators

This subsection is devoted to describe the variation operators. For the crossover operator, we propose a new operator, while for the mutation operator mutation we adopt the well-known Cheng and Church's algorithm (CC algorithm) [8]. Figure 2 describes the main idea of our crossover operator. As previously noted, each parent chromosome is composed of two parts: (1) the genes (rows) and (2) conditions (columns). For the genes, a two-point crossover is applied by randomly generating two Gene Cutting Points (GCP1 = 2 and GCP2 = 5 for our example) and then exchange the chromosomes' sub-parts falling between the cutting points. The choice of the two-point crossover operator is justified

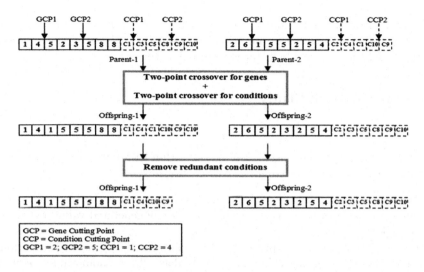

Fig. 2. A two-point crossover operator for variable-size chromosomes. The number of genes is fixed while the number of conditions is variable. Redundant conditions are removed from the offspring solutions in order to obtain meaningful generations.

by the ability of this kind of crossover to vary any part from the chromosome whatever its position is [8], which is not the case for the single-point crossover where extreme chromosomes' components cannot be exchanged; therefore causing a lack of diversity. For the conditions parts, we use a two-point crossover version that takes into account the number of conditions which is variable from one chromosome to another. First, we count the number of conditions of each parent chromosome and then determine the least one. After that, we generate two Condition Cutting Points (CCP1 = 1 and CCP2 = 4 in our example) between one and the minimum number of conditions among the parents chromosomes, which is equal to five in our example.

For the mutation operator, we use the CC algorithm [8], which is one of the most famous algorithms in the field of bi-clustering. This algorithm takes as input a single bi-cluster, usually corresponding to the whole data set, and iteratively removes some rows and columns with the aim to minimize the *MSR* value. Once the latter is less than or equal to a user-specified threshold δ, the algorithm reinserts rows and columns that do not belong to the bi-cluster until obtaining an *MSR* value above δ. The CC algorithm could not directly be applied to our chromosome encoding and hence an adaptation step is required. Indeed, each chromosome in Bi-MOCK corresponds to a set of bi-clusters having the same conditions. In fact, the chromosome is decoded to obtain a set of bi-clusters. Since the latter has the conditions, we cannot apply the CC algorithm on the columns. Thus, the mutation is performed by applying the first two steps of the CC algorithm only on rows (genes) in order to preserve the shared set of conditions among the chromosome's bi-clusters. The δ threshold rule is used in a controlled manner in our mutation operator in order to limit the computational cost allowed for the mutation. For instance, if the δ threshold condition is not met after a certain number of mutation iterations (*MaxMutIter*), the process is stopped and the current bi-cluster is returned as the output.

4 Experimental Study

4.1 Data Sets

In order to assess the performance of our algorithm Bi-MOCK, we perform a set of experiments using three commonly used data sets:

- *Yeast* (Yeast cell cycle) [9]: It contains 2884 genes, 17 conditions, and 0.07% missing data. The latter is replaced by randomly generated values between 0 and 800.
- *Human* (Human B-cell expression) [9]: It contains 4026 genes, 96 conditions, and 12.3% missing data. The latter is replaced by random values between −800 and 800.
- *Colon* (Colon cancer) [10]: It contains 2000 genes and 62 conditions. It does not contain missing data.

4.2 Performance Metrics

The following metrics are used to measure the performance of each bi-clustering algorithm:

- *Average MSR*: It corresponds to the average of the *MSR* values of δ-biclusters obtained by a particular algorithm.
- *Row Variance (RV)*: A bi-cluster characterized by a high value of *RV* contains genes that present large changes in their expression values under different conditions. The higher the RV value is, the better the quality of the obtained bi-cluster is. A high *RV* is desirable to escape from the insignificant constant bi-clusters. The *RV* of a bi-cluster $B(I', J')$ is analytically expressed as follows:

$$RV(B(I', J')) = \frac{\sum_{i \in I', j \in J'} (w_{ij} - w_{iJ'})^2}{|I'||J'|}, \tag{8}$$

where w_{ij} is the element of the ith row and j^{th} column and $w_{iJ'}$ is the mean of the ith row, which is expressed as follows:

$$w_{iJ'} = \frac{1}{J'} \sum_{j \in J'} w_{ij}. \tag{9}$$

- *Average RV*: It corresponds to the mean of the *RV* values of δ-biclusters generated by a particular algorithm.
- *Average Bi-cluster Size*: It corresponds to the average of the sizes of bi-clusters obtained by a particular algorithm.

4.3   Competitor Algorithms and Parameter Settings

In this experimental study, we compare our Bi-MOCK algorithm against three existing bi-clustering algorithms that are: (1) The CC algorithm [8] which is a greedy algorithm that has been described in Sect. 3.3, (2) HMOBI [11] which is the result of the hybridization of the MOEA MOBI and the Dominance-based Multi-objective Local Search (DMLS), and (3) SMOB [12] which is an evolutionary algorithm that optimizes the weighted sum of three objectives: *MSR*, *Size*, and *RV*. It is important to note that each of these three algorithms encodes a single bi-cluster in each solution, which is not the case of Bi-MOCK where multiple bi-clusters are encoded in each solution (individual). Table 1 presents the parameter settings of the different algorithms for comparison. The parameters were settled in a way that ensures the fairness of comparisons.

Table 1. Parameter settings

Common parameters		Settings
All algorithms	δ for Yeast dataset	300
	δ for Human dataset	1200
	δ for Colon dataset	500
	Stopping criterion	25000 evaluations
Specific parameters		
BI-MOCK	*EP* size = 200; *IP* size = 100; *MaxMutIter* = 10 Crossover rate = 0.7; Mutation rate = 0.3	
HMOBI	Population size = 200 Crossover rate = 0.5; Mutation rate = 0.4	
SMOB	Population size = 200 Crossover rate = 0.85; Mutation rate = 0.2	

4.4   Obtained Results

Table 2 summarizes the obtained comparative results according to the considered metrics with a statistical significance of 5% based on the Kruskal-Wallis test. In fact, for each algorithm, we consider the best 100 obtained bi-clusters having the smallest *MSR* values that are less than the pre-defined threshold δ. Based on this table, we notice that Bi-MOCK outperforms the three other algorithms in terms of the average *MSR* and average *Size* metrics. Bi-MOCK has provided bi-clusters having the greatest sizes and the smallest residues for all data sets. The bi-clusters are biologically relevant since Bi-MOCK has the highest average *RV* values for the three considered data sets. A high RV value means that there are coherent up and down regulations in the obtained bi-clusters, which indicates their interestingness and allows escaping from trivial bi-clusters having constant values.

Table 2. Median metrics' values of Bi-MOCK, CC, HMOBI, and SMOB over 30 independent runs, where in each run each metric value is computed based on bi-clusters respecting the δ threshold provided by each algorithm. Best statistically significant values (with $\alpha = 0.05$) are shown in bold.

Dataset	Algorithm	Avg. MSR	Avg. size	Avg. RV	Avg. genes	Avg. conditions
Yeast dataset	Bi-MOCK	**203.56**	**15530.19**	**1307.42**	1451.42	10.7
	CC	204.29	2015.52	801.27	166.71	12.09
	HMOBI	299.6	7827.86	789.12	759.25	10.31
	SMOB	206.17	415.06	678.53	27.28	15.25
Human dataset	Bi-MOCK	**821.93**	**75637.88**	**4113.05**	1524.65	49.61
	CC	850.04	6595.89	2144.14	269.22	24.5
	HMOBI	1199.9	47356.56	2231.04	759.37	62.6
	SMOB	1019.16	505.41	3169.19	11.6	43.57
Colon dataset	Bi-MOCK	**461.35**	**36075.86**	**2711.65**	1213.45	29.73
	CC	464.21	2988.1	2300.89	156.2	19.13
	HMOBI	482.15	19503.22	2288.43	748.11	26.07
	SMOB	472.01	632.05	2265.8	15.45	40.91

Based on the last two columns of Table 2, we observe that Bi-MOCK calibrates the compromise well between the number of genes and the number of conditions. In fact, the *size* of a particular bi-cluster gets larger with the increase of the number of genes and/or the number of conditions, and vice versa. Thus, in order to preserve an interesting *MSR* value, a bi-clustering algorithm should be well-balanced between the number of genes and the number of conditions. For example, for Yeast data set, Bi-MOCK has the highest average number of genes and the second-smallest average number of conditions, while presenting the best average *MSR* value, the best average *Size*, and also the best average *RV* value. A similar observation could be seen for the Colon data set.

The outperformance of Bi-MOCK over the three tested algorithms can be explained by two reasons. On the one hand, a single solution in Bi-MOCK encodes more than one bi-cluster having the same conditions instead of a single bi-cluster. This allows Bi-MOCK to explore the search space more efficiently. Indeed, the evaluation of a single solution in Bi-MOCK allows visiting several bi-clusters in the search space, while the evaluation of a single solution in the other algorithms allows visiting only a single bi-cluster in the search space. This fact makes Bi-MOCK able to summarize the huge search space of the bi-clustering problem. On the other hand, Bi-MOCK minimizes the average *MSR* and maximizes the average *Size*, where the average value is computed over the set of bi-clusters encoded in the corresponding solution. Since bi-clusters belonging to the same solution have exactly the same conditions, the optimization process over a single solution is similar to a one-dimensional clustering of the genes based on the fixed conditions (features). This makes Bi-MOCK able to well-partition the genes based on the fixed conditions, thereby finding good candidate bi-clusters for the fixed set of conditions. As conditions vary from one solution to another through genetic operations, Bi-MOCK has the ability to explore the genes and conditions with the aim to find the largest bi-clusters with the smallest MSR values.

5 Conclusion and Future Work

In this paper, we have proposed a new multi-objective bi-clustering algorithm, called Bi-MOCK, which is an extended version of the multi-objective clustering algorithm MOCK. Bi-MOCK inherits the main feature of MOCK, which is the ability to determine the number of bi-clusters automatically. Thanks to our newly proposed solution encoding scheme and variation operators and the optimization of average MSR and average bi-cluster $Size$, Bi-MOCK has been shown to outperform two evolutionary bi-clustering algorithms (HMOBI and SMOB) in addition to the well-known greedy CC algorithm. This result can be mainly explained by the ability of Bi-MOCK to efficiently sample the search space by encoding several bi-clusters in each population individual (solution), which is not the case in the other three algorithms. As for future work, it would be interesting to further assess the performance of Bi-MOCK against other algorithms using other data sets which have diversified characteristics. Moreover, we intend to develop an adaptive version of Bi-MOCK with the aim to reduce its sensitivity to the parameter setting.

References

1. Madeira, S.C., Oliveira, A.L.: Biclustering algorithms for biological data analysis: a survey. IEEE/ACM Trans. Comput. Biol. Bioinform. **1**(1), 24–45 (2004)
2. Kasim, A., Shkedy, Z., Kaiser, S., Hochreiter, S., Talloen, W.: Applied biclustering methods for big and high dimensional data using R (2016). ISBN 9781482208238
3. Freitas, A.V., Ayadi, W., Elloumi, M., Oliveira, J., Oliveira, J., Hao, J.-K.: A survey on biclustering of gene expression data. In: Elloumi, M., Zomaya, A.Y. (eds.) Biological Knowledge Discovery Handbook: Preprocessing, Mining, and Postprocessing of Biological Data. Wiley, Hoboken (2013). doi:10.1002/9781118617151.ch25
4. Handl, J., Knowles, J.D.: An evolutionary approach to multiobjective clustering. IEEE Trans. Evol. Comput. **11**(1), 56–76 (2007)
5. Mitra, S., Banka, H.: Multi-objective evolutionary biclustering of gene expression data. Pattern Recogn. **39**(12), 2464–2477 (2006)
6. Handl, J., Knowles, J.: Exploiting the trade-off—the benefits of multiple objectives in data clustering. In: Coello Coello, C.A., Hernández Aguirre, A., Zitzler, E. (eds.) EMO 2005. LNCS, vol. 3410, pp. 547–560. Springer, Heidelberg (2005). doi:10.1007/978-3-540-31880-4_38
7. Bechikh, S., Ben Said, L., Ghedira, K.: Negotiating decision makers' reference points for group preference-based evolutionary multi-objective optimization. In: International Conference on Hybrid Intelligent Systems, pp. 377–382 (2011)
8. Cheng, Y., Church, G.M.: Biclustering of expression data. In: International Conference on Intelligent Systems for Molecular Biology, pp. 93–103 (2000)

9. Yeast and Human datasets. http://arep.med.harvard.edu/biclustering/
10. Colon data set. http://genomics-pubs.princeton.edu/oncology/affydata/index.html
11. Seridi, K., Jourdan, L., Talbi, E.G.: Using multiobjective optimization for biclustering microarray data. Appl. Soft Comput. **33**(1), 239–249 (2015)
12. Divina, F., Aguilar-Ruiz, J.S.: A multi-objective approach to discover biclusters in microarray data. In: International Conference on Genetic and Evolutionary Computation (GECCO 2007), pp. 385–392 (2007)

A Transferable Framework: Classification and Visualization of MOOC Discussion Threads

Lin Feng, Guochao Liu, Sen Luo, and Shenglan Liu[✉]

Springer-Verlag, Computer Science Editorial,
Tiergartenstr. 17, 69121 Heidelberg, Germany
fenglin@dlut.edu.cn, guochao_liu@foxmail.com, heyrosen@163.com,
liusl@mail.dlut.edu.cn

Abstract. Analysis of Massive Open Online Course (MOOC) forums data often use natural language processing (NLP) technology to extract keywords from discussions content as features. However, discussions in different course forums vary significantly, so the analyzed results obtained on these specific forums are not easily applied to other irrelevant forums. Besides, a lot of discussion threads are not related to the course in MOOCs forums. To address above problems, we analyze about 100,000 discussion threads from the forums of 60 MOOCs offered by Coursera, and design many features related to user interaction ways in different sub-forums. This work proposes a transferable framework to classify MOOC discussion threads using these features. The classification framework is not sensitive to the subjects and the forum discussions, so the classification model can be used directly by other course forums without being trained again. Experiments show that the average classification performance with Area Under ROC Curve (AUC) is 0.8. This work also gives the methods to remove noisy discussion threads and visulize the interactive characteristic of MOOC forum threads using dimensionality reduction technology.

Keywords: MOOC · Forum discussion threads classification · Dimensionality reduction · Interaction features

1 Introduction

MOOC is a novel education model emerging in recent years, more than 58 million users get education through this way before February 2017. Discussion forums in MOOC offer the only venue for communication between students and instructors [1], it determines the quality of course interaction.

Discussion thread in MOOC forum is a question or advice proposed by one user who wants to get an answer or responses from other users. Because the number of users belong to a course is too large, the forums are full of various information relevant and irrelevant to the course. Course staff may be unable to adequately track the forum threads to find all issues that need a resolution. So

© Springer International Publishing AG 2017
D. Liu et al. (Eds.): ICONIP 2017, Part IV, LNCS 10637, pp. 377–384, 2017.
https://doi.org/10.1007/978-3-319-70093-9_39

classifying the threads to reasonable categories automatically is very important to reduce the works of course staff and improve MOOC forums service quality.

Classifying threads in MOOC forums can use NLP technology to extract key words features from user discussions content [2–5]. These text features are strong correlation to the threads discussions content. But this method has many problems, e.g. the classification model trained on some specific course forums is not easily applied to other irrelevant courses forum directly.

Another idea to classify threads is to use user behavior data to design features, and combining these features with key words can get a better classification performance [6–9]. But training a transferable model should only use features that are not related to course subject and discussions content.

In this paper, we propose a framework to classify threads in MOOC forums through only using the interaction features extracted from user behavior data, thus the classification model can be used in other unrelated course forums directly. Our work gives the following insights and contributions to MOOC data analysis:

- We design some interaction ways features extracted from user behavior data for the supervised classification of the threads.
- We put forward a method to remove noisy labels of the threads in MOOC forums.
- We first introduce dimensionality reduction techniques to visualize the interactive characteristic of MOOC forum threads.

2   Analysis and Visualization of Threads

2.1   Analyze the Interaction Ways of Threads

In this section, the difference of interaction ways associated to threads in different sub forum will be analyzed. Interaction ways indicate a variety of behaviors which forum user showed. For example, participating in the thread discussion, posting on the thread etc. We analyze some typical interaction features from the dataset offered by Rossi [10], the analysis results are shown in Table 1 n_{stu}, n_{sta}, n_{ins}, n_v, r_{avg}, t_{avg} are defined in Sect. 3.

From Table 1, we can know average number of course instructors (n_{ins}) participated in the threads of Assignments sub forums is 4 times more than Meetups sub forum. Average number of course staffs (n_{sta}) is 5.5 times more than Meetups. Average number of votes (n_v) is 1.7 times more than Meetups. But for the average number of student user (n_{stu}) is only 0.5 times as many as the Meetups sub forum. More interesting differences can be found in Table 1.

These differences are very consistent with assumptions about the interaction ways in different sub forum threads. For instance, most of threads associated to Assignments sub forum are about course assignments, so course staffs and instructors incline to participate in the discussion threads. And assignment threads are some problems that many students will encounter, so the thread will

Table 1. Interaction features of different sub forum threads

Sub forum	n_{stu}	n_{sta}	n_{ins}	n_v	r_{avg}	t_{avg}
Gen. Discc.	6.143	0.209	0.072	2.504	0.209	0.322
Assignments	5.378	0.417	0.090	2.412	0.620	0.476
Meetups	10.279	0.075	0.026	1.452	1.472	0.143
Logistics	4.034	0.368	0.138	2.453	1.178	0.373
Lectures	6.751	0.280	0.126	4.040	0.926	0.364
Feedback	3.952	0.580	0.096	1.535	1.032	0.398

get more browsing times, and votes than Meetups. However, for the threads of Meetups sub forum, most of them are about making friends, so the student users more intend to participate in the discussion than Assignments, and the reason of other differences between them are also obvious.

2.2 Visual Interaction Ways Characteristics of Threads

Dimensionality reduction techniques can be used to visual data distribution in a lower-dimensional space. Principal component analysis (PCA) and linear discriminant analysis (LDA) are classical dimensionality reduction method often used in the research [10]. PCA is a unsupervised dimensionality and often used to eliminate the correlations between features. LDA is a supervised dimensionality method and it is very suitable for dimensionality reduction of classification problems.

Because the labels of threads is marked directly by the sub forum, so the labels for many threads are wrong. In this section, we propose a method to detect these wrong label threads based on the PCA and One-Class SVM method, the specific operations are as follows.

1. Use PCA to reduce the threads dimension. We retain 85% variance of the thread data. This step not only can eliminate the correlation among the features we proposed, but also can eliminate the noise data distribution on some unimportant features.
2. Use One-Class SVM method to detect the noisy outliers of threads. We use radial basis function (RBF) kernel, and parameter nu is set to 0.05.

The detection result of Outliers of threads in different sub forums is shown in Fig. 1.

In order to visualize the data distribution in General Discussion, Assignments, and Meetups sub forum, LDA method is used to reduce dimensions of thread interaction features to 2, and plot it on a 2-dimensional plane. The result is shown in Fig. 2.

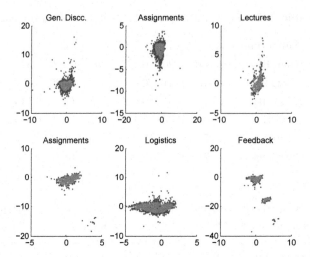

Fig. 1. Outliers distributions in different sub forums (Blue spots indicate outliers) (Color figure online)

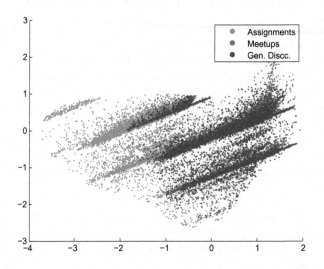

Fig. 2. Data distribution of threads in different sub forums

3   Interaction Ways Features

We designed 18 interaction ways features related to 4 different aspects to the threads (structure, underlying social network, popularity, quality). Compared with text features extracted by NLP technology, the dimensionality of features we used is far lower, and does not require a priori knowledge of the course to selected key words features manually. Features related to interaction ways of threads are as follows.

1. **Thread structure** (Features related to structural aspects associated to a discussion thread)
   - **Number of posts** (n_p): number of posts in a thread.
   - **Number of comments** (n_c): number of comments in a thread.
   - **Day of week** (*week_day*): the day of week for the first post of the thread.
   - **Relative time** (t_{rel}): the relative time for the first post of the thread. We normalize and quantize the post time to [0, 1] by the following formula:

$$t_{rel} = \frac{t - t_{start}}{t_{end} - t_{start}} \tag{1}$$

   Here, t is post time of thread, t_{start} is course start time, t_{end} is course end time.
   - **Average post time** (t_{avg}): the average post time for all posts of the threads.

$$t_{avg} = \frac{1}{n_p} \sum_{i=1}^{n_p} t_{rel}^i \tag{2}$$

2. **Underlying social network** (Features related to the user participation and interaction of the threads)
   - **Number of registered student users** (n_{stu}): the number of registered student users participated in this threads.
   - **Number of anonymous users** (n_{ano}): the number of anonymous ID users participated in this threads.
   - **Number of instructors** (n_{ins}): the number of instructors and teach assistants participated in the threads.
   - **Number of course staffs** (n_{sta}): the number of course staffs and course technical support belong to the threads.
   - **User chain** (u_{chain}): Boolean feature to indicate that at least 3 users participated in the discussion threads and post more than 1 post each in the thread.
3. **Popularity measures** (Features related to the popularity of the threads)
   - **Number of views** (n_v): the number of views of the thread.
   - **Average response time** (r_{avg}): the average response time for all posts for the thread.
   - **Average number of posts** (p_{avg}): the average number of posts for all users participated in the thread.
   - **Average number of comments** (c_{avg}): the average number of comments for all the users participated in the thread.
4. **Quality measures** (Features related to the quality of the threads)
   - **Sum of votes** (v_{add}): sum of votes in the threads.
   - **Maximum votes** (v_{max}): the maximum votes number of the threads for all the posts (best post) belong to the threads.
   - **Most popular user role** (v_r): the user role corresponding to the best post. In this paper, we divided user roles as registered student users, anonymous users, course staff, and course instructors.
   - **Best post order** (v_o): the post order for the best post of the thread. The first post order is 1, the last post order is the number of the posts for this thread.

4 Experiments and Results

Our motivation is to build a classification framework which can be used directly in another course forum, so we design features related to interaction ways of thread replaced the key words features. In order to verify our classification framework, we randomly select 40 courses (these courses are all taught by English) forum data as training dataset, and use the other 20 courses (8 courses are taught by non-English) forum data as test dataset. The number of threads in training dataset account for about 70% of the total threads number. The noisy labels of threads are removed by using the method presented in Sect. 4.2.

This paper use one-vs-rest method to build multiple classifier. Because the thread number distribution is very unbalanced, so training classification model on training dataset directly will get a bad performance for the small samples [11]. In order to solve the problem, we use synthetic minority oversampling tech (SMOTE) to produce samples.

K-nearest neighbor classifier (KNN), Logistic regression (LR), and support vector machine (SVM) classifiers were selected to train classification model. And using AUC index to measure classification performance. The result is shown in Table 2.

Table 2. Classifier performance for different sub forum threads

Sub forum	Classifier	AUC	
		Without SMOTE	With SMOTE
Gen. Discc.	KNN	0.749	0.773
	LR	0.750	0.760
	SVM	0.753	0.774
Assignments	KNN	0.749	0.773
	LR	0.750	0.760
	SVM	0.753	0.774
Meetups	KNN	0.749	0.773
	LR	0.750	0.760
	SVM	0.753	0.774
Logistics	KNN	0.749	0.773
	LR	0.750	0.760
	SVM	0.753	0.774
Lectures	KNN	0.749	0.773
	LR	0.750	0.760
	SVM	0.753	0.774
Feedback	KNN	0.749	0.773
	LR	0.750	0.760
	SVM	0.753	0.774

Table 3. Comparison of classification performance

Sub forum	AUC	
	Rossi	Our
Gen. Discc.	0.600	0.720
Assignments	0.679	0.756
Meetups	0.917	0.918
Logistics	0.635	0.753
Lectures	0.611	0.686
Feedback	0.660	0.780

In order to compare features designed in this paper and the features proposed by Rossi, we construct training set and test set according to the way offered by Rossi, and using the SVM (Linear kernel) to train classifier. The results are shown in Table 3.

As we can see from the Table 3, our classification performance is 12% higher than Rossi. Thus our interaction features can measure the characters of threads better than Rossi.

5   Conclusion

This paper propose a transferable framework to classify the discussion threads of MOOCs. Compared to the classification model trained by key words features extracted from the discussion content, our framework can be used in other irrelevant courses forum directly. Moreover, interaction features selection does not require a priori knowledge of the course like key words.

The conclusion that interaction ways features can indeed describe threads very well is proved through analyzing about 100000 threads data offered by Rossi. A method for removing noisy labels is proposed based on PCA and One-Class SVM algorithm. And we also utilize dimensionality reduction techniques to visual noise labels of threads and the character of threads in different sub forums on a 2-dimensional plane.

The classification performance of our framework for 6 classes threads is at least 0.72, for the discussion threads in Assignments and Meetups forum, AUC is more than 0.85. Compare the work of Rossi, our performance gained an average increase of 12%.

Acknowledgments. This work was supported by MOE Research Center for Online Education, and Foundation of LiaoNing Educational Committee (Grant no. 2016YB121, 201602151)

References

1. Jiang, Z., Zhang, Y., Liu, C.: Influence analysis by heterogeneous network in MOOC forums: what can we discover? In: International Educational Data Mining Society (2015)
2. Ravi, S., Kim, J.: Profiling student interactions in threaded discussions with speech act classifiers. Front. Artif. Intell. Appl. **158**, 357 (2007)
3. Lin, F.R., Hsieh, L.S., Chuang, F.T.: Discovering genres of online discussion threads via text mining. J. Comput. Educ. **52**, 481–495 (2009)
4. Hecking, T., Chounta, I.A., Hoppe, H.U.: Analysis of user roles and the emergence of themes in discussion forums. In: 2th Network Intelligence Conference, pp. 114–121. IEEE Press, New York (2015)
5. Atapattu, T., Falkner, K., Tarmazdi, H.: Topic-wise classification of MOOC discussions: a visual analytics approach. In: 9th International Conference on Educational Data Mining, pp. 276–281. ERIC Press, United States (2016)
6. Song, J., Zhang, Y., Duan, K.: TOLA: topic-oriented learning assistant based on cyber-physical system and big data. J. Future Gener. Comput. Syst. **75**, 200–205 (2017)
7. Chaturvedi, S., Goldwasser, D., Daume III, H.: Predicting instructor's intervention in MOOC forums. In: 52th Annual Meeting of the Association for Computational Linguistics, pp. 1501–1511. ACL Press, Baltimore (2014)
8. Rossi, L.A., Gnawali, O.: Language independent analysis and classification of discussion threads in Coursera MOOC forums. In: 15th IEEE International Conference on Information Reuse and Integration, pp. 654–661. IEEE Press, New York (2014)
9. Anonymized Coursera Discussion Threads Dataset. http://github.com/elleros/courseraforumsJuly2014
10. Liu, S., Feng, L., Qiao, H.: Scatter balance: an angle-based supervised dimensionality reduction. IEEE Trans. Neural Netw. Learn. Syst. **26**, 277–289 (2015)
11. He, H., Garcia, E.A.: Learning from imbalanced data. IEEE Trans. Knowl. Data Eng. **21**, 1263–1284 (2009)

A Simple Convolutional Transfer Neural Networks in Vision Tasks

Wenlei Wu[1,2], Zhaohang Lin[1], Xinghao Ding[2], and Yue Huang[2(✉)]

[1] Tencent Computer Systems Company Limited, Shenzhen 518000, China
[2] Department of Communication Engineering, Xiamen University,
Xiamen 361005, China
yhuang2010@xmu.edu.cn

Abstract. Convolutional neural networks (ConvNets) is multi-stages trainable architecture that can learn invariant features in many vision tasks. Real-world applications of ConvNets are always limited by strong requirements of expensive and time-consuming labels generating in each specified task, so the challenges can be summarized as that labeled data is scarce while unlabeled data is abundant. The traditional ConvNets does not consider any information hidden in the large-scale unlabeled data. In this work, a very simple convolutional transfer neural networks (CTNN) has been proposed to address the challenges by introducing the idea of unsupervised transfer learning to ConvNets. We propose our model with LeNet5, one of the simplest model in ConvNets, where an efficient unsupervised reconstruction based pre-training strategy has been introduced to kernel training from both labeled and unlabeled data, or from both training and testing data. The contribution of the proposed model is that it can fully use all the data, including training and testing simultaneously, thus the performances can be improved when the labeled training data is insufficient. Widely used hand-written dataset MNIST, together with two retinal vessel datasets, DRIVE and STARE, are employed to validate the proposed work. The classification experiments results have demonstrated that the proposed CTNN is able to reduce the requirement of sufficient labeled training samples in real-world applications.

Keywords: Convolutional neural networks · Transfer learning · Unsupervised pre-training · PCA

1 Introduction

ConvNets aims to learn features with a hierarchical neural networks whose convolutional layers alternate with subsampling layers, reminiscent of simple and complex cells in the primary visual cortex of the brain [1,2]. Higher order features can be directly extracted based on the stacked trainable stages in ConvNets using repeating convolutions, non-linear mapping and max-pooling the large segmented region from the previous step. Finally, object recognition is implemented as the final layer in ConvNets with a supervised back-propagation (BP) neural network classifier [3].

© Springer International Publishing AG 2017
D. Liu et al. (Eds.): ICONIP 2017, Part IV, LNCS 10637, pp. 385–392, 2017.
https://doi.org/10.1007/978-3-319-70093-9_40

As for the typical structure of ConvNets, back-propagation is used to train the entire system in a supervised fashion. ConvNets finds successes in many recognition problems in computer vision, e.g. handwritten recognition, face recognition, license plate detection and even obstacle avoidance for off-load mobile robots. [4–7]. Many improved ConvNets models are still pursuiting for a better feature extraction in different classification tasks [8–10].

However, traditional ConvNets is always trained with large-scale labeled data for each specified task in its success. Therefore, challenges still remain in the real-world applications of Big Data, where (1) it is labor-intensive and time-consuming for generate large-scale data, (2) training samples and testing samples are assumed to follow the same distribution. Therefore, repeatable labeling and model training is still required in each specified task. This can be explained that traditional ConvNets belongs to isolated learning, which does not consider related information from data across different domains and from related tasks.

Source domain and target domain are two basic definitions in transfer learning. The theory assumes to transfer knowledge from the source domain (auxiliary domain) to the target domain. The knowledge transfer would greatly improve the performance of learning by avoiding much expensive data-labeling efforts [11]. Feature-based transfer is one of the major categories in transfer learning, which is supposed to find a good feature representation that reduces the differences between the source and the target domains, and the error of classification and regression models [11]. Considered that ConvNets has been well-known for its strong ability of extracting high-level vision features, so it is very nature to introduce the idea of transfer learning to ConvNets to generate the strong representative feature from insufficient labeled data.

In our proposed work, a very simple convolutional transfer neural networks (CTNN) is proposed, where a unsupervised reconstruction based pre-training strategy has been introduced to improve LeNet5, one of the most simplest model in ConvNets. The proposed model is able to mine the representative patterns hidden in the large-scale of data without considering their labels, thus the information can be transferred to the target domain to reduce the strong requirement of labeled data. Meanwhile, the proposed model also generates better performances on some extreme cases of insufficient labeled data, e.g. cross-data recognition tasks.

2   Related Works

The proposed model improves the ConvNets with LeNet5, as shown in Fig. 1. As LeNet5 has been widely-used, we will have a very brief description of the architecture. It contains two convolutional layers $C1$ and $C3$, two pooling layers $S2$ and $S4$, and two fully connection layers $F5$ and $F6$. Based on convolving the input image with different kernels (or weights), several feature maps can be generated in layer $C1$, denoted as $C1_i$, i = 1:N, where N is the number of kernel.

$$C1_i = (sigmoid\,(w_i \otimes x)) \tag{1}$$

where x is the original image with size $M_x \times M_y$, $\otimes$ is the convolutional operation, w_i corresponds to the ith convolutional kernel and the non-linear mapping $sigmoid(x)$ is defined as $f(x) = 1/(1 + e^{-x})$. The filter w_i is randomly initialized at first, and then is trained with a well-known BP network [12,13]. Each feature map in $S2$ is obtained by a pooling operation called max-pooling which is performed on the corresponding feature map in layer $C1$. The convolution and max-pooling procedures in layer $C3$ and layer $S4$ are the same as in layer $C1$ and layer $S2$. For final recognition, local features from the input image are combined by the subsequent layers in order to obtain higher order features after last max-pooling layer S4. Such high-order features are eventually encoded into a 1-D vector, which is then categorized by a BP neural networks classifier in the last layer of the ConvNets structure.

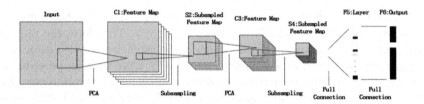

Fig. 1. Architecture of LeNet5.

3   Proposed Protocol

As described in the previous subsection, kernels w_i in Eq. (1) are usually estimate with large-scale of labeled data with BP after random initialization. When the labeled data is insufficient, the estimation becomes a ill-posed problem. In the proposed model, we want to reduce the illness by introducing some knowledge from related data or tasks in an unsupervised manner. Source domain and target domains are two definitions in transfer learning. The basic idea of transfer learning is to transfer some knowledge from source domain to the tasks of interest in the target domain. So the question is how to define the source domain and what kind of knowledge we are going to transfer to the target domain from the source domain. The proposed model is based on the observation that although the labeled data is scarce, however, the unlabeled data is still abundant. The proposed model assumes that there are many representative patterns hidden in the large-scale of data, although they are unlabeled. The patterns may have the contributions for enhancing the recognition task in the target domain. So we tackle the problem by combining all available data, including labeled and unlabeled data, or including training and testing data, as the source domain at first; and then the representative patterns are mined without supervision as transferred knowledge. We will give a more detail description of implementation with LeNet 5.

Each convolution layer in LeNet5 is implemented as follows: there are M training images, the ith feature maps in j^{th} training image is defined as:

$$y_j^i = w_i \otimes x_j \tag{2}$$

where $\otimes$ denotes the 2D convolution, $w_i \in R^{l_x \times l_y}$ is the ith kernel, and $x_j \in R^{s_x \times s_y}$ denotes the jth training image; the boundary of x_j is zero-padded before convolution so as to make $y_j^i \in R^{s_x \times s_y}$ have the same size of x_j. Let $x_j^v \in R^{s_x s_y \times 1}$ be an image x_j in vectorized form, let $P_l \in R^{l_x l_y \times s_x s_y}$ be the lth patch extraction matrix, it extracts a vectorized $l_x \times l_y$ patch from the image x_j^v as $x_j^l = P_l x_j^v \in R^{l_x l_y \times 1}$ for $l = 1, 2, ..., s_x s_y$. Therefore, Eq. (2) can be rewritten as

$$Y_j^i = W_i^T X_j \tag{3}$$

where $Y_j^i \in R^{1 \times s_x s_y}$ is the vectorized form of y_j^i, $W_i \in R^{l_x l_y \times 1}$ is the vectorized form of w_i, and $X_j = [x_j^1, x_j^2, ..., x_j^{s_x s_y}] \in R^{l_x l_y \times s_x s_y}$. Therefore, the ill-posed problem of estimating convolutional weights with insufficient labeled data, turns to be seeking for a good initial value of $\{W_i\}_{i=1}^N$ by a pre-training procedure.

In the proposed model, the source domain includes all the available data without considering their labels. The data could be coming from both training and testing data. Since there is no label information in the step, the pre-training step is unsupervised. By combining all the available images together, a new matrix is defined as $X = [X_1, X_2, ...X_j, ...X_M] \in R^{l_x l_y \times M s_x s_y}$. The pre-training aims to find a set of kernels/basis $\{W_i\}_{i=1}^N$ that are able to reconstruct the X with minimum reconstruction error. Principle component analysis (PCA) is one of the most efficient solutions to the problem. Therefore, the pre-trained kernels are the first N eigenvectors of the matrix $X X^T$. It should be emphasized that the number of kernels N can be determined adaptively using a simple thresholding operations. If the power ratio of the first N_p components to X is larger than a given threshold, where N_p is defined as the number of features maps in the layers. The unsupervised learning in proposed protocol can use data from all available domains, thus can enhance the performances with insufficient labeled data by sharing the representation across all domains. In addition, there is no fine-tune in the proposed model anymore, thus the training procedure could be accelerated.

4   Experiments

Handwritten dataset MNIST and two retinal vessel datasets DRIVE and STARE are employed for validating the proposed work. MNIST is used to validate the proposed model with insufficient labeled data, while DRIVE and STARE are used to evaluate the model with a more challenged case, cross-dataset recognition. In the latter one, the model can be trained in one dataset, and then tested in the other one. All the experiments are implemented in Matlab 2015 on a normal PC with Intel-I3 CPU and 4G RAM. Datasets STARE and DRIVE are two independent retinal vessel segmentation datasets generated from two diabetic retinopathy screening programs in the Netherlands and the United States. Manually segmentation in both STARE and DRIVE are provided together with the original images as groundtruth.

4.1   Experiments on MNIST

MNIST is considered as a popular benchmark for ConvNets. Here we use MNIST basic for validation. In the implementation of LeNet5, the kernel size is 5×5, number of feature maps in first layer and second layer are set to be 16 and 32 respectively, same as mentioned in [4]. The pooling strategy is averaging. To simulate the fact in many applications in big Data that the size of labeled data is much smaller than that of unlabeled data, the proposed method is also validated on data with decreasing labeled training samples size. As in Fig. 2, proposed method has the ability of being more robust than ConvNets model in classification with decreasing percent of labeled data, especially when the labeled data size is very small (10%). This can be explained that the pre-training extract the features based on reconstruction instead of classification; therefore, all the samples, including labeled and unlabeled, can be fully used in the pre-training in the proposed method. As for comparisons, only labeled samples are used to train the deep networks in the LeNet5 model, thus the performances depends strongly on the labeled training samples size, and will decrease dramatically with the decreasing size of labeled samples.

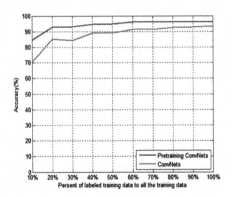

Fig. 2. Accuracy comparisons between ConvNets and proposed with decreasing labeled data at MNIST basic.

4.2   Experiments on DRIVE and STARE

We also evaluate our method on a more challenge task, cross-dataset recognition with two public available datasets for retinal vessel segmentation. DRIVE includes 40 eye-fundus images, 20 images act for training and the other 20 images act for testing [14]. Each image was captured at a resolution of 768×584 pixels. A single binary image of manual segmentation of the vasculature is provided as the ground-truth with each training image. The STARE dataset [15,16] contains 20 colored eye-fundus images, each of them is digitalized with 700×605 pixels. Ten of the images are normal cases and the remaining 10 images show pathological signs, which we reserve as control cases. Ground-truth labeling is also available. The mask of field of view (FOV) for each image is generated as a binary image, where a white circular with a diameter of approximately 650 pixels is shown in the middle of the image. Recent works in retinal vessel segmentation cast the segmentation task into a binary pattern recognition task, where each pixel is assigned a label of 'non-vessel' or 'vessel'.

The experiment is designed to simulate the case where labeled training data is coming from another independent dataset. A sample in training or testing is defined as an image patch for each pixel in the image as the N-connected neighborhood of a given pixel. In the implementation, the kernel size is 5×5, number of feature maps in first layer and second layer are set to be 8 and 8 respectively. Given that some images in STARE show pathological signs, the validation is performed with those images, with only 50,000 labeled samples in DRIVE. In feature extraction with proposed CTNN, 50,000 unlabeled pixels are randomly selected from the STARE dataset, so the procedure of feature extraction can still be trained with 100,000 pixels, from the combination of all the labeled training samples and unlabeled testing samples. The ConvNets model (LeNet 5) is then trained with 50,000 labeled samples from DRIVE. The cross-data validation is implemented for all images in STARE. As shown in Table 1, it can be observed that our proposed method outperforms typical CNN with higher value of both accuracy and area under the curve (AUC) in cross-dataset validation. This can be explained by noting that all the data, including labeled and unlabeled training data, are utilized in the adaptive feature extraction of our method.

Table 1. Cross-dataset validation: trained with DRIVE and tested in STARE

Training dataset	Average accuracy	Average AUC
ConvNets(LeNet5)	0.9110	0.9291
Proposed(CTNN)	0.9381	0.9413

5 Conclusions

A simple convolutional transfer neural networks (CTNN) has been proposed to address the challenges in real-world applications. Inspired by the idea of transfer learning, the tedious kernel training is replaced by a reconstruction-based pre-training strategy to fully use all the available data, including labeled and unlabeled. The proposed work is validated on MNIST and two retinal vessel datasets. The experiments demonstrated that the proposed model outperforms typical ConvNets in the case of insufficient labeled data, and in the more challenged cross-validation task as well.

Acknowledgments. This work was supported in part by the National Natural Science Foundation of China under Grants 8167176681301278, 61172179, 61103121, 61571382, and 61571005, in part by the Guangdong Natural Science Foundation under Grant 2015A030313007, in part by the Fundamental Research Funds for the Central Universities under Grants 20720160075, 20720150169 and 20720150093, in part by the National Natural Science Foundation of Fujian Province, China 2017J01126, in part by the CCF-Tencent research fund.

References

1. LeCun, Y., Bottou, L., Bengio, Y., Haffner, P.: Gradient-based learning applied to document recognition. Proc. IEEE. **86**(11), 2278–2324 (1998)
2. Hubel, D.H., Wiesel, T.N.: Receptive fields of single neurones in the cats striate cortex. J. Physiol. **148**(3), 574–591 (1959)
3. Lecun, Y., Boser, B., Denker, J.S., Henderson, D., Howard, R.E., Hubbard, W., Jackel, L.D.: Handwritten digit recognition with a back-propagation network. Neural Netw. Curr. Appl. Chappman Hall **86**(11), 2278–2324 (1992)
4. Jarrett, K., Kavukeuoglu, K., Ranzato, M.A., Lecun, Y.: What is the best multistage architecture for object recognition? In: IEEE International Conference on Computer Vision, pp. 2146–2153 (2009)
5. Lawrence, S., Giles, C.L., Tsoi, A.C.: Face recognition: a convolutional neural-network approach. IEEE Trans. Neural Netw. **8**(1), 98–113 (1997)
6. Frome, A., Cheung, G., Abdulkader, A., Zennaro, M., Wu, B.: Large-scale privacy protection in street level imagery. IEEE Int. Conf. Comput. Vis. **1**(2), 2373–2380 (2009)
7. Farabet, C., Couprie, C., Najman, L., Lecun, Y.: Learning hierarchical features for scene labeling. IEEE Trans. Pattern Anal. Mach.Intell. **35**, 1915–1929 (2013)
8. Lecun, Y., Kavukcuoglu, K., Farabet, C.: Convolutional networks and applications in vision. In: IEEE International Conference on Computer Vision pp, pp. 253–256 (2010)
9. Soltau, H., Sano, G., Sainath, T.N.: Joint training of convolutional and non-convolutional neural networks. In: International Conference on Acoustics, Speech and Signal Processing, pp. 5572–5576 (2014)
10. Sainath, T.N., Mohamed, A.R., Kingsbury, B., Ramabhardran, B.: Deep convolutional neural networks for LVCSR. In: International Conference on Acoustics, Speech and Signal Processing, pp. 8614–8618 (2014)
11. Pan, S.J., Yang, Q.: A survey on transfer learning. IEEE Trans. Knowl. Data Eng. **22**, 1345–1359 (2010)
12. Pieer, S., Lecun, Y.: Traffic sign recognition with multi-scale convolutional networks. In: International Joint Conference on Neural Networks, pp. 2809–2813 (2011)
13. Ji, S., Xu, W., Yang, M., Yu, K.: 3D convolutional neural networks for human action recognition. IEEE Trans. Pattern Anal. Mach. Intell. **35**(1), 221–231 (2013)

14. Staal, J., Abramoff, M.D., Niemeijer, M., Viergever, M.A., van Ginneken, B.: Ridge-based vessel segmentation in color images of the retina. IEEE Trans. Med. Imaging **23**, 501–509 (2004)
15. Hoover, A., Goldbaum, M.: Locating the optic nerve in a retinal image using the fuzzy convergence of the blood vessels. IEEE Trans. Med. Imaging **22**, 951–958 (2003)
16. Hoover, A.D., Kouznetsova, V., Goldbaum, M.: Locating blood vessels in retinal images by piecewise threshold probing of a matched filter response. IEEE Trans. Med. Imaging **19**, 203–210 (2000)

Dissimilarity-Based Sequential Backward Feature Selection Algorithm for Fault Diagnosis

Yangtao Xue, Li Zhang[✉], and Bangjun Wang

School of Computer Science and Technology and Joint International Research
Laboratory of Machine Learning and Neuromorphic Computing,
Soochow University, Suzhou 215006, China
zhangliml@suda.edu.cn

Abstract. The aim of feature selection applied to fault diagnosis is
to select an optimal feature subset that is relevant to the faults. The
optimal feature subset with fewer features contains more discriminative information which can improve the performance of fault diagnosis models. A novel sequential backward feature selection method based
on dissimilarity is proposed to detect the difference of features between
normal and fault data. The proposed feature selection method can be
used to find relevant features with fault. Furthermore, the fault diagnosis model combines the proposed feature selection method with support vector machine. Experimental results on a chemical process indicate
that the proposed feature selection method is useful and superior in fault
diagnosis.

Keywords: Feature selection · Dissimilarity · Fault diagnosis · Support
vector machine

1 Introduction

Enormous collected process data promotes the wide researches and applications
of the technology of pattern recognition in fault diagnosis. The high-efficient
and powerful generalization enables support vector machine (SVM) as a great
diversity of application for fault diagnosis [1,2]. The faults were found with the
effectiveness of wavelet-based features for fault diagnosis using support vector
machines (SVM) and proximal support vector machines (PSVM) in [1]. However,
the collected industrial data consists of possibly irrelevant or redundant features,
which usually results in the poor performance of fault diagnosis model. It is
essential to extract the significant and valuable features.

Feature selection as an important aspect of data process, has been introduced
in fault diagnosis models [3–5]. While relevant feature with faults are essential
in determining where faults occur and need to be recovered [6]. Feature selection
methods contain the original information not like feature extraction methods
which project the original data into a new feature space, so that the feature
selection method is more advantageous for fault diagnosis. In addition, these
methods not only decrease the computational load, but also improve accuracy.

© Springer International Publishing AG 2017
D. Liu et al. (Eds.): ICONIP 2017, Part IV, LNCS 10637, pp. 393–401, 2017.
https://doi.org/10.1007/978-3-319-70093-9_41

Considering computationally complexity and implementation, the filter method is a good choice. This kind of method evaluates features by some feature ranking algorithm, which is independent of the subsequent classifier. Feature ranking algorithms select features with various measures of the general characteristic, such as data variance, distance estimate, and correlation estimate [7,8]. The filter method highlights the difference between two different classes when selecting discriminative features for practical applications. Generally, the filter method is faster and can be used as a pre-processing step to reduce dimensionality and avoid overfitting [9]. In [10], Fisher score (Fscore) measures the discrimination of two class data, and the importance of features is ranked by Fscore. In [11], Pearson correlation coefficient (Pearson) is used to measure the correlation between a feature and a class, and as a ranking criterion for feature selection. In [12], Relief is a feature weight-based algorithm that updates the relevant weight values based on Euclidean distance between the selected instance and the two nearest instances of the same and opposite class.

Therefore, feature selection is introduced into the fault diagnosis model which is built in SVM. In [13], a dissimilarity index is introduced to detect a change of operating condition, and on the basis of the idea, the dissimilarity (DISSIM) method is proposed in [13]. Inspired by DISSIM, the dissimilarity of process data can also be utilized for feature ranking which measures the difference of two-class data with different feature subsets. The data with different features have different distribution, and the important features have significant influence on the distribution of dataset. The features selected by the proposed method make great contributions to the difference of two data sets. The proposed method avoid measuring the relationship between features. Compared with the linear feature selection methods, such as Fscore and Pearson, the proposed method has an advantage of addressing nonlinear and complex data set. In addition, the method contains the information of all samples which can reduce the influence of outliers.

The remainder of the paper is structured as follows. Section 2 introduces the proposed feature selection based on dissimilarity for fault diagnosis. The experiments are conducted and their results are presented in Sect. 3. Lastly, Sect. 4 presents the conclusion.

2   Dissimilarity-Based Sequential Backward Feature Selection

The dissimilarity index can detect a change of the distribution of data that can be taken as a measurement to distinguish the difference between different features. Compared with other filter methods, the importance of features is measured by the dissimilarity. That is to say, the feature ranking is determined by the dissimilarity indexes of different feature subsets. The dissimilarity index and the proposed feature selection method are described as shown below.

2.1 Dissimilarity Index

The original concept of dissimilarity is used for classification. In [15], the dissimilarity is proposed on the basis of Karhunen-Loeve expansion to apply to feature selection and ordering. There are two-class data matrices $\mathbf{X}_1 = [\mathbf{x}_{11}, \mathbf{x}_{12}, \ldots, \mathbf{x}_{1d}] \in \mathbb{R}^{N_1}$ and $\mathbf{X}_2 = [\mathbf{x}_{21}, \mathbf{x}_{22}, \ldots, \mathbf{x}_{2d}] \in \mathbb{R}^{N_2}$, where d is the number of features, N_1 and N_2 are the sample number of $\mathbf{X}_1$ and $\mathbf{X}_2$, respectively. The mixture covariance matrix $\mathbf{R}$ of the two-class data can be calculated as:

$$\mathbf{R} = \frac{1}{N-1}\mathbf{X}_1^T\mathbf{X}_1 + \frac{1}{N-1}\mathbf{X}_2^T\mathbf{X}_2 \tag{1}$$

where $N = N_1 + N_2$ is the number of total samples.

By applying Eigendecomposition to $\mathbf{R}$, we can obtain the eigenvalue matrix $\mathbf{\Lambda}$ and the eigenvector matrix $\mathbf{P}$:

$$\mathbf{RP} = \mathbf{P\Lambda} \tag{2}$$

The transformation matrix $\mathbf{W}$ is given by

$$\mathbf{W} = \mathbf{P\Lambda}^{-\frac{1}{2}} \tag{3}$$

Then, $\mathbf{X}_i$ can be transformed into $\mathbf{Y}_i$, $i = 1, 2$:

$$\mathbf{Y}_i = \sqrt{\frac{N_i - 1}{N - 1}}\mathbf{X}_i\mathbf{W} \tag{4}$$

The covariance matrices $\mathbf{A}_i$ of the transformed data matrices $\mathbf{Y}_i$ can be obtained by

$$\mathbf{A}_i = \frac{1}{N_i - 1}\mathbf{Y}_i^T\mathbf{Y}_i \tag{5}$$

The relationship of covariance matrices of $\mathbf{Y}_i$ is given by

$$\mathbf{A}_1 + \mathbf{A}_2 = \mathbf{I} \tag{6}$$

which can easily proved so that $\mathbf{A}_1$ and $\mathbf{A}_2$ have same eigenvalues but oppositely ordered.

Therefore, it is sufficient that we perform Eigendecomposition on only one matrix, $\mathbf{A}_1$ or $\mathbf{A}_2$. Without loss of generality, $\mathbf{A}_1$ is adopted. Namely,

$$\mathbf{A}_1\mathbf{P}_0 = \mathbf{\Lambda}_0\mathbf{P}_0 \tag{7}$$

where $\mathbf{\Lambda}_0$ is the eigenvalue matrix with the jth eigenvalue λ_j, and $\mathbf{P}_0$ is an orthogonal matrix consisting of eigenvectors.

Finally, a dissimilarity index D is computed to evaluate the difference between $\mathbf{X}_1$ and $\mathbf{X}_2$.

$$D = DS(\mathbf{X}_1, \mathbf{X}_2) = \frac{1}{d}\sum_{j=1}^{d}(\lambda_j - 0.5)^2 \tag{8}$$

When two-class data are quite different from each other, D is larger and near one. On the contrary, if the data is similar to each other, D should approach to zero.

2.2 Sequential Backward Feature Selection Algorithm

The dissimilarity can indicate the difference between two datasets. If we delete one feature from both original datasets and compute the dissimilarity between the new formed datasets which not contain the deleted feature, we can find the dissimilarity difference between the corresponding feature. The smaller the dissimilarity difference is, the less important the corresponding feature. In other words, this feature makes no difference between the normal and fault data. To rank features, the search strategy of the proposed feature selection adopts sequential backward selection [14].

The candidate feature subsets are generating by eliminating one feature at a time. That is to say, the dissimilarity of candidate feature subsets is to be computed as

$$D^{(-k)} = DS(\mathbf{X}_1^{(-k)}, \mathbf{X}_2^{(-k)}), k = 1, \cdots, d \tag{9}$$

where $(-k)$ means that the kth feature is removed, the candidate feature matrices are $\mathbf{X}_i^{(-k)} = [\mathbf{x}_{i1}, \ldots, \mathbf{x}_{i(k-1)}, \mathbf{x}_{i(k+1)}, \ldots, \mathbf{x}_{id}] \in \mathbb{R}^{N_i}$, $i = 1, 2$. Then, the importance of features is determined the difference between D and $D^{(-k)}$. The ranking criterion is to find the kth feature with the smallest value of the dissimilarity difference and remove it from the feature set. Namely,

$$\min_{k=1,\cdots,d} \left| D - D^{(-k)} \right| \tag{10}$$

where $|.|$ denotes the absolute value.

In each iteration of the algorithm, that the kth feature with the smallest dissimilarity difference is deleted from the candidate feature set. The ranked feature set is ranked by the importance of features in an descending order. The feature ranking list is determined by the dissimilarities of two data sets with different feature subsets, the detailed algorithm is shown in Algorithm 1.

The optimal feature subset is determined when the F1-measure (FM) value of the classifier is maximum. FM is defined by

$$FM = \frac{2 \times Presision \times Recall}{Presision + Recall} \tag{11}$$

where $Presision = \frac{T1}{T1+F1}$ and $Recall = \frac{T1}{T1+F2}$, $F1$ is the number of false normal points, $T1$ is the number of true normal samples, and $F2$ is the number of false fault samples. The greater value of F1-measure indicates that the feature subset is the more useful.

3 Experiments

The Tennessee Eastman (TE) process is a chemical industrial process and has been widely used for evaluating the fault diagnosis models. The TE process has 52 features: 11 control variables, 22 continuous process measures and 19 compositions. All the data is available at http://brahms.scs.uiuc.edu. The training

Algorithm 1. Dissimilarity-based sequential backward feature selection (DSBFS) algorithm

---

Input: A data set $\mathbf{X}_1 = [\mathbf{x}_{11}, \mathbf{x}_{12}, \ldots, \mathbf{x}_{1d}] \in \mathbb{R}^{N_1 \times d}$ and a fault set $\mathbf{X}_2 = [\mathbf{x}_{21}, \mathbf{x}_{22}, \ldots, \mathbf{x}_{2d}] \in \mathbb{R}^{N_2 \times d}$.

Output: The optimal feature subset F.

1. Initialize: Let the ranked feature list $R = \emptyset$, the candidate feature subset $S = [1, 2, \ldots, d]$, the optimal feature subset $F = \emptyset$, and $n = |S|$.
2. Normalize $\mathbf{X}_1$ and $\mathbf{X}_2$: Let $\mathbf{X}_1$ be the reference dataset, and find the mean and variance of $\mathbf{X}_1$. Then, normalize $\mathbf{X}_1$ to be zero-mean and unit-variance, and normalize $\mathbf{X}_2$ according to the mean and variance of $\mathbf{X}_1$.
3. Compute the dissimilarity of two normalized data matrixes D according to (8).
4. Rank features:
    (a). **While** S is not empty
    (b).   **For** $k = 1, 2, \ldots, n$ **Do**
    (c).   Construct candidate data matrices $\mathbf{X}_1^{(-k)}$ and $\mathbf{X}_2^{(-k)}$ by removing the kth feature from S.
    (d).   Compute the feature dissimilarity D^{-k} for $\mathbf{X}_1^{(-k)}$ and $\mathbf{X}_2^{(-k)}$ according to (9).
    (e).   **End For**
    (f). Find feature q with the smallest dissimilarity difference by the criterion (10)
    (g). Update the ranked feature list $R \leftarrow R \bigcup q$, the candidate feature subset $S \leftarrow S \setminus q$, and $n = |S|$.
    (h). **End While**
5. Determine the optimal feature subset F.
    (a). Partition the whole dataset into training and validation subsets.
    (b). **For** $p = 1, \cdots, d$ **Do**
    (c).   Use the training subset with features $1, 2, \ldots, p$ in the ranked list R to train a classifier on the training set.
    (d).   Compute the F1-measure score on the validation subset with features $1, 2, \ldots, p$. Denote this value be s_p.
    (e). **End For**
6. Find $q = \arg\max_{p=1,\cdots,d} s_p$. Then return the optimal feature subset $F = \{j_1, \cdots, j_q\}$.

---

set contains 500 normal samples which are taken as the reference data and 480 fault data. In each test set, there are 960 normal test samples, and 960 fault test data consisting of 160 normal and 800 abnormal ones. Considering the realistic process, we choose SVM with the RBF kernel where the RBF kernel parameter is determined by using the method in [16] as the subsequent classifier. Let the regular parameter $C = 1$ in SVM model which is an empirical value.

Several feature selection methods including Fisher score (Fscore), Pearson correlation measure (Pearson) and Relief algorithms [8] are conducted by comparison on the ability of selecting discriminant features. The optimal feature subset is determined by 10-fold cross validation for all compared methods. In this paper, we list the results of Fault 2, 14, 19 and 21 for analysis.

Figure 1 displays the distribution of Fault 14 with the top two selected feature using four methods. It is clear that the fault data points are far away from the region which the normal data ones are gathered, while the normal data points are overlapped with the fault ones when using other methods. In addition, the proposed method have less selected features in Table 1. DSBFS has great superior in selecting related features than other three methods.

Five fault diagnosis models are used in this experiment, including SVM, Fscore+SVM, Pearson+SVM, Relief+SVM, and DSBFS+SVM. The results are

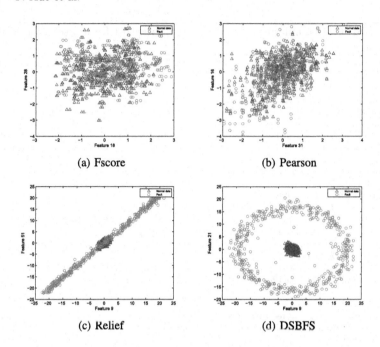

Fig. 1. Distribution of Fault 14 data with top-two features obtained by different methods: (a) Fscore, (b) Pearson, (c) Relief, and (d) DSBFS

Table 1. The number of selected features using four methods

Fault	Fscore	Pearson	Relief	DSBFS
2	18	16	11	6
14	51	28	3	2
19	25	45	30	8
21	33	46	35	1

Table 2. F1-measure of different fault diagnosis models

Fault	SVM	Fscore + SVM	Pearson + SVM	Relief + SVM	DSBFS + SVM
2	99.25%	99.03%	97.62%	99.14%	99.25%
14	99.78%	99.79%	99.95%	100%	100%
19	76.12%	81.35%	81.17%	75.66%	87.45%
21	42.46%	42.07%	37.92%	60.64%	98.88%

listed in Table 2. We can see that SVM with all features also have good performance in Fault 2 and 14, which indicates the SVM can effectively diagnose fault data on the TE process. The fault diagnosis model combining feature selection methods with SVM decreases the time of online diagnosis and increases the

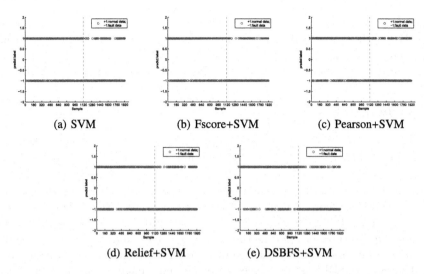

Fig. 2. Monitoring performance for Fault 19 using different models: (a) SVM, (b) Fscore+SVM, (c) Pearson+SVM, (d) Relief+SVM, and (e) DSBFS+SVM.

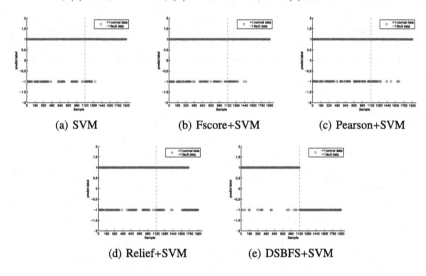

Fig. 3. Monitoring performance for Fault 21 using different models: (a) SVM, (b) Fscore+SVM, (c) Pearson+SVM, (d) Relief+SVM, and (e) DSBFS+SVM.

accuracy of classification. Compared with the performance of other methods for Fault 19 and 21, the selected features using DSBFS which are beneficial in improving precision and efficiency in classification of fault diagnosis.

The monitoring performance of Fault 19 and 21 is shown in Figs. 2 and 3 in which the first 1120 samples are normal, and the subsequent samples are faulty. For Fault 19, the performance of DSBFS+SVM in diagnosing fault data using is little better than other methods, while in diagnosing normal data is obviously the best. For Fault 21, the fault data can be correctly distinguished from the normal data. The proposed model has better performance than others, meanwhile the fault cause can be detected according to the selected features using DSBFS.

4 Conclusion

The idea of dissimilarity of different feature subsets can be taken as the measurement of feature importance. The dissimilarity-based sequential backward selection (DSBFS) method is proposed to select relevant features to accurately distinguish two different class data. The effectiveness of DSBFS is validated compared with Fisher score, Pearson correlation measure and Relief algorithms. For the ability on selecting features, DSBFS is superior to the others on the TE process. The feature selection method based on dissimilarity optimizes classification accuracy and simplify the training model by reducing more dimensionality. Therefore, DSBFS is introduced into the fault diagnosis model to improve the classification performance. The proposed fault diagnosis model combining DSBFS and SVM is more favorable to address realistic process data.

Acknowledgments. This work was supported in part by the National Natural Science Foundation of China under Grant No. 61373093, by the Natural Science Foundation of Jiangsu Province of China under Grant No. BK20140008, and by the Soochow Scholar Project.

References

1. Saravanan, N., Kumar Siddabattuni, V.N.S., Ramachandran, K.I.: A comparative study on classification of features by SVM and PSVM extracted using morlet wavelet for fault diagnosis of spur bevel gear box. Expert Syst. Appl. **35**(3), 1351–1366 (2008)
2. Cheng, J., Yu, D., Tang, J., Yang, Y.: Application of SVM and SVD technique based on emd to the fault diagnosis of the rotating machinery. Shock Vibr. **16**(1), 89–98 (2013)
3. Liu, C., Jiang, D., Yang, W.: Global geometric similarity scheme for feature selection in fault diagnosis. Expert Syst. Appl. **41**(8), 3585–3595 (2014)
4. Wang, L., Yu, J.: Fault feature selection based on modified binary PSO with mutation and its application in chemical process fault diagnosis. In: Wang, L., Chen, K., Ong, Y.S. (eds.) ICNC 2005. LNCS, vol. 3612, pp. 832–840. Springer, Heidelberg (2005). doi:10.1007/11539902_102
5. Lu, L., Yan, J., Silva, C.W.D.: Dominant feature selection for the fault diagnosis of rotary machines using modified genetic algorithm and empirical mode decomposition. J. Sound Vibr. **344**, 464–483 (2015)
6. Sulaiman, M.A., Labadin, J.: Feature selection based on mutual information. In: International Conference on It in Asia, pp. 1–6. IEEE Press, New York (2015)
7. Lei, Y., He, Z., Zi, Y., Chen, X.: New clustering algorithm-based fault diagnosis using compensation distance evaluation technique. Mech. Syst. Sig. Process. **22**(2), 419–435 (2008)

8. Zhang, K., Li, Y., Scarf, P., Ball, A.: Feature selection for high-dimensional machinery fault diagnosis data using multiple models and radial basis function networks. Neurocomputing **74**(17), 2941–2952 (2011)

9. Chen, F.L., Li, F.C.: Combination of feature selection approaches with SVM in credit scoring. Expert Syst. Appl. Int. J. **37**(7), 4902–4909 (2010)

10. Chen, Y.W., Lin, C.J.: Combining SVMs with various feature selection strategies. In: Guyon, I., Nikravesh, M., Gunn, S., Zadeh, L.A. (eds.) Feature Extraction. SFSC, vol. 207, pp. 315–324. Springer, Heidelberg (2006). doi:10.1007/978-3-540-35488-8_13

11. Guyon, I., Elisseeff, A.: An introduction to variable and feature selection. JMLR.org (2003)

12. Dash, M., Liu, H.: Feature Selection for Classification. IOS Press, Amsterdam (1997)

13. Kano, M., Hasebe, S., Hashimoto, I., Ohno, H.: Statistical process monitoring based on dissimilarity of process data. AIChE J. **48**(6), 1231–1240 (2002)

14. Aha, D.W., Bankert, R.L.: A comparative evaluation of sequential feature selection algorithms. In: Fisher, D., Lenz, H.J. (eds.) Learning from Data. LNS, vol. 112, pp. 199–206. Springer, New York (1996). doi:10.1007/978-1-4612-2404-4_19

15. Fukunaga, K., Koontz, W.L.G.: Application of the Karhunen-Loeve expansion to feature selection and ordering. IEEE Trans. Comput. **C–19**(4), 311–318 (1970)

16. Zhang, L., Zhou, W.D., Chang, P.C., Liu, J., Yan, Z., Wang, T., Li, P.Z.: Kernel sparse representation-based classifier. Neural Process. Lett. **60**(1), 1684–1695 (2016)

Online Chaotic Time Series Prediction Based on Square Root Kalman Filter Extreme Learning Machine

Shoubo Feng, Meiling Xu, and Min Han[(✉)]

Faculty of Electronic Information and Electrical Engineering,
Dalian University of Technology, Dalian 116024, Liaoning, China
minhan@dlut.edu.cn

Abstract. In this paper, we proposed a novel neural network prediction model based on extreme learning machine for online chaotic time series prediction problems. The model is characterized by robustness and generalization. The initial weights are initialized by orthogonal matrix to improve the generalization performance and the output weights are updated by square root Kalman filter. The convergence of the algorithm is proved by Lyapunov stability theorem. Simulations based on artificial and real-life data sets demonstrate the effectiveness of the proposed model.

Keywords: Time series prediction · Square root Kalman filter · Extreme learning machine · Convergence analysis

1 Introduction

Time series is the sampling data in the chronological order from dynamic systems. In practice, most systems are chaotic such as meteorological, hydrology and financial systems and it is difficult to model chaotic systems. Thence, time series prediction has been one of the most important research areas. So far, autoregressive moving average model, autoregressive integrated moving average model, and artificial intelligence methods such as artificial neural network [1–4] and deep neural networks [5] has been used for time series prediction. Extreme learning machine [6] is a theory to initialize the neural network, and is used in time series prediction [2], but it cannot be used in online prediction applications. Online sequential extreme learning machine (OSELM) [3] can be used in online prediction mission but it is sensitive to noises and unstable in prediction applications. Modified online ELM [4,7] has been proposed and used in time series prediction applications. Kalman filter is a classical dynamic system state estimation algorithm under linear gaussian state space model assumption, which could capture the state of system [8]. While in practical application, there are numerical stability problems due to the round-off error of filtering-error covariance matrix. Generally, square root filter method are used to ensure the stability of the filter and could achieve better numerical performance. In this paper,

© Springer International Publishing AG 2017
D. Liu et al. (Eds.): ICONIP 2017, Part IV, LNCS 10637, pp. 402–409, 2017.
https://doi.org/10.1007/978-3-319-70093-9_42

we proposed a robust model named square root Kalman filter extreme learning machine, abbreviated as SRKF-ELM, for online prediction problems to overcome the weakness of OS-ELM. In this model, input weights and biases are initialized by random orthogonal matrices and the output weights are pre-trained by least square method with regularization term. With the arrival of stream data, the output weights are updated by square root kalman filter to capture the state of chaotic systems.

The paper is organized as follows. In Sect. 2, we introduce the algorithm of SRKF-ELM. Section 3 provides the proof of convergence. The experiment results of chaotic time series prediction are given in Sect. 4. The study is concluded in Sect. 5.

2   Square Root Kalman Filter Extreme Learning Machine

The single hidden layer feedforward neural network is used to extract characteristics of chaotic data and approximate the chaotic system states. The nonlinear mapping relationship between input and output is as follows:

$$\mathbf{y}_{k+\eta} = f(\mathbf{W}_{in}\mathbf{x}_k + \mathbf{b})^T \mathbf{w}_k \tag{1}$$

where $\mathbf{W}_{in}$ is the input weights of neural network, $\mathbf{b}$ is the biases of hidden layers, they are all set as orthogonal matrix to enhance the generalization performance [6]. $\mathbf{x}_k$ is the input of neural networks, $\mathbf{w}_k$ is the output weight of the neural networks. f is the activation function, normally set as sigmoid function. $\mathbf{y}_{k+\eta}$ is the predicting value, η is prediction horizon.

2.1   Pre-training Stage

In Pre-training stage, the output weights is obtained by extreme learning machine. ELM tends to minimize not only the training error but also the L2-norm of the output weights in order to obtain generalization performance. The details are as follows, where $\mathbf{H}$ is the output matrix of hidden layer

$$\mathbf{H} = \begin{bmatrix} f(\mathbf{W}_{in}\mathbf{x}_1 + \mathbf{b}) \\ \vdots \\ f(\mathbf{W}_{in}\mathbf{x}_N + \mathbf{b}) \end{bmatrix}. \tag{2}$$

$\mathbf{T}$ is the target matrix.

$$\mathbf{T} = \begin{bmatrix} \mathbf{t}_1^T \\ \vdots \\ \mathbf{t}_N^T \end{bmatrix}. \tag{3}$$

$\mathbf{w}_0$ is the output weights, and C is the regularization coefficient. According to the ridge regression theory [6], the solution performs stable and has better generalization behavior in real-life applications.

$$\mathbf{w}_0 = \mathbf{H}^T \left(\frac{1}{C} + \mathbf{H}\mathbf{H}^T \right)^{-1} \mathbf{T}. \tag{4}$$

2.2   Online Training Stage

The online training process of output weight $\mathbf{w}_k$ in Eq. (1) could be represented by a discrete state space model without controlling inputs [8]:

$$\mathbf{w}_{k+1} = \mathbf{A}_{k+1,k}\mathbf{w}_k + \omega_k$$
$$\mathbf{y}_k = \mathbf{H}_k\mathbf{w}_k + \mathbf{v}_k \tag{5}$$

where $\mathbf{w}_k \in \mathbf{R}^n$ is the state of the system, $\mathbf{w}_0$ is the initial value computed by pre-training stage (2)–(4), $\mathbf{w}_{k+1}$ is the predicting state, $\mathbf{y}_k$ is observation value. $\mathbf{A}_{k+1,k}$ is the state transition matrix between time k and $k+1$, set as identity matrix in the model. $\mathbf{H}_k$ is observation matrix, which reflects the mapping between $\mathbf{w}_k$ and $\mathbf{y}_k$. The process noise ω_k and observation noise $\mathbf{v}_k$ are additive, independent and identically distributed gaussian noises. $\mathbf{Q}_{\omega,k}$ and $\mathbf{Q}_{v,k}$ are covariance matrix of ω_k and $\mathbf{v}_k$.

In practical application, the covariance matrix $\mathbf{P}$ in Kalman filter tends to get into numerical accuracy trouble. Therefore, we propagate $\mathbf{P}$ in the form of square root to improve the robustness of the model. Kalman gain matrix $\mathbf{G}_k$ and covariance matrix $\mathbf{P}_{k|k}$ are defined as:

$$\mathbf{G}_k = \mathbf{P}_{k|k-1}\mathbf{H}_k^T\mathbf{R}_k^{-1} \tag{6}$$
$$\mathbf{P}_{k|k} = \mathbf{P}_{k|k-1} - \mathbf{G}_k\mathbf{H}_k\mathbf{P}_{k|k-1} \tag{7}$$
$$\mathbf{R}_k = \mathbf{H}_k\mathbf{P}_{k|k-1}\mathbf{H}_k^T + \mathbf{Q}_{v,k} \tag{8}$$

and according to Eqs. (6)–(8), it yields

$$\mathbf{P}_{k|k} = \mathbf{P}_{k|k-1} - \mathbf{P}_{k|k-1}\mathbf{H}_k^T\mathbf{R}_k^{-1}\mathbf{H}_k\mathbf{P}_{k|k-1}. \tag{9}$$

Rewriting Eqs. (6)–(9) in the form of matrix and performing Cholesky factorization to it:

$$\mathbf{M} = \begin{bmatrix} \mathbf{Q}_{v,k} + \mathbf{H}_k\mathbf{P}_{k|k-1}\mathbf{H}_k^T & \mathbf{H}_k\mathbf{P}_{k|k-1} \\ \mathbf{P}_{k|k-1}\mathbf{H}_k^T & \mathbf{P}_{k|k-1} \end{bmatrix}$$
$$= \begin{bmatrix} \mathbf{Q}_{v,k}^{1/2} & \mathbf{H}_k\mathbf{P}_{k|k-1}^{1/2} \\ 0 & \mathbf{P}_{k|k-1}^{1/2} \end{bmatrix} \begin{bmatrix} \mathbf{Q}_{v,k}^{1/2} & 0^T \\ \mathbf{P}_{k|k-1}^{1/2}\mathbf{H}_k^T & \mathbf{P}_{k|k-1}^{1/2} \end{bmatrix} \tag{10}$$

and Eq. (10) could be written as $\mathbf{M} = \mathbf{X}\mathbf{X}^T$, which satisfies the condition of matrix factorization lemma. Then there must be a matrix $\mathbf{Y}_k$ and an orthogonal matrix $\mathbf{\Theta}_k$ that $\mathbf{Y}_k = \mathbf{X}\mathbf{\Theta}_k$, which is given by:

$$\begin{bmatrix} \mathbf{Q}_{v,k}^{1/2} & \mathbf{H}_k\mathbf{P}_{k|k-1}^{1/2} \\ 0 & \mathbf{P}_{k|k-1}^{1/2} \end{bmatrix} \mathbf{\Theta}_k = \begin{bmatrix} \mathbf{Y}_{11,k} & 0^T \\ \mathbf{Y}_{21,k} & \mathbf{Y}_{22,k} \end{bmatrix}. \tag{11}$$

And $\mathbf{\Theta}_n$ could be derived by QR factorization or Givens rotations. According to the orthogonal character, (11) could be rewritten as $\mathbf{X}\mathbf{X}^T = \mathbf{Y}\mathbf{Y}^T$. The partitioned lower triangular matrix in (12) is obtained by multiplying $\boldsymbol{\Theta}_n$ and the left matrix:

$$\begin{bmatrix} \mathbf{Q}_{v,k}^{1/2} & \mathbf{B}_k\mathbf{P}_{k|k-1}^{1/2} \\ \mathbf{0} & \mathbf{P}_{k|k-1}^{1/2} \end{bmatrix} \boldsymbol{\Theta}_k = \begin{bmatrix} \mathbf{R}_k^{1/2} & \mathbf{0}^T \\ \mathbf{G}_k\mathbf{R}_k^{1/2} & \mathbf{P}_{k|k}^{1/2} \end{bmatrix}. \tag{12}$$

Three elements on the right side of Eq. (12) are the target matrices. $\mathbf{R}_k^{1/2}$ is the square-root form of covariance matrix of the innovations process. $\mathbf{G}_k\mathbf{R}_k^{1/2}$ is used to compute Kalman gain. $\mathbf{P}_{k|k}^{1/2}$ represents the square-root form of the filtering-error covariance matrix.

Calculate the optimal estimation at time k:

$$\mathbf{G}_k = \begin{bmatrix} \mathbf{G}_k\mathbf{R}_k^{1/2} \end{bmatrix} \begin{bmatrix} \mathbf{R}_k^{1/2} \end{bmatrix}^{-1} \tag{13}$$

$$\mathbf{w}_{k|k} = \mathbf{w}_{k|k-1} + \mathbf{G}_k(\mathbf{y}_k - \mathbf{H}_k\mathbf{w}_{k|k-1}) \tag{14}$$

$$\mathbf{P}_{k|k} = \mathbf{P}_{k|k}^{1/2}\begin{bmatrix}\mathbf{P}_{k|k}^{1/2}\end{bmatrix}^T \tag{15}$$

where $\mathbf{w}_{k|k}$ is the optimal estimation of output weights, $\mathbf{P}_{k|k}$ is filtering-error covariance matrix at time k.

The state prediction of square root Kalman filter are shown as follows

$$\begin{aligned} \mathbf{w}_{k+1|k} &= \mathbf{A}_{k+1|k}\mathbf{w}_{k|k} \\ \mathbf{P}_{k+1|k} &= \begin{bmatrix} \mathbf{A}_{k+1|k}\mathbf{P}_{k|k}^{1/2} & \mathbf{Q}_{\omega,k}^{1/2} \end{bmatrix} \begin{bmatrix} \mathbf{P}_{k|k}^{1/2}\mathbf{A}_{k+1|k}^T \\ \mathbf{Q}_{\omega,k}^{T/2} \end{bmatrix}. \end{aligned} \tag{16}$$

3   Convergence of SRKF-ELM

In this section, we provide the convergence analysis of online training stage [1,9,10]. For the sake of convenience, $\hat{\mathbf{w}}_{k+1}$ represents $\mathbf{w}_{k+1|k+1}$, and $\mathbf{P}_k$ represents $\mathbf{P}_{k|k}$.

Define the estimation error of output weights at time $k + 1$:

$$\tilde{\mathbf{w}}_{k+1} = \mathbf{w}^* - \hat{\mathbf{w}}_{k+1} \tag{17}$$

where $\mathbf{w}^*$ is the optimal solution of output weights, $\hat{\mathbf{w}}_{k+1}$ is the estimation at time $k + 1$, then define the Lyapunov function:

$$V_{k+1} = \tilde{\mathbf{w}}_{k+1}^T\mathbf{P}_{k+1}^{-1}\tilde{\mathbf{w}}_{k+1}. \tag{18}$$

The convergence of SRKF-ELM could be proved only if $\lim_{k\to\infty} V_{k+1} = 0$. First, we prove $\{V_{k+1}\}$ is a decreasing order.

From Eq. (14) we have

$$\hat{\mathbf{w}}_{k+1} = \hat{\mathbf{w}}_k + \mathbf{G}_{k+1}\mathbf{e}_{k+1}. \tag{19}$$

Minus $\mathbf{w}^*$ on both sides, it yields

$$\tilde{\mathbf{w}}_{k+1} = \tilde{\mathbf{w}}_k + \mathbf{G}_{k+1}\mathbf{e}_{k+1} \tag{20}$$

Substituting (6)–(9) into (20), it yields

$$\mathbf{K}_{k+1} = \mathbf{P}_{k+1}\mathbf{H}_{k+1}\mathbf{R}_{k+1}^{-1} \tag{21}$$

$$\mathbf{P}_{k+1}^{-1} = \mathbf{P}_k^{-1} + \mathbf{H}_{k+1}\mathbf{R}_{k+1}^{-1}\mathbf{H}_{k+1}^T. \tag{22}$$

Substituting (19)–(22) into (18), it yields

$$\begin{aligned}
V_{k+1} &= (\tilde{\mathbf{w}}_k - \mathbf{P}_{k+1}\mathbf{H}_{k+1}\mathbf{R}_{k+1}^{-1}\mathbf{e}_{k+1})^T\mathbf{P}_{k+1}^{-1}(\tilde{\mathbf{w}}_k - \mathbf{P}_{k+1}\mathbf{H}_{k+1}\mathbf{R}_{k+1}^{-1}\mathbf{e}_{k+1}) \\
&= V_k + \mathbf{e}_{k+1}^T\mathbf{R}_{k+1}^{-1}(\mathbf{H}_{k+1}^T\mathbf{P}_k\mathbf{H}_{k+1}(\mathbf{R}_{k+1} + \mathbf{H}_{k+1}^T\mathbf{P}_k\mathbf{H}_{k+1})^{-1} - \mathbf{I})\mathbf{e}_{k+1} \tag{23}
\end{aligned}$$

Usually, we set $\mathbf{R}_{k+1} = \sigma^2\mathbf{I}$. V_k is a decreasing order only if $\varphi = \mathbf{H}_{k+1}^T\mathbf{P}_k\mathbf{H}_{k+1}(\mathbf{R}_{k+1} + \mathbf{H}_{k+1}^T\mathbf{P}_k\mathbf{H}_{k+1})^{-1} - \mathbf{I}$ is a negative matrix, and it obviously satisfies. According to the bounded monotonic principle, we have

$$\lim_{k \to \infty} V_{k+1} = V. \tag{24}$$

Lemma 1: For the proposed algorithm, given positive real number σ_1, σ_2 and positive integer n, for $\forall k > n$, if

$$0 < k\sigma_1 \leq \lambda(\mathbf{P}_k^{-1} - \mathbf{P}_0^{-1}) \leq k\sigma_2 \tag{25}$$

then

$$\lim_{k \to \infty} \lambda_{\min}(\mathbf{P}_k^{-1}) = \infty \tag{26}$$

$$\text{Sup} \lim_{k \to \infty} \frac{\lambda_{\max}(\mathbf{P}_k^{-1})}{\lambda_{\min}(\mathbf{P}_k^{-1})} < \infty. \tag{27}$$

In the hidden layer with sigmoid as the activation function, we have $\|\mathbf{H}_k\|_\infty < 1$, then the right side of (25) holds. When n is big enough, for $\forall k > n$,

$$\Omega = \mathbf{P}_k^{-1} - \mathbf{P}_0^{-1} = \mathbf{H}(1, k+1)^T\mathbf{R}(1, k+1)\mathbf{H}(1, k+1) \tag{28}$$

is non-singularly, it yields the left side of (25). According to Lemma 1, we have

$$\frac{V_k}{\text{tr}(\mathbf{P}_k^{-1})} \geq \frac{\lambda_{\min}(\mathbf{P}_k^{-1})\tilde{\mathbf{w}}_k^T\tilde{\mathbf{w}}_k}{d\lambda_{\max}(\mathbf{P}_k^{-1})} \geq 0 \tag{29}$$

$$\lim_{k \to \infty} \text{tr}(\mathbf{P}_k^{-1}) > \lim_{k \to \infty} \lambda_{\min}(\mathbf{P}_k^{-1}) = \infty \tag{30}$$

$$\lim_{k \to \infty} \frac{V_k}{\text{tr}(\mathbf{P}_k^{-1})} = \lim_{k \to \infty} \frac{\lambda_{\min}(\mathbf{P}_k^{-1})\tilde{\mathbf{w}}_k^T\tilde{\mathbf{w}}_k}{d\lambda_{\max}(\mathbf{P}_k^{-1})}. \tag{31}$$

According to (26), (31), it yields

$$\lim_{k \to \infty} \tilde{\mathbf{w}}_k = 0. \tag{32}$$

Therefore, the convergence of SRKF-ELM has been proved.   □

4   Experimental Evaluation

Simulations were performed on MATLAB 2014a, on a 64 bit Windows 7 systems, using an Intel core i3 CPU and 8 GB of RAM. We validate the proposed model on 3 artificial data sets and 3 real-life applications.

In the simulation of artificial data sets, there are 3000 samples in each data set and the first 1000 samples are used for pre-training. According to Takens' embedding theorem, the delayed times and embedding dimensions of Lorenz x and y time series are set as $\mathbf{m} = [2, 1]$, $\tau = [8, 8]$, the parameter of Rossler time series are set as $\mathbf{m} = [3, 5, 7]$, $\tau = [19, 13, 12]$, and the parameter of Mackey-Glass time series are set as $m = 17$, $\tau = 4$. The performance comparison on artificial data set is shown in Table 1. It points that although SRKF-ELM spend longer time training the model, it outperforms ELM and OS-ELM owing to the generalization performance of pre-training and the robustness of square root Kalman filter algorithm. In the mean time, we present experiments about predictive steps and hidden nodes to evaluate the performance of SRKF-ELM. Predictive steps are set from 1 to 20, and hidden nodes are set as $[0, 5, \ldots, 100]$, parameter C in pre-training stage is set as 10^6. Figure 1 shows the relationship between hidden nodes and prediction steps. It illustrates that SRKF-ELM performs best when the number of hidden nodes is around 30 regardless of the predictive steps.

Table 1. One-step-ahead performance on artificial data sets (mean of 50 times)

	ELM		OS-ELM		SRKF-ELM	
	RMSE	Time	RMSE	Time	RMSE	Time
Lorenz	3.42e−3	0.0462 s	1.67e−3	0.0505 s	**1.27e−3**	0.4615 s
Rossler	6.46e−3	0.1104 s	1.15e−2	0.1173 s	**2.08e−3**	0.4533 s
Mackey-Glass	4.46e−4	0.1257 s	4.87e−4	0.1345 s	**6.83e−5**	0.4549 s

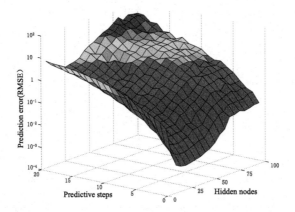

Fig. 1. The relationship between hidden nodes and predictive steps

We validate the SRKF-ELM in meteorological system applications to demonstrate the robustness of SRKF-ELM. The Melbourne weather data set consists of daily maximum temperature, daily minimum temperature and daily rainfall, including 3652 sampling points, from 1981–1990 (https://datamarket. com/data/). The delayed times and embedding dimensions are set as $\mathbf{m} = [8, 5, 3]$, $\tau = [1, 3, 3]$. And the prediction target is daily maximum temperature. The monthly runoff time series of Yangtze river data set includes 1380 sampling points, from 1865–1979. The delayed times and embedding dimensions are set as $m = 3$, $\tau = 4$. Shanghai monthly weather data set consists of monthly mean temperature and El Nino 3.4 index, including 774 samples, from 1950–2015, and the delayed times and embedding dimensions are set as $\mathbf{m} = [6, 5]$, $\tau = [4, 3]$. And El Nino 3.4 index is the prediction target. In all experiments, the first 500 samples are used for pre-training.

Table 2 shows the comparison between ELM, OS-ELM and SRKF-ELM. In the simulation of Melbourne weather data set, OS-ELM is so unstable that at some points the prediction error is infinite. Simultaneously, the predictive results of monthly runoff of Yangtze River are shown in Fig. 2. At 53th and 60th points, the OS-ELM failed to make the right predictions, but SRKF-ELM goes well. According to Table 2 and Fig. 2, it indicates that the SRKF-ELM is unsensitive to noises, and the generalization performance is better than OS-ELM.

Table 2. One-step-ahead performance in real-life applications (mean of 50 times)

RMSE (NRMSE)	ELM	OS-ELM	SRKF-ELM
Monthly runoff of Yangtze river	6.57e+3 (0.1040)	6.60e+3 (0.104)	**6.56e+3 (0.1040)**
Melbourne daily weather	**1.20 (0.0419)**	-	1.25 (0.0443)
ShangHai monthly weather	0.224 (0.0489)	0.236 (0.0515)	**0.217 (0.0473)**

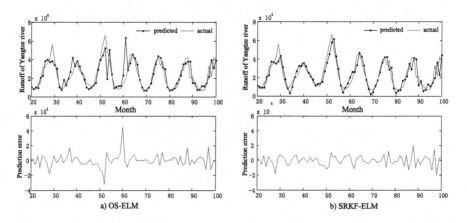

Fig. 2. The comparison of robustness between OS-ELM and SRKF-ELM on the monthly runoff time series of Yangtze river.

5   Conclusion

In this paper, we proposed a novel robust model for chaotic time series prediction. The model is initialized by orthogonal matrix and pre-trained by extreme learning machine, and square root Kalman filter is used to updated the output weights to obtain stable performance. The SRKF-ELM has following advantages: (1) after pre-trained by extreme learning machine, SRKF-ELM can adaptively tune weights when new samples arrive. (2) SRKF-ELM is more robust and performs similar or better on noisy data than classical algorithms. In the future, we will focus on the time series prediction application of deep extreme learning machine based on square root Kalman filter.

Acknowledgment. This paper is supported by the National Natural Science Foundation of China (61374154) and the Fundamental Research Funds for the Central Universities (DUT16RC(3)123).

References

1. Han, M., Wang, Y.N.: Multivariate time series online predictor with Kalman filter trained reservoir. Acta Automat. Sin. **36**(1), 169–173 (2010)
2. Lian, C., Zeng, Z.G., Yao, W., et al.: Ensemble of extreme learning machine for landslide displacement prediction based on time series analysis. Neural Comput. Appl. **24**(1), 99–107 (2014)
3. Ye, Y., Squartini, S., Piazza, F.: Online sequential extreme learning machine in nonstationary environments. Neurocomputing **116**, 94–101 (2013)
4. Wang, X.Y., Han, M.: Improved extreme learning machine for multivariate time series online sequential prediction. Eng. Appl. Artif. Intell. **40**, 28–36 (2015)
5. Zhang, C.Y., Chen, C.L.P., Gan, M., et al.: Predictive deep Boltzmann machine for multiperiod wind speed forecasting. IEEE Trans. Sustain. Energy **6**(4), 1416–1425 (2015)
6. Kasun, L.L.C., Zhou, H., Huang, G.B., et al.: Representational learning with ELMs for big data. IEEE Intell. Syst. **28**(6), 31–34 (2013)
7. Rong, H.J., Huang, G.B., Sundararajan, N., et al.: Online sequential fuzzy extreme learning machine for function approximation and classification problems. IEEE Trans. Syst. Man Cybern. Part B (Cybern.) **39**(4), 1067–1072 (2009)
8. Haykin, S.: Neural Networks and Learning Machines, 3rd edn. Prentice-Hall, Cambridge (2008)
9. Michel, A.N., Hou, L., Liu, D.R.: Stability of Dynamical Systems: Continuous, Discontinuous, and Discrete Systems. Springer Science Business Media, Boston (2007). doi:10.1007/978-0-8176-4649-3
10. Li, Y., Tan, Y., Dong, R., et al.: State estimation of macromotion positioning tables based on switching Kalman filter. IEEE Trans. Control Syst. Technol. **25**(3), 1076–1083 (2017)

Automatic Detection of Epileptic Seizures Based on Entropies and Extreme Learning Machine

Xiaolin Cheng, Meiling Xu, and Min Han[(✉)]

Faculty of Electronic Information Electrical Engineer,
Dalian University of Technology, Dalian 116024, China
minhan@dlut.edu.cn

Abstract. Epilepsy is a common neurological disease, and it is usually judged based on EEG signals. Automatic detection and classification of epileptic EEG gradually get more and more attention. In this work, we adopt two-step program to implement the automatic classification. Three entropies (approximate entropy, sample entropy, permutation entropy) are extracted as features to prepare for classifying. Then extreme learning machine is utilized to realize feature classification. Experimental results on Bonn epilepsy EEG dataset indicate that the proposed method is capable of recognizing normal, pre-ictal and ictal EEG with an accuracy of 99.31%, which is helpful for doctors to diagnose epilepsy disease.

Keywords: Entropy · Extreme learning machine · EEG signal classification

1 Introduction

Epilepsy disease is a common disease at nervous system, which affects about 1–2% of the population around the world. And there will appear 2.4 million new cases around the world yearly [1]. The direct reason of epileptic seizure is paradoxical discharge of the brain, and it has the characteristics of suddenness and transient dysfunction of the brain. Electroencephalogram (EEG) is the comprehensive reflection of the state of the electricity activity of cerebral tissue and brain function status. EEG signal is used for epileptic disease diagnosis, and automatic epileptic EEG signal classification is gradually becoming the focus of the study of many scholars [2]. Automatic epileptic EEG signal classification includes two parts: features are extracted from EEG signals firstly, and then the feature information is input into the classifier for classification. The classification results are used to analyze diseases for example epilepsy [3].

At present, main measures of EEG feature extraction include linear methods and nonlinear methods. Linear methods are based on linear models, including wavelet transform, autoregressive model and so on. Nonlinear methods are developed from nonlinear dynamics methods, which analyze signals according to nonlinear characteristics, including correlation dimension, entropy and so on. The performance of classifiers determines whether the classification result is good or not. Many machine

© Springer International Publishing AG 2017
D. Liu et al. (Eds.): ICONIP 2017, Part IV, LNCS 10637, pp. 410–418, 2017.
https://doi.org/10.1007/978-3-319-70093-9_43

learning models have been used in EEG signal classification, such as relevance vector machine (RVM) [4], support vector machine (SVM), Extreme learning machine (ELM) and so on.

This paper puts forward a technique for automatic detection of epileptic seizures based on entropies and extreme learning machine. Four entropies, approximate entropy, sample entropy, permutation entropy and fuzzy entropy, are extracted and then combined as different feature sets. Choose the best combination, and fed it to four different classifiers: ELM, SVM, K-nearest neighbor (KNN) and decision tree (DT). Then according to the evaluation indexes, choose the optimal combination.

The organization of this paper is as follows. Feature extraction method and classifiers will be described in Sects. 2 and 3, respectively. In Sect. 4, experiments are implemented for EEG classification. Finally, the study is concluded in Sect. 5.

2 Feature Extraction

Entropy is a kind of nonlinear dynamic indexes. In the field of epilepsy detection research, entropies serve as high-efficiency feature extraction methods.

2.1 Approximate Entropy

Approximate entropy is an estimation tool for evaluating the regularity of time series, and presents it with a nonnegative number. The more irregular the signal is, the greater the approximate entropy value is. The approximate entropy is calculated as follows.

Step 1. Embed the signal $X = [x(1), x(2), \ldots x(N)]$ into an m-dimension subspace, $X_m(i) = [x(i), x(i+1), \ldots, x(i+m-1)]$, $1 \leq i \leq N - m + 1$, m is chosen as 2 in this research.

Step 2. Calculate the scaling range $r = g \times SD$, g is the coefficients of tolerance and set as 0.2 in this work, SD is the standard deviation of the time series X.

Step 3. Define $d[X_m(i), X_m(j)]$ between $X_m(i)$ and $X_m(j)$ is the biggest absolute difference in corresponding element, $d[X_m(i), X_m(j)] = \max\limits_{k=0,\ldots,m-1} (|x(i+k) - x(j+k)|)$.

Step 4. Record the numbers of $d[X_m(i), X_m(j)] < r$ as C_i, and $C_i^m(r) = \frac{1}{N-m+1} C_i$.

Step 5. Define $\phi^m(r) = \frac{1}{N-m+1} \sum\limits_{i=1}^{N-m+1} \ln C_i^m(r)$.

Step 6. Increase the embedded dimension to $m + 1$, and calculate $\phi^{m+1}(r)$.

Step 7. The approximate entropy of the signal is calculated by

$$\text{ApEn}(m, r) = \lim_{N \to \infty} [\phi^m(r) - \phi^{m+1}(r)] \tag{1}$$

2.2   Sample Entropy

Sample entropy is expressing the complexity of a time series [5]. We can get the sample entropy estimation of a short data reliably, and it has good anti-interference ability and the ability to resist noise. Sample entropy algorithm is introduced as follows.

Step 1. Embed the signal X into an m-dimension subspace.
Step 2. Calculate the distance $d[X_m(i), X_m(j)]$.
Step 3. Define the similar tolerance $r = K \times SD$, K is defined as 0.2. Record all numbers of $d[X_m(i), X_m(j)] < r$ as B_i, then calculate $B_i^m(r) = \frac{1}{N-m}B_i$.
Step 4. Average all i, save as $B^m(r) = \frac{1}{N-m+1} \sum_{i=1}^{N-m+1} B_i^m(r)$.
Step 5. Calculate $B^{m+1}(r)$.
Step 6. Therefore, the value of the sample entropy can be calculated by

$$\text{SampEn}(m, r, N) = \lim_{N \to \infty} \left\{ -\ln \left[\frac{B^{m+1}(r)}{B^m(r)} \right] \right\} \tag{2}$$

2.3   Permutation Entropy

Permutation entropy is put forward by Bandt and Pompe and first used as a feature of EEG signal by Nicoletta and Georgiou [6]. It's an evaluation pattern of the complexity of time series, which will increase in the wake of the growth of irregularity. The calculation process of the permutation entropy is as follows.

Step 1. Embed the signal X into an m-dimension subspace, m is chosen as 3.
Step 2. Rank the elements of $X_m(i)$ according to the order since the childhood.
Step 3. For a certain m, confirm the $m!$ patterns.
Step 4. Count the number of subsequences corresponding each pattern, and calculate the ratio between each number and $N - m + 1$ as p_j.
Step 5. On the basis of the Shannon entropy, the permutation entropy of data X is

$$H_p(m) = -\sum_{j=1}^{J} p_j \ln(p_j) \tag{3}$$

where J is the total number of permutation patterns and $J \leq m!$.

2.4   Fuzzy Entropy

Compared to the sample entropy algorithm who is on the strength of two-valued function, the values of fuzzy entropy have continuity and smoothness [7]. The fuzzy entropy algorithm is introduced in detail as follows.

Step 1. Embed the signal X into an m-dimension subspace.

$$X_i^m = \{x(i), x(i+1), \cdots x(i+m-1)\} - x_0(i), \quad i = 1, 2, \cdots N - m + 1,$$
$$x_0(i) = \frac{1}{m} \sum_{j=0}^{m-1} x(i+j) \tag{4}$$

Step 2. Define the distance $d_{ij}^m = \max\limits_{k \in (0, m-1)} \{|(x(i+k) - x_0(i)) - (x(j+k)$ $- x_0(j))|\}, j \neq i.$

Step 3. Determine similar tolerance r, then define $D_{ij}^m = \exp\left(-\left(d_{ij}^m\right)^n / r\right).$

Step 4. Define the function $\phi^m(n, r) = \frac{1}{N-m} \sum\limits_{i=1}^{N-m} \left[\frac{1}{N-m-1} \sum\limits_{j=1, j \neq i}^{N-m} D_{ij}^m\right].$

Step 5. Calculate $\phi^{m+1}(n, r)$.

Step 6. The fuzzy entropy of the given signal can be calculated by

$$\text{FuzzyEn}(m, n, r, N) = \ln \phi^m(n, r) - \ln \phi^{m+1}(n, r) \tag{5}$$

In this work, the similar tolerance r is taken as 0.2 times standard deviation of data X, m and n are set to 3 and 2, respectively.

3 Classifier

Features obtained will be fed into four classifiers orderly. In this work, we adopt four classifiers which are widespread used in various classification studies.

3.1 Extreme Learning Machine

ELM proposed in recent years is a kind of feed forward neural network, which has been successfully employed in time series prediction, data classification, etc. In the process of training, the input weight vectors and hidden layer bias of extreme learning machine do not need to adjust, which overcomes the shortcomings in the traditional neural network, such as slow training speed, and easily falling into local optimum.

3.2 Support Vector Machine

SVM is a commonly used classifier in machine learning and a kind of kernel-based methods [8]. The main idea of SVM classifier is mapping the input data into a feature space, which is realized by kernel functions. We employ RBF and adopt cross-validation technique in the training process. It can find the best parameters automatically.

3.3   K-Nearest Neighbor Classifier

KNN is a nonparametric method and based on statistics. The working principle of KNN is that for a given new test data, which class it belongs to is decided by the most common class among its k nearest neighbors in feature space. In addition, KNN is sensitive to the dimension of input sample. It has good property when the dimension is low.

3.4   Decision Tree

DT is generally employed in working out classification problem in the field of machine learning and data mining [9]. It can handle missing data and has robustness to noise. Currently, the common algorithms for decision tree include CART, ID3, C4.5 and so on. Among them, the C4.5 algorithm is most popular and employed in this work.

4   Experimental Results

4.1   Data Prepare and Classification Evaluation Criteria

The EEG data of epilepsy research center at the University of Bonn contain five subsets (A-E), and each subset includes 100 section of EEG signals. And each section of EEG signals has 4097 data points. The sets of A, D, E are chosen for the experiment. Subset A is obtained from healthy volunteers. Subsets D records the activity of seizure free intervals. Subsets E is seizure activity. To construct more sample data, we cut the original signals and get a series of time sequences length of 1024, and each time series is regarded as a sample. Then we get a total of 1200 samples (400 in each category). 300 samples are chosen randomly for training in each category, and the rest for testing.

In order to compare the classification performance of different methods, we choose 8 evaluation indexes, TN (True Negative, number of non-seizure (normal condition) classified as non-seizure), FN (False Negative, number of seizure (pre-ictal or ictal condition) classified as non-seizure), TP (True Positive, number of seizure classified as seizure), FP (False Positive, number of non-seizure classified as seizure), Sensitivity (TP/(TP + FN)), Specificity (TN/(TN + FP)), PPV (Positive predictive value, TP/(TP + FP)), Accuracy (((TP + TN)/(TP + FN + TN + FP))) [10].

4.2   Simulation Results

In this work, we do four different groups of classification experiments. They are A\D, A\E, D\E, A\D\E, respectively. All the results show next are the mean values of the 50 times simulations.

Firstly, calculate approximate entropy, sample entropy, permutation entropy and fuzzy entropy of A, D, E signal sets. The obtained boxplot are shown in Fig. 1. We can easily find approximate entropy of group A is very different from group D and group E, as well as sample entropy. Hence they are conducive to distinguish normal EEG from pre-ictal EEG or epileptic EEG. Permutation entropy of group E is greatly distinguishing from group A and group D, as well as fuzzy entropy. Hence, they are efficient measures of classifying epileptic EEG against pre-ictal EEG or normal EEG signals.

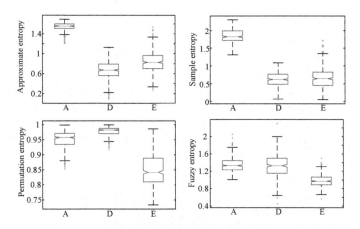

Fig. 1. Boxplot of groups A, D, E.

Table 1 expresses the performance comparison of feature extraction methods based on different classification tasks. The feature extraction methods are autoregressive model (AR), wavelet transform (WT), wavelet packet decomposition (WPD), FPS_En (the combination of fuzzy entropy, permutation entropy and sample entropy), AFS_En (the combination of approximate entropy, fuzzy entropy and sample entropy), APS_En (the combination of approximate entropy, permutation entropy and sample entropy), respectively. The classifier used here is ELM. It is obvious that classification accuracies based on APS_En are always highest in the four groups of classification experiments. This indicates APS_En can extract more abundant and overall information of original signals than other feature extraction methods. In addition, the classification accuracies of A\D and A\E are 100%, which means by using the features extracted by APS_En method, we can distinguish normal EEG signals and pre-ictal EEG signals with 100% accuracy, as well as normal EEG signals and ictal EEG signals.

The features extracted are fed into ELM, SVM, KNN and DT, respectively. Figure 2 is the histogram of A/D/E classification accuracies based on different feature extraction methods and classifiers. We can see that the features obtained by APS_En method always get the highest classification accuracy, even though with different classifiers, which illustrates APS_En method possesses the best performance in feature extraction. In addition, from the perspective of classifier, KNN gets the lowest classification accuracy under WT, WPD, FPS_En feature extraction methods, and DT gets the lowest classification accuracy under AR, AFS_En and APS_En feature extraction methods. ELM and SVM performs well under all kinds of feature extraction methods, which means they have strong robustness. On the whole, ELM is superior to the other three classifiers under different feature extraction methods.

Table 2 shows classification performance evaluation results of A/D/E classification tasks based on different feature extraction methods, and the classifier used in this contrast experiment is ELM. It's clear that features extracted based on APS_En method get the best classification performance, the 8 evaluation indexes all get the highest values. It proves that the data get by APS_En can extract overall effective information of EEG signals and is advantageous to automatic classification of the classifier.

Table 1. Comparison of feature extraction methods based on different classification tasks.

Feature extraction	A/D	A/E	D/E	A/D/E
AR	97.83%	98.89%	94.88%	95.49%
WT	92.58%	92.01%	88.77%	91.34%
WPD	97.40%	99.38%	98.69%	96.92%
FPS_En	100%	99.00%	95.37%	97.20%
AFS_En	100%	99.51%	97.56%	97.64%
APS_En	**100%**	**100%**	**98.98%**	**99.31%**

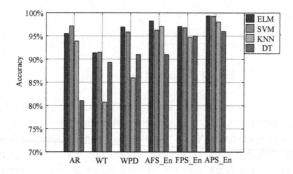

Fig. 2. The histogram of classification accuracies based on different feature extraction methods and classifiers.

Table 2. Classification performance evaluation based on different feature extraction methods.

	TN	FN	TP	FP	Sensitivity	Specificity	PPV	Accuracy
AR	99.95	13.48	186.52	0.05	99.98%	88.14%	93.26%	95.49%
WT	94.16	17.78	182.22	5.84	96.90%	84.19%	91.11%	92.13%
WPD	98.37	7.59	192.41	1.63	99.16%	92.85%	96.20%	96.73%
FPS_En	99.96	8.36	191.64	0.04	99.98%	92.29%	95.82%	97.20%
AFS_En	100	7.08	192.92	0	100%	93.39%	96.46%	97.64%
APS_En	**100**	**2.2**	**197.8**	**0**	**100%**	**97.89%**	**99.28%**	**99.31%**

Table 3 expresses the classification performance of different classifiers in the A/D/E classification tasks, and the feature extraction method used here is APS_En. On the whole, the ELM and SVM perform well than KNN and DT under the 8 evaluation indexes. It's obvious that among these classifiers, ELM is the best while DT is the worst, and the values of sensitivity, PPV and accuracy of ELM are higher than the other three classifiers. The TN value of ELM is 100, and the FP value is 0, that means ELM gets 100% accuracy in the task of recognizing normal EEG signals from normal condition, pre-ictal condition and ictal condition. This is a great help to identify whether the patient is suffering from epilepsy in clinical practice. And it also gets 99.31% accuracy in identifying the three kinds of EEG signals to their correct categories, which is a very reliable precision that can help doctors make clinical decisions.

Table 3. Classification performance evaluation of different classifiers

	TN	FN	TP	FP	Sensitivity	Specificity	PPV	Accuracy
ELM	**100**	**2.2**	**197.8**	**0**	**100%**	**97.89%**	**99.28%**	**99.31%**
SVM	99.8	2	198	0.2	99.90%	98.05%	99.00%	99.27%
KNN	100	6	194	0	100%	94.37%	97.00%	98.00%
DT	99.6	10.6	189.4	0.4	99.79%	90.49%	94.70%	96.33%

Based on the above results and analysis, APS_En is determined as the feature extraction method, and ELM is chosen as the classifier in this work. The proposed method gets the classification accuracy of 99.31% and experimental variance of 6.39E-07 in real epileptic EEG datasets (A\D\E), which shows the method has strong credibility in practical application, and it can accurately accomplish the classification of normal, pre-ictal and epileptic EEG signals, which can help the doctor for auxiliary decision-making of diagnosing the patient condition.

5 Conclusion

This paper realizes automatic classification of epileptic seizures by means of a combination entropies method and ELM classifier. Four entropies are combined as different features, APS_En who possesses the best superiority of identification ability of various states EEG is chosen as the feature extraction method. Among ELM, SVM, KNN and DT, the ELM classifier possesses the greatest satisfactory accuracy and stability of the classification. Experiment results and analysis prove that the proposed method for EEG signals classification is effective and efficient, which is helpful for doctors to diagnose epilepsy disease.

Acknowledgments. This work was supported by the project (61374154) of the National Nature Science Foundation of China and the Fundamental Research Funds for the Central Universities (DUT16RC (3)123).

References

1. Zhou, W., Liu, Y., Yuan, Q., Li, X.: Epileptic seizure detection using lacunarity and bayesian linear discriminant analysis in intracranial EEG. IEEE Trans. Biomed. Eng. **60**(12), 3375–3381 (2013)
2. Coito, A., Michel, C.M., Mierlo, P.: Directed functional brain connectivity based on EEG source imaging: methodology and application to temporal lobe epilepsy. IEEE Trans. Biomed. Eng. **63**(12), 2619–2628 (2016)
3. Salam, M.T., Velazquez, J.L.P., Genov, R.: Seizure suppression efficacy of closed-loop versus open-loop deep brain stimulation in a rodent model of epilepsy. IEEE Trans. Neural Syst. Rehabil. Eng. **24**(6), 710–719 (2016)
4. Han, M., Sun, L.: EEG signal classification for epilepsy diagnosis based on AR model and RVM. In: IEEE ICICIP, pp. 134–139 (2010)

5. Zhang, C., Wang, H., Fu, R.: Automated detection of driver fatigue based on entropy and complexity measures. IEEE Trans. Intel. Transp. Syst. **15**(1), 168–177 (2014)
6. Nicolaou, N., Georgiou, J.: Detection of epileptic electroencephalogram based on permutation entropy and support vector machines. Expert Syst. Appl. **39**(1), 202–209 (2012)
7. Cao, Z., Lin, C.T.: Inherent fuzzy entropy for the improvement of EEG complexity evaluation. IEEE Trans. Fuzzy Syst. (2017). doi:10.1109/TFUZZ.2017.2666789
8. Ghamisi, P., Couceiro, M.S., Benediktsson, J.A.: A novel feature selection approach based on FODPSO and SVM. IEEE Trans. Geosci. Remote Sens. **53**(5), 2935–2947 (2015)
9. Jindal, A., Dua, A., Kaur, K.: Decision tree and SVM-based data analytics for theft detection in smart grid. IEEE Trans. Industr. Inf. **12**(3), 1005–1016 (2016)
10. Acharya, U.R., Molinari, F., Sree, S.V., Chattopadhyay, S., Ng, K.H., Suri, J.S.: Automated diagnosis of epileptic EEG using entropies. Biomed. Signal Process. Control **7**(4), 401–408 (2012)

Community Detection in Networks by Using Multiobjective Membrane Algorithm

Chuang Liu[✉], Linan Fan, Liangjie Li, Zhou Liu, Xiang Dai, and Wei Gao

School of Information Engineering, Shenyang University, Liaoning 110044, China
chuang.liu@mail.dlut.edu.cn

Abstract. This paper introduces a multi-objective optimization idea to solve the community detection. First, the problem of community detection is transformed into complex multi-objective optimization problem. Second, an evolutionary multi-objective membrane algorithm is proposed for discovering community structure. Finally, the proposed algorithm is conducted on the synthetic networks, and the experimental results demonstrate that our algorithm is effective and promising, and it can detect communities more accurately compared with PSO and GSA.

Keywords: Community detection · Evolutionary multiobjective optimization · Multiobjective membrane algorithm

1 Introduction

Recently, the analysis of complex networks from real-world systems has attracted more and more attention. The community structure is the basic theory of the important characteristics for the analysis of complex networks, such as the Internet of things, the big data, the social network and so on. It has a very close relationship with the nodes within the same structure, and is relatively sparse relationship between the nodes within the different structure [1]. Revealing the community structure of the network, including more deeply understanding the network function, finding the network potential model, predicting network behavior and so on, is of great significance [2].

At present, the common method is to convert the problem of community detection into optimization problem [3]. Evolutionary algorithms are more and more applied to the field, but most of the algorithms are to optimize a single target. Although the single-objective community partitioning algorithm has been successful in theory and application, there are some problems such as high complexity and limitation of solution [4]. In order to solve these problems, a natural way is to divide the community problem as a multi-objective optimization problem.

This paper studies a multi-objective membrane algorithm for finding community structures in complex networks. Firstly, two objective functions including modularity Q values and normalized mutual information are selected, which can identify the group of nodes with close internal relations and sparsely connected

© Springer International Publishing AG 2017
D. Liu et al. (Eds.): ICONIP 2017, Part IV, LNCS 10637, pp. 419–428, 2017.
https://doi.org/10.1007/978-3-319-70093-9_44

with each other. Secondly, after the multi-objective membrane algorithm returns a non-dominated solution of a set of two objective functions, the most appropriate solution is selected by means of modularity and normalization mutual information. The algorithm can discover the hierarchical structure of the network in which the deeper solution of a larger number of communities is included in the solution with a smaller number of communities. The number of communities automatically depends on the objective function for better trade-offs. Finally, the experimental comparison between the simulation network shows that the algorithm can successfully discover the complex network community structure, and has certain competitiveness compared with other algorithms at present.

2   Related Work

2.1   Membrane System

As a new branch of natural computing, membrane computing is a distributed and parallel computing theory by imitating the structure and functioning of the living cell, which is proposed by Păun from the European Academy of Sciences in 1998 [5]. It is an abstract computational idea or model, which is inspired by the structure and functioning of living cells, such as processing chemical compounds in tissues or higher order structures. The device computational model of membrane computing is called as membrane system or P system. A membrane system, as we will see later on, is a distributed and parallel theoretical computing device and whose aim is mimicking the inner mechanism of the living cell [6,7]. Most of these membrane systems could tackle specific problems (i.e., optimization problems) in a feasible time. At presents, applications of the membrane system have been expanded to various fields, e.g., medicine, biology, linguistics and computer science [8–11].

In a Membrane System, three basic elements consists of a membrane structure, multisets of symbol-objects, and reaction rules. A membrane structure is a hierarchically arranged set of membranes, as shown in Fig. 1.

Figure 1 shows the generic structure of a membrane system. The concept description of the membrane system is elaborated in the following sections. The hierarchical structure denotes that a skin membrane contains several membranes. The external membrane is usually called the skin membrane. And the several internal membranes are corresponded to the membranes and the elementary membrane. Each membrane determines a compartment, called a region. If a membrane has not any other membrane in its region, it is said to be elementary membrane.

The region of a membrane may contain some multisets of symbol-objects abstracted by molecular compounds and reaction rules inspired by other biological processes. The reaction rules may evolve multisets of symbol-objects and move them from a region to a neighboring one.

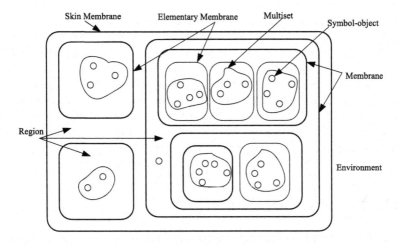

Fig. 1. The generic structure of membrane computing.

2.2   Membrane Algorithm

The membrane system in optimization is called as membrane algorithm. Membrane algorithm provides the opportunity to apply evolutionary algorithms or swarm intelligence in a parallel and distributed environment [12]. The study on membrane algorithm has been an increasing interest, most of them will be discussed as follows. Nishida proposed a compound membrane algorithm to solve the traveling salesman problem [13]. Zhang et al. proposed a membrane-inspired approximate algorithm for traveling salesman problems, which implemented the rules of Ant Colony Optimization in membrane systems. Huang et al. presented an optimization algorithm inspired by the membrane system to solve optimization problems [14]. Zhang et al. proposed a hybrid algorithm based on the quantum-inspired evolutionary approach and membrane systems to solve a well-known combinatorial optimization the knapsack problem [15]. Huang et al. presented a dynamic multi-objective optimization algorithm [16], which is inspired by membrane systems. Simulation results verify the effectiveness of the algorithm. Buiu et al. proposed a membrane controller based on membrane systems for mobile robots [17]. Liu et al. presented a novel algorithm based on the membrane system for solving multi-objective optimization problems. The proposed algorithm could quickly obtain the approximate Pareto front and satisfy the requirement of diversity of Pareto front [18]. Zhang et al. proposes a novel way to design a membrane system for directly obtaining the approximate solutions of combinatorial optimization problems without the aid of evolutionary operators [19]. Extensive experiments on knapsack problems have been reported to experimentally prove the viability and effectiveness of the proposed neural system. Liu and Fan proposes an evolutionary membrane algorithm to solve the dynamic or uncertain optimization problems. Experimental study is conducted based on the moving peaks benchmark. The results indicate the proposed algorithm is effective to solve the dynamic or uncertain optimization problems [20]. Xiao et al. proposed an improved dynamic membrane evolutionary algorithm based on particle swarm optimization and differential solution to solve constrained engineering design problems. The simulation experiments show that the proposed algorithm outperforms other state-of-the-art-algorithms [21]. Liu and Fan proposed a hybrid algorithm to solve the single objective real-parameter numerical optimization problems. The proposed algorithm consists of tissue membrane systems and the evolution strategy with covariance matrix adaptation algorithm. Numerical results show that the proposed algorithm has a very good performance on thirty benchmark functions on the CEC14 test suite [22].

3   The Proposed Algorithm

The complex network community detection can be treated as an optimization problem. Considering the advantages of multi-objective optimization algorithm in dealing with the problem, the complex network community detection is regarded as a multi-objective optimization problem, and two objective functions are constructed. The optimal compromise of the function is used to solve the optimal partitioning result.

The description of the proposed multiobjective membrane algorithm is given as follows. The proposed algorithm is based on the membrane system which consists of a membrane structure, multisets of symbol-objects, reaction rules. The network structure includes several of elementary membranes. Each elementary membrane has its own symbol-objects. These symbol-objects can be evolved by invoking the related reaction rules. After the appointed generation, these elementary membranes communicate with each other by using symbol-objects through protein channels. In the proposed algorithm, the symbol-objects in each elementary membrane represent the candidate partition of networks. The reaction rules are designed to evolve symbol-objects in different elementary membranes. Reaction rules consists of evolutionary operators and communication rules of membrane systems. These features are very useful to develop a new multiobjective membrane algorithm to improve its solving performance. Additionally, we also use an archive (or elitism) strategy that stores the Pareto optimal symbol-objects found from the different elementary membranes. Such archive strategy is an elitist mechanism most commonly adopted in MOEAs, and it helps symbol-objects to move towards the true Pareto front of MOPs.

The flowchart of the proposed multiobjecitve membrane algorithm is pictorially shown as Fig. 2.

3.1   Objects

In membrane systems, an object is an abstract representation of the atom, molecule and other chemical substance. For the community detection problem, the object represents a partition of a complex network. The object is encoded by

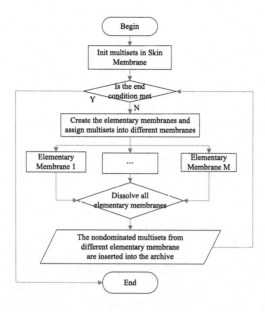

Fig. 2. The flowchart of MOMA

using string-based representation schema. According to the number of the communities of the networks, each object can be expressed as a set of real integer values.

3.2 Objective Functions

The first objective function is modularity and its function is shown in Eq. 1. The modularity Q values can be calculated according to the difference between the actual fraction of edges within communities and the value obtained by experimental algorithms. A higher modularity Q value has, a better community structure is attained by a network partition of a given network. If Q equals to 1, a network partition represents a good community structure.

$$Q = \sum_{i=1}^{N_c} (\frac{l_i}{2m} - (\frac{d_i}{2m})^2) \tag{1}$$

where l_i is the total number of links in the community i, m represents the edges of the networks, N_c represents the number of communities of the networks, d_i is the total degree of vertices in the community i.

In addition to the objective function, normalized mutual information (NMI) is a similarity measure estimating the similarity between the detected partitions and the true ones. A higher NMI value represents a greater similarity between two partitions. If NMI takes its maximum value which is equal to 1, all communities obtained by the experimental algorithms are identical to all real communities. In the following experiment, NMI is used to evaluate the results between the true partition and the partition obtained by experimental algorithms. The definition of NMI(A, B) is shown in Eq. 2.

$$NMI(A,B) = \frac{-2\sum_{i=1}^{C_A}\sum_{j=1}^{C_B} D_{ij}\log(\frac{D_{ij}N}{D_i \cdot D_j})}{\sum_{i=1}^{C_A} D_i \cdot \log(\frac{D_i}{N}) + \sum_{j=1}^{C_B} D_j \cdot log(\frac{D_j}{N})} \tag{2}$$

where A and B are partitions of a network, and C_A represents the number of communities in A while C_B denotes that of B. D is a confusion matrix, and $D_{i,j}$ stands for the number of nodes in community i of A that also appear in community j of B. N is the number of elements. D_i is the sum over row i of D while D_j is the sum of elements in column j.

3.3   Membrane Structure

The structure in MOMA consists of several elementary membranes. In the experiments, the number of elementary membranes is set to 4. Each elementary membrane can be seen as an evolution unit. In the proposed algorithm, the initialization of symbol-objects is implemented in each elementary membrane. The initialized symbol-objects are evolved in the inner region of own elementary membrane according to executing the reacting rules. After executing several generation, some good symbol-objects can be generated by executing reaction rules in the different elementary membranes. These good symbol-objects are sent into the common environments of all elementary membranes. The best object can be found by comparing the fitness of these symbol-objects. The worst object in each elementary membrane is replaced with the best object from the common environments, which helps other symbol-objects to move toward the direction of the global optimal partition of networks.

The structure of MOMA is benefit to improve the search efficiency of the proposed algorithm, which is suit to solve the community detection problems.

3.4   Reaction Rules

The rewrite_rule (Given a specific rule, $[s_j \rightarrow s'_j]i$) is applied to evolve the symbol-objects which could be found by using select_rule in the i-th elementary membrane. Under its role, a new symbol-object may be produced on the basis of the current symbol-objects. It can improve the algorithm in the aspect of exploiting and exploring the capability of the solution space.

Two rewriting rules, which simulate the process of the molecular motion, are designed to evolve the symbol-objects. The rewriting rules, including Eqs. 3 and 4, incorporate the Gaussian distribution function because the randomness of the molecular motion can be described by the Gaussian process. The detail of the rewriting rules can be described as follows:

$$S_{i,j} = r * (2 * S_{a,j} - S_{i,j}) + (1 - r) * S_{g,j} \tag{3}$$
$$S_{i,j} = (1 + 2 * r) * S_{g,j} - 2 * r * S_{i,j}) - S_{a,j} \tag{4}$$

where $S_{i,j}$ is the j-th dimension of the i-th symbol-object. r is a Gaussian distribution function. $S_{a,j}$ is the j-th dimension of the balanced symbol-object. $S_{g,j}$ is the j-th dimension of the intermediate symbol-object.

Communication rules not only send the best object from the different elementary membranes to the common environments of all elementary membranes, but send the common best object to the different elementary membranes. The rules is benefit to share the information of candidate solutions among elementary membranes.

3.5 Archive Strategy

An archive with the fixed size in the Skin Membrane is implemented in MOMA to record the best symbol-objects found from the different elementary membranes. And these symbol-objects in the archive serve as potential candidate solutions. After each iterative loop was executed, symbol-objects from the elementary membrane are compared with symbol-objects in the archive. If the symbol-object is not dominated by any symbol-objects in the archive, they are inserted into the archive. Similarly, if any symbol-objects in the archive is dominated by a new symbol-object, it is replaced by the new symbol-object. When the size of the archive exceeds the threshold (its value is set to 100 in the experiments), the mechanism of the crowding distance will be employed to remove the symbol-objects having the minimum of crowding distance. Finally, the archive is updated and the multiset consisted of the symbol-objects are formed in the next generation. The archive can improve the convergence speed to the known Pareto front and increase the number of the excellent symbol-objects.

4 Experiments

The performance of the proposed algorithm is evaluated on the synthetic network datasets. In simulation, some compared algorithms are chosen, including GSA [23], and PSO [24]. Section 4.1 will describe the artificial network datasets and the experimental condition when the simulation is run. Section 4.2 will discuss the experimental results on different algorithms.

4.1 Artificial Network Datasets

The artificial network datasets, which were proposed by Girvan and Newman, have been widely used to evaluate the performance of evolutionary algorithms in detecting network community structure [25]. The network randomly generated has 128 vertices, which is partitioned by four communities each with 32 vertices. Each vertex has an average degree of 16, $Z_{in} + Z_{out} = 16$. Z_{in} is a internal degree from a vertex connected to other vertex in the same community. Z_{out} is a external degree from a vertex connected to other vertex in the other community. With the Z_{out} increased, the community structure of the network become more and more noisy and fuzzy. In this experiment, the range of Z_{out} is from 1 to 10.

All experiments were run in Windows 7 enterprise version under the hardware environment of Intel Pentium dual-core 2.93 GHZ and 16 GB RAM. The proposed algorithm is implemented using matlab2015. The maximum evaluated times is set to 100 on each run. To evaluate the statistical performance of algorithms and reduce statistical errors, each network is evaluated independently at 20 times. Moreover, five statistical metrics are designed, such as Mean, Std, Worst, Best, and Num. These metrics can be employed to evaluate the solving performance of these various algorithms.

4.2   Comparison of the Proposed Algorithm with Other Algorithms

In this section, the proposed algorithms are compared with other algorithms for artificial networks. In the Newman benchmark of artificial networks, each network includes 128 nodes put into 4 communities with 32 nodes each. With the increase of Z_{out}, the true community detection of the generated networks is more hard to be found by GSA [23], PSO [24], respectively. In particular, with Z_{out} reached to 10, the proposed MOMA are also difficult to find the true community detection.

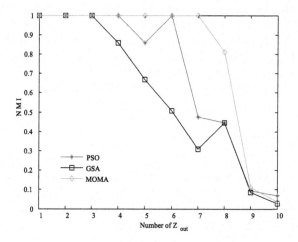

Fig. 3. Comparison our algorithm with the other experimental algorithms on the Newman benchmark in terms of the NMI.

Figure 3 shows the results of NMI attained by each algorithm. As we can see, the proposed algorithm outperforms the other algorithms in terms of NMI on the Newman benchmark. Especially, when the number of z_{out} is greater than 3, the proposed algorithm has better NMI than GSA. When the number of z_{out} is greater than 4, the proposed algorithm has better NMI than PSO. Therefore, the superiority of the propose algorithm becomes more and more significant when the number of z_{out} becomes larger and larger.

5   Conclusion

In this paper, we propose a multiobjective membrane algorithm called MOMA, to simultaneously optimize the modularity values and normalized mutual information, for solving the community detection problems which do not require any prior knowledge about the number of communities in complex networks. The simultaneous optimization of these two contradictory objects returns a set of tradeoff solutions between the objectives. Each object of MOMA represents a candidate partition of a complex network. After some generation of each membrane is executed, the optimal object are sent to the common environment, which achieve the information exchange from the different membranes. The experimental results show MOMA is effective and efficient at detecting community structure in complex networks. Therefore, the proposed algorithm is a viable method to solve community detection.

Acknowledgments. This project was supported by Shenyang Science and Technology Program (Grant No. 17-175-3-00).

References

1. Gong, M., Fu, B., Jiao, L., Du, H.: Memetic algorithm for community detection in networks. Phys. Rev. E **84**(5), 056101 (2011)
2. Cai, Q., Ma, L., Gong, M., Tian, D.: A survey on network community detection based on evolutionary computation. Int. J. Bio-inspired Comput. **8**(2), 84–98 (2016)
3. Atay, Y., Koc, I., Babaoglu, I., Kodaz, H.: Community detection from biological and social networks: a comparative analysis of metaheuristic algorithms. Appl. Soft Comput. **50**, 194–211 (2017)
4. Pizzuti, C.: A multiobjective genetic algorithm to find communities in complex networks. IEEE Trans. Evol. Comput. **16**(3), 418–430 (2012)
5. Paun, G.: Computing with membranes. J. Comput. Syst. Sci. **61**(1), 108–143 (2000)
6. Paun, G., Rozenberg, G.: A guide to membrane computing. Theoret. Comput. Sci. **287**(1), 73–100 (2002)
7. Paun, G., Rozenberg, G., Salomaa, A.: The Oxford Handbook of Membrane Computing. Oxford University Press, Inc., Oxford (2010)
8. Ciobanu, G., Paun, G., Paerez-Jimaenez, M.J.: Applications of Membrane Computing. Springer, Heidelberg (2006). doi:10.1007/3-540-29937-8
9. Cecilia, J.M., Garcia, J.M., Guerrero, G.D., Martinez-del Amor, M.A., Perez-Jimenez, M.J., Ujaldon, M.: The GPU on the simulation of cellular computing models. Soft. Comput. **16**(2), 231–246 (2012)
10. Pan, L., Martin-Vide, C.: Solving multidimensional 0–1 knapsack problem by P systems with input and active membranes. J. Parallel Distrib. Comput. **65**(12), 1578–1584 (2005)
11. Liu, C., Chen, D., Wan, F.: Multiobjective learning algorithm based on membrane systems for optimizing the parameters of extreme learning machine. Optik - Int. J. Light Electron Opt. **127**(4), 1909–1917 (2015)
12. Singh, G., Deep, K., Nagar, A.K.: Cell-like P-systems based on rules of particle swarm optimization. Appl. Math. Comput. **246**, 546–560 (2014)

13. Nishida, T.Y.: An approximate algorithm for NP-complete optimization problems exploiting P systems. In: Proceedings of Brainstorming Workshop on Uncertainty in Membrane Computing, pp. 185–192 (2004)

14. Zhang, Y., Huang, L.: A variant of P systems for optimization. Neurocomputing **72**(4), 1355–1360 (2009)

15. Zhang, G., Gheorghe, M., Wu, C.: A quantum-inspired evolutionary algorithm based on P systems for knapsack problem. Fundam. Inform. **87**(1), 93–116 (2008)

16. Huang, L., Suh, I.H., Abraham, A.: Dynamic multi-objective optimization based on membrane computing for control of time-varying unstable plants. Inf. Sci. **181**(11), 2370–2391 (2011)

17. Buiu, C., Vasile, C., Arsene, O.: Development of membrane controllers for mobile robots. Inf. Sci. **187**, 33–51 (2012)

18. Liu, C., Han, M., Wang, X.: A multi-objective evolutionary algorithm based on membrane systems. In: 2011 Fourth International Workshop on Advanced Computational Intelligence (IWACI), pp. 103–109. IEEE (2011)

19. Zhang, G., Rong, H., Neri, F., Perez-Jimenez, M.J.: An optimization spiking neural P system for approximately solving combinatorial optimization problems. Int. J. Neural Syst. **24**(05), 1440006 (2014)

20. Liu, C., Fan, L.: Evolutionary algorithm based on dynamical structure of membrane systems in uncertain environments. Int. J. Biomath. **9**(02), 1650017 (2016)

21. Xiao, J., He, J.J., Chen, P., Niu, Y.Y.: An improved dynamic membrane evolutionary algorithm for constrained engineering design problems. Natural Comput. 1–11 (2016)

22. Liu, C., Fan, L.: A hybrid evolutionary algorithm based on tissue membrane systems and CMA-ES for solving numerical optimization problems. Knowl.-Based Syst. **105**, 38–47 (2016)

23. Rashedi, E., Nezamabadi-Pour, H., Saryazdi, S.: GSA: a gravitational search algorithm. Inf. Sci. **179**(13), 2232–2248 (2009)

24. Liang, J.J., Qin, A.K., Suganthan, P.N., Baskar, S.: Comprehensive learning particle swarm optimizer for global optimization of multimodal functions. IEEE Trans. Evol. Comput. **10**(3), 281–295 (2006)

25. Girvan, M., Newman, M.E.: Community structure in social and biological networks. Proc. Nat. Acad. Sci. **99**(12), 7821–7826 (2002)

Double-Coding Density Sensitive Hashing

Xiaoliang Tang$^{(\boxtimes)}$, Xing Wang, Di Jia, Weidong Song,
and Xiangfu Meng

Liaoning Technical University, Huludao 125105, China
xiaoliang@mail.dlut.edu.cn

Abstract. This paper proposes a double-coding density sensitive hashing (DCDSH) method. DCDSH accomplishes approximate nearest neighbor (ANN) search tasks based on its double coding scheme. First, DCDSH generates real-valued hash codes by projecting objects along the principle hyper-planes. These hyper-planes are determined by principle distributions and geometric structures of data set. Second, DCDSH derives binary hash codes based on these real-valued hash codes. Real-valued hash codes can avoid undesirable partition of objects in low density areas and effectively improve representation capability and discriminating power. Binary codes contribute to query speed owing to the low complexity for computing hamming distance. DCDSH integrates the advantages of these two kinds of hash codes. Experimental results on large scale high dimensional data show that the proposed DCDSH exhibits superior performance compared to several state-of-the-art hashing methods.

Keywords: Double coding · Density sensitive hashing · Real-valued codes

1 First Section

Finding nearest neighbors is a fundamental step in many machine learning algorithms such as kernel density estimation, spectral clustering, manifold learning, and semi-supervised learning [1–3]. Exhaustively comparing the query with each sample in the database is infeasible because linear complexity is not scalable in practical settings. Besides the scalability issue, most searching tasks also suffer from the curse of dimensionality. Therefore, beyond the infeasibility of exhaustive search, storage of the original data also becomes a critical bottleneck [4].

To bypass the difficulty of finding exact query answers in high-dimensional space, the approximate version of the problem, called the Approximate Nearest Neighbor (ANN) search, has attracted extensive studies. During the past decade, hashing based ANN search methods [4] received considerable attentions.

Different hashing algorithms partition feature space based on different criterion. One of the most popular hashing algorithms is Locality Sensitive Hashing (LSH) [5]. LSH is fundamentally based on random projection, i.e., it uses random projection to partition feature space. LSH needs to use many random vectors to generate the hash tables with long codes, leading to a large storage space and a high computational cost.

Different from all the existing random projection based hashing methods, DSH tries to utilize the geometric structure of the data to guide the projections (hash tables)

© Springer International Publishing AG 2017
D. Liu et al. (Eds.): ICONIP 2017, Part IV, LNCS 10637, pp. 429–438, 2017.
https://doi.org/10.1007/978-3-319-70093-9_45

selection. Specifically, DSH uses *k-means* to roughly partition the data set into k groups. Then for each pair of adjacent groups, DSH generates one projection vector which can well split the two corresponding groups. From all the generated projections, DSH select the final ones according to the maximum entropy principle, in order to maximize the information provided by each bit.

Although DSH achieves superior performance over state-of-the-art hashing methods, there still exists some further development space for DSH. Binary hash codes can significantly reduce computational complexity in the query process. However, the representation capability of binary hash codes is still much lower than original feature vectors. The low representation capability of binary hash codes is the limitation of query accuracy not only for DSH but also for many other kinds of hashing methods that utilize binary hash codes. On the other hand, the projections of DSH [6] are obtained based on *k-means* method which reveals structural information of data set in real-valued features. If we introduce real-valued codes into DSH, the discriminating power will be improved.

In order to further improve the performance of hashing based ANN methods, we propose a novel hashing method called double-coding density sensitive hashing (DCDSH) method for effective high dimensional nearest neighbor search. Our proposed method DCDSH can be regarded as an extension of DSH. Different from DSH and other existing hashing methods, DCDSH relies on a double coding scheme. Specifically, first, DCDSH generates real-valued hash codes by projecting objects along the hyper-planes that are determined by principle distributions and geometric structures; second, DCDSH derives binary hash codes based on these real-valued hash codes. Experimental results show the superior performance of the proposed DCDSH algorithm over the existing state-of-the-art approaches.

The remainder of this paper is organized as follows. Section 2 presents the proposed DCDSH method. Section 3 provides the experimental results that compared our algorithm with the state-of-the-art hashing methods on two real world large scale data sets. Conclusion is provided in Sect. 4.

2 Double-Coding Density Sensitive Hashing (DCDSH)

2.1 Training Process of DCDSH

Let $x \in R^d$ denote the d-dimensional object, and $\mathcal{X} = [x_1, x_2, \ldots, x_n] \in R^{d \times n}$ denote a collection of n objects. Implement *k-means* quantization on the n objects in $\mathcal{X}$. Specifically, we stop the *k-means* after t iterations, where t is usually a small number (5 is enough). In real applications, the cluster number k is set as $k = \alpha L$, where $\alpha(>1)$ is a parameter and L denotes the length of hash codes. After *k-means* quantization, we obtain the quantization results, i.e., k cluster groups $\mathcal{S}_1, \ldots, \mathcal{S}_k$ and their center vectors $\mu_1, \ldots, \mu_k$. Furthermore, we compute the size rate of each cluster as

$$v_j = |\mathcal{S}_j| / \sum_{i=1}^{k} |\mathcal{S}_i|, j = 1, \ldots, k.$$

Let $N_r(\mu_i)$ denote the set of r-nearest neighbors $(r = 1, 2, \ldots)$ of μ_i, i.e., $N_r(\mu_i) = \left\{ \mu_j \middle| \mu_j \text{ is one of the } r \text{ nearest clusters of } \mu_i \right\}$. Then the median plane between μ_i and its r-adjacent cluster μ_j, can be defined by

$$\left(x - \left(\mu_i + \mu_j \right)/2 \right)^T \left(\mu_i - \mu_j \right) = 0, \ \mu_j \in N_r(\mu_i) \tag{1}$$

Let $w = \mu_i - \mu_j$ and $b = \left(\mu_i + \mu_j \right)^T \left(\mu_i - \mu_j \right)/2$, we rewrite the median plane as

$$w^T x - b = 0 \tag{2}$$

where w denotes the normal vector, b denotes the intercept of the median plane.

For the k clusters, the previous step can generate around $kr/2$ median planes. Since $k = \alpha L$, the number of median planes is about $\alpha r L > 2L$ [6]. From the information theoretic point of view, good binary codes should maximize the information/entropy provided by each bit [6]. Using maximum entropy principle, a binary bit that gives balanced partitioning of the data points provides maximum information. We utilize the strategy in [6] to compute the approximate entropy of each median hyper-plane:

$$\vartheta_m = -P_m^+ \log P_m^+ - P_m^- \log P_m^-, \ m = 1, \ldots, M \tag{3}$$

where ϑ_m denotes the entropy of hyperplane $w_m^T x - b = 0$, M denotes the number of hyper-planes, P_m^+ and P_m^- are computed as follows

$$P_m^+ = \sum_{i \in \mathcal{I}_m^+} v_i \quad \text{and} \quad P_m^- = \sum_{i \in \mathcal{I}_m^-} v_i \tag{4}$$

where v_i denotes the size rate of cluster i $(i = 1, \ldots, M)$, $\mathcal{I}_m^+$ and $\mathcal{I}_m^-$ denote the sets of cluster indices defined by

$$\mathcal{I}_m^+ = \left\{ i \middle| w_m^T \mu_i - b_m \geq 0, \ i = 1, \ldots, k \right\} \quad \text{and} \quad \mathcal{I}_m^- = \left\{ i \middle| w_m^T \mu_i - b_m < 0, \ i = 1, \ldots, k \right\} \tag{5}$$

Sort the approximate entropy ϑ_m (see Eq. (3)), $m = 1, \ldots, M$, in descending order and select the top L hyper-planes from them as principle median planes. Let $\varphi_\ell(x) = w_\ell^T x - b_\ell$, $\ell = 1, \ldots, L$, denote L principle hyper-planes, we compute the real-valued hash codes for x as

$$g_\ell(x) = \delta(\varphi_\ell(x)) \exp(\varphi_\ell(x)) / \sum_{s=1}^{L} \exp(\varphi_s(x)), \ \ell = 1, \ldots, L \tag{6}$$

where $\delta(x) = \begin{cases} 1 & x \geq 0 \\ -1 & \text{otherwise} \end{cases}$

Based on real-valued hash codes, $g_\ell(x)$, we compute L-bit binary hash codes, $h_\ell(x)$, for x as

$$h_\ell(x) = \begin{cases} 1 & g_\ell(x) \geq 0 \\ 0 & \text{otherwise} \end{cases}, \ell = 1, \ldots, L \tag{7}$$

Based on Eqs. (6) and (7), we obtain double-coding density sensitive hash codes w.r.t. object x, as $(G(x), H(x))$ where $G(x) = [g_1(x), \ldots, g_L(x)]$ and $H(x) = [h_1(x), \ldots, h_L(x)]$. Table 1 shows the training process of DCDSH.

Table 1. Training process of DCDSH

Algorithm 1 Training process of double-coding density-sensitive hashing

---

Input

   n training objects $x_1, x_2, \ldots, x_n \in R^d$, hash code length L, the parameter α controlling the cluster number, the iteration number t in the k-means, the number of nearest neighbors r

Output

   The hyperplane models: $\{w_\ell, b_\ell\}_{\ell=1}^{L}$, double-coding density sensitive hash codes for all training objects: $(G(x_i), H(x_i)), i = 1, \ldots, n$.

Steps

   1. Implement k-means with t iterations to generate αL clusters with center vectors $\mu_1, \mu_2, \ldots, \mu_{\alpha L}$;

   2. Build the median plane between each pair of r-adjacent clusters based on Eq.(1) or Eq.(2), and obtain normal vector and the intercept of each median plane, $\{w_m, b_m\}_{m=1}^{M}$, where M is the total number of median planes;

   3. Compute the approximate entropy of each median hyperplane using Eq.(3), and then select L median hyper-planes with largest approximate entropy values from all median hyper-planes, i.e, $\{w_\ell, b_\ell\}_{\ell=1}^{L} \subset \{w_m, b_m\}_{\ell=1}^{L}$;

   4. Generate the real-valued hash codes for all training objects based on the selected L hyper-planes based on Eq.(6), $G(x) = \left[g_1(x), \ldots, g_L(x)\right]$;

   5. Compute binary hash codes for training objects based on Eq.(7), $H(x) = \left[h_1(x), \ldots, h_L(x)\right]$.

2.2   Query Process of DCDSH

Suppose we implement $\mathcal{K}$-NN search task ($\mathcal{K}$ is an integer larger or equal to 1), $q \in R^d$ denote the query object. Compute the real-valued and binary density sensitive hash codes based on Eqs. (6) and (7), respectively. The query process of DCDSH for q involves two steps. First, from all training objects, select $\lceil \beta \mathcal{K} \rceil$ nearest training objects according to their Hamming distances of binary hash codes to the query object q, where $\beta > 1$ is a coefficient and $\lceil x \rceil$ denotes the least integer greater than or equal to x.

Second, from the selected $\lceil \beta \mathcal{K} \rceil$ objects (see previous step), select $\mathcal{K}$ nearest objects according to their Euclidean distance of real-valued hash codes to the query object q. Table 2 shows the query process of DCDSH.

Table 2. Query process of double-coding density-sensitive hashing

---

Algorithm 2 Training process of double-coding density-sensitive hashing

---

Input

Query object q, the hyperplane models $\{w_\ell, b_\ell\}_{\ell=1}^{L}$, double-coding density sensitive hash codes for all training objects $(G(x_i), H(x_i))$, $i = 1, \ldots, n$, the number of nearest training objects $\mathcal{K}$, β the coefficient controlling the size of candidates on binary hash codes.

Output

$\mathcal{K}$ nearest training objects to the query object

Steps

1. Compute real-valued hash code of query object q, i.e., $G(q) = [g_1(q), \ldots, g_L(q)]$, using Eq.(6) with hyperplane models $\{w_\ell, b_\ell\}_{\ell=1}^{L}$;

2. Compute binary hash code of query object q, i.e., $H(q) = [h_1(q), \ldots, h_L(q)]$, using Eq.(7) with $G(q)$;

3. Generate initial set of candidates from training objects via binary hash codes:

$$\mathcal{Z}(q) = \left\{ x_j \, \middle| \, \begin{array}{l} hamm\_d\big(H(x_j), H(q)\big) \leq \text{the } \beta\mathcal{K}\text{-th smallest hamming} \\ \text{distance between } x_i \text{ and } q, i = 1, \ldots, n \end{array} \right\}$$

where $hamm\_d(\cdot)$ denotes the hamming distance between two binary vectors.

4. Select $\mathcal{K}$ nearest objects from initial set of candidates via real-valued hash codes:

$$\left\{ x_j \, \middle| \, x_j \in \mathcal{Z}(q) \text{ and } \left\| G(x_j) - G(q) \right\|_2^2 \leq \text{the } \mathcal{K}\text{-th smallest Euclidean distance betwee } x_j \text{ and } q \right\}$$

In the first step, the DCDSH algorithm generates an initial set of candidates through hamming embedding (binary hash codes) which can shrink the range of candidates and significantly reduce the computational complexity; in the second step, the DCDSH algorithm determines the $\mathcal{K}$ nearest objects from the initial set of candidates using real-valued hash codes which effectively deals with the curse of dimensionality and improves search accuracy.

2.3 Computational Complexity Analysis of DCDSH

Given n training objects with d dimensionality, the computational complexity of DCDSH in the training process (Table 1) mainly consists of five parts. First, implementing *k-means* with t iterations to generate αL clusters costs $O(\alpha Ltnd)$; second, building the median plane between each pair of r-adjacent clusters costs $O(\alpha^2 L^2(d+r))$; third, computing the approximate entropy of each median hyperplane costs $O(\alpha^2 L^2 dr)$; fourth, selecting L median hyper-planes with largest approximate entropy values from them costs $O(\alpha Lr \log (\alpha Lr))$; fifth, real-valued and binary hash codes can be obtained in $O(Lnd)$. Generally, $\alpha Lr \ll n$, the computational complexity of DCDSH is mainly determined by k-means clustering $O(\alpha Ltnd)$.

The computational complexity of DCDSH is comparable with that of DSH. The main reason is that binary codes of both DSH and DCDSH (in both training stage and query stage) are obtained based on real-valued codes and there is no additional computations in the training stage of DCDSH. In the query process, given a query object, DCDSH needs $O(Ld)$ to compress the query object into a real-valued hash code and its corresponding binary code, which is the same as the complexity of DSH and LSH.

3 Experiments

In this section, we evaluate the performance of DCDSH algorithm on the high dimensional nearest neighbor search task. Four state-of-the-art hashing algorithms for high dimensional nearest neighbor search, LSH [5], DSH [6], QALSH [7] and are compared in our experiments. Three large scale real-world data sets, GIST1M, SIFT1M and Flickr1M, are used in our experiments. GIST1M contains one million GIST features and each feature is represented by a 960-dim vector. SIFT1M contains one million SIFT features and each feature is represented by a 128-dim vector. Flickr1M contains one million GIST features and each feature is represented by a 512-dim vector. The publicly available links of the three data sets can be found in [6].

We use the same criteria as in [5–7]. For each data set, randomly select 1000 data points as the queries and use the remaining to form the gallery database; a returned point is considered to be a true neighbor if it lies in the top two percentile points closest (measured by the Euclidian distance in the original space) to the query. We evaluate retrieval performance by the commonly used metric precision-recall curve.

QALSH [7] is implemented in C++ while other algorithms are mainly implemented in Matlab [5, 6]. The difference of implementation speed of different languages should be considered. Therefore, for fair comparison, we use a metric *Speedup_Ratio* to represent the improvement effectiveness for query speed as follows

$$Speedup\_Ratio = \frac{\text{Brute-force search time in the original feature space}}{\text{Search time using DCDSH algorithm}}$$

Speedup_Ratio can avoid the influence of different languages.

In the training process of DCDSH, there are three key parameters (i.e., α the parameter controlling the cluster number, t the number of iterations in the k-means, and r the parameter for r-nearest neighbors) which are the same as those of DSH [6]. In the query process, there is one key parameter (i.e., β the coefficient controlling the size of selected base objects in the initial query based on binary hash codes). In the training process, for fair comparison, we employ the same values of the three parameters (i.e., $\alpha = 1.5$, $t = 3$ and $r = 3$) with 64 bit hash codes as reported in [6]. In the query process, β is selected according to Mean Average Precision (MAP).

Figure 1(a) shows how the *Speedup_Ratio* of DCDSH varies as β changes at 64 bits w.r.t. GIST1M data set. Figure 1(a) indicates that as β increases, the *Speedup_Ratio* of DCDSH decreases. The main reason is that when β becomes larger, the number of initial candidates rises which raises the query time of real-valued hash codes, and then the *Speedup_Ratio* decreases. Figure 1(b) shows how the MAP of DCDSH varies as β changes at 64 bit hash code length w.r.t. GIST1M data set. Figure 1(b) indicates that when β becomes larger than some value, the performance MAP of DCDSH will remain stable and consistent. Although the initial set of candidates extends as β increases, the final query results usually lie in a relatively fixed range of candidates. No matter how large the range is, the final query results will not change significantly. For the other two data sets, Flickr1M and SIFT1M, we obtain similar trends. Based on the performance analyses we empirically select $\beta = 4$.

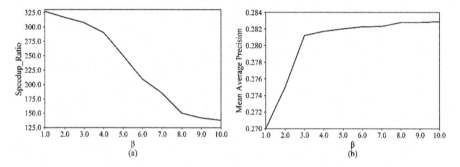

Fig. 1. Speedup_Ratio (a) and MAP (b) of DCDSH vs. β at 64 bits w.r.t. GIST1M data set.

Figure 2(a), (b) and (c) show precision-recall curves of DCDSH and other three algorithms (i.e., LSH [5], DSH [6], QALSH [7]) on data sets (a) GIST1M (b) SIFT1M and (c) Flickr1M at 64 bit hash code length, respectively. Figure 2(a), (b) and (c) indicate that the proposed DCDSH outperforms other three algorithms in terms of retrieval performance. The main reason is that DCDSH not only considers density distribution of data set but also utilizes real-valued hash codes to represent more characteristics than binary hash codes.

Figure 3(a), (b) and (c) show the *Speedup-Ratios* of the four algorithms on data sets GIST1M, SIFT1M and Flickr1M (at 64 bit hash code length), respectively. Figure 3(a), (b) and (c) indicate that the proposed DCDSH achieves better speedup effectiveness than QALSH and LSH. DCDSH approaches comparable speed with DCH. The main reason is that the candidate set is much smaller than the whole data base and the query time consumed by real-valued hash codes can be ignored.

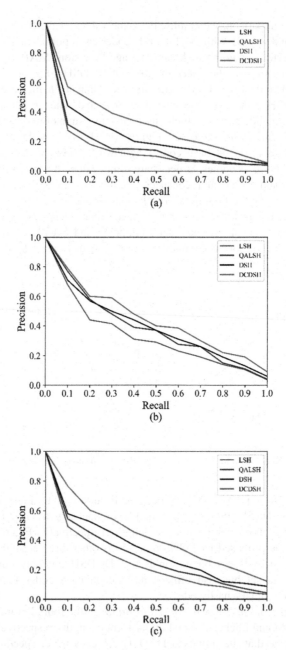

Fig. 2. Precision-recall curves of four algorithms (LSH, DSH, DCDSH and QALSH) on (a) GIST1M, (b) SIFT1M and (c) Flickr1M at 64 bit hash code length.

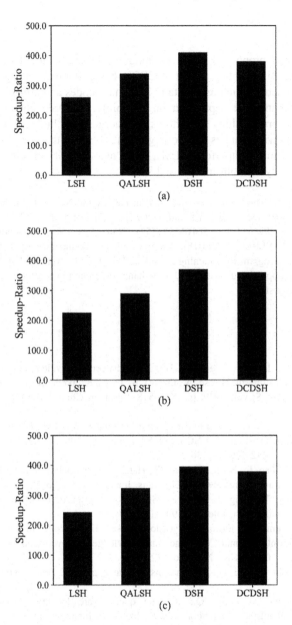

Fig. 3. *Speedup-Ratios* of four algorithms on (a) GIST1M, (b) SIFT1M and (c) Flickr1M at 64 bit hash code length.

4   Conclusion

In this paper, we propose a novel hashing algorithm, called DCDSH, for high dimensional nearest neighbor search. DCDSH has a double coding scheme which explores the geometric structure of data to build hash codes and utilizes the real-valued has codes to enhance the representation capability. DCDSH not only inherits the advantage of general hashing algorithms in query speed but also improves query performance. Empirical studies on two large data sets show that the proposed DCDSH algorithm significantly outperforms LSH, DSH and QALSH in terms of query performance, especially in high-dimensional space.

Acknowledgements. This work was supported in part by the National Natural Science Foundation of China (61401185, 61601213, and 61402212), China Postdoctoral Science Foundation Funded Project (2017M611252, 2016M591452), Natural Science Foundation of Liaoning General Project (LJYL017, LJYL018), Natural Science Foundation of Liaoning Province (2015020098), and Program for Liaoning Excellent Talents in University (LJQ2015045). All of these supports are appreciated. We would also thank the anonymous referees for their detailed comments and suggestions.

References

1. Ma, C., Gu, Y., Liu, W., Yang, J., He, X.: Unsupervised video hashing by exploiting spatio-temporal feature. In: Hirose, A., Ozawa, S., Doya, K., Ikeda, K., Lee, M., Liu, D. (eds.) ICONIP 2016. LNCS, vol. 9949, pp. 511–518. Springer, Cham (2016). doi:10.1007/978-3-319-46675-0_56

2. Zhao, K., Liu, D., Lu, H.: Local linear spectral hashing. In: Lee, M., Hirose, A., Hou, Z.-G., Kil, R.M. (eds.) ICONIP 2013. LNCS, vol. 8228, pp. 283–290. Springer, Heidelberg (2013). doi:10.1007/978-3-642-42051-1_36

3. Liu, L., Lin, Z., Shao, L., Shen, F., Ding, G., Han, J.: Sequential discrete hashing for scalable cross-modality similarity retrieval. IEEE Trans. Image Process. 26(1), 107–118 (2017)

4. Wang, J., Zhang, T., Song, J., Sebe, N., Shen, H.T.: A survey on learning to hash. IEEE Trans. Pattern Anal. Mach. Intell. PP(99), 1 (2017)

5. Dasgupta, A., Kumar, R., Sarlos, T.: Fast locality-sensitive hashing. In: 17th Proceedings of ACM SIGKDD International Conference on Knowledge Discovery and Data Mining (KDD 2011), pp. 1073–1081. ACM, New York (2011)

6. Jin, Z., Li, C., Lin, Y., Cai, D.: Density sensitive hashing. IEEE Trans. Cybern. 44(8), 1362–1371 (2014)

7. Huang, Q., Feng, J., Fang, Q.: Reverse query-aware locality-sensitive hashing for high-dimensional furthest neighbor search. In: 33rd International Conference on Data Engineering (ICDE), pp. 167–170. IEEE, San Diego, CA (2017)

Improving Shape Retrieval by Fusing Generalized Mean First-Passage Time

Danchen Zheng$^{(\boxtimes)}$, Wangshu Liu, and Hanxing Wang

Faculty of Electronic Information Electrical Engineering,
Dalian University of Technology, 2 Linggong Lu, Dalian, China
dcjeong@dlut.edu.cn, {liuwangshu,whx064}@mail.dlut.edu.cn

Abstract. In recent years, many efforts have been made to fuse different similarity measures for robust shape retrieval. In this paper, we firstly propose generalized mean first-passage time (GMFPT) that extends the mean first-passage time (MFPT) to the general form. Instead of focusing on the propagation of similarity information, GMFPT is introduced to improve pairwise shape distances, which denotes the mean time-steps for the transition from one state to a set of states. Through a semi-supervised learning framework, an iterative approach with a time-invariant state space is further proposed to fusing multiple distance measures, and the relative objects on the geodesic paths can be gradually and explicitly retrieved. The experimental results on different databases demonstrate that shape retrieval results can be effectively improved by the proposed method.

Keywords: Shape retrieval · Pairwise distance · Generalized mean first-passage time · Fusion of measures

1 Introduction

Shape-based retrieval plays an important role in computer vision with many applications, such as shape recognition and image retrieval. Traditionally, the shape retrieval problem is solved by analyzing pairwise distance, which measures the similarity between two elements. It seems to be an obvious statement that the more similar two shapes are, the smaller distance is measured [19]. However, traditional pairwise retrieval researches ignore the structure of the underlying data manifold, which may lead to incorrect retrieval results.

In the past decade, context-sensitive similarities are proposed to capture the geometry of the underlying manifold within database instances, and it can be considered as a post-processing procedure for improving the shape retrieval results obtained from pairwise distances [3]. Bai et al. [5] firstly introduced a context-sensitive shape similarity on the basis of label propagation (LP), and the relative shapes can be effectively obtained according to the updated similarities after a limited number of iterations. And since then, many approaches are proposed to improve the shape retrieval results by employing the contextual

© Springer International Publishing AG 2017
D. Liu et al. (Eds.): ICONIP 2017, Part IV, LNCS 10637, pp. 439–448, 2017.
https://doi.org/10.1007/978-3-319-70093-9_46

information, such as modified mutual kNN graph (mkNN) [11], meta descriptor (MD) [9], shortest path propagation (SSP) [18], and tensor product graph (TPG) [20]. Donoser et al. [8] revisited the state-of-the-art approaches of diffusion process for improving retrieval performance, and a generic framework is further introduced.

For recent rears, in consideration of different measures focusing on different emphasis, many studies prefer improving the retrieval results with multiple measures, which are complementary to each other. Co-transduction is proposed to fuse distance measures through a semi-supervised learning framework in [4], and it utilizes two or three metrics to help each other for better retrieval performance. Wang et al. [17] presented a novel approach to fuse multiple metrics through diffusion process on the basis of an unsupervised framework. In [2], a feature vector called sparse contextual activation (SCA) is proposed and applied to fuse multiple metrics for highly effective re-ranking. Moreover, multiple distance measures fusion has been applied in various visual fields, such as visual tracking [22], salient object detection [13], and image retrieval [21] etc.

In this paper, different from previous methods, our approach concentrates on the average time spent on propagating similarity information. Based on this idea, generalized mean first-passage time (GMFPT) is firstly introduced as the updated distance from unknowns to the queries. Then, we improve the shape retrieval results by fusing GMFPTs obtained from different measures through a semi-supervised framework. Comparing to previous algorithms updating similarities by implicitly capturing the shortest paths, our approach gradually searching related objects within certain ranges dynamically determined according to the query objects. The shape retrieval experiments on well-known data sets demonstrates that the proposed approach performs better than state-of-the-art methods based on the same baseline methods.

2   Distance Learning by GMFPT

2.1   Generalized Mean First-Passage Time

We firstly give the basic concept of the generalized mean first-passage time. Given a set of objects with N elements, the corresponding state space can be constructed as $S = \{1, 2, \cdots, N\}$. We have the transition matrix corresponding to the state space defined as $\mathbf{P} = [p_{ij}]$ derived by normalizing the weighted matrix $\mathbf{W} = [w_{ij}]_{N \times N}$ in row wise, where p_{ij} is (one step) state transition probability and $i, j \in S$. And then, a homogeneous discrete-time Markov chain $X_0, X_1, \cdots, X_n$ can be defined, and we have

$$
\begin{aligned}
&\Pr\left(X_{n+1} = j | X_0 = i_0, X_1 = i_1, \cdots, X_n = i\right) \\
&= \Pr\left(X_{n+1} = j | X_n = i\right) \\
&= \Pr\left(X_1 = j | X_0 = i\right) \\
&= p_{ij}.
\end{aligned}
\tag{1}
$$

For $i \in S$ and a set $L \subset S$, the generalized mean first-passage time from i to L is defined as

$$\eta_{iL} = \mathrm{E}\left(T_L | X_0 = i\right) = \sum_{n=1}^{\infty} n\mathrm{Pr}\left(T_L = n | X_0 = i\right), \tag{2}$$

where $T_L = \min\{n > 0 : X_n \in L\}$ represents the number of transfer times when a particle first reach the set L. Given a random walk starting from i, η_{iL} can be viewed as the expected number of time-steps for reaching any state in L for the first time. In particular, if L only contains one state j, we have the mean first-passage time η_{ij} instead of η_{iL} as follows

$$\eta_{ij} = \mathrm{E}\left(T_j | X_0 = i\right) = \sum_{n=1}^{\infty} n\mathrm{Pr}\left(T_j = n | X_0 = i\right). \tag{3}$$

Compared to other distance learning approaches, GMFPT concentrates on the cost of time for the propagations of similarity information. Previous methods mostly prefer learning a new distance by collectively propagating the measures, which depend the idea that the similarity information spreads quickly along geodesic paths. Instead of focusing on the processes of diffusing, we are interested in the time-steps taken to accomplish the state transitions between states on the data manifold. According to GMFPTs, the state transition between the within-class objects could be accomplished in a few time-steps via other relative states on the shortest paths. Meanwhile, it takes much more time-steps to carry out the state transition between the vertices from various categories.

2.2 Calculation of GMFPT

For convenience, we introduce conditional probability as follows

$$f_{iL}^{(n)} = \mathrm{Pr}\left(X_n \in L, X_m \notin L, 1 \leq m \leq n - 1 | X_0 = i\right). \tag{4}$$

$f_{iL}^{(n)}$ indicates the probability of a particle shifts from i to L with n steps. According to Eq. 4, we have $f_{iL}^{(n)} = \sum_{j \notin L} p_{ij} f_{jL}^{(n-1)}$, and η_{ij} can be further expressed as follows

$$\eta_{iL} = \sum_{n=1}^{\infty} n f_{iL}^{(n)} = \sum_{j \notin L} p_{ij} \eta_{jL} + \sum_{n=1}^{\infty} n f_{iL}^{(n-1)}. \tag{5}$$

We divide S into two subsets $U = \{1, \cdots, M\}$ with M elements and $L = \{M + 1, \cdots, N\}$ with $N - M$ elements. The transition matrix $\mathbf{P}$ can be written as the following partitioned matrix

$$\mathbf{P} = \begin{bmatrix} \mathbf{P}_{UU} & \mathbf{P}_{UL} \\ \mathbf{P}_{LU} & \mathbf{P}_{LL} \end{bmatrix}, \tag{6}$$

where $\mathbf{P}_{UU} = [p_{ij}]$, $\mathbf{P}_{UL} = [p_{ik}]$, $\mathbf{P}_{LU} = [p_{lj}]$, $\mathbf{P}_{LL} = [p_{lk}]$, $1 \le i \le M$, $1 \le j \le M$, $M+1 \le k \le N$, and $M+1 \le l \le N$. With $S = U + L$, we have $f_{iL}^{(n)} = \sum_{j=1}^{M} p_{ij} f_{jL}^{(n-1)}$ and $f_{iL}^{(1)} = \sum_{j=M+1}^{N} p_{ij}$, which can be expressed as

$$\mathbf{P}_{UU} \begin{bmatrix} f_{1L}^{(n-1)} \\ \vdots \\ f_{ML}^{(n-1)} \end{bmatrix} = \begin{bmatrix} f_{1L}^{(n)} \\ \vdots \\ f_{ML}^{(n)} \end{bmatrix}, \quad \mathbf{P}_{UL} \times \mathbf{1}_{N-M} = \begin{bmatrix} f_{1N}^{(1)} \\ \vdots \\ f_{MN}^{(1)} \end{bmatrix}. \tag{7}$$

Then, we have

$$\begin{bmatrix} \sum_{n=1}^{\infty} f_{1L}^{(n)} \\ \vdots \\ \sum_{n=1}^{\infty} f_{ML}^{(n)} \end{bmatrix} = \left(\sum_{k=0}^{\infty} (\mathbf{P}_{UU})^k \right) \mathbf{P}_{UL} \times \mathbf{1}_{N-M}, \tag{8}$$

where $\mathbf{1}_k$ is a k-dimensional vector with each element having the value of 1.

With S divided into two subspaces U and L, Eq. 5 can be written as $\eta_{iL} = \sum_{j=1}^{M} p_{ij} \eta_{jL} + \sum_{n=1}^{\infty} f_{iL}^{(n)}$, and we have

$$\boldsymbol{\eta}_L = \mathbf{P}_{UU} \boldsymbol{\eta}_L + \begin{bmatrix} \sum_{n=1}^{\infty} f_{1L}^{(n)} \\ \vdots \\ \sum_{n=1}^{\infty} f_{ML}^{(n)} \end{bmatrix}, \tag{9}$$

where $\boldsymbol{\eta}_L = [\eta_{1L}, \cdots, \eta_{ML}]^{\mathrm{T}}$. And then, we have

$$(\mathbf{I} - \mathbf{P}_{UU}) \boldsymbol{\eta}_L = \left(\sum_{k=0}^{\infty} (\mathbf{P}_{UU})^k \right) \mathbf{P}_{UL} \times \mathbf{1}_{N-M}. \tag{10}$$

From Eq. 10 it can be seen that if the matrix $\mathbf{I} - \mathbf{P}_{UU}$ is invertible, $\boldsymbol{\eta}_L$ can be obtained by directly solving the linear equation.

For a large enough h which determines $p_{ij}^{(h)} > 0$, we define q_{ij} as the entry of $(\mathbf{P}_{UU})^h$, and have

$$q_{ij} = \sum_{k_1=1}^{M} \cdots \sum_{k_{h-1}=1}^{M} p_{ik_1} \cdots p_{k_{h-1}j} \le \sum_{k_1=1}^{N} \cdots \sum_{k_{h-1}=1}^{N} p_{ik_1} \cdots p_{k_{h-1}j} = p_{ij}^{(h)}, \tag{11}$$

where $1 \le i, j \le M$. If $\forall p_{ij}^{(h)} > 0$, then we have $\sum_{j=1}^{M} p_{ij}^{(h)} > 0$, and the inequalities can be obtained as follows

$$\sum_{j=1}^{M} q_{ij} \le \sum_{j=1}^{M} p_{ij}^{(h)} = 1 - \sum_{j=M+1}^{N} p_{ij}^{(h)} < 1, \tag{12}$$

$$\left\| (\mathbf{P}_{UU})^h \right\|_\infty = \max_{1 \le i \le M} \sum_{j=1}^{M} |q_{ij}| < 1. \tag{13}$$

Given $\rho(\cdot)$ as the spectral radius of matrix, we have

$$\rho(\mathbf{P}_{UU}) \le \left\| (\mathbf{P}_{UU})^h \right\|_\infty^{\frac{1}{h}} < 1. \tag{14}$$

Therefore, the matrix $\mathbf{I} - \mathbf{P}_{UU}$ is invertible, and $\sum_{k=0}^{\infty} (\mathbf{P}_{UU})^k = (\mathbf{I} - \mathbf{P}_{UU})^{-1}$. As $\sum_{j=1}^{N} p_{ij} = 1$, we have $(\mathbf{I} - \mathbf{P}_{UU})^{-1} \mathbf{P}_{UL} \mathbf{1}_{N-M} = \mathbf{1}_M$. Therefore, the solution of the linear equations Eq. 10 is unique, and $\boldsymbol{\eta}_L$ can be obtained as follows

$$\boldsymbol{\eta}_L = (\mathbf{I} - \mathbf{P}_{UU})^{-1} \mathbf{1}_M. \tag{15}$$

3   Improving Retrieval by Fusing GMFPTs

3.1   Construction of Weighted Matrix

As the GMFPTs are obtained based on a state transition matrix $\mathbf{P}$, we firstly discuss the construction of $\mathbf{W}$ with the pairwise shape distances. Given d_{ij} as the pairwise distance, the corresponding similarity sim_{ij} is defined with the Gaussian kernel as

$$sim_{ij} = \exp\left(-\frac{d_{ij}^2}{\sigma_{ij}^2} \right), \tag{16}$$

where σ_{ij} is the kernel width of the Gaussian kernel function. We use $knn_d(i)$ to denote the set of K_d nearest neighbors of i based on pairwise distance, and σ_{ij} can be expressed as follows

$$\sigma_{ij} = \frac{\alpha}{2K_d} \left(\sum_{i \in knn_d(j)} d_{ij} + \sum_{j \in knn_d(i)} d_{ij} \right). \tag{17}$$

K_d and α are two hyper-parameters that are decided empirically. Specifically, the choice of K_d is depend on the distribution of samples from the database, and K_d ranges from 10 to 20 in this paper. On the other hand, α is mainly determined by the distance measure, and it ranges from 0.3 to 0.4.

To remove noisy edges and keep meaningful connections, $\mathbf{W} = [w_{ij}]$ is constructed based on reciprocal kNN (also named mutual kNN) [16] as

$$w_{ij} = \begin{cases} sim_{ij} & i \in knn_s(j), j \in knn_s(i) \\ 0 & \text{others} \end{cases}, \tag{18}$$

where $knn_s(j)$ denotes the set of K_s nearest neighbors of j based on sim_{ij}. For the convenience of parameter setting, we set K_s close to K_d empirically in the following experiments.

3.2   Iterative Approach Based on GMFPTs

To effectively gain the relative objects with different distance measures, we introduce an iterative approach to retrieve the within-class objects gradually and explicitly. L_t and U_t are used to denote the set of queries and the set of unlabeled objects respectively in iteration t. With $L_0 = \{N\}$ and $U_0 = \{1, 2, \cdots, N - 1\}$, each iteration consists of two procedures: calculating $\boldsymbol{\eta}_{L_t}$, and updating $L_{t+1} = L_t + \Delta_t$, where $\Delta_t = \arg\min_\lambda (\boldsymbol{\eta}_{L_t})$. And $\arg\min_\lambda (\boldsymbol{\eta}_{L_t})$ denotes the samples corresponding to the λ minimum elements from $\boldsymbol{\eta}_{L_t}$, where the iteration step-size λ is set to a small value empirically. To fuse various distance measures, for the first procedure of the iteration approach, the GMFPTs based on different distances are calculated alternately. In this way, the structure of the underlying data manifold viewed from different perspectives can be explicitly captured and effectively integrated. After T iterations, the retrieval result is obtained according to the order in which the objects are labeled and added to L_T.

In many cases, a data set contains lots of elements, the calculation of inverse matrix is time-consuming. As only top λ samples nearby the queries are added to L_t in each iteration, we construct a time-variant S_t with a small number of examples instead of the state space composed of all the elements. With this idea, we define the time-variant S_t as follows

$$S_t = L_t + U_t = L_t + R_t + O_t, \tag{19}$$

where $U_t = R_t + O_t$, $R_t = \{j : w_{ij} > 0, i \in L_t, j \notin L_t\}$ and $O_t = \{k : w_{jk} > 0, j \in R_t, k \notin L_t + R_t\}$.

In most cases, the graph based on S/S_t is composed of connected components. With C_t defined as the union of the components containing the elements from L_t, we have the meaningless $\eta_{iL} \rightarrow \infty$ for $U_t \cap C_t = \varnothing$, where $i \in U_t$. To solve this problem, after all the elements from the components containing the queries are retrieved, we employ a temporary weighted matrix $\mathbf{W}'$ instead of $\mathbf{W}$ in next iteration. And $\mathbf{W}'$ connect all the components together with the largest similarities between two elements from different components. Given N_Q as the number of objects expected to be retrieved, the proposed algorithm (co-GMFPT) with K distance measures is summarized as the pseudocode of Algorithm 1, and it can be proved that $\mathbf{I} - \mathbf{P}_{UU}^t$ is always invertible.

4   Experimental Results

To verify the effectiveness of the proposed method, our experiments are carried out on two well-known shape databases: Tari-1000 data set [1] and MPEG-7 data set [12]. With each shape represented by 100 sample points, shape context (SC) [6] and inner distance shape context (IDSC) [14] are selected as the baseline methods. In the following experiments, we focus on fusing two different measures together, and set $\lambda = 3$.

Algorithm 1. co-GMFPT

Input: $\{\mathbf{W}_k : 1 \leq k \leq K\}$, $L_0 = \{N\}$, $t = 0$, and $i = 1$.
Output: L_T
1 **while** $\#L_t < N_Q$ **do**
2   **if** $\#L^t = \#C^t$ **then**
3   create S^t by Eq. 19 with $\mathbf{W}'_i$;

4   **else**
5   create S^t by Eq. 19 with $\mathbf{W}_i$;

6   create $\mathbf{P}^t = [p_{ij}^t]$ based on S^t;
7     $\eta_{L^t} = (\mathbf{I} - \mathbf{P}_{UU}^t)^{-1} \times \mathbf{1}$;
8   **if** $\#U^t \leq \lambda$ **then**
9        $L^{t+1} = L^t + U^t$;

10   **else**
11       $L^{t+1} = L^t + \Delta^t$, $\Delta^t = \arg\min_\lambda (\eta_{L^t})$;
12    $t = t + 1$;
13   **if** $i \geq K$ **then**
14       $i = 1$;

15   **else**
16       $i = i + 1$;

4.1 Tari-1000 Shape Database

Tari-1000 shape data set [1] consists of 1000 silhouette images grouped into 50 classes, and each contains 20 different shapes. The retrieval score is employed to evaluate the performances of different methods on this data set. With each object viewed as the query, the number of correct retrievals in the top 20 ranks is counted, including the self-match. And the retrieval score is defined as the ratio of the retrieved relative objects over the total number of relative objects.

The baseline retrieval scores are 88.01% for SC and 90.43% for IDSC. In this experiment, we set $\alpha = 0.38$, $K_d = 18$ and $K_s = 20$. Table 1 gives the retrieval scores of different methods on Tari-1000 database. The proposed approach achieves the retrieval scores of 99.54%, which is superior to the other state-of-the-art methods. And the baseline results are greatly and effectively improved by our approaches.

4.2 MPEG-7 Shape Database

In this section, we further test the proposed method on a widely-used MPEG-7 database [12]. A total of 1400 silhouette images are included in the MPEG-7 data set, which grouped into 70 classes with each class containing 20 different shapes. Bull's eye score is used to evaluate the performance. Each sample in the database is viewed as query in turn and compared with all the shapes. Then the number of objects belonging to the same category among the top 40 similar shapes is

Table 1. Comparison of retrieval results on Tari-1000 shape database

Baseline	Algorithm	Bull's eye score
IDSC [14]	-	90.43%
SC [6]	-	88.01%
IDSC+SC	DM [7]	89.03%
IDSC+SC	SSO [10]	95.43%
IDSC+SC	Co-Transduction [4]	98.59%
IDSC+SC	**Co-GMFPT**	**99.54%**

counted. And the bulls eye score represents the ratio of the total number of samples belonging to the same category to the possible number.

The bull's eye scores of the baselines: SC and IDSC are 86.80% and 85.40%. We set $\alpha = 0.38$, $K_d = 18$ and $K_s = 20$ for this experiment. Table 2 shows the retrieval results on MPEG-7 database. As can be can seen, by taking SC and IDSC as the baselines, our method achieves the bull's eye score of 99.14%, which is superior to other methods based on the same pairwise distances. The retrieval results are significantly improved by taking the proposed methods as the post-processing step.

Table 2. Comparison of retrieval results on MPEG-7 shape database

Baseline	Algorithm	Bull's eye score
IDSC	-	85.40%
SC	-	86.80%
IDSC+SC	DM [7]	92.07%
IDSC+SC	SSO [10]	97.64%
IDSC+SC	Co-Transduction [4]	97.72%
IDSC+SC	LCDM [15]	98.84%
IDSC+SC	SCA [2]	99.01%
IDSC+SC	**Co-GMFPT**	**99.14%**

5   Conclusion

This paper proposed co-GMFPT method for improving shape retrieval based on GMFPT. In this paper, GMFPT as an extension of MFPT is introduced as the updated distance from target objects to the queries. With a time-invariant state space, An iterative approach is proposed to fuse multiple pairwise distances as the post-processing step through a semi-supervised learning framework. The geodesic paths can be effectively captured and the relative objects can be accurately searched. The experimental results on different data sets indicate the effectiveness of the proposed algorithm for shape retrieval.

Acknowledgement. This research is supported by the project (DUT14RC(3)128) of Fundamental Research Funds for the Central Universities.

References

1. Aslan, C., Erdem, A., Erdem, E., Tari, S.: Disconnected skeleton: shape at its absolute scale. IEEE Trans. Pattern Anal. Mach. Intell. **30**(12), 2188–2203 (2008)
2. Bai, S., Bai, X.: Sparse contextual activation for efficient visual re-ranking. IEEE Trans. Image Process. **25**(3), 1056–1069 (2016)
3. Bai, S., Sun, S., Bai, X., Zhang, Z., Tian, Q.: Smooth neighborhood structure mining on multiple affinity graphs with applications to context-sensitive similarity. In: Leibe, B., Matas, J., Sebe, N., Welling, M. (eds.) ECCV 2016. LNCS, vol. 9906, pp. 592–608. Springer, Cham (2016). doi:10.1007/978-3-319-46475-6_37
4. Bai, X., Wang, B., Yao, C., Liu, W., Tu, Z.: Co-transduction for shape retrieval. IEEE Trans. Image Processing **21**(5), 2747–2757 (2012)
5. Bai, X., Yang, X., Latecki, L.J., Liu, W., Tu, Z.: Learning context-sensitive shape similarity by graph transduction. IEEE Trans. Pattern Anal. Mach. Intell. **32**(5), 861–874 (2010)
6. Belongie, S., Malik, J., Puzicha, J.: Shape matching and object recognition using shape contexts. IEEE Trans. Pattern Anal. Mach. Intell. **24**(4), 509–522 (2002)
7. Coifman, R.R., Lafon, S.: Diffusion maps. Appl. Comput. Harmonic Anal. **21**(1), 5–30 (2006)
8. Donoser, M., Bischof, H.: Diffusion processes for retrieval revisited. In: IEEE Conference on Computer Vision and Pattern Recognition, pp. 1320–1327 (2013)
9. Egozi, A., Keller, Y., Guterman, H.: Improving shape retrieval by spectral matching and meta similarity. IEEE Trans. Image Process. **19**(5), 1319–1327 (2010)
10. Jiang, J., Wang, B., Tu, Z.: Unsupervised metric learning by self-smoothing operator. In: IEEE International Conference on Computer Vision, pp. 794–801 (2011)
11. Kontschieder, P., Donoser, M., Bischof, H.: Beyond pairwise shape similarity analysis. In: Zha, H., Taniguchi, R., Maybank, S. (eds.) ACCV 2009. LNCS, vol. 5996, pp. 655–666. Springer, Heidelberg (2010). doi:10.1007/978-3-642-12297-2_63
12. Latecki, L.J., Lakamper, R., Eckhardt, T.: Shape descriptors for non-rigid shapes with a single closed contour. In: IEEE Conference on Computer Vision and Pattern Recognition, pp. 424–429 (2000)
13. Li, H., Lu, H., Lin, Z., Shen, X., Price, B.: Inner and inter label propagation: salient object detection in the wild. IEEE Trans. Image Process.: Publ. IEEE Sig. Process. Soc. **24**(10), 3176–86 (2015)
14. Ling, H., Jacobs, D.W.: Shape classification using the inner-distance. IEEE Trans. Pattern Anal. Mach. Intell. **29**(2), 286–299 (2007)
15. Luo, L., Shen, C., Zhang, C., van den Hengel, A.: Shape similarity analysis by self-tuning locally constrained mixed-diffusion. IEEE Trans. Multimedia **15**(5), 1174–1183 (2013)
16. Guimarães Pedronette, D.C., Penatti, O.A.B., Torres, R.D.S.: Unsupervised manifold learning using reciprocal KNN graphs in image re-ranking and rank aggregation tasks. Image Vis. Comput. **32**(2), 120–130 (2014)

17. Wang, B., Jiang, J., Wang, W., Zhou, Z.H., Tu, Z.: Unsupervised metric fusion by cross diffusion. In: IEEE Conference on Computer Vision and Pattern Recognition (2012)
18. Wang, J., Li, Y., Bai, X., Zhang, Y., Wang, C., Tang, N.: Learning context-sensitive similarity by shortest path propagation. Pattern Recogn. 44(10C11), 2367–2374 (2011)
19. Yang, X., Bai, X., Latecki, L.J., Tu, Z.: Improving shape retrieval by learning graph transduction. In: Forsyth, D., Torr, P., Zisserman, A. (eds.) ECCV 2008. LNCS, vol. 5305, pp. 788–801. Springer, Heidelberg (2008). doi:10.1007/978-3-540-88693-8_58
20. Yang, X., Prasad, L., Latecki, L.J.: Affinity learning with diffusion on tensor product graph. IEEE Trans. Pattern Anal. Mach. Intell. 35(1), 28–38 (2013)
21. Zhang, S., Yang, M., Cour, T., Yu, K., Metaxas, D.N.: Query specific rank fusion for image retrieval. IEEE Trans. Pattern Anal. Mach. Intell. 37(4), 803–815 (2015)
22. Zhou, Y., Bai, X., Liu, W., Latecki, L.J.: Similarity fusion for visual tracking. Int. J. Comput. Vis. 118(3), 337–363 (2016)

Complex-Valued Neural Networks for Wave-Based Realization of Reservoir Computing

Akira Hirose[1(⊠)], Seiji Takeda[2], Toshiyuki Yamane[2], Daiju Nakano[2],
Shigeru Nakagawa[2], Ryosho Nakane[1], and Gouhei Tanaka[1]

[1] Social Cooperation Program on Energy Efficient Information Processing (EEIP)
and Department of Electrical Engineering and Information Systems,
The University of Tokyo, Tokyo, Japan
ahirose@ee.t.u-tokyo.ac.jp, nakane@cryst.t.u-tokyo.ac.jp,
gouhei@sat.t.u-tokyo.ac.jp
[2] IBM Research - Tokyo, Tokyo, Japan
{seijitkd,tyamane,dnakano,snakagaw}@jp.ibm.com

Abstract. In this paper, we discuss the significance of complex-valued
neural-network (CVNN) framework in energy-efficient neural networks,
in particular in wave-based reservoir networks. Physical-wave reservoir
networks are highly enhanced by CVNNs. From this viewpoint, we also
compare the features of reservoir computing and other architectures.

Keywords: Neural hardware · Complex-valued neural networks
(CVNN)

1 Introduction

The main function of modern artificial intelligence (AI) is the extraction and utilization of correlations among diverse events. Some of found correlations lead to explicit expression of causality and/or generation of new statistical information. Storing correlations is also the most basic function of a neuron microscopically as well as a neuron network macroscopically. This is the fundamental origin of why neural networks play the principal role in recent AI.

However, some serious problems have arisen in these years since AI systems are realized as neuro-based software on von-Neumann type hardware. That is, a large amount of energy is consumed in a system to deal with large-scale data for processing and learning with deep learning or other methods. In addition, edge nodes in sensor networks also exhibit large energy consumption in a total system. Saving the energy is a seriously pressing issue.

Reservoir computing, or reservoir neural networks, is the key technology to solve this problem in relation to both the hardware and software [1,2]. A reservoir network is formally a perceptron that contains recurrent connections among the hidden neurons [1,3–5]. The recurrence enables the network to deal with time-sequential data processing. This is in contrast with perceptrons suitable only for such tasks as function approximation and classification. However, the

D. Liu et al. (Eds.): ICONIP 2017, Part IV, LNCS 10637, pp. 449–456, 2017.
https://doi.org/10.1007/978-3-319-70093-9_47

advantages of reservoir networks extend wider. That is, the combination of delays and nonlinaerity realizes not only the time-sequential processing but also more general processing.

This paper considers reservoir neural networks realized by physical waves to clarify the usefulness of complex-valued neural networks (CVNNs). Generally speaking, CVNNs have great advantages in wave-based hardware [6–12]. But they have advantages also in processing information generated by wave phenomena such as radar imaging and computer generated holography [13–16]. In this paper, we discuss the realization of reservoir neural networks with the learning and self-organization in the framework of CVNN.[1]

2   Two Serious Key Issues in Hardware

Computing hardware, not only in neural networks but also in von Neumann-type computers, has always held the following serious problems regardless of which direction a technology goes. One is the wiring explosion, and the other is the variability in miniaturization.

2.1   Wiring Explosion

The most serious issue is the increase of the wiring amount. In particular in neural networks, it grows at an exponential rate along the increase of the network size, resulting in fatal infabricability. Furthermore, the increase of the total wiring length leads directly to the rise of electric power used to charge and discharge the wires. Energy saving definitely requires a solution of wiring explosion.

2.2   Element Variability

High integration requires the downscaling of element devices. Ultimate downscaling of such a level that an element is composed of only several tens of atoms brings about relatively huge variability in the electronic characteristics of the element devices. The variability disables us from fabricating a circuit precisely. This fact suggests strongly the increasing importance of neural adaptability in information processing systems in the near future. This is the other of the most serious problems in general though this limitation is extremely serious in von Neumann-type computers.

3   Pattern/Symbol Representation and Connection Amount

Figure 1 maps various information processing architectures in the space of information representation and connection amount. The horizontal axis shows the

---

[1] This paper concentrates upon a long-span perspective of reservoir networks with CVNNs. Detailed dynamics of CVNNs are given in literature such as Ref. [17].

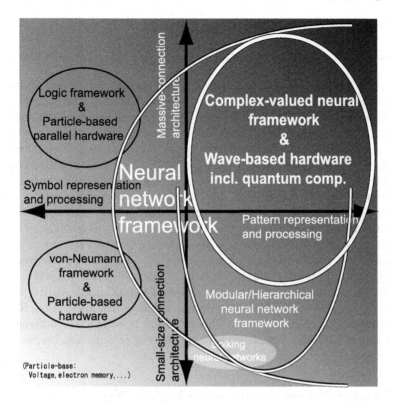

Fig. 1. Conceptual diagram showing the information-processing frameworks and hardware architectures mapped in the coordinates of the degree of pattern-/symbol-information representations and processing modes as well as the degree of massiveness in wiring.

pattern/symbol representations and processing while the vertical one represents the massiveness of connections.

Von Neumann-type computers are grounded in symbol representation and symbol processing. Though neural networks are basically founded on pattern representation and pattern processing, a group of networks such as self-organizing maps and hourglass-type networks provides functions of symbolization. Then the neural networks extend also to symbol representation/processing areas in Fig. 1.

Vertically, on the other hand, conventional hardware is placed as small-size connection architecture. The widely used von Neumann-type computers are consequently placed in the bottom-left area. Upper-left area shows those with ultimate parallelization with pipelines, asynchronization and other special implementation. Located in bottom right are normal and hierarchical neural networks as well as spiking networks.

In contrast to these architectures, we have the possibility of upper areas in Fig. 1, that is, neural networks or von Neumann systems with massive connections, by use of waves. In addition, we can cope with high variability with the learning and/or self-organization dynamics when we employ neural networks. This is a great merit of wave-based neural networks.

4   CVNNs in Reservoir Computing

4.1   Wiring Avoidance and Neuro-Dynamics Realization by Waves

There are two aspects in the use of waves in information processing devices and systems. One is the interconnection. For years, optical fiber interconnection has been widely used very effectively for high-speed information transmission. However, such a use requires optical-fiber "wiring" anyway. But in these years, many people are interested in free-space lightwave or electromagnetic-wave connections again because of their high flexibility. Being free from cables is also an advantage. The cableless flexibility sometimes overcomes communication overheads.

The other one is wave-based information processing in neural networks [6–10,18,19]. Its specific feature is to deal with not only amplitude but also phase information in the complex amplitude of lightwave, electromagnetic wave, sonic wave. In this sense, CVNN is a coherent neural network.

In old days, phase information was introduced also in "parametron" [20]. However, the parametron could not fully utilize the phase property and its dynamics since parametron system is a digital circuit manipulating symbols such as "bit." This point, besides the incompatibility with integration, has been a heavy drawback of parametron. This situation is completely different from that of neural networks, which is based on pattern information representation and processing.

4.2   Complex-Valued Neural Networks (CVNNs)

CVNN is a framework to deal with complex amplitude [17,21–28]. They extend their applicable fields mainly in electronics such as coherent imaging [13–15], channel prediction in multi-path mobile communications to treat complicated electromagnetic field [16], lightwave information processing systems [6,7], in particular in adaptive processing of phase information [8] and lightwave frequency-multiplexed information processing [9,10], lightwave processing without physical wire interconnections [19,29–31], as well as quantum computing [32].

The most important advantage of a CVNN lies in the superior generalization ability in application to processing of wave-originating information and wave-based neural hardware [33]. The merit is significant also in wave-based reservoir computing.

4.3   Reservoir Neural Networks

Figure 2 illustrates the schematic construction of a reservoir neural network. Input signals fed at the input terminals are input to the so-called reservoir where many neurons are connected one another with recurrence. Output neurons receive signals from the reservoir to synthesize output signals adaptively. Learning and/or self-organization occurs only in this output layer, not in the reservoir, just like perceptron. High variety in the reservoir connections generates signal components useful for building desired output in the output layer.

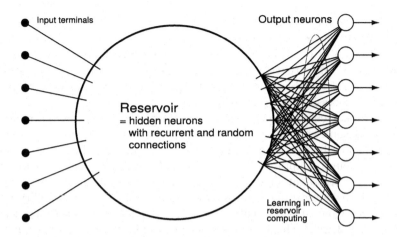

Fig. 2. Schematic construction diagram of a reservoir computing module

Table 1. Reservoir networks and recurrent neural networks

Learning in hidden connections	Dynamics	Merits
Yes	Fully recurrent network	High accuracy
No	Reservoir computing	Energy efficient, fast in learning

Table 1 compares reservoir networks with fully recurrent networks. In these years, many groups are working on reservoir computing theories [2] including echo-satte networks [3,4] as well as physical implementation using lightwave or other physical entities [1,34]. The decision not to introduce learning in the reservoir leads to a great advantage of energy efficiency. This strategy may also result in lower accuracy in theory. However, we have many applications in the real society where a rough but quick solution is more useful than accurate but energy- and time-consuming processing.

When we deal fully with wave information, we actually work on its amplitude and phase. We have to introduce the natural metric involved in the complex amplitude for use of physical wave nature into the neural dynamics. Nonliearity should also be in harmony with the amplitude-phase construction so that it works as a meaningful nonlinear function. Such a neural framework is the CVNN. Its importance is significant even in reservoir computing where only the output layer neurons learn and/or self-oragnize. Note that, when we deal with polarization additionally, we should employ/combine quaternion neural networks, an extended framework of CVNNs [15,35,36].

4.4 Comparisons of Various Neural Network Hardware

Conventional computers represent information physically by whether one or more static particles (electrons) exist or not. In this sense, they use baseband (static) physical representation. This strategy is suitable for realizing a memory using electrons and representing information as digital symbols (bits). In particular, the ease in function updates by rewriting software programs has been worthy to note. This situation applies also to neural network implementation by using field programmable gate arrays (FPGAs) or general purpose computing on graphics processing unit (GPGPU).

Wave implementation of neural connections for adaptive learning and processing is somewhat in contrast. Though it is true that analog use sometimes results in a limited accuracy in each element, this is not a serious problem because the neural adaptability compensates this weakness in the operation as a system. We would rather utilize the great merit of analog use, namely the continuity in response, that realizes diverse and flexible learning ability in the physical level. Contrarily, in digital systems, dynamics such as learning need to be written as software. This is because the metric of bit information is completely separate from the physical metric such as voltage, resulting in the necessity for human beings to assign information and/or meaning of a bit by writing software.

Pulse neural networks, or spiking neural networks, may be located between these two architectures mentioned above. They have a set of merits such as the fact that a multiplication product is obtained simply as a time-domain average of a series of pulses multiplied by another sequence of pulses. However, because of the baseband circuit structure, they have a shortcoming of charge-and-discharge energy consumption just like conventional digital neural networks. In addition, the multiple-pulse information representation requires a frequency bandwidth wider than what is needed intrinsically for the information representation, resulting in larger power consumption.

Consequently, it is clear that wave-based neural networks hold a great advantage in particular in realization of energy-efficient neural-network hardware. It is also found that the framework of CVNNs plays an important role more and more in neural hardware in the next generation.

5　Conclusion

We discussed the significance of CVNNs in reservoir computing, in particular in wave-based reservoir computing hardware. Reservoir neural hardware has a great advantage in energy-efficient information processing. Reservoir networks utilizing lightwaves or other physical waves will make progress based on the CVNN framework.

References

1. Takeda, S., Nakano, D., Yamane, T., Tanaka, G., Nakane, R., Hirose, A., Nakagawa, S.: Photonic reservoir computing based on laser dynamics with external feedback. In: Hirose, A., Ozawa, S., Doya, K., Ikeda, K., Lee, M., Liu, D. (eds.) ICONIP 2016. LNCS, vol. 9947, pp. 222–230. Springer, Cham (2016). doi:10.1007/978-3-319-46687-3_24

2. Yamane, T., Takeda, S., Nakano, D., Tanaka, G., Nakane, R., Nakagawa, S., Hirose, A.: Dynamics of reservoir computing at the edge of stability. In: Hirose, A., Ozawa, S., Doya, K., Ikeda, K., Lee, M., Liu, D. (eds.) ICONIP 2016. LNCS, vol. 9947, pp. 205–212. Springer, Cham (2016). doi:10.1007/978-3-319-46687-3_22
3. Tanaka, G., Nakane, R., Yamane, T., Nakano, D., Takeda, S., Nakagawa, S., Hirose, A.: Exploiting Heterogeneous units for reservoir computing with simple architecture. In: Hirose, A., Ozawa, S., Doya, K., Ikeda, K., Lee, M., Liu, D. (eds.) ICONIP 2016. LNCS, vol. 9947, pp. 187–194. Springer, Cham (2016). doi:10.1007/978-3-319-46687-3_20
4. Mori, R., Tanaka, G., Nakane, R., Hirose, A., Aihara, K.: Computational performance of echo state networks with dynamic synapses. In: Hirose, A., Ozawa, S., Doya, K., Ikeda, K., Lee, M., Liu, D. (eds.) ICONIP 2016. LNCS, vol. 9947, pp. 264–271. Springer, Cham (2016). doi:10.1007/978-3-319-46687-3_29
5. Yamane, T., Katayama, Y., Nakane, R., Tanaka, G., Nakano, D.: Wave-based reservoir computing by synchronization of coupled oscillators. In: Arik, S., Huang, T., Lai, W.K., Liu, Q. (eds.) ICONIP 2015. LNCS, vol. 9491, pp. 198–205. Springer, Cham (2015). doi:10.1007/978-3-319-26555-1_23
6. Hirose, A., Eckmiller, R.: Proposal of frequency-domain multiplexing in optical neural networks. Neurocomputing **10**(2), 197–204 (1996)
7. Hirose, A., Eckmiller, R.: Coherent optical neural networks that have optical-frequency-controlled behavior and generalization ability in the frequency domain. Appl. Opt. **35**(5), 836–843 (1996)
8. Kawata, S., Hirose, A.: Coherent optical neural network that learns desirable phase values in frequency domain by using multiple optical-path differences. Opt. Lett. **28**(24), 2524–2526 (2003)
9. Kawata, S., Hirose, A.: Frequency-multiplexed logic circuit based on a coherent optical neural network. Appl. Opt. **44**(19), 4053–4059 (2005)
10. Kawata, S., Hirose, A.: Frequency-multiplexing ability of complex-valued Hebbian learning in logic gates. Int. J. Neural Syst. **12**(1), 43–51 (2008)
11. Tanizawa, K., Hirose, A.: Performance analysis of steepest-descent-based feedback control of tunable dispersion compensator for adaptive dispersion compensation in all-optical dynamic routing networks. IEEE/OSA J. Lightwave Technol. **25**(4), 1086–1094 (2007)
12. Tanizawa, K., Hirose, A.: Fast tracking algorithm for adaptive compensation of high-speed PMD variation caused by SOP change in milliseconds. IEEE Photonics Technol. Lett. **21**(3), 140–142 (2009)
13. Hara, T., Hirose, A.: Adaptive plastic-landmine visualizing radar system: effects of aperture synthesis and feature-vector dimension reduction. IEICE Trans. Electron. **E88–C**(12), 2282–2288 (2005)
14. Suksmono, A.B., Hirose, A.: Interferometric sar image restoration using Monte-Carlo metropolis method. IEEE Trans. Sig. Process. **50**(2), 290–298 (2002)
15. Shang, F., Hirose, A.: Quaternion neural-network-based PolSAR land classification in poincare-sphere-parameter space. IEEE Trans. Geosci. Remote Sens. **52**(9), 5693–5703 (2014)
16. Ding, T., Hirose, A.: Fading channel prediction based on combination of complex-valued neural networks and chirp Z-transform. IEEE Trans. Neural Netw. Learn. Syst. **25**(9), 1686–1695 (2014)
17. Hirose, A.: Complex-Valued Neural Networks, 2nd edn. Springer, Heidelberg (2012)
18. Hirose, A., Eckmiller, R.: Behavior control of coherent-type neural networks by carrier-frequency modulation. IEEE Trans. Neural Netw. **7**(4), 1032–1034 (1996)

19. Hirose, A., Kiuchi, M.: Coherent optical associative memory system that processes complex-amplitude information. IEEE Photon. Tech. Lett. **12**(5), 564–566 (2000)
20. Goto, E.: The parametron - a new circuit element which utilizes non-linear reactors. Paper of Technical Group of Electronic Computers and Nonlinear Theory, IECE (1954, in Japanese)
21. Hirose, A.: Complex-Valued Neural Networks: Theories and Applications. Innovative Intelligence, vol. 5. World Scientific Publishing, Singapore (2003)
22. Hirose, A. (ed.): Complex-Valued Neural Networks: Advances and Applications. IEEE Press Series on Computational Intelligence. IEEE Press and Wiley, New Jersey (2013)
23. Mandic, D.P., Goh, V.S.L.: Complex Valued Nonlinear Adaptive Filters - Noncircularity, Widely Linear and Neural Models. Wiley, Hoboken (2009)
24. Adali, T., Haykin, S.: Adaptive Signal Processing: Next Generation Solutions. Wiley-IEEE Press, New Jersey (2010)
25. Aizenberg, I.: Complex-Valued Neural Networks with Multi-Valued Neurons. Studies in Computational Intelligence. Springer, Heidelberg (2011)
26. Nitta, T.: Complex-Valued Neural Networks: Utilizing High-Dimensional Parameters. Information Science Reference, Pennsylvania (2009)
27. Bayro-Corrochano, E.: Geometric Computing for Wavelet Transforms, Robot Vision, Learning, Control and Action. Springer, Heidelberg (2010)
28. Suresh, S., Sundararajan, N., Savitha, R.: Supervised Learning with Complex-valued Neural Networks. Springer, Heidelberg (2013)
29. Hirose, A., Higo, T., Tanizawa, K.: Efficient generation of holographic movies with frame interpolation using a coherent neural network. IEICE Electron. Expr. **3**(19), 417–423 (2006)
30. Tay, C.S., Tanizawa, K., Hirose, A.: Error reduction in holographic movies using a hybrid learning method in coherent neural networks. Appl. Opt. **47**(28), 5221–5228 (2008)
31. Takeda, M., Kirihara, S., Miyamoto, Y., Sakoda, K., Honda, K.: Localization of electromagnetic waves in three dimensional fractal cavities. Phys. Rev. Lett. **92**, 093902 (2004)
32. Ono, A., Sato, S., Kinjo, M., Nakajima, K.: Study on the performance of neuromorphic adiabatic quantum computation algorithms. In: International Joint Conference on Neural Networks (IJCNN) 2008, Hong Kong, Nakajima, pp. 2508–2512, June 2008
33. Hirose, A., Yoshida, S.: Generalization characteristics of complex-valued feedforward neural networks in relation to signal coherence. IEEE Trans. Neural Netw. Learn. Syst. **23**, 541–551 (2012)
34. Antonik, P., Duport, F., Smerieri, A., Hermans, M., Haelterman, M., Massar, S.: Online training of an opto-electronic reservoir computer. In: Arik, S., Huang, T., Lai, W.K., Liu, Q. (eds.) ICONIP 2015. LNCS, vol. 9490, pp. 233–240. Springer, Cham (2015). doi:10.1007/978-3-319-26535-3_27
35. Matsui, N., Isokawa, T., Kusamichi, H., Peper, F., Nishimura, H.: Quaternion neural network with geometrical operators. J. Intell. Fuzzy Syst. **15**, 149–164 (2004)
36. Takizawa, Y., Shang, F., Hirose, A.: Adaptive land classification and new class generation by unsupervised double-stage learning in poincare sphere space for polarimetric synthetic aperture radars. Neurocomputing **248**, 3–10 (2017)

Waveform Classification by Memristive Reservoir Computing

Gouhei Tanaka[1(✉)], Ryosho Nakane[1], Toshiyuki Yamane[2], Seiji Takeda[2],
Daiju Nakano[2], Shigeru Nakagawa[2], and Akira Hirose[1]

[1] Graduate School of Engineering, The University of Tokyo, Tokyo 113-8656, Japan
gouhei@sat.t.u-tokyo.ac.jp, nakane@cryst.t.u-tokyo.ac.jp,
ahirose@ee.t.u-tokyo.ac.jp
[2] IBM Research – Tokyo, Kawasaki, Kanagawa 212-0032, Japan
{tyamane,seijitkd,dnakano,snakagw}@jp.ibm.com

Abstract. Reservoir computing is one of the computational frameworks based on recurrent neural networks for learning sequential data. We study the memristive reservoir computing where a network of memristors, instead of recurrent neural networks, provides a nonlinear mapping from input sequential signals to high-dimensional spatiotemporal dynamics. First we formulate the circuit equations of the memristive networks and describe the simulation methods. Then we use the memristive reservoir computing for solving a waveform classification problem. We demonstrate how the classification ability depends on the number of reservoir outputs and the variability of the memristive elements. Our methods are useful for finding a better architecture of the memristive reservoir under the inevitable element variability when implemented with nano/microscale devices.

Keywords: Reservoir computing · Recurrent networks · Memristors · Pattern classification · Energy efficiency

1 Introduction

Reservoir computing is a computational framework originating from neural information processing, particularly suited for sequential data processing [1,2]. The concept of reservoir computing was presented as a unified framework including the echo state network [3,4] and the liquid state machine [5], both of which are variants of recurrent neural networks. A reservoir computing system consists of a *reservoir* part and a *readout* part: the reservoir part converts input sequential signals into high-dimensional spatiotemporal patterns; the readout part extracts useful information on the input signals from the spatiotemporal patterns. This system is available for computational tasks such as sequential pattern recognition and prediction. The main characteristic of reservoir computing is that only the connection weights in the readout part are adjusted by a simple learning method, which enables to save the cost for learning compared with classical recurrent neural networks.

© Springer International Publishing AG 2017
D. Liu et al. (Eds.): ICONIP 2017, Part IV, LNCS 10637, pp. 457–465, 2017.
https://doi.org/10.1007/978-3-319-70093-9_48

In general, the reservoir should have some specific properties for effective computation. To reflect time correlations of input signals in the output signals, the reservoir needs to generate history-dependent dynamics. At the same time, a fading memory property is important for avoiding a long-lasting memory causing a loss of variety in the output signals. If the reservoir system is contracting from the dynamical systems viewpoint, then different input patterns are mapped into similar output signals and thus a differentiation of them from the output signals is difficult. In contrast, if the reservoir system is expanding, then similar input signals are transformed into very different output signals and no consistent behavior required for computation is expected. In this way, an appropriate design of the reservoir is significant for realizing high computational performance.

Toward device implementation of reservoir computing systems, there are many attempts to construct the reservoir using physical systems, instead of sparse recurrent neural networks. One of such physical systems is the memristive network, i.e. a network of memristors. The memristor is a two-terminal passive circuit element which has a changeable conductance depending on the history of the electric charge that passed through the element [6]. The existence of memristor was first predicted by Chua [7] and after a long time the electric device has been established [8]. The memristor with scalar state was generalized to memristive systems with vector state, which are mathematically defined in Ref. [9]. The history-dependent property of the memristor is available to mimic synaptic facilitation and depression in biological neurons, and therefore, expected to be used for neuronal units in neuromorphic circuits [10]. On the other hand, memristor-based networks [11–15] and atomic switch networks with memristive property [16–18] have been applied to reservoir computing. Most of these studies use the network of memristor elements, but the explicit formulation of circuit equations has not been given so far. In addition, the computational ability of memristive reservoir computing has yet to be fully clarified. There remain many factors which would affect the computational performance of memristive reservoir computing.

One of the concerns in nano/micro-scale device implementation of memristive networks is the effect of the variability of individual memristive elements on the computational performance [12, 19]. Motivated by this issue, we investigate the memristive reservoir computing for sequential pattern recognition in this study. First, we construct the circuit equations of memristive networks and describe the simulation method. Then, we apply the networks to a waveform classification problem. We demonstrate that the classification ability depends on the number of reservoir outputs and the element variability.

2   Methods

2.1   Single Memristor Model

The single memristor changes its resistance with time when a voltage input is applied. The mathematical model of a general memristive system is described as follows [8]:

$$v(t) = R(w(t), i(t))i(t), \tag{1}$$

$$\frac{dw(t)}{dt} = f(w(t), i(t)), \tag{2}$$

where $v(t)$ and $i(t)$ represent the voltage and the current at time t, respectively, $w(t)$ represents the internal state variable which evolves with time, and R represents the resistance depending on the internal variable. The function f governing the intrinsic V-I characteristic of the memristor is determined by modelling ionic drifts [20].

In the linear drift model [8], it is assumed that the memristor consists of the doped region with a small resistance R_{off} and the undoped region with a large resistance R_{on}. The length of the device is denoted by D and the length of the doped region is denoted by w. Then, the resistance of the memristor (memristance) M is given by

$$M(w) = \left(R_{\text{on}} \frac{w}{D} + R_{\text{off}} \left(1 - \frac{w}{D} \right) \right). \tag{3}$$

The equations for the linear drift model is described as follows [8]:

$$v(t) = M(w(t))i(t), \tag{4}$$

$$\frac{dw(t)}{dt} = \mu_v \frac{R_{\text{on}}}{D} i(t), \tag{5}$$

where μ_v represents the average ion mobility. The conductance of the memristor (memductance) W is calculated as follows [21]:

$$W(\Phi) = \frac{dq(\Phi)}{d\Phi} = \frac{1}{\sqrt{M(w(0))^2 - 2a\Phi}}, \tag{6}$$

where Φ denotes the magnetic flux, q denotes the charge, and $a = \mu_v R_{\text{on}}(R_{\text{off}} - R_{\text{on}})/D^2$.

2.2 Memristive Networks

We consider a reservoir computing system with a network of M memristors as illustrated in Fig. 1. The number of nodes connecting memristors is denoted by N. The number of voltage sources is denoted by S. By regarding the nodes as vertices and the memristor branches as edges, the connectivity of memristors is represented as a directional graph, which is described by an $N \times M$ incidence matrix E_m: if branch m ($m = 1, \ldots, M$) connects the starting node n_s and the end node n_e, then $E_m(n_s, m) = -1$, $E_m(n_e, m) = 1$, and $E_m(n, m) = 0$ for $n \neq n_s, n_e$. Similarly, the connectivity of voltage sources is represented by an $N \times S$ matrix E_s. We denote the vectors of node voltages and node fluxes by $\mathbf{v}_n = (v_1, \ldots, v_N)^\top$ and $\mathbf{\Phi}_n = (\Phi_1, \ldots, \Phi_N)^\top$, respectively. Similarly, the vectors of source voltages and currents are denoted by $\mathbf{V}_s = (V_1, \ldots, V_S)^\top$ and $\mathbf{I}_s = (I_1, \ldots, I_S)^\top$, respectively.

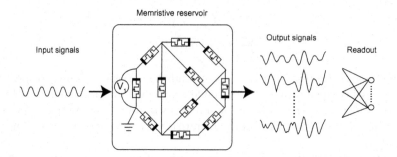

Fig. 1. Schematic illustration of memristive reservoir compputing.

We use the new modified nodal analysis [22] to formulate the circuit equations governing the dynamics of the memristive network. The system of the memristive network is given by the differential-algebraic equations (DAEs) as follows:

$$\frac{d}{dt}E_m q(E_m^\top \mathbf{\Phi}_n(t)) + E_s \mathbf{I}_s(t) = \mathbf{0}, \tag{7}$$

$$\frac{d\mathbf{\Phi}_n(t)}{dt} - \mathbf{v}_n(t) = \mathbf{0}, \tag{8}$$

$$E_s^\top \mathbf{v}_n(t) - \mathbf{V}_s(t) = \mathbf{0}, \tag{9}$$

where Eq. (7) represents the Kirchhoff's law, Eq. (8) corresponds to the Faraday's law, and Eq. (9) indicates the constraint that the source voltage is the same as the voltage difference between the end nodes. Using the first-order linearization and the memductance in Eq. (6), we can rewrite Eq. (7) as follows:

$$E_m W(E_m^\top \mathbf{\Phi}_n) E_m^\top \frac{d\mathbf{\Phi}_n}{dt} + E_s \mathbf{I}_s = \mathbf{0}. \tag{10}$$

The DAEs in Eqs. (8)–(10) are numerically integrated using the solver ode15i operating on the software package Matlab 2016 [23].

2.3 Computational Task

We apply the memristive reservoir computing system to the waveform classification problem [24] which aims to distinguish between sinusoidal and triangular waves. The classification of wave patterns is important as a fundamental task for more practical applications such as speech recognition [2] as well as for developing wave-based reservoir computing [25, 26]. We generate 50 waveform data with different frequencies for each type and toally obtain 100 waveform data as shown in Fig. 2. We use 25 sinusoidal and 25 triangular waveform data for training and use the remaining data for testing. The waveform data is given as the voltage source $V_1(t)$ in the memristive reservoir as illustrated in Fig. 1. The kth waveform signal in each group is represented as $V_1(t) = V_0 \sin(\omega_k t)$ or $V_1(t) = V_0 \cdot \text{triangular}(\omega_k t)$. The voltage amplitude is fixed at constant and the

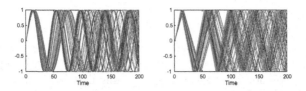

Fig. 2. Examples of sinusoidal (left) and triangular (right) waveforms.

angular frequency of the kth waveform is set at $\omega_k = \omega_0(1 + r_k)$ where r_k is randomly taken from the uniform distribution with range $[-r, r]$.

The output sequential signals of the reservoir, represented by an L-dimensional vector $\mathbf{x}(t) = (x_1(t), \ldots, x_L(t))^\top$, are given as the inputs to the two output neurons in the readout part. The output neurons convert the sum of weighted inputs to the network output state $\mathbf{y}(t) = (y_1, y_2)^\top$ by the sigmoid function as follows:

$$y_i(t) = (1 + \exp(-h_i(t)))^{-1}, \quad \text{for } i = 1, 2, \tag{11}$$

where $h_i(t) = \sum_{j=1}^{L} w_{ij} x_j(t)$ is the weighted sum of inputs. The teacher outputs are given by $\mathbf{d}(t) = (d_1(t), d_2(t))^\top = (1, 0)^\top$ for sinusoidal waves and $\mathbf{d}(t) = (d_1(t), d_2(t))^\top = (0, 1)^\top$ for triangular waves. In the training phase, the connection weights w_{ij} are determined so as to minimize the difference between the network outputs and the teacher outputs, using the least-square method. In the testing phase, the network outputs are computed from the reservoir output for the testing data and the obtained weights in the training phase. We evaluate the classification accuracy in the testing phase. The parameter values in the numerical experiments are listed in Table 1. To take into account the variability of individual memristors, we set $R_{on} = \bar{R}_{on}(1 + u_k)$, $R_{off} = \bar{R}_{off}(1 + u_k)$, and $w_0 = 0.5D(1 + u_k)$ for the kth memristor, where u_k is randomly taken from the uniform distribution with range $[-u, u]$.

Table 1. Parameters

Parameter	Nominal values	Meaning
$\bar{R}_{on}$	3.33×10^7 $[\Omega]$	Resistance of doped region
$\bar{R}_{off}$	3.33×10^{10} $[\Omega]$	Resistance of undoped region
D	1.0×10^{-8} $[m]$	Length of the device
μ_v	2.5×10^{-6} $[m^2 s^{-1} V^{-1}]$	Average ion mobility
V_0	1.0 $[V]$	Voltage amplitude
ω_0	1.0×10^8 $[rad/s]$	Angular frequency
r	0.2	Range of the waveform frequency

3   Results

The memristive reservoir transforms the input waveform patterns into spatiotemporal patterns. Figure 3 shows the time courses of the branch currents of the
memristor reservoir for the sinusoidal and triangular waves. We use randomly
chosen L branch currents as the reservoir outputs. The amplitudes are different
but the phases are synchronized. This means that the low-dimensional input pattern is converted to the high-dimensional spatiotemporal patterns. Our purpose
is to well differentiate between these responses to the two waveform patterns in
the readout part.

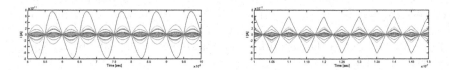

Fig. 3. The output signals (branch currents) of the reservoir for a sinusoidal input
(left) and a triangular input (right).

In the waveform classification task, the number of vertices is fixed at $N = 20$.
First, we consider a simple ring structure with $M = 20$ for the memristive reservoir [19,27]. Figure 4(a) shows the classification accuracy plotted against the
number of reservoir outputs, L. The five symbols correspond to the different values of the variability parameter u. When the single reservoir output is used, the
accuracy is 50%, which is the same as that in the case without reservoir. When
two or more than two reservoir outputs are used, the accuracy is highly improved.
In the case of $u = 0$ for which all the memristors are identical, the perfect classification is achieved. As the variability u increases, the accuracy is degraded.
However, the variability of memristor elements is unavoidable when they are

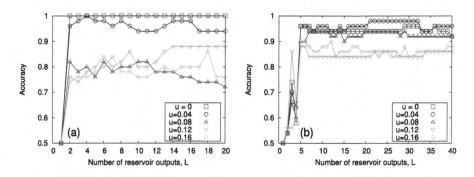

Fig. 4. Classification accuracy for the memristive networks. (a) The ring topology. (b)
The small-world topology.

implemented with nano/micro-scale devices. Therefore, we need to improve the accuracy under large element variability for practical purposes.

We change the network topology by adding random connections to the ring topology and obtain a memristive network with $M = 40$. This is similar to the small-world network architecture [28]. By this modification, the symmetry and regularity of the network topology is broken and more rich nonlinear spatiotemporal patterns are expected to arise. Figure 4(b) shows the result of the classification test. We notice that the accuracy is decreased for the low variability cases but increased for the large variability cases. This result suggests that one of the approaches for improving the classification ability of the memristive reservoir computing is to appropriately design the network architecture.

4   Conclusions

We have developed the reservoir computing system using memristive networks for sequential pattern classification. We have shown that a high classification performance is achieved with the ring topology network if the variability of memristors is zero or low but not if the variability is high. By incorporating random connections into the ring network, we have demonstrated that the classification accuracy under the element variability is improved. Our results are useful to realize variability-tolerant memristive reservoir computing. Future works include the analysis of nonlinear dynamics of the memristive networks, particularly focusing on how the network topology, the element variability, and the input signals influence the properties of the spatiotemporal dynamics and the performance in computational tasks. After fully understanding the memristive reservoir computing system in the model studies, the practical conditions can be considered for hardware implementation of the memristive network device.

Acknowledgments. This work was partially supported by JSPS KAKENHI Grant Number 16K00326 (GT).

References

1. Schrauwen, B., Verstraeten, D., Van Campenhout, J.: An overview of reservoir computing: theory, applications and implementations. In: Proceedings of the 15th European Symposium on Artificial Neural Networks, pp. 471–482 (2007)
2. Verstraeten, D., Schrauwen, B., d'Haene, M., Stroobandt, D.: An experimental unification of reservoir computing methods. Neural Netw. **20**(3), 391–403 (2007)
3. Jaeger, H.: The "echo state" approach to analysing and training recurrent neural networks-with an erratum note. Bonn, Germany: German National Research Center for Information Technology GMD Technical report 148, 34 (2001)
4. Jaeger, H.: Tutorial on training recurrent neural networks, covering BPPT, RTRL, EKF and the "echo state network" approach. GMD-Forschungszentrum Informationstechnik (2002)
5. Maass, W., Natschlager, T., Markram, H.: Real-time computing without stable states: a new framework for neural computation based on perturbations. Neural Comput. **14**(11), 2531–2560 (2002)

6. Di Ventra, M., Pershin, Y.V., Chua, L.O.: Circuit elements with memory: memristors, memcapacitors, and meminductors. Proc. IEEE **97**(10), 1717–1724 (2009)
7. Chua, L.: Memristor-the missing circuit element. IEEE Trans. Circ. Theory **18**(5), 507–519 (1971)
8. Strukov, D.B., Snider, G.S., Stewart, D.R., Williams, R.S.: The missing memristor found. Nature **453**(7191), 80–83 (2008)
9. Chua, L.O., Kang, S.M.: Memristive devices and systems. Proc. IEEE **64**(2), 209–223 (1976)
10. Chang, T., Yang, Y., Lu, W.: Building neuromorphic circuits with memristive devices. IEEE Circ. Syst. Mag. **13**(2), 56–73 (2013)
11. Kulkarni, M.S., Teuscher, C.: Memristor-based reservoir computing. In: 2012 IEEE/ACM International Symposium on Nanoscale Architectures (NANOARCH), pp. 226–232 (2012)
12. Bürger, J., Teuscher, C.: Variation-tolerant computing with memristive reservoirs. In: Proceedings of the 2013 IEEE/ACM International Symposium on Nanoscale Architectures, pp. 1–6. IEEE Press (2013)
13. Bürger, J., Goudarzi, A., Stefanovic, D., Teuscher, C.: Hierarchical composition of memristive networks for real-time computing. In: 2015 IEEE/ACM International Symposium on Nanoscale Architectures (NANOARCH), pp. 33–38. IEEE (2015)
14. Burger, J., Goudarzi, A., Stefanovic, D., Teuscher, C.: Computational capacity and energy consumption of complex resistive switch networks. AIMS Mater. Sci. **2**(4), 530–545 (2015)
15. Merkel, C., Saleh, Q., Donahue, C., Kudithipudi, D.: Memristive reservoir computing architecture for epileptic seizure detection. Procedia Comput. Sci. **41**, 249–254 (2014)
16. Stieg, A.Z., Avizienis, A.V., Sillin, H.O., Martin-Olmos, C., Aono, M., Gimzewski, J.K.: Emergent criticality in complex turing B-type atomic switch networks. Adv. Mater. **24**(2), 286–293 (2012)
17. Sillin, H.O., Aguilera, R., Shieh, H.H., Avizienis, A.V., Aono, M., Stieg, A.Z., Gimzewski, J.K.: A theoretical and experimental study of neuromorphic atomic switch networks for reservoir computing. Nanotechnology **24**(38), 384004 (2013)
18. Stieg, A.Z., Avizienis, A.V., Sillin, H.O., Aguilera, R., Shieh, H.-H., Martin-Olmos, C., Sandouk, E.J., Aono, M., Gimzewski, J.K.: Self-organization and emergence of dynamical structures in neuromorphic atomic switch networks. In: Adamatzky, A., Chua, L. (eds.) Memristor Networks, pp. 173–209. Springer, Cham (2014). doi:10.1007/978-3-319-02630-5_10
19. Tanaka, G., Nakane, R., Yamane, T., Nakano, D., Takeda, S., Nakagawa, S., Hirose, A.: Exploiting heterogeneous units for reservoir computing with simple architecture. In: Hirose, A., Ozawa, S., Doya, K., Ikeda, K., Lee, M., Liu, D. (eds.) ICONIP 2016. LNCS, vol. 9947, pp. 187–194. Springer, Cham (2016). doi:10.1007/978-3-319-46687-3_20
20. Joglekar, Y.N., Wolf, S.J.: The elusive memristor: properties of basic electrical circuits. Eur. J. Phys. **30**(4), 661 (2009)
21. McDonald, N.R., Pino, R.E., Rozwood, P.J., Wysocki, B.T.: Analysis of dynamic linear and non-linear memristor device models for emerging neuromorphic computing hardware design. In: The 2010 International Joint Conference on Neural Networks (IJCNN), pp. 1–5. IEEE (2010)
22. Fei, W., Yu, H., Zhang, W., Yeo, K.S.: Design exploration of hybrid CMOS and memristor circuit by new modified nodal analysis. IEEE Trans. Very Large Scale Integr. (VLSI) Syst. **20**(6), 1012–1025 (2012)

23. MATLAB: version 9.0 (R2016a). The MathWorks Inc., Natick, Massachusetts (2016)
24. Takeda, S., Nakano, D., Yamane, T., Tanaka, G., Nakane, R., Hirose, A., Nakagawa, S.: Photonic reservoir computing based on laser dynamics with external feedback. In: Hirose, A., Ozawa, S., Doya, K., Ikeda, K., Lee, M., Liu, D. (eds.) ICONIP 2016. LNCS, vol. 9947, pp. 222–230. Springer, Cham (2016). doi:10.1007/978-3-319-46687-3_24
25. Katayama, Y., Yamane, T., Nakano, D., Nakane, R., Tanaka, G.: Wave-based neuromorphic computing framework for brain-like energy efficiency and integration. IEEE Trans. Nanotechnol. **15**(5), 762–769 (2016)
26. Yamane, T., Katayama, Y., Nakane, R., Tanaka, G., Nakano, D.: Wave-based reservoir computing by synchronization of coupled oscillators. In: Arik, S., Huang, T., Lai, W.K., Liu, Q. (eds.) ICONIP 2015. LNCS, vol. 9491, pp. 198–205. Springer, Cham (2015). doi:10.1007/978-3-319-26555-1_23
27. Rodan, A., Tino, P.: Minimum complexity echo state network. IEEE Trans. Neural Netw. **22**(1), 131–144 (2011)
28. Watts, D.J., Strogatz, S.H.: Collective dynamics of "small-world" networks. Nature **393**(6684), 440–442 (1998)

A Preliminary Approach to Semi-supervised Learning in Convolutional Neural Networks Applying "Sleep-Wake" Cycles

Mikel Elkano[1,2(✉)], Humberto Bustince[1,2], and Andrew Paplinski[3]

[1] Department of Automatics and Computation, Public University of Navarre, Campus Arrosadia s/n, 31006 Pamplona, Spain
{mikel.elkano,bustince}@unavarra.es
[2] Institute of Smart Cities, Public University of Navarre, Campus Arrosadia s/n, 31006 Pamplona, Spain
[3] Monash University, 25 Exhibition Walk, Clayton, Melbourne 3800, Australia
andrew.paplinski@monash.edu

Abstract. The scarcity of labeled data has limited the capacity of convolutional neural networks (CNNs) until not long ago and still represents a serious problem in a number of image processing applications. Unsupervised methods have been shown to perform well in feature extraction and clustering tasks, but further investigation on unsupervised solutions for CNNs is needed. In this work, we propose a bio-inspired methodology that applies a deep generative model to help the CNN take advantage of unlabeled data and improve its classification performance. Inspired by the human "sleep-wake cycles", the proposed method divides the learning process into sleep and waking periods. During the waking period, both the generative model and the CNN learn from real training data simultaneously. When sleep begins, none of the networks receive real data and the generative model creates a synthetic dataset from which the CNN learns. The experimental results showed that the generative model was able to teach the CNN and improve its classification performance.

Keywords: Semi-supervised learning · Sleep-wake cycles · Variational autoencoders · Convolutional neural networks · Generative models · Deep learning

1 Introduction

Deep learning has revolutionized the field of machine learning in the last decade. Among existing techniques, convolutional neural networks (CNNs)[8] have been shown to be the best performing approach in image processing [4,7]. These networks are based on bio-inspired architectures that capture important characteristics of the mammalian visual system, such as hierarchical organization and receptive fields. However, CNNs require vast amounts of training data due to the large number of model parameters that need to be adjusted. Moreover, this

© Springer International Publishing AG 2017
D. Liu et al. (Eds.): ICONIP 2017, Part IV, LNCS 10637, pp. 466–474, 2017.
https://doi.org/10.1007/978-3-319-70093-9_49

drawback is even more accentuated by the scarcity of labeled data. Although a number of unsupervised solutions have been proposed to alleviate this problem [11,12,14], further research is needed to improve supervised classification performance using unlabeled data [12,14].

In this work, we propose the synergy of deep generative models and CNNs to improve supervised learning using unsupervised techniques. This combination has been motivated by the so-called *sleep-wake cycles* and the interaction between the hippocampus and the neocortex that takes place in the human memory consolidation process. It is now well established that sleep plays a key role in human memory performance by stabilizing memory traces and protecting them against interference [1,2,13]. According to [2], one of the functions of dreams might also be to create a virtual environment in which the human brain reinforces and tests certain behaviors. During the memory consolidation process, the hippocampus is responsible for the acquisition and integration of new information that will then be transferred to widespread high-order neocortical areas [3]. McClelland et al. suggest that after the initial acquisition, the hippocampal system serves as a teacher to the neocortex, allowing for the reinstatement of representations of past events in the neocortex [10]. In this manner, this information may be gradually acquired by the cortical system via interleaved learning.

The synergy proposed in this work takes the aforementioned concepts and puts them all together to train a variational autoencoder (VAE) [5] that allows supervised CNNs to take advantage of unlabeled data. Therefore, our model comprises two different neural networks: the VAE and the supervised CNN. The way in which the VAE helps the CNN to deal with mostly unlabeled datasets is (vaguely) inspired by the human sleep-wake cycles. During the waking period, both networks learn from real training data simultaneously. When sleep begins, none of the networks receive real data and the VAE creates a synthetic (virtual) dataset from which the CNN learns. During sleep, only the CNN modifies the parameters of the model. As the VAE becomes more reliable, the amount of synthetic data used by the CNN increases. In this manner, the model will become more confident in its own internal representation as it evolves. We speculate that the VAE would be acting as a "hippocampal system" that helps the CNN ("neocortical visual areas") to reinforce the patterns received during the waking period. Results obtained from experiments carried out on MNIST handwritten digit recognition dataset show the effectiveness of this preliminary approach.

The paper is organized as follows. Section 2 briefly describes the generative model used in this work, i.e., the variational autoencoder. The proposed synergy of variational autoencoders and convolutional networks is presented in Sect. 3, while Sect. 4 shows the effectiveness of this method on MNIST dataset. Finally, we conclude the paper in Sect. 5 and suggest a possible future line to extend this methodology to deeper architectures and larger datasets.

2   Preliminaries: Variational Autoencoders

Variational autoencoders (VAEs) [5] have become one of the most popular frameworks for building generative models. The reason behind their success is a fast backpropagation-based learning process which does not need strong assumptions. The word *autoencoders* comes from the fact that the neural network built by this technique is composed of an *encoder* and a *decoder*. An autoencoder is a neural network that tries to build an approximate copy of its input that resembles the training data. To this end, the encoder learns a low-dimensional *code* or *internal representation* z of the input x, while the decoder is responsible for reconstructing the original data from this internal code. Autoencoders allow one to extract useful properties from training data.

In the case of VAEs, the hidden code z represents a probability distribution that is learned during training, instead of single values. Therefore, the encoder becomes a variational inference network that maps the data to the distribution of the hidden code ($q_\phi(z|x)$), and the decoder becomes a generative network that maps the hidden code back to the distribution of the data ($p_\theta(x|z)$). In this manner, the data generation process starts by sampling z from its distribution. More specifically, VAEs assume that samples of z can be drawn from a simple distribution, i.e., $z \sim N(0, I)$, where I is the identity matrix. This is reasonable because any distribution in d dimensions can be generated by taking d variables that are normally distributed and mapping them through a sufficiently complicated function, such as a Multi-Layer Perceptron (MLP).

In this work, we use a semi-supervised VAE introduced by Kingma et al. [6] that is able to learn from both unlabeled and labeled data. This method is a combination of two different VAEs:

- M1 model: provides a low-dimensional latent representation z_1 of the original data using the following generative model:

$$p(z_1) = N(z_1|0, I); \qquad p_\theta(x|z_1) = f(x; z_1, \theta), \qquad (1)$$

where $f(x; z_1, \theta)$ is a suitable likelihood function (e.g., a Gaussian or Bernoulli distribution) whose probabilities are formed by a non-linear transformation, with parameters θ, of a set of latent variables z_1. This non-linear transformation is given by a deep neural network.

- M2 model: describes the data as being generated by a latent class variable y plus a continuous latent variable z_2 as follows:

$$p(y) = Cat(y|\pi); \qquad p(z_2) = N(z_2|0, I); \qquad p_\theta(x|y, z_2) = f(x; y, z_2, \theta), \quad (2)$$

where $Cat(y|\pi)$ is the multinomial distribution, class labels y are treated as latent variables if no class label is available, and the input data x is given by the latent representation z_1 provided by M1. When y is unobserved, the inferred posterior distribution $p_\theta(y|x)$ predicts the class label, performing classification as inference.

3   Proposal: Semi-supervised Learning Based on the Synergy of Variational Autoencoders and Convolutional Neural Networks

The semi-supervised methodology proposed in this work consists in the interaction of two different deep learning models: variational autoencoders (VAEs) and supervised convolutional neural networks (CNNs). More specifically, our proposal is based on a sequence of "sleep-wake cycles" in which a VAE serves as a teacher to the CNN. This scheme allows the supervised CNN to take advantage of the internal representation created by the VAE from unlabeled data.

The learning process of our model applies the following procedure:

1. The VAE is first trained using the whole dataset (which typically contains a small amount of labeled data) in order to obtain a robust internal representation of the data.
2. Both the VAE and the CNN evolve (learn) simultaneously throughout a sequence of "sleep-wake cycles". Each of these cycles is composed of "sleep" and "waking" periods:
   - Wake: the supervised CNN is trained using only labeled data from the training set. Simultaneously, the VAE learns from the training set using both labeled and unlabeled data.
   - Sleep: the CNN does not receive real data anymore (in this cycle). Instead, it is trained using synthetic data generated and labeled by the VAE, which does not carry out any learning process during sleep (only the CNN learns during this period). We refer to these synthetic samples as "dreams". In order to generate new data, the encoders of M1 and M2 are dropped from the computations, since only the reconstruction path is needed for this process. Instead of having a real image as the input of the network, the decoder of M2 receives a vector sampled from a normal distribution directly and generates the input of the layer z_1 of M1. Finally, the decoder of M1 builds a new image from z_1. In this work, the values taken by the variable y during the generation process are not based on any previously learned parameter. Instead, the VAE builds the same number of samples for all labels in each cycle. It is worth noting that dreams vary from one cycle to another to prevent the CNN from overfitting. This variation is given by a Gaussian diffusion process defined by the following expression:

$$df = \sqrt{(1-\gamma)} * \varepsilon_1 + \sqrt{\gamma} * \varepsilon_2$$
$$input\_z_2 = \varepsilon_1 + s * (df - \varepsilon_1), \tag{3}$$

where $\varepsilon_1 \sim N(0, I)$ and $\varepsilon_2 \sim N(0, I)$, $\gamma \in \mathbb{R}$ ranges from 0 to 1, $s \in \mathbb{R}$ sets the smoothing factor, and $input\_z_2$ is the input of the layer z_2 of M2. Both γ and s control the trajectory variations of $input\_z_2$, which determines how dreams vary from one cycle to another. In this work we set γ to 0.8 and s to 1.

As the VAE becomes more reliable, the amount of synthetic data used by the CNN increases. In this manner, the model will become more confident in its own internal representation as it evolves. We must remark that the amount of synthetic data can be kept constant if the CNN varies the weight assigned to synthetic samples, obtaining a similar scenario. A diagram of the proposed methodology is shown in Fig. 1.

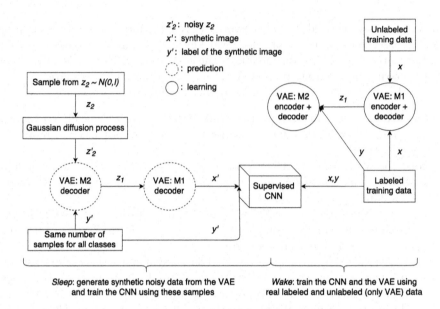

Fig. 1. Proposed sleep-wake cycle.

From a bio-inspired point of view, we suggest the following scheme. As described, the VAE is responsible for the construction of a robust internal representation of the real data received during the waking period, while the CNN focuses on maximizing discrimination capabilities. Therefore, there are two different neural circuits that are specialized in different tasks. We speculate that the VAE can be vaguely serving as a "hippocampal system" that might reinforce certain patterns in neocortical areas that would be represented by the CNN, which would be acting as a dedicated visual system. This reinforcement comes from a virtual environment created by the VAE during sleep [2], where the CNN learns from data generated from the input representation of past events [10]. In this manner, the VAE would be responsible for the acquisition of new information coming from unlabeled data and the subsequent reinstatement of these patterns in the visual system (CNN).

In comparison with unified objective functions [14] and self-training techniques [12], the advantage of training two separate neural networks (VAE and CNN) is that each network is specialized in a different task. As a consequence, this scheme can be applied with a wide variety of deep generative models and convolutional neural networks, and thus the capacity of each network is not limited by the method. The aforementioned properties allow us to build biologically inspired deep models that might capture certain interactions between human neural circuits.

4 Experimental Results on MNIST Handwritten Digits Dataset

We tested the effectiveness of our method using the MNIST database of handwritten digits [9]. The architecture and hyperparameters used throughout the experiments are the following:

- Variational autoencoder (VAE): we used the code published by Kingma at GitHub (https://github.com/dpkingma/nips14-ssl), which is written in Theano. Both M1 and M2 models were built considering the configuration recommended by the authors [6]. For M1 we used a 50-dimensional latent variable z. The Multi-Layer Perceptrons (MLPs) of the generative and inference models were composed of two hidden layers, each with 600 hidden units, using softplus $log(1 + e^x)$ activation functions. M2 also used 50-dimensional z and softplus activation functions, but in this case the MLPs had one hidden layer, each with 500 hidden units. The likelihood functions for $p_\theta(x|z_1)$ and $p_\theta(x|y, z_2)$ were given by Bernoulli and Gaussian distributions, respectively.
- Convolutional neural network (CNN): we used a simple CNN with two convolutional layers and two fully-connected layers. The convolutional layers had 3×3 receptive fields, 2×2 max-pooling, and 20 and 50 filters, respectively. The fully-connected layers were composed of 512 and 10 units. Dropout ratio was set to 20% and 50% in convolutional and fully-connected layers, respectively. All layers applied ReLU activation functions, except for the output layer, which used a softmax non-linearity.

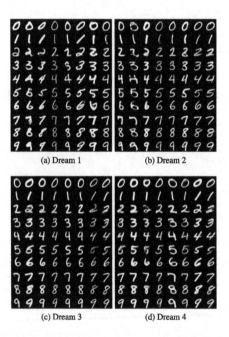

Fig. 2. Synthetic data generated by the VAE during sleep.

Both models were trained with the Adam optimizer using the following default settings: $\alpha = 0.001$, $\beta_1 = 0.9$, $\beta_2 = 0.999$, and $\varepsilon = 10^{-8}$.

For the experiment, a random subset of the original MNIST dataset is treated as unlabeled data by discarding its labels. Regarding sleep-wake cycles, the VAE and the CNN ran for 10 and 20 epochs in each cycle, respectively, for a total of 20 cycles. Note that the VAE ran for less epochs than the CNN, since it does not perform any learning process during sleep. The number of epochs for the initial training of the VAE was set to 300. In each cycle, the number of synthetic images created by the VAE increases according to the following equation:

$$n_s(i) = n_s(i-1) + \frac{(n_l * 4 - n_s(0))}{c} \quad \text{for all } i \in \mathbb{N} \quad with \quad i > 0, \quad (4)$$

where $n_s(i)$ is the number of synthetic images in the cycle i, $n_s(0)$ is the number of synthetic images in the initial cycle, n_l is the number of labels, and c is the number of cycles. Due to the small amount of labeled data, the batch size for the CNN was set to 20 in all cases except for the 50-label dataset, where it was set to 10.

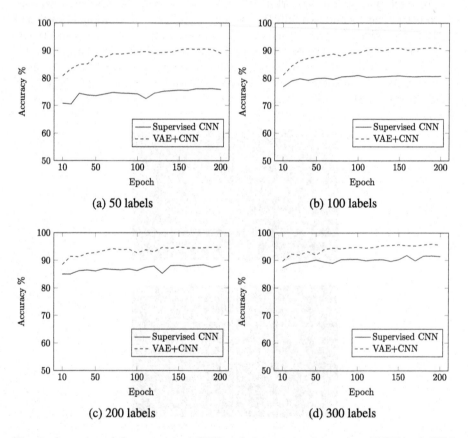

Fig. 3. Accuracy of the supervised CNN and the proposed synergy of VAE and CNN on MNIST dataset.

Figure 2 shows a subset of synthetic samples generated by the VAE during sleep. As we can observe, the model was able to construct new images from its internal representation, modifying the trajectory of digits in each cycle to prevent the CNN from overfitting.

In order to assess whether the sleep period was beneficial for the CNN, we ran our method (VAE+CNN) along with a supervised CNN in isolation (baseline). According to Fig. 3, when we used 50, 100, 200, and 300 labels our approach yielded 13%, 10%, 6%, and 4% improvement, respectively, after 20 sleep-wake cycles. These results suggest that the rate of improvement increased as the proportion of labeled data decreased, and thus the CNN was able to take advantage of the data generated by the VAE to improve its classification performance. Moreover, the plots in Fig. 3 show that the CNN of our method kept learning throughout the sleep-wake cycles, while the baseline CNN stopped improving its performance in a few epochs. This behavior suggests that the proposed methodology could be an interesting solution to improve incremental or online learning in CNNs, since the VAE and the CNN are trained simultaneously and the interaction between both neural networks takes place gradually.

5   Discussion

In this work, we have proposed a bio-inspired methodology to improve semi-supervised learning in convolutional neural networks (CNNs) using variational autoencoders (VAEs). In order to improve the classification performance of the CNN with the knowledge extracted by the VAE from unlabeled data, our model runs a sequence of "sleep-wake cycles" composed of sleep and waking periods. These cycles define the way in which both networks interact with each other and allow the CNN to learn from the VAE. During the waking period, both networks are simultaneously trained using real training data. When sleep begins, none of the networks receive real data and the VAE creates a number of synthetic noisy images from which the CNN learns. Based on this procedure, we speculate that the function of the VAE would be twofold: (1) during the waking period, it acts as a "hippocampal system" that is responsible for acquiring unlabeled data and building an internal representation that integrates new information [3]; (2) during sleep, it serves as a teacher to the CNN (that would represent high-level neocortical areas) by creating a virtual environment in which the CNN reinforces the patterns received in the waking period [2] and reinstates the representations of past stimuli [10]. The experiments carried out on MNIST dataset show that the CNN was able to learn from images created by the VAE. More specifically, the classification performance was improved by up to 13% over the purely supervised CNN. The advantage of our approach over simpler semi-supervised solutions is that one could apply any type of generative model or convolutional neural network. Consequently, the usage of deeper models should allow our method to tackle more complex problems, since each network would specialize in either "hippocampal" or "visual" tasks.

However, this work is a preliminary approach to the proposed methodology and further experiments are needed to assess its performance on large-scale datasets. Future work involves adding mechanisms that allow the VAE to learn how to create "useful dreams" for the CNN. This could be done by adding an extra branch to the decoder of M2, which would specialize in learning the distribution from which "useful dreams" are drawn. In this context, "useful dreams" might be the set of synthetic images that are generated with a high confidence level and force the CNN to make doubtful predictions. The learning process would consist in maximizing the norm of $p_\theta(y|x')$ (which implies confident labeling), x' being the synthetic image, while minimizing the norm of the CNN's output vector (which implies a doubtful prediction). This branch would be active only during sleep (replacing the original branch of the decoder) and would not affect the learning process of the VAE in the waking period.

Acknowledgments. This work has been partially supported by the Spanish Ministry of Science and Technology under the project TIN2016-77356-P (AEI/FEDER, UE).

References

1. Diekelmann, S., Born, J.: The memory function of sleep. Nat. Rev. Neurosci. **11**(2), 114–126 (2010)
2. Franklin, M.S., Zyphur, M.J.: The role of dreams in the evolution of the human mind. Evol. Psychol. **3**(1), 59–78 (2005)
3. Geib, B.R., Stanley, M.L., Dennis, N.A., Woldorff, M.G., Cabeza, R.: From hippocampus to whole-brain: the role of integrative processing in episodic memory retrieval. Hum. Brain Mapp. **38**(4), 2242–2259 (2017)
4. He, K., Zhang, X., Ren, S., Sun, J.: Deep residual learning for image recognition, pp. 770–778, January 2016
5. Kingma, D.P., Welling, M.: Auto-encoding variational Bayes. In: ICLR (2014)
6. Kingma, D., Rezende, D., Mohamed, S., Welling, M.: Semi-supervised learning with deep generative models. NIPS **4**, 3581–3589 (2014)
7. Krizhevsky, A., Sutskever, I., Hinton, G.: ImageNet classification with deep convolutional neural networks, vol. 2, pp. 1097–1105 (2012)
8. Lecun, Y., Bottou, L., Bengio, Y., Haffner, P.: Gradient-based learning applied to document recognition. Proc. IEEE **86**(11), 2278–2324 (1998)
9. LeCun, Y., Cortes, C.: The MNIST database of handwritten digits (1998)
10. McClelland, J.L., McNaughton, B.L., O'Reilly, R.C.: Why there are complementary learning systems in the hippocampus and neocortex: insights from the successes and failures of connectionist models of learning and memory. Psychol. Rev. **102**(3), 419–457 (1995)
11. Rasmus, A., Valpola, H., Honkala, M., Berglund, M., Raiko, T.: Semi-supervised learning with ladder networks. In: NIPS, pp. 3546–3554, January 2015
12. Shinozaki, T.: Semi-supervised learning for convolutional neural networks using mild supervisory signals. In: Hirose, A., Ozawa, S., Doya, K., Ikeda, K., Lee, M., Liu, D. (eds.) ICONIP 2016. LNCS, vol. 9950, pp. 381–388. Springer, Cham (2016). doi:10.1007/978-3-319-46681-1_46
13. Wamsley, E.J.: Dreaming and offline memory consolidation. Curr. Neurol. Neurosci. Rep. **14**(3), 433 (2014)
14. Zhang, Y., Lee, K., Lee, H.: Augmenting supervised neural networks with unsupervised objectives for large-scale image classification, vol. 2, pp. 939–957 (2016)

Deep Reinforcement Learning: From Q-Learning to Deep Q-Learning

Fuxiao Tan[1]([✉]), Pengfei Yan[2], and Xinping Guan[3]

[1] School of Computer and Information Engineering, Fuyang Normal University,
Fuyang 236037, Anhui, China
fuxiaotan@gmail.com
[2] School of Automation and Electrical Engineering,
University of Science and Technology Beijing, Beijing 100083, China
pengfei.yan.ia@foxmail.com
[3] School of Electronic Information and Electrical Engineering,
Shanghai Jiao Tong University, Shanghai 200240, China
xpguan@sjtu.edu.cn

Abstract. As the two hottest branches of machine learning, deep learning and reinforcement learning both play a vital role in the field of artificial intelligence. Combining deep learning with reinforcement learning, deep reinforcement learning is a method of artificial intelligence that is much closer to human learning. As one of the most basic algorithms for reinforcement learning, Q-learning is a discrete strategic learning algorithm that uses a reasonable strategy to generate an action. According to the rewards and the next state generated by the interaction of the action and the environment, optimal Q-function can be obtained. Furthermore, based on Q-learning and convolutional neural networks, the deep Q-learning with experience replay is developed in this paper. To ensure the convergence of value function, a discount factor is involved in the value function. The temporal difference method is introduced to training the Q-function or value function. At last, a detailed procedure is proposed to implement deep reinforcement learning.

Keywords: Deep reinforcement learning · Q-learning · Deep Q-learning · Convolutional neural networks

1 Introduction

In 2006, the concept of deep learning (DL) was proposed by Hinton et al., which was originated from the research of artificial neural networks [1]. Deep learning is a new field in the study of machine learning. Its motivation lies in the establishment and simulation of the human brain by artificial neural networks, which mimics the human brain's mechanism to interpret data, in the fields of computer vision, image processing, speech and audio [2].

The important factors for the success of deep learning include the strong fitting and generalization ability of the model, the high-density of computing

© Springer International Publishing AG 2017
D. Liu et al. (Eds.): ICONIP 2017, Part IV, LNCS 10637, pp. 475–483, 2017.
https://doi.org/10.1007/978-3-319-70093-9_50

power, and the huge amount of training data. Based on the nonlinear operation of a large number of neurons, the deep neural network can acquire a strong fitting and generalization ability. Using a high-density computing device such as a GPU, the network model can be trained on a vast amount of training data (million level) in an acceptable amount of time (days) [3].

Machine learning is a multidisciplinary and interdisciplinary subject, which involves many subjects such as probability theory, statistics, approximation theory, convex analysis and computational complexity theory. Machine learning is a specialized discipline, which investigates how the computer simulates or realizes human learning behavior to acquire new knowledge or skills and reorganizes existing knowledge structures to continually improve their performance. Machine learning is the core of artificial intelligence, which is also the fundamental approach to make computers intelligent [6].

In the field of machine learning, based on the theories of Markov decision processes and temporal difference learning, reinforcement learning is still facing many difficulties and challenges [7], such as high dimensional approximate optimal strategy of continuous space, the stability of the learning algorithms, structured hierarchical optimization, reward function design, and so on [8].

1.1   Deep Learning and Neural Networks

The concept of deep learning is originated from the research of artificial neural networks (ANNs). The mathematical model of ANNs is made of layers of neurons [9]. ANNs are distributed parallel information processing algorithms, which are used to simulate the behavior of animal neurons. Depending on the complexity of the system, ANNs realize the purpose of processing information by adjusting the interconnection between large numbers of internal nodes.

The so-called deep learning is the neural network composed of multi-layer neurons to approximate the function of machine learning. The structure of deep learning is multilayer perceptron with multiple hidden layers. By combining low-level features to form more abstract high-level representations of attribute categories or features, deep learning can discover the distributed characteristics of data [10].

1.2   Deep Reinforcement Learning

The hottest branches of machine learning are deep learning and reinforcement learning. Reinforcing learning (RL) is a process of continuous decision-making [11]. The characteristics of RL is not giving any data annotations, but just providing a return function that determines the results of the current state (such as "good" or "bad"). In the essence of mathematics, it is a Markov decision process [12]. The ultimate goal of reinforcement learning is to obtain the expectation of optimal overall return function in the decision making process [13].

Reinforcement learning and deep learning are two very different algorithms of machine learning, but the deep learning algorithm can be used to implement reinforcement learning, which forms deep reinforcement learning (DRL).

Combining reinforcement learning with deep learning, it is possible to find an agent that can solve any human level task. Reinforcement learning defines the goal of optimization, while deep learning gives the mechanism of operation, which is the approach to characterize the problem and the way to solve the problem [14]. Therefore, deep reinforcement learning has a kind of ability to solve many complex problems. Thus, it will present more intelligent behaviors. So, deep reinforcement learning can be understood as part of the reinforcement learning achieved by deep learning. In essence, DRL can be considered as a learning and thinking algorithm [15].

2   Reinforcement Learning

Combining the perceptive ability of deep learning with the decision-making ability of reinforcement learning, deep reinforcement learning is a kind of artificial intelligence method which is closer to the human learning modes.

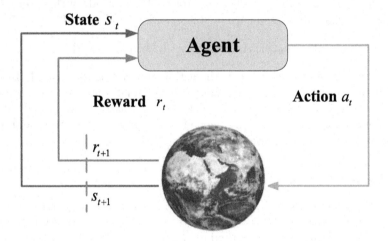

Fig. 1. The agent-environment interaction in reinforcement learning.

In Fig. 1, the agent and environment interact at each discrete time steps of a sequence, $t = 0, 1, 2, 3, \cdots$. At each time step t, the agent receives some representation of the environment's state $s_t \in S$, where S is the set of possible states and the action is $a_t \in A(s_t)$, where $A(s_t)$ is the set of actions available in state s_t. At time $t + 1$, as a consequence of its actions, the agent receives a numerical reward $r_{t+1} \in R$ and finds itself in a new state s_{t+1}.

At each time step, the agent implements a mapping from state to probabilities of each possible action. This mapping is called the agent's policy and is denote as π_t, where $\pi_t(s, a)$ is the probability that $a_t = a$ if $s_t = s$. Reinforcement learning methods specify how the agent changes its policy as a result of its experience. The agent's goal is to maximize the total amount of reward received over the long term.

2.1   Markov Decision Process

In general, Markov decision process (MDP) is denoted as a tuple $\{A, S, R, P\}$, where S is the state space of the environment, A is the action space of the agent, $R : S \times A \times S \to \Re$ is a reward function, and $P : S \times A \times S \to [0, 1]$ is a state transition function. In Fig. 1, for the RL structure, we define a function $\pi : S \times A \to \Pr(A)$ as a stochastic stationary policy of MDP, where $\Pr(A)$ is a probability distribution in the action space. So, the stationary policy π directly maps states to distributions over the action space, which is denoted as $a_t = \pi(s_t), t \geq 0$.

In sequential making, RL can be modeled as a MDP. At each time step t, the agent receives a state s_t and selects an action a_t from action space A, following a policy $\pi(a_t \mid s_t)$, i.e., mapping from state s_t to action a_t. Then, it receives a scalar reward r_t, and transitions to the next state s_{t+1}, according to the model, reward function $R(s, a)$ and state transition probability $P(s_{t+1} \mid s_t, a_t)$.

The set of states and actions, together with rules for transition from one state to another, make up a Markov decision process. One episode of this process (e.g. one game) forms a finite sequence of states, actions and rewards:

$$\{(s_0, a_0, r_1), (s_1, a_1, r_2), \cdots, (s_{t-1}, a_{t-1}, r_t)\} \tag{1}$$

where s_t represents the state, a_t is the action and r_{t+1} is the reward after performing the action. The episode ends with terminal state s_t (e.g. game over screen). A Markov decision process relies on the Markov assumption, that the probability of the next state s_{t+1} depends only on current state s_t and action a_t, but not on preceding states or actions.

2.2   Value Function

According to the theory of Markov decision process, the total reward for one episode can be calculated: $R = r_1 + r_2 + \cdots + r_n$. Thus, the total future reward from time t onward can be expressed as: $R_t = r_t + r_{t+1} + \cdots + r_{t+n}$.

If the same actions are performed and the same rewards next time are also got, the total future reward will be diverged. Therefore, the above rewards are substituted by discounted future reward:

$$R_t = r_t + \gamma r_{t+1} + \gamma^2 r_{t+2} + \cdots + \gamma^{n-t} r_n \tag{2}$$

where γ is the discount factor between 0 and 1. It is easy to see, that discounted future reward at time step t can be expressed in terms of the same thing at time step $t + 1$:

$$R_t = r_t + \gamma(r_{t+1} + \gamma(r_{t+2} + \cdots)) = r_t + \gamma R_{t+1} = \sum_{k=0}^{\infty} \gamma^k r_{t+k} \tag{3}$$

A good strategy for an agent would be to always choose an action that maximizes the (discounted) future reward.

Along the state trajectories by policy π, let J_π be the expected total rewards. Without loss of generality, the goal of reinforcement learning is to estimate an optimal policy π^* which satisfies

$$J_{\pi^*} = \max_\pi J_\pi = \max_\pi E_\pi \left[\sum_{t=0}^{\infty} \gamma^t r_t \right] \tag{4}$$

where r_t is the reward at time-step t, γ is a discount factor, and $E_\pi[\cdot]$ is the expectation with respect to the policy π and the state.

For a stationary policy π, in this paper, the state value function is defined as $V^\pi(s) - E_\pi \left[\sum_{t=0}^{\infty} \gamma^t r_t \mid s_0 = s \right]$. Thus, the optimal value function is defined as $V^{\pi^*}(s) = E_{\pi^*} \left[\sum_{t=0}^{\infty} \gamma^t r_t \mid s_0 = s \right]$.

To facilitate policy improvement, we defines the state-action value function $Q^\pi(s, a)$ as

$$Q^\pi(s, a) = E_\pi \left[\sum_{t=0}^{\infty} \gamma^t r_t \mid s_0 = s, a_0 = a \right] \tag{5}$$

and the optimal state-action value function

$$Q^{\pi^*}(s, a) = E_{\pi^*} \left[\sum_{t=0}^{\infty} \gamma^t r_t \mid s_0 = s, a_0 = a \right] \tag{6}$$

According to the Bellman optimality

$$V^\pi(s) = \sum \pi(a \mid s) E\left[R_{t+1} + \gamma V(s_{t+1}) \mid S_t = s \right] \tag{7}$$

the optimal value function satisfies

$$V^*(s) = E\left[R_{t+1} + \gamma \max^\pi V(s_{t+1}) \mid S_t = s \right] \tag{8}$$

and

$$Q^*(s, a) = E\left[R_{t+1} + \gamma \max_{a'} Q(s_{t+1}, a') \mid S_t = s, A_t = a \right] \tag{9}$$

where $R(s, a)$ is the expected reward received after taking action a in state s and s' is the successive state of s.

Thus, the optimal state-action value function is $Q^*(s, a) = \max_\pi Q^\pi(s, a)$. So, the optimal policy is computed by

$$\pi^*(s) = \arg \max_a Q^*(s, a) \tag{10}$$

In Fig. 1, after estimating the optimal state-action value function by $\tilde{Q}^*(s, a)$, a near-optimal policy can be obtained as

$$\tilde{\pi}^*(s) = \arg \max_a \tilde{Q}^*(s, a). \tag{11}$$

2.3   Q Learning

Temporal difference (TD) learning is a central idea in RL. It learns value function $V(s)$ directly from experience with the TD error, with bootstrapping, in a model-free, online, and fully incremental way. The update rule is

$$V(s_t) \leftarrow V(s_t) + \alpha \left[r_t + \gamma V(s_{t+1}) - V(s_t) \right] \tag{12}$$

where α is a learning rate. The $[r_t + \gamma V(s_{t+1}) - V(s_t)]$ is called TD error.

Similarly, Q-learning learns action value function with the update rule

$$Q(s_t, a_t) \leftarrow Q(s_t, a_t) + \alpha \left[r + \gamma \max_{a_{t+1}} Q(s_{t+1}, a_{t+1}) - Q(s_t, a_t) \right] \tag{13}$$

Q-learning is an off-policy control method [16].

3   Deep Reinforcement Learning

3.1   Deep Neural Network

We obtain deep reinforcement learning (deep RL) methods when we use deep neural networks to approximate any of the following component of reinforcement learning: value function, $V(s, \theta)$ or $Q(s, a; \theta)$, policy $\pi(a| s; \theta)$, and model (state transition and reward). Where the parameters θ are the weights in deep neural networks.

In the DeepMind paper [17], in order to obtain the four last screen images, there need resize them to 84×84 and convert to grayscale with 256 gray levels. Thus, there are $256^{84 \times 84 \times 4} \approx 10^{67970}$ possible game states. This means that there need 10^{67970} rows in the Q-table, and the number is a lot bigger than the number of atoms in the known universe.

Since many states (pixel combinations) do not occur, you can actually use a sparse table to include the states that are being accessed. Even so, many states are still rarely accessed. Maybe it will take lifetime of the universe for the Q-table to converge. Therefore, in algorithmic designing, the idealized scenario is to have a good guess for Q-values that has not yet been met.

To solve the above problems, it requires deep learning to take part in. The main role of neural networks is to derive features from highly structured data. Therefore, Q-function can be represented by neural network, which means the state and action as input and the corresponding Q-value as the output. In general, this is actually a classical convolutional neural network with three convolutional layers and followed by two fully connected layers.

3.2   Experience Replay

By now we have an idea how to estimate the future reward in each state using Q-learning and approximate the Q-function using a convolutional neural network. However, the main problem is that approximation of Q-values using non-linear functions is not very stable. So, it requires some tricks to guarantee the convergence of the algorithm.

In deep learning, the widely-used of trick is experience replay. In algorithm execution, all experience $\{s, a, r, s'\}$ are stored in a replay memory. When conducting deep Q network training, the random minibatches from replay memory will replace the most recent transition. This will break the similarity of subsequent training samples, otherwise it is easy to direct the network into one of a local minimum. Furthermore, the experience replay makes the training task more similar to the usual supervised learning, which also simplifies the program's debug and algorithm testing.

3.3 Exploration-Exploitation

Reinforcement learning is a kind of a trial-and-error learning method. In the beginning, agent is not clear that the environment and the ways of working. The agent don't know what kind of action (behavior) is right, what kind is wrong. Therefore, agent needs to find a good policy from the experience of trial and trial to gain more reward in this process.

Therefore, in deep reinforcement learning, there is a tradeoff between Exploration and Exploitation. Exploration will give up some known information of reward and try a few new choice. Namely, under some kind of states, algorithm may have to learn how to choose the action for larger reward, but it doesn't make the same choice every time. Perhaps another choice that hasn't been tried will make the reward larger, that is the agent's goal to explore more information about environment. Moreover, exploitation refers to maximizing reward according to known information of agent.

3.4 Deep Q-Networks

To approximate the value function $Q(s, a; \theta)$ with parameters θ (that is, weights), a deep convolutional neural network can be applied in the designing of DRL. To carry out experience replay at each time t, the agent's experiences $e_t = (s_t, a_t, r_t, s_{t+1})$ is stored into a dat set $D_t = \{e_1, \cdots, e_t\}$.

The deep Q-learning updating at iteration i uses the following loss function

$$L_i(\theta_i) = E_{(s,a,r,s')}\left[(y_i - Q(s, a; \theta_i))^2\right] \tag{14}$$

with

$$y_i = r + \gamma \max_{a'} Q(s', a'; \theta_i^-) \tag{15}$$

where θ_i^- are the network parameters used to compute the target at iteration i.

Without any generalization, it is common to use a function approximate to estimate the action-valued function

$$Q(s, a; \theta) \approx Q^*(s, a) \tag{16}$$

Differentiating the loss function with respect to the parameters, the gradient is

$$\nabla_{\theta_i} L(\theta_i) = E_{s,a,r,s'} \left[\left(r + \gamma \max_{a'} Q\left(s', a', \theta_i^-\right) - Q\left(s, a; \theta_i\right) \right) \nabla_{\theta_i} Q\left(s, a; \theta_i\right) \right] \tag{17}$$

From above, the deep Q-learning algorithm is as follows [5].

Algorithm 1. Deep Q-learning with experience replay.

Require:

Initialize replay memory D to capacity N

Initialize action-value function Q with random weights θ

Initialize the target action-value function $\hat{Q}$ with weights $\theta^- = \theta$

Ensure:

1: **for** episode $= 1$ to M **do**

2:    initialize sequence $s_1 = \{x_1\}$ and preprocessed sequence $\phi_1 = \phi(s_1)$;

3:    **for** $t = 1$ to T **do**

4:        With probability ε select a random action a_t;

5:        Otherwise select $a_t = \arg\max_a Q(\phi(s_t), a; \theta)$;

6:        Execute action a_t in emulator and observe reward r_t and image x_{t+1};

7:        Set $s_{t+1} = s_t, a_t, x_{t+1}$ and preprocess $\phi_{t+1} = \phi(s_{t+1})$;

8:        Store transition $(\phi_t, a_t, r_t, \phi_{t+1})$ in D;

9:        Sample random minibatch of transitions $(\phi_j, a_j, r_j, \phi_{j+1})$ from D;

10:       Set $y_j = r_j$ if episode terminates at step $j + 1$, otherwise $y_j = r_j + \gamma \max_{a'} \hat{Q}(\phi_{j+1}, a'; \theta^-)$;

11:       Perform a gradient descent step on $(y_j - Q(\phi_j, a_j; \theta))^2$ with respect to the network parameter θ;

12:       Every C steps reset $\hat{Q} = Q$, i.e., $\theta^- = \theta$;

13:    **end for**

14: **end for**

---

4   Conclusion

Combining deep learning with reinforcement learning, deep reinforcement learning is a method of artificial intelligence that approaches the human way of thinking. Furthermore, combining the deep convolution neural network with Q-learning, the algorithm of deep Q-learning is investigated in the paper.

There need to be pointed out that the essence of reinforcement learning is a markov decision process, reinforcement learning gives only a return function, which is the decision in the risk of a particular condition to perform an action (which may be a good result, or probably worst). However, RL performance depends on the extraction technology of artificial features, and deep learning just makes up for this shortcoming. The deep reinforcement learning algorithm is mainly aimed at the optimization problems of discrete-time systems. However, because there are continuous action space and continuous state space in the practical application, it also limits the application range of deep reinforcement learning.

Acknowledgments. This work was supported by the National Natural Science Foundation of China under Grant 61673117.

References

1. Hinton, G.E., Osindero, S., Teh, Y.W.: A fast learning algorithm for deep belief nets. Neural Comput. **18**(7), 1527–1554 (2006)
2. LeCun, Y., Bengio, Y., Hinton, G.: Deep learning. Nature **521**(7553), 436–444 (2015)
3. Silver, D., Huang, A., Maddison, C.J., Guez, A., Sifre, L., van den Driessche, G., Schrittwieser, J., Antonoglou, I., Panneershelvam, V., Lanctot, M., Dieleman, S., Grewe, D., Nham, J., Kalchbrenner, N., Sutskever, I., Lillicrap, T., Leach, M., Kavukcuoglu, K., Graepel, T., Hassabis, D.: Mastering the game of go with deep neural networks and tree search. Nature **529**(7587), 484–489 (2016)
4. Hinton, G.E., Salakhutdinov, R.R.: Reducing the dimensionality of data with neural networks. Science **313**(5786), 504–507 (2006)
5. Mnih, V., Kavukcuoglu, K., Silver, D., Rusu, A.A., Veness, J., Bellemare, M.G., Graves, A., Riedmiller, M., Fidjeland, A.K., Ostrovski, G., Petersen, S., Beattie, C., Sadik, A., Antonoglou, I., King, H., Kumaran, D., Wierstra, D., Legg, S., Hassabis, D.: Human-level control through deep reinforcement learning. Nature **518**(7540), 529–533 (2015)
6. Bengio, Y., Courville, A., Vincent, P.: Representation learning: a review and new perspectives. IEEE Trans. Pattern Anal. Mach. Intell. **35**(8), 1798–1828 (2013)
7. Werbos, P.J.: Approximate dynamic programming for realtime control and neural modeling. In: Handbook of Intelligent Control: Neural, Fuzzy, and Adaptive Approaches. Van Nostrand Reinhold, New York (1992)
8. Sutton, R.S., Barto, A.G.: Reinforcement Learning: An Introduction. MIT Press, Cambridge (1998)
9. Liu, D.R., Wang, D., Wang, F.Y., Li, H.L., Yang, X.: Neural-network-based Online HJB solution for optimal robust guaranteed cost control of continuous-time uncertain nonlinear systems. IEEE Trans. Cybern. **44**(12), 2834–2847 (2014)
10. Schmidhuber, J.: Deep learning in neural networks: an overview. Neural Netw. **61**, 85–117 (2015)
11. Liu, D.R., Wang, D., Li, H.: Decentralized stabilization for a class of continuous-time nonlinear interconnected systems using online learning optimal control approach. IEEE Trans. Neural Netw. Learn. Syst. **25**(2), 418–428 (2014)
12. Wei, Q.L., Liu, D.R., Lin, H.: Value iteration adaptive dynamic programming for optimal control of discrete-time nonlinear systems. IEEE Trans. Cybern. **46**(3), 840–853 (2015)
13. Liu, D.R., Wei, Q.L.: Policy iteration adaptive dynamic programming algorithm for discrete-time nonlinear systems. IEEE Trans. Neural Netw. Learn. Syst. **25**(3), 621–634 (2014)
14. Kaelbling, L.P., Littman, M.L., Moore, A.W.: Reinforcement learning: a survey. J. Artif. Intell. Res. **4**, 237–285 (1996)
15. Arel, I., Rose, D.C., Karnowski, T.P.: Deep machine learning - a new Frontier in artifical intelligence research. IEEE Comput. Intell. Mag. **5**(4), 13–18 (2010)
16. Watkins, C.J.H., Dayan, P.: Technical note: Q-learning. Mach. Learn. **8**(3–4), 279–292 (1992)
17. Mnih, V., Kavukcuoglu, K., Silver, D., Graves, A., Antonoglou, I., Wierstra, D., Riedmiller, M.: Playing atari with deep reinforcement learning. In: NIPS Deep Learning Workshop, arxiv preprint arXiv:1312.5602 (2013)

Origami Folding Sequence Generation Using Discrete Particle Swarm Optimization

Ha-Duong Bui[1]([✉]), Sungmoon Jeong[1], Nak Young Chong[1,2], and Matthew Mason[2]

[1] School of Information Science,
Japan Advanced Institute of Science and Technology,
Nomi, Ishikawa 923-1211, Japan
{bhduong,jeongsm,nakyoung}@jaist.ac.jp
[2] School of Computer Science and Robotics Institute, Carnegie Mellon University,
Pittsburgh, PA 15213, USA
matt.mason@cs.cmu.edu

Abstract. This paper proposes a novel approach to automating origami or paper folding. The folding problem is formulated as a combinatorial optimization problem to automatically find feasible folding sequences toward the desired shape from a generic crease pattern, minimizing the dissimilarity between the current and desired origami shapes. Specifically, we present a discrete particle swarm optimization algorithm, which can take advantage of the classical particle swarm optimization algorithm in a discrete folding action space. Through extensive numerical experiments, we have shown that the proposed approach can generate an optimum origami folding sequence by iteratively minimizing the Hausdorff distance, a dissimilarity metric between two geometric shapes. Moreover, an in-house origami simulator is newly developed to visualize the sequence of origami folding.

Keywords: Mathematical origami · Folding sequence generation · Combinatorial optimization problem · Particle swarm optimization

1 Introduction

Many innovative products, such as bendable electronics, deployable solar array, foldable paper lithium-ion battery, are inspired by origami, the art of paper folding. From the 1930s, problems related to the folding and unfolding have attracted a large attention in the computational geometry community [10]. Recent advances in computing environment allow researchers to study complex folding systems that include programmable matters [8,11] and folding machines [3,7]. However, few researches have addressed the following issues: (1) how to find a crease pattern of a desired origami shape without going through the unfolding process, (2) how to generate a folding sequence to create a desired shape from a generic crease pattern.

© Springer International Publishing AG 2017
D. Liu et al. (Eds.): ICONIP 2017, Part IV, LNCS 10637, pp. 484–493, 2017.
https://doi.org/10.1007/978-3-319-70093-9_51

Three main areas of origami in computational geometry, related to folding problems are origami simulator, folding sequence generation, and multiple object folding from a single sheet, respectively. *Freeform Origami* [14–16] is a well-known folding simulator providing users with an environment to interact with a virtual paper. Users are able to fold papers and interactively modify crease patterns, and generate crease patterns for polyhedra. Using the quadrilateral mesh information of the 3D input shape, *Freeform Origami* first unfolds the input shape to get the candidate crease pattern which is then simultaneously folded and controlled through affine transformations. This process is iterated until the desired shape is achieved. Another interesting research performed by Akitaya et al. [1] is how to generate folding sequences of flat-foldable origami. To accomplish this goal, the framework builds a new graph-like data structure called the extended crease pattern. This data is constructed using the input crease pattern by which the input shape is unfolded. The folding sequence can then be found by inverting the unfolding process. However, users are often requested to decide the next step in the unfolding sequence, because multiple outcomes are possible from the extended crease pattern. Therefore, it is considered a semi-autonomous system. An et al. [2] studied how the 3D shapes can be transformed using the programmable matter. They designed a programmable sheet with a set of hinges. Multiple shapes can be constructed and transformed using the programmable sheet, without considering any folding action sequences. In the above-mentioned literatures, origami problems and their solution approaches are mainly studied from a mathematical perspective. It is important to note that crease patterns were folded into desired shapes, but the crease patterns could only be obtained through the unfolding process of the desired shapes.

In this paper, we aim to present an algorithmic framework for automating the folding process. We show how to find the folding sequence from a generic crease pattern. A generic crease pattern could possibly folded into multiple shapes. Our approach tries to find an optimal solution through the sequential combination of appropriate folding actions. Specifically, in our problem definition, the inputs are a generic crease pattern and a desired shape, and the output is a folding sequence. Our proposed algorithm can create folded shapes from the given generic crease pattern, maximizing the similarity with the desired shape. We formulate our problem as a combinatorial optimization problem (**COP**) and employ a discrete particle swarm optimization (**DPSO**) algorithm. The proposed approach will be described in detail in the following sections.

2   Methodology

2.1   Problem Formulation

Problem Definition. Figure 1 shows a diagram of the proposed folding automation system. The input to our system consists of

1. A square sheet of paper, *OriginalPaper*
2. A set of n predefined creases, $ActionSet = \{Action_1, Action_2, \ldots, Action_n\}$

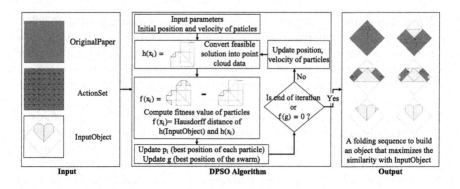

Fig. 1. Overview of the proposed approach (Color figure online)

3. A desired folded shape, *InputObject*.

All of the following are outputs from the proposed system.

1. An optimal folding sequence g from the given **generic** crease pattern
2. A final folded shape *OutputObject* achieved by applying the folding sequence g to *OriginalPaper*
3. The similarity score between *InputObject* and *OutputObject*.

Feasible Solution. We define a nominal folding action called $Action_0$ causing the current shape to remain unchanged. Adding the $Action_0$ to the *ActionSet*, our new *ActionSet* becomes

$$ActionSet = \{Action_0, Action_1, Action_2, \ldots, Action_n\}$$

A folding process is a list of actions chosen out of *ActionSet*, meaning that the list of actions are sequentially applied to the original paper to be folded into an object shape.

From the above description, a **feasible solution** x is defined as a folding process with a fixed length n given by

$$\begin{cases} x = (x_1, x_2, \ldots, x_n) \\ x_i \in ActionSet, 1 \leq i \leq n \end{cases}$$

Search Space. We define a search space A which is a set of all feasible solutions. Typically, A is a $n - dimensional$ integer lattice, denoted by $\mathbb{Z}^n$, which is the lattice in the Euclidean space $\mathbb{R}^n$. Because each candidate solution x is a vector of n elements, and we also have $n+1$ candidates for each element, then the set A has a cardinality (the total number of feasible solutions) of $(n+1)^n$. The size of the search space A increases exponentially with the number of folding actions. Figure 2 illustrates the search space A and the feasible solutions with $n = 2$ and $n = 3$, respectively.

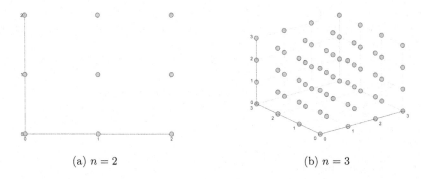

(a) $n = 2$ (b) $n = 3$

Fig. 2. Examples of search space. (a) With $n = 2$, 9 feasible solutions exist. (b) With $n = 3$, 64 feasible solutions exist. The green dots are feasible solutions. (Color figure online)

Objective Function. Let $h(x)$ be a function that converts a feasible solution x into an object shape (Sect. 2.3). Also let $f(x) : \mathbb{Z}^n \to \mathbb{R}$ be a function that assesses the degree of similarity between $h(x)$ and *InputObject*. A lower value of $f(x)$ indicates a high degree of similarity with *InputObject*, and vice versa. The function f is the objective function to be minimized. We can now formulate our problem as an optimization problem in the following way:

Given $f : A \to \mathbb{R}$ from A to the real number,

Seek an element g in A such that $f(g) \leq f(x)$ for all x in A.

2.2 Problem Optimization: DPSO Algorithm

We choose DPSO as a meta-heuristics algorithm that can provide an approximate solution to our problem. A basic variant of the PSO [9] algorithm contains a list of feasible solutions called particles which are members of a swarm. These particles are initially given their positions and velocities randomly. Each particle employs the objective function to evaluate its position. It also uses its velocity to move to new positions. Here the moving function should be carefully designed by users. Moreover, (1) the particle's current direction, (2) the particle's best-known position, as well as (3) the swarm's best-known position, are combined to update the particle's velocity. When a particle discovers a new best position, it will communicate and update the entire swarm. The process of updating and relocation of the swarm is repeated, and by doing so, it is expected, but not guaranteed, that an optimal solution will eventually be explored.

As having described, obviously, the most important things in PSO algorithm are the particle's velocity and position. The behavior of particles is directly affected by the velocity. In order to move, three information are processed that are (1) particle's current direction, (2) particle's previous best position and (3) swarm's best-known position [4]. In this research, we redefined these concepts and mathematical operations. This discretization method helps adjust the classical PSO's properties and make it capable of searching in the discrete search space of our problem.

The above description can be formalized by the following moving functions,

$$\begin{cases} v_i^{t+1} \leftarrow \omega \otimes v_i^t \oplus \varphi_p r_p \otimes (p_i^t \ominus x_i^t) \oplus \varphi_g r_g \otimes (g^t \ominus x_i^t) \\ x_i^{t+1} \leftarrow x_i^t \oplus v_i^{t+1}, \end{cases} \tag{1}$$

where

- v_i^t velocity at time step t of particle i
- x_i^t position at time step t of particle i
- p_i^t particle i's best known position at time step t
- g^t swarm's best known position at time step t
- $\omega, \varphi_p r_p, \varphi_g r_g \sim U(0,1)$ social/cognitive confidence coefficients
- $a \otimes b = a \times b \mod (n+1)$
- $a \oplus b = a + b \mod (n+1)$
- $a \ominus b = a - b \mod (n+1)$.

The pseudo code of DPSO is introduced in Algorithm 1.

Algorithm 1. Discrete Particle Swarm Optimization

1: **for all** particle i **do**
2:      $x_i \leftarrow InitializeParticle(n)$
3:      **for all** dimension d **do**
4:          $v_{i,d} \sim U\{0,n\}$
5: **while** termination condition not reached **do**
6:      **for all** particle i **do**
7:          Pick random numbers $r_p, r_g \sim U(0,1)$
8:          Update particle's velocity and position using Eq. (1)
9:          Evaluate the fitness $f(x_i)$
10:          **if** $f(x_i) < f(p_i)$ **then**
11:              $p_i \leftarrow x_i$
12:              **if** $f(p_i) < f(g)$ **then**
13:                  $g \leftarrow p_i$
14: g is the proposed solution of the algorithm

2.3   Converting Folding Sequences into Object Shapes

In order to convert a feasible solution x into an object shape (function $h(x)$ in Sect. 2.1) and visualize the proposed sequence of folding, we develop an in-house origami simulator. This system was originally presented by Miyazaki et al. [12]. We can construct flat folding origami shapes from generic crease patterns. Specifically, we use faces, edges, and vertices to describe the state of the folded paper. Each flat sheet of paper is represented by a face. A face contains multiple edges. Moreover, each edge has two vertices. A folded paper is a list of faces.

Based on the idea that each crease will separate the origami plane into two half-planes, we construct a binary tree structure, in which the root is the *OriginalPaper*. Each node of the tree is a face and stores the information about the relative position of the plane with the original plane. When applying a folding action, we traverse from the root and find all the leaves that contain the crease, make these leaves become parent nodes, and create two new children nodes (as two new faces which are created by the folding action). Obviously, two nodes with the same parent will share one common edge. Finally, we combine all the leaves of the binary tree to get the final shape.

An example is shown in Fig. 3, where we apply the folding actions in the following order: $Action_1 \rightarrow Action_2 \rightarrow Action_3$. E1, E2, E3, E4, and E5 are the common edges between faces. When we combine all the leaves F5, F6, F7, F8, F9, and F10, we can get the object shape. The states of the origami shape after each folding action are shown in Table 1.

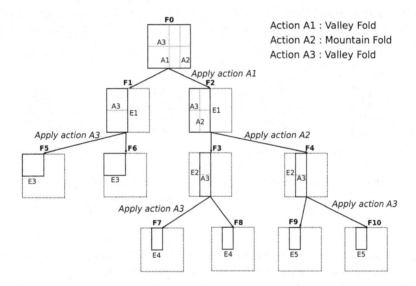

Fig. 3. Example of converting a feasible solution into an object shape

We also implement a function to calculate the similarity among folded origami papers using the point cloud data (function $f(x)$ in Sect. 2.1). With the PCL library [13], we convert the object shapes into the point cloud data based on the information about faces, edges, and vertices of shapes. We normalize the point cloud data in size and density. Then, the Hausdorff distance [6] between two object shapes can be calculated. The lower the Hausdorff distance, the higher the similarity between two shapes, and vice versa.

Table 1. The states of the origami shape after each folding action in Fig. 3

Applied action	Leaf nodes	Parent nodes	Common edges
Begin	$\{F_0\}$	$\emptyset$	
$Action_1$	$\{F_1, F_2\}$	$\{F_0\}$	$\{F_1 \cap F_2 = E_1,$
$Action_2$	$\{F_1, F_3, F_4\}$	$\{F_0, F_2\}$	$F_3 \cap F_4 = E_2, F_5 \cap F_6 = E_3,$
$Action_3$	$\{F_5, F_6, F_7, F_8, F_9, F_{10}\}$	$\{F_0, F_1, F_2, F_3, F_4\}$	$F_7 \cap F_8 = E_4, F_9 \cap F_{10} = E_5\}$

3   Experimental Evaluation

3.1   Experimental Set-Up

Original Origami Paper. The original paper used in our experiments is a flat square paper with a size of 60×60 *units*. The front side of the paper is blue and the back side is white.

Predefined Crease Patterns. We use four predefined crease patterns in our experiments as shown in Fig. 4. Each crease represents a valley fold or a mountain fold.

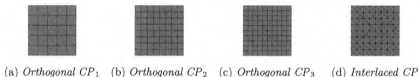

(a) *Orthogonal CP_1*   (b) *Orthogonal CP_2*   (c) *Orthogonal CP_3*   (d) *Interlaced CP*

Fig. 4. Crease patterns used in experiments (Color figure online)

Experimental Objective Models. Using the same folding simulation was introduced in Sect. 2.3, we fold the desired shapes shown in Fig. 5. In the figures, the black dashed lines represent the original paper.

DPSO Parameters. Referring to *Standard Particle Swarm Optimization* [5], we choose the parameters of DPSO algorithm as follows:

- The swarm size $S = 35 + B(10, 0.5)$. $B(10, 0.5)$ is a binomial distribution, where 10 is the number of trials and 0.5 is success probability in each trial.
- The maximum number of iterations is 100
- $\omega \simeq 0.721$
- $\varphi_p = 1$
- $\varphi_g = 1$.

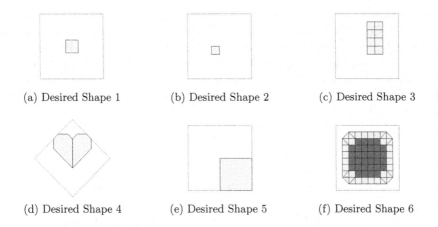

(a) Desired Shape 1   (b) Desired Shape 2   (c) Desired Shape 3

(d) Desired Shape 4   (e) Desired Shape 5   (f) Desired Shape 6

Fig. 5. Desired shapes used in experiments (Color figure online)

Table 2. Results of experiments

	Predefined crease pattern	Objective model	Haursdoff distance			Iteration			Avg. runtime (ms)
			Min	Max	Avg	Min	Max	Avg	
E1	Orthogonal CP_1	Desired Shape 1	0.0	0.0	0.0	1	4	1.06	838.24
E2	Orthogonal CP_2	Desired Shape 2	0.0	0.0	0.0	1	90	18.33	35064.85
E3	Orthogonal CP_2	Desired Shape 3	0.0	3.75	0.07	1	100	20.50	33893.63
E4	Interlaced CP	Desired Shape 4	0.0	5.32	2.98	1	100	73.42	45533.28
E5	Orthogonal CP_1	Desired Shape 5	8.5	8.50	8.50	100	100	100	78594.77
E6	Orthogonal CP_2	Desired Shape 6	3.3	16.37	6.0	100	100	100	164159.70

3.2   Experimental Results

We report the experimental results obtained from the 6 test sets in Table 2.

In the experiments E1–E4, the folding sequence of the desire shape is a subset of *ActionSet*, which means we can fold the desire shape from the given predefined crease pattern. It was confirm that if the desired shape is built using a set of creases chosen out of the generic input crease pattern, then our method can always find the optimal solution.

Furthermore, in the experiments E1–E4, different types of shapes (square, rectangular and heart shape) are employed to evaluate our system. As expected, our experiments prove that with a suitable generic crease pattern, we can create arbitrary shapes. Moreover, our system also provides an efficient way to transform shapes.

To evaluate the cases where the desired shapes are not built from the folding actions of *ActionSet*, we perform the experiments E5 and E6. In the experiment E5, the desired shape in Fig. 5e is a square with a size of $\frac{1}{4}OriginalPaper$. Obviously, we cannot construct any shape that is the same as the desired shape from the *Orthogonal CP_1*. In Fig. 6a, the proposed solution of our algorithm for the experiment E5 is shown. Because attempting to get 0.0 in Hausdorff distance is impossible, the shapes that are most similar to the desired shape are presented. Correspondingly, in the experiment E6, the desired shape is neither a rectangle nor a square but an octagon. We need to find a folding sequence to build it from the *Orthogonal CP₇*. The shape of the optimal solution has been obtained and introduced in Fig. 6b. To summarize, our approach is quite successful in finding the minimum Hausdorff distance between the desired shape and the feasible solutions.

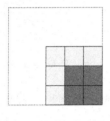   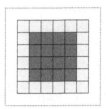

(a) Output Shape of E5   (b) Output Shape of E6

Fig. 6. Output shapes of experiments E5 and E6 (Color figure online)

From the test cases, we see that the computational time increases with the number of creases. There are two reasons for it. First, the calculation time of the folding simulation is dependent on the size of the *ActionSet*. The more folding actions we apply to the origami paper, the longer time the program needs to build the object shape. Secondly, as we discussed in Sect. 2.1, the search space exponentially grows with the input size. Therefore, the computational costs for arriving at feasible solutions in the search space A increase accordingly as the input size increases.

4   Conclusion

We presented a combinatorial optimization approach to automating origami. The DPSO algorithm was employed to solve the folding problem. Moreover, an in-house origami simulator was newly developed to perform extensive numerical experiments. Our method was verified to show an efficient technique to generate feasible folding sequences in an autonomous way. We have confirmed that, with a properly prepared generic crease pattern, various types of origami shapes can be folded easily with the proposed method.

References

1. Akitaya, H.A., Mitani, J., Kanamori, Y., Fukui, Y.: Generating folding sequences from crease patterns of flat-foldable origami. In: ACM SIGGRAPH 2013 Posters, SIGGRAPH 2013, p. 20:1. ACM, New York (2013). http://doi.acm.org/10.1145/2503385.2503407

2. An, B., Benbernou, N., Demaine, E.D., Rus, D.: Planning to fold multiple objects from a single self-folding sheet. Robotica **29**(1), 87–102 (2011)
3. Balkcom, D.J., Mason, M.T.: Robotic origami folding. Int. J. Robot. Res. **27**(5), 613–627 (2008). doi:10.1177/0278364908090235
4. Clerc, M.: Discrete particle swarm optimization, illustrated by the traveling salesman problem. In: Onwubolu, G.C., Babu, B.V. (eds.) New Optimization Techniques in Engineering, pp. 219–239. Springer, Heidelberg (2004). doi:10.1007/978-3-540-39930-8_8
5. Clerc, M.: Standard Particle Swarm Optimisation, 15 p., September 2012. https://hal.archives-ouvertes.fr/hal-00764996
6. Deza, M.M., Deza, E.: Encyclopedia of distances. In: Deza, M.M., Deza, E. (eds.) Encyclopedia of Distances, pp. 1–583. Springer, Heidelberg (2009). doi:10.1007/978-3-642-00234-2_1
7. Felton, S., Tolley, M., Demaine, E., Rus, D., Wood, R.: A method for building self-folding machines. Science **345**(6197), 644–646 (2014). http://science.sciencemag.org/content/345/6197/644
8. Hawkes, E., An, B., Benbernou, N.M., Tanaka, H., Kim, S., Demaine, E.D., Rus, D., Wood, R.J.: Programmable matter by folding. Proc. Natl. Acad. Sci. **107**(28), 12441–12445 (2010). http://www.pnas.org/content/107/28/12441.abstract
9. Kennedy, J., Eberhart, R.: Particle swarm optimization. In: 1995 Proceedings of the IEEE International Conference on Neural Networks, vol. 4, pp. 1942–1948, November 1995
10. Margalit, F.: Akira Yoshizawa, 94, modern origami master. N. Y. Times (2005)
11. Miyashita, S., Guitron, S., Ludersdorfer, M., Sung, C.R., Rus, D.: An untethered miniature origami robot that self-folds, walks, swims, and degrades. In: 2015 IEEE International Conference on Robotics and Automation (ICRA), pp. 1490–1496, May 2015
12. Miyazaki, S., Yasuda, T., Yokoi, S., Toriwaki, J.: An origami playing simulator in the virtual space. J. Vis. Comput. Animat. **7**(1), 25–42 (1996). doi:10.1002/(SICI)1099-1778(199601)7:1⟨25::AID-VIS134⟩3.0.CO;2-V
13. Rusu, R.B., Cousins, S.: 3D is here: point cloud library (PCL). In: 2011 IEEE International Conference on Robotics and Automation, pp. 1–4, May 2011
14. Tachi, T.: Freeform rigid-foldable structure using bidirectionally flat-foldable planar quadrilateral mesh. In: Ceccato, C., Hesselgren, L., Pauly, M., Pottmann, H., Wallner, J. (eds.) Advances in Architectural Geometry, pp. 87–102. Springer Vienna, Vienna (2010). doi:10.1007/978-3-7091-0309-8_6
15. Tachi, T.: Freeform variations of origami. J. Geom. Graph. **14**(2), 203–215 (2010)
16. Tachi, T.: Freeform origami (2010–2016). www.tsg.ne.jp/TT/software/

CACO-LD: Parallel Continuous Ant Colony Optimization with Linear Decrease Strategy for Solving CNOP

Shijin Yuan, Yunyi Chen, and Bin Mu[(✉)]

School of Software Engineering, Tongji University, Shanghai 201804, China
yuanshijin2003@163.com, cyy_lele@outlook.com, binmu@tongji.edu.cn

Abstract. Increasing intelligence algorithms have been applied to solve conditional nonlinear optimal perturbation (CNOP), which is proposed to study the predictability of numerical weather and climate prediction. Currently, swarm intelligence algorithms have much lower stability and efficiency than single individual intelligence algorithms, and the validity of CNOP (in terms of CNOP magnitude and CNOP pattern) obtained by swarm intelligence algorithms is not as good as that obtained by single individual intelligence algorithms. In this paper, we propose an improved parallel swarm intelligence algorithm, continuous ant colony optimization with linear decrease strategy (CACO-LD), to solve CNOP. To verify the validity of the CACO-LD, we apply it to study EI Niño-Southern Oscillation (ENSO) event with Zebiak-Cane (ZC) model. Experimental results show that the CACO-LD can achieve better CNOP magnitude, better CNOP pattern with much higher stability than the modified artificial bee colony algorithm (MABC) and the continuous tabu search algorithm with sine maps and staged strategy (CTS-SS), which respectively are the latest best swarm intelligence algorithm and single individual intelligence algorithm for solving CNOP. Moreover, when using 32 processes, the parallel CACO-LD runs 3.9 times faster than the parallel MABC, and is competitive with the parallel CTS-SS in efficiency.

Keywords: CNOP · Ant colony optimization · MPI · Zebiak-Cane model

1 Introduction

Conditional nonlinear optimal perturbation (CNOP) [1] was proposed to study the predictability of numerical weather and climate prediction. CNOP is the initial perturbation that will most easily evolve into a weather or climate event, and it has been successfully applied to study EI Niño-Southern Oscillation (ENSO) [2–4], spring predictability barrier of ENSO events [3,4], targeted observations for typhoon [5], and the Kuroshio large meander [6], etc.

Solving CNOP can be regarded as an optimization problem in nonlinear system. The most popular method for solving CNOP is the adjoint-based method

© Springer International Publishing AG 2017
D. Liu et al. (Eds.): ICONIP 2017, Part IV, LNCS 10637, pp. 494–503, 2017.
https://doi.org/10.1007/978-3-319-70093-9_52

(ADJ-CNOP) [7], which requires gradient information provided by corresponding adjoint models of numerical models. However, the adjoint model is hard to develop and many modern numerical models don't have corresponding adjoint models, which limit the application of CNOP.

To avoid those problems, increasing intelligence algorithms (IAs) have been applied to solve CNOP because they are free of gradient information and easy to parallelize. Those IAs for solving CNOP can be divided into two categories, which are swarm IAs and single individual IAs. The parallel sensitive area selection-based particle swarm optimization algorithm (SASPSO) [8], the PCA based particle swarm optimization (PPSO) [9] and the modified artificial bee colony algorithm (MABC) [10] are examples of swarm IAs. While the principal components-based great deluge method (PCGD) [11], the dynamic step size sphere gap transferring algorithm (DSGT) [12], and the continuous tabu search algorithm with sine maps and staged strategy (CTS-SS) [13] belong to the category of single individual IAs. Recent studies show that both types of IAs can obtain similar CNOP to that obtained by the adjoint-based method, which is the benchmark. However, swarm IAs have much lower stability and efficiency than single individual IAs, and the validity of CNOP obtained by swarm IAs is not as good as that obtained by single individual IAs.

In this paper, we propose a parallel continuous ant colony optimization with linear decrease strategy (CACO-LD), which is an improved swarm intelligence algorithm based on ant colony optimization for continuous domain (ACO_R) [14], to solve CNOP. To accelerate the computational speed, we parallelize the CACO-LD with message passing interface (MPI) technique. For verifying the validity of CACO-LD, we apply it to solve CNOP in Zebiak-Cane (ZC) model [15] for studying the predictability of ENSO event, and compare it with the adjoint-based method, MABC and CTS-SS. Experimental results show that the CACO-LD is an effective and efficient method in such research.

The remainder of this paper is organized as follows. Section 2 demonstrates the model and methodology. Details of parallel CACO-LD for solving CNOP are introduced in Sect. 3. Experiments and results analysis are shown in Sect. 4. Finally, this paper ends with conclusions and future works in Sect. 5.

2   Model and Methodology

2.1   CNOP

CNOP represents an initial perturbation imposed on the initial state of model under certain physical constraint, which will cause the largest nonlinear evolution at prediction time. An initial perturbation u_0^* that subjects to constraint condition $\|u_0^*\| \leq \delta$ is called CNOP, if and only if

$$J(u_0^*) = \max_{\|u_0\| \leq \delta} J(u_0), \tag{1}$$

where $\|\cdot\|$ denotes the Euclidean norm, δ is the constraint radius of u_0^*, and function J is computed by (2).

$$J(u_0) = \|M_{T_0 \to T}(U_0 + u_0) - M_{T_0 \to T}(U_0)\|, \tag{2}$$

where u_0 represents the initial perturbation superposed on the initial basic state U_0, and $M_{T_0 \rightarrow T}$ denotes the nonlinear propagator of a nonlinear model from initial time T_0 to prediction time T. Thus, J describes the nonlinear evolution of initial perturbation u_0. Function J is usually regarded as the adaptive function and its value is called fitness value in optimization algorithms.

For convenience of computation, (1) is converted to a minimum problem:

$$f(u_0) = -J(u_0) \tag{3}$$

$$J(u_0^*) = \min_{\|u_0\| \leq \delta} -J(u_0) = \min_{\|u_0\| \leq \delta} f(u_0), \tag{4}$$

where f is regarded as the objective function.

2.2   Zebiak-Cane Model

ZC model is a mesoscale atmosphere-ocean coupled model covering the scope of the Tropical Pacific, and it has been widely used to study the predictability and dynamics of ENSO event [2–4] since it successfully predicted the ENSO event of 1986–1987 [16]. In ZC model, sea surface temperature anomaly (SSTA) and thermocline height anomaly (THA) are the variables used for studying ENSO event. Since we divide the region of SSTA and THA into 20×27 grids after removing unused marginal area, SSTA and THA can be regarded as two (20×27)-dimensional matrixes. We reshape them to two 540 (20×27)-dimensional vectors, so that the u_0 in (1) is denoted as a 1080 (540×2)-dimensional perturbation vector that consists of SSTA and THA. Hence, solving CNOP in ZC model is a 1080-dimensional optimization problem.

3   Parallel CACO-LD for Solving CNOP

3.1   CACO-LD and Its Parallelization

Through a mass of experiments, we find that ACO_R sometimes is trapped into local optimum and show some instability in our study. To improve the stability, we design a linear decrease strategy and propose CACO-LD based on ACO_R.

For an n-dimensional problem, CACO-LD uses a candidate table to keep track of k candidates s_l $(l = 1, \ldots, k)$:

$$s_l = \{s_l^1, \ldots, s_l^n\}, \tag{5}$$

where s_l^i $(i = 1, \ldots, n)$ denotes the ith variable of s_l. CACO-LD incrementally constructs new candidates based on Gaussian kernel probability density function $G^i(x)$, which is defined as the weighted sum of k Gaussian functions $g_l^i(x)$ as (6) shows:

$$G^i(x) = \sum_{l=1}^{k} w_l g_l^i(x), \tag{6}$$

Algorithm 1. CACO-LD

Input : N_{ant} $(N_{ant} > k)$ ants
Output: The best candidate (CNOP)

1 **Initialize candidate table:**
Initialize every ant and calculate the corresponding objective function value;
Select the best k ants as candidates, and store candidates in the candidate table;
/* *maxIter*: the maximum number of iterations   */
for *curIter* = 1 to *maxIter* do

2 | **Rank the candidates:**
Rank the candidates according to their quality (objective function value);
Compute the weight w_l associated to the lth candidate by (7, 8):

$$w_l = \frac{1}{qk\sqrt{2\pi}} e^{-\frac{(l-1)^2}{2q^2k^2}} \tag{7}$$

/* q_{max} (q_{min}): the upper (lower) bounds of q */

$$q = q_{max} - \frac{q_{max} - q_{min}}{maxIter} \cdot curIter . \tag{8}$$

3 | **Construct new candidates:**
for *curAnt* = 1 to N_{ant} do
Choosing the lth Gaussian functions with probability p_l:

$$p_l = \frac{w_l}{\sum_{r=1}^{k} w_r} . \tag{9}$$

/* Suppose the lth Gaussian functions are chosen, that is, g_l^i
is chosen for dimension i $(i = 1,\ldots,n)$ */
for i = 1 to n do
The mean of g_l^i is set as s_l^i;
The standard deviation σ_l^i of g_l^i is obtained by (10, 11):

$$\sigma_l^i = \xi \sum_{r=1}^{k} \frac{|s_r^i - s_l^i|}{k-1} \tag{10}$$

/* ξ_{max} (ξ_{min}): the upper (lower) bounds of ξ */

$$\xi = \xi_{max} - \frac{\xi_{max} - \xi_{min}}{maxIter} \cdot curIter . \tag{11}$$

Sampling g_l^i for the ith dimension;

/* The samples of all dimensions comprise a new candidate */
4 | Calculate the objective function value of the new candidate;

5 **Update candidate table:**
Add the set of new candidates and remove the same number of worst candidates;
Update the best candidate, then go to step 2;

---

where w_l is the weight of $g_l^i(x)$ $(l = 1, \ldots, k)$. CACO-LD makes $G^i(x)$ correspond to the ith $(i = 1, \ldots, n)$ dimension, thus there are n Gaussian kernel probability density functions in total. The algorithm description is shown in Algorithm 1.

In Algorithm 1, the linear decrease strategy used in (8) and (11) makes a difference. Since w_l in (7) follows a Gaussian distribution $w_l \sim \mathcal{N}(1, (qk)^2)$, when q is small, the top-ranked candidates are endowed with much larger weights than others. As q decreases linearly, the distribution of the weights becomes more uniform. Furthermore, as ξ decreases linearly, CACO-LD fine-tunes the search range to better explore the global optimum with higher convergence speed.

During the computation of CNOP, the objective function will be called a great many times, especially in step 4, and the objective function contains the computation of nonlinear propagator as shown in (3), which is the most time-consuming part for solving CNOP. To improve the efficiency of CACO-LD, we parallelize the computation of the objective function values with MPI technique. Suppose that we allocate N processes initially, one process is designated as the master process while other $N - 1$ processes are regarded as slave processes when the MPI program runs. The master process divides the ants into $N - 1$ groups and distributes each group to one slave process. The slave processes calculate the objective function values of the ants assigned to them in parallel, and send the results back after finishing their computation. Finally, the master process collects the results from the slave processes.

3.2 Solving CNOP with Parallel CACO-LD

The dimension of the solution space in ZC model is relatively high, in order to further improve the efficiency of solving CNOP, we reduce the dimension to a lower one before applying parallel CACO-LD to solve CNOP. The framework of solving CNOP with parallel CACO-LD is shown in Fig. 1.

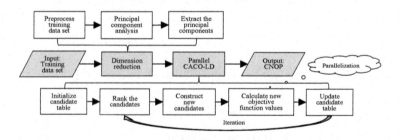

Fig. 1. The framework of solving CNOP with parallel CACO-LD.

In the framework, we firstly run the numerical model to export the training data set, then the dimension reduction is implemented. In the process of dimension reduction, the training data set is preprocessed, then the principal component analysis (PCA) method [9] is adopted to extract the principal components (PCs), so that the 1080-dimensional solution space is converted to a lower-dimensional feature space denoted by the PCs. Finally, we apply the parallel CACO-LD, which has been demonstrated in Sect. 3.1, to solve CNOP in the feature space and output CNOP.

4   Experiments and Results

All the experiments run on a Lenovo ThinkServer RD430 with Intel Xeon CPU E5-2450, 32 cores and 128G RAM. To verify the validity of CACO-LD, we apply it to solve CNOP in ZC model. The optimization time starts from each month in a year and the time span is 9 months. We compare CACO-LD with ADJ-CNOP, MABC, CTS-SS and ACO_R. To ensure the fairness of comparison, all the IAs in the experiments are based on PCA and parallelized by MPI. After lots of experiments, all the parameters involved in solving CNOP with the CACO-LD are listed in Table 1.

Table 1. Parameter setting of CACO-LD for solving CNOP.

Name	Meaning	Value
N_{ant}	The number of ants	150
$maxIter$	The maximum iteration number	60
PCs	The number of principal components	80
k	The number of candidates stored in candidate table	40
q_{max}	The upper bounds of q	0.2
q_{min}	The lower bounds of q	0.1
ξ_{max}	The upper bounds of ξ	0.7
ξ_{min}	The lower bounds of ξ	0.6

4.1   Validity Analysis

Generally, CNOP magnitude and CNOP pattern are used to evaluate the validity of CNOP. CNOP magnitude measures the fitness value of CNOP, the larger the better. Whats more, the year-round tendency of CNOP magnitude should be in accord with the ADJ-CNOP. CNOP patterns (including the patterns of SSTA and THA) reveal the CNOP spatial structures, which are of the most importance. It's a precursor of ENSO event if the SSTA component presents a zonal dipole with negative anomalies in the central tropical Pacific and positive anomalies in the eastern tropical Pacific, meanwhile, the THA component shows a uniform deepening across the whole equatorial Pacific.

CNOP Magnitude. For each initial month, every method is executed 60 times to solve CNOP, and the best CNOP magnitude is selected for comparison, as shown in Fig. 2.

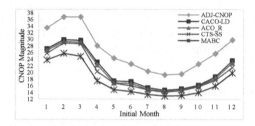

Fig. 2. CNOP magnitude of each month in a year obtained by five methods.

In Fig. 2, the CNOP magnitude of all IAs conforms to the same tendency with that of ADJ-CNOP. The CNOP magnitude obtained by CACO-LD is much larger than MABC, a bit larger than CTS-SS and ACO_R, but smaller than ADJ-CNOP. The reason is that the dimension reduction method applied in the IAs inevitably results in some loss of information, which is acceptable as long as the CNOP patterns are correctly captured. Hence, CACO-LD is better than MABC, CTS-SS and ACO_R in the aspect of CNOP magnitude.

CNOP Pattern. Since the CNOP magnitude of each method has the same tendency, we select the CNOP patterns whose gap of CNOP magnitude between CACO-LD and ADJ-CNOP is the largest for comparison, as shown in Fig. 3. The initial month is March.

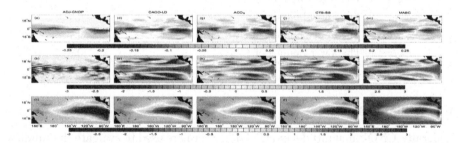

Fig. 3. CNOP patterns obtained by ADJ-CNOP, CACO-LD, ACO_R, CTS-SS and MABC from left to right. Three rows from top to bottom denote the patterns of SSTA, THA and the corresponding evolution of SSTA after 9 months respectively. The initial month is March. (Color figure online)

In Fig. 3, we can observe that the SSTA and THA obtained by every method show the spatial patterns that can be regarded as the precursor of ENSO event, and the evolutions of SSTAs after 9 months conform to the character of real ENSO event. The major difference is that the CACO-LD obtains the second largest and darkest red areas after ADJ-CNOP. Besides, the red area of THA obtained by ADJ-CNOP looks discontinuous, while the others are smooth.

The reason is that CNOPs of all the other four methods are calculated in feature space, thereby losing some accuracy.

Compared with other IAs, CACO-LD captures closer CNOP pattern to that of ADJ-CNOP, which is in accord with the conclusion of the CNOP magnitude. Hence, CACO-LD performs best among the four IAs in validity.

4.2 Stability Analysis

It's significant to analyze the stability of CACO-LD since stochasticity is the natural character of IAs. In the stability experiments, we run CACO-LD, ACO_R, CTS-SS and MABC for 60 times respectively to compute the standard deviation of CNOP magnitude obtained by each method, as shown in Fig. 4. The initial month is still March.

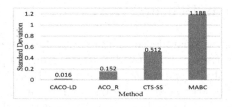

Fig. 4. The standard deviations of the CNOP magnitude obtained by four methods.

In Fig. 4, CACO-LD has the highest stability and its corresponding standard deviation is only 0.016, much smaller than that of ACO_R (0.152), CTS-SS (0.512) and MABC (1.188). Similar standard deviations can be observed for other months, which are not listed here due to the limited space.

4.3 Efficiency Analysis

In order to illustrate the efficiency of parallel CACO-LD, we compare it with parallel MABC, parallel CTS-SS and ADJ-CNOP (with 15 initial guess fields). Figure 5 shows the time consumption of four methods when the process number varies.

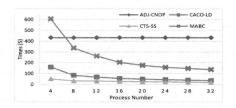

Fig. 5. Time consumption of four methods.

In Fig. 5, with the increase of process number, all the IAs keep a decline trend in time consumption, while the time consumption of ADJ-CNOP stays constant because ADJ-CNOP cannot be parallelized. When 32 processes are used for parallelization, CACO-LD consumes only 35.1 s (with a speedup of 24.4 times) and is 3.9 times faster than MABC (135 s), 12.3 times faster than ADJ-CNOP (432 s), only 16.4 s slower than CTS-SS (18.7 s). The reason why CACO-LD is still a bit slower than CTS-SS is that CACO-LD is a swarm IA and has much more times of calling objective function than CTS-SS, which is a single individual IA, according to the parameter setting in this paper and that in [13].

We can observe that the time consumption of CACO-LD is always much less than that of ADJ-CNOP and MABC, and gets closer and closer to that of CTS-SS as more processes are available. Hence, CACO-LD thoroughly defeats ADJ-CNOP and MABC, and is competitive with CTS-SS in efficiency.

5   Conclusions and Future Works

In this paper, we propose a parallel continuous ant colony optimization algorithm with linear decrease strategy (CACO-LD), which is an improved swarm intelligence algorithm, to solve CNOP in ZC model for studying ENSO event. Taking ADJ-CNOP as benchmark, we compare CACO-LD with MABC and CTS-SS, which respectively are the latest best swarm intelligence algorithm and single individual intelligence algorithm for solving CNOP. Experimental results show that CACO-LD obtains the best CNOP magnitude, the best CNOP pattern and the highest stability among CACO-LD, CTS-SS and MABC. Whats more, the parallel CACO-LD shows overwhelming superiority compared with the parallel MABC, and is competitive with the parallel CTS-SS in efficiency. In conclusion, CACO-LD is the best swarm intelligence algorithm for solving CNOP till now.

In the future, we will implement CACO-LD to more complex numerical model that has much higher dimension, such as the fifth-generation Penn State/NCAR mesoscale model (MM5) [17] and the weather research and forecasting model (WRF) [18].

Acknowledgements. This work is funded by the National Natural Science Foundation of China (Grant No. 41405097).

References

1. Mu, M., Duan, W., Wang, B.: Conditional nonlinear optimal perturbation and its applications. Nonlinear Process. Geophys. **10**(6), 493–501 (2003)
2. Duan, W., Yu, Y., Hui, X., Peng, Z.: Behaviors of nonlinearities modulating the El Niño events induced by optimal precursory disturbances. Clim. Dyn. **40**(5–6), 1399–1413 (2012)
3. Yu, L., Mu, M., Yu, Y.: Role of parameter errors in the spring predictability barrier for enso events in the Zebiak-Cane model. Adv. Atmos. Sci. **31**(3), 647 (2014)

4. Duan, W., Mu, M.: Dynamics of nonlinear error growth and the spring predictability barrier for El Niño predictions. In: Climate Change: Multidecadal and Beyond, p. 81 (2015)
5. Qin, X., Duan, W., Mu, M.: Conditions under which CNOP sensitivity is valid for tropical cyclone adaptive observations. Quart. J. R. Meteorol. Soc. **139**(675), 1544–1554 (2013)
6. Wang, Q., Ma, L., Xu, Q.: Optimal precursor of the transition from kuroshio large meander to straight path. Chin. J. Oceanol. Limnol. **31**(5), 1153–1161 (2013)
7. Hui, X., Wansuo, D., Jianchao, W.: The tangent linear model and adjoint of a coupled ocean-atmosphere model and its application to the predictability of ENSO. In: 2006 IEEE International Symposium on Geoscience and Remote Sensing, pp. 640–643. IEEE (2006)
8. Yuan, S., Ji, F., Yan, J., Mu, B.: A parallel sensitive area selection-based particle swarm optimization algorithm for fast solving CNOP. In: Arik, S., Huang, T., Lai, W.K., Liu, Q. (eds.) ICONIP 2015. LNCS, vol. 9490, pp. 71–78. Springer, Cham (2015). doi:10.1007/978-3-319-26535-3_9
9. Mu, B., Wen, S., Yuan, S., Li, H.: PPSO: PCA based particle swarm optimization for solving conditional nonlinear optimal perturbation. Comput. Geosci. **83**, 65–71 (2015)
10. Ren, J., Yuan, S., Mu, B.: Parallel modified artificial bee colony algorithm for solving conditional nonlinear optimal perturbation. In: IEEE International Conference on High Performance Computing and Communications; IEEE International Conference on Smart City; IEEE International Conference on Data Science and Systems, pp. 333–340 (2016)
11. Wen, S., Yuan, S., Mu, B., Li, H., Ren, J.: PCGD: principal components-based great deluge method for solving CNOP. In: 2015 IEEE Congress on Evolutionary Computation (CEC), pp. 1513–1520. IEEE (2015)
12. Yuan, S., Yan, J., Mu, B., Li, H.: Parallel dynamic step size sphere-gap transferring algorithm for solving conditional nonlinear optimal perturbation. In: 2015 IEEE 17th International Conference on High Performance Computing and Communications (HPCC), 2015 IEEE 7th International Symposium on Cyberspace Safety and Security (CSS), 2015 IEEE 12th International Conferen on Embedded Software and Systems (ICESS), pp. 559–565. IEEE (2015)
13. Yuan, S., Qian, Y., Mu, B.: Paralleled continuous tabu search algorithm with sine maps and staged strategy for solving CNOP. In: Wang, G., Zomaya, A., Perez, G.M., Li, K. (eds.) ICA3PP 2015. LNCS, vol. 9530, pp. 281–294. Springer, Cham (2015). doi:10.1007/978-3-319-27137-8_22
14. Socha, K., Dorigo, M.: Ant colony optimization for continuous domains. Eur. J. Oper. Res. **185**(3), 1155–1173 (2008)
15. Zebiak, S.E., Cane, M.A.: A model El Niño-southern oscillation. Water Resour. Res. **115**(5906), 2262–2278 (1987)
16. Yu, Y., Duan, W., Xu, H., Mu, M., et al.: Dynamics of nonlinear error growth and season-dependent predictability of El Niño events in the Zebiak-Cane model. Quart. J. Roy. Meteorol. Soc. **135**(645), 2146 (2009)
17. Grell, G.A., Dudhia, J., Stauffer, D.R., et al.: A description of the fifth-generation Penn State/NCAR Mesoscale Model (MM5) (1994)
18. Skamarock, W.C., Klemp, J.B., Dudhia, J.: Prototypes for the WRF (weather research and forecasting) model. Preprints, Ninth Conference Mesoscale Processes, pp. J11–J15. Amer. Meteorol. Soc. Fort Lauderdale, FL (2001)

New Decrease-and-Conquer Strategies for the Dynamic Genetic Algorithm for Server Consolidation

Chanipa Sonklin, Maolin Tang$^{(\boxtimes)}$, and Yu-Chu Tian

School of Electrical Engineering and Computer Science,
Queensland University of Technology, 2 George Street,
Brisbane, QLD 4001, Australia
chanipa.sonklin@hdr.qut.edu.au, {m.tang,y.tian}@qut.edu.au

Abstract. The energy consumption in a data center is a big issue as it is responsible for about half of the operational cost of the data centres. Thus, it is desirable to reduce the energy consumption in data centre. One of the most effective ways of cutting the energy consumption in a data centre is through server consolidation, which can be modelled as a virtual machine placement problem. Since virtual machines in a data centre may come and go at any time, the virtual machine placement problem is a dynamic one. As a result, a decrease-and-conquer dynamic genetic algorithm has been proposed for the dynamic virtual machine placement problem. The decrease-and-conquer strategy plays a very important role in the dynamic genetic algorithm as it directly affects the performance of the dynamic genetic algorithm. In this paper we propose three new decrease-and-conquer strategies and conduct an empirical study of the three new decrease-and-conquer strategies as well as the existing one being used in the decrease-and-conquer genetic algorithm. Through the empirical study we find one of the decrease-and-conquer strategy, namely new first-fit decreasing, is significantly better than the existing decrease-and-conquer strategy.

Keywords: Decrease and conquer · Server consolidation · Virtual machine placement · Dynamic optimisation · Genetic algorithm

1 Introduction

The energy consumption of data centres has been increasing dramatically due to the rapid growth of cloud computing as data centres are the infrastructure of cloud computing. In the United States, the energy consumption in the data center continuously increase to 140 Billion kWh by 2020 and this increasing can bring about 150 million tons of carbon pollution every year [5]. In addition, the increasing energy consumption is leading to the increase in the operational costs of data centres. Thus, it is desirable to cut the energy consumption of data centres.

D. Liu et al. (Eds.): ICONIP 2017, Part IV, LNCS 10637, pp. 504–512, 2017.
https://doi.org/10.1007/978-3-319-70093-9_53

One of the most effective ways of cutting the energy consumption is *server consolidation*, which can be transformed into two consecutive problems: *Virtual Machine (VM) placement problem* [1,4] and *live migration of multiple VMs problem* [2]. The VM placement problem, by its nature, is an optimisation problem. Thus, Genetic Algorithms (GAs) have been applied to the VM placement problem [4,6–8]. Since GAs are global search algorithms, they can find an optimal or near-optimal solution for the VM placement problem. However, their computation times are generally long. As a result, the VM placement may not be updated in time.

In order to deal with the problem, we have proposed a decrease-and-conquer dynamic GA recently [3]. Rather than considering all the PMs and all the VMs in the data centre, the decrease-and-conquer dynamic GA adopts a decrease-and-conquer strategy to select a small set of PMs and a small set of VMs when finding a new optimal or near-optimal VM placement. Experimental results have shown that the decrease-and-conquer dynamic GA can generate a VM placement solution that is the same with or similar to the VM placement solution considering all the PMs and all the VMs with significantly shorter computation time and incurring less VM migrations.

The decrease-and-conquer strategy plays a very important role in the dynamic genetic algorithm as it directly affects the performance of the dynamic genetic algorithm. In this paper we will propose three new decrease-and-conquer strategies and conduct an empirical study of the three new decrease-and-conquer strategies as well as the existing one being used in the decrease-and-conquer genetic algorithm.

The remaining of this paper is as follows: Sect. 2 presents three new decrease-and-conquer strategies and an existing decrease-and-conquer strategy used in the dynamic genetic algorithm; Sect. 3 is an empirical study of the four decrease-and-conquer strategies; Finally, Sect. 4 concludes athis research work.

2   Decrease-and-Conquer Strategies

The decrease-and-conquer strategy in the decrease-and-conquer dynamic GA for the server consolidation problem is used to reduce the search space of the server consolidation problem. The size of the search space of the server consolidation problem is determined by the number of PMs and the number of VMs that need to be considered by the dynamic GA. Generally speaking, the smaller the number of PMs and the smaller the number of VMs, the less the computation time of the genetic algorithm is. In addition, reducing the number of PMs and the number of VMs contributes to reduction of the number of VM migration.

A big challenge in designing a decrease-and-conquer strategy for the dynamic genetic algorithm is how to guarantee the reduced search space covers at least one optimal or near-optimal solution. In the following we will present four decrease-and-conquer strategies: Best-Fit (BF), Worst-Fit (WF), Original First-Fit Decrease (OFFD) and New First-Fit Decrease (NFFD). OFFD is the decrease-and-conquer strategy that is currently being used in the dynamic genetic algorithm for the server consolidation problem and the other three are new ones.

2.1   Best Fit

The basic idea behind the Best Fit (BF) decrease-and-conquer strategy is to find a PM among all the PMs for each of the new VMs such that the energy consumption increase after placing the VM on the new VM is minimal. Then, it only selects those PMs that have new VMs placed and all the old and new VMs that are placed on these selected PMs. In this way, the BF decrease-and-conquer strategy can make sure there exists an optimal or near-optimal solution, which is the solution found by the BF decrease-and-conquer strategy, in the search space.

2.2   Worst Fit

The Worst Fit (WF) decrease-and-conquer strategy selects a PM among all the PMs for each of the new VMs such that after placing the new VM on that PM the PM has the most resources (CPU and main memory) left, which is measured by $W(pm_i)$ as below:

$$W(pm_i) = \frac{V_{n_j}^{cpu}}{pm_i^{cpu} - pm_i^{cu}} + \frac{V_{n_j}^{mem}}{pm_i^{mem} - pm_i^{mu}} \tag{1}$$

where $V_{n_j}^{cpu}$ and $V_{n_j}^{mem}$ are the CPU and memory requirements of new V_{n_j} respectively; pm_i^{cpu} and pm_i^{mem} are the CPU and memory capacities of pm_i respectively; and pm_i^{cu} and pm_i^{mu} are the current CPU and memory of pm_i respectively.

Then, it only selects those PMs that have new VMs placed and all the old and new VMs that are placed on these selected PMs. In this way, the WF decrease-and-conquer strategy can also make sure there exists an optimal or near-optimal solution, which is the solution found by the WF decrease-and-conquer strategy, in the search space. The VMs to be selected are all those VMs on those selected PMs, including the old and new VMs.

2.3   Original First Fit Decrease (OFFD)

The Original First Fit Decrease (OFFD) decrease-and-conquer strategy was proposed in our previous work [3]. It sorts all the PMs in the reverse order by their energy efficiency and then from the first PM to the last PM in the sorted PMs to find the first PM that a VM can be placed for all the new VMs. It selects from the first PM to the last PM on which has a new VM placed. The VMs to be selected are all those VMs on those selected PMs, including the old and new VMs.

2.4   New First Fit Decrease (NFFD)

Similar to the OFFD decrease-and-conquer strategy, the New First Fit Decrease (NFFD) strategy also sorts all the PMs in the reverse order by their energy efficiency, and then for each of the new VMs the NFFD strategy finds the first active PM in the sorted PMs which can accommodate the new VM. If such an active PM cannot be found, then the NFFD strategy finds the first inactive PM in the sorted PMs which can accommodate the new VM. A PM is called *active PM*, if there exists at least one VM that has already been placed on it; otherwise, it is called *inactive PM*. Finally, the NFFD strategy selects all the PMs where a new VM is placed and all the VMs on these selected PMs, including existing and new VMs.

3 Empirical Study

This section conducts an empirical study of the four decrease-and-conquer strategies in the previous section.

3.1 Experimental Design

First of all, we implement four dynamic genetic algorithms for server consolidation using the four different decrease-and-conquer strategies. These dynamic genetic algorithms are called BF-based dynamic GA, WF-based dynamic GA, OFFD-based dynamic GA and NFFD-based dynamic GA. Then, we use a program to randomly generate ten dynamic server consolidation problems of different sizes.

The CPU capacity of those PMs is a randomly generated number of multiple of 5000 between 5000 and 25000, inclusive; the main memory capacity of those PMs is a random value between 0.7 and 1.3 times of its CPU capacity. Similarly, the CPU requirement of those VMs is randomly selected from between 300 and 3000, inclusive; and the main memory requirement of those VMs is a random value between 0.7 and 1.3 times of its CPU requirement.

The time duration for all the randomly generated test problems is 24 hours. During the period there are 10 times at which up to three new VM come and/or go. Table 1 shows the number of PMs and the initial number of VMs in each of the randomly generated test problems.

Table 1. Characteristics of test problems

Test problem	VM (#)	PM (#)
1	100	20
2	200	40
3	300	60
4	400	80
5	500	100
6	600	120
7	700	140
8	800	160
9	900	180
10	1000	200

For each of the test problems, we use the four dynamic GAs to solve it. Considering the stochastic nature of the dynamic GAs, we repeat each of the experiments for 30 times and use the average of the 30 runs as the result. For all the dynamic GAs, their population size is 200, the crossover probability and mutation probability are 0.75 and 0.15, respectively, and the termination condition is "no improvement in the best solution for 100 generations".

3.2   Experimental Results

Tables 2, 3, 4, and 5 show the experimental results about the BF-based dynamic GA, the WF-based dynamic GA, the OFFD-based dynamic GA, and the NFFD-based dynamic GA for the 10 test problems, respectively. The experimental

Table 2. The performance of the BF-based dynamic GA

Test problem	Energy consumption (kWh)	VM migrations (#)	Computation time (sec)
1	71.49	15	0.68
2	140.45	19	1.19
3	212.04	17	1.56
4	316.94	17	1.98
5	375.35	24	2.54
6	458.35	19	2.94
7	517.01	16	3.35
8	599.83	22	4.38
9	652.03	28	5.22
10	762.63	20	5.48

Table 3. The performance of the WF-based dynamic GA

Test problem	Energy consumption (kWh)	VM migrations (#)	Computation time (sec)
1	71.54	13	0.67
2	141.02	17	1.06
3	212.81	15	1.55
4	317.77	11	1.93
5	375.21	16	2.32
6	457.48	15	2.70
7	516.48	11	3.10
8	598.70	15	3.70
9	651.44	15	4.70
10	762.70	14	5.17

Table 4. The performance of the OFFD-based dynamic GA

Test problem	Energy consumption (kWh)	VM migrations (#)	Computation time (sec)
1	72.42	73	0.83
2	140.38	163	1.64
3	212.62	231	3.64
4	318.40	229	3.34
5	375.12	234	4.46
6	459.13	276	4.00
7	517.44	358	5.31
8	599.27	487	7.05
9	653.06	408	7.75
10	762.41	617	9.41

Table 5. The performance of the NFFD-based dynamic GA

Test problem	Energy consumption (kWh)	VM migrations (#)	Computation time (sec)
1	72.15	11	0.64
2	141.41	14	1.12
3	212.81	11	1.49
4	319.30	7	1.81
5	375.25	6	2.25
6	459.18	7	2.67
7	518.98	7	3.13
8	599.13	9	3.68
9	653.23	10	4.71
10	763.30	8	5.05

results include the quality of the solutions, measured by energy consumption and number of VM migrations, and computation times.

Figures 1, 2, and 3 are the comparisons between those four dynamic GAs in terms of total energy consumption, number of VM migrations, and computation time, respectively.

It can be shown from Fig. 1 that the total energy consumptions of all the dynamic GAs are very close to each other for all the 10 test problems. However, it can be shown from Figs. 2 and 3 that the number of VM migrations incurred by the solutions generated by the dynamic GAs using the three new decrease-and-conquer strategies is less than the number of VM migrations incurred by the solutions generated by the dynamic GA using the original decrease-and-conquer

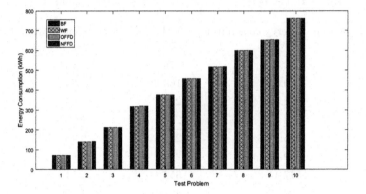

Fig. 1. The energy consumption comparisons

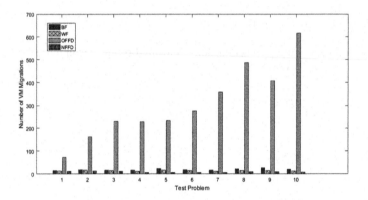

Fig. 2. The number of VM migration comparisons

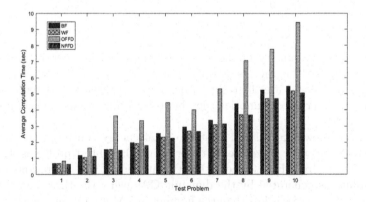

Fig. 3. The computation time comparisons

strategy and the computation time of those dynamic GAs using new decrease-and-conquer strategies is shorter than that of the dynamic GA using original decrease-and-conquer strategy.

It can be concluded that overall all the three new decrease-and-conquer strategies are better than the original one, and that NFFD is the best of all the three new decrease-and-conquer strategies.

4   Conclusion

The decrease-and-conquer strategy plays an important role in the decrease-and-conquer genetic algorithm for the server consolidation problem as it directly affects the performance of the decrease-and-conquer algorithm. In this paper we have proposed three new decrease-and-conquer strategies (BF, WF and NFFD) for the dynamic genetic algorithm for the server consolidation problem, which is a major contribution of this paper.

In addition, we have conducted an empirical study of the three new decrease-and-conquer strategies (FB, WF and NFFD) and an existing decrease-and-conquer strategy (OFFD). The empirical study has revealed that one of the new decrease-and-conquer strategies, NFFD, outperforms the existing decrease-and-conquer strategy, OFFD. Thus, the current decrease-and-conquer strategy (OFFD) that is being used in the dynamic genetic algorithm can be replaced with NFFD to develop a new dynamic genetic algorithm for the server consolidation problem, which will have better performance than the current dynamic genetic algorithm for the server consolidation problem. This is another major contribution of this paper.

In the future, we will study how to further improve the performance of the dynamic genetic algorithm.

References

1. Dong, J., Jin, X., Wang, H., Li, Y., Zhang, P., Cheng, S.: Energy-saving virtual machine placement in cloud data centers. In: 2013 13th IEEE/ACM International Symposium on Cluster, Cloud and Grid Computing (CCGrid), pp. 618–624, May 2013
2. Sarker, T.K., Tang, M.: Performance-driven live migration of multiple virtual machines in datacenters. In: IEEE International Conference on Granular Computing, pp. 253–258 (2013)
3. Sonklin, C., Tang, M., Tian, Y.C.: A decrease-and-conquer genetic algorithm for energy efficient virtual machine placement in data centers. In: IEEE International Conference on Industrial Informatics. IEEE Press, July 2017, in press
4. Tang, M., Pan, S.: A hybrid genetic algorithm for the energy-efficient virtual machine placement problem in data centers. Neural Process. Lett. $41(2)$, 211–221 (2015)
5. Whitney, J., Delforge, P.: Data center efficiency assessment-scaling up energy efficiency across the data center industry: evaluating key drivers and barriers. NRDC and Anthesis, Rep. IP: 14–08 (2014)

6. Wu, G., Tang, M., Tian, Y.-C., Li, W.: Energy-efficient virtual machine placement in data centers by genetic algorithm. In: Huang, T., Zeng, Z., Li, C., Leung, C.S. (eds.) ICONIP 2012. LNCS, vol. 7665, pp. 315–323. Springer, Heidelberg (2012). doi:10.1007/978-3-642-34487-9_39
7. Wu, Y., Tang, M., Fraser, W.: A simulated annealing algorithm for energy efficient virtual machine placement. In: 2012 IEEE International Conference on Systems, Man, and Cybernetics (SMC), pp. 1245–1250, October 2012
8. Xu, J., Fortes, J.A.B.: Multi-objective virtual machine placement in virtualized data center environments. In: Green Computing and Communications (GreenCom), 2010 IEEE/ACM International Conference on International Conference on Cyber, Physical and Social Computing (CPSCom), pp. 179–188, December 2010

Feature Extraction for the Identification of Two-Class Mechanical Stability Test of Natural Rubber Latex

Weng Kin Lai[1,3]($\boxtimes$), Kee Sum Chan[1], Chee Seng Chan[2],
Kam Meng Goh[1], and Jee Keen Raymond Wong[1]

[1] Tunku Abdul Rahman University College, Kuala Lumpur, Malaysia
laiwk@acd.tarc.edu.my
[2] Faculty of Computer Science and Information Technology,
University of Malaya, Kuala Lumpur, Malaysia
[3] Swinburne University of Technology, Sarawak Campus, Kuching, Malaysia

Abstract. Rubber latex concentrate is a popular raw material widely used for making many common household and industrial products. As its quality is not consistent due to either, the source, weather, storage time, etc. there is a need to be able to measure its quality. A common measure of its quality is the *mechanical stability*, which is defined as the time at the first onset of flocculation when the latex is subjected to physical stress. Currently, the assessment is performed manually by trained personnel, closely adhering to the specifications defined by the ISO 35 standard mechanical stability test that is widely adopted by the rubber industry. Nevertheless, there is some level of subjectivity involved as the test heavily depends on the human eyesight as well as the technician's experience. In this paper, we proposed a new feature set for a computer vision-based mechanical stability classification system that is based on the current standard test. We investigated this with several features as well as a new feature set that is based on the particle size. These were classified with a feedforward neural network. Experimental results demonstrated that the proposed system was able to provide good classification accuracies for this two-class MST problem.

Keywords: Colloids · Latex · Quality testing · Mechanical stability test · Feature extraction · Neural networks

1 Introduction

Natural rubber (NR) latex is a complex, renewable biosynthesis polymer extracted from the *Hevea Brasiliensis* tree. It has many unique characteristics such as high strength, excellent elasticity and processability, making it an appropriate raw material for products such as surgical gloves, tyres, etc. [1]. Clearly, the quality of the raw NR latex which will have a major impact on the manufacturing processes, as well as the performance of the end-product, is of major industrial and economic concern. Unfortunately, the quality of natural rubber latex is not within human control because it is affected by the weather, altitude of the plantation, drainage of the soil, age and health of

© Springer International Publishing AG 2017
D. Liu et al. (Eds.): ICONIP 2017, Part IV, LNCS 10637, pp. 513–522, 2017.
https://doi.org/10.1007/978-3-319-70093-9_54

the tree [2]. Moreover, when the raw latex is transported to the processing facility, it is usually subjected to large mechanical forces which can significantly destabilise the NR latex. This would be in addition to other destabilization effects which are commonly introduced during the manufacturing processes. With so many different factors that can easily modify the quality of the NR latex, it may not be surprising to learn that there is a real need for a standard quality assessment in order to ensure that the batch of NR latex is suitable for a specific application [2].

The mechanical stability test is commonly used to measure the ability of the latex concentrate to withstand the destabilization forces. The widely accepted standard for the latex's mechanical stability is the one specified by the American Society for Testing & Materials ASTM D1076 and International Organization of Standardization ISO 35. The time that elapses before the first visible signs of flocculation to occur is its Mechanical Stability Time (MST). The conventional and common method relies on the experience and analytical skills of the trained human operator to determine this. This is where measurement of the size of the dispersed particles is useful as it can provide an indication of the onset of flocculation. Figure 1 shows several images from the two different rubber latex concentrates that have been subjected to mechanical stress taken at different times, i.e. pre-MST (a and b) and at MST (c and d). The pre-MST images in Fig. 1(a and b) do not really show any significant signs of flocculation as compared to the later images (c and d), where the larger particles can now be easily seen.

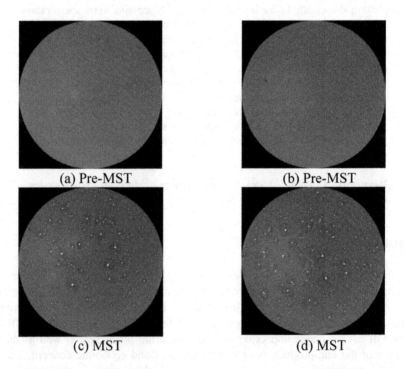

(a) Pre-MST (b) Pre-MST

(c) MST (d) MST

Fig. 1. Mechanical Stability Test (MST) – pre-flocculation and flocculation

In this paper, we proposed a computer vision-based mechanical stability classification system to help minimise the potential for biasness in the current standard test. We investigated this with a new set of features that is based on the count of the particle size. The rest of the paper is organized as follows. Section 2 describes some of the prior work which measured the quality of NR latex. Section 3 describes the methodology used to collect the MST images at a rubber latex processing facility and key details of the features used. Results of the feedforward neural network in recognizing the MST classes for each of the features investigated are also reported here. Finally, Sect. 4 concludes the paper and suggests some areas for further work.

2   Prior Work

The tests to determine the MST of NR latex that is commonly adopted by the industry follows the American Society for Testing & Materials 'ASTM D-1076 and International Organization of Standardization's ISO 35 standards. ASTM D-1076 specifies the standard requirements for the first grade concentrated NR latex of either Category 1, centrifuged Hevea natural latex preserved with ammonia only or by formaldehyde followed by ammonia [3, 4]. However, both these standards are based on qualitative methods that are determined manually [5]. A common approach, based on palm-of-the-hand, determines the MST by dipping a clean glass rod into the test beaker to extract a drop of NR latex that has been mechanically stirred at 14,000 revolutions per minute (rpm) by a latex test machine (LTM), with samples taken at regular time intervals and gently spread on the palm of the hand. The quality is computed based on the first appearance of flocculum with the MST expressed as the number of seconds elapsed from the start of the test to this end point. This is the standard procedure described by both ASTM D-1076 as well as ISO 35 [3, 4].

However, there is also a variation to the ISO 35 standard, adopted by the rubber industry that is based on the *dispersibility-in-water* to determine the end point of the NR latex. This method does not require any direct physical contact between the operator and the colloidal liquid. Similar to the palm-of-the-hand technique, a pointed rod is used to pick up a small drop of latex in the test beaker which is then immediately dispersed onto a petri dish filled with distilled water. When the human operator detects signs of flocculation, the latex would deemed to have reached its MST end-point. Hence, there is an element of subjectivity in such an approach [6].

Another common method that is widely used in industry to identify particle size is by mechanical sieving [7] but clearly this only works for hard objects, for example, rice grains, sand particles, iron ore, etc. Muller and Muller [8] measured the light intensity fluctuations by the number of photons emitted from a laser light source at a set sampling time, to determine the particle size of microemulsions and other highly dispersed systems. This approach makes use of a combination of fast electronics and an optimal correlator design. Vega et al. [9] adopted a more sophisticated approach, using mullti-angle dynamic light scattering from a laser source to compute the particle size. Their contribution comes in the form of a novel autocorrelation measurement function that estimates the weighting coefficients. Others have used specialized particle size measuring equipment that are generally suitable for coarse aggregates but these are known to produce results which are inconsistent and generally inaccurate [10, 11]. Mora et al. [12] investigated the use of digital image processing (DIP) techniques to compute the particle size distribution. They investigated this on three different types of aggregates. A size correction factor was used to convert the particle sizes measured by DIP to equivalent square sieve sizes so that comparison between the DIP and mechanical sieving results can be made. Such a DIP approach was found be fast, convenient, versatile, and accurate for particle size analysis. They extended their work with DIP to measure the flakiness and elongation characteristics of such coarse aggregates, consisting of various different rock types and sizes [13]. Tobias Andersson had investigated machine vision to produce accurate estimates of particle sizes, primarily focusing on industrial material, for example crushed limestone particles which can vary from 20 mm to 40 mm that were transported on conveyor belts [7]. The accuracy was validated with classifiers based on both nominal logistic regression and discriminant analysis with satisfactory accuracies. Barreto et al. [14] also investigated the use of computer vision to determine the size of microstructures, based on the proposed methodology that enables a clear definition of separation between grain and the boundary regions. Their focus was on the size of the microstructures whereas the MST test only looks at the first sign of flocculation which manifest itself in the form of larger latex particles. Nevertheless, a common approach is to extract the key features which were then used by the classifier to identify the relevant structures accurately.

3   Experimental Set-Up and Feature Selection

3.1   Preparing NR Latex for MST Testing

The standard approach defined in the ISO 35 for the mechanical stability tests starts by preparing about 80 g (Fig. 2(a)) of the NR latex concentrate diluted with an ammonia solution to about 55% total solids (Fig. 2(b)). This mixture is then immersed into a warm water bath (Fig. 2(c)) to ensure the mixture is maintained at a constant temperature of 36-37 deg. C. The diluted and warmed latex is quickly strained through a stainless sieve to remove any impurities before being agitated with the LTM (Fig. 2(d)) which shears the liquid at a speed of 14,000 rpm $\pm$ 200 rpm. The stirring continues with samples of the mixture taken at regular intervals for analysis until the MST point is reached [5]. A pointed rod is used to pick up a small drop of latex from the LTM (Fig. 2(e)). Each sample was subsequently dropped into a petri-dish containing distilled water (Fig. 2(f)) and photographs of the latex samples in the petri-dish were taken with a standard digital single-lens reflex (DSLR) camera fitted with a 60 mm f2.8 macro lens. This is repeated until signs of mechanical destabilization, i.e. large number of obvious coagulum is observed. The whole process involving taking samples at regular intervals to identify the MST can be quite time consuming as it takes an average of 30 min to complete. The actual physical mechanical stress test of the NR latex takes up approximately 50% of this time. Hence, the data collected would involve two basic classes, viz., pre-MST and MST. In the 50 laboratory experiments, we collected a total of 487 high-resolution images comprising of 400 and 87 pre-MST and MST respectively. To maintain a class balance, we used two samples from each of the pre-MST and MST images, resulting in a total of 200 images from these 50 experiments.

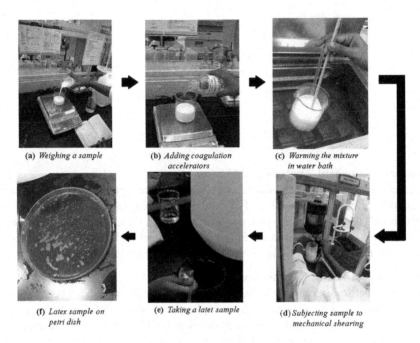

(a) *Weighing a sample*   (b) *Adding coagulation accelerators*   (c) *Warming the mixture in water bath*

(f) *Latex sample on petri dish*   (e) *Taking a latet sample*   (d) *Subjecting sample to mechanical shearing*

Fig. 2. Process of MST testing

3.2   Feature Extraction

Histogram of Particle Size (HOPS)

We proposed a new feature, *Histogram Of Particle Size* (HOPS) to assist in the classification of the MST classes. Computing the particle size for each of these images is quite straightforward but the size of the particles obtained range from particles of just a few pixels to as much as more than 7,000 – even though there are not many of such large latex particles.

The median of the particle size is 16 with a standard deviation, σ of 47. Based on this, we set the size of each bin to be 37, primarily to provide a better resolution of the features, resulting in a total of 200 features each. There is usually a significantly large amount of small particles, tapering to a very small number at the larger extremes, as shown in Fig. 3.

Nevertheless these will have to be normalised before they can be classified. An intuitive normalization scheme to linearly transform for each and every particle into the range of values from 0.0 to 1.0 was investigated. Figure 4 shows the distribution of the particles of various sizes for this and the sigmoid transforms (with medians of 355 and 3,000).

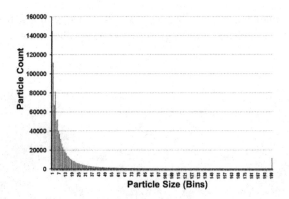

Fig. 3. Histogram of particle size

Principal Component Analysis (PCA)

PCA is a useful statistical technique that has found application in fields such as pattern recognition [15], face recognition [16] and image compression [17, 18], and is a common technique for finding patterns in data of high dimension. This is especially relevant if we treat each and every image as a vector of high dimensions. Hence the higher dimensionality HOPS can be reduced to a lower dimension with PCA.

Scale Invariant Feature Transform (SIFT)

Local photometric descriptors computed for key areas in an image have been proven to be very useful in a variety of different applications. Many different techniques for extracting local image features have been developed [19] ranging from the simple and intuitive to the more sophisticated and powerful. Lowe [20] proposed a scale invariant feature transform (SIFT) which combines a scale invariant region detector that generates a set of descriptors based on the gradient distribution in the detector regions. Each descriptor is represented by a 3D histogram of gradient locations and orientations, with each weighted by the gradient magnitude. Due to the quantization of these gradient locations and orientations, *Lowe* had shown that the descriptors were robust to small geometric distortions as well as small errors in the selected regions of interest. Furthermore, the descriptors are normalised by the square root of the sum of the squared components to achieve illumination invariance. However, there will be a significant variation in the regions of interest which will result in a different number of detectors. For a range of different images, there will be a different number of descriptors with some having more than 2,400. While we have used the smallest number of detectors nevertheless, we are still left with 384 descriptors - which is still very much larger than the set of features used for the other schemes. With such a large set of features it will increase the amount of computation needed to train the system to identify the MST class for each image - in addition to the need to identify the optimal MLP architecture. PCA (principal component analysis) had been widely used to reduce the dimensionality by identifying the more relevant descriptors instead. With PCA, we managed to reduce the 384 features to just 15 identified by PCA.

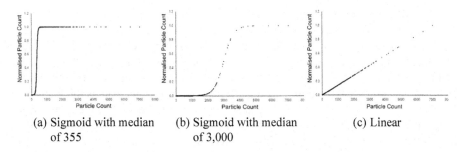

(a) Sigmoid with median   (b) Sigmoid with median   (c) Linear
        of 355   of 3,000

Fig. 4. Data normalisation

Classification

The data generated from the two-class MST were then classified. The multi layer perceptron (MLP) feedforward neural network had been used extensively in many different areas with good results. Zabidi et al. [21] used the MLP to identify the infant cries with Asphyxia, while Catalan et al. [22] reduced the dimension features with MLP. Other applications include mineral prospectivity mapping [23], identification of stress in reading [24], classification of Fish Ectoparasite Genus Gyrodactylus SEM images [25], box-office success of motion pictures [26], data mining analysis [27], phoneme recognition and academic emotion detection [28].

Our multilayer perceptron neural network (MLP) consists of the standard topology [29] i.e. an input layer with one hidden layer and an output layer. Each layer has a set of nodes with the nodes from different layers connected in a feed-forward manner to the nodes of other layers. The neural network maintains a set of connection weights used to adjust the inputs as it propagates through the network towards the output layer. During the training with the backpropagation [30, 31] learning algorithm, the weights were adjusted to minimize the misclassification error of the outputs between the desired and actual values. Like most supervised classification algorithms, MLP is sensitive to the training data. Hence, to test the effectiveness of the MLP in recognising the two classes of the MST and not be subjected to such sensitivities, we repeated the test for different training-test data set pairs. The data was randomly shuffled a number of times to produce 5,000 sets of data partitioned approximately in a 70%-30% mix of training and test data sets respectively. After training with learning rate of 0.15 and momentum of 0.05 was completed, the remaining 30% of the data were then used to verify its accuracy in predicting the correct MST class. Several feature schemes were studied and the performance of the MLP in accurately identifying the correct MST class are shown in Table 1.

Table 1. Classification accuracy for various feature sets (70%–30% partition)

	Recognition accuracy					
	PCA 15	**Sigmoid (median 3000)**	**Sigmoid (median 355)**	**Linear 10_5**	**Linear 15_0**	**SIFT +PCA**
Maximum	91.67	100.00	100.00	100.00	100.00	68.33
Average	73.41	93.29	91.56	97.47	92.03	50.04
Minimum	50.00	80.00	73.33	86.67	50.00	26.67

The MLP was able to correctly identify the correct MST class with an average accuracy of 73.41% for the set of features computed from the PCA of the histogram of particle sizes, using the top 15 components. Furthermore, the MLP was able to achieve an average accuracy of just slightly over 50% for the set of 15 SIFT-PCA features. With the HOPS features, we investigated both the sigmoid as well as linear transforms. Here, we used the first 10 features and the remainder 5 from the end as the feature set for each image, making a total of 15 features. Finally, we also studied the performance of just the first 15 components from the HOPS data set that had been processed with a linear transformation. Overall, the best results were with the linear transformation of the proposed new features, using the 10-5 combination that managed to achieve an average accuracy of 97.47%.

4   Conclusions

The quality of natural rubber latex is known to be affected by mechanical influences which can occur at almost every stage in its manufacturing process, starting from the pumping to the processing of the latex. Hence there is a need to know the quality of any given rubber latex sample and this is where the MST was introduced to satisfy this requirement. However there is some level of subjectivity involved as the test heavily depends on the naked eye as well as the human operator's experience. In this paper, we have described a computer vision-based mechanical stability classification system using a novel set of size-based features. In the first set, the features consist of the finer and smaller particles size but we do know that as the mechanical shearing of the NR latex continues during the mechanical stability test, larger particles would start to form. Samples were then taken at regular intervals and the size of the particles were monitored. The features for each sample were extracted and several types of features were investigated. Moreover, we have also proposed another set of features that is based on both the small particles (10) as well as the larger ones (5). Each set of features were tested on the multi layer perceptron (MLP) feedforward neural network. The experimental results have shown significantly good recognition accuracies for this two-class MST problem, with an average accuracy of 97% with the new HOPS feature set. Even though encouraging results have been achieved with this MLP classification system on a new feature set for NR latex, we would like to extend this work by investigating how the nodes in the hidden layer may affect the performance of the MLP classifier in correctly identifying the MST classes.

Acknowledgements. The authors are grateful to Ming Chieng TAN for her assistance in collecting the rubber latex MST images as well as *Jaya Kumar Veellu*, Chief Chemist of Sime Darby R&D for giving us access to the rubber latex testing facility. The work reported here was partially funded by the Malaysian Ministry of Education's (MOE) *Fundamental Research Grant Scheme* (FRGS/1/2014/TK04/TARUC/02/1).

References

1. Malaysian Investment Development Authority. http://www.mida.gov.my/env3/index.php?page=rubber-based-industries

2. Official Website of the Malaysian Rubber Export Promotion Council. http://www.mrepc. com/industry/industry.php
3. Dawson, H.G.: Mechanical stability test for Hevea Latex. Anal. Chem. **21**(9), 1068–1071 (1949)
4. Maron, S.H., Ulebitch, I.N.: Mechanical stability test for rubber latices. Anal. Chem. **25**(7), 1087–1091 (1953)
5. Akmal, M.K., Othman, A., Mansor, M.N.: Invention of RRIM MST tester for quantitative method of mechanical stability time (MST) of natural rubber latex concentrates. In: Proceedings of the Plastics and Rubber Institute Malaysia (PRIM) Annual Polymer Technology Seminar 2013 (2013)
6. Amran, M.K.A., Mansor, M.N., Othman, A.: Method of quantitative measurement of mechanical stability time (MST) of latex suspensions and the apparatus for use in the method. In: International Patent WO 2012158015 A1 (2012)
7. Andersson, T.: Estimating particle size distributions based on machine vision. Doctoral thesis, Department of Computer Science and Electrical Engineering, Lulea University of Technology, Sweden (2010)
8. Muller, B.W., Muller, R.H.: Particle size analysis of latex suspensions and microemulsions by photon correlation spectroscopy. J. Pharm. Sci. **73**(7), 915–918 (1984)
9. Vega, J.R., Gugliotta, L.M., Gonzalez, V.D.G., Meira, G.R.: Latex particle size distribution by dynamic light scattering: novel data processing for Multiangle Measurements. J. Colloid Interface Sci. **261**(1), 74–81 (2003)
10. Etzler, F.M., Sanderson, M.S.: Particle size analysis: a comparative study of various methods. Part. Part. Syst. Charact. **12**(5), 217–224 (1995)
11. Monnier, O., Klein, J.P., Ratsimba, B., Hoff, C.: Particle size determination by laser reflection: methodology and problems. Part. Part. Syst. Charact. **13**(1), 10–17 (1996)
12. Mora, C.F., Kwan, A.K.H., Chan, H.C.: Particle size distribution analysis of coarse aggregate using digital image processing. Cem. Concr. Res. **28**(6), 921–932 (1998)
13. Kwan, A.K.H., Mora, C.F., Chan, H.C.: Particle shape analysis of coarse aggregate using digital image processing. Cem. Concr. Res. **29**(9), 1403–1410 (1999)
14. Bareto, H.P., Vcillalobos, I.R.T., Magdeleno, J.J.R., Navarro, A.M.H., Hernandez, L.A.M., Guerro, F.M.: Automatic grain size determination in microstructures using image prepocessing. Measurement **46**, 249–258 (2013)
15. Segreto, T., Simeone, A., Teti, R.: Principal component analysis for feature extraction and NN pattern recognition in sensor monitoring of chip form during turning. CIRP J. Manufact. Sci. Technol. **7**(3), 202–209 (2014)
16. Turk, M., Pentland, A.: Eigenfaces for recognition. J. Cogn. Neurosci. **3**(1), 71–86 (1991)
17. Mahendran, M., Jayavathi, S.D.: Compression of hyperspectral images using PCA with lifting transform. In: Proceedings of the International Conference on Emerging Engineering Trends and Science (ICEETS 2016), pp 68–73 (2016)
18. Lim, S., Sohn, K., Lee, C.: Principal component analysis for compression of hyperspectral images. In: Proceedings of the IEEE International Geoscience and Remote Sensing Symposium, (IGARSS 2001), Sydney, Australia, pp 97–99 (2001)
19. Mikolajczyk, K., Schmid, C.: A performance evaluation of local descriptors. IEEE Trans. Pattern Anal. Mach. Intell. **27**(10), 1615–1630 (2005)
20. Lowe, D.G.: Object recognition from local scale-invariant features. In: Proceedings of the Seventh IEEE International Conference on Computer Vision, Kerkyra, pp. 1150–1157 (1999)
21. Zabidi, A., Khuan, L.Y., Mansor, W., Yassin, I.M., Sahak, R.: Classification of infant cries with asphyxia using multilayer perceptron neural network. Proc. Second Int. Conf. Comput. Eng. Appl. **1**, 204–208 (2010)

22. Catalan, J.A., Jin, J.S., Gedeon, T.D.: Reducing the dimensions of texture features for image retrieval using multi-layer neural networks. J. Pattern Anal. Appl. **2**(2), 196–203 (1999)
23. Brown, W, Gedeon, T.D., Barnes, R.: The use of a multilayer feedforward neural network for mineral prospectivity mapping. In: Proceedings 6th International Conference on Neural Information Processing (ICONIP 1999), Perth, pp. 160–165 (1999)
24. Sharma, N., Gedeon, T.: Artificial neural network classification models for stress in reading. In: Proceedings of 19th International Conference on Neural Information Processing 2012 (ICONIP 2012), pp. 388–395 (2012)
25. Ali, R., Jiang, B., Man, M., Hussain, A., Luo, B.: Classification of fish ectoparasite genus gyrodactylus SEM images using ASM and complex network model. In: Proceedings of the 21st International Conference on Neural Information Processing 2014 (ICONIP 2014), pp. 103–110 (2014)
26. Sharda, R., Delen, D.: Predicting box-office success of motion pictures with neural networks. Expert Syst. Appl. **30**(2), 243–254 (2006)
27. Eftekharian, E., Khatami, A., Khosravi, A., Nahavandi, S.: Data mining analysis of an urban tunnel pressure drop based on CFD data. In: Proceedings of the 22nd International Conference on Neural Information Processing 2015 (ICONIP 2015), pp. 128–135 (2015)
28. Azcarraga, A., Talavera, A., Azcarraga, J.: Gender-specific classifiers in phoneme recognition and academic emotion detection. In: Proceedings of the 23rd International Conference on Neural Information Processing 2016 (ICONIP 2016), pp 497–504 (2016)
29. Rosenblatt, F.: The perceptron: a probabilistic model for information storage and organization in the brain. Psychol. Rev. **65**(6), 386–408 (1958)
30. Rosenblatt, F.: Principles of neurodynamics: perceptrons and the theory of brain mechanisms. In: Spartan Books, Washington DC (1961)
31. Russell, S.J., Norvig, P.: Artificial Intelligence: A Modern Approach, 3rd (edn.). Pearson Education Inc., London (2010)

Neural Data Analysis

Evolutionary Modularity Optimization Clustering of Neuronal Spike Trains

Chaojie Yu[1], Yuquan Zhu[1], Yuqing Song[1], and Hu Lu[2]([✉])

[1] School of Computer Science and Communication Engineering,
Jiangsu University, 301 Xuefu Road, Zhenjiang 212003, China
[2] School of Computer Science, Fudan University, 825 Zhangheng Road,
Shanghai 201203, China
luhu@ujs.edu.cn

Abstract. We propose a method for automatic evolutionary clustering of multi neuronal spike trains on the basis of community detection in complex networks. We use a genetic algorithm for optimization to maximize the modularity for community partitioning and then automatically determine the number of clusters hidden in the multi neuronal spike trains. The number of clusters does not need to be specified in advance. Compared with the traditional graph partitioning method, the genetic evolutionary modularity optimization clustering algorithm can obtain the maximum value of modularity and, determine the number of communities. We evaluate the performance of this method on surrogate spike train datasets with ground truth. The results obtained showed improvement. We then apply this proposed method to raw real spike trains. We obtain a larger value for modularity and the results. This finding suggests that the proposed method can be used to detect the hidden firing pattern.

Keywords: Spike trains · Modularity · Genetic algorithm

1 Introduction

Understanding the process of encoding information on neuronal populations in decision making is important in systems neuroscience research. Synfire chains and coordinated firing between neurons are regarded as important principles of information transmission between cortical neurons. With the development of recording methods in recent years, such as multi-channel arrays and 2-photon imaging, researchers have obtained the firing of individual neurons at the micro level. The spikes of hundreds of individual neurons can be recorded using multi-channel electrodes. These discrete spike sequences at different time points are referred to as neuronal spike trains [1]. Although the theory generally states that neural information processing is composed of clusters of neuronal populations; however, the method of organizing these neurons and forming clusters between neurons, or whether special structures are present to form similar firing patterns need to be determined. Analytical tools for multiple neuronal firings

© Springer International Publishing AG 2017
D. Liu et al. (Eds.): ICONIP 2017, Part IV, LNCS 10637, pp. 525–532, 2017.
https://doi.org/10.1007/978-3-319-70093-9_55

are currently lacking. The clustering algorithm is the most efficient method of analyzing neuronal firing patterns [2].

Clustering algorithm is an unsupervised learning method. The traditional clustering algorithms, such as k-means [2] and spectral clustering algorithms [3] are widely used in various data analyses, including medical image data and biological information analyses. A clustering algorithm is currently applied to analyze neuronal spike trains and to help recognize the inner structures and interesting patterns of spike trains. Fellous et al. (2004) introduced the fuzzy k-means algorithm to identify the precise temporal pattern of spikes from trial to trial of a single neuron. The spectral clustering algorithm performs better than the k-means algorithm and is used to identify groups of synchronized spike trains, which can measure synchronous activity. These clustering algorithms need to specify a predefined number of clusters before the clustering algorithm is implemented. While the true recorded spike trains cannot contain a transcendental number of clusters as a value, the result obtained by clustering is subjective. Humphries [4] proposed a new algorithm, based on the community detection method, which regards the functional connectivity between spike trains as a functional network by maximizing the modularity of the network, automatically determining the number of communities. Several neuronal spike trains are divided into groups. In a study of complex networks, Newman et al. proposed the modularity Q, which became an important evaluation index [5]. The determination of modularity is regarded as an external optimization problem. Given that value distribution is a non-monotonic function, the traditional spectral partitioning method cannot be used to obtain the global optimal Q value [6], and optimizing the Q value is an NP-hard problem. Thus, other methods need to be developed [7].

In the present, we propose a new method based on the evolutionary optimization of the modularity Q to cluster spike trains into groups. We first compute for the similarity of pairwise neuronal spike trains. On the basis of the similarity matrix of all spike trains, we construct a neuronal functional network and then use the genetic algorithm to maximize the Q value of this network. This method does not require specifying the number of groups before partitioning. According to the maximum Q, this method can automatically determine the number of clusters in spike trains, without inputting any parameters. The proposed method can be applied in the analysis of original spike trains without a priori knowledge. Compared with the existing pattern of spike train pattern clustering, the method proposed in the current study show a better performance.

2   Materials and Methods

2.1   Surrogate Data Set

To evaluate the performance of the proposed algorithm compared with those of other algorithms, we first tested the experiments on surrogate data sets, which was created by Fellous (available from http://cnl.salk.Edu/fellous/data/ JN2004data/data.html). Each data set contained a certain number of neuronal spike trains and simulated neuronal firing at different time points. Different firing patterns of spike trains were contained in each data set with different numbers of clusters e.gs., 2, 3, and 5 clustersthus representing a similar firing pattern. Certain extra spikes and jitter spikes were added to each data set. The data could be used to evaluate the performance of the algorithms because it was specified for the number and type of patterns.

2.2 Data on Multi-electron Recording Spike Trains

In addition, the proposed method is applied in real-life public spike trains data sets, referred to as CRCNS.org ret-1 data, which can be obtained from sharing website the National Science Foundation-funded based collaborative research in computational neuroscience data sharing website (http://crcns.org/). The data sets provided by Harvard University record the neuronal activities in the isolated retina by 61 electrodes. We chose the data set containing the maximum number of neurons (including 37 neurons) from the 16 original visual cortex data sets. The initial 100 s recordings of spike trains were used, which contained all 37 neuronal spike trains. We then divided the recording time into 10 time windows, with the all neuronal spike trains in one window considered as a trial. Ten trial data sets were obtained.

2.3 Method

The recording of the action potential firing of neurons by the multi-electrode arrays can be regarded as a discrete time point process. Therefore, the spike train data can be considered as special time series data set. Clustering of analyzing data is typically divided into 2 steps. First, the similarity between them is calculated according to the characteristics of pairwise spike trains. In total, we obtain $N(N-1)/2$, and a correlation matrix $R(N*N)$ is constructed by considering only the functional connectivity among N neurons, where $R(i, j)$ represents the correlation between the ith neuronal spike train and the jth neuronal spike train. Finally, various clustering algorithms can be applied, such as k-means and spectral clustering, to partition the correlation matrix R into different clusters.

Various methods have been proposed to calculate the similarity between pairwise spike trains [8], mainly including binning spike trains with time windows and binless spike trains [9]. The results can be used to analyze the synchronized neuron firings and clustering of patterns. For the time window to bin spike trains, the size of the windows should be specified to calculate the number of spike trains within the window with high subjectivity. Humphries applied 2 methods and compared them in a study; the method without binning spike trains performed better. To compare the experimental results with the results of other algorithms, a preprocessing method similar to that used by Humphries was adopted in the current study. First, each neuronal spike train was converted by a Gaussian function to obtain the standard waveform from the discrete time series, resulting in a continuous vector representation of the spike train. The cosine of the angle between pairs of vectors was then considered as the similarity of pairwise spike trains.

$$s_{ij} = \frac{g_i \cdot g_j}{\|g_i\| \cdot \|g_j\|} \tag{1}$$

where g_i and g_j represent the vectors transformed from the ith and jth neuronal spike train by Gaussian transformation, respectively. S ranges between 0 and 1. A larger value for S indicates a higher correlation between 2 neurons, exhibiting synchronization. The size of the time windows does not have to be specified in advance. In addition, this method can be applied to any types of neuronal firing pattern. We then used $R_{ij} = S_{ij}$ as the correlation matrix of N neuronal spike trains. The resulting matrix R is a symmetric matrix, where $R_{ij} = R_{ji}$. On the basis of the matrix R, some clustering methods can be used [4]. Similar to other studies, the present study applies fuzzy c-means algorithm, along with modular detection based on eigenvector decomposition [4]. The former method requires that the number of clusters be specified in advance. The latter method can identify the optimal number of clusters via the maximization of modularity Q.

In the current study, we propose a new spike trains clustering method based on genetic algorithm to optimize modularity Q, which was presented by Newman for the community structure of weighted networks,

$$Q = \frac{1}{2m} \sum_{ij} \left(A_{ij} - \frac{k_i k_j}{2m} \right) \delta \left(C_i, C_j \right) \tag{2}$$

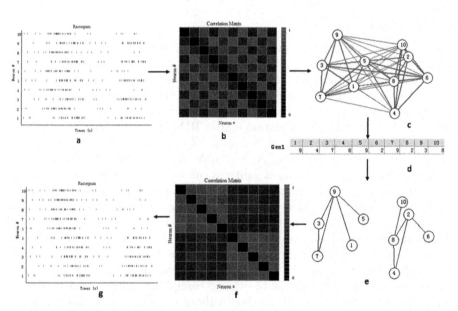

Fig. 1. Process of the proposed method. (a) 10 original neuronal spike trains. (b) Similarity matrix of 10 neuronal spike trains constructed based on Eq. 1. (c) Functional network generated by 10 neuronal spike trains (bolder lines indicate higher correlations). (d) Chromosome generated according to (b). (e) Two communities partitioned using the method based on optimizing modularity Q. (f) Ordered similarity matrix according to the 2 communities. (g) Ordered 10 spike trains. (Ten neuronal spike trains can be divided into 2 communities by the method proposed in this study.

The value of Q lies in $[-0.5, 1]$. Larger values of Q represent the stronger community in the network. In this sense, community detection is transformed into finding the optimal value of Q.

The genetic algorithm exhibits a satisfactory performance in searching for extreme values; thus, the modularity function Q of the community structure was used as the objective function of the genetic algorithm in iterative optimization. Compared with the traditional community partition methods, the proposed method can obtain the global optimal value of Q [10], avoiding the local optimal values. In the genetic algorithm, a series of operations should be conducted, including encoding community structures, chromosome crossover, chromosome mutation, and decoding chromosomes into community structures. Four kinds of operation were applied in this study as they were proposed in the study.

In the iterative process, the chromosome corresponding to the largest values of modularity Q was preserved as the clustering result of multi-neuronal spike trains until the completion of the iterative process. The whole process of evolutionary modularity optimization clustering of neuronal spike trains is presented in Fig. 1.

3   Experiment Results

3.1   Experiment Evaluation

The method was tested on 2 sets of surrogate data sets and 1 real spike train data set. To evaluate the performance of the algorithm, we first analyze artificial data sets with spike train patterns known in advance to test the detection algorithm. These artificial data sets were set as clustering labels during their creation; thus, an external evaluation can be used as a standard to measure the performances of different community partitioning algorithms in these data sets. Mutual information is a commonly used method. If the community structure and the number of clusters of a network are known in advance, normalized mutual information (NMI) can be applied to measure the similarity of the detected community structure and the original community structure. Mutual information as an evaluation function has been widely used in the community structure. If A and B represent the original partition and the detected partition, respectively, then C is the confusion matrix, where C_{ij} is the number of nodes in the ith group that are also in the jth group of B. NMI is calculated as follows:

$$NMI(A, B) = \frac{-2\sum_{i=1}^{C_A}\sum_{j=1}^{C_B} C_{ij} log(\frac{C_{ij}N}{C_i C_j})}{\sum_{i=1}^{C_A} C_i log(\frac{C_i}{N}) + \sum_{i=1}^{C_B} C_j log(\frac{C_j}{N})} \qquad (3)$$

NMI ranges from 0 to 1. The larger value represents the detected community structures that are more consistent with the original partition, indicating satisfactory results.

3.2   Surrogate Data Sets

Certain data sets were selected to evaluate the proposed method with other existing clustering methods for spike train patterns, including the spike trains of 20 trials in 2 groups. These methods include our proposed evolutionary modularity optimization clustering of neuronal spike trains based on genetic algorithm (**A1**); the community detection method based on spectral partition (**A2**); the spike train clustering algorithms based on fuzzy k-means (**A3**); and the spike train algorithm based on traditional spectral clustering (**A4**). We calculated the modularity Q, the number of clusters of firing patterns (N), and NMI according to the clustering results. We used 2 datasets, K-E-J, with K = 3 containing 3 * 35 trials, E = 1, ..., 10 containing E extra spikes, and J = 1, ..., 10 containing J jitter spikes. The results are shown in Tables 1 and 2.

Table 1. The results of four clustering methods on data1

Data1	A1			A2			A3			A4		
	Q	N	NMI	Q	N	NMI	Q	N	NMI	Q	N	NMI
3-1-1	0.3023	3	1	0.3023	3	1	0.3023	3	1	0.3023	3	1
3-1-2	0.2789	3	1	0.2788	3	0.9605	0.2789	3	1	0.0588	3	0.2081
3-1-3	0.4659	3	1	0.3822	6	0.8507	0.4669	3	1	0.0060	3	0.0055
3-1-4	0.4280	3	1	0.3935	4	0.9414	0.4280	3	1	0.0126	3	0.0379
3-1-5	0.4198	3	1	0.3533	4	0.8798	0.4198	3	1	0.01	3	0.0114
3-1-6	0.3903	3	0.9330	0.3594	2	0.7337	0.2995	3	0.5929	0.0125	3	0.0455
3-1-7	0.3966	3	0.8339	0.3196	5	0.6583	0.3757	3	0.6849	0.0226	3	0.0433
3-1-8	0.2200	3	0.6105	0.2073	3	0.5002	0.2108	3	0.4203	0.0253	3	0.0333
3-1-9	0.1562	8	0.3504	0.1513	7	0.5308	0.1627	3	0.1333	0.0179	3	0.0730
3-1-10	0.1447	13	0.3359	0.1558	6	0.2277	0.1508	3	0.1002	0.0178	3	0.0201

Table 2. The results of four clustering methods on data2

Data2	A1			A2			A3			A4		
	Q	N	NMI	Q	N	NMI	Q	N	NMI	Q	N	NMI
3-1-1	0.3023	3	1	0.3023	3	1	0.3023	3	1	0.3023	3	1
3-2-1	0.3415	3	0.9605	0.2727	4	0.7520	0.3401	3	1	0.0711	3	0.3339
3-3-1	0.2390	3	1	0.2195	2	0.7337	0.2390	3	1	0.1663	3	0.479
3-4-1	0.2413	3	1	0.1440	5	0.5195	0.2413	3	1	0.0646	3	0.2281
3-5-1	0.1073	3	0.7648	0.0556	4	0.2602	0.1138	3	1	0.0550	3	0.3813
3-6-1	0.0908	3	0.7123	0.0582	2	0.4311	0.0955	3	0.9213	0.0107	3	0.0799
3-7-1	0.0381	4	0.3460	0.0361	4	0.1864	0.0529	3	1	0.0258	3	0.2963
3-8-1	0.0237	8	0.1959	0.0248	3	0.0796	0.0347	3	0.9211	0.0078	3	0.1811
3-9-1	0.0173	8	0.1135	0.0220	4	0.0396	0.0268	3	0.9605	0.0025	3	0.0427
3-10-1	0.0121	16	0.2053	0.0158	3	0.1234	0.163	3	0.7821	3.6276e−04	3	0.0296

Table 2: N is the detected the number of clusters. A3 and A4 cannot automatically detect the number of clusters, requiring the number of clusters before clustering. Thus, we set the number of clusters in these methods as the original cluster number. From the experimental results, the algorithm proposed in this study can obtain larger values for modularity Q and NMI. This result indicates that the method is effective in spike train pattern clustering.

3.3 Multi-electron Recording Data Sets

We performed experiments in ret-1 data with 10 trials of spike trains. Considering the number of clusters in the real data sets, the algorithms that can identify the number of pattern clusters automatically in real data sets need to be applied. Fuzzy k-means and spectral clustering require the number of clusters in advance; otherwise, the algorithms cannot be executed. Therefore, the most difficult problem related to clustering algorithms is that the number of pattern clusters in spike trains cannot be identified automatically, thereby limiting their practical application. In this study, we showed the results of the proposed method and spike train community detection method (A1 and A2 methods). We also cannot use NMI to evaluate the performance of the clustering algorithm in real data sets. The results of the experiments are presented in Table 3. The experimental results indicate that compared with A2, the proposed method obtained larger values for modularity Q, demonstrating that the clustering results have stronger modularity. The proposed method in this study obtained more pattern clusters, compared with the A2 method. This result shows that the proposed method can identify smaller modules.

Table 3. The results of two methods

Trial		1	2	3	4	5	6	7	8	9	10
A1	Q	0.0262	0.0358	0.0503	0.0591	0.0703	0.0932	0.0459	0.0437	0.0654	0.0665
	N	6	6	5	4	5	6	7	5	6	5
A2	Q	0.0166	0.0299	0.0525	0.0486	0.0565	0.0813	0.0450	0.0375	0.0649	0.0621
	N	2	2	3	4	2	3	3	2	4	2

4 Discussion and Conclusion

Clustering of spike trains is important in identifying multi-neuronal firing patterns. In this study, we proposed a clustering algorithm for spike trains on the basis of optimizing modularity Q without specifying the number of clusters in advance. It can identify the number of clusters by optimizing the modularity Q and detecting the number of firing patterns and the types of firing automatically; therefore, it can be used in real data sets of spike trains. The algorithm shows a wide range of applicability.

Determining the firing patterns of spike trains is not sufficient. The bigger challenge is to analyze the significance of these patterns, which will be the aim of our future studies.

Acknowledgments. This study was supported by the National Natural Science Foundation of China (Project No. 61375122 and Project No. 61572239), China Postdoctoral Science Foundation (Project No. 2014M551324). Scientific Research Foundation for Advanced Talents of Jiangsu University (Project No. 14JDG040).

References

1. Stevenson, I.H., Kording, K.P.: How advances in neural recording affect data analysis. Nat. Neurosci. **14**(2), 139–142 (2011)
2. Hartigan, J.A.: Cluster Algorithms. Wiley, New York (1975)
3. Ng, A., Jordan, M., Weiss, Y.: On spectral clustering: analysis and an algorithm. In: Advances in Neural Information Precessing Systems 14 (2001)
4. Humphries, M.D.: Spike-train communities: finding groups of similar spike trains. J. Neurosci. **31**(6), 2321–2336 (2011)
5. Newman, M.E.: Fast algorithm for detecting community structure in networks. Phys. Rev. E **69**(6), 066133 (2004)
6. Newman, M.E.: Modularity and community structure in networks. Proc. Nat. Acad. Sci. **103**(23), 8577–8582 (2006)
7. Hu, L., Hui, W.: Detecting community structure in networks based on community coefficients. Physica A Stat. Mech. Appl. **391**, 6156–6164 (2012)
8. Dauwels, J., Vialatte, F., Weber, T., Cichocki, A.: On similarity measures for spike trains. In: Köppen, M., Kasabov, N., Coghill, G. (eds.) ICONIP 2008. LNCS, vol. 5506, pp. 177–185. Springer, Heidelberg (2009). doi:10.1007/978-3-642-02490-0_22
9. Paiva, A.R., Park, I., Prncipe, J.C.: A comparison of binless spike train measures. Neural Comput. Appl. **19**(3), 405–419 (2010)
10. Duch, J., Arenas, A.: Community detection in complex networks using extremal optimization. Phys. Rev. E **72**(2), 027104 (2005)

Identifying Gender Differences in Multimodal Emotion Recognition Using Bimodal Deep AutoEncoder

Xue Yan[1], Wei-Long Zheng[1], Wei Liu[1], and Bao-Liang Lu[1,2,3(✉)]

[1] Department of Computer Science and Engineering,
Center for Brain-like Computing and Machine Intelligence,
Shanghai Jiao Tong University, 800 Dong Chuan Road, Shanghai 200240, China
{yanxue_10085,weilong,liuwei-albert,bllu}@sjtu.edu.cn
[2] Key Laboratory of Shanghai Education Commission for Intelligent Interaction
and Cognitive Engineering, Shanghai Jiao Tong University,
800 Dong Chuan Road, Shanghai 200240, China
[3] Brain Science and Technology Research Center, Shanghai Jiao Tong University,
800 Dong Chuan Road, Shanghai 200240, China

Abstract. This paper mainly focuses on investigating the differences between males and females in emotion recognition using electroencephalography (EEG) and eye movement data. Four basic emotions are considered, namely happy, sad, fearful and neutral. The Bimodal Deep AutoEncoder (BDAE) and the fuzzy-integral-based method are applied to fuse EEG and eye movement data. Our experimental results indicate that gender differences do exist in neural patterns for emotion recognition; eye movement data is not as good as EEG data for examining gender differences in emotion recognition; the activation of the brains for females is generally lower than that for males in most bands and brain areas especially for fearful emotions. According to the confusion matrix, we observe that the fearful emotion is more diverse among women compared with men, and men behave more diversely on the sad emotion compared with women. Additionally, individual differences in fear are more pronounced than other three emotions for females.

Keywords: EEG · Eye movement data · Emotion · BDAE · Gender differences

1 Introduction

Gender differences in many aspects such as temperaments, cognition and social behavior have been widely studied. Whether sexes are different or not has been systematically considered in a wide range of psychological aspects. However, gender differences are continually attracting people's interests because many observations should be verified.

One of the observations, that females are more emotional than males, is a prevailing acknowledgment among gender differences. In this paper, we intend to

© Springer International Publishing AG 2017
D. Liu et al. (Eds.): ICONIP 2017, Part IV, LNCS 10637, pp. 533–542, 2017.
https://doi.org/10.1007/978-3-319-70093-9_56

study gender differences in emotion recognition using EEG and eye movement data. Our previous work has found that the fusion on both the feature level and the decision level of EEG data and eye movement data can improve the accuracy of the emotion recognition model [10]. Moreover, the complementarity of EEG and eye movement data to the emotion recognition model indicates that using the combination of both data is a more appropriate method for emotion recognition [4].

Weiss *et al.* pointed out that in the neuropsychological processes, men showed significantly more activation in parietal areas, while women showed significantly stronger right frontal activation [9]. Therefore, different responses in emotion processes are supposed to occur in males and females' brain areas. In our previous work, we have indicated that neural patterns are distinct in different emotions [11], there exist some gender differences in three emotions (happy, sad and neutral) in EEG patterns [12]. However, whether there is difference in fear, a very important emotional state explored in animals in neuroscience field, between women and men in EEG and eye movement data is an open question. The goal of this paper is to investigate the gender differences of EEG data, eye movement data and their combination in emotion recognition using BDAE, and whether there are different brain responses between males and females to EEG patterns of four emotions. Moreover, individual differences in males and females for recognizing four emotions are also discussed.

2   Methodology

2.1   Feature Extraction and Feature Smoothing

In this paper, several eye movement parameters are applied according to our previous study [4]. Differential Entropy (DE) features of the pupil diameter using short-term Fourier transform (STFT) in four frequency bands (0–0.2 Hz, 0.2–0.4 Hz, 0.4–0.6 Hz and 0.6–1 Hz), as well as the mean and standard deviation features in X and Y axes are computed. Moreover, the mean and standard deviation of the pupil dispersion in X and Y axes, eye saccade in duration and amplitude and fixation are used as well. Besides, nine event statistics are also included. The total dimension of eye movement features is 31.

For preprocessing, EEG data is filtered between 1 Hz and 75 Hz and sampled down to 200 Hz to remove artifacts and to reduce computation. STFT with a 4-s-long window and no overlapping Hanning window is employed to calculate the DE features of EEG data. EEG data in each channel are filtered into five frequency bands (δ: 1–3 Hz, θ: 4–7 Hz, α: 8–13 Hz, β: 14–30 Hz and γ: 31–50 Hz) [1]. There are 62 channels used in the electrode cap, which means the dimension of EEG features is 310. To filter rapid fluctuations, the linear dynamic system is employed for feature smoothing [8].

2.2   Feature Fusion and Model Combination

In this paper, we apply two modality fuision strategies to combine EEG and eye movement data: the fuzzy-integral-based method [6] and the Bimodal Deep

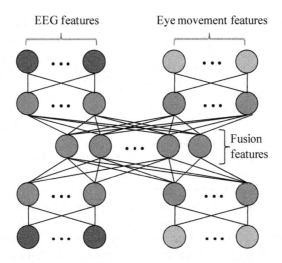

Fig. 1. The architecture of BDAE for fusing EEG and eye movement features. The high-level shared features of BDAE are used finally as the input of SVM.

AutoEncoder (BDAE) method which is shown in Fig. 1. There are three layers in BDAE model and the hyperparameters are determined by cross-validation.

BDAE is formed by stacking Restricted Boltzmann Machines (RBMs) and contrastive divergence (CD) algorithm [3] is used to train Bernoulli RBM in this paper. RBM consists of hidden nodes and visible nodes. The energy function of visible nodes and hidden nodes is defined as below:

$$E(v, h; \theta) = -\sum_{i=1}^{M}\sum_{j=1}^{N} W_{ij}v_ih_j - \sum_{i=1}^{M} b_iv_i - \sum_{j=1}^{N} a_jh_j \tag{1}$$

where visible nodes $v \in \{0, 1\}^M$, hidden nodes $h \in \{0, 1\}^N$ and $\theta = \{a, b, W\}$ are parameters, a_j and b_i are bias of visible nodes and hidden nodes, respectively, and W_{ij} is the weight between visible and hidden layers. Then, the joint distribution of visible nodes and hidden nodes can be calculated from the energy function:

$$p(v, h; \theta) = \frac{exp\left(E\left(v, h; \theta\right)\right)}{\sum_v \sum_h exp\left(E\left(v, h; \theta\right)\right)} \tag{2}$$

Next, the derivative of log-likelihood with respect to W can be computed:

$$\frac{1}{N}\sum_{i=1}^{N} \frac{\partial logp\left(v_n; \theta\right)}{\partial W_{ij}} = E_{P_{data}}\left[v_ih_j\right] - E_{P_{model}}\left[v_ih_j\right] \tag{3}$$

2.3 Classification

The classifier of all models after parameter regulation is linear SVM with soft margin. Two training strategies are developed: the *different-gender strategy* where the training data and the testing data come from different genders, and the *same-gender strategy* where both data come from the same gender. In both methods, the testing data holds only one subject's data and the training data is all the rest data with the same or the opposite gender.

3 Experiment Design

A total of 16 subjects (8 females) aged between 18 and 28 participated the experiments for three times. All the subjects were healthy, right-handed, had sufficient sleep with normal or corrected-to-normal vision and were told the harmlessness and the goal of the experiment.

At the start of each trial, the textual description of the following movie clip was presented for 5 s, the clip evoking a single emotion was presented for about 4 min and then the self-assessment stage lasted for 45 s for subjects to assess whether the corresponding emotions are evoked. In the end, 15 s were left for subjects to relax.

Movie clips used as stimuli were evaluated and selected for experiments. 20 persons were asked to score the clips according to the degree that emotions were evoked and finally, 72 clips (24 for one experiment) were chosen on the basis of scores. There are twenty-four movie clips (6 clips per emotion) for one experiment in total. In terms of avoiding the influence of the movie order and similar movies, the movie clips in each experiment were shuffled randomly and no two same clips were used.

During the experiment, the ESI NeuroScan System with a 62-channel electrode cap and the SMI ETG eye tracking glasses were used to collect EEG data with 1000 Hz sampling rate and eye movement data, respectively. Each subject conducted experiments for three times to avoid individual deviations. The experiments were carried out when subjects were in good mental states and they were familiar with the procedure. Feedback forms collected after the experiment indicated that certain emotions were successfully evoked.

4 Results and Discussions

4.1 Gender Differences in Different Data

Using single EEG data. Figure 2(a) illustrates two models using EEG data. The upper one is female models: a female's data is used as the testing and two strategies are employed to train the model. The lower one is male models. One way analysis of variance (ANOVA) is used in both models. Under female models $[F(1, 46) = 8.72, p = 0.0049]$ and male models $[F(1, 46) = 9.7, p = 0.004]$, gender differences have highly significant influence on the accuracy changes. Furthermore, the average accuracy is higher when using the model trained and tested by the same gender for both models. These experimental results indicate that there does exist some gender differences in neural patterns for emotion recognition.

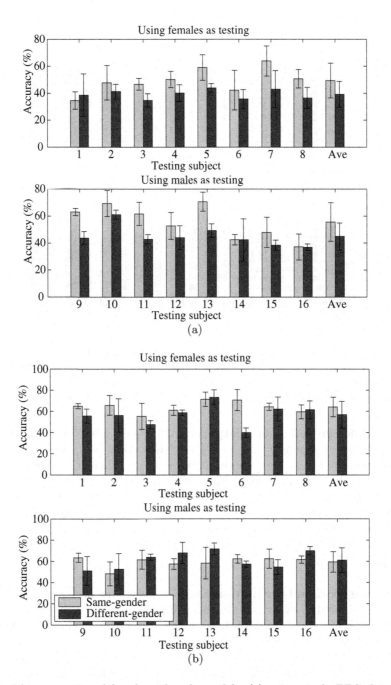

Fig. 2. The accuracies of female and male models: (a) using single EEG data, and (b) using single eye movement data. The upper figure represents female models using females as testing data. Two training strategies are applied for each model.

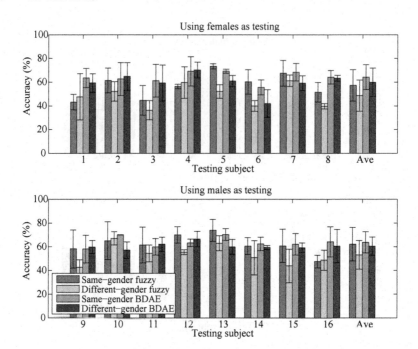

Fig. 3. The accuracies of female and male models using both EEG and eye movement data. The same-gender strategy and the different-gender strategy are used. Each strategy is separately used fuzzy integral and the BDAE algorithm to compare effects. The upper figure represents female models using females as testing data.

Using single eye movement data. As for female models $[F(1, 46) = 5.02, p = 0.0299]$, the average performance in Fig. 2(b) shows that using the same gender's data to train can obtain higher accuracies than using different gender's data. On the other hand, gender has no significant effect on the accuracy of the subjects for male models $[F(1, 46) = 0.29, p = 0.5949]$. These indicate that when using single eye movement data, no significant gender differences were found when testing males' data, but differences were more pronounced in women, which implies that eye movement patterns are not as obvious as EEG patterns on gender differences in emotion recognition.

Combining EEG and eye movements. Two fusion methods and two training strategies are used in two models in this part. The fuzzy-integral-based method is to fuse the output of two classifiers on the decision level while the BDAE is to fuse EEG and eye movement features on the feature level.

The experimental results shown in Fig. 3 demonstrate that when using BDAE, average accuracies for females and males training are 64.26% and 59.88% in the female models, respectively, and as for the male models, average accuracies for males and females are 63.77% and 60.56%, respectively. These performances are all better than the fuzzy method. When using the fuzzy-integral-based method, genders have a significant influence on classification accuracies for both male models $[F(1, 46) = 5.35, p = 0.0253]$ and female models $[F(1, 46) = 4.97, p = 0.0308]$. However, no significant influences on the accuracy results have been shown when the BDAE method is applied to male models $[F(1, 46) = 1.73, p = 0.1947]$ and female models $[F(1, 46) = 1.75, p = 0.1928]$. These results indicate that when using two kinds of data, whether genders have influences on classification accuracies is affected by the fusion strategy.

On the content of our present discussion, EEG data is the most suitable data for studying the gender differences in emotions. To examine specific differences,

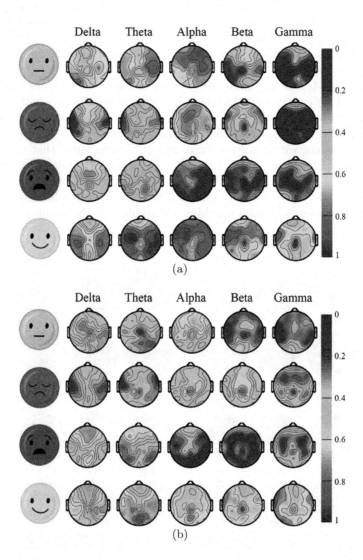

Fig. 4. The brain topographic mapping of females and males: (a) female's pattern, and (b) male's pattern.

we studied gender differences with respect to two cases using single EEG data: one is to focus on neural patterns, and another is to focus on different emotions.

4.2   Gender Differences in Neural Patterns

To identify neural patterns of two genders, the brain topographic mapping of four emotions across five frequencies is given in Fig. 4. DE features are normalized between 0 and 1 to represent the neural patterns of subjects. Common patterns exist for two genders. The temporal lobe activates the most under happy emotions on gamma band, and there are lower responses in the occipital lobe under fearful emotions than neutral and sad emotions on the alpha band. Neutral patterns hold more activation on the parietal and frontal lobe in the alpha band than sad patterns.

Figure 5 illustrates the difference between females and males in neural patterns. Except for the temporal lobe in theta band under neutral emotion and the frontal lobe in alpha band under happy emotion, most brain areas under four emotions across five frequencies band activate more in males than in females, which is consistent with results of Imaizumi et al. who found that under emotional tasks, males show greatly stronger activation in certain areas [2]. This phenomenon is more obvious in fear. The whole brain of men response more under fear in all frequency band than women. Schienle et al. also observed that men emerge greater activation watching fearful pictures than women [7].

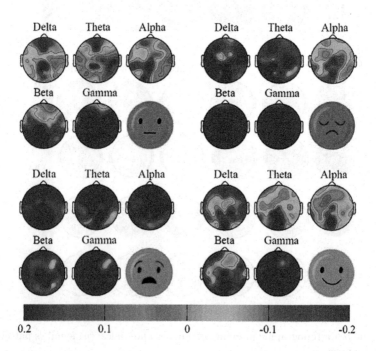

Fig. 5. The difference between the average neural pattern of females and males.

4.3 Gender Differences in Different Emotions

In this section, the same-gender training strategy is used to explore gender differences in emotion recognition for males and females. The confusion matrix of EEG data is calculated as listed in Table 1, where the first row and column of Table 1 mean true labels and predicted labels, respectively.

Table 1. The confusion matrix of EEG data for two models

Female models	Neutral	Sad	Fearful	Happy	Male models	Neutral	Sad	Fearful	Happy
Neutral	**0.53**	0.19	0.09	0.19	Neutral	**0.56**	0.14	0.16	0.14
Sad	0.20	**0.52**	0.17	0.11	Sad	0.18	**0.43**	0.20	0.19
Fearful	0.20	0.26	**0.38**	0.16	Fearful	0.15	0.16	**0.49**	0.20
Happy	0.26	0.12	0.12	**0.50**	Happy	0.19	0.14	0.17	**0.50**

On account of the use of the same-gender training strategy, bold numbers in Table 1 mean abilities to use existing data to predict new data, which indicates that as for neutral and happy emotions, the abilities are similar for males and females. The main difference lies in the other two emotions where women share less fearful emotion patterns among females compared with men and men share less sad emotion patterns among males compared with women. Other researches also point out that women perform more fearful emotions during their lifetimes than men [5].

Besides, a significantly low accuracy of the fear emotion in female models, with the number of 38%, means females share more individual differences in the fearful emotion compared with other emotions among women.

5 Conclusions

In this paper, we have investigated the gender differences in multimodal emotion recognition using Bimodal Deep AutoEncoder. From our experiment results, we have obtained the following observations: (1) Gender differences do exist in neural patterns and the same sex has more similar emotion patterns than the opposite sex. (2) Eye movement data is not as obvious as EEG data in discussing gender differences in emotion recognition. (3) Females' responses are generally lower than males' in most bands and brain areas for four emotion patterns, especially for the fearful emotion. (4) Females have more individual differences in the fear emotion among four emotions, and males differ more in sad emotions compared with females while females differ more in fearful emotions compared with males.

Acknowledgments. This work was supported in part by grants from the National Key Research and Development Program of China (Grant No. 2017YFB1002501), the National Natural Science Foundation of China (Grant No. 61673266), the Major

Basic Research Program of Shanghai Science and Technology Committee (Grant No. 15JC1400103), ZBYY-MOE Joint Funding (Grant No. 6141A02022604), and the Technology Research and Development Program of China Railway Corporation (Grant No. 2016Z003-B).

References

1. Duan, R.N., Zhu, J.Y., Lu, B.L.: Differential entropy feature for EEG-based emotion classification. In: 6th International IEEE/EMBS Conference on Neural Engineering, vol. 8588, pp. 81–84 (2013)
2. Imaizumi, S., Homma, M., Ozawa, Y., Maruishi, M., Muranaka, H.: Gender differences in emotional prosody processing-an fMRI study. Psychologia **47**(2), 113–124 (2004)
3. Liu, W., Zheng, W.-L., Lu, B.-L.: Emotion recognition using multimodal deep learning. In: Hirose, A., Ozawa, S., Doya, K., Ikeda, K., Lee, M., Liu, D. (eds.) ICONIP 2016. LNCS, vol. 9948, pp. 521–529. Springer, Cham (2016). doi:10.1007/978-3-319-46672-9_58
4. Lu, Y., Zheng, W.L., Li, B., Lu, B.L.: Combining eye movements and EEG to enhance emotion recognition. In: International Conference on Artificial Intelligence, pp. 1170–1176 (2015)
5. Mclean, C.P., Anderson, E.R.: Brave men and timid women? A review of the gender differences in fear and anxiety. Clin. Psychol. Rev. **29**(6), 496–505 (2009)
6. Sugeno, M.: Theory of fuzzy integrals and its applications. Ph.D. thesis. Tokyo Institute of Technology (1974)
7. Schienle, A., Schäfer, A., Stark, R., Walter, B., Vaitl, D.: Gender differences in the processing of disgust- and fear-inducing pictures: an fMRI study. NeuroReport **16**(3), 277–80 (2005)
8. Wang, X.W., Nie, D., Lu, B.L.: Emotional state classification from EEG data using machine learning approach. Neurocomputing **129**(4), 94–106 (2014)
9. Weiss, E., Siedentopf, C.M., Hofer, A., Deisenhammer, E.A., Hoptman, M.J., Kremser, C., Golaszewski, S., Felber, S., Fleischhacker, W.W., Delazer, M.: Sex differences in brain activation pattern during a visuospatial cognitive task: a functional magnetic resonance imaging study in healthy volunteers. Neurosci. Lett. **344**(3), 169–72 (2003)
10. Zheng, W.L., Dong, B.N., Lu, B.L.: Multimodal emotion recognition using EEG and eye tracking data. In: 36th Annual International Conference of the IEEE Engineering in Medicine and Biology Society, pp. 5040–5043. IEEE (2014)
11. Zheng, W.L., Zhu, J.Y., Lu, B.L.: Identifying stable patterns over time for emotion recognition from EEG. IEEE Trans. Affect. Comput. **99**, 1 (2017)
12. Zhu, J.-Y., Zheng, W.-L., Lu, B.-L.: Cross-subject and cross-gender emotion classification from EEG. In: Jaffray, D.A. (ed.) World Congress on Medical Physics and Biomedical Engineering, June 7-12, 2015, Toronto, Canada. IP, vol. 51, pp. 1188–1191. Springer, Cham (2015). doi:10.1007/978-3-319-19387-8_288

EEG-Based Sleep Quality Evaluation with Deep Transfer Learning

Xing-Zan Zhang[1], Wei-Long Zheng[1], and Bao-Liang Lu[1,2,3]($\boxtimes$)

[1] Department of Computer Science and Engineering,
Center for Brain-Like Computing and Machine Intelligence,
Shanghai Jiao Tong University, 800 Dong Chuan Road, Shanghai 200240, China
zxzsprinkle@hotmail.com, {weilong,bllu}@sjtu.edu.cn
[2] Key Laboratory of Shanghai Education Commission for Intelligent Interaction
and Cognitive Engineering, Shanghai Jiao Tong University,
800 Dong Chuan Road, Shanghai 200240, China
[3] Brain Science and Technology Research Center, Shanghai Jiao Tong University,
800 Dong Chuan Road, Shanghai 200240, China

Abstract. In this paper, we propose a subject-independent approach
with deep transfer learning to evaluate the last-night sleep quality using
EEG data. To reduce the intrinsic cross-subject differences of EEG data
and background noise variations during signal acquisition, we adopt two
classes of transfer learning methods to build subject-independent classifiers. One is to find a subspace by matrix decomposition and regularization theory, and the other is to learn the common shared structure
with the deep autoencoder. The experimental results demonstrate that
deep transfer learning model achieves the mean classification accuracy of
82.16% in comparison with the baseline SVM (65.74%) and outperforms
other transfer learning methods. Our experimental results also indicate
that the neural patterns of different sleep quality are discriminative and
stable: the delta responses increase, the alpha responses decrease when
sleep is partially deprived, and the neural patterns of 4-h sleep and 6-h
sleep are more similar compared with 8-h sleep.

Keywords: Sleep quality · EEG · Neural pattern · Deep transfer
learning

1 Introduction

Sleep quality evaluation has remarkable value in both scientific research and
practical applications. Sufficient sleep is of great importance to human daily life,
and the study of sleep mechanisms is an important part of brain science. An
objective and effective measurement of sleep quality is quite valuable in transportation, medicine, health care, and neuroscience. For example, the tiredness of
the drivers due to insufficient sleep imposes a severe threat to the public safety
in the transportation industry.

© Springer International Publishing AG 2017
D. Liu et al. (Eds.): ICONIP 2017, Part IV, LNCS 10637, pp. 543–552, 2017.
https://doi.org/10.1007/978-3-319-70093-9_57

Methods like polysomnography, actigraphy and smart bands have been shown as efficient approaches for evaluating sleep quality. However, these approaches require the subjects to wear equipments such as EEG cap, eye sensors, nose sensors, elastic belt sensors during the whole sleep procedure. The whole-process signal acquisition requirement limits their feasibility in real-world applications.

In this paper, we propose an objective EEG-based approach to classify last-night sleep quality into three categories: poor, normal and good, which excludes whole-process physiological signal acquisition and expert knowledge. Rather than building subject-specific models, which require the collection of labeled data for each subject and thus is time-consuming and unfeasible in practice, we build a subject-independent model and then make inference on the new subjects. The performances of conventional algorithms, when applied to the cross-subject classification tasks, are unsatisfactory because of the intrinsic cross-subject differences and background noise variations. The cornerstone assumption of traditional machine learning methods is that the training data and test data are identically distributed, which is seldom satisfied if not never in sleep quality evaluation due to the variability of cross-subject and cross-session. Our previous work [10] only considers the total sleep time in the experiment setup and the subject-dependent evaluations. In this paper, we refine the experiments by taking deep sleep into consideration and apply transfer learning methods to deal with the cross-subject variations.

Transfer learning approaches have been proved to have the capability to reduce the differences of EEG data across subjects and sessions recently [11,12]. In this paper, we explore two categories of subject-to-subject transfer learning. One is to find a subspace in Reproducing Kernel Hilbert Space (RKHS) in which the EEG data distributions of different subjects are drawn closer when mapped into this subspace. TCA [7] and ARRLS [5] belongs to this class. The other is to learn the common shared, higher-level structure underlying different categories of sleep quality among different subjects while eliminate the influences of background noise with deep learning, which refers to TLDA [13].

2   Transfer Learning Methods

2.1   TCA-Based Subject Transfer

Transfer Component Analysis (TCA) [7] aims to find a set of transfer components across different subjects in a RKHS. In the new space spanned by the transfer components, the data distributions of different subjects are drawn closer, while the data variance properties within each subject are preserved. TCA assumes that there is a kernel function that can simultaneously adapt marginal distribution and conditional distribution. Under this assumption, the transfer components are found by minimizing the Maximum Mean Discrepancy (MMD) between training subject and test subject.

2.2 ARRLS-Based Subject Transfer

Adaptation Regularization based Transfer Learning using Regularized Least Squares (ARRLS) [5] simultaneously optimizes the structural risk, the joint distribution and the manifold consistency of two subjects based on the structural risk minimization principle and the regularization theory.

Suppose that $f = \mathbf{w}^T \phi(x)$ is the prediction function where $\mathbf{w}$ is the classifier parameters and $\phi : \mathcal{X} \mapsto \mathcal{H}$ is the kernel function that projects the original feature space into a RKHS space $\mathcal{H}_K$. The prediction function f is learnt by

$$f = \min_{f \in \mathcal{H}_K} \sum_{i=1}^{n} \ell(f(x_i), y_i) + \sigma\|f\|_K^2 + \lambda D_{f,K}(J_s, J_t) + \delta M_{f,K}(P_s, P_t). \quad (1)$$

where K is the kernel function induced by $\phi(\cdot)$, $\sum_{i=1}^{n} \ell(f(x_i), y_i) + \sigma\|f\|_K^2$ denotes the structural risk minimization of training subject, and $D_{f,K}(J_s, J_t)$ represents the minimization term of marginal distribution and conditional distribution. ARRLS measures the marginal distribution difference with MMD as same as TCA. Since there are no labels in the test subject, the conditional distribution adaption is achieved by the trick of pseudo target labels. The manifold regularization term $M_{f,K}(P_s, P_t)$ is computed by normalized graph Laplacian matrix. And σ, λ and δ are corresponding regularization parameters.

2.3 Transfer Learning with Deep Autoencoders

Deep autoencoders are effective and efficient in learning robust and higher-level representing features. Besides the optimization of individual reconstruction error as the ordinary autoencoders, Transfer Learning with Deep Autoencoders (TLDA) [13] learns a common feature representation shared by the training subject and test subject by explicitly minimizing the symmetrized Kullback-Leibler (KL) divergence, which is a non-symmetric measure of the divergence between two probability distributions, of the two subjects.

The framework of TLDA is shown in Fig. 1, which contains two encoding layers and two decoding layers. The first encoding layer is used to learn robust and high-level features of original EEG features and the second encoding layer encodes the hidden features into labels, which is in fact an classifier based on the softmax regression model. Given nonlinear activation function f (we adopt sigmoid function in our study), for $r \in \{s, t\}$, $\xi_i^r = f(W_1 x_i^r + b_1)$ is the hidden representations, $z_i^r = f(W_2 \xi_i^r + b_2)$ is the encoded labels, and $\hat{\xi}_i^r = f(W_2' z_i^r + b_2')$, $\hat{x}_i^r = f(W_1' \hat{\xi}_i^r + b_1')$ are the corresponding reconstructions of ξ_i^r and x_i^r.

The objective function to be minimized in TLDA can be formalized as

$$J = J_r(x, \hat{x}) + \alpha\Gamma(\xi_s, \xi_t) + \beta L(\Theta, \xi^s) + \gamma\Omega(W, b, W', b'), \quad (2)$$

where α, β and γ are the trade-off parameters to balance between the four different optimization terms.

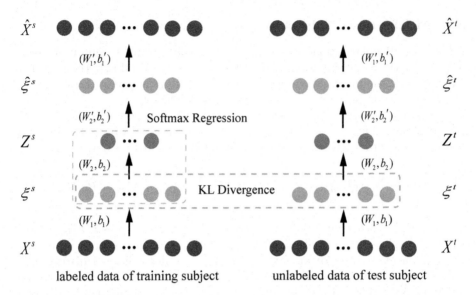

Fig. 1. The framework of TLDA. The encoding and decoding weights are shared by both the training and test subjects. We draw individual networks for each subject to better illustrate the idea that the distributions of two subjects are enforced to be similar in the hidden feature space by minimizing KL divergence.

The first term in Eq. (2) is the reconstruction error for both training and test subjects, which can be calculated as $J_r(x, \hat{x}) = \sum_{r \in \{s,t\}} \sum_{i=1}^{n_r} ||x_i^r - \hat{x}_i^r||^2$.

The second term measures the symmetrized KL divergence among data distributions of two subjects in the embedded space.

$$\Gamma(\xi_s, \xi_t) = D_{KL}(E_s||E_t) + D_{KL}(E_t||E_s) = E_s ln(\frac{E_s}{E_t}) + E_t ln(\frac{E_t}{E_s})$$

$$E_s = \frac{E_s'}{\Sigma E_s'}, E_s' = \frac{1}{n_s} \sum_{i=1}^{n_s} \xi_i^s, E_t = \frac{E_t'}{\Sigma E_t'}, E_t' = \frac{1}{n_t} \sum_{i=1}^{n_t} \xi_i^t. \tag{3}$$

The third term denotes the loss function of the final softmax regression classifier. $L(\Theta, \xi^s) = -\frac{1}{n_s} \sum_{i=1}^{n_s} \sum_{j=1}^{c} sgn\{y_i^s = j\} log \frac{e\Theta_j^T \xi_i^s}{\Sigma_{l=1}^{c} e\Theta_j^T \xi_i^s}$, where Θ_j^T is the j-th row of W_2, sgn is the indicator function.

The last term is the regularization on the complexity of the model parameters which can be formulated as $\Omega(W, B, W', b') = ||W_1||^2 + ||b_1||^2 + ||W_2||^2 + ||b_2||^2 + ||W_1'||^2 + ||b_1'||^2 + ||W_2'||^2 + ||b_2'||^2$.

After taking the partial derivatives of the objective Eq. 2 with respect to W_1, b_1, W_2, b_2, W_1', b_1', W_2', b_2', we apply the the gradient descent methods to calculate the final weight matrices.

3   Experiments

3.1   Deep Sleep Time

According to the findings of National Sleep Foundation (NSF), 8-h sleep presents a high sleep quality for adults people, and a sleep time of less than 4 h means an awful sleep quality [4]. Among whole night sleep procedure, the deep sleep part is the period that human brain gets a full rest, thus is the main factor that counts for sleep quality [2].

Based on the results of sleep medicine, we take 4-h sleep, 6-h sleep and 8-h sleep with increasing deep sleep time as poor, normal and good in terms of the sleep quality in our study. And the sleep time and wake up time for three experiments are 3:00–7:00, 1:00–7:00 and 23:00–7:00, respectively.

3.2   Subjects and EEG Data Acquisition

Ten graduate and undergraduate students (six males and four females, age range: 21–26, mean: 23.57, std: 1.62) with self-reported healthy conditions and regular daily routines participate in the experiments. Each subject performs 4-h, 6-h

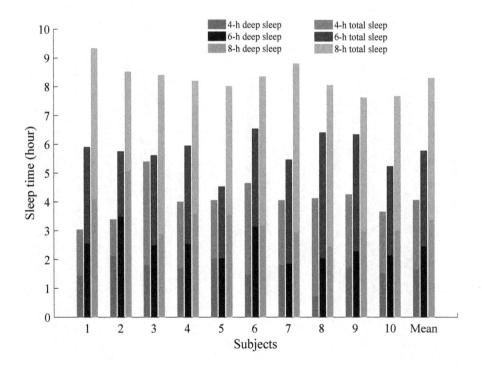

Fig. 2. The total sleep time and deep sleep time of all 10 subjects recorded by the smart bands.

and 8-h experiments wearing smart bands[1] with an interval of two days and declines coffee, drugs, alcohols and other medicines that may disturb human sleep during the experiments. Figure 2 shows the total sleep time and deep sleep time of all subjects recorded by the smart bands. The EEG signals are recorded for 30 min in each experiment with a 62-channel electrode cap according to the international 10–20 system using the ESI NeuroScan system at a sampling rate of 1000 Hz. During the data acquisition procedures, the subjects are required to stare at a green dot on the screen and count numbers to keep a peaceful state. We used the mean signals of all 62 electrodes as the reference.

3.3   Feature Extraction

The raw EEG data are first down-sampled to 200 Hz and fed to a bandpass filter (0–50 Hz). Then the Infomax [1] denoising algorithms is applied to eliminate the noise and artifacts. We adopt DE feature [8] which is calculated in five frequency bands (delta: 1–3 Hz, theta: 4–7 Hz, alpha: 8–13 Hz, beta: 14–30 Hz, gamma: 31–50 Hz) using a Short-Time Fourier Transform with 200-point windows. Therefore, a 310 dimensional DE vector is extracted in each second. The DE features are further smoothed by the LDS algorithm [9] and normalized

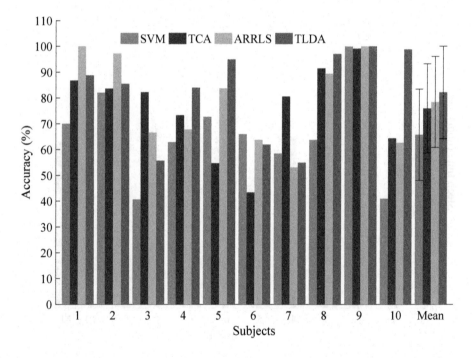

Fig. 3. Accuracy comparation of SVM, TCA, ARRLS and TLDA for each subject.

---

[1] We use Mi band 2, the website is https://www.mi.com/shouhuan2.

between 0 and 1 before traininging classifiers. Finally, there are 1800 data samples for each experiment, 5400 data samples for each subject, and 54000 data samples for all subjects.

3.4 Classifier Traininging Details

A leave-one-subject-out cross validation scheme is applied for the evaluation. Each time, the 5400 samples from one subject without labels are treated as test data and the 48600 samples from the rest 9 subjects with labels are treated as training data. For TCA and ARRLS, it is impracticable to include all the available data due to limits of memory and time cost. Therefore, we randomly select 1/5 samples from 9 subjects (9720) as the traininging data each time.

We use SVM with linear kernel and $C = 0.01$ as the baseline. In TCA, $\mu = 1$ and the optimal dimension is 30 by searching from 5 to 100 with step 5. We adopt line search rather than grid search to avoid tremendous computation in ARRLS

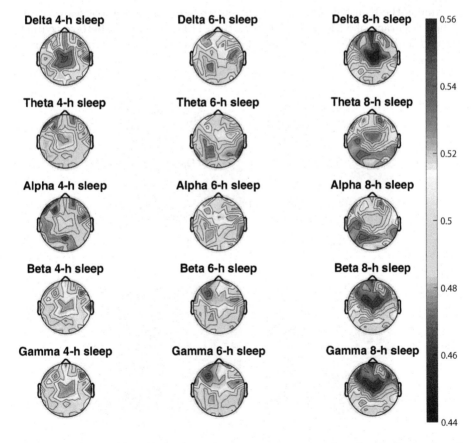

Fig. 4. Neural patterns of three kinds of sleep quality: 4-h (poor), 6-h (normal) and 8-h (good).

and TLDA. The best configuration for ARRLS is $p = 10$, $\sigma = 0.1$, $\lambda = 10$, $\gamma = 1$, and linear kernel. In TLDA, since the objective function is not convex, we first run Sparse Auto-Encoder (SAE) on training and test data, and initialize the weight matrices with the output of SAE to achieve a better local optimal solution. The optimal parameters are $k = 10$, $\alpha = 5$, $\beta = 1$ and $\gamma = 10^{-7}$.

4   Results and Discussion

The classification accuracies of baseline SVM, TCA, ARRLS and TLDA for all 10 subjects are shown in Fig. 3. Accuracy means and standard deviations are 65.74%, 75.96%, 78.44%, 82.16% and 17.69%, 17.23%, 17.61%, 17.91%, respectively. Firstly, all three transfer methods have promotions compared with the baseline SVM. The promotions verify that transfer learning methods are effective at promoting generic classifier with the capability of capturing the underlying common structure shared by different subjects while eliminating sleep quality unrelated noise.

Among the transfer methods, TLDA outperforms other approaches with the highest accuracy of 82.16%. The reasons that TLDA achieves a better accuracy are two folds. The first one is that TLDA learns the mapping functions and

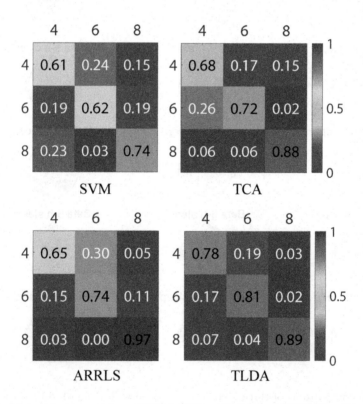

Fig. 5. The confusion matrices of SVM, TCA, ARRLS and TLDA for all subjects.

hidden features from the traininging and test data sets with impressive fitting capability, while TCA and ARRLS find the projection function with pre-defined kernels, which is seldom optimal for the data sets we are dealing with. The second reason lies in that TLDA learns the new features and classifiers altogether, while TCA does them in two steps separately.

The DE pattern comparison for three kinds of sleep quality on five bands of all subjects are shown in Fig. 4. The experimental results present that neural structures of different categories of sleep quality are discriminative and stable: the delta responses increase and the alpha responses decrease when sleep is partially deprived. Our findings are in accord with the studies from Borbely *et al.* [3] and Lorenzo *et al.* [6].

Figure 5 presents the confusion matrices. By examining the accuracies of different categories of sleep quality, we find that 8-h sleep can be more effectively recognized with the comparatively high accuracy of 89%, while 4-h and 6-h sleep are more inclined to intertwine with each other with misclassification rates of 19% and 17%, respectively, in TLDA. This observation is consistent in SVM, TCA and ARRLS, which indicates that the neural patterns of 4-h and 6-h sleep are more similar. These findings can again be verified in Fig. 4 which shows that the DE patterns of 4-h sleep and 6-h sleep are much closer than the patterns of 8-h sleep.

5   Conclusion

In this paper, we have adopted deep transfer learning approaches to build a subject-independent model to classify three kinds of sleep quality: poor, normal and good. Among the four evaluated methods (SVM, TCA, ARRLS and TLDA), the TLDA algorithm achieves the best performance with mean accuracy of 82.16%. Our experimental results demonstrate that the neural structures of different sleep quality among people are discriminative and stable: the energy of delta bands has an increasing trend while the alpha responses are depressed when sleep is partially deprived. The results also indicate that the neural patterns of 4-h sleep and 6-h sleep are more similar compared with 8-h sleep.

Acknowledgments. This work was supported in part by grants from the National Key Research and Development Program of China (Grant No. 2017YFB1002501), the National Natural Science Foundation of China (Grant No. 61673266), the Major Basic Research Program of Shanghai Science and Technology Committee (Grant No. 15JC1400103), ZBYY-MOE Joint Funding (Grant No. 6141A02022604), and the Technology Research and Development Program of China Railway Corporation (Grant No. 2016Z003-B).

References

1. Bell, A.J., Sejnowski, T.J.: An information-maximization approach to blind separation and blind deconvolution. Neural Comput. **7**(6), 1129–1159 (1995)

2. Berry, R.B.: Fundamentals of Sleep Medicine. Elsevier Health Sciences, Amsterdam (2011)
3. Borbely, A.A., Baumann, F., Brandeis, D., Strauch, I., Lehmann, D.: Sleep deprivation: effect on sleep stages and EEG power density in man. Electroencephalogr. Clin. Neurophysiol. **51**(5), 483 (1981)
4. Hirshkowitz, M., Whiton, K., Albert, S.M., Alessi, C., Bruni, O., DonCarlos, L., Hazen, N., Herman, J., Katz, E.S., Kheirandish-Gozal, L., et al.: National Sleep Foundation's sleep time duration recommendations: methodology and results summary. Sleep Health **1**(1), 40–43 (2015)
5. Long, M., Wang, J., Ding, G., Pan, S.J., Philip, S.Y.: Adaptation regularization: a general framework for transfer learning. IEEE Trans. Knowl. Data Eng. **26**(5), 1076–1089 (2014)
6. Lorenzo, I., Ramos, J., Arce, C., Guevara, M.A., Corsi-Cabrera, M.: Effect of total sleep deprivation on reaction time and waking EEG activity in man. Sleep **18**(5), 346–354 (1995)
7. Pan, S.J., Tsang, I.W., Kwok, J.T., Yang, Q.: Domain adaptation via transfer component analysis. IEEE Trans. Neural Netw. **22**(2), 199–210 (2011)
8. Shi, L.C., Jiao, Y.Y., Lu, B.L.: Differential entropy feature for EEG-based vigilance estimation. In: 35th Annual International Conference of the IEEE Engineering in Medicine and Biology Society, pp. 6627–6630. IEEE (2013)
9. Shi, L.C., Lu, B.L.: Off-line and on-line vigilance estimation based on linear dynamical system and manifold learning. In: 32nd Annual International Conference of the IEEE Engineering in Medicine and Biology Society, pp. 6587–6590. IEEE (2010)
10. Wang, L.L., Zheng, W.L., Ma, H.W., Lu, B.L.: Measuring sleep quality from EEG with machine learning approaches. In: International Joint Conference on Neural Networks, pp. 905–912. IEEE (2016)
11. Zhang, Y.-Q., Zheng, W.-L., Lu, B.-L.: Transfer components between subjects for EEG-based driving fatigue detection. In: Arik, S., Huang, T., Lai, W.K., Liu, Q. (eds.) ICONIP 2015. LNCS, vol. 9492, pp. 61–68. Springer, Cham (2015). doi:10.1007/978-3-319-26561-2_8
12. Zheng, W.L., Lu, B.L.: Personalizing EEG-based affective models with transfer learning. In: Proceedings of the 25th International Joint Conference on Artificial Intelligence, pp. 2732–3738 (2016)
13. Zhuang, F., Cheng, X., Luo, P., Pan, S.J., He, Q.: Supervised representation learning: transfer learning with deep autoencoders. In: Proceedings of the 24th International Joint Conference on Artificial Intelligence, pp. 4119–4125 (2015)

A Stochastic Neural Firing Generated at a Hopf Bifurcation and Its Biological Relevance

Huijie Shang[1], Rongbin Xu[1], Dong Wang[1,2(✉)], Jin Zhou[1,2], and Shiyuan Han[1,2]

[1] School of Information Science and Engineering,
University of Jinan, Jinan 250022, China
ise_wangd@ujn.edu.cn
[2] Shandong Provincial Key Laboratory of Network Based Intelligent
Computing, University of Jinan, Jinan 250022, China

Abstract. The integer multiple firing patterns, generated in the rabbit depressor baroreceptors under the different static blood pressure, were observed between the resting state and the periodic firing and were characterized to be stochastic but not chaotic by a series of nonlinear time series estimations. These patterns exhibited very similar characteristics to those observed in the experimental neural pacemaker. Using $I_{na,p} + I_K$ models with dynamics of a supercritical Hopf bifurcation, we successfully simulated the bifurcation process of firing patterns and observed the induction of the integer multiple firing patterns by adding noise. The results strongly suggest that the integer multiple firing rhythms generated by rabbit baroreceptors result from the interplay between noise and the system's dynamics. Because of the important normal physiological function of baroreceptors, the biological significance of noise and the noise-induced firing rhythms at a Hopf bifurcation is interesting to be addressed.

Keywords: Stochastic neural firing pattern · Hopf bifurcation · Blood pressure · Baroreceptors

1 Introduction

With great progresses in nonlinear science and neuroscience during recent years, the dynamic characteristics of many neural firing activities have been revealed by analytical results of different firing rhythms observed both in theoretical model and biological experiment [1–5]. Some important nonlinear phenomena, such as bifurcation, chaos and stochastic oscillation, were approved to play key roles in neural information processing and have explicit biological relevance [6–8]. However, some others were not completely clear. A number of noise-induced stochastic neural firing rhythms, which were termed as the integer multiple firing patterns, were reported in previous studies [6, 9, 10]. Such non-periodic firing patterns were identified to be stochastic by means of nonlinear time series measures and suggested to be generated at a Hopf bifurcation via interplay between noise and the system's dynamics. A series of such patterns were experimentally discovered by using an experimental neural pacemaker,

© Springer International Publishing AG 2017
D. Liu et al. (Eds.): ICONIP 2017, Part IV, LNCS 10637, pp. 553–562, 2017.
https://doi.org/10.1007/978-3-319-70093-9_58

which was produced by chronic nerve injury [11]. However, the injured nerve does not encode normal sensory information, and thus, the biological relevance of the integer multiple firing patterns are unclear.

Baroreceptors are sensory nerve terminals of blood pressure. They generate neural firing trains to encode and input the changes of blood pressure into the central nervous system. Their biological function is fundamentally important in the maintenance of a normal blood pressure level [12]. To our best knowledge, experimental demonstration of the integer multiple firing patterns generated by baroreceptors is seldom reported. In the present investigation, we briefly reported integer multiple firing patterns observed in vitro experiments by adjusting the static pressure in the blood vessel. A series of nonlinear time series analysis and theoretical simulation were carried on to study these firing patterns. The results supported the proposed ideas and provided new insights into the complex ongoing activities of baroreceptors working in physiological context.

In this paper, we applied the simple nearest-neighbor of nonlinear prediction to evaluate the deterministic property of the integer multiples interspike intervals (ISIs) series, while surrogate data method was employed to enhance the reliability of the normalized prediction error (*NPE*). Besides, an entropy measures which could reflect the confusion degree of symbolic time series to evaluate the discrete ISI series, and Kasper-Shuster method which was used to calculate the Lemple-Ziv complexity of the binary series of transformed from the original ISI series, were also applied. The algorithms of all the nonlinear time series analysis methods were particularly described in the previous study [6].

2   Experimental Model and Results

Our experiments were performed on the rabbit depressor baroreceptors. Adult male New Zealand rabbit, weighing 2–2.5 kg, was anesthetized with urethane intravenously, 1 g/kg, and additional doses were given in the course of the experiment if required. The unit of common carotid arteries, subclavian arteries and aortic arch, including the depressor nerve, was dissociated and perfused continuously with 34 °C Kreb's solution. The static blood pressure in artery was controlled by the velocity of flow in the physiological fluid and recorded with a cannula pressure transducer connected to an ML221 amplifier from arteria carotis. The depressor nerve, about 2 cm long, was isolated and marinated in an oil pool. A thin bundle of depressor fibers was separated, and its afferent firing trains were induced by means of a fine platinum electrode with a nearby reference and connected to a bioelectrical amplifier. The static blood pressure and spike trains of individual fibers were recorded simultaneously with a Powerlab system (ASInstruments, Sydney, Australia) with a sampling frequency of 10.0 kHz. The time intervals between the maximal values of the successive spikes were recorded seriatim as interspike interval series [3].

Our previous works studied three types of firing patterns with the changes of blood pressure, which were observed in vitro experimental procedure, but no further analysis was given [4]. This paper will only report integer multiple firing patterns observed on 13 depressor baroreceptors in vitro experiment.

A typical example of such firing patterns was observed when the static blood pressure was stabilized at 120 mmHg. As shown in Fig. 1(a), the spike trains of bursting exhibited obvious non-periodic. The simple analysis of ISI series suggested that it had classical characteristics of the integer multiple firing patterns. First, the stable

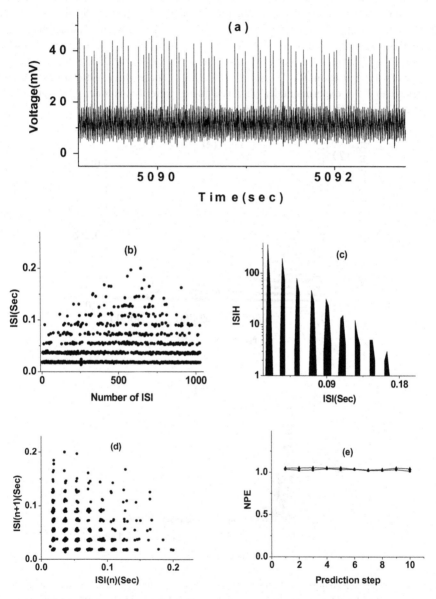

Fig. 1. Integer multiple firing pattern generated by rabbit baroreceptors when static blood pressure was stabilized at 120 mmHg in the experiment: (a) Spike trains; (b) Interspike intervals (ISIs) of spike trains; (c) The histogram of ISI series; (d) The first return map of ISI series; (e) NPE of raw and surrogate data of ISI series. Where, circle, raw data; triangle, surrogate data.

ISI series showed multiple layers, where the first layer was corresponding to a basic ISI, as shown in Fig. 1(b). Second, the peaks in ISIH mostly located at integer multiples of a basic ISI and the amplitudes of the peaks decayed approximately exponentially, as shown in Fig. 1(c). And third, the first return map of ISI series was a lattice-like structure, as shown in Fig. 1(d). By nonlinear time series estimations, this non-periodic firing pattern was identified to be stochastic, but not chaotic. The results of *NPE* showed that it could not be predictable. Compared to 50 realizations of surrogate data, *NPE* values of raw ISI series in all prediction steps are nearly equal to 1.0, as shown in Fig. 1(e). The entropy values of ISI series is 0.8255, and the result of complexity computation for ISI series is 0.9528. These three estimations strongly suggested that the experimentally observed firing pattern was stochastic, rather than chaotic. The analysis on the other cases obtained very similar results.

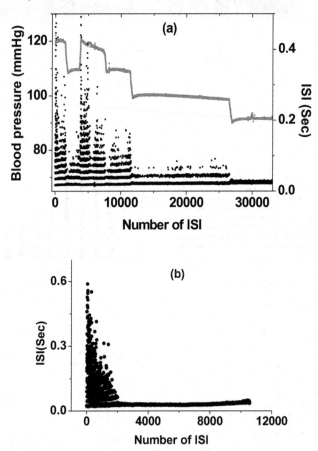

Fig. 2. (a) A gradual change of ISI series from resting to periodic firing with static blood pressure changing (red dots) in the experimental baroreceptors. Integer multiple firing patterns were observed between the resting state and the periodic firing. (b) A gradual change of ISI series from resting to periodic firing with extracellular calcium concentration changing in the experimental neural pacemaker. Integer multiple firing patterns were also observed. (Color figure online)

In order to find out the dynamic mechanism of the integer multiple firing patterns observed in rabbit baroreceptors, we adjusted the static pressure and discovered a gradual change from resting to periodic firing in the firing activities of the barore-ceptors. Interestingly, the integer multiple firing patterns were observed between the resting state and the periodic firing, as shown in Fig. 2(a). Then we compared these ones with those previously observed in the experimental neural pacemaker, as shown in Fig. 2(b), and discovered very similar dynamic characteristics.

3 Theoretical Model

The previous studies suggested that integer multiple firing patterns were tended to be generated at a Hopf bifurcation via interplay between noise and the system's dynamics. So as to know whether such patterns in baroreceptors have the same dynamics, the $I_{Na,p} + I_K$ model was employed to numerically simulate the firing activities of rabbit baroreceptors under different parameter conditions. This model consists of a fast Na^+ current and a relatively slower K^+ current. Its deterministic form contains the following two simultaneous differential equations:

$$C\dot{V} = I - g_L(V - E_L) - g_{Na}m_\infty(V)(V - E_{Na}) - g_Kn(V - E_K) \qquad (1)$$

$$\dot{n} = (n_\infty(V) - n)/\tau(V) \qquad (2)$$

The variables and parameters were detailed introduced in descriptions [13]. Besides, the stochastic $I_{Na,p} + I_K$ model, in which a Gaussian white noise $\xi(t)$ was directly added to the right hand of Eq. (1), was also studied here. The stochastic factor possesses the statistical properties as $<\xi(t)> = 0$; $<\xi(t), \xi(t')> = 2D\delta(t - t')$; where D is the noise density and δ is the Dirac δ-function.

In both deterministic and stochastic model, E_k is choosen as bifurcation parameter and others are as follow: $C = 1.0$; $E_L = -78.0$; $g_L = 8.0$; $g_{Na} = 20.0$; $g_K = 10.0$; $V_m = -20.0$; $V_n = -45.0$; $K_m = 15.0$; $K_n = 5.0$; $\tau(V) = 8.0$; $E_{Na} = 60.0$; $I = 3.0$. Models are solved by Mannelle numerical integrate method proposed by Mannella and Palleschi [14], which integration time step is 10^{-3} s. Upstrokes of the voltage reached the amplitude of -25.0 mV are counted as spikes.

In the deterministic $I_{Na,p} + I_K$ model, $E_k \approx -88.2162$ is a supercritical Hopf bifurcation point of the equilibrium point. When $-89 < E_k < -88.2162$, the stable behavior of $I_{Na,p} + I_K$ model is at a stable focus, corresponding to the polarized resting state. When $-88.2162 < E_k < -87$, the stable behavior of $I_{Na,p} + I_K$ model is on a stable period 1 limit cycle, corresponding to period 1 firing. There is no other patterns except rest and periodic firing when E_k was from -89 to -87 in the deterministic model, as shown in Fig. 3.

However, in the stochastic model, when noise density $D = 0.5$, a gradual change from resting to periodic firing was observed and we found integer multiple firing patterns between the resting state and the periodic firing, as shown in Fig. 4(a). Taking $E_k = -88.45$ for example, which is near supercritical Hopf bifurcation point in the deterministic model, we obtained the obviously non-periodic spike trains, as shown in

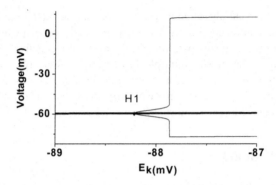

Fig. 3. Structures of the supercritical Hopf bifurcation of the equilibrium point ($E_k \approx -88.2162$) and the generation of limit cycles in deterministic $I_{Na,p} + I_K$ modelas parameter E_k is varied from -89 to -87. H1 is a supercritical Hopf bifurcation point. The bold solid line corresponds to stable focus. The upper (lower) thin solid line corresponds to maximal (minimal) amplitude of the stable period 1 limit cycle.

Fig. 4(b). Its ISI series exhibited similar multimodal characteristics to those in the experiment, as shown in Fig. 4(c), (d) and (e). The stochastic characteristics were also validated by a series of nonlinear time series estimations, including the results of nonlinear prediction, as shown in Fig. 4(f), entropy value (0.8665) and complexity computation (0.9822) for ISI series. All the results proved that the stochastic $I_{Na,p} + I_K$ model could well simulate the observed integer multiple pattern and the bifurcation from rest to period firing in experimental baroreceptors.

The degree of coherence (β) [11] and signal-to-noise ratio (*SNR*) of the peak of power spectrum have been frequently used to quantify the phenomenon of coherence resonance [12], which were also employed to verify the effect of coherence resonance with different noise density in this investigation. According to the method in [15], the ISI series corresponding to $E_k = -88.45$ in the stochastic $I_{Na,p} + I_K$ model was converted to a time series made from standard pulses whose amplitude was $V_0 = 30$ mV and duration was $\Delta t = 10$ ms. Each pulse represented an event of the original firing series. For each noise density D, 100 realizations were used to generate spectrum average of pulses series re-sampled at 500 Hz. For each realization, 1024-point Hanning-windowed fast Fourier transforms (FFT) were computed. Then, the power spectrum could be obtained as shown in Fig. 5(a). *SNR* and β could be defined and computed with respect to the dominant spectral peak followed as the previous studies [16, 17]. The amplitude changes of *SNR* and β with different noise density D reflected the signature of coherence resonance: The value rises to a maximum at some optimal noise density, and then decreases, as shown in Fig. 5(b) and (c). As there was no external stimulation, we concluded that these two alternation firing patterns resulted from the interplay of the noise and the system, and might exactly be induced by the effect of autonomous stochastic resonance (ASR).

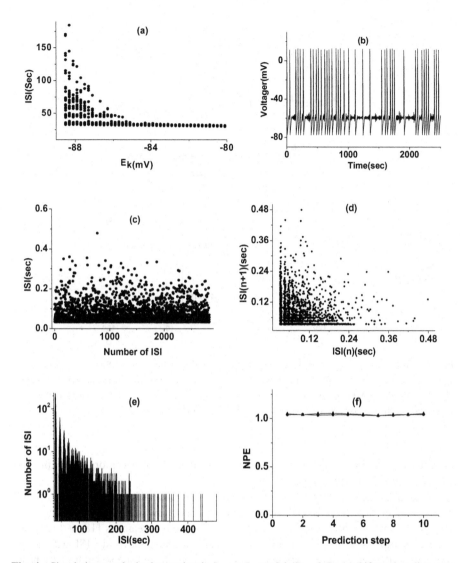

Fig. 4. Simulation results in the stochastic $I_{Na,p} + I_K$ model ($D = 0.5$): (a) Bifurcation diagrams with respect to ISI series when the parameter is from -88.5 to -79.9; (b), (c), (d), (e) were all obtained when $E_K = -88.45$, showing integer multiple firing pattern (b) Spike trains; (c) ISI series of spike trains; (d) The first return map of ISI series; (e) The histogram of ISI series; (f) NPE of raw and surroagate data of ISI series. Where, circle, raw data; triangle, surrogate data.

In this paper, with rabbit depressor baroreceptors which were forced by static blood pressure, we observed stochastic integer multiple firing patterns between the resting state and the periodic firing and found it had as similar dynamics as that in the experimental neural pacemaker. In theoretical studies, with a neural model with a supercritical Hopf bifurcation, we successfully simulated the bifurcation process of the

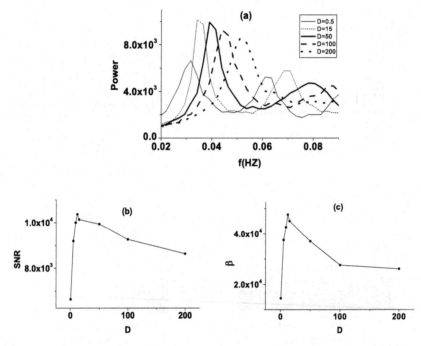

Fig. 5. Simulation results in the stochastic $I_{Na,p} + I_K$ model when $E_k = -88.45$: (a) The power spectrum of trains occurred with different noise density D; (b) Relationship between *SNR* value, obtained by the power spectrum, with different noise density D; (c) Relationship between β value, obtained by the power spectrum, with different noise density D.

integer multiple firing patterns observed in experimental baroreceptors and explained its generation induced by noise.

Stochastic bursting generated near a supercritical Hopf bifurcation has been found in many theoretical models with or without external stimulation and their biological significance has not been very clear so far. The previous researches [18, 19] reported integer multiple spiking caused by external signal where the basic ISI was corresponding to the period of the external periodic stimulation, and suggested it might be of help to encode the information under the external stimuli, for example, variations in temperature and/or electric field. In the experimental neural pacemaker, we also found integer multiple spiking without external signal. However, experimental neural pacemaker, which is produced by chronic nerve injury, does not encode normal sensory information. Baroreceptors are believed to respond uniformly, acting like one transducer. The appearance of paradoxical bursting during blood pressure elevation indicates that the physiological range of blood pressure spans the bifurcation scenario of some baroreceptors. The new finding in this letter implies the existence of integer multiple spiking could encode information under normal physiological environment without external signal as well as other firing patterns and have significant roles during regulating blood pressure.

In realistic biological systems, including the rabbit baroreceptors, intrinsic noise is inevitably produced from perturbations and thermal dynamic fluctuations. The variation of blood pressure is also unstable. The biological significance of noise and the noise-induced firing rhythms is an interesting question to be addressed.

Acknowledgment. This research was supported by the National Key Research And Development Program of China (No. 2016YFC0106000), the Natural Science Foundation of China (Grant No. 61302128), and the Youth Science and Technology Star Program of Jinan City (201406003), although supported by NSFC (Grant Nos. 61573166, 61572230, 61671220, 61640218), the Natural Science Foundation of Shandong Province (ZR2013FL002), the Shandong Distinguished Middle-aged and Young Scientist Encourage and Reward Foundation, China (Grant No. ZR2016FB14), the Project of Shandong Province Higher Educational Science and Technology Program, China (Grant Nos. J16LN07, J16LBO6, J17KA047), the Shandong Province Key Research and Development Program, China (Grant No. 2016GGX101022).

References

1. Bashkirtseva, I., Ryashko, L., Slepukhina, E.: Order and chaos in the stochastic Hindmarsh-Rose model of the neuron bursting. Nonlinear Dyn. **82**(1–2), 919–932 (2015)
2. Xing, J.L., Hu, S.J., Yang, J.: Electrophysiological features of neurons in the mesencephalic trigeminal nuclei. Neuro-Signals (22), 79–91 (2015)
3. Abbasi, S., Abbasi, A., Sarbaz, Y., Janahmadi, M.: Power spectral density analysis of Purkinje cell tonic and burst firing patterns from a rat model of ataxia and riluzole treated. Basic Clin. Neurosci. **8**(1), 61–68 (2017)
4. Lu, T., Wade, K., Hong, H., Sanchez, J.T.: Ion channel mechanisms underlying frequency-firing patterns of the avian nucleus magnocellularis: a computational model. Channels (3), 1–15 (2017)
5. Li, Y., Gu, H.: The distinct stochastic and deterministic dynamics between period-adding and period-doubling bifurcations of neural bursting patterns. Nonlinear Dyn. **87**(4), 2541–2562 (2017)
6. Bashkirtseva, I., Ryashko, L.: Stochastic sensitivity analysis of noise-induced order-chaos transitions in discrete-time systems with tangent and crisis bifurcations. Physica A: Stat. Mech. Appl. (467), 573–584 (2017)
7. Zhao, Z., Jia, B., Gu, H.: Bifurcations and enhancement of neuronal firing induced by negative feedback. Nonlinear Dyn. **86**(3), 1–12 (2016)
8. Yang, M.H., Liu, Z.Q., Li, L., Xu, Y.L., Liu, H.J., Gu, H.G.: Identifying distinct stochastic dynamics from chaos: a study on multimodal neural firing patterns. Int. J. Bifurc. Chaos (19), 453–485 (2009)
9. Jia, B., Gu, H.G.: Identifying type I excitability using dynamics of stochastic neural firing patterns. Cogn. Neurodyn. (6), 485–497 (2012)
10. Hu, G., Ditzinger, T., Ning, C.Z.: Stochastic resonance without external periodic force. Phys. Rev. Lett. (71), 807–810 (1993)
11. Gu, H.G., Zhang, H.M., Wei, C.L., Yang, M.H., Liu, Z.Q., Ren, W.: Coherence resonance induced stochastic neural firing at a saddle-node bifurcation. Int. J. Mod. Phys. B (25), 3977–3986 (2011)
12. Azevedo, R.T., Garfinkel, S.N., Critchley, H.D., Tsakiris, M.: Cardiac afferent activity modulates the expression of racial stereotypes. Nat. Commun. (8), 13854 (2017)

13. Izhikevich, E.M., Edelman, G.M.: Large-scale model of mammalian thalamocortical systems. Proc. Natl. Acad. Sci. U.S.A. **105**(9), 3593–3598 (2008)
14. Mannella, R., Palleschi, V.V.: Mean first-passage time in a bistable system driven by strongly correlated noise: introduction of a fluctuating potential. Phys. Rev. A **39**(7), 3751–3753 (1989)
15. Huaguang, G., Zhiguo, Z., Bing, J., Shenggen, C.: Dynamics of on-off neural firing patterns and stochastic effects near a sub-critical Hopf bifurcation. Plos One **10**(4), e0121028 (2015)
16. Xing, J.L., Hu, S.J., Xu, H., Han, S., Wan, Y.H.: Subthreshold membrane oscillations underlying integer multiples firing from injured sensory neurons. NeuroReport (12), 1311–1313 (2011)
17. Chay, T.R.: Chaos in a three-variable modle of an excitable cell. Physica D Nonlinear Phenom. **16**(2), 233–242 (1985)
18. Sancristóbal, B., Rebollo, B., Boada, P., Sanchezvives, M.V., Garciaojalvo, J.: Collective stochastic coherence in recurrent neuronal networks. Nat. Phys. (12), 881–888 (2016)
19. Lai, Y.C., Park, K.: Noise-sensitive measure for stochastic resonance in biological oscillators. Math. Biosci. Eng. **3**(4), 583–602 (2006)

Functional Connectivity Analysis of EEG in AD Patients with Normalized Permutation Index

Lihui Cai[1], Jiang Wang[1], Ruofan Wang[2(✉)], Bin Deng[1], Haitao Yu[1], and Xile Wei[1]

[1] School of Electrical and Information Engineering, Tianjin University, Tianjin 300072, China
{clhfio,jiangwang,dengbin,htyu,xilewei}@tju.edu.cn
[2] School of Information Technology Engineering, Tianjin University of Technology and Education, Tianjin 300222, China
wangrf@tju.edu.cn

Abstract. In this work, we proposed Normalized Permutation Index (NPI) to analysis the functional connectivity of EEG from human brain with Alzheimer's disease. NPI is modified method of permutation disalignment index based on permutation entropy, and can be used for the functional network analysis. The simulation analysis of NPI is first performed and the results show that NPI could effectively estimate the coupling strength with high sensitivity. Then NPI is applied to the synchronization and network analysis of AD brain. It can be observed that the functional connectivity in AD brain is weakened in most channel pairs, and the network properties are also altered with decreased global and local efficiency. These preliminary results demonstrate that NPI could be used to provide a new biomarker for AD pathology.

Keywords: EEG · Alzheimer's disease · Functional connectivity · Normalized Permutation Index

1 Introduction

Alzheimer's disease (AD), the most prevalent form of dementia, is a disabling neuro-degenerative disorder that affects mainly the older [1, 2]. In recent guidelines concerning AD, the accumulation of amyloid in the brain is considered to be the pathophysiological markers, which can be detected by positron emission tomography (PET) maps, magnetic resonance imaging (MRI) and so on [3]. However, these neuroimaging techniques are not widely applied in the diagnose of AD considering their high costs and invasion property. EEG, as a cheap and non-invasive neuroimaging technique, has proved to be effective in the neurophysiological assessment of AD [4–6]. In AD patients, EEG abnormalities arise on account of the degeneration of synapses and death of neurons, which may induce a functional brain disconnection syndrome between some cortical areas [7]. Therefore, investigation of brain functional connectivity seems a promising method to provide effective biomarkers for AD.

© Springer International Publishing AG 2017
D. Liu et al. (Eds.): ICONIP 2017, Part IV, LNCS 10637, pp. 563–571, 2017.
https://doi.org/10.1007/978-3-319-70093-9_59

Previous studies demonstrated that AD patients show perturbations in EEG synchrony and altered network properties [8–12]. For instance, Stam et al. found that the synchronization is weakened in beta band for AD patients compared with the healthy controls [8]. Further, they studied EEG based networks and found that AD networks have longer shortest path length than the healthy controls in beta band [9]. Tahaei et al. also showed the destruction of functional brain networks in early AD with a decrease in synchronizability [10]. Various linear and nonlinear methods have been used to compute the connectivity strength between two network nodes from EEG [13–16]. Pearson correlation coefficient, for example, is a time-domain connectivity measure to evaluate the similarity between two nodes. Coherence is another linear method which aims to measure the similarity of frequency distribution in two EEG series. Recently, Mammone et al. proposed a novel measure called permutation disalignment index (PDI) to estimate the brain connectivity [17]. This index is inversely proportional to the coupling strength and not within the range [0, 1], thus not suitable for the network analysis.

In this work, we make a modification to PDI and proposed Normalized Permutation Index (NPI) to analysis the functional connectivity in AD brain. First, the simulation systems are used to test its ability in estimating the coupling strength. Then this approach is applied to the characterization of synchronization and graph theoretical analysis of AD networks.

2 Materials and Methods

2.1 Experiment Design and EEG Recordings

Thirty subjects are recruited in this study and divided into two groups. (a) Fifteen right-handed patients with a diagnosis of probable AD (age: 74–78 years old; nine females and six males). The Mini-Mental-Status examination (MMSE) scores are ranged from 12 to 15. (b) Fifteen healthy age-matched subjects (age: 70–76 years old; ten females and five males) are served as controls whose MMSE scores are ranged from 28 to 30. During the experiment, the subjects were seated in a semi-dark quiet room and stay awake with eyes closed. The data collection lasts more than 10 min for each subject.

The EEG was recorded by a Symtop amplifier according to the international 10–20 system at a sampling frequency of 1024 Hz. The linked earlobe A1 and A2 are used as a reference, and the 16 Ag-AgCl scalp electrodes are channels Fp1, Fp2, F3, F4, C3, C4, P3, P4, O1, O2, F7, F8, T3, T4, T5, T6, as shown in Fig. 1(A), and the real EEG signals recorded are shown in Fig. 1(B). EEG signals are band-pass filtered at 0.5–30 Hz to eliminate the effect of high frequency noises. Before the analysis of filtered EEG, the artifacts caused by eye movement or other visible disturbances were manually labeled by the expert and then removed. In this study, a 8s artifact-free signal was selected for each subject.

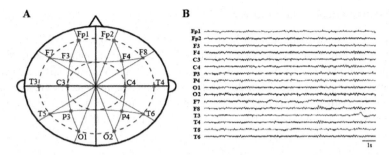

Fig. 1. Electrode positions on the brain (A) and an example of 16-channnal EEG signals recorded from one AD patient (B).

2.2   Normalized Permutation Index

Given an N-point time series u, it can be mapped into m-dimension space, thus a set of vectors are formed as $X_i = \{x(i), x(i+L), \ldots, x(i+(m-1)L)\}$ with i ranged from 1 to $N - (m-1)L$, where m is the embedding dimension and L is the time delay. Next, arrange X_i in the incremental order $[x_{i+(j_1-1)L} \leq x_{i+(j_2-1)L} \leq \cdots \leq x_{i+(j_m-1)L}]$, and define $\pi = [j_1, j_2, \ldots, j_m]$ as the ordinal pattern (also called motif). There will be $m!$ motifs for time series μ and each vector can be mapped into one of the $m!$ motifs. Obviously, the $m!$ motifs can be also obtained for time series v with the same data length. Let $f_{uv}(\pi_j)$ represent the simultaneous occurrence frequency of each motif π_j in μ and v, the simultaneous occurrence rate is defined as

$$p_{uv}(\pi_j) = \frac{f_{uv}(\pi_j)}{N - (m-1)L} \tag{1}$$

Then the permutation index is computed as

$$PI(u, v) = -\sum_{j=1}^{m!} p_{uv}(\pi_j) \log(p_{uv}(\pi_j)) \tag{2}$$

The strong coupling of u and v will lead to a large PI value, as the highly coupled time series show the same motifs with a high probability. When u and v are completely coupled and the simultaneous occurrence rate is equal for each motif, the PI will reach its maximum $\log(m!)$. Therefore, PI can be normalized as

$$NPI(u, v) = -\sum_{j=1}^{m!} p_{uv}(\pi_j) \log(p_{uv}(\pi_j)) / \log(m!) \tag{3}$$

Since a large embedding dimension m will significantly increase the computational burden for NPI, m is set to 3 (corresponds to 6 motifs) in this study.

2.3   Graph Theoretical Approach

A nonparametric technique named Minimum Connected Component (MCC) is applied to extract the binary networks. MCC is a special connected spanning subgraph and defined as follows. For an undirected weighted graph with N nodes, any link between two nodes is removed first. Then the strongest weight is considered to create a binary link and the corresponding nodes are marked. Next, the second strongest weight is considered and so on. This procedure is continued until all the nodes are visited (without isolated nodes) [18].

To investigate the network properties (such as information segregation and integration) of AD brain, global efficiency and local efficiency are calculated. Global efficiency is the inverse of its average shortest path length and is defined as

$$gE = \frac{1}{N(N-1)} \sum_{i,j,i \neq j} \frac{1}{d_{ij}} \tag{4}$$

where d_{ij} is the shortest path length between node i and j. Local efficiency of node k is defined as

$$locE(k) = \frac{1}{N_{G_k}(N_{G_k}-1)} \sum_{i,j \in G_k} \frac{1}{d_{ij}} \tag{5}$$

where G_k is the subgraph constituted by the neighbor nodes of node k and N_{G_k} is the number of nodes in G_k. Local efficiency of the network is computed as

$$locE = \frac{1}{N} \sum_k locE(k) \tag{6}$$

To assess whether AD and the control group have significant difference for synchronization and network metrics, one way ANOVA is used. The results with $p < 0.01$ indicates significant difference.

3   Results

3.1   Henon Systems Analysis with NPI

In this work, we first investigated the dependence of NPI on the coupling strength between simulated time series x and y, which is resulted from two unidirectionally coupled Henon systems X and Y. The Henon systems could be defined as

$$\begin{cases} X : x_{n+1} = 1.4 - x_n^2 + b_x x_{n-1} \\ Y : y_{n+1} = 1.4 - [cx_n + (1-c)y_n]y_n + b_y y_{n-1} \end{cases} \tag{7}$$

where the coupling strength c is ranged from 0 to 1 with 0 indicating no coupling and 1 indicating complete coupling. For identical systems, bx and by are both set to 0.3, whereas they are set to 0.3 and 0.1 for nonidentical systems. X and Y are initialized randomly in the range [0, 1] and computed with c increased from 0 to 1 (step = 0.1). All the simulation is repeated 20 times for each coupling strength.

Figure 2 shows the performance of NPI in estimating the coupling strength for coupled Henon systems. NPI exhibits an increasing trend as c increases from 0 to 0.7 and then tends to be stable when $c > 0.7$ for identical systems, while it increases monotonically for nonidentical systems, implying that NPI could detect the difference between two kinds of systems. For $0 < c < 0.2$, a larger fluctuation is observed for NPI with $L = 1, 2$ compared with NPI with $L = 3$. Therefore, $L = 3$ is selected for further analysis. Note that NPI is always within the range [0, 1] as c increases. These observations indicate that NPI can be applied to estimate the synchronization strength and construct functional networks for real EEG series.

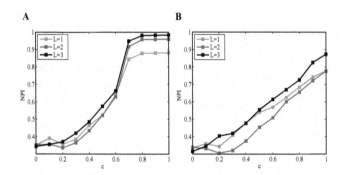

Fig. 2. NPI as a function of coupling strength c with different time delay L for coupled identical Henon systems (A) and nonidentical Henon system (B).

3.2 EEG Analysis for AD with NPI

We calculated NPI of each pair-wise EEG channels to explore the synchronization differences between AD and the control group. A 16×16 connectivity matrix is obtained for each group (averaged across subjects in the group) shown in Figs. 3A and B. For AD group, NPI is mainly within the range [0.37, 0.41]; while for the control group, it is within the range [0.39, 0.43]. The connectivity strength for AD group is weakened in most of channel pairs. In addition, One way ANOVA is applied to investigate the group difference of mean synchronization strength, as shown in Fig. 3C. It is observed that the connectivity strength in AD group is significantly larger than that in the control group. We Further explored the influence of sampling frequency on NPI with the original EEG signals are downsampled to 512 Hz and 256 Hz. Significant group difference is found for all the three sampling frequency, though NPI decreases when the sampling frequency is reduced, which indicates that the sampling frequency has little effect on detecting the difference between patients and the controls.

We further calculated the mean synchronization strength in each channel for AD and the control group and the results are shown in Fig. 4. The mean strength in central (channel C3 and C4) and parietal region (channel P3 and P4) is markedly stronger than

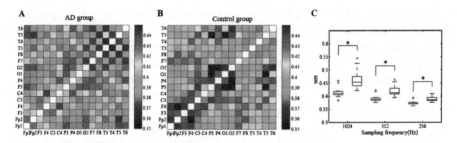

Fig. 3. Differences of connectivity matrices computed by NPI between AD and the control group. Connectivity matrices are shown for AD (A) and the control group (B) (averaged across subjects in the group). (C) the boxplot of mean connectivity strength averaged across the matrix for both groups. * indicates significant group difference ($p < 0.01$).

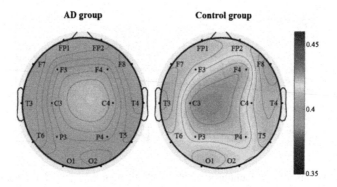

Fig. 4. Mean synchronization strength in each channel for AD and the control group.

that in other brain regions for both groups. However, the mean strength in AD group is smaller than the control one in each channel. Moreover, the largest mean strength is observed in different hemisphere for two groups (C3 channel for AD group and C4 channel for the control group). Table 1 presented the statistical analysis results of mean synchronization strength in each channel for both groups. Significant group difference was found in most channels, which are mainly located in the frontal (i.e. channel FP1 and Fp2) and temporal regions (i.e. channel T3 and T4).

Finally, MCC method is applied to construct binary functional networks based on the connectivity matrices. We used graph theory metrics to gain further insight into the network properties of the AD brain. Figure 5 shows the global efficiency (gE) and local efficiency (locE) for both groups. Both gE and locE are significantly higher in AD group than the control group, indicating decreased capacity of information transmission in AD brain (see Fig. 5A). Figure 5B shows the scatterplot of gE and locE for all the patients and controls. It can be found that the two network metrics can discriminate patients from the controls with little a high accuracy. These observations confirm that NPI can be applied to detect the abnormalities of network structure for AD brain.

Table 1. Range of mean synchronization strength in each channel for AD and the control group with the results of ANOVA test (denoted by p values).

Channels	AD (mean ± std)	CON (mean ± std)	p-value
FP1	0.390 ± 0.010	0.415 ± 0.011	**<0.001**
FP2	0.393 ± 0.009	0.417 ± 0.009	**<0.001**
F3	0.400 ± 0.005	0.416 ± 0.018	0.013
F4	0.404 ± 0.005	0.425 ± 0.012	**<0.001**
C3	0.402 ± 0.007	0.429 ± 0.014	**<0.001**
C4	0.408 ± 0.006	0.414 ± 0.016	0.356
P3	0.400 ± 0.007	0.422 ± 0.017	**<0.001**
P4	0.403 ± 0.007	0.423 ± 0.019	**0.002**
O1	0.403 ± 0.007	0.417 ± 0.016	0.056
O2	0.398 ± 0.006	0.422 ± 0.015	**<0.001**
F7	0.387 ± 0.005	0.414 ± 0.010	**<0.001**
F8	0.391 ± 0.007	0.401 ± 0.011	0.034
T3	0.387 ± 0.010	0.404 ± 0.017	**0.001**
T4	0.392 ± 0.006	0.415 ± 0.008	**<0.001**
T5	0.391 ± 0.011	0.422 ± 0.015	**<0.001**
T6	0.399 ± 0.006	0.417 ± 0.019	0.053

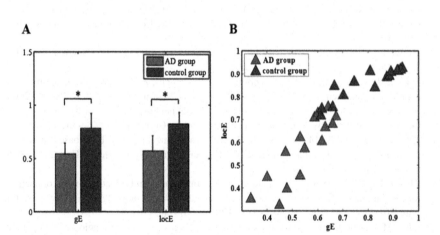

Fig. 5. Network metrics for AD and the control group. (A) statistical analysis of global efficiency (gE) and local efficiency (locE) for both groups. (B) scatterplot of global and local efficiency for patients and controls. * indicates significant group difference ($p < 0.01$).

4   Conclusion

In this paper, NPI is proposed to estimate the functional connectivity strength between different brain regions and construct functional networks. The ability of NPI in estimating coupling strength is verified through the simulated systems. In addition, NPI approach proves to be effective in detecting the abnormalities of AD brain, with weakened synchronization and altered network properties observed. We infer that NPI may not only facilitate our study of the functional alteration in AD brain, but also could be used as a useful tool to gain insights into other neurologic disorders.

Acknowledgments. This work was supported by National Natural Science Foundation of China (NO. 61601331).

References

1. Dauwels, J., Vialatte, F., Cichocki, A.: Diagnosis of Alzheimer's disease from EEG signals: where are we standing? Curr. Alzheimer Res. **7**(6), 487–505 (2010)
2. Mattson, M.P.: Pathways towards and away from Alzheimer's disease. Nature **430**(7000), 631–639 (2004)
3. Brayne, C.: A population perspective on the IWG-2 research diagnostic criteria for Alzheimer's disease. Lancet Neurol. **13**(6), 532–534 (2014)
4. Cao, Y.Z., Cai, L.H., Wang, J., Wang, R.F., Yu, H.T., Cao, Y.B., Liu, J.: Characterization of complexity in the electroencephalograph activity of Alzheimer's disease based on fuzzy entropy. Chaos **25**(8), 083116 (2015)
5. Adeli, H., Ghosh-Dastidar, S., Dadmehr, N.: Alzheimer's disease: models of computation and analysis of EEGs. Clin. EEG Neurosci. **36**(3), 131–140 (2005)
6. Barzegaran, E., van Damme, B., Meuli, R., Knyazeva, M.G.: Perception-related EEG is more sensitive to Alzheimer's disease effects than resting EEG. Neurobiol. Aging **43**, 129–139 (2016)
7. Jeong, J.: EEG dynamics in patients with Alzheimer's disease. Clin. Neurophysiol. **115**, 1490–1505 (2004)
8. Stam, C.J., van der Made, Y., Pijnenburg, Y.A., Scheltens, P.: EEG synchronization in mild cognitive impairment and Alzheimer's disease. Acta Neurol. Scand. **108**(2), 90–96 (2003)
9. Stam, C.J., Jones, B.F., Nolte, G., Breakspear, M., Scheltens, P.: Small-world networks and functional connectivity in Alzheimer's disease. Cereb. Cortex **17**(1), 92–99 (2007)
10. Tahaei, M.S., Jalili, M., Knyazeva, M.G.: Synchronizability of EEG-based functional networks in early Alzheimer's disease. IEEE Trans. Neural Syst. Rehabil. Eng. **20**(5), 636–641 (2012)
11. Blinowska, K.J., Rakowski, F., Kaminski, M.: Functional and effective brain connectivity for discrimination between Alzheimer's patients and healthy individuals: a study on resting state EEG rhythms. Clin. Neurophysiol. **128**(4), 667–680 (2017)
12. Jalili, M.: Graph theoretical analysis of Alzheimer's disease: discrimination of AD patients from healthy subjects. Inf. Sci. **384**, 145–156 (2016)
13. Stam, C.J., van Dijk, B.W.: Synchronization likelihood: an unbiased measure of generalized synchronization in multivariate datasets. Physica D **163**, 236–251 (2002)
14. Knyazeva, M.G., Jalili, M., Brioschi, A.: Topography of EEG multivariate phase synchronization in early Alzheimer's disease. Neurobiol. Aging **31**(7), 1132–1144 (2010)
15. Zalesky, A., Fornito, A., Bullmore, E.: On the use of correlation as a measure of network connectivity. NeuroImage **60**, 2096–2106 (2012)
16. Jalili, M.: Functional brain networks: does the choice of dependency estimator and binarization method matter? Sci. Rep. **6**, 29780 (2016)

17. Mammone, N., Bonanno, L., Salvo, S., Marino, S., Bramanti, P., Bramanti, A., Morabito, F. C.: Permutation disalignment index as an indirect, EEG-based, measure of brain connectivity in MCI and AD patients. Int. J. Neural Syst. **27**(5), 1750020 (2017)
18. Vijayalakshmi, R., Nandagopal, D., Dasari, N., Cocks, B., Dahal, N., Thilaga, M.: Minimum connected component – a novel approach to detection of cognitive load induced changes in functional brain networks. Neurocomputing **170**(C), 15–31 (2015)

Emotion Annotation Using Hierarchical Aligned Cluster Analysis

Wei-Ye Zhao[1], Sheng Fang[1], Ting Ji[1], Qian Ji[1], Wei-Long Zheng[1], and Bao-Liang Lu[1,2,3(✉)]

[1] Department of Computer Science and Engineering,
Center for Brain-like Computing and Machine Intelligence, Shanghai, China
{andylaw,weilong}@sjtu.edu.cn
[2] Key Laboratory of Shanghai Education Commission for Intelligent
Interaction and Cognitive Engineering, Shanghai, China
[3] Brain Science and Technology Research Center,
Shanghai Jiao Tong University, Shanghai, China
bllu@sjtu.edu.cn

Abstract. The correctness of annotation is quite important in supervised learning, especially in electroencephalography(EEG)-based emotion recognition. The conventional EEG annotations for emotion recognition are based on the feedback like questionnaires about emotion elicitation from subjects. However, these methods are subjective and divorced from experiment data, which lead to inaccurate annotations. In this paper, we pose the problem of annotation optimization as temporal clustering one. We mainly explore two types of clustering algorithms: aligned clustering analysis (ACA) and hierarchical aligned clustering analysis (HACA). We compare the performance of questionnaire-based, ACA-based, HACA-based annotation on a public EEG dataset called SEED. The experimental results demonstrate that our proposed ACA-based and HACA-based annotation achieve an accuracy improvement of 2.59% and 4.53% in average, respectively, which shows their effectiveness for emotion recognition.

Keywords: Neural data analysis · Time series analysis · EEG annotations · Emotion recognition

1 Introduction

Emotion is a subjective, conscious experience when people are faced with internal or external stimuli, and it is crucial for natural communication, decision making and human-machine interface. Emotion recognition can be performed through facial movement, voice, speech, text, and physiological signals [5,9]. Among these approaches, emotion recognition from electroencephalography (EEG) has attracted increasing interest [12,14,15].

In many supervised learning tasks, accurate training data annotations are the key factors to obtain ideal learning performance [4]. It was shown that the

© Springer International Publishing AG 2017
D. Liu et al. (Eds.): ICONIP 2017, Part IV, LNCS 10637, pp. 572–580, 2017.
https://doi.org/10.1007/978-3-319-70093-9_60

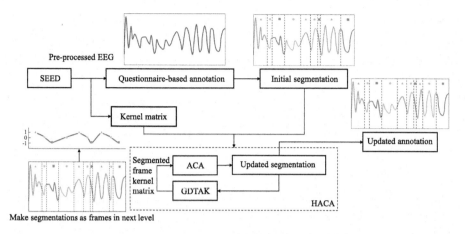

Fig. 1. The main processing steps to optimize questionnaire-based annotations with ACA and HACA.

presence of mislabeled training data, even on a very small scale, can deteriorate the performance of classifiers in a broad range of classification problems [3,8] and inaccurate annotations are even more harmful than noisy training features [18]. However, obtaining annotations is error-prone since the process is inherently subjective [1]. For example, EEG signals are often annotated according to testers' evaluation of their emotion states in affective brain-computer interfaces [7].

Two categories of methods to eliminate mislabeled training examples have been widely studied recently: designing noise robust supervised learning algorithm and filtering mislabeled data [4]. Some classification methods in the presence of label noise were proposed [2]. As for filtering mislabeled data, a simple approach is to remove low-quality data in preprocessing stage, whereas such strategy might remove useful instances. Guan and colleagues proposed a nearest neighbor editing method to remove noisy samples from the training dataset [3].

To our best knowledge, there are limited studies reported in the literature dealing for emotion annotation optimization using temporal clustering algorithms. In our paper, we explore aligned clustering analysis (ACA) [16] and hierarchical aligned clustering analysis (HACA) [17]. HACA is a hierarchical extension of ACA using generalized dynamic time alignment kernel (GDTAK). The performance are evaluated on a public EEG dataset named SEED, which gives EEG annotations according to test subjects' evaluation on their emotional states (questionnaire-based annotations). The process for optimizing questionnaire-based EEG annotations is shown in Fig. 1.

2   Methods

2.1   Dynamic Time Alignment Kernel

Temporal clustering method utilizes a distance metric that is capable of matching time series points even for series of different length. Shimodaira *et al.* [11] proposed Dynamic Time Alignment Kernel (DTAK) as an efficient metric between time sequences. Given two sequences $X = [x_1, \ldots, x_{n_x}]$ and $Y = [y_1, \ldots, y_{n_y}]$, and their distance kernel matrix $K \in \mathbb{R}^{n_x \times n_y}$ ($\kappa_{i,j} = \exp\left(-\frac{\|x_i - y_j\|^2}{2\sigma^2}\right)$), DTAK is defined by recursively computing the similarity between the two time sequences.

$$\tau(X, Y) = \frac{\omega_{n_x, n_y}}{n_x + n_y}, \omega_{i,j} = \max \begin{cases} \omega_{i-1,j} + \kappa_{i,j} \\ \omega_{i-1,j-1} + 2\kappa_{i,j} \\ \omega_{i,j-1} + \kappa_{i,j} \end{cases} \quad (1)$$

2.2   Aligned Clustering Analysis

Aligned Clustering Analysis [16] extends kernel k-means for temporal clustering. Given an EEG sequence $X \in \mathbb{R}^{dim \times n_x}$, instead of minimizing the sum of distances, ACA minimizes the energy function:

$$J_{aca}(G, s) = \sum_{c=1}^{k} \sum_{i=1}^{m} g_{ci} \underbrace{\|\psi(X_{[s_i, s_{i+1}]}) - z_c\|^2}_{dist_\psi^2(Y_i, z_c)} = \|[\psi(Y_1), \ldots, \psi(Y_m)] - ZG\|_F^2$$

$$\text{s.t.} \quad G^T 1_k = 1_m,$$

$$(2)$$

where $\psi(\cdot)$ denotes a mapping of the sequence into a feature space, $G \in \{0,1\}^{k \times m}$ is the EEG segment indicator matrix with k classes and m segments ($g_{ci} = 1$ only when segment i belongs to class c), $s \in \mathbb{R}^{m+1}$ is a vector contains start and end position of each EEG segment ($s_{i+1} - s_i \in [1, n_{max}]$), $Y_i = X_{[s_i, s_{i+1}]}$ denotes an EEG segment, and z_c is the geometric centroid for class c.

With the ability to handle variable length features and DTAK to calculate the distance between the segment and the class centroid, ACA is suitable for EEG sequence analysis. An effective algorithm called Dynamic Programming Search (DPSearch) [17] is proposed to minimize ACA energy function. Process of using ACA to optimize questionnaire-based annotations is described in Algorithm 1.

2.3   Hierarchical Aligned Clustering Analysis

In this section, we introduce HACA optimization (see Algorithm 2), which extends ACA with hierarchical implementation [17] to optimize questionnaire-based annotations. The core idea is using ACA to find an optimal segmentation in a smaller segment length constraint n_{max1} in the first level, then propagating the solution to the second level by treating segments in the first level as temporal series frames. After calculating the "frame" kernel matrix in the second level, ACA is applied in the second level to gain the segmentation of longer temporal scale in the constraint of n_{max2}. The generalized dynamic time alignment kernel (GDTAK) is used for calculating kernel matrix based on segmented frames.

Note that both ACA and HACA share the same energy function minimizing algorithm, ACA minimizes distance among sampling points, whereas HACA

Algorithm 1. ACA Optimization

ACA Optimization $(X, label)$;

Parameter: EEG segment length constraint n_{max} and number of emotion
classes k

Input   : EEG sequence X and questionnaire-based EEG annotations $label$

Output   : Optimized EEG annotations $label_{new}$

Construct indicator matrix G and segment vector s;

Construct kernel matrix K from distance matrix of X;

$s_{new} \leftarrow s$;

do

   | $s \leftarrow s_{new}$;

   | Use $DPSearch(G, s, K)$ to obtain s_{new} and G_{new};

while $s_{new} \neq s$;

$m + 1 \leftarrow length(s_{new})$;

for $i = 1$ *to* m **do**

   | **for** $j = s_{new}(i)$ *to* $s_{new}(i+1)$ **do**

      | Create $label_{new}(j) \leftarrow G(i)$;

   | **end**

end

Algorithm 2. HACA Optimization

HACA Optimization $(X, label)$;

Parameter: EEG segment length constraint in 1st level n_{max1}, EEG segment
length constraint in 2nd level n_{max2} and number of emotion
classes k

Input   : EEG sequence X and questionnaire-based EEG annotations $label$

Output   : Optimized EEG annotations $label_{new}$

Construct indicator matrix G and segment vector s;

Construct kernel matrix K from distance matrix of X;

Use ACA to optimize segmentation in first level: $(G, s) \leftarrow ACA(G, s, K)$;

Construct segmented frame kernel matrix $T \leftarrow GDTAK(K, s)$;

for $i = 1$ *to* $length(s) - 1$ **do**

   | Create $s_{hierarchy}(i) \leftarrow i$, $G_{hierarchy} \leftarrow G_j$;

end

Create $s_{hierarchy}(length(s)) \leftarrow length(s)$;

Use ACA to optimize segmentation in second level:

$(G_{hierarchy}, s_{hierarchy}) \leftarrow ACA(G_{hierarchy}, s_{hierarchy}, T)$;

for $j = 1$ *to* $length(s_{hierarchy}) - 1$ **do**

   | Create $s_{new}(j) \leftarrow s(s_{hierarchy}(j))$, $G_{new}(j) \leftarrow G_{hierarchy}(j)$;

end

Create $s_{new}(length(s_{hierarchy})) \leftarrow s(length(s))$;

Create $label_{new}$ based on G_{new} and s_{new};

---

minimizes distance among ACA optimized segments. Therefore, HACA replace DTAK with GDTAK. As for more details about ACA and HACA algorithms, please refer to the previous study [17].

3 Experiment and Result

3.1 EEG Dataset

We evaluate the performance of these approaches on a public dataset, SEED dataset[1][13], which consists of stimuli and EEG data. There are 15 emotional film clips which elicit three emotions: positive, neutral and negative. For each session, a 5-s hint for starting is given before clip and a 45-s self-assessment and a 15-s rest after. There are totally 45 subjects participating in the experiments, who are required to elicit their own corresponding emotions while watching the clips. EEG data are recorded with a 62-electrode cap according to the international 10–20 system using ESI Neuroscan system.

3.2 Data Preprocessing and Feature Extraction

For data preprocessing, we apply bandpass filter between 1 Hz and 75 Hz to process raw EEG data. 62-channel EEG signals are further down-sampled to 200 Hz to reduce the computational complexity. Then EEG features are calculated using short-term Fourier transform from pre-processed EEG segments with non-overlapping 1-s time window.

For feature extraction, differential entropy (DE) features as EEG features are used for emotion recognition [13], which shows superior performance when compared to the conventional power spectral density features. For a fixed length EEG segment, DE is equivalent to the logarithm energy spectrum in a certain frequency band [10]. Therefore, DE features can be calculated in five frequency bands: delta (1–4 Hz), theta (4–8 Hz), alpha (8–14 Hz), beta (14–31 Hz), and gamma (31–50 Hz). The total dimension of a 62-channel EEG segment is 310.

3.3 Questionnaire-Based Annotations Optimization using HACA

In this section, we carry out experiments to evaluate the effectiveness of ACA (HACA) for optimizing questionnaire-based annotations. There is problem that if the annotations are optimized by temporal clustering methods, relabeled classes have higher similarity within each class, classification on these EEG data could be easier, which will cause inaccurate results. In order to avoid this problem, we use first 9 sessions for training and rest 6 sessions for testing. Only training data are clustered by ACA and HACA, and testing data annotations remain questionnaire-based to compare the performance of different annotation methods. Unlike typical clustering problem which determines the intrinsic grouping in a collection of unlabeled data, we make use of questionnaire-based annotations initialization to rectify the mislabeled annotations. Figure 2 shows that HACA annotation reaches the peak accuracy when $n_{max1} = 15$ and $n_{max2} = 4$.

Figure 3 shows the average accuracies of questionnaire-based (original), ACA-based, HACA-based annotation methods for total 45 experiments. The

---

[1] http://bcmi.sjtu.edu.cn/~seed/index.html.

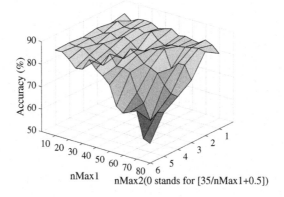

Fig. 2. Classification accuracy surface of HACA method using questionnaire-based initiation in 1st level and 1st level annotation optimization-based initiation in 2nd level.

questionnaire-based annotation achieves the moderate performance with an average accuracy of only 85.13%. HACA outperforms the other two approaches with an average accuracy of 89.70%.

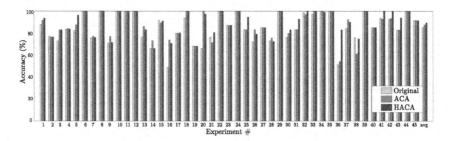

Fig. 3. Accuracies of questionnaire-based annotation (original), ACA-based annotation, HACA-based annotation on 45 experiments.

In supervised learning, the classification accuracy of mislabeled testing data will be aberrantly low [6], based on this observation, for investigating the relationship between mislabeled samples and classification accuracy, we employ 5-fold cross validation, which uses EEG sequences from 3 consecutive emotional film clips as testing data and the rest from 12 film clips as training data to compute the average classification accuracy (*curve Red*) of every sampling point over 45 experiments. The annotation change ratios (*curve Blue*) of every sampling point over 45 experiments are calculated after HACA annotation optimization. As shown in Fig. 4, the sampling points with low classification accuracy have high annotation change ratio. This comparison result shows HACA annotation optimization can efficiently rectify inaccurate annotations of EEG data.

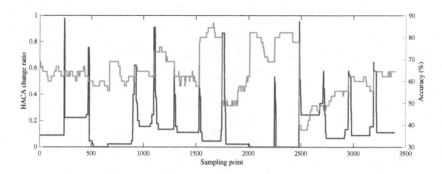

Fig. 4. Comparison between annotation change ratio and average classification accuracy. (Color figure online)

Figure 5 shows part of EEG annotations of 45 experiments using HACA, which indicates annotation variations across different experiments. To be specific, the classification accuracy of Experiment 16 and Experiment 20 improves from 48.99% to 70.59% and from 66.04% to 97.49%, respectively, which demonstrate the effectiveness of HACA annotation optimization compared with conventional questionnaire-based annotation.

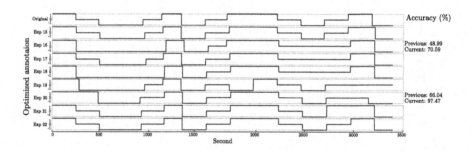

Fig. 5. Part of HACA EEG annotation optimization results of 45 experiments.

The average change ratios for three labels are shown in Table 1, in which negative emotions are often mislabeled as neutral (10%) and vice versa (10%).

Table 1. Average label change ratio of 45 experiments for HACA annotation

Rectified	Original		
	Positive	Neutral	Negative
Positive	0.90	0.03	0.04
Neutral	0.05	0.87	0.10
Negative	0.05	0.10	0.86

These results are consistent with previous finding that negative emotion is often confused with neutral emotion for EEG [7]. In summary, the experimental results demonstrate the efficiency of HACA approach for optimizing questionnaire-based EEG annotations.

4   Conclusion and Future Work

In this paper, we have adopted two temporal clustering methods, ACA and HACA to optimize the original questionnaire-based annotations. The experimental results have shown that ACA-based annotation achieves 2.59% average accuracy improvement and HACA-based annotation performs better with a 4.53% improvement. The comparison of HACA change ratio and performance improvement over original classification accuracy shows our annotation methods rectify inaccurate annotations.

Our future work will focus on testing performance of ACA and HACA annotation methods in larger real-world data sets, and comparing them with other existing works on mislead sampling filtering.

Acknowledgments. This work was supported in part by grants from the National Key Research and Development Program of China (Grant No. 2017YFB1002501), the National Natural Science Foundation of China (Grant No. 61673266), the Major Basic Research Program of Shanghai Science and Technology Committee (Grant No. 15JC1400103), ZBYY-MOE Joint Funding (Grant No. 6141A02022604), and the Technology Research and Development Program of China Railway Corporation (Grant No. 2016Z003-B).

References

1. Brodley, C.E., Friedl, M.A.: Identifying mislabeled training data. J. Artif. Intell. Res. **11**, 131–167 (1999)
2. Frénay, B., Verleysen, M.: Classification in the presence of label noise: a survey. IEEE Trans. Neural Netw. Learn. Syst. **25**(5), 845–869 (2014)
3. Guan, D., Yuan, W., Lee, Y.K., Lee, S.: Nearest neighbor editing aided by unlabeled data. Inf. Sci. **179**(13), 2273–2282 (2009)
4. Guan, D., Yuan, W., Ma, T., Khattak, A.M., Chow, F.: Cost-sensitive elimination of mislabeled training data. Inf. Sci. **402**, 170–181 (2017)
5. Kim, K.H., Bang, S.W., Kim, S.R.: Emotion recognition system using short-term monitoring of physiological signals. Med. Biol. Eng. Comput. **42**(3), 419–427 (2004)
6. Lam, C.P., Stork, D.G.: Evaluating classifiers by means of test data with noisy labels. In: IJCAI, pp. 513–518 (2003)
7. Lu, Y., Zheng, W.L., Li, B., Lu, B.L.: Combining eye movements and EEG to enhance emotion recognition. In: IJCAI, pp. 1170–1176 (2015)
8. Sáez, J.A., Galar, M., Luengo, J., Herrera, F.: A first study on decomposition strategies with data with class noise using decision trees. In: Corchado, E., Snášel, V., Abraham, A., Woźniak, M., Graña, M., Cho, S.-B. (eds.) HAIS 2012. LNCS, vol. 7209, pp. 25–35. Springer, Heidelberg (2012). doi:10.1007/978-3-642-28931-6_3

9. Schuller, B., Rigoll, G., Lang, M.: Hidden Markov model-based speech emotion recognition. In: Proceedings of the 2003 International Conference on Multimedia and Expo, vol. 1, p. I–401. IEEE (2003)
10. Shi, L.C., Jiao, Y.Y., Lu, B.L.: Differential entropy feature for EEG-based vigilance estimation. In: 35th Annual International Conference of the IEEE, Engineering in Medicine and Biology Society, pp. 6627–6630. IEEE (2013)
11. Shimodaira, H., Noma, K.I., Nakai, M., Sagayama, S., et al.: Dynamic time-alignment kernel in support vector machine. In: NIPS, vol. 2, pp. 921–928 (2001)
12. Wang, X.W., Nie, D., Lu, B.L.: Emotional state classification from EEG data using machine learning approach. Neurocomputing **129**, 94–106 (2014)
13. Zheng, W.L., Lu, B.L.: Investigating critical frequency bands and channels for EEG-based emotion recognition with deep neural networks. IEEE Trans. Auton. Ment. Dev. **7**(3), 162–175 (2015)
14. Zheng, W.L., Zhu, J.Y., Lu, B.L.: Identifying stable patterns over time for emotion recognition from EEG. IEEE Trans. Affect. Comput. (2017). doi:10.1109/TAFFC.2017.2712143
15. Zheng, W.L., Zhu, J.Y., Peng, Y., Lu, B.L.: EEG-based emotion classification using deep belief networks. In: IEEE International Conference on Multimedia and Expo, pp. 1–6. IEEE (2014)
16. Zhou, F., De la Torre, F., Hodgins, J.K.: Aligned cluster analysis for temporal segmentation of human motion. In: 8th IEEE International Conference on Automatic Face & Gesture Recognition, pp. 1–7. IEEE (2008)
17. Zhou, F., De la Torre, F., Hodgins, J.K.: Hierarchical aligned cluster analysis for temporal clustering of human motion. IEEE Trans. Pattern Anal. Mach. Intell. **35**(3), 582–596 (2013)
18. Zhu, X., Wu, X.: Class noise vs. attribute noise: a quantitative study. Artif. Intell. Rev. **22**(3), 177–210 (2004)

Identify Non-fatigue State to Fatigue State Using Causality Measure During Game Play

Yuying Zhu[1], Yi-Ning Wu[2($\boxtimes$)], Hui Su[1], Sanqing Hu[1], Tong Cao[1],
Jianhai Zhang[1], and Yu Cao[2]

[1] College of Computer Science, Hangzhou Dianzi University,
Hangzhou 310018, Zhejiang, China
xjjhzyy@126.com, jhzhang@hdu.edu.cn
[2] Department of Physical Therapy, University of Massachusetts Lowell,
Lowell, MA 01854, USA
yining_wu@uml.edu, ycao@cs.uml.edu

Abstract. In this paper, Granger causality (GC) and New causality (NC) analysis methods are applied in frequency domain to reveal causality changes from non-fatigue state to fatigue state with EEG signals during video game-playing. EEG signals were recorded while a subject was playing video-games. Results show that fatiguing phenomenon was observed in 15 subjects using NC in [20, 30] Hz while only 13 subjects were identified with GC for comparison. The NC further showed the bi-directional causality changes between the two hemispheres during unilateral forearm movements. We noticed that half of the subjects had predominant active hemisphere while the other half showed the opposite, especially to the ones with higher fatigue level. The findings demonstrate that the NC method is better than the GC to reveal causal influence between homologous motor areas of active and inactive hemispheres in this study.

Keywords: Causality · EEG · Fatigue state · Frequency domain · Neurorehabilitation

1 Introduction

Fatigue is attributed to either a build-up of metabolic variables within the muscle fibers (peripheral fatigue) or a decline in the efficacy of motor neurons (central fatigue). Both of them contribute to decline motor performance in people with and without neurological disorders [1,2]. As fatigue begins to accumulate, motor neurons become less receptive to synaptic input, which impairs the function of the muscle. Therefore fatigue has been a concern during neurorehabilitation that involves high intensity with few rest periods or even the boredom and stress [3]. When the high repetition of movement is required to engage neural plasticity during neurorehabilitation or motor skill learning, excess practice can cause fatigue that might be detrimental (causing a longer period to recover and interrupting the neurorehabilitaiton routine). Due to the less desirable effects of

© Springer International Publishing AG 2017
D. Liu et al. (Eds.): ICONIP 2017, Part IV, LNCS 10637, pp. 581–588, 2017.
https://doi.org/10.1007/978-3-319-70093-9_61

fatigue on motor performance, it is valuable to find a reliable way to monitor the development of fatigue and to determine a distinct threshold of fatigue during neurorehabilitation.

In the last few years, determining the dynamic relationships between EEG signals and progression of mental fatigue has been a very important research topic [4,5]. And some studies have shown relationships between EEG signals and muscle fatigue [6,7]. Furthermore, the phenomenon that hemiplegic persons often show associated movements (contralateral homologous movement occurs while a person intends to move one side of body), which is also seen in healthy people during fatigue. Therefore examining the causal influences between two hemispheres might provide an opportunity to detect fatigue development more intuitively in nuerorehabilitation. Granger causality (GC) [8,9] is one of the most popular approaches which has been adopted in neuroscience [10]. However, it is unable to reveal the true causality and only suitable for examining pairwise combination. Hu et al. [11] addressed the limitations of the Granger-like causality methods and proposed New Causality (NC) that had been validated the accuracy and rationales in neuroscience [12].

Above all, the existing causality analysis techniques [13] may provide a possible way to employ EEG signals for fatigue determination during neurorehabilitation. In this paper, we use GC and NC to reveal the fatigue state of subjects during repetitive movements and compare the results of GC and NC followed by more detailed comparisons in the frequency domain.

2   Granger Causality (GC) and New Causality (NC)

We consider two stochastic time series to be jointly stationary. Individually, under fairly general conditions, each time series admits an autoregressive representation and their joint representations are described as

$$
\begin{cases}
X_{1,t} = \sum_{j=1}^{m} a_{11,j} X_{1,t-j} + \sum_{j=1}^{m} a_{12,j} X_{2,t-j} + \eta_{1,t} \\
X_{2,t} = \sum_{j=1}^{m} a_{21,j} X_{1,t-j} + \sum_{j=1}^{m} a_{22,j} X_{2,t-j} + \eta_{2,t}
\end{cases}
\tag{1}
$$

where $t = 0, 1, \cdots, N$, the noise terms are uncorrelated over time (that is, let $\theta_{k,t} = \eta_{1,t-k}$ and $\xi_{k,t} = \eta_{2,t-k}, k = 1, 2, \cdots m'$, then $E[\theta_i \theta_j] = E[\xi_i \xi_j] = E[\theta_i \eta_2] = E[\xi_i \eta_1] = 0$ where $E[\cdot]$ is the expectation value of a variable, $i,j = 1, 2, \cdots, m', i \neq j$), ϵ_i and η_i have zero means and variances of $\sigma_{\epsilon_i}^2$, and $\sigma_{\eta_i}^2, i = 1, 2$. The covariance between η_1 and η_2 is defined by $\sigma_{\eta_1 \eta_2} = cov(\eta_1, \eta_2)$ [11]. For a practical system, a general approach for determining the order of the MVAR model is the AIC–Akaike Information Criterion [14].

2.1   GC in Frequency Domain

Granger causal influence from X_2 to X_1 in frequency domain is defined by $I_{X_2 \to X_1}(f) =$

$$- \ln\left(1 - \frac{(\sigma_{\eta_2}{}^2 - \frac{\sigma_{\eta_1 \eta_2}{}^2}{\sigma_{\eta_1}{}^2})| H_{12}(f) |^2}{S_{X_1 X_1}}\right) \in [0, +\infty) \qquad (2)$$

where $H_{12}(f) = \frac{1}{det(A)}\overline{a}_{12}(f), \overline{a}_{hl}(f) = -\sum\limits_{j=1}^{m} a_{hl,j} e^{-i2\pi fk}, h, l = 1, 2, h \neq l.$

Similarly, we define Granger causal influence from X_1 to X_2 by $I_{X_1 \to X_2}(f) =$

$$- \ln\left(1 - \frac{(\sigma_{\eta_1}{}^2 - \frac{\sigma_{\eta_1 \eta_2}{}^2}{\sigma_{\eta_2}{}^2})| H_{21}(f) |^2}{S_{X_2 X_2}}\right) \in [0, +\infty) \qquad (3)$$

where $H_{21}(f) = \frac{1}{det(A)}\overline{a}_{21}(f).$

2.2   NC in Frequency Domain

Implementing Fourier transformation on both sides of (1) leads to

$$\begin{cases} X_1(f) = a_{11}(f)X_1(f) + a_{12}(f)X_2(f) + \eta_1(f) \\ X_2(f) = a_{21}(f)X_1(f) + a_{22}(f)X_2(f) + \eta_2(f) \end{cases} \qquad (4)$$

where $a_{lj}(f) = \sum\limits_{k=1}^{m} a_{lj,k} e^{-i2\pi fk}, i = \sqrt{-1}, l, j = 1, 2.$

From (5), one can see that contributions to $X_1(f)$ include $a_{11}(f)X_1(f)$, $a_{12}(f)X_2(f)$ and noise term $\eta_1(f)$. So NC from X_2 to X_1 in frequency domain is defined as $N_{X_2 \to X_1}(f) =$

$$\frac{| a_{12}(f) |^2 S_{X_2 X_2}(f)}{| a_{11}(f) |^2 S_{X_1 X_1}(f) + | a_{12}(f) |^2 S_{X_2 X_2}(f) + \sigma_{\eta_1}^2} \qquad (5)$$

Similarly, NC from X_1 to X_2 in frequency domain is defined as $N_{X_1 \to X_2}(f) =$

$$\frac{| a_{21}(f) |^2 S_{X_1 X_1}(f)}{| a_{21}(f) |^2 S_{X_1 X_1}(f) + | a_{22}(f) |^2 S_{X_2 X_2}(f) + \sigma_{\eta_2}^2} \qquad (6)$$

3   Experimental Method

This study was approved by the institutional review board of University of Massachusetts Lowell, MA, USA. We designed an experiment to record electroencephalography (EEG) signals of non-fatigue and fatigue states. Twenty subjects were recruited in the study including ten typically developing children (six females and four males, 10 ± 2.4 years old) and ten healthy adults (five females and five males, 51.1 ± 16 years old). We marked ten children as C01 to C10, and ten adults as A01 to A10.

During the experiment, We had subjects play several video-games by rotating the forearm repetitively with the non-dominant hand. Subjects were seated in front of the Forearm-Intelli Stretcher comfortably without moving or tilting the head and the forearm was kept slight shoulder flexion ($\sim 30°$) and elbow flexion ($90 \sim 100°$). We recorded EEG sampled at 500 Hz with the wireless EEG headset device. The EEG signals of each game-playing were recorded as one session which lasted three minutes. The record was stopped once the subject felt tired or the tenth session ended [15]. At the end of each session, we asked the subject to score the fatigue level using the Borg perceived exertion scale [16], which demonstrated each subject developed certain level of fatigue at the end of the experiment.

In this study, we analyzed the data as follows: firstly, we calculated GC and NC in [20, 30]Hz between C3 and C4, and compared the strength changes of NC with GC from the non-fatigue state to fatigue state of each subject. Secondly, we compared the strength of causal influence from the active hemisphere to its contralateral hemisphere. The active hemisphere here refers to the hemisphere contralateral to the moving arm while the contralateral hemisphere is used to describe the hemisphere contralateral to the active hemisphere. To investigate the trends of changing of each subject between two hemispheres, we calculated $NC_{active \to contra}$ and $NC_{contra \to active}$. $NC_{active \to contra}$ is the influence strength of active hemisphere to the contralateral hemisphere and vise versa. For all analyzed EEG data, the GC and NC values of each session were calculated under a significance threshold $p = 0.05$ (Wilcoxon signed rank test). After applying AIC to determine the choices of m, we got $m = 10$ as the order of the estimated MVAR models.

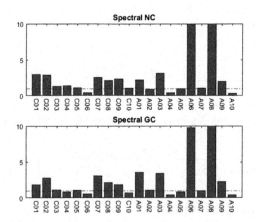

Fig. 1. Ratios of non-fatigue NC and GC to fatigue NC and GC respectively. NC on the interval [20, 30] Hz reflects 75% of subjects showing decreased causal influence from the active hemisphere to the contralateral hemisphere when fatigue developed (ratio value <1). GC reflects 65% of subjects showing decreased causal influence from the active hemisphere to the contralateral hemisphere. Moreover, there is no statistically significant decrease in GC.

4   Results

The NC and GC within [20, 30] Hz were further analyzed and the results are shown as follow. Figure 1 demonstrates the strength of NC and GC changes from the non-fatigue state to fatigue state of each subject. The changes of NC reveals 15 among 20 subjects (15/20 = 75%), which including 6 adults (6/10 = 60%) and 9 children (9/10 = 90%). The NC strength from the active hemisphere to the contralateral hemisphere within [20, 30] Hz was significantly higher in the

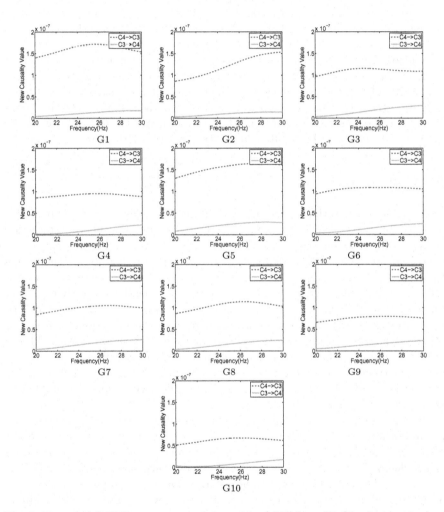

Fig. 2. Pattern 1.1: $NC_{active \rightarrow contra}$ of the subject C07 from G1 (the first session) to G10 (the last session). One can see that (i) in the interval [20, 30] Hz the dotted curve is above the solid curve in each session, this demonstrates that the causal influence from C4 to C3 is always larger than that from C3 to C4, and (ii) the area under the dotted curve in the interval [20, 30] Hz decreased toward the end of the experiment, while the area under the solid curve did not change much.

non-fatigue state than in the fatigue state (p = 0.014). In contrast, the changes of GC reveals 13 among 20 subjects (13/20 = 65%), which including 6 adults (6/10 = 60%) and 7 children (7/10 = 90%). However, the GC strength can't show significant changes (p = 0.145). Furthermore, two major patterns with two sub-patterns about the inter-hemispheric interaction were observed among the subjects by contrasting $NC_{active \to contra}$ and $NC_{contra \to active}$. Here we choose one of the sub-patterns (Pattern 1.1) to show our results.

Pattern 1.1. Dominant $NC_{active \to contra}$ with non-reversed pattern: The area under the $NC_{active \to contra}$ curve is larger than $NC_{contra \to active}$ in every session. Subjects C02, C03, C06, C07, C10, A03, and A07 are the representatives. The exemplary NC changes of C07 is shown in Fig. 2. Subject C07 played the video-games with left hand, therefore the electrode of active hemisphere is channel C4. One can see that in the interval [20, 30] Hz the dotted curve is above the solid curve in each session, which demonstrates that the causal influence from the active hemisphere to the contralateral hemisphere is always larger than the other way around. Particularly, the area under the dotted curve in the interval [20, 30] Hz is the smallest in the last session (G10).

5 Conclusions

In this study, we find that the subject felt fatigue in the last session can be better revealed by the NC method. And we notice the interhemispheric interaction changes which demonstrate that the feasibility of using NC method can reveal causal influence between C3 and C4 in our experiments. In addition, the results we observed further support the currently accumulating evidences that fatigue related to movement is not restricted to the peripheral factors that prevent force generation of muscles, but also the supraspinal structures especially the motor cortex [17,18]. Insufficient neural drive from the supraspinal center was observed under motor fatigue state [19]. The insufficient neural drive could also alter the inhibition from the active motor area to the homologous motor area of the contralateral hemisphere which has shown in the transcranial magnetic stimulation study [20]. Our results are in concert with what Matsuura and Ogata demonstrated in their study.

In conclusion, our paper demonstrates the feasibility of using EEG to monitor the motor fatigue while performing repetitive forearm movements. More specifically, utilizing NC method in frequency domains provides us the information of interhemispheric interaction during aforementioned movement. These information can be potentially used to adjust the training parameters and dosages of neurorehabilitation such as robotic-aided therapy or high intensity training in the individual with neurological disorder to prevent the motor fatigue and enhance the effectiveness of movement treatments. Furthermore, utilizing fatigue to guide neurorehabilitation may shed a light on determining the optimal dosage for intensive practice but not yet available.

Acknowledgments. This work was supported in part by the National Natural Science Foundation of China under Grant 61473110 and Grant 61633010, in part by the International Science and Technology Cooperation Program of China under Grant 2014DFG12570, and in part by the U.S. National Science Foundation under Grant 1156639 and Grant 1229213.

References

1. Allen, D.G., Westerblad, H.: Role of phosphate and calcium stores in muscle fatigue. J. Physiol. **536**(3), 657–665 (2001)
2. Abd-Elfattah, H.M., Abdelazeim, F.H., Elshennawy, S.: Physical and cognitive consequences of fatigue: a review. J. Adv. Res. **6**(3), 351–358 (2015)
3. Kacem, A., Ftaiti, F., Chamari, K., Dogui, M., Grélot, L., Tabka, Z.: EEG-related changes to fatigue during intense exercise in the heat in sedentary women. Health **6**(11), 1277–1285 (2014)
4. Jap, B.T., Lal, S., Fischer, P., Bekiaris, E.: Using EEG spectral components to assess algorithms for detecting fatigue. Expert Syst. Appl. **36**(2), 2352–2359 (2009)
5. Chen, C., Li, K., Wu, Q., Wang, H., Qian, Z., Sudlow, G.: EEG-based detection and evaluation of fatigue caused by watching 3DTV. Displays **34**(2), 81–88 (2013)
6. Yang, Q., Fang, Y., Sun, C.K., Siemionow, V., Ranganathan, V.K., Khoshknabi, D., et al.: Weakening of functional corticomuscular coupling during muscle fatigue. Brain Res. **1250**, 101–112 (2009)
7. Yang, Q., Siemionow, V., Yao, W., Sahgal, V., Yue, G.H.: Single-trial EEG-EMG coherence analysis reveals muscle fatigue-related progressive alterations in corticomuscular coupling. IEEE Trans. Neural Syst. Rehabil. Eng. **18**(2), 97–106 (2010)
8. Wiener, N.: The theory of prediction. Mod. Math. Eng. **1**, 125–139 (1956)
9. Granger, C.W.: Investigating causal relations by econometric models and crosss pectral methods. Econom.: J. Econom. Soc. **37**(3), 424–438 (1969)
10. Ding, M., Chen, Y., Bressler, S.L.: Granger causality: basic theory and application to neuroscience. In: Handbook of Time Series Analysis: Recent Theoretical Developments and Applications, p. 437 (2006)
11. Hu, S., Dai, G., Worrell, G.A., Dai, Q., Liang, H.: Causality analysis of neural connectivity: critical examination of existing methods and advances of new methods. IEEE Trans. Neural Netw. **22**(6), 829–844 (2011)
12. Hu, S., Jia, X., Zhang, J., Kong, W., Cao, Y.: Shortcomings/limitations of blockwise granger causality and advances of blockwise new causality. IEEE Trans. Neural Netw. Learn. Syst. **27**(12), 2588–2601 (2016)
13. Granger, C.W.: Some recent development in a concept of causality. J. Econom. **39**(1–2), 199–211 (1988)
14. Akaike, H.: A new look at the statistical model identification. IEEE Trans. Autom. Control **19**(6), 716–723 (1974)
15. Wu, Y.N., Hwang, M., Ren, Y., Gaebler-Spira, D., Zhang, L.Q.: Combined passive stretching and active movement rehabilitation of lower-limb impairments in children with cerebral palsy using a portable robot. Neurorehabilitation Neural Repair **25**(4), 378–385 (2011)
16. Borg, G.: Borgs Perceived Exertion and Pain Scales. Human Kinetics, Champaign (1998)
17. Taylor, J.L., Todd, G., Gandevia, S.C.: Evidence for a supraspinal contribution to human muscle fatigue. Clin. Exp. Pharmacol. Physiol. **33**(4), 400–405 (2006)

18. Hou, L.J., Song, Z., Pan, Z.J., Cheng, J.L., Yu, Y., Wang, J.: Decreased activation of subcortical brain areas in the motor fatigue state: an fMRI study. Front. Psychol. **7** (2016). Article 1154

19. Amann, M., Dempsey, J.A.: Locomotor muscle fatigue modifies central motor drive in healthy humans and imposes a limitation to exercise performance. J. Physiol. **586**(1), 161–173 (2008)

20. Matsuura, R., Ogata, T.: Effects of fatiguing unilateral plantar flexions on corticospinal and transcallosal inhibition in the primary motor hand area. J. Physiol. Anthropol. **34**(1), 4 (2015)

A Graph Theory Analysis on Distinguishing EEG-Based Brain Death and Coma

Gaochao Cui[1(✉)], Li Zhu[2,4], Qibin Zhao[3], Jianting Cao[1],
and Andrzej Cichocki[2]

[1] Saitama Institute of Technology, Fusaiji 1690, Fukaya,
Saitama 3690293, Japan
e4001hbx@sit.ac.jp
[2] Brain Science Institute, RIKEN, Hirosawa 2-1, Wakoshi,
Saitama 3510198, Japan
[3] Advanced Intelligence Project, RIKEN, 1-4-1 Nihonbashi,
Chuo-ku, Tokyo 1030027, Japan
[4] Department of Cognitive Science, Xiamen University,
No. 422, Siming South Road, Xiamen 361005, Fujian, China

Abstract. Electroencephalogram (EEG) is always used to diagnosis the patients consciousness clinically because it is safe and easy to be record from patients. The aim of this paper is to analysis the relations between each channel in order to find out the brain network of brain death and coma patients particularity. In this paper, we use 10 adult patients' EEG data to calculate the partial directed coherence (PDC) and build the average brain network for the two groups' data after t-test based on the PDC results. Results showed that, these two clinical data are at most difference in the network parameters of degree, centrality and cluster coefficient as the threshold of PDC is set of 0.3. The time-varying connectivity could lead to better understanding of non-symmetric relations between different EEG channels and application in prediction of patients in brain death or coma state.

Keywords: Brain death · Coma · EEG · Cross coherence · PDC · Brain network

1 Introduction

Nowadays, EEG is quite an important tool in clinical practice to do the brain related disease diagnosis [1,2]. Machine learning methods have been used into EEG analysis [3,4]. So to have a good use of signal processing and statistics methods into clinical EEG analysis is quite important because more potential knowledge could be found out through advanced data processing methods which could be applied into adding clinical criterions.

Brain death is taken as the standard into termination of a patient's life so the fast and accuracy diagnosis of brain death is very important in clinics [5]. As the deep coma clinical expressions are similar to brain death so there are several

© Springer International Publishing AG 2017
D. Liu et al. (Eds.): ICONIP 2017, Part IV, LNCS 10637, pp. 589–595, 2017.
https://doi.org/10.1007/978-3-319-70093-9_62

wrong diagnosis cases in recent years [6]. The main reasons include that the diagnosis standard is not clear and the limitation criterions for brain death. So that, more effective methods should be used on differentiating the brain death and coma.

Brain can be seen as a complex system because hundreds of millions of nerve cells are connected to each other. There are different abnormal connectivity in certain brain diseases such as Alzheimer, depressive disorder, autism and so on [7–9].

Based on the related work, the dynamic connectivity for brain death and coma EEG analysis would be a good way to find out the features for these two kinds of patients' data. Cross coherence is the method to estimate the connectivity between two channels varying time. Moreover, the brain network analysis based on graph theory is an important means [10]. In the brain network, the different connectivity ways show different flowing roads within different information. It would be quite significant to explore the information transformation mechanisms among different brain areas for brain death and coma patients network structure based on graph theory.

The structure of this paper includes the signal preprocessing, cross coherence analysis and the brain network based on graph theory construction. We wish to have some useful analysis results to aid the diagnosis in clinics, improving the speed and accuracy of distinguishing these two diseases.

2   Methods

2.1   Brain Network Construction

PDC Algorithm. PDC is proposed to express Granger Causality in a new way based on MVAR (multivariate autoregressive) model. Normalized $PDC_{x_i \to x_j}$ accounts for the proportion between x_i flowing to x_j and signals outgoing from x_i. $PDC_{x_i \to x_j}$ is nearer 0 and then it means no much connection while its value is greater than 0.1, which is regarded as the two channels connected to each other [11]. Specifically, the algorithm will be discussed below [12,13]. A time-varying N-variate AR process of order p could be expressed in formula 1.

$$
\begin{bmatrix} X_1 \\ X_2 \\ \cdot \\ \cdot \\ \cdot \\ X_k \end{bmatrix} = \sum_{i=1}^{p} A_r \begin{bmatrix} x_1(n-i) \\ x_2(n-i) \\ \cdot \\ \cdot \\ \cdot \\ x_k(n-i) \end{bmatrix} + \begin{bmatrix} \mu_1(t) \\ \mu_2(t) \\ \cdot \\ \cdot \\ \cdot \\ \mu_k(t) \end{bmatrix}
\tag{1}
$$

where μ is the noise vector and $\sum_{i=1}^{p} A_r$ is given by formula 2.

$$
A_r(n) = \begin{bmatrix} a_{11r} & \cdots & a_{1kr} \\ a_{21r} & \cdots & a_{2kr} \\ \vdots & \ddots & \vdots \\ a_{k1r} & \cdots & a_{kkr} \end{bmatrix}
\tag{2}
$$

Here, we process the brain death EEG data of 6 channels $k = 6$ so first the AR model of EEG in time domain should be solved. And then do the FFT on formula 2. The time-varying version of partial directed coherence is defined as formula 3,

$$PDC_{x_i \to x_j}(f) = \frac{A_{i,j}(f)}{\sqrt{a_j^H(f)a_j(f)}} \qquad (3)$$

where $a_j(f)$ is the jth column of the matrix $A(f)$.

$PDC_{x_i \to x_j}(f)$ is normalized where the higher value (taking 0.3 as the border) in a certain frequency band represents the linear influence from channel i to channel j. As a result, we will get a connectivity matrix (6 multiple 6) to test the information flow strength and direction between two channels in EEG data.

2.2   Brain Network Based on Graph Theory

In our paper, we take the 6 channels as nodes and the PDC value as the edge and direction to build the brain topological graph network for brain death. Through network map, we can analyze the characteristics for coma and brain death groups and then to locate the abnormal connectivity in brain death.

In the graph, the connectivity between channels is illustrated by connectivity matrix. We have 6 nodes in the network so the size of adjacency matrix is $6 * 6$. Suppose the adjacency matrix is B and b_{ij} is the element of B. If $b_{ij} = 1$ and then there is connection between node i and j. Else $b_{ij} = 0$ and then there is no connection from node i to j. So the constructed brain network based on PDC value is directed weighted network. But some of the parameters of network would be transferred into two-value one to be analyzed and calculated. We use three topological graph parameters to do quantificational estimation.

(1) Degree: for our network graph, degree has in-degree and out-degree respectively. In-degree represents the number of edges importing into the certain node; out-degree shows the number of edges out of the node. The value degree could be used as the evaluation of the node importance level in the network.

(2) Clustering coefficient: the clustering degree could be measured by clustering coefficient, showing the probability of the connectivity between each node adjacency edge. The formulas 4 and 5 is taken as the calculated function for node i and the average whole network.

$$L_i = \frac{2connectede_i}{k_i(k_i - 1)} \qquad (4)$$

$$L = \frac{\sum_{i \in V} L_i}{N} \qquad (5)$$

(3) Betweenness centrality: the role and importance could be described by betweenness centrality. The value of it is bigger, the more important of the

node in the network (named core node). The solved function is formula 6. The ε_{jk} means the number of optimal path from node j to node k.

$$N(i) = \sum_{j \neq I \neq k \in G} \frac{\varepsilon_{jk}(i)}{\varepsilon_{jk}},$$ (6)

$$\varepsilon_{jk}(i), j \xrightarrow{i} k$$

In our paper, the specific steps for brain network construction of these two groups data based on PDC value are given as follows. There are 10 adults patients EEG data for the two groups. Each patient has a corresponding PDC matrix $6 * 6$. Set different threshold values $\omega_0 (\omega_{min} \leq \omega_0 \leq \omega_{max})$ for PDC, where $\omega_{min} = 0.1$ and $\omega_{max} = 1.79$. The single sampling t-test is applied to determine every group functional connectivity and then the final brain network structure will be fixed.

Suppose the sample mean value is $\bar{\omega}$ and the zero hypothesis is given as follows:

$$H_0 : \omega \leq \omega_0, \ H_1 : \omega > \omega_0$$

The test statistic function is given in formula 7.

$$t = \frac{\bar{\omega} - \omega_0}{\sqrt{\frac{s^2}{n}}}$$ (7)

If t is in the reject domain and then the connectivity strength between each channel is bigger than ω_0, which means that there is connection between the corresponding nodes in the brain network; else there is no connection.

3   Experiments and Results

3.1   Experiment Data Recording and Preprocessing

The EEG data in our study were collected from one Grade A of level III hospital in Shanghai, China. The EEG data were directly from the ICU clinics so the noise

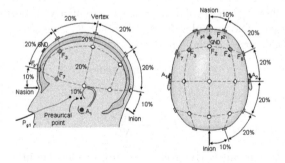

Fig. 1. Channel layout

would be high. The recording equipment is portable NeuroScan ESI-64 system. The electrode layout includes 6 channels such as Fp1, Fp2, F7, F3, F4 and F8 based on 10/20 system. A total of 10 patients (5 brain death and 5 coma) (Fig. 1).

The signal sampling is 1000 Hz and filtering ranges from 0.5 to 100 Hz. The procedure of noise remove includes: (1) the raw EEG is taken A1 and A2 as the references. (2) 0.5–30 Hz band pass to remove high frequency interference.

3.2 Partial Directed Coherence Result

Figure 2 shows the PDC results of the brain death group and coma group. In the result of coma group (Fig. 2b), the connectivity between channels are notably higher than brain death group (Fig. 2a). Especially in Fp1 and Fp2 of coma group, both of these two channels have strong connectivity with other channels.

(a) Brain death

(b) Coma

Fig. 2. PDC for brain death group and coma group.

3.3 Brain Network Connection Results

As the aforementioned, we use the three parameters to do quantity description and then we find that the difference between the two groups is the most clear at $\omega_0 = 0.3$. The difference description for the two groups is taken at $\omega_0 = 0.3$ below.

Table 1. In-degree of each node in brain network for two groups.

Groups	Channel locations					
	Fpl	Fp2	F7	F3	F4	F8
Brain death	4	3	2	3	0	4
Coma	1	5	3	4	2	3

Table 2. Out-degree of each node in brain network for two groups.

Groups	Channel locations					
	Fpl	Fp2	F7	F3	F4	F8
Brain death	3	1	2	2	1	2
Coma	0	2	4	2	2	3

Distribution of Degree. In-degree represents for the information importing into one certain node, taking as 'causality'; out-degree represents for the information flowing out of one node, taking as 'effect'. Table 1 is the distribution of in-degree and Table 2 is distribution of out-degree of the two brain networks.

Clustering Coefficient. The average clustering coefficient stands for the clustering degree of the whole network. If the value is higher and then the operation efficiency of corresponding network is higher. The mean clustering coefficient of average brain network for brain death and coma groups (6 nodes) is 0.775 and 0.851. It is clearly that the network efficiency of brain death is much lower than coma.

Betweenness Centrality. The constructed average brain network for the two groups based on t-test $p \leq 0.05$ is shown in Fig. 3 and the betweenness centrality important position node is labeled. There are 3 nodes (F7, F3 and F8) for brain death group and 4 nodes (FP1, F7, F3 and F8) for coma group. From the result, we can see that the brain death is more dispersed than coma, which means the clustering degree is decreasing for brain death and the importance of core node is dropping.

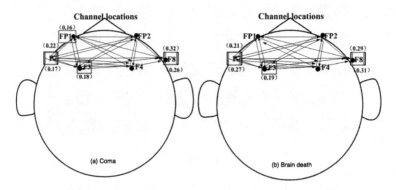

Fig. 3. Brain network connection for the brain death group and coma group.

4   Conclusion

In our paper, we analyze the dynamic connectivity for brain death and coma EEG data, build the brain network based on graph theory and compare the brain connectivity features for each group to differentiate the two groups. The graph theory analysis shows that the in-degree is more clustering than out-degree. The operation efficiency is decreasing in brain death than coma group. In summary, compared with the brain death and coma brain network, there are quite difference in these two groups EEG data. The core nodes in these two groups show variance which means the transferring in these two data. To have a deep learning in this topic will be helpful to find out the abnormal of connectivity in these two patients brain.

Acknowledgements. This work was supported in part by KAKENHI (15K15955 and 15H04002).

References

1. Szurhaj, W., Lamblin, M.D., Kaminska, A., Sediri, H.: EEG guidelines in the diagnosis of brain death. Clin. Neurophysiol. **45**(1), 97–104 (2015)
2. Malter, M.P., Bahrenberg, C., Niehusmann, P., Elger, C.E., Surges, R.: Features of scalp EEG in unilateral mesial temporal lobe epilepsy due to hippocampal sclerosis: determining factors and predictive value for epilepsy surgery. Clin. Neurophysiol. **127**(2), 1081–1087 (2016)
3. Brinkmann, B.H., Patterson, E.E., Vite, C., Vasoli, V.M., Crepeau, D., Stead, M., Howbert, J.J., Cherkassky, V., Wagenaar, J.B., Litt, B., Worrell, G.A.: Correction: forecasting seizures using bivariate intracranial EEG measures and SVM in naturally occurring canine epilepsy. PLoS ONE, **10** (2016)
4. Wang, X.W., Nie, D., Lu, B.L.: Emotional state classification from EEG data using machine learning approach. Neurocomputing **129**(10), 94–106 (2014)
5. Ad Hoc Committee of the Harvard Medical School to Examine the Definition of Brain Death. A definition of irreversible coma. JAMA **205**, 337–340 (1968)
6. Big Data for Healthcare. In: Chinese Academy of Social Sciences (2016)
7. Zhou, Y., Dougherty, J.H., Hubner, K.F., Bai, B., Cannon, R.L., Hutson, R.K.: Abnormal connectivity in the posterior cingulate and hippocampus in early Alzheimer's disease and mild cognitive impairment. Alzheimer's Dementia **4**(4), 265–270 (2008)
8. Zhang, J., Wang, J., Wu, Q., Kuang, W., Huang, X., He, Y., Gong, Q.: Disrupted brain connectivity networks in drug-naive, first episode major depressive disorder. Biol. Psychiatry **70**(4), 334–342 (2011)
9. Sam, W.: Distortions and disconnections: disrupted brain connectivity in autism. Brain Cogn. **75**(1), 18–28 (2011)
10. Bullmore, E., Sporns, O.: Complex brain networks: graph theoretical analysis of structural and functional systems. Nat. Rev. Neurosci. **10**, 186–198 (2009)
11. Sameshima, K., Baccala, L.A.: Using partial directed coherence to describe neuronal ensemble interactions. J. Neurosci. Methods **94**(1), 93–103 (1999)
12. Yasumasa Takahashi, D., Antonio Baccal, L., Sameshima, K.: Connectivity inference between neural structures via partial directed coherence. J. Appl. Stat. **34**(10), 1259–1273 (2007)
13. Youssofzadeh, V., Prasad, G., Naeem, M., Wong-Lin, K.: Temporal information of directed casual connectivity in multi-trial ERP data using Partial Granger Causality. Neuroinformatics **14**(1), 99–120 (2016)

EEG Comparison Between Normal and Developmental Disorder in Perception and Imitation of Facial Expressions with the NeuCube

Yuma Omori[1(✉)], Hideaki Kawano[1], Akinori Seo[1], Zohreh Gholami Doborjeh[2], Nikola Kasabov[2], and Maryam Gholami Doborjeh[2]

[1] Kyushu Insititute of Techonology, Kitakyushu 804-8550, Japan
omori.yuma780@mail.kyutech.jp, kawano@ecs.kyutech.ac.jp
[2] Knowledge Engineering and Discovery Research Institute,
Auckland University of Technology, Auckland 1142, New Zealand
zohreh.gholamidoborjeh@aut.ac.nz

Abstract. This paper is a feasibility study of using the NeuCube spiking neural network (SNN) architecture for modeling EEG brain data related to perceiving versus mimicking facial expressions. We collected EEG patterns during perception and imitation of facial expressions for each emotion. Comparing the collected data in perceiving and mimicking facial expressions, EEG patterns were very similar. This fact suggests that it seems that there are mirror neurons on facial expression in the human brain. Recently, some studies have been reported that the mirror neuron system does not work well in the case of subjects with brain disorders. In this study, we calculated differences between EEG patterns when we perceived facial expressions and mimicking facial expressions for healthy people and developmental disorders.

Keywords: EEG data · SNN · Mirror neuron system · Developmental disorders

1 Introduction

Facial expression is a fundamental tool in human communication. Understanding the facial expression effects on a third person is of a crucial importance to develop a comprehensive communication. Neuropsychological studies reported that communications through facial expressions are highly related to the Mirror Neuron System (MNS). MNS principle has been introduced in 1990s by Rizzolatti when he discovered similar areas of the brain became activated when a monkey performed an action and when a monkey observed the same action performed by another [1]. The MNS in human were also confirmed by an experiment using functional magnetic resonance imaging (fMRI) data [2]. Different facial expressions of emotion have different effects on the human brain activity.

D. Liu et al. (Eds.): ICONIP 2017, Part IV, LNCS 10637, pp. 596–601, 2017.
https://doi.org/10.1007/978-3-319-70093-9_63

The brain processes of perceiving an emotional facial expression and mimicking expression of the same emotion are spatio-temporal processes. The analysis of collecting Spatio-Temporal Brain Data (STBD) related to these processes could reveal personal characteristics or abnormalities that would lead to a better understanding of the brain processes related to the MNS. This can be achieved only if the models created from the STBD can capture both spatio and temporal components from this data. Despite of the rich literature on the problem, such models still do not exist.

Recently, a brain-inspired Spiking Neural Network (SNN) architecture, called NeuCube [4–6], has been proposed to capture both the time and the space characteristics of STBD, such as EEG, fMRI, DTI, etc. In contrast to traditional statistical analysis methods that deal with static vector-based data, the NeuCube has been successfully shown to be a rich platform for STBD mapping, learning, classification and visualization [7–9].

In this paper, we examined differences in brain activity patterns between healthy people and developmental disabilities by calculating the difference EEG data of facial expression task (both perceiving and mimicking) in two kinds of emotional faces (anger, happiness). The models allow for a detail understanding on the problem.

2   The NeuCube Spiking Neural Network Architecture

The NeuCube architecture [4] consists of: an input encoding module; a 3D recurrent SNN reservoir/cube (SNNc); an evolving SNN classifier. The encoding module converts continuous data streams into discrete spike trains. As one implementation, a Threshold Based Representation (TBR) algorithm is used for encoding. The NeuCube is trained in two learning stages. The first stage is unsupervised learning based on spike-timing-dependent synaptic plasticity (STDP) learning [10] in the SNNc. The STDP learning is applied to adjust the connection weights in the SNNc according to the spatiotemporal relations between input data variables. The second stage is a supervised learning that aims at learning the class information associated with each training sample. The dynamic evolving SNNs (deSNNs) [11] is employed as an output classifier. In this study, the NeuCube is used for modeling and learning of the case study EEG data corresponding to different facial expressions.

3   The Case Study STBD: EEG Data Evoked by Facial Expression

The subjects were 11 Japanese adult males, 10 healthy person and 1 development disabled in the case study of the facial expression task. As facial stimuli, JACFEE collection [12] was used, consisting of 56 color photographs of 56 different individuals. Each individual illustrates one of the two different emotions, i.e. anger, happiness. The collection is equally divided into male and female populations (28 males, 28 females).

During the experiments, subjects were wearing an EEG headset (Emotive EPOC+) which consists of 14 electrodes with the sampling rate of 128 Hz and the bandwidth is between 0.2 and 45 Hz.

The EEG data was recorded while the subjects were performing two different facial expression tasks. During the first presentation, subjects were instructed to perceive different facial expression images shown on a screen, and in the second presentation they were asked to mimic the facial expression images. We used five patterns of facial expression images per emotion in these experiments.

Each facial expression image was exposed for 5 s followed by randomly 5 to 10 s inter stimulus interval (ISI) as shown in Fig. 1.

Exposition: 5 seconds   Inter stimulus interval (ISI)
                                       5-10 seconds

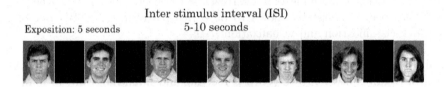

Fig. 1. The facial expression-related task: the order of emotion expressions is alternation of anger and happiness. Each subject watched 10 images during an experiment.

4   Analysis of the Spatiotemporal Connectivity in a Trained SNNc of a NeuCube Model

A 3D brain-like SNNc is created to map the Talairach brain template of 1471 spiking neurons [13,14]. The spatio-temporal data of EEG channels were encoded into spike trains and entered to the SNNc via 14 input neurons which spatial locations in the SNNc correspond to the 10–20 system location of the same channels on the scalp. The SNNc is initialized with the use of the "small world" connectivity [4].

We input EEG data obtained from five patterns of facial expressions images into one SNNc, and created a model for each subject. Table 1 shows that Neu-Cube parameter values used in the simulations.

Table 1. NeuCube parameter values used in the simulations.

Parameter	Value
TBR	0.5
Small world connectivity distance	2.5
STDP rate	0.01
Training iteration	1
Training time length	0.2

During the unsupervised STDP learning, the SNNc connectivity evolves with respect to the spike transmission between neurons. Stronger neuronal connection between two neurons means stronger information (spikes) exchanged between them.

Table 2 shows the numerical differences of EEG data between imitation and perception in 2 facial expressions (ANGRY and HAPPY) from each subject. The definitions of the L1-difference D_{L1} and the L2-difference D_{L2} are shown in Eqs. (1) and (2).

$$D_{L1} = \sum_i^N |w_i^{perceiving} - w_i^{mimicking}|, \tag{1}$$

$$D_{L2} = \sum_i^N (w_i^{perceiving} - w_i^{mimicking})^2, \tag{2}$$

where w_i represents weight parameter between neurons in SNNc.

Table 2. The difference between facial expressions (perceiving and mimicking) of 2 kinds of emotion (Angry and Happy).

Subject	D_{L1}		D_{L2}	
	ANGRY	HAPPY	ANGRY	HAPPY
A (Developmental disorder)	1897.8	1894.9	86.2	86.2
B (healthy)	1889.8	1880.9	85.6	85.2
C (healthy)	1890.7	1881.8	85.4	85.2
D (healthy)	1893.3	1874.1	85.5	84.5
E (healthy)	1888.7	1884.9	85.9	84.8
F (healthy)	1888.6	1881.4	85.1	85.4
G (healthy)	1885.4	1875.8	85.1	84.7
H (healthy)	1886.1	1892.4	85.2	85.4
I (healthy)	1886.1	1886.9	85.2	85.6
J (healthy)	1879.8	1872.9	84.6	84.5
K (healthy)	1890.5	1894.2	85.9	85.7
AVG of healthy	1887.9	1882.5	85.5	85.1
STDDV of healthy	3.75	7.24	0.39	0.45

As shown in Table 2, the difference in the developmental disorder is higher than the one in the healthy subjects. Especially, L1-difference in ANGRY and L2-difference in HAPPY show a significant difference between a developmental disorder and healthy subjects. Indeed the number of samples in the experiment is quite small, but we believe that this fact implicates a possibility to use the difference between weight connections learnt by the NeuCube as an index to evaluate a kind of social ability.

5   Conclusion

In this paper, we used the NeuCube architecture of SNN [4] for mapping and learning of EEG data recorded from subjects when they were performing a facial expression-related task. From Table 2, it was found that the person with developmental disability has a larger difference between EEG data of perception and imitation than healthy people. This finding can prove the principle of the mirror neurons in the human brain. This is only the first study in this respect. Further studies will require more subject data to be collected for a more models developed before the proposed method is used for cognitive studies and medical practice.

References

1. Gallese, V., Fadiga, L., Fogassi, L., Rizzolatti, G.: Action recognition in thepremotor cortex. Brain **119**, 593–609 (1996)
2. Lacoboni, M., Woods, R.P., Brass, M., Bekkering, H., Mazziotta, J.C., Rizzolatti, G.: Cortical mechanisms of human imitation. Science **186**, 2526–2528 (1999)
3. Binkofski, F., Buccino, G., Seitz, R.J., Rizzolatti, G., Freund, H.-J.: Afrontoparietal circuit for object manipulation in man: evidence from an fMRIstudy. Eur. J. Neurosci. **11**, 3276–3286 (1999)
4. Kasabov, N.: NeuCube: a spiking neural network architecture for mapping, learning and understanding of spatio-temporal brain data. Neural Netw. **52**, 62–76 (2014)
5. Tu, E., Kasabov, N., Yang, J.: Mapping temporal variables into the NeuCube for improved pattern recognition, predictive modelling and understanding of stream data. In: IEEE Transactions on Neural Networks and Learning Systems, pp. 1–13. IEEE Press, New York (2016)
6. Kasabov, N., Scott, E., Tu, E., Marks, S., Sengupta, N., Capecci, E.: Evolvingspatio- temporal data machines based on the NeuCube neuromorphic framework: design methodology and selected applications. Neural Netw. **78**, 1–14 (2016)
7. Doborjeh, M.G., Capecci, E., Kasabov, N.: Classification and segmentation of fMRI spatio-temporal brain data with a neucube evolving spiking neural network model. In: IIEEE International Symposium on Circuits and Systems, pp. 73–80. IEEE Press, Melbourne (2014)
8. Doberjeh, M.G., Wang, G., Kasabov, N., Kydd, R., Russell, B.R.: A Neucube-Spiking neural network model for the study of dynamic brain activities during a GO/NO GO task: a case study on using EEG data of healthy vs addiction vs treated subjects. IEEE Trans. Biomed. Eng. **63**, 1830–1841 (2016)
9. Doborjeh, M.G., Kasabov, N.: Dynamic 3D clustering of spatio-temporal brain data in the NeuCube spiking neural network architecture on a case study of fMRI data. In: Arik, S., Huang, T., Lai, W.K., Liu, Q. (eds.) ICONIP 2015. LNCS, vol. 9492, pp. 191–198. Springer, Cham (2015). doi:10.1007/978-3-319-26561-2_23
10. Song, S., Miller, K.D., Abbott, L.F.: Competitive Hebbian learning throughspike-timing-dependent synaptic plasticity. Nat. Neurosci. **3**, 919–926 (2000)
11. Kasabov, N., Dhoble, K., Nuntalid, N., Indiveri, G.: Dynamic evolving spiking neural networks for on-line spatio-and spectro-temporal pattern recognition. Neural Netw. **41**, 188–201 (2013)

12. Matsumoto, D., Ekman, P.: Japanese and Caucasian facial expressions of emotion (IACFEE) [Slides]. Intercultural and Emotion Research Laboratory, Department of Psychology, San Francisco State University, San Francisco (1988)
13. Talairach, J., Tournoux, P.: Co-planar Stereotaxic Atlas of the Human Brain: 3- Dimensional Proportional System: An Approach to Cerebral Imaging. Thieme Medical Publishers, New York (1988)
14. Koessler, L., Maillard, L., Benhadid, A., Vignal, J.P., Felblinger, J., Vespignani, H., Braun, M.: Automated cortical projection of EEG sensors: anatomical correlation via the international 10–10 system. Neuroimage **46**, 64–72 (2009)
15. Alfano, K.M., Cimino, C.R.: Alteration of expected hemispheric asymmetries: valence and arousal effects in neuropsychological models of emotion. Brain Cogn. **66**, 213–220 (2008)
16. Kawano, H., Seo, A., Doborjeh, Z.G., Kasabov, N., Doborjeh, M.G.: Analysis of similarity and differences in brain activities between perception and production of facial expressions using EEG data and the NeuCube spiking neural network architecture. In: Hirose, A., Ozawa, S., Doya, K., Ikeda, K., Lee, M., Liu, D. (eds.) ICONIP 2016. LNCS, vol. 9950, pp. 221–227. Springer, Cham (2016). doi:10.1007/978-3-319-46681-1_27

Testing and Understanding Second-Order Statistics of Spike Patterns Using Spike Shuffling Methods

Zedong Bi[1] and Changsong Zhou[1,2,3,4(✉)]

[1] HKBU Shenzhen Institute of Research and Continuing Education,
Shenzhen, China
zedong.bi@outlook.com, cszhou@hkbu.edu.hk
[2] Department of Physics, Hong Kong Baptist University,
Kowloon Tong, Hong Kong
[3] Centre for Nonlinear Studies, and Beijing-Hong Kong-Singapore Joint Centre
for Nonlinear and Complex Systems (Hong Kong), Institute of Computational and
Theoretical Studies, Hong Kong Baptist University, Kowloon Tong, Hong Kong
[4] Beijing Computational Science Research Center, Beijing, China

Abstract. We introduce a framework of spike shuffling methods to test
the significance and understand the biological meanings of the second-
order statistics of spike patterns recorded in experiments or simulations.
In this framework, each method is to evidently alter a specific pattern statistics, leaving the other statistics unchanged. We then use this
method to understand the contribution of different second-order statistics to the variance of synaptic changes induced by the spike patterns self-
organized by an integrate-and-fire (LIF) neuronal network under STDP
and synaptic homeostasis. We find that burstiness/regularity and heterogeneity of cross-correlations are important to determine the variance
of synaptic changes under asynchronous states, while heterogeneity of
cross-correlations is the main factor to cause the variance of synaptic
changes when the network moves into strong synchronous states.

Keywords: Spike shuffling methods · Spike pattern statistics

1 Introduction

Neuronal spike trains contain rich statistical structures, which may play important roles on information encoding [1,2], memory embedding [3,4] and neural
development [5,6]. Spike shuffling method provides a tool to test the significance
or understand the biological effects of these statistical structures (see e.g. [7,8]).
The basic idea is to shuffle the recorded spike patterns using some method, evidently changing a specific pattern statistic while keeping others unchanged, then
compare the statistics or the biological effects (such as the encoded information
or the plasticity process) of these two patterns before and after being shuffled.

Shuffling spikes can be used not only to analyze spike patterns, but also to
generate spike patterns from recorded ones. Among many methods to analyze

© Springer International Publishing AG 2017
D. Liu et al. (Eds.): ICONIP 2017, Part IV, LNCS 10637, pp. 602–612, 2017.
https://doi.org/10.1007/978-3-319-70093-9_64

and generate spike patterns (see e.g., [9–12]), spike shuffling method is model-free and non-parametric. This means that the analysis results from this method is irrelevant to the model to be used, and the generated spike patterns can keep the flavor of the recorded pattern to the most extent, except the pattern statistic we choose to alter.

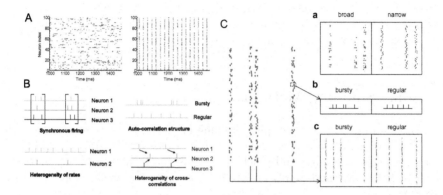

Fig. 1. Schematic of the spike patterns statistics we study in this paper. (A) Examples of asynchronous (left) and synchronous (right) spike patterns. (B) The four types of spike pattern structure we study in this paper, see text for explanation. (C) The three types of auto-correlation structure we consider under synchronous firing. Panel a: The broadness of the distribution of the spike numbers a neuron fires in different synchronous events, denoted as $\text{AT}_{SpikeNum}$. Note that in the left subplot, a neuron fires quite different number of spikes during different synchronous events; while in the right panel, the spike numbers of a neuron during different synchronous events are almost the same. Panel b: The burstiness/regularity of the pieces of spike trains within synchronous events, denoted as $\text{AT}_{WithinEvent}$. Panel c: The burstiness/regularity of the occurrence of synchronous events, denoted as AT_{events}.

In this paper, we study how several second-order statistics of spike patterns may be changed by a number of spike shuffling methods, and provide a systematic solution to tease apart the effect of a statistic from a recorded spike pattern. We consider two types of spike patterns: asynchronous pattern (Fig. 1A, left), in which the population firing rate fluctuates only weakly with time; and synchronous pattern (Fig. 1A, right), in which spikes spurt in short time windows, forming synchronous events. For both types of patterns, we explicitly consider four types of second-order statistics (Fig. 1B). First is *synchronous firing*, which typically means the spurt of firing activity of a population; in this paper, it also represents the time fluctuation of the population rate in asynchronous patterns. Second is *auto-correlation structure*, which reflects the burstiness/regularity of the spike trains. For asynchronous patterns, auto-correlation structure may be quantified by, say, coefficient of variance. For synchronous patterns, we consider three types of auto-correlation structure to reflect the burstiness/regularity in the context of synchronous firing (Fig. 1C). Third is *heterogeneity of rates*, which means that the time-averaged firing rates are different for different neurons. Fourth is *heterogeneity of cross-correlations*, which means that the cross-correlations between different neuronal pairs are different.

This paper is organized as follows. First, we list out the spike shuffling methods we consider, and discuss how each of them changes spike pattern statistics. Second, we provide a systematic solution on how to use these shuffling methods to tease apart different spike pattern statistics. Third, we exemplify the usage of this method by considering how the statistics of the spike pattern generated by a LIF neuronal network may influence the variance of synaptic changes under spike-timing dependent plasticity (STDP) and synaptic homeostasis.

2   Spike Shuffling Methods

In this section, we explain the spike shuffling methods we propose to treat spike patterns, and discuss how each of them changes spike pattern statistics. The statistics of the asynchronous and synchronous spike patterns are sharply different, so some of these methods apply only to asynchronous (or synchronous) patterns (see Fig. 2).

2.1   Spike Shuffling Methods for both Asynchronous and Synchronous Patterns

(1) Whole-train Swap (WTS)

This method randomly shuffles the neuronal indexes of spike trains in the pattern. For example, if we denote $\mathcal{T}_a$ to be the spike train of the ath neuron, then the whole spike pattern can be denoted as a set of neuron-train pairs $\{(a, \mathcal{T}_a), (b, \mathcal{T}_b), (c, \mathcal{T}_c), \cdots\}$; then after WTS, the spike pattern may become $\{(a, \mathcal{T}_c), (b, \mathcal{T}_a), (c, \mathcal{T}_b), \cdots\}$. WTS keeps all the statistics of a spike pattern, but destroys the possible correlation between the spike trains and the structure of the underlying neuronal network. To get rid of this pattern-network coupling thereby focusing on the influence solely from spike pattern statistics (see more discussions in Sect. 3), we suggest to treat all the recorded spike patterns by WTS before any other shuffling method.

(2) Spike-time Rescaling (STR)

STR is realized by first ordering all the M spikes in the pattern, then setting the time of the ith spike at iT/M. By definition, this shuffling method flattens population firing rate, while conserving the time-averaged firing rate of each neuron. As it keeps the order of spikes, the burstiness/regularity of spike trains and cross-correlations between spike trains in the original pattern can be, to some extent, kept in the pattern after shuffling, especially if the rate fluctuation in the original spike pattern is weak.

(3) Neuron Re-choosing (NRC)

In this method, each spike in the pattern is assigned to a randomly selected neuron, keeping spike time unchanged. When the population size is large, this

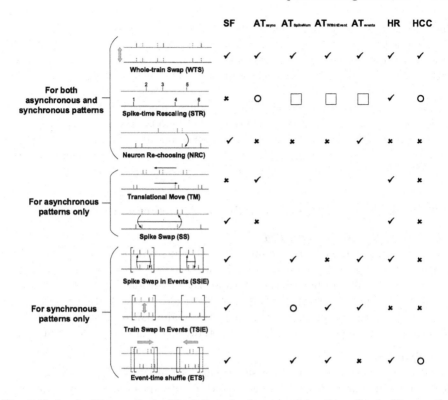

Fig. 2. Spike shuffling methods. The leftmost column explains the spike shuffling methods we propose for asynchronous and synchronous spike patterns, and the rest columns show their influence onto synchronous firing (SF), auto-correlation structure (AT), heterogeneity of rates (HR) and heterogeneity of cross-correlations (HCC). We consider three types of auto-correlation structure for synchronous patterns $AT_{SpikeNum}$, $AT_{WithinEvent}$ and AT_{events} (see Fig. 1C), and the auto-correlation structure under asynchronous states are shortly represented by AT_{async}. ✓ means that a shuffling method keeps a pattern structure unchanged; × means that a shuffling method *destroys* a pattern structure. Here, "destroy" has a sense of "completely randomize". For SF, "destroy" means that there is no time fluctuation of population firing rate. For AT_{async}, it means that the spike train of the ath neuron can be regarded as an inhomogeneous Poisson process with rate $r_a(t) = r_a x(t)$, with r_a being the time-averaged firing rate, and $x(t)$ being the same for all the neurons. For $AT_{SpikeNum}$, it means that the spike numbers of the neurons within a synchronous event follows Poisson distribution of parameter p, with p being the mean spike number per neuron within the synchronous event. For $AT_{WithinEvent}$, it means that the spike train of the ath neuron can be regarded as an inhomogeneous Poisson process with rate $r_a(t) = r_a x(t)$, with r_a being the time-averaged firing rate, and $x(t)$ being the same for all the neurons. For AT_{events}, it means that the occurrence of synchronous events can be regarded as a Poisson process. For HR, it means that the time-averaged firing rates of all the neurons are the same. For HCC, it means that the unit cross-correlations $C_{ab}(\tau) = \frac{\langle r_a(t) r_b(t+\tau) \rangle}{\langle r_a(t) \rangle \langle r_b(t) \rangle}$ (with $r_a(t)$ representing the firing rate of the ath neuron) are the same for different neuronal pairs. ○ means that a shuffling method may change, but does not "completely randomizes", a pattern structure. Note that STR completely flattens the rate fluctuation with time, so $AT_{SpikeNum}$, $AT_{WithinEvent}$ and AT_{events} are not applicable to the synchronous spike patterns after STR (indicated by the squares in the figure).

method makes all the neurons to fire as Poisson processes with equal time-dependent firing rate (i.e. $r_a(t) = r_b(t)$ for two different neurons a and b). NRC destroys auto-correlation structure, heterogeneity of cross-correlations and heterogeneity of firing rates, but keeps the time fluctuation of population firing rate.

Note that in synchronous patterns, NRC also changes the distribution of the spike numbers a neuron fires in different synchronous events (i.e. $AT_{SpikeNum}$). When the population size is large, this distribution after NRC is a Poisson distribution with parameter p, with p being mean spike number per neuron within a synchronous event (i.e. the strength of the synchronous event).

2.2   Spike Shuffling Methods only for Asynchronous States

(1) Translational Move (TM)

In this method, each spike train is translationally moved by a random displacement, and periodic boundary condition is used to deal with the spikes which are moved out of the boundaries of time. By definition, TM keeps the auto-correlation structure of spike trains, and the time-averaged firing rate of each neuron. It flattens the cross-correlations between any pair of spike trains, thereby destroying both synchronous firing and heterogeneity of cross-correlations.

(2) Spike Swap (SS):

The idea of this method is to swap pairs of randomly chosen spikes of different neurons many times. A spike pattern can be denoted as a set of number pairs $\{(a, t_1), (b, t_2), (c, t_3), \cdots\}$, with $a, b, c, \cdots$ being neuronal indexes, and t_1, $t_2, t_3, \cdots$ being spike times. SS shuffles the order of the first fields of these number pairs, so that the spike pattern after SS may be $\{(b, t_1), (c, t_2), (a, t_3), \cdots\}$. SS does not change the spike number of a neuron, but randomizes the occurrence of these spikes. When the size of the network is large, the probability that a neuron fires near time t is proportional to the global firing rate at t, and the times of different spikes are independent with each other. Because of this, SS makes the spike trains Poisson processes, and the rate fluctuations of all these spike trains are simultaneously time-modulated: i.e., the firing rate of the ath spike train can be written as $r_a(t) = r_a x(t)$, with r_a being the time-averaged firing rate, and $x(t)$ being the same for all the spike trains. By definition, SS destroys auto-correlation structure and heterogeneity of cross-correlations, but keeps the time fluctuation of population firing rate and the time-averaged firing rate of each neuron.

2.3   Spike Shuffling Methods only for Synchronous States

(1) Spike Swap in Events (SSiE)

The idea of SSiE is to swap pairs of randomly chosen spikes of different neurons many times, with each pair of chosen spikes being within the same synchronous event. It can be understood as implementing SS (see Sect. 2.2) onto each synchronous event in the spike pattern. SSiE keeps the spike number of a neuron within a synchronous event; and the same as SS, SSiE also makes the neurons to fire like Poisson processes when the size of the neuronal population is large, and the rate fluctuation of all the spike trains are simultaneously time-modulated. By definition, SSiE destroys heterogeneity of cross-correlations and $AT_{\text{WithinEvent}}$.

(2) Train Swap in Events (TSiE)

The idea of TSiE is to swap the pieces of spike trains of randomly selected neurons pairs in the same synchronous event many times, and the pieces of spike pattern in different synchronous events are shuffled independently. By definition, TSiE destroys heterogeneity of rates and heterogeneity of cross-correlations, but keeps $AT_{\text{WithinEvent}}$. The spike number distribution per neuron per synchronous event for the whole neuronal population is kept unchanged under TSiE, but the distribution for a single neuron in different synchronous events is changed dramatically.

(3) Event-time Shuffle (ETS)

The idea of ETS is that all the spikes within the same synchronous events are translationally moved by a random displacement, at the same time (1) avoiding the overlapping of different synchronous events, and (2) keeping the order of synchronous events unchanged. Technically, this is realized by first randomly selecting N_{event} points in the duration $[0, T - \sum_{i=1}^{N_{events}} T_i]$ (with N_{event} being the number of the synchronous events, T being the time duration of the spike pattern, and T_i being the duration of the ith synchronous event), then set the beginning time of the jth synchronous event at $x_j + \sum_{i=1}^{j-1} T_i$ (with x_j being the jth selected points). ETS mainly changes AT_{event}, but may also change heterogeneity of cross-correlations by changing the cross-correlations between spikes in adjacent synchronous events.

3   Using Spike Shuffling Methods to Tease Apart Different Spike Pattern Statistics

As we mentioned in the Introduction, the basic idea behind the spike-shuffling methodology is to shuffle the spike pattern in some way, dramatically changing one pattern structure while keeping the other structures unchanged, and then compare the effects of the patterns before and after shuffling, thereby under-standing how the specific pattern structure influences this effect. However, there are several problems to be solved before this methodology gets its life. Here, we discuss these problems and provide our solutions.

The first problem is the pattern-network coupling. Spike activity depends on the neuronal interactions in a network, so it is not surprising that the spike pattern statistics may be coupled with the network structure. When we shuffle a spike pattern, we may not only change the statistics of the spike pattern itself, but also destroy this coupling. Therefore, if we want to focus on the effect from the spike pattern statistics, we may need to first tease apart the influence from this pattern-network coupling. To this end, we propose a shuffling method called Whole-train Swap (WTS, see Sect. 2.1), and also suggest to first treat the recorded spike patterns using this method to remove the pattern-network coupling before using other methods to study the effect of the spike pattern statistics.

Another problem is that from Fig. 2, a shuffling method may simultaneously destroy more than one aspects of pattern structure. Therefore, we must carefully design the order to implement them when trying to understand the influence from each aspect of pattern structure. For example, to understand the effect of population rate fluctuation with time in asynchronous patterns, we would like to treat the patterns using TM. However, from Fig. 2, TM destroys not only synchronous firing but also heterogeneity of cross-correlations, so to investigate the effect of synchronous firing, we must implement TM onto the spike patterns whose heterogeneity of cross-correlations are already destroyed. Therefore, we first treat the spike patterns using SS, thereby destroying heterogeneity of cross-correlations, and then investigate the change of the efficacy variability after further implementing TM. This is the basic idea we used to design our research.

Table 1. The spike patterns that we suggest to compare to understand the effects of different spike pattern statistics. Abbreviations are the same as those used in Fig. 2.

	Spike pattern statistics		Spike patterns to be compared
Asynchronous patterns	SF		P_{SS} vs P_{SS+TM}
	HCC		P_{STR} vs P_{STR+TM}
	AT		P_{TM} vs P_{TM+SS}
	HR		P_{TM+SS} vs P_{TM+NRC}
Synchronous patterns	SF		P_{SSiE} vs $P_{SSiE+STR}$
	HR		P_{SSiE} vs $P_{SSiE+TSiE}$
	AT	$AT_{SpikeNum}$	$P_{TSiE+SSiE}$ vs $P_{TSiE+NRC}$
		$AT_{WithinEvent}$	P_{TSiE} vs $P_{TSiE+SSiE}$
		AT_{events}	$P_{TSiE+SSiE}$ vs $P_{TSiE+SSiE+ETS}$
	HCC		P_{WTS} vs P_{SSiE}

We outline the spike patterns we suggest to compare to understand each aspect of spike pattern statistics in Table 1. For the convenience of the following discussions, we denote $P_{S_1+S_2+\cdots}$ to be the spike patterns sequentially shuffled by methods $S_1, S_2, \ldots$ from the spike patterns firstly shuffled by WTS, and denote P_{WTS} to be the patterns shuffled by WTS.

4 Understanding the Contribution of Spike Pattern Statistics to Variability of Synaptic Changes Under STDP Using Spike Shuffling Methods

As an example of the spike-shuffling methodology introduced above, here, we use it to investigate how different spike pattern statistics influence the variability of synaptic changes under STDP and synaptic homeostasis.

We simulated a network of excitatory-inhibitory conductance-based LIF neurons, randomly connected with probability 0.2. We kept the time constant of the excitatory synaptic input at 4 ms, and changed the decaying time scale τ_d^I of the inhibitory synaptic inputs from 3 ms to 14 ms, taking integer values. When $3\,\text{ms} \leq \tau_d^I \leq 6\,\text{ms}$, the network works in asynchronous state (Fig. 1A, left); when $7\,\text{ms} \leq \tau_d^I \leq 14\,\text{ms}$, it works in synchronous state (Fig. 1A, right), with the number of spikes per synchronous event getting more and period getting longer with τ_d^I. Excitatory-excitatory synapses were changed according to a spike-timing plasticity (STDP) rule, and synaptic homeostasis was implemented so that the mean excitatory synaptic input to each excitatory neuron was fixed during training (see [13] for details of the simulation method).

We are interested in the variance of synaptic changes during a plasticity process: $\text{Var}_{ab}(\Delta w_{ab})$, with Δw_{ab} being the change of the synapse from neuron b to a. As explained in [13], this variance may be important for understanding the biological meaning of spike pattern statistics during memory embedding and neural development.

In reality, the dynamics of a plastic network co-evolve with the synaptic weights. To only investigate the influence of the network dynamics onto the variance of synaptic changes without worrying about the feedback to network dynamics from synaptic changes, we take the following strategy: We first record the spike patterns of the excitatory population of the LIF network, then shuffle the recorded spike patterns using different methods to change different statistical features, and at last evolve the E-E links under STDP and synaptic homeostasis when the excitatory population are supposed to fire according to the recorded or shuffled spike patterns. By comparing the statistics of the patterns as well as the variance of synaptic changes under the patterns before and after implementing a spike shuffling method, we can gain understanding on how different aspects of the pattern structure may influence the variance.

For asynchronous states, we sequentially shuffled the original spike pattern by WTS, STR, TM, SS and NRC, and then observed the variance of synaptic changes in the resulting spike patterns (Fig. 3A). The changes of the variances after WTS, TM, SS and NRC respectively manifest the contributions from pattern-network coupling, heterogeneity of cross-correlations, auto-correlation structure, and heterogeneity of rates. The contribution from population rate fluctuation with time can be understood by comparing P_{SS} and P_{SS+TM}. We found that under our parameter values, heterogeneity of cross-correlations and auto-correlation structure are the two major sources of the efficacy variability; and auto-correlation structure contributes most to the increase of the efficacy variability with τ_d^I (Fig. 3A).

For synchronous states, we first used WTS to destroy the pattern-network coupling, and found that this coupling hardly influences the variance of synaptic changes (Fig. 3B). Then, we used TSiE to destroy heterogeneity of rates and heterogeneity of cross-correlations, and found that the variance is reduced by more than 10 times (Fig. 3B). This suggests that these two statistics are the main sources of the variance. To compare the contributions from these two sources, we compared the variance under P_{WTS} with that under P_{SSiE}, which destroys heterogeneity of cross-correlations but keeps heterogeneity of rates.

We found that the variance under P_{SSiE} gets its maximum around $\tau_d^I = 9\,\mathrm{ms}$; the variance under P_{WTS} is smaller than that under P_{SSiE} when τ_d^I is small, but tends to monotonically increase with τ_d^I (Fig. 3B). These observations suggest that heterogeneity of rates contributes most to the variance when τ_d^I is small; but when τ_d^I increases, heterogeneity of cross-correlation gradually becomes the dominating factor to the variance.

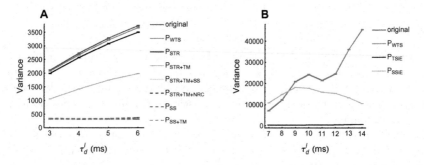

Fig. 3. Understanding the contributions to the variance of synaptic changes from different spike pattern statistics under STDP and the spike patterns of the LIF network. (A) The variance of synaptic changes under the original and shuffled asynchronous patterns. Note that the variances under $P_{STR+TM+SS}$, $P_{STR+TM+NRC}$ and P_{SS+TM} are strongly overlapped. (B) The variance under the original and shuffled synchronous patterns. Note that the results for the original pattern and P_{WTS} are strongly overlapped.

5   Discussion

In this paper, we introduce a non-parametric spike-shuffling methodological framework to systematically tease apart different second-order statistics from the recorded spike patterns, both in asynchronous and synchronous spike patterns. This helps to test the significance or understand the effects of these statistics respectively. We then exemplify this method by investigating the influence of different second-order statistics onto the variance of synaptic changes during STDP and synaptic homeostasis under spike patterns self-organized by a LIF neuronal network. We found that burstiness/regularity and heterogeneity of cross-correlations are important to determine the variance of synaptic changes under asynchronous states, while heterogeneity of cross-correlations is the main factor to cause this variance when the network moves into strong synchronous states.

From Fig. 2, some methods simultaneously strongly alter more than one pattern statistics. As a result, before we study the influence of a pattern statistics $\mathcal{A}$ using a shuffling method, we have to first treat the spike pattern using other shuffling methods to randomize other pattern statistics $\mathcal{B}$ thereby nullifying their influences. Therefore, an implicit assumption under this spike-shuffling framework is that this randomization of $\mathcal{B}$ does not prohibit us from understanding the significance or the effect of $\mathcal{A}$. We need to keep this assumption in mind, and if necessary, use other method to double-check our results (see [14] for double-checking our understanding on the variance of synaptic changes using spike-generating models). Additionally, from this point, we can see that model-based spike generating models [11,12] and model-free spike shuffling methods both have their pros and cons. This inspires us that developing a hybrid method may be an interesting research direction to generate biologically realistic while statistically controllable spike patterns. For example, one may use some algorithm to perturb an experimentally recorded spike pattern, changing a specific statistic to the desirable value while minimizing the perturbation.

Acknowledgments. CZ is partially supported by Hong Kong Baptist University (HKBU) Strategic Development Fund, NSFC-RGC Joint Research Scheme (Grant No. HKUST/NSFC/12-13/01) and NSFC (Grant No. 11275027). ZB is supported by HKBU Faculty of Science.

References

1. Averbeck, B.B., Latham, P.E., Pouget, A.: Neural correlations, population coding and computation. Nat. Rev. Neurosci. **7**, 358–366 (2006)
2. Brette, R.: Computing with neural synchrony. PLoS Comput. Biol. **8**, e1002561 (2012)
3. Jutras, M.J., Fries, P., Buffalo, E.A.: Gamma-band synchronization in the macaque hippocampus and memory formation. J. Neurosci. **29**, 12521–12531 (2009)
4. Yamamoto, J., Suh, J., Takeuchi, D., Tonegawa, S.: Successful execution of working memory linked to synchronized high-frequency gamma oscillations. Cell **157**, 845–857 (2014)
5. Clause, A., Kim, G., Sonntag, M., Weisz, C.J., Vetter, D.E., Rubsamen, R., et al.: The precise temporal pattern of prehearing spontaneous activity is necessary for tonotopic map refinement. Neuron **82**, 822–835 (2014)
6. Xu, H., Furman, M., Mineur, Y.S., Chen, H., King, S.L., Zenisek, D., et al.: An instructive role for patterned spontaneous retinal activity in mouse visual map development. Neuron **70**, 1115–1127 (2011)
7. Ji, D., Wilson, M.A.: Coordinated memory replay in the visual cortex and hippocampus during sleep. Nat. Neurosci. **10**, 100–107 (2007)
8. Nirenberg, S., Latham, P.E.: Decoding neuronal spike trains: how important are correlations? Proc. Natl. Acad. Sci. USA **100**, 7348–7353 (2003)

9. Prentice, J.S., Marre, O., Ioffe, M.L., Loback, A.R., Tkačik, G., Berry II, M.J.: Error-robust modes of the retinal population code. PLoS Comput. Biol. **12**, e1005148 (2016)

10. Calabrese, A., Schumacher, J.W., Schneider, D.M., Paninski, L., Woolley, S.M.N.: A generalized linear model for estimating spectrotemporal receptive fields from responses to natural sounds. PLoS One **6**, e16104 (2011)

11. Macke, J.H., Berens, P., Ecker, A.S., Tolias, A.S., Bethge, M.: Generating spike trains with specified correlation coefficients. Neural Comput. **21**, 397–423 (2009)

12. Krumin, M., Shoham, S.: Generation of spike trains with controlled auto- and cross-correlation functions. Neural Comput. **21**, 1642–1664 (2009)

13. Bi, Z., Zhou, C.: Spike pattern structure influences synaptic efficacy variability under STDP and synaptic homeostasis. II: spike shuffling methods on LIF networks. Front. Comput. Neurosci. **10**, 83 (2016)

14. Bi, Z., Zhou, C.: Spike pattern structure influences synaptic efficacy variability under STDP and synaptic homeostasis. I: spike generating models on converging motifs. Front. Comput. Neurosci. **10**, 14 (2016)

Self-connection of Thalamic Reticular Nucleus Modulating Absence Seizures

Daqing Guo[1,2(✉)], Mingming Chen[1], Yang Xia[1,2], and Dezhong Yao[1,2]

[1] Key Laboratory for NeuroInformation of Ministry of Education,
School of Life Science and Technology, University of Electronic Science
and Technology of China, Chengdu 610054, China
dqguo@uestc.edu.cn

[2] Center for Information in BioMedicine,
University of Electronic Science and Technology of China,
Chengdu 610054, China

Abstract. Accumulating evidence has suggested that the corticothalamic system not only underlies the onset of absence seizures, but also provides functional roles in controlling absence seizures. However, few studies are involved in the roles of self-connection of thalamic reticular nucleus (TRN) in modulating absence seizures. To this end, we employ a biophysically based corticothalamic network mean-field model to explore these potential control mechanisms. We find that the inhibitory projection from the TRN to specific relay nuclei of thalamus (SRN) can shape the self-connection of TRN controlling absence seizures. Under certain condition, the self-connection of TRN can bidirectionally control absence seizures, which increasing or decreasing the coupling strength of the self-connection of TRN could successfully suppress absence seizures. These findings might provide a new perspective to understand the treatment of absence epilepsy.

Keywords: Absence seizures · Spike and wave discharges · Thalamic reticular nucleus · Self-connection · Mean-field model

1 Introduction

Absence seizures, recognized as the classical characteristics of absence epilepsy, abruptly start and terminate, during which bilaterally synchronous 2–4 Hz spike and wave discharges (SWDs) can be easily observed on the electroencephalogram (EEG), especially in childhood absence epilepsy patients [1,2]. In the past two decades, a growing body of evidence has suggested that the genesis of absence seizures is closely associated with abnormal excessive neural oscillations within corticothalamic network [3–5]. Based on these findings, it is reasonable to speculate that appropriately tuning interactions between cerebral cortex and thalamus to abate these abnormal neural oscillations might provide a manner to suppress absence seizures.

© Springer International Publishing AG 2017
D. Liu et al. (Eds.): ICONIP 2017, Part IV, LNCS 10637, pp. 613–621, 2017.
https://doi.org/10.1007/978-3-319-70093-9_65

Both the inside and outside regulations of corticothalamic network have been suggested to successfully terminate absence seizures [6–13]. For example, in terms of the inside modulation, previous animal experimental findings suggested that both thalamic stimulation and closed-loop optogenetic control of thalamus could interrupt epileptic seizures [9,10]. Recently, a computational model study also demonstrated that the thalamic feed-forward inhibition contributes to suppressing absence seizures [13]. Unlike these inside modulations, several external nuclei of corticothalamic network, such as basal ganglia and cerebellar nuclei, also have been suggested to play crucial roles in controlling absence seizures [6–8,11,12,14]. Particularly, computational simulations suggested that the basal ganglia not only could bidirectionally control absence seizures through the direct nigro-thalamic pathway, but also could suppress absence seizures via the pallido-cortical pathway [11,12]. Taken together, both animal experiments and computational simulations suggest that appropriately tuning abnormal interactions within the corticothalamic network could effectively prevent absence seizures.

Anatomically, the thalamic reticular nucleus (TRN) lies like a shell between cerebral cortex and specific relay nuclei of thalamus (SRN), and entirely composes GABAergic neurons. Note that the TRN-SRN pathway has been demonstrated to contribute to modulating absence seizures by the feed-forward or feedback inhibition for SRN neurons [13,15]. Actually, besides the two types of inhibition, the self-inhibition of TRN neurons might also participate in suppressing absence seizures. However, few studies are involved in the potential functional roles of self-connection of TRN in controlling absence seizures [15–18]. Although past animal experimental studies have suggested that loss of recurrent inhibition in TRN in β_3 knockout mice might evoke absence epilepsy [18], how this recurrent inhibition participates in suppressing absence seizures is still poorly understood. Therefore, in this study, we employ a biophysically based corticothalamic network mean-field model to investigate the potential control roles played by the self-connection of TRN. We find that the self-connection of TRN could bidirectionally control absence seizures under certain condition, which might provide some new insights into the treatment of absence epilepsy.

2   Methods and Analysis

2.1   Computational Model

Similar to previous studies [11–13], we use the main epileptogenic factor that the slow kinetics of $GABA_B$ receptors in SRN to induce absence seizures. Here, we modify the corticothalamic network by inducing the self-connection of TRN, as shown in Fig. 1. Notably, four neural populations contained in the corticothalamic network are excitatory pyramidal neurons (E, e), inhibitory interneurons (I, i), TRN (r) and SRN (s). The connections between neural populations are primarily mediated by glutamate (red arrow lines) and GABA (blue lines with round dots), which the solid and dashed blue lines represent the $GABA_A$ and $GABA_B$ mediated inhibitory connections, respectively. The external inputs to the SRN (ϕ_n) show the non-specific thalamic inputs.

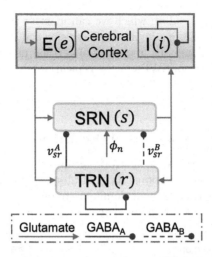

Fig. 1. The framework of the corticothalamic network. In the cerebral cortex, we mainly focus on excitatory pyramidal neurons (E, e) and inhibitory interneurons (I, i). The thalamus consists of thalamic reticular nucleus (TRN, r) and specific relay nuclei (SRN, s). Notably, the red arrow lines show the excitatory connections mediated by glutamate, and the blue lines with round dots depict the inhibitory connections mediated by GABA, including GABA$_A$ (solid blue lines) and GABA$_B$ (the dashed blue line). External inputs to the SRN (ϕ_n) represent the non-specific thalamic inputs (Color figure online)

To describe the neural dynamics of corticothalamic network, we employ the mean-field model proposed by Robinson and his colleagues [19–21], in which for a given neural population a, three main state variables that the mean membrane potential V_a, mean firing rate Q_a and presynaptic activity ϕ_a are considered. Accordingly, we build the corticothalamic network mean-field model, as shown in the Appendix.

It is worth noting that we set $v_{se} = 2.4$ mV s and $\tau = 50$ ms to evoke absence seizures in this model, which is consistent with previous studies [13]. But for more details about the model parameter values and the related physiological significances, we refer the reader to previous literature [11–13, 19–22].

2.2 Data Analysis

In the current study, we use the standard fourth-order Runge-Kutta method with fixed step length of 0.05 ms to solve the corticothalamic network mean-field model, and all model simulations are performed on the MATLAB software (Mathworks, Natick, MA). Here, two primary data analysis techniques adapted from previous studies have been applied to analyze cortical neural oscillations [11–13]. One is the bifurcation analysis that based on the stable local minimum and maximum values of cortical neural oscillations ϕ_e in a single oscillatory period. According to the number of extreme values and amplitude of ϕ_e, we can easily determine the dynamical states of cortical oscillations. The other one is the frequency analysis. We use the Fast Fourier transform to calculate the power spectral density of ϕ_e, based on which we can identify the dominant frequency that corresponding to the maximum of power spectral density. Combining the dynamical states and dominant frequency of cortical neural oscillations ϕ_e, we can outline the boundary of SWDs in the typical range 2–4 Hz.

3   Results

3.1   Dynamical States of Cortical Oscillations

In the modified corticothalamic network mean-field model, we first examine whether the model could reproduce typical absence seizures. To this end, we calculate the dynamical states and dominant frequency of cortical oscillations ϕ_e in the two-dimensional parameter space $(-v_{sr}^{A,B}, -v_{rr})$ (Fig. 2). We find that

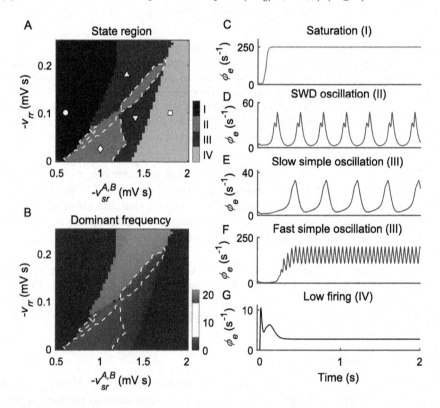

Fig. 2. Dynamical states of cortical oscillations. (A) Four different dynamical states of cortical oscillations distributed in the two-dimensional parameter space $(-v_{sr}^{A,B}, -v_{rr})$ are saturation state (I), SWD oscillation state (II), simple oscillation state (III) including slow and fast simple oscillations, and low firing state (IV). (B) The distribution of dominant frequency of cortical oscillations in the parameter space $(-v_{sr}^{A,B}, -v_{rr})$. Note that the regions of the typical 2–4 Hz SWDs are surrounded by white dashed lines in (A) and (B). (C-G) The time series of cortical oscillations (ϕ_e) and the corresponding parameter values are shown in (A) with different white symbols that "○" (I), "◇" (II), "△" and "▽" (III), and "□" (IV).

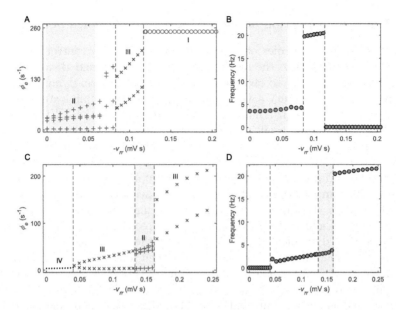

Fig. 3. Roles of self-connection of TRN in modulating absence seizures. (A) The dynamical state analysis of cortical oscillations ϕ_e with $-v_{sr}^{A,B} = 1.0$ mV s. Gradually increasing the coupling strength of self-connection of TRN ($-v_{rr}$) could push cortical oscillations ϕ_e from the SWD oscillation state (II) to simple oscillation state (III) and to saturation state (I). (B) The dominant frequency of cortical oscillations ϕ_e that corresponding to those in (A) changes with the increase of $-v_{rr}$. (C, D) Same as in (A) and (B), but with $-v_{sr}^{A,B} = 1.46$ mV s. The dynamical state analysis (C) and dominant frequency (D) of cortical oscillations ϕ_e changes with the enhancement of $-v_{rr}$. Notably, all gray shaded regions show the range of typical 2–4 Hz SWDs

four dynamical states that the saturation state (I), SWD oscillation state (II), simple oscillation state (III) and low firing state (IV) are distributed in this parameter space (Fig. 2A). Correspondingly, the distribution of the dominant frequency of ϕ_e for each dynamical state can be clearly observed (Fig. 2B). Furthermore, we outline the boundary of regions of the typical 2–4 Hz SWDs (as shown in Fig. 2A and B, the areas surrounded by white dashed lines), for which are often observed in real patients. These findings suggest that the model can replicate absence seizures, as well as other brain dynamical states. Moreover, the coupling strength of self-connection of TRN ($-v_{rr}$) can affect the transition between different dynamical states. Additionally, to show the firing mode of each dynamical state, we select five groups of parameters and calculate the corresponding cortical oscillations ϕ_e (Fig. 2C–G). Interestingly, we find that the strength of $-v_{rr}$ has a significant influence on the dominant frequency of cortical simple oscillation state (III), which relatively strong and weak $-v_{rr}$ are corresponding to fast and slow simple oscillations, respectively (Fig. 2E and F). For this case, we speculate that strong $-v_{rr}$ might weaken the effect of GABA$_B$ in the TRN-SRN pathway.

3.2   The Self-connection of TRN Modulating Absence Seizures

To further uncover the roles of self-connection of TRN in controlling absence seizures, and according to the results obtained in Fig. 2A and B, we select two different values of $-v_{sr}^{A,B}$ to observe the transition of cortical dynamical states. First, we set $-v_{sr}^{A,B}=1.0$ mV s. Gradually enhancing the coupling strength $-v_{rr}$, we find that the cortical dynamical state switches from the SWD oscillation state (II) to the simple oscillation state (III), and to the saturation state (I) (Fig. 3A). Correspondingly, the dominant frequency of ϕ_e smoothly change with the increase of $-v_{rr}$, but significantly increase in the simple oscillation state (III) (Fig. 3B). These findings suggest that appropriately increase the coupling strength $-v_{rr}$ can effectively interrupt absence seizures. Then, we change the value of $-v_{sr}^{A,B}$ to 1.46 mV s. Similarly, we calculate the transition of cortical dynamical states and corresponding dominant frequency. In this case, we find that the cortical dynamical state switches from the low firing state (IV) to the slow simple oscillation state (III) at first, and then to the SWD oscillation state (II), and next to the fast simple oscillation state (III) (Fig. 3C). A sudden jump of the dominant frequency can be observed when the cortical dynamical state is in the fast simple oscillation state (III) (Fig. 3D). Obviously, these findings suggest that there exists a bidirectional control region for the coupling strength $-v_{rr}$, in which appropriately enhancing or decreasing the coupling strength $-v_{rr}$ can achieve the control of absence seizures, implying that $-v_{rr}$ can bidirectionally modulate absence seizures under this condition.

4   Conclusions

In the present study, we found that the self-connection of TRN plays crucial roles in modulating absence seizures. Particularly, the coupling strength of TRN-SRN pathway that mediated by GABA$_A$ and GABA$_B$ can shape the self-connection of TRN suppressing absence seizures. Moreover, the coupling strength $-v_{rr}$ has a significant effect on the dominant frequency of cortical oscillations in the simple oscillation state. Importantly, under certain condition, enhancing or decreasing the coupling strength $-v_{rr}$ can terminate absence seizures, implying that the self-connection of TRN could bidirectionally control absence seizures. It should be noted that the corticothalamic network mean-field model makes specific predictions for the functional roles of recurrent inhibition of TRN in suppressing absence seizures that are testable in animal models of absence epilepsy. Therefore, these findings provide a new perspective to design animal experiments, as well as some novel insights into the treatment of absence epilepsy.

Acknowledgments. This study was supported by the National Natural Science Foundation of China (Nos. 81571770, 61527815, 81401484 and 81330032).

Appendix: The Corticothalamic Network Mean-Field Model

To obtain the corticothalamic network mean-field model, we assume that absence seizures are involved in the whole brain, so that the spatial effect is ignored in this model. Also, based on the assumption that intracortical connections are proportional to the synapses involved, we have $V_i = V_e$ and $Q_i = Q_e$. Note that both of these assumptions are in line with previous studies [11–13, 20, 21]. Accordingly, the corticothalamic network mean-field equations are given in the following:

$$\frac{d\phi_e(t)}{dt} = \dot{\phi}_e(t) \tag{1}$$

$$\frac{d\dot{\phi}_e(t)}{dt} = \gamma_e^2 \left\{ -\phi_e(t) + F[V_e(t)] \right\} - 2\gamma_e \dot{\phi}_e(t) \tag{2}$$

$$\frac{dV_e(t)}{dt} = \dot{V}_e(t) \tag{3}$$

$$\frac{d\dot{V}_e(t)}{dt} = \alpha\beta \left\{ -V_e(t) + v_{ee}\phi_e(t) + v_{ei}F[V_e(t)] + v_{es}F[V_s(t)] \right\} - (\alpha + \beta)\dot{V}_e(t) \tag{4}$$

$$\frac{dV_r(t)}{dt} = \dot{V}_r(t) \tag{5}$$

$$\frac{d\dot{V}_r(t)}{dt} = \alpha\beta \left\{ -V_r(t) + v_{re}\phi_e(t) + v_{rs}F[V_s(t)] + v_{rr}F[V_r(t)] \right\} - (\alpha + \beta)\dot{V}_r(t) \tag{6}$$

$$\frac{dV_s(t)}{dt} = \dot{V}_s(t) \tag{7}$$

$$D_s = \alpha\beta \left\{ -V_s(t) + v_{se}\phi_e(t) + v_{sr}^A F[V_r(t)] + v_{sr}^B F[V_r(t - \tau)] + \phi_n \right\} \tag{8}$$

$$\frac{d\dot{V}_s(t)}{dt} = D_s - (\alpha + \beta)\dot{V}_s(t) \tag{9}$$

Here, ϕ_a ($a = e, i, r, s$) shows the propagating axonal fields of the neural population a, v_{ab} describes the coupling strength from the neural population b to the neural population a. The inverses of α and β represent the decay and rise time constants, respectively. γ_e governs the temporal damping rate of cortical excitatory pulses. V_a denotes the mean membrane potential, through which the mean firing rate Q_a can be obtained with the sigmoid function $F[V_a(t)]$ [11, 12, 20] given by

$$Q_a(t) \equiv F[V_a(t)] = \frac{Q_a^{max}}{1 + \exp[-\frac{\pi}{\sqrt{3}} \frac{(V_a(t) - \theta_a)}{\sigma}]} \tag{10}$$

where Q_a^{max} denotes the maximum firing rate, θ_a shows the mean firing threshold, and σ is the standard deviation of the mean firing threshold.

References

1. Crunelli, V., Leresche, N.: Childhood absence epilepsy genes, channels, neurons and networks. Nat. Rev. Neurosci. **3**(5), 371–382 (2002)
2. Moshé, S.L., Perucca, E., Ryvlin, P., Tomson, T.: Epilepsy: new advances. Lancet **385**(9971), 884–898 (2015)
3. Danober, L., Deransart, C., Depaulis, A., Vergnes, M., Marescaux, C.: Pathophysiological mechanisms of genetic absence epilepsy in the rat. Prog. Neurobiol. **55**(1), 27–57 (1998)
4. Blumenfeld, H.: Cellular and network mechanisms of spike-wave seizures. Epilepsia **46**, 21–33 (2005)
5. Avoli, M.: A brief history on the oscillating roles of thalamus and cortex in absence seizures. Epilepsia **53**(5), 779–789 (2012)
6. Deransart, C., Vercueil, L., Marescaux, C., Depaulis, A.: The role of basal ganglia in the control of generalized absence seizures. Epilepsy Res. **32**(1–2), 213–223 (1998)
7. Paz, J.T., Chavez, M., Saillet, S., Deniau, J.M., Charpier, S.: Activity of ventral medial thalamic neurons during absence seizures and modulation of cortical paroxysms by the nigrothalamic pathway. J. Neurosci. **27**(4), 929–941 (2007)
8. Luo, C., Li, Q., Xia, Y., Lei, X., Xue, K., Yao, Z., Lai, Y., Martínez-Montes, E., Liao, W., Zhou, D., Valdes-Sosa, P.A., Gong, Q., Yao, D.: Resting state basal ganglia network in idiopathic generalized epilepsy. Hum. Brain Mapp. **33**(6), 1279–1294 (2012)
9. Lüttjohann, A., van Luijtelaar, G.: Thalamic stimulation in absence epilepsy. Epilepsy Res. **106**(1–2), 136–145 (2013)
10. Paz, J.T., Davidson, T.J., Frechette, E.S., Delord, B., Parada, I., Peng, K., Deisseroth, K., Huguenard, J.R.: Closed-loop optogenetic control of thalamus as a tool for interrupting seizures after cortical injury. Nat. Neurosci. **16**(1), 64–70 (2013)
11. Chen, M., Guo, D., Wang, T., Jing, W., Xia, Y., Xu, P., Luo, C., Valdes-Sosa, P.A., Yao, D.: Bidirectional control of absence seizures by the basal ganglia: a computational evidence. PLoS Comput. Biol. **10**(3), 1–17 (2014)
12. Chen, M., Guo, D., Li, M., Ma, T., Wu, S., Ma, J., Cui, Y., Xia, Y., Xu, P., Yao, D.: Critical roles of the direct gabaergic pallido-cortical pathway in controlling absence seizures. PLoS Comput. Biol. **11**(10), 1–23 (2015)
13. Chen, M., Guo, D., Xia, Y., Yao, D.: Control of absence seizures by the thalamic feed-forward inhibition. Front. Comput. Neurosci. **11**, 31 (2017)
14. Kros, L., Eelkman Rooda, O.H.J., Spanke, J.K., Alva, P., van Dongen, M.N., Karapatis, A., Tolner, E.A., Strydis, C., Davey, N., Winkelman, B.H.J., Negrello, M., Serdijn, W.A., Steuber, V., van den Maagdenberg, A.M.J.M., De Zeeuw, C.I., Hoebeek, F.E.: Cerebellar output controls generalized spike-and-wave discharge occurrence. Ann. Neurol. **77**(6), 1027–1049 (2015)
15. Paz, J.T., Huguenard, J.R.: Microcircuits and their interactions in epilepsy: is the focus out of focus? Nat. Neurosci. **18**(3), 351–359 (2015)
16. Huguenard, J.R., Prince, D.A.: Clonazepam suppresses gabab-mediated inhibition in thalamic relay neurons through effects in nucleus reticularis. J. Neurophysiol. **71**(6), 2576–2581 (1994)
17. Kim, U., Sanchez-Vives, M.V., McCormick, D.A.: Functional dynamics of gabaergic inhibition in the thalamus. Science **278**(5335), 130–134 (1997)
18. Huntsman, M.M., Porcello, D.M., Homanics, G.E., DeLorey, T.M., Huguenard, J.R.: Reciprocal inhibitory connections and network synchrony in the mammalian thalamus. Science **283**(5401), 541–543 (1999)

19. Robinson, P.A., Rennie, C.J., Wright, J.J.: Propagation and stability of waves of electrical activity in the cerebral cortex. Phys. Rev. E **56**, 826–840 (1997)
20. Robinson, P.A., Rennie, C.J., Rowe, D.L.: Dynamics of large-scale brain activity in normal arousal states and epileptic seizures. Phys. Rev. E **65**, 041924 (2002)
21. Breakspear, M., Roberts, J.A., Terry, J.R., Rodrigues, S., Mahant, N., Robinson, P.A.: A unifying explanation of primary generalized seizures through nonlinear brain modeling and bifurcation analysis. Cereb. Cortex **16**(9), 1296–1313 (2006)
22. Marten, F., Rodrigues, S., Benjamin, O., Richardson, M.P., Terry, J.R.: Onset of polyspike complexes in a mean-field model of human electroencephalography and its application to absence epilepsy. Philos. Trans. R. Soc. Lond. A: Math. Phys. Eng. Sci. **367**(1891), 1145–1161 (2009)

Learning a Continuous Attractor Neural Network from Real Images

Xiaolong Zou[1,2], Zilong Ji[2], Xiao Liu[2], Yuanyuan Mi[3], K.Y. Michael Wong[4], and Si Wu[2(✉)]

[1] School of Systems Science, Beijing Normal University, Beijing 100875, China
[2] State Key Laboratory of Cognitive Neuroscience and Learning,
IDG/McGovern Institute for Brain Research, Beijing Normal University,
Beijing 100875, China
wusi@bnu.edu.cn
[3] Brain Science Center, Institute of Basic Medical Sciences, Beijing 100850, China
[4] Department of Physics, Hong Kong University of Science and Technology,
Kowloon, Hong Kong

Abstract. Continuous attractor neural networks (CANNs) have been widely used as a canonical model for neural information representation. It remains, however, unclear how the neural system acquires such a network structure in practice. In the present study, we propose a biological plausible scheme for the neural system to learn a CANN from real images. The scheme contains two key issues. One is to generate high-level representations of objects, such that the correlation between neural representations reflects the sematic relationship between objects. We adopt a deep neural network trained by a large number of natural images to achieve this goal. The other is to learn correlated memory patterns in a recurrent neural network. We adopt a modified Hebb rule, which encodes the correlation between neural representations into the connection form of the network. We carry out a number of experiments to demonstrate that when the presented images are linked by a continuous feature, the neural system learns a CANN successfully, in term of that these images are stored as a continuous family of stationary states of the network, forming a sub-manifold of low energy in the network state space.

Keywords: Continuous attractor neural network · Correlated patterns · Modified Hebb rule · Deep neural network

1 Introduction

The brain performs computation via dynamics of neural circuits formed by a large number of neurons. A neural system can reliably retrieve the stored information even when external inputs are noisy. This can be viewed as attractor computation mathematically, that is, the network dynamics enables the neural

---

X. Zou and Z. Ji—Equal contribution.

D. Liu et al. (Eds.): ICONIP 2017, Part IV, LNCS 10637, pp. 622–631, 2017.
https://doi.org/10.1007/978-3-319-70093-9_66

system to reach to the same stationary state (attractor), once an external input falls into the basin of attraction. Hopfield network is such a representative attractor model, which has been applied successfully to explain associative memory [1].

The experimental and theoretical studies have shown that there exists another form of attractor model, called continuous attractor neural networks (CANNs) [2–5]. CANNs have been successfully applied to describe the encoding of a number of continuous features in neural systems, such as orientation, moving direction, head direction, and spatial location of objects (see [6] and references therein). The key property of a CANN is that it holds a continuous family of stationary states, which form an *approximately* flat sub-manifold of low energy in the network state space; whereas, in the Hopfield network, attractor states are isolated with each other with high-energy barriers.

Despite the importance of CANN being widely recognized in the field [2], the mechanism of how the neural system acquires such a structure remains largely unknown. For certain features, such as orientation tuning in V1, it is known that the neural system gains a rough structure from heredity and continually refines it after birth based on visual experiences. However, for many other less directly perceivable features, in particular, for those varying with computational tasks, the neural system needs to develop a CANN in a data-dependent way. For instances, in the monkey experiments, Logothetis et al. found that after training, neurons in the inferior temporal cortex can form a CANN to encode the view angle of an object [7]; and Mante et al. found that after training, neurons in PFC can form a CANN to encode the evidence values of different sensory cues [8]. It is far from clear how the neural system develops CANNs in those on-line computations.

To learn a CANN, it faces two fundamental challenges technically. One is to generate proper neural representations. In effect, the unsupervised Hebb rule, or its variations, is to convert the overlap between neural representations (the correlation) into the strengths of neuronal connections, such that the similarity/dissimilarity between objects is encoded in the network structure. Thus, it is crucial that the overlap between neural representations reflects the sematic, rather than a superficial (e.g., different spatial locations), relationship between objects. The other challenge is to learn correlated representation patterns. Using the conventional Hebb rule, the Hopfield model can learn statistically independent memory patterns, but once the patterns are correlated, the performance of the Hopfield model is deteriorated dramatically.

In this study, we propose a biologically plausible scheme to learn a CANN from real images. Firstly, mimicking the dorsal visual pathway, we adopt a deep neural network trained by a large number of nature images to generate neural representations of objects. It has been observed that the neural representations generated by a deep neural network (i.e., the activities in the representation layer preceding the reading-out layer) can explain a large portion of statistical characteristics of the real neural data recorded in the higher visual cortexes [9], indicating that the representations generated by deep learning capture some sematic information of objects. Secondly, mimicking the hippocampus, we adopt a modified Hebb rule to construct neuronal connections based on the representations of a deep neural network. This modified Hebb rule encodes the correlations between objects' representations into the structure of the network. We carry out experiments, which demonstrate that if the presented images are linked by a continuous feature, a CANN is learned successfully.

2 Learning Correlated Patterns

To start, we introduce how to store correlated patterns in a recurrent network, presuming that the proper neural representations of objects have been generated already (to be discussed in Sect. 4). It is well known that the conventional Hebb rule can only learn statistically uncorrelated patterns, resulting in the classical Hopfield model [1]. For correlated patterns, their associated stationary states will merge into only a few attractor states. To solve this issue, several strategies have been proposed. Specifically, Kroff and Treves [10] proposed a popularity-based approach, whose idea is reduce the learning ratio of popular neurons (i.e., those neurons which are involved in many memory patterns). This method works well only under the condition that neuronal activities in a memory pattern are statistically independent, which is hard to be satisfied in practice (Fig. 1). Blumenfeld et al. [11] proposed a novelty-based method, which dynamically adjusts the learning ratio relying on the difference between the newly arriving pattern and those already stored in the network. This novelty-based learning idea explains several interesting neurobiological experiments [12,13]. However, the way of modifying the learn ratio in this approach is empirical and can not guarantee that it works well always (Fig. 1).

Here, we adopt an orthogonal Hebb rule to learn correlated patterns [14]. This method works well and is also biologically plausible. Its basic idea is that when

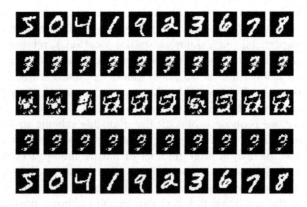

Fig. 1. Retrieval performances of different learning rules. From top to bottom: the original ten digital numbers, the result of the conventional Hebb rule, the result of the popularity-based method, the result of novelty-based method, and the result of the orthogonal Hebb rule.

a pattern ξ^{p+1} is presented, the network first orthogonalizes it with respect to the stored patterns, and then learns the orthogonalized pattern η^{p+1}. The latter is calculated to be,

$$\eta^{p+1} = \xi^{p+1} - \sum_{\mu=1}^{p} \hat{\eta}^{\mu} \hat{\eta}^{\mu} \xi^{p+1}, \tag{1}$$

where $\hat{\eta}^{\mu} = \eta^{\mu}/|\eta^{\mu}|$ is the normalization of η^{μ}.

According to the Hebb rule, the connection strength between neurons i and j is set to be,

$$W_{ij} = \sum_{\mu=1}^{P} (\hat{\eta}_i^{\mu} \hat{\eta}_j^{\mu} - \delta_{ij} \hat{\eta}_i^{\mu} \hat{\eta}_i^{\mu}) \tag{2}$$

where P is the total number of patterns, and δ_{ij} is Kronecker delta function, i.e., $\delta_{ij} = 1$ for $i = j$ and otherwise $\delta_{ij} = 0$.

Denote $\mathbf{S} = \{S_i\}$, for $i = 1, 2, \ldots, N$, to be the network state, with S_i taking values of ± 1. The network dynamics is given by

$$S_i(t+1) = sign(\sum_j W_{ij} S_j), \tag{3}$$

and the energy function of the network state is calculated to be

$$E = -\frac{1}{2} \mathbf{S}^T \mathbf{W} \mathbf{S}. \tag{4}$$

Note that the network dynamics contains a novelty-detection process. Suppose a pattern ξ^{P+1} is already learned as the pattern ξ^v, for $v < P + 1$, in the network. When this pattern is presented, we have

$$
\begin{aligned}
\mathbf{W}\xi^{p+1} &= \mathbf{W}\xi^v, \\
&= \sum_{\mu=1}^{v-1} \hat{\eta}^{\mu} \hat{\eta}^{\mu,T} \xi^v + \hat{\eta}^v \hat{\eta}^{v,T} \xi^v + \sum_{\mu=v+1}^{P} \hat{\eta}^{\mu} \hat{\eta}^{\mu,T} \xi^v - O(\frac{P}{N})\xi^v, \\
&= \xi^v - \eta^v + \eta^v - O(\frac{P}{N})\xi^v, \\
&= [1 - O(\frac{P}{N})]\xi^v.
\end{aligned} \tag{5}
$$

where,

$$\mathbf{W}\xi^v = \sum_{\mu=1}^{P} \hat{\eta}^{\mu} \hat{\eta}^{\mu,T} \xi^v - O(\frac{P}{N})\xi^v \tag{6}$$

since $(\hat{\eta}_i)^2$ is of the order of $1/N$.

Thus, according to Eq. (3), this pattern is recognized by the network as learned previously.

Although the above orthogonal Hebb rule may appear to be quite mathematic, its fundamental idea can be implemented in the neural system.

For instances, the orthogonal operation may be implemented in the dentate gyrus of the hippocampus, where pattern separation is known to be performed, the novelty detection may be implemented in CA3 [15], and the normalization operation is the canonical neural computation [16]. Figure 1 shows the results of applying different methods to learn the highly correlated handwriting numbers. We see that only the orthogonal Hebb rule works well.

3 Learning a CANN from Continuous Morphed Patterns

3.1 A CANN of Discrete Dynamics

Before learning real images, we first try synthetic morphed patterns. As illustrated in Fig. 2A, these morphed patterns are constructed by shifting an active patch (taking values of ones, and all other units are zeros) step by step along the network. We select a set of continuously morphed patterns as the memories to be stored in the network, and use the orthogonal Hebb rule to construct the neuronal recurrent connections. The network dynamics follows Eq. (3). After learning, the energies of the network states given by Eq. (4) are shown in Fig. 2B. We see that those continuously morphed patterns are successfully learned as attractors (local minima) of the network. Furthermore, these attractors form a relatively flat valley of low energy in the network state space, and with more morphed patterns stored, the valley becomes more flat (see the inset in Fig. 2B). We also find that the learned neuronal recurrent connections are shift-invariant along the morphing parameter (Fig. 2C), a key property of CANNs. Overall, the conclusion is that when continuously morphed patterns are presented, a CANN is learned by the orthogonal Hebb rule.

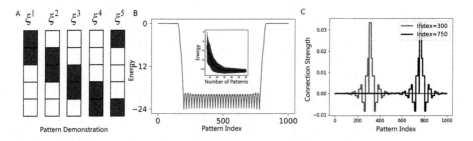

Fig. 2. (A) Illustrating the construction of morphed patterns. The network size $N = 5$ and the patch size $L = 2$. (B) The energies of all morphed patterns in the learned CANN. The network size $N = 1000$. Totally 1000 morphed patterns are constructed, with the patch size $L = 50$, and the patterns are indexed from 1 to 1000. We choose those patterns indexed from 200 to 800 to be the memory patterns. The inset shows the relationship between the energy of the valley and the number of stored patterns. The black line represents the mean of the valley energy averaged over different realizations of selecting stored patterns, and the green shadow denotes the standard deviation. (C) The learned recurrent connections from two neurons at different locations, which are translation-invariant. (Color figure online)

3.2   A CANN of Continuous Dynamics

To be biologically more plausible, we extend the above discrete model to be a continuous one, which is written as,

$$\tau\frac{dV_i}{dt} = -V_i + \sum_j W_{ij}g(V_i) + I_i, \tag{7}$$

where V_i denotes the state of the ith neuron, τ the time constant, I_i the external input, and $g(V_i) = \frac{1}{1+e^{-\alpha*V_i}}$ is a nonlinear function.

Consider that the network learns a set of continuously morphed (shift) gaussian bumps (Fig. 3A). We construct the neuronal connections **W** according to the orthogonal Hebb rule, which is translation-invariant among neurons (Fig. 3B). Figure 3C shows the attractors held by the network after learning, which form a one-dimensional manifold embedded in the high-dimensional space, i.e., a CANN is learned. Note that there are distortions between the training patterns and the attractors held by the network (Fig. 3A), nevertheless, one-to-one correspondences between them are established (Fig. 3C), meaning that the information of training patterns is acquired by the network. We can also diminish these distortions by applying the stationary states of the network as training patterns iteratively.

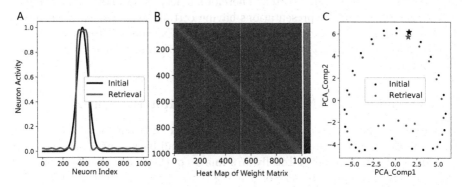

Fig. 3. (A) A training gaussian-shaped pattern (black) and the corresponding attractor learned by the network (red). (B) The translation-invariant recurrent connections between neurons after a set of continuously morphed gaussian bumps are learned. (C) The one-to-one correspondences between the training patterns and the attractors learned by the network. Using the first two dominant PCA components, it is shown that the attractors of the network form a one-dimensional manifold. The parameters: $\tau = 1, \alpha = 8$. (Color figure online)

4   Learning a CANN from Real Images

4.1   Generating Suitable Neural Representations

In the above study, we have presumed that the proper neural representations of objects are already available. In practice, this is not a trivial issue. Essentially, an unsupervised Hebb rule is to convert the correlation (overlap) between neural representations into the connection form between neurons, such that the network encodes not only objects but also their relationships. Thus, it is crucial that the neural representations of objects are organized, in term of that their correlation reflects a sematic, rather than a superficial, relationship between objects. In other words, we expect that the closeness between two neural representations should reflect that the two objects they represent are similar in categorization, rather than that the two objects are close in spatial locations. If the raw images of objects are used as the neural representations, then the network will not learn a meaningful result: the network will treat the same object at different locations to be different, whereas treat different objects at the same location to be similar. So, how does the neural system generate proper representations of objects?

Recently, the model of deep neural networks, which mainly mimics the feed-forward, hierarchical information processing in the dorsal visual pathway, has made great success in object recognition. Interestingly, it has been observed that in spite of the simple structure, a deep neural network pre-trained by a large number of natural images generates object representations (the neural activity preceding the read-out layer), which capture some statistical characteristics of neural responses in the higher visual cortex. It is known that the neural representations in the higher visual cortex are much more abstract than raw images and those in the early visual cortex. Therefore, a deep neural network provides a way to generate neural representations having certain-level of sematic meaning. In the below, we adopt a deep neural network (DNN) adapted from the Oxford VGG CNN-S network [17] trained by ImageNet to generate neural representations for objects.

4.2   Learning a CANN from Continuously Rotating Chairs

We collected a dataset of rotating chairs in an indoor environment. The telescopic chair was taken by a digital camera with a white wall background, and the chair was rotated from $0°$ to $360°$. Totally, 679 images of the chair were obtained, and they were cropped into the size of 224×224 to fit the input of the DNN. No other preprocessing to the inputs was done except normalization to $[0, 1]$. The neural representations generated by the DNN were then used as the memory patterns to construct a CANN. Figure 4A shows that indeed a CANN is learned, and the representations of the rotating chairs form a subspace of attractors.

To further confirm that the network does have the CANN structure, we carry out the mental rotation experiment. Mental rotation is an important characteristic of CANNs, an property coming from the flat subspace of low-energy in a CANN, and was observed in the experiment [18]. Mimicking the experimental setting, we set the network state to be stable initially at an angle, and then apply an external input pointing to a target angle. Under the drive of the external input, the network state changes from the initial to the target position. We record how the network state evolves during this process by measuring the instant overlap between the network state and all attractor states. Interestingly, we observe that if the target angle is not far away from the initial angle, the network state (a bump) rotates smoothly from the initial angle to the target one, displaying the mental rotation behavior as observed in the experiment (Fig. 4B).

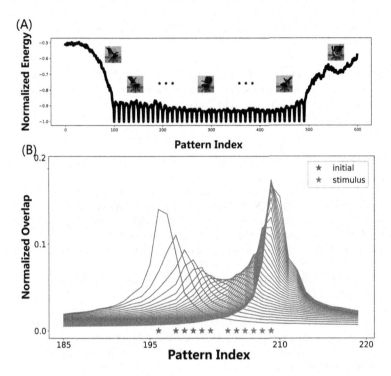

Fig. 4. (A) A CANN learned from continuously rotating chairs. The attractor corresponding to different rotating angles form a subspace of low energy. (B) Mental rotation in the CANN. Starting from an initial angle, the network state (bump) rotates smoothly to the target angle where the stimulus is applied. The parameters are the same as in Fig. 3.

5   Conclusion

The canonical neural representation model of CANNs is receiving increasing attention in both neuroscience and brain-inspired computation. In the present study, we explore the long-standing, unresolved question in the field, i.e., how the neural system acquires the CANN structure in practice. We focus on solving two technical challenges, namely, how to generate neural representations of sematic meaning and how to store correlated patterns in a neural network. For the former, we consider that a deep neural network, which mimics the dorsal visual pathway, generates suitable neural representations. For the latter, we show that the orthogonal Hebb rule, which may be achieved in the hippocampus, can learn correlated patterns efficiently. Tested on synthetic data and real images of rotating chairs, we demonstrate that our method learns CANNs successfully. We also test others real images and confirm that our method works very well (data not shown). Future research will look into extending the proposed approach to object tracking in real scene, as it has been shown that a CANN incorporating some synaptic mechanisms exhibits intrinsic mobility and produces different tracking behaviors [19].

Acknowledgments. This work was supported by BMSTC (Beijing municipal science and technology commission) under grant No: Z161100000216143 (SW), Z171100000117007 (DHW&YYM). The National Natural Science Foundation of China (31371109), National Key Basic Research Program of China (2014CB846101).

References

1. Hopfield, J.J.: Neural networks and physical systems with emergent collective computational abilities. Proc. Nat. Acad. Sci. **79**, 2554–2558 (1982)
2. Kim, S.S., Rouault, H., Druckmann, S., Jayaraman, V.: Ring attractor dynamics in the Drosophila central brain. Science **356**, 849–853 (2017)
3. Seelig, J.D., Jayaraman, V.: Neural dynamics for landmark orientation and angular path integration. Nature **521**, 186–191 (2015)
4. Amari, S.I.: Dynamics of pattern formation in lateral-inhibition type neural fields. Biol. Cybern. **27**, 77–87 (1977)
5. Zhang, K.: Representation of spatial orientation by the intrinsic dynamics of the head-direction cell ensemble: a theory. J. Neurosci. **16**, 2112–2126 (1996)
6. Wu, S., Wong, K.M., Fung, C.A., Mi, Y., Zhang, W.: Continuous attractor neural networks: candidate of a canonical model for neural information representation. F1000Research, **5** (2016)
7. Yoon, K., Buice, M.A., Barry, C., Hayman, R., Burgess, N., Fiete, I.R.: Specific evidence of low-dimensional continuous attractor dynamics in grid cells. Nat. Neurosci. **16**, 1077–1084 (2013)
8. Mante, V., Sussillo, D., Shenoy, K.V., Newsome, W.T.: Context-dependent computation by recurrent dynamics in prefrontal cortex. Nature **503**, 78 (2013)
9. Yamins, D.L., Hong, H., Cadieu, C.F., Solomon, E.A., Seibert, D., DiCarlo, J.J.: Performance-optimized hierarchical models predict neural responses in higher visual cortex. Proc. Nat. Acad. Sci. **111**, 8619–8624 (2014)
10. Kropff, E., Treves, A.: Uninformative memories will prevail: the storage of correlated representations and its consequences. HFSP J. **1**, 249–262 (2007)
11. Blumenfeld, B., Preminger, S., Sagi, D., Tsodyks, M.: Dynamics of memory representations in networks with novelty-facilitated synaptic plasticity. Neuron **52**, 383–394 (2006)
12. Leutgeb, J.K., Leutgeb, S., Treves, A., et al.: Progressive transformation of hippocampal neuronal representations in morphed environments. Neuron **48**, 345–358 (2005)
13. Wills, T.J., Lever, C., Cacucci, F., Burgess, N., O'keefe, J.: Attractor dynamics in the hippocampal representation of the local environment. Science **308**, 873–876 (2005)

14. Srivastava, V., Sampath, S., Parker, D.J.: Overcoming catastrophic interference in connectionist networks using gram-schmidt orthogonalization. PLoS One **9**, e105619 (2014)
15. Kumaran, D., Hassabis, D., McClelland, J.L.: What learning systems do intelligent agents need? Complementary learning systems theory updated. Trends Cogn. Sci. **20**, 512–534 (2016)
16. Carandini, M., Heeger, D.J.: Normalization as a canonical neural computation. Nat. Rev. Neurosci. **13**, 51–62 (2012)
17. Simonyan, K., Zisserman, A.: Very deep convolutional networks for large-scale image recognition. arXiv preprint arXiv:1409.1556 (2014)
18. Georgopoulos, A.P., Taira, M., Lukashin, A.: Cognitive neurophysiology of the motor cortex. Science **260**, 47–52 (1993). New York then Washington
19. Mi, Y., Fung, C.A., Wong, K.M., Wu, S.: Spike frequency adaptation implements anticipative tracking in continuous attractor neural networks. In: Advances in Neural Information Processing Systems, pp. 505–513 (2014)

Active Prediction in Dynamical Systems

Chun-Chung Chen[1(✉)], Kevin Sean Chen[1,2], and C.K. Chan[1,3]

[1] Institute of Physics, Academia Sinica, Taipei, Taiwan, Republic of China
cjj@phys.sinica.edu.tw
[2] Department of Life Science, National Taiwan University,
Taipei, Taiwan, Republic of China
[3] Department of Physics and Center for Complex Systems,
National Central University, Chungli, Taiwan, Republic of China

Abstract. Using a hidden Markov model (HMM) that describes the position of a damped stochastic harmonic oscillator as a stimulus input to a data processing system, we consider the optimal response of the system when it is targeted to predict the coming stimulus at a time shift later. We quantify the predictive behavior of the system by calculating the mutual information (MI) between the response and the stimulus of the system. For a passive sensor, the MI typically peaks at a negative time shift considering the processing delay of the system. Using an iterative approach of maximum likelihood for the predictive response, we show that the MI can peak at a positive time shift, which signifies the functional behavior of active prediction. We find the phenomena of active prediction in bullfrog retinas capable of producing omitted stimulus response under periodic pulse stimuli, by subjecting the retina to the same HMM signals encoded in the pulse interval. We confirm that active prediction requires some hidden information to be recovered and utilized from the observation of past stimulus by replacing the HMM with a Ornstein–Uhlenbeck process, which is strictly Markovian, and showing that no active prediction can be observed.

Keywords: Retina · Mutual information · Predictive dynamics · Omitted stimulus response · Stochastic process

1 Introduction

Biological systems are built to provide functions that help the continuation of the organisms. An important function for the neural systems in animals is to predict future conditions of their environment so the animals can anticipate coming events and react accordingly to increase their chance of survival. An example of such anticipation is the omitted stimulus response (OSR) which has been observed in lives as simple as amoeba [1] or even organs such as retina in animals [2]. In the OSR phenomena of retina, the periodicity information of the input stimulus is retained by the retina and a well-timed response is produced right after the periodic stimulus is removed. Such is a very simple

© Springer International Publishing AG 2017
D. Liu et al. (Eds.): ICONIP 2017, Part IV, LNCS 10637, pp. 632–638, 2017.
https://doi.org/10.1007/978-3-319-70093-9_67

case of anticipation and it has been shown that the function of producing well-timed OSR can be realized with an adaptive FitzHugh–Nagumo excitable and oscillatory system [3].

Naturally, it is desirable to quantify the predictive properties of retina through the study of OSR. However, while a strictly periodic stimulus carries minimal information rate, it is difficult to identify or even produce OSR when there are fluctuations in the inter-pulse intervals. Furthermore, it is unclear how to differentiate behaviors of the systems that are acting as a passive sensor or recorder from that are actively predicting coming events. To quantify the predictive properties of a data processing system, Bialek and Tishby introduced the idea of predictive information based on the mutual information (MI) between the momentary output of the system and stimulus input at a time shift later [4,5]. This idea was applied to describe the response of a retina to a stimulus in the form of a stochastic moving bar [6]. The retina was shown to provide predictive information at near optimal level under a constrain of limited memory capacity.

In this paper, we consider the hidden Markov model (HMM) that controls the stochastic moving bar in [6] and quantify how well such a stimulus can be predicted by an idealized system. We show that by using the hidden variable of the HMM, one can actively produce signals that is optimized to match the stimulus at a targeted time in the future. By encoding the same signal in the pulse intervals to the retina in a setup that can produce OSR, we show that the retina can perform a similar active prediction of coming signals when the information rate of the stimulus is low [7]. We propose that such active prediction is only possible with the help of some hidden information such as that in the HMM. This proposal is checked in a modification of the retina experiment where the HMM is replaced by an Ornstein–Uhlenbeck (OU) process [8], which has no hidden information, while maintaining the mean, correlation time, and standard deviation of the input signal, and no active predictions can be observed.

2   Predicting Stochastic Signal with Hidden Variable

We consider a discrete time sequence signal $\{\tau_i\}$ from a hidden Markov model following the idea of [6], which describes a damped harmonic oscillator driven by a noise. The generation of τ_i is described by the equations,

$$\tau_{i+1} = \tau_i + v_i \Delta \tag{1}$$

$$v_{i+1} = (1 - \Gamma \Delta) v_i - \omega^2 \tau_i \Delta + \xi_i \sqrt{D \Delta} \tag{2}$$

where the hidden variable v is the change rate for the observable τ; ξ is a unit Gaussian noise with zero mean and $D = 2$ controls the amplitude of the noise term. We fix the iteration step size Δ at $1/60$ s and keep $\Gamma/(2\omega)$ at 1.06 so that the oscillator is slightly over damped. Figure 1 shows a typical input sequence of $\{\tau_i\}$ generated by the HMM.

Imagine a smart agent who has been observing the sequence for a very long time. It must be able to recover the dynamic Eqs. (1) and (2) as well as all

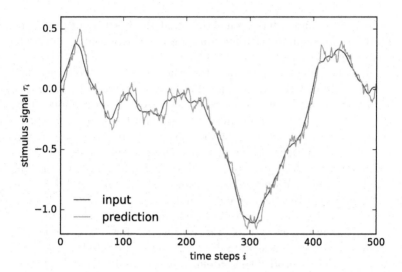

Fig. 1. Stimulus input generated by the dynamic Eqs. (1) and (2) compared with the prediction from $n = 10$ steps earlier using Eqs. (3)–(5)

the parameters used to generate the sequence. For a prediction of the stimulus at n-th step in the future, the agent can simply iterate the dynamics (1) and (2) for n steps to obtain τ_{t+n} where t is the current time. The only missing information for such iterations is the actual value of the noise ξ at each step. Nonetheless, noting that the distribution of the noise ξ can also be obtained from past observations, the agent can choose to use the most probable value $\xi_i = 0$ at each step in performing the iterations as described below.

With the observations τ_{i-1} and τ_i, we can derive the value v_{i-1} as

$$v_{i-1} = \frac{1}{\Delta}\left(\tau_i - \tau_{i-1}\right). \tag{3}$$

We then estimate the value of v_i, assuming the most probable value of ξ_{i-1}, namely, zero:

$$\tilde{v}_i = (1 - \Gamma\Delta)\,v_{i-1} - \omega^2\tau_{i-1}\Delta \tag{4}$$

where the tilde over a symbol denotes an estimation. The next τ_{i+1} can thus be estimated by

$$\tilde{\tau}_{i+1} = \tau_i + \tilde{v}_i\Delta. \tag{5}$$

For a prediction targeted at n steps in the future, the iterations (3)–(5) are repeated n times to obtain $\tilde{\tau}_{i+n}$. The result of the prediction for 10 steps in the future is compared in Fig. 1 with the stimulus input at the targeted time. As shown in Fig. 1, the predictive response has greater fluctuations than the actual stimulus. This likely follows the fact that predictions are based on trends which can overshoot and be corrected by new observations. Similar behavior can be found, for example, in a financial market, where the derivative securities, which are speculative in nature, are generally more volatile than the corresponding securities.

We calculate the mutual information between the predictive response actively produced by the idealized system and the stimulus input as a function of the time shift between the two signals, for different numbers of targeted time steps n into the future. The results are shown in Fig. 2, where we can see the MI peak moves towards the positive δt direction as the system actively predicts further into the future. From these results, we can also see the peaks of the MI are generally above the auto-mutual information curve. This indicates that the predictive output of the system is more informative of the future stimulus than the signal itself at these time shifts. This is only possible when the hidden information can be recovered from the history of past stimulus and utilized by the system in producing the predictive responses.

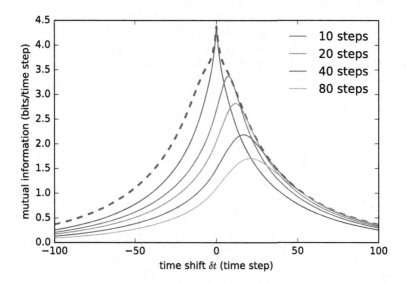

Fig. 2. Mutual information as functions of time shift for active predictive responses produced by an idealized system for different number of targeted time steps. The blue curve shows the auto-mutual information which can correspond to a "prediction" at 0 step ahead using the input signal itself as the output. The dashed green line is the mutual information between the full internal state (v, τ) of the HMM and the presented input τ, which represents the upper bound of the mutual information any processing system can have with the input signal (Color figure online)

In Fig. 2, we also calculate the mutual information between the full internal state of the HMM, which includes both v and τ, and the stimulus input τ of the system at different time shift. Since the full internal state of an HMM is all that is relevant for producing the next state of the observable τ, this represents an upper bound of the mutual information the output of any system can have with the stimulus. From our calculation shown in Fig. 2, we see the MI of the active predictions using the iterative method described above closely approaches the upper bound at the time shifts that the responses are targeted to predict. On the other hand, while the peak position δt_p of the MI curve becomes more positive when targeted time steps n becomes larger, its movement stalls significant with increasing n. Therefore, while the position of the MI peak is an indication of an active prediction, it does not faithfully reflect the target of such prediction.

3   Active Prediction in Retina

To study the predictive behavior of a retina, our experiment is similar to that of Schwartz et al. [2,9] for the study of OSR except that retinas from bullfrogs are used in our setup as detailed in [7]. Instead of using a periodic pulse train as stimulus, we use it as a carrier and modulate the signal τ_i generated by the HMM dynamics (1) and (2) in the variation of the pulse intervals s_i as follows: After $\{\tau_i\}$ is generated, the signal is rescaled so to have a standard deviation of 20 ms. An offset around 200 ms is also added to $\{\tau_i\}$ to obtain the desired mean $\langle s \rangle$ so to keep the system operating near the dynamics range of OSR. Beside the value of the Γ parameter in the HMM, the correlation time of s_i is also affected by the rescaling as well as the offset process. The values of correlation time as shown in Fig. 3 are measured retroactively by computing the decay time of the autocorrelation function of the pulse intervals.

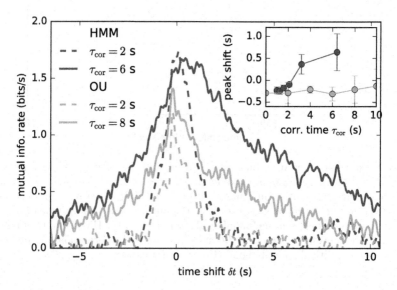

Fig. 3. Mutual information curves between input and output from the retina under pulse stimulus with stochastic intervals generated from hidden Markov model (blue) and Ornstein–Uhlenbeck process (green) with different correlation time as indicated. The Inset shows the peak positions of the curves as a function of correlation time of the stimulus with error bars indicating variations across recorded channels (Color figure online)

Beside the stimulus generated by the HMM, we also subject the retina to the stimulus generated by a discrete OU process, following the dynamics

$$\tau_{i+1} = \tau_i - \frac{1}{T}\tau_i \Delta + \xi_i \sqrt{D\Delta} \tag{6}$$

where D and Δ are identical in values to that in the HMM described above, while the parameter T is used to control the correlation time of the stimulus generated by the OU process. The same scaling and offset procedures are performed on the OU interval sequences to ensure that they have the same mean and variance as that from the HMM. We see in Fig. 3 the MI peaks at $\delta t_p < 0$ for both kinds of the stimulus inputs when the correlation time is short, e.g., $\tau_{\mathrm{cor}} = 2$, which corresponds to high information rate from the input. For low information rate, or long correlation time, $\tau_{\mathrm{cor}} \gtrsim 3$, the peak position of the MI shifts to $\delta t_p > 0$ in the case of the HMM, indicating the behavior of active prediction. On the other hand, the MI peak for the OU stimulus remains at the same $\delta t_p < 0$ indicative of a processing delay that can be expected for a passive sensor.

4   Discussion

As shown above, using the hidden information recovered from past observations of an HMM stimulus, one can actively produce responses that are optimally informative of the stimulus at some targeted time in the future. Such active prediction can be more informative of the stimulus for the targeted time than a perfect sensor that faithfully copies the input to the output. It is long realized that some biological systems such as retinas are doing more than a sensor in processing the input signal, for example, in producing the OSR. Here, we propose to quantify the active prediction of the system by considering the mutual information between the input and output at different time shifts as suggested by Bialek and Tishby [4].

We identify the functional behavior of active prediction when the peak of the MI curve moves to a positive time shift indicating the instant output of the system is most informative of the stimulus input at another instance in the future. There are two key components to this behavior: the retention of information from the past stimulus and the computation to filter out information that is not pertinent to the stimulus of the targeted time. The former allows the recovery of hidden information that are not directly observable by the system. The end point of this recovery is the full internal state of the upstream system, e.g., the HMM, that generates this stimulus. This full internal state has an MI curve with the stimulus as shown by the green dashed line in Fig. 2 for the specific HMM we consider here. Without further constraints, such as the information bottleneck [10], on the processing system, this limit can be simply approached by a system that records everything and outputs everything.

The second key point of the active prediction is the computation, or the filtering of information. It can tell the intention of the system. In our case, it is the number of time steps n to the future targeted by the predictive behavior of our idealized system. We note that while for small n, the peak of MI is close to the target, for large n, the peak position is actually a significant underestimate of n. The target n is best estimated by the point where the MI curves approach the upper bound as shown by the green dashed line in Fig. 2. Unfortunately, this upper bound of MI is not readily available in an experimental system and real biological systems are likely not optimally predictive of their stimulus.

Finally, we show that a system can only produce active predictions for stimulus that is actively predictable, that is, there is some hidden information that can be recovered from past observations and used in bettering the prediction. For the OU stimulus input, while there is nonzero predictive information in the sense defined in [4], the peak of MI remains at a lag $\delta t < 0$, and the response produced by the retina is never actively predictive.

5 Conclusion

In this paper, we introduced the concept of active prediction, which can set apart some information processing systems from passive sensors. We showed how such functional behavior can be identified through the calculation of mutual information between stimulus and response. And, we provided evidence of such predictive behavior in a bull frog retina.

References

1. Saigusa, T., Tero, A., Nakagaki, T., Kuramoto, Y.: Amoebae anticipate periodic events. Phys. Rev. Lett. **100**(1), 018101 (2008)
2. Schwartz, G., Harris, R., Shrom, D., Berry, M.J.: Detection and prediction of periodic patterns by the retina. Nat. Neurosci. **10**(5), 552–554 (2007)
3. Yang, Y.J., Chen, C.C., Lai, P.Y., Chan, C.K.: Adaptive synchronization and anticipatory dynamical systems. Phys. Rev. E **92**(3), 030701 (2015)
4. Bialek, W., Tishby, N.: Predictive information (1999). arXiv:cond-mat/9902341
5. Rubin, J., Ulanovsky, N., Nelken, I., Tishby, N.: The representation of prediction error in auditory cortex. PLoS Comput. Biol. **12**(8), e1005058 (2016)
6. Palmer, S.E., Marre, O., Berry, M.J., Bialek, W.: Predictive information in a sensory population. Proc. Natl. Acad. Sci. **112**(22), 6908–6913 (2015)
7. Chen, K.S., Chen, C.C., Chan, C.K.: Characterization of predictive behavior of a retina by mutual information. Front. Comput. Neurosci. **11**, 66 (2017)
8. Uhlenbeck, G.E., Ornstein, L.S.: On the theory of the brownian motion. Phys. Rev. **36**(5), 823–841 (1930)
9. Schwartz, G., Berry, M.J.: Sophisticated temporal pattern recognition in retinal ganglion cells. J. Neurophysiol. **99**(4), 1787–1798 (2008)
10. Tishby, N., Pereira, F.C., Bialek, W.: The information bottleneck method (2000). arXiv:physics/0004057

A Biophysical Model of the Early Olfactory System of Honeybees

Ho Ka Chan[(⊠)] and Thomas Nowotny[(⊠)]

School of Engineering and Informatics, University of Sussex,
Falmer, Brighton BN1 9QJ, UK
{hc338, t.nowotny}@sussex.ac.uk

Abstract. Experimental measurements often can only provide limited data from an animal's sensory system. In addition, they exhibit large trial-to-trial and animal-to-animal variability. These limitations pose challenges to building mathematical models intended to make biologically relevant predictions. Here, we present a mathematical model of the early olfactory system of honeybees aiming to overcome these limitations. The model generates olfactory response patterns which conform to the statistics derived from experimental data for a variety of their properties. This allows considering the full dimensionality of the sensory input space as well as avoiding overfitting the underlying data sets. Several known biological mechanisms, including processes of chemical binding and activation of receptors, and spike generation and transmission in the antennal lobe network, are incorporated in the model at a minimal level. It can therefore be used to study how experimentally observed phenomena are shaped by these underlying biophysical processes. We verified that our model can replicate some key experimental findings that were not used when building it. Given appropriate data, our model can be generalized to the early olfactory systems of other insects. It hence provides a possible framework for future numerical and analytical studies of olfactory processing in insects.

Keywords: Insect olfaction · Honeybees · Model · Biophysical

1 Introduction

To develop a quantitative understanding of animals' sensory systems, researchers build mathematical models based on experimental data. Unfortunately, due to limitations in experimental techniques, these data are necessarily noisy and incomplete. For example, in honeybees, responses of only around 30 of a total of 160 known types of olfactory receptor neurons (ORNs) can routinely be measured [1, 2]. A typical modelling approach is to create reduced models using incomplete data. An example is the olfactory model in [3], which contains only the ORNs and corresponding glomeruli in the antennal lobe for which the data are available. However, it is unclear whether such models sufficiently relate to the biological systems they aim to describe, since the scaling of noise, synaptic efficacies and finite size network effects may change the dynamics of the system significantly. In addition, many numerical models create response patterns by directly fitting data from experiments in which specific properties

© Springer International Publishing AG 2017
D. Liu et al. (Eds.): ICONIP 2017, Part IV, LNCS 10637, pp. 639–647, 2017.
https://doi.org/10.1007/978-3-319-70093-9_68

of a subset of the systems are measured [3, 4]. It cannot be expected that the predictions from these models would reflect other properties of the observed subsystem or any characteristics of the remainder of the system. It would also be unlikely for these models to be consistent with unrelated experimental data not used to build them. Yet other models are purely phenomenological [5]. In this case it is difficult to address how the processing and coding of stimuli are implemented biologically. It is, therefore, highly desirable to develop statistical models that both consider inputs representative of the full sensory input space of the animals to guard against over-fitting to limited data and at the same time incorporate relevant underlying biophysical processes to allow relating the model back to biology.

In this work, we illustrate a method for building full-size models of animals' sensory systems that extrapolates inputs from a limited subset of available experimental observations using the example of the early olfactory system of honeybees. The resulting model comprises the full number of 160 different types of ORNs as well as local neurons (LNs) and projection neurons (PNs) organized in 160 corresponding glomeruli. The ORN response patterns are generated using a set of ordinary differential equations describing the binding and activation of receptors closely related to the actual biological processes [3, 4], while the response of PNs are determined by network input from ORNs and LNs using a simple rate model derived from the leaky integrate-and-fire model.

The remainder of the paper is organized as follows. In Sect. 2, we describe in details how our model was built. In Sect. 3, we show that our model reproduces key features of ORN and PN responses to continuous stimuli (3.1) and short pulses (3.2) observed in experimental work that was not considered when building the model [9, 10, 13, 14]. In Sect. 4, we discuss the strength and limitations of our model and our plans for future work.

2   Methods

In our model, responses from the same type of ORNs, LNs or PNs are approximated by their ensemble average. We therefore use a single unit to represent all units of the same type. In this report, we refer to each type of a certain entity by its representatives (e.g. 20 types of ORNs will be referred to as 20 ORNs).

2.1   Asymptotic Response of Model ORNs for Time-Invariant Odor Inputs at High Concentration

Asymptotic responses of 28 ORNs for time-invariant odor inputs at high concentration to 16 different odors have been measured using calcium imaging of glomeruli with bath-applied Ca^{2+} dyes [1]. We adopted these responses directly to form the responses of the first 28 ORNs in our model. We then generated the responses of the remaining 132 ORNs to the same 16 odors using a method inspired by [6]. The response patterns were generated from a combination of previously generated responses, including those from [1], and noise. The parameters were chosen such that the statistical distribution of the pairwise correlations of ORNs across odours in the generated responses matches that of the 28 ORNs adopted from data. The generated responses were then rescaled such that the mean and the variance of the responses for all receptor-odour combinations, and the mean and the variance of the variance of the responses across odours for each ORN also match.

ORN responses to chemically similar odours are correlated [7]. In our model, such correlations are quantified using the normalized Euclidean distances d_{ij} between the response vectors of 2 odours i and j, denoted by x_i and x_j, as in [7].

$$d_{ij} = \sqrt{\frac{\sum_k (x_{ik} - x_{jk})^2}{N}}, \tag{1}$$

where N is the total number of odours in our input space and the subscript k labels the different ORNs.

The responses of all previously generated ORNs were then iteratively tuned so that the Euclidean distance matrix d for the generated response patterns matches that calculated from the experimental data. The tuning processes are designed to cause insignificant changes to the statistical quantities calibrated previously.

2.2 Time Series Response at Other Concentrations

The time series response to a single odour stimulus was generated using a set of ordinary differential equations describing the binding and activation of receptors as in [3]:

$$\begin{cases} \dot{r} = k_{-1}r_b - k_1 r c^n \\ \dot{r}_b = k_1 r c^n - k_{-1}r_b + k_{-2}r_b^* - k_2 r_b \\ \dot{r}_b^* = k_2 r_b - k_{-2}r_b^* \\ r + r_b + r_b^* = r_0 \end{cases}, \tag{2}$$

where k_1 (k_{-1}) and k_2 (k_{-2}) are the (un)binding constants and (de)activation constants respectively, c is the concentration of the odor, n describes the effects of the transduction cascade, and r, r_b and r_b^* are the 'effective concentration' of free, bound and activated receptors such that r_b^* is proportional to the excitatory conductance of the ORN. The sum of the number of receptors in different states is equal to the total number of available receptors, r_0, as described by the last equation.

Denoting r_b^* as the receptor response, the equilibrium response-dose relationship can be described by Hill curves when a time-invariant stimulus is applied [4]. n, k_1 (k_{-1}) and k_2 (k_{-2}) are partially constrained by the parameters in the Hill curves, which are statistically sampled in accordance to experimental observations in [8]. To deal with the remaining degrees of freedom, we took into account the typical timescale of dynamics in AL responses measured experimentally [9, 10].

2.3 Obtaining the Instantaneous Firing Rate of Neurons

To obtain the firing rate of the ORNs from its input conductances (which are assumed to be proportional to r_b^*; see 2.2), the dynamics of a neuron is approximated by the conductance-based leaky integrate-and-fire model with adaptation as shown in (3).

$$\tau_{eff}(t)\frac{dV}{dt} = -V + RI_{eff}(t) - RI_{adapt}(t)$$

$$\tau_{adapt}\frac{dI_{adapt}(t)}{dt} = -I_{adapt}(t) \tag{3}$$

$$I_{adapt} = I_{adapt}^{max} \text{ at } t = t_f$$

Detailed descriptions of the parameters in (3) can be found in [11]. We then adopted the adiabatic approximation by considering the input to be quasi-time-invariant on the time scale of neuronal firing, such that $\tau_{eff}(t)$ and $I_{eff}(t)$ are taken to be constant. With the additional assumption of noise-free input and setting $t_f = 0$, the membrane potential before the next firing event can be obtained analytically as follows:

$$V = V_{reset}e^{\frac{-t}{\tau_{eff}}} + I_{eff}\left(1 - e^{\frac{-t}{\tau_{eff}}}\right) - \frac{\tau_{adapt}I_{adapt}^{max}}{\tau_{adapt} - \tau_{eff}}\left(e^{\frac{-t}{\tau_{adapt}}} - e^{\frac{-t}{\tau_{eff}}}\right), \tag{4}$$

where V_{reset} is the reset potential after the neuron has fired. The instantaneous firing rate of the neuron can then be obtained using:

$$v = \frac{1}{t_{thres}} + t_{refract}, \tag{5}$$

where t_{thres} is the time when $V = V_{th}$, which is to be obtained numerically, and $t_{refract}$ is the absolute refractory period. Note that in (4) and (5), we have set $R = 1$ by absorbing it into I_{adapt}^{max} and other variables.

We chose $I_{adapt}^{max} = I_{adapt}^{base}\sqrt{r_b^*}$ for ORNs, and $I_{adapt}^{max} = I_{adapt}^{base}\sqrt{v_{pre}}$ for PNs and LNs, where I_{adapt}^{base} is a constant and v_{pre} is the firing rate of the corresponding units in the previous iteration. However, qualitatively similar results can be obtained by assuming I_{adapt}^{max} to be constant (results not shown).

2.4   Generating PN Responses

In our model, ORNs provide excitatory input to PNs and LNs. Both receive excitatory input from the ORNs of their own glomerulus, with uniform connectivity, as well as inhibitory input from LNs of all other glomeruli. This is illustrated in Fig. 1.

To be consistent with the findings in [12], for any pair of glomeruli i and j, the connectivity between the PN in glomerulus i and the LN in glomerulus j is weighted by w_{ij}, which is based on the correlations ρ_{ij} between the corresponding ORN responses across different odours,

$$w_{ij} = (1 - \delta_{ij})\left[w_0 + H(\rho_{ij}) \times \rho_{ij}w_{corr}\right], \tag{6}$$

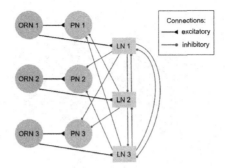

Fig. 1. Illustration of part of the model AL network.

where δ_{ij} is the Kronecker delta function, H is the Heaviside step function, w_0 and w_{corr} are normal distributed random variables. ρ_{ij} is the Pearson correlation between the conductance of ORNs i and j as obtained in 2.1, across odours.

$$\rho_{ij} = \frac{\text{Cov}\left(x_i^T, x_j^T\right)}{\sigma(x_i^T)\sigma\left(x_j^T\right)} \tag{7}$$

The firing rate of LNs and PNs are calculated by (4) and (5). The calculations are iterated several times to allow the system to settle into a steady state. This effectively assumes that any oscillations in PN activity are negligible and thus ignored.

3 Results

3.1 Response to Continuous Stimuli

Using the methods described in Sect. 2, we built a model that can generate asymptotic and time series responses of all 160 glomeruli to 16 different odours. We first tested the ORN responses generated by our model with continuous stimuli. As an example, the asymptotic ORN responses to 1-hexanol at two different concentrations is shown in Fig. 2. At low concentration, most ORNs are quiescent; while at high concentration, almost all ORNs are activated to some degree.

We next compared the results of PN responses obtained from our model to Ca^{2+} imaging data with back-filled PNs [13]. Previous studies [2] have shown that unlike ORNs, which responses almost always increase with dose, PN responses display a variety of relationships with dose due to inhibition from LNs. In Fig. 3, we divide the response-dose relationships into 4 different types: "inactivated" where PNs show no or very weak responses to stimuli at any dose, "decrease" ("increase") where responses decrease (increase) with dose and "other" where responses are independent of or display non-monotonic relationships with dose, and show that the statistical distribution of each type of response-dose relationship observed in the model PNs matches

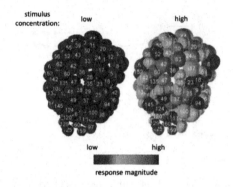

Fig. 2. The model predicted asymptotic ORN responses to 1-hexanol at low ($c = 10^{-4}M$, left) and high ($c = 10^{-1}M$, right) concentrations.

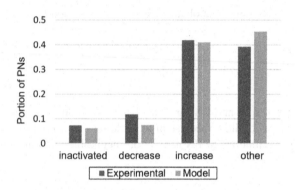

Fig. 3. Statistical distribution of different kinds of response-dose relationships observed in model PN responses (right bars) and experimental PN data from [7] (left bars).

very well to experimental PN data in [13]. This suggests that our network model can reasonably capture the effects of LN inhibition.

Finally, we compared the ORN and PN responses from our model. Figure 4 (left) shows the probability distribution of pairwise correlations between experimental and model ORN and PN responses across odours. While the model ORN responses are highly correlated as in the corresponding experimental ORN data, the model PN responses are mostly uncorrelated, with the peak of the probability around zero correlation. Our model results for both ORNs and PNs show a good fit to their experimental counterparts in [1, 13] qualitatively, even though the model was only fitted to ORN data. Please note that the 'model responses' correspond to firing rate, which may not map directly to Ca^{2+} imaging data. Figure 4 (right) shows that the correlation between our model ORN and PN response patterns to different odours centred around 0.7, which agrees with experimental data [14], which shows that the correlation between ORN and AL activity is around 0.6–0.7 (Fig. 5).

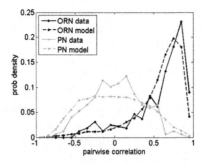

Fig. 4. Statistical distribution for pairwise correlations for different ORNs ($c = 1$) and PNs ($c = 0.1$) across response patterns, observed in Ca^{2+} imaging experiments [1, 13] and our model.

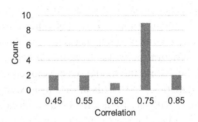

Fig. 5. Statistical distribution of the pairwise correlations between the overall ORN and PN response for different odour stimuli.

3.2 Response to Short Pulses

In addition to continuous stimuli, we further tested our model with stimuli consisting of short pulses. Figure 6 shows the average ORN responses to 1-hexanol to square pulses of 2 ms with different inter-pulse intervals and compares them to those measured by electro-antennogram recordings [9]. The model responses exhibit all relevant features

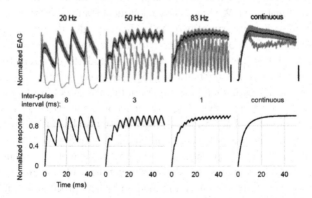

Fig. 6. ORN responses to pulsed and constant stimuli (top, red line) as measured by electro-antennogram recordings [9] (top, black line) and the average normalized ORN responses to similar stimuli (1-hexanol at concentration 0.1 M in square pulses of 2 ms) predicted by our model (bottom). (Color figure Online)

observed in the experimental data except that the time scale of response latency is smaller. This can be explained by the lack of temporal filtering of input conductance to output spiking in our rate model.

4 Discussion

In this work, we demonstrated how we built a mathematical model of the early olfactory system of honeybees using severely limited data. We were able to generate ORN response patterns using a very simple biophysical model of receptors and generate responses of other parts of the system using a network model mimicking the widely accepted structure of honeybees' antennal lobe. The response to different types of stimuli predicted by our model matches that obtained from experimental measurements very well on the statistical level. Our methods make full use of our knowledge about honeybees' olfactory system and require extremely small amounts of computation. They are general enough such that, given appropriate data, they could be readily applied to model olfactory systems of other insects, or, with slight modifications to the receptor dynamics and the neural network, to model other sensory systems.

Unlike model in previous work, the model we developed here is not directly fitted to data, but to their overall statistics [3, 4]. This allows us to study the statistical responses to different types of stimuli, with the trade-off that a generated model glomerulus may not correspond to any particular glomerulus in a honeybee. On the other hand, it allows us to model the full number of 160 glomeruli, even though data normally are only acquired from around 30. Biologically, responses of a sensory unit to a stimulus may differ greatly among individual animals due to factors like genetic heterogeneity and past learning experience. Moreover, many coding strategies are believed to be based on ensemble behaviours [2, 3, 14]. Therefore, we believe that reproducing the statistics of observed data, rather than detailed measurements of individual cells, is more useful when modelling sensory systems of animals in most circumstances.

In this work, we used a simple firing rate model to describe the input-output relationship of neurons. A major benefit is that it is analytically tractable and drastically reduces computational costs. We have shown that most features in experimental findings can be reproduced with such a simplified model. In the model, the major approximations are the adiabatic approximation in the firing rate and the absence of input noise. The lack of temporal filters for output spiking in the model leads to an overestimation in the sensitivity of neurons to input fluctuations [15]. However, in this context, such effects are alleviated by the additional temporal filters in the process of binding and activation of receptors, as described in (3), which smoothen I_{eff} to some extent even if the stimulus intensity fluctuates rapidly. The absence of input noise also has little consequence since most of the neurons are mean-driven. Mathematical formalisms [15, 16] have been developed such that the mean firing rate can be computed without having to make the above-mentioned approximations, but this is beyond the scope of this work.

Our future work will involve analysing the biophysical processes which give rise to the above results [17]. We also aim to modify (3) to study the olfactory response to mixtures. Preliminary results [17] suggest that receptor dynamics account for differences in olfactory processing between complex mixtures and simpler compounds, which may lead to more efficient and robust coding for complex mixtures.

References

1. Galizia, C.G., Sachse, S., Rappert, A., Menzel, R.: The glomerular code for odor representation is species specific in the honeybee Apis mellifera. Nat. Neurosci. 2(5), 473–478 (1999)
2. Sachse, S., Galizia, C.G.: The coding of odour-intensity in the honeybee antennal lobe: local computation optimizes odour representation. Eur. J. Neurosci. 18(8), 2119–2132 (2003)
3. Nowotny, T., Stierle, J.S., Galizia, C.G., Szyszka, P.: Data-driven honeybee antennal lobe model suggests how stimulus-onset asynchrony can aid odour segregation. Brain Res. 1536, 119–134 (2013)
4. Rospars, J.-P., Lansky, P., Chaput, M., Duchamp-Viret, P.: Competitive and noncompetitive odorant interactions in the early neural coding of odorant mixtures. J. Neurosci. 28(10), 2659–2666 (2008)
5. Luo, S.X., Axel, R., Abbott, L.F.: Generating sparse and selective third-order responses in the olfactory system of the fly. Proc. Natl. Acad. Sci. USA 107(23), 10713–10718 (2010)
6. Haenicke, J.: Modeling insect inspired mechanisms of neural and behavioral plasticity. [PhD thesis]. Berlin: Freie Universität Berlin (2015)
7. Carcaud, J., Hill, T., Giurfa, M., Sandoz, J.: Differential coding by two olfactory subsystems in the honeybee brain. J. Neurophysiol. 108, 1106–1121 (2012)
8. Grémiaux, A., Nowotny, T., Martinez, D., Lucas, P., Rospars, J.-P.: Modelling the signal delivered by a population of first-order neurons in a moth olfactory system. Brain Res. 1434, 123–135 (2012)
9. Szyszka, P., Gerkin, R.C., Galizia, C.G., Smith, B.H.: High-speed odor transduction and pulse tracking by insect olfactory receptor neurons. Proc. Natl. Acad. Sci. USA 111(47), 16925–16930 (2014)
10. Szyszka, P., Stierle, J.S., Biergans, S., Galizia, C.G.: The speed of smell: odor-object segregation within milliseconds. PLoS One 7(4), 4–7 (2012)
11. Chan, H.K., Yang, D.-P., Zhou, C., Nowotny, T.: Burst firing enhances neural output correlation. Front. Comput. Neurosci. 10(May), 1–12 (2016)
12. Linster, C., Sachse, S., Galizia, C.G.: Computational modeling suggests that response properties rather than spatial position determine connectivity between olfactory glomeruli. J. Neurophysiol. 93(6), 3410–3417 (2005)
13. Ditzen, M.: Odor concentration and identity coding in the antennal lobe of the honeybee Apis mellifera. [PhD thesis]. Berlin: Freie Universität Berlin (2005)
14. Deisig, N., Giurfa, M., Sandoz, J.C., Deisig, N., Giurfa, M., Sandoz, J.C.: Antennal lobe processing increases separability of odor mixture representations in the honeybee. J. Physiol. 103, 2185–2194 (2010)
15. Ostojic, S., Brunel, N.: From spiking neuron models to linear-nonlinear models. PLoS Comput. Biol. 7(1), 1–16 (2011)
16. Chan, H.K., Nowotny, T.: Firing probability for a noisy leaky integrate-and-fire neuron receiving arbitrary external currents. In: Poster Presented at the 3rd International Conference on Mathematical Neuroscience, Boulder, CO, USA (2017)
17. Chan, H.K., Nowotny, T.: Mixture are more salient stimuli in olfaction. bioRxiv 163238 (2017). doi:https://doi.org/10.1101/163238

The Dynamics of Bimodular Continuous Attractor Neural Networks with Moving Stimuli

Min Yan[1(✉)], Wen-Hao Zhang[2], He Wang[1], and K.Y. Michael Wong[1]

[1] Department of Physics, Hong Kong University of Science and Technology,
Hong Kong SAR, People's Republic of China
{myanaa,hwangaa,phkywong}@ust.hk
[2] Center for the Neural Basis of Cognition, Carnegie Mellon University,
Pittsburgh, USA
wenhaoz1@andrew.cmu.edu

Abstract. The single-layer continuous attractor neural network (CANN) model has been applied successfully to describe the tracking of moving stimuli of a single modality. Experimental evidence shows that stimuli of different modalities interact with each other in the neural system. To study these interaction effects, we generalize the single-module structure to a bimodular one. We found that when there is one static stimulus in one module and a moving one in the other, the network have very different behaviours depending on whether the inter-modular couplings are excitatory or inhibitory. We further compare the model with experimental observations that illustrate the interactions between two sensory modalities, such as the motion-bounce Illusion. Agreement between model and experimental results can be obtained for appropriate choice of parameters.

Keywords: Continuous attractor neural networks · Multisensory information processing · Motion-Bounce Illusion

1 Introduction

Our brain performs computations based on the structure of the brain network and the inputs. Various neural network models have been proposed to illustrate different functions and properties of the brain. In recent years, the continuous attractor neural network (CANN) model has gained wide attention due to its ability to use a continuous family of neuronal states to represent a continuum of information and features [1,2]. This endows the model with the capacity to track moving stimuli continuously, which provides an intuitive and practical way to study the functions of the brain in processing evolving information. Based on the CANN model, we can use localized neuronal activities to mimic visual, auditory or vestibular signals (stimulus), and elucidate network responses to static or moving stimuli. There have already been extensive studies about the CANN

© Springer International Publishing AG 2017
D. Liu et al. (Eds.): ICONIP 2017, Part IV, LNCS 10637, pp. 648–657, 2017.
https://doi.org/10.1007/978-3-319-70093-9_69

model, and many properties of the model have been revealed [3,4]. In particular, the CANN model is an effective tool to study the dynamics of tracking moving stimuli.

However, our brain receives sensory inputs in multiple modalities, such as visual, auditory, olfactory and vestibular inputs. These inputs are integrated to form a comprehensive picture of the environment. Hence it is interesting to study how different inputs interact with each other in the neural system. The brain is organized in modules, with each module playing certain functional role [5]. Most of the studies on CANN were based on the structure of a single module, but a modular network architecture for processing multisensory inputs was proposed [6]. The dynamics of processing motional inputs in this modular network remains unclear. In this paper we generalize the CANN model to a two-module structure, each receiving input of a modality. The structure seems to be more complicated compared with that of a single module, but the network is able to process inputs from the two modalities separately. We will present results describing in detail the effects of various parameter settings, including different couplings between the two layers (excitatory or inhibitory), the properties of the two inputs (static or moving), and the velocity of the moving stimulus. When processing a moving stimulus, the tracking dynamics of the network will have a diverse behavior under the influence of the neuronal couplings, moving speed, and the strength of the external input.

The two-module CANN model enables us to compare our theoretical predictions with experiments that explore the interaction of different sensory modalities, e.g., sensory illusions [7–9]. In this paper we consider the visual-auditory sensory illusion experiment [10,11], namely, the Motion-Bounce Illusion, and simulate the experiment in our two-module CANN model. Agreement between model and experimental results can be obtained for appropriate choice of parameters.

2 Formulation

We first introduce a single-module network processing a one-dimensional stimulus which is encoded by a population of neurons. The stimulus can be thought of as the position of an object, head direction or other continuously distributed information. We use $U(x,t)$ to denote the synaptic input at time t to the neurons whose preferred stimulus is x, ranging from 0 to 2π. The dynamics of $U(x,t)$ is [3,4]:

$$\tau \frac{\partial U(x,t)}{\partial t} = I_{ext}(x,t) + \rho \int_{-\infty}^{\infty} J(x,x')r(x',t)dx' - U(x,t), \tag{1}$$

where τ is a time constant, typically of the order of 1 ms, which controls the rate at which the synaptic input relaxes to the total input of the neuron. The function $I_{ext}(x,t)$ denotes the external input to the network at time t and position x, and ρ is the density of neurons. The coupling between neurons is denoted as $J(x,x')$:

$$J(x, x') = \frac{J_0}{\sqrt{2\pi}a} \exp\left[-\frac{(x-x')^2}{2a^2}\right], \tag{2}$$

where a is the interaction range. Note that the coupling is translationally invariant, that is, dependent on x and x' only through their displacement $x - x'$. This is important for the network to support a continuous family of attractors. The firing rate $r(x,t)$ at x and t is given by:

$$r(x,t) = \frac{[U(x,t)]_+^2}{1 + k\rho \int_{-\infty}^{\infty} dx'[U(x',t)]_+^2}, \tag{3}$$

in which $[U]_+ \equiv \max(U, 0)$, and k is the global inhibition that controls the extent at which the firing rate saturates. To simplify the analysis, we adopt the following rescaled parameters: $\tilde{U} = \rho J_0 U, \tilde{I}_{ext} = \rho J_0 I_{ext}, \tilde{r} = (\rho J_0)^2 r, \tilde{k} = \frac{8\sqrt{2\pi}ak}{\rho J_0^2}$. So Eqs. (1)–(3) can be rewritten as:

$$\tau\frac{\partial \tilde{U}(x,t)}{\partial t} = \int_{-\infty}^{\infty} \frac{1}{\sqrt{2\pi}a} \exp\left[-\frac{(x-x')^2}{2a^2}\right] \tilde{r}(x',t)dx' + \tilde{I}_{ext}(x,t) - \tilde{U}(x,t), \tag{4}$$

$$\tilde{r}(x,t) = \frac{[\tilde{U}(x,t)]_+^2}{1 + \frac{\tilde{k}}{8\sqrt{2\pi}a}\int_{-\infty}^{\infty} dx'[\tilde{U}(x',t)]_+^2}, \tag{5}$$

Next, we extend the model to a bimodular structure [12], which processes inputs from two modalities, such as visual and auditory senses. Experimental evidence shows that different modalities interact with each other in the process of the information encoding [15]. So we add couplings between the two modules. The dynamics of the two-module CANN model is (For convenience, hereafter we will use U, r, k, I to denote the rescaled variables $\tilde{U}, \tilde{r}, \tilde{k}$ and $\tilde{I}$, respectively):

$$\tau\frac{\partial U_1(x,t)}{\partial t} = -U_1(x,t) + \omega_{11} \int_{-\infty}^{\infty} dx' \frac{1}{\sqrt{2\pi}a} \exp\left[-\frac{(x-x')^2}{2a^2}\right] r_1(x',t)$$

$$+ \omega_{12} \int_{-\infty}^{\infty} dx' \frac{1}{\sqrt{2\pi}a} \exp\left[-\frac{(x-x')^2}{2a^2}\right] r_2(x',t) + I_{1ext}(x,t),$$

$$\tau\frac{\partial U_2(x,t)}{\partial t} = -U_2(x,t) + \omega_{22} \int_{-\infty}^{\infty} dx' \frac{1}{\sqrt{2\pi}a} \exp\left[-\frac{(x-x')^2}{2a^2}\right] r_2(x',t)$$

$$+ \omega_{21} \int_{-\infty}^{\infty} dx' \frac{1}{\sqrt{2\pi}a} \exp\left[-\frac{(x-x')^2}{2a^2}\right] r_1(x',t) + I_{2ext}(x,t), \tag{6}$$

in which ω_{11} (ω_{22}) indicates the coupling strength between neurons within the first (second) module. ω_{12} and ω_{21} denote the couplings from module 2 to module 1 and from module 1 to module 2 respectively. As shown in Eq. (6), besides the reciprocal couplings within the same module, we also add the couplings between different modules. In both modules, there are two external inputs respectively, which are assumed to be independent of each other.

3   Network Dynamics

Since each module is a CANN, we set the reciprocal couplings within the same layer (ω_{11} and ω_{22}) are positive, which denote excitatory couplings. Below we consider the cases that the couplings between the distinct layers (ω_{12} and ω_{21}) can range from inhibitory to excitatory. We focus on the situation in which the external stimulus in one module is moving while the input in the other module is static. As shown in Fig. 1 for the case that the couplings between the two modules are one excitatory and one inhibitory, and that the inputs have the same strengths, the response of the module at the receiving end of the inhibitory inter-modular coupling is suppressed by the other module, and quickly decays to silence. On the other hand, the response of the second module is strong and stable due to the excitatory coupling from the other module.

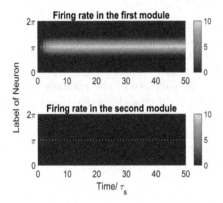

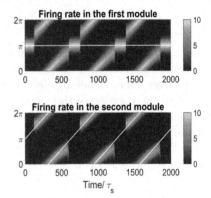

Fig. 1. The firing rate patterns under static stimuli. $I_{ext1} = I_{ext2} = 0.8$. $\omega_{12} = 0.8$, $\omega_{21} = -0.2$.

Fig. 2. The firing rate patterns under one slow moving stimulus. $I_{ext1} = 1.6$, $I_{ext2} = 0.8$. $\omega_{12} = 0.5$, $\omega_{21} = -0.2$.

In Fig. 2, the input in the second module (the module receiving the inhibitory inter-modular coupling) is a slow moving stimulus, whose trajectory is indicated by the white dash lines. The input of the first module (the one receiving the excitatory inter-modular coupling) is still static. In this case, the dynamics of the modules show very different behaviors. Suppose again that the modules are in the silent state for $t < 0$ and the external inputs are applied for $t \geq 0$, with the two inputs located at the same position at $t = 0$ and input 2 moving away thereafter. In the first module which receives excitatory input from the second module, the firing rate builds up rapidly and strongly from the beginning with the assistance of the second module. When the two inputs are more separated later, the response in the first module starts to track the moving input instead of its own. When the moving stimulus approaches the position of the static input 1 again after moving almost a cycle, the static input is reinforced again and the response becomes static once more.

The response in the second module which receives inhibitory inter-modular input from the first module is different. Initially its response is inhibited by the first module. However, when input 2 moves away, input 1 is no longer interacting with response 2 effectively, and the moving response begins to build up. Nevertheless, due to the inhibition from module 1, response 2 is not very strong.

From the comparison of Figs. 1 and 2, it can be seen that there is a competition between the two external stimuli. Hence our next step is to find out the appropriate settings for which the network will track the moving stimulus, and for which it will track the static stimulus, and for which both modulus will track its own stimulus respectively without being interfered with each other. To address these questions, we refer to the modules with the moving and static inputs as modules m and s respectively. To monitor the tracking behavior of module m, we introduce the variance measures for module m, given by

$$\sigma_s^2 = \langle(x_m(t) - v_s \cdot t)^2\rangle_t - \langle x_m(t) - v_s \cdot t\rangle_t^2$$
$$\sigma_m^2 = \langle(x_m(t) - v_m \cdot t)^2\rangle_t - \langle x_m(t) - v_m \cdot t\rangle_t^2, \tag{7}$$

where $x_m(t)$ denotes the center of mass of the moving firing rate profile, v_m (v_s) represents the velocities of the two inputs, with $v_s = 0$, and $\langle\cdots\rangle$ represents average over time. Hence σ_m^2 (σ_s^2) represents the variance of the response with respect to the position of the moving (static) input in module m. By comparing the magnitudes of the two variances, we can get an understanding about the tracking behaviors. When σ_m^2 is less than σ_s^2, the response is tracking its own input. When σ_s^2 is less than σ_m^2, it is tracking the static input. The tracking behavior in module s can be studied similarly.

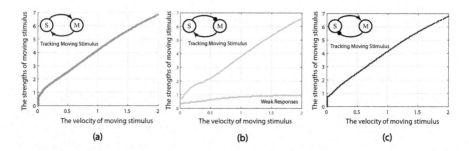

Fig. 3. The phase diagram of the tracking behaviors module with the moving stimulus. (a) $\omega_{ms} = 0.1, \omega_{sm} = 0.1$; (b) $\omega_{ms} = -0.1, \omega_{sm} = 0.1$; (c) $\omega_{ms} = 0.1, \omega_{sm} = -0.1$.

Figure 3 shows the phase diagram of the tracking dynamics. We tested three kinds of couplings between the two modules. The '+' denotes excitatory couplings and '−' represents inhibitory ones. The strengths of all the couplings are the same and the input strength of the static stimulus is fixed at $I_{exts} = 0.7$. In all 3 cases, the module is not able to track the moving stimulus when the input strength of the moving stimulus is weak. When the input strength of the moving stimulus increases, the module starts to track the moving stimulus. For the two cases that module s excites module m, the phase boundaries are very similar. For the case that module s inhibits module m, the position of the phase boundary is slightly lower, but in addition, there exists a weak response when the input strength is very weak, indicating that module m cannot sustain a stable and strong response.

The transitions between these phases also depend on the velocity of the moving stimulus. When the moving stimulus has a high speed, it is more difficult for module m to support a moving response. Hence the transition from static response to moving response takes place at larger stimulus strength. Similarly, for the case that module s inhibits module m, it is more difficult for module m to support a static response, and the transition from the weak response phase to the static response phase occurs at larger stimulus strength.

3.1 Sensory Illusion

Understanding the dynamics of the network in the presence of a moving and a static stimulus is of great importance for us to study real-life multisensory phenomena. There have been extensive studies on the integration of multisensory signals, such as visual-auditory [17,20], visual-vestibular [14,16], and others [13,15,18,19,21,24]. Here we consider a sensory illusion experiment called the 'Motion-Bounce Illusion' [9] involving visual and auditory perceptions [10]. As shown in Fig. 4, the subject observes on the screen two balls located at points A and B initially. When the experiment starts, the two balls begin to move towards the center point O simultaneously with the same speed. In test 1, the two balls meet each other at point O and keep their original motion directions and velocities after meeting, and finally arrive at the destinations at the same time, which are indicated by the two deep blue balls. No auditory input is applied when the balls meet. The majority of observers reported that they only perceive the two balls streaming through each other rather than colliding and changing their course of motion. In test 2, the motions of the balls are the same, but a "tink" sound is introduced when the two balls meet at point O. In this test a considerable fraction of observers reported that they perceived the two balls bounced off

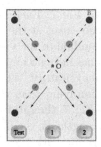

Fig. 4. Motion-Bounce Illusion. (Color figure online)

each other instead of streaming through, and the trajectories of the balls become ')⟨' rather than 'X'.

To simulate these results in our model, we use two modules to represent the visual and auditory modalities respectively. We use a moving stimulus with two peaks approaching each other to represent the visual input of the balls, and a momentary static stimulus to represent the auditory input. Since the processing time of the auditory input is shorter than that of the visual input, we vary the timing and duration of the auditory stimulus to represent the sound "tink". We introduce two reference patterns to quantify the statistical weight of "streaming through" (meaning two balls overlap with each other so that an observer can see only one ball) and "bouncing off" (meaning two balls bounce off so that an observer sees two balls) as illustrated in Figs. 5 and 6 respectively. This enables us to calculate the 'bouncing ratio (BR)' defined by

$$BR = \frac{M_B}{M_S + M_B},\qquad(8)$$

where M_S and M_B are the overlaps with the reference patterns in Figs. 5 and 6 respectively. A low value of BR means that more observers see no illusions, while a high value of BR implies more observers feel the 'bouncing off' illusions.

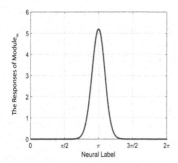

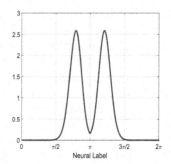

Fig. 5. The 'reference pattern' for streaming through obtained from point O in the absence of the "tink".

Fig. 6. The 'reference pattern' for bouncing off obtained when the central minimum is 3% of the two maxima.

We summarize our results in Fig. 7. Each point in the figure denotes the BR value when the two visual cues meet at the center point. In the simulations the coupling from the auditory module to the visual module (denoted as J_{VA}) is inhibitory. Nevertheless, visual cues have no inhibition effects on the audition, so the couplings J_{AV}) are set to be excitatory. When the strength of the auditory input increases, BR increases, implying that it is more likely to observe the 'bouncing illusion'. Therefore in Fig. 7, the ratio of the sensory illusion increases. When the inhibitory coupling strengthens, the illusion becomes more significant.

We compare the modeling results with the 'Motion-Bounce Illusion' experiment [7–10], in which the bouncing ratio increases by around 80% when compared with the case without auditory inputs. In the simulation, the changes made

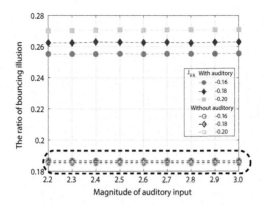

Fig. 7. The BR under different couplings and input strengths settings. The box enclosed by the dashed line represents the simulation results without auditory inputs.

by the auditory input are around 50%, which is comparable to the experimental result. Hence we have a better and deeper understanding of the relation between theory and experiment [22,23,25].

4   Conclusions

We have generalized the single-module CANN model to two modules, which are able to process inputs with two modalities. The couplings between the modules, the input strengths, and the moving speed of the stimulus play important role in determining the dynamics of the network. We observed that excitatory inter-modular couplings result in attraction and enhancement of the response profile, whereas inhibitory couplings lead to repulsion and suppression. Competition between stimuli results in a rich spectrum of behaviors. The phase diagram in Fig. 3 reveals that response to moving stimuli may be silenced by inhibitory inter-modular input if the input strength is weak, or overridden by static stimuli if its input strength is larger but still not sufficiently large. Stimuli moving too fast may be missed by the system and became dominated by static stimuli.

Our two-module CANN model is relevant to the 'Motion-Bounce Illusion' experiment that demonstrates the interaction between visual and auditory stimuli. The enhancement of the bouncing off illusion when a tink sound is introduced is consistent with a phenomenological inhibitory coupling from the auditory module to the visual module, as shown in Fig. 7.

The processing of multisensory signals is an important issue in modeling the functions of the brain as well as in technological applications of neural computation. It is commonly accepted that excitatory couplings between modules are useful when signals from different channels are correlated, and inhibitory couplings are useful when they are uncorrelated or anti-correlated [6,26]. The existence of the so-called congruent and opposite cells when visual and vestibular signals are combined in the monkeys brain [16] shows that the response of the neural system to signals with different disparities can be rather diverse. Recent work shows that the network structure to achieve Bayes-optimal performance involves both excitatory or inhibitory couplings depending on the prior distribution of signals [27]. While most of these studies focus on the steady-state behavior of the neural system, our work shows that dynamical and temporal aspects are also important, and transient behaviors of the neural system may also be useful in conveying inter-modular information. Experiments based on temporal illusions can also be designed to deepen our understanding of multisensory information processing.

Acknowledgments. This work is supported by grants from the Research Grants Council of Hong Kong (grant numbers N_HKUST606/12, 605813 and 16322616).

References

1. Camperi, M., Wang, X.J.: A model of visuospatial working memory in prefrontal cortex: recurrent network and cellular bistability. J. Comput. Neurosci. **5**(4), 383–405 (1998)
2. Wu, S., Hamaguchi, K., Amari, S.I.: Dynamics and computation of continuous attractors. Neural Comput. **20**(4), 994–1025 (2008)
3. Fung, C.C.A., Wong, K.Y.M., Wu, S.: A moving bump in a continuous manifold: a comprehensive study of the tracking dynamics of continuous attractor neural networks. Neural Comput. **22**(3), 752–792 (2010)
4. Fung, C.C.A., Wong, K.Y.M., Wang, H., Wu, S.: Dynamical synapses enhance neural information processing: gracefulness, accuracy, and mobility. Neural Comput. **24**(5), 1147–1185 (2012)
5. Zhou, C.S., Zemanov, L., Zamora, G., Hilgetag, C.C., Kurths, J.: Hierarchical organization unveiled by functional connectivity in complex brain networks. Phys. Rev. Lett. **97**(23), 238103 (2006)
6. Zhang, W.H., Chen, A., Rasch, M.J., Wu, S.: Decentralized multisensory information integration in neural systems. J. Neurosci. **36**(2), 532–547 (2016)
7. Watanabe, K.: Crossmodal interaction in humans. Doctoral dissertation, California Institute of Technology (2001)
8. Shimojo, S., Shams, L.: Sensory modalities are not separate modalities: plasticity and interactions. Curr. Opin. Neurobiol. **11**(4), 505–509 (2001)
9. Watkins, S., Shams, L., Tanaka, S., Haynes, J.D., Rees, G.: Sound alters activity in human V1 in association with illusory visual perception. Neuroimage **31**(3), 1247–1256 (2006)
10. Sekuler, R., Sekuler, A.B., Lau, R.: Sound alters visual motion perception. Nature **385**(6614), 308 (1997)
11. Jaekl, P.M., Harris, L.R.: Auditoryvisual temporal integration measured by shifts in perceived temporal location. Neurosci. Lett. **417**(3), 219–224 (2007)
12. Zhang, W.H., Wu, S.: Neural information processing with feedback modulations. Neural Comput. **24**(7), 1695–1721 (2012)
13. Ernst, M.O., Banks, M.S.: Humans integrate visual and haptic information in a statistically optimal fashion. Nature **415**(6870), 429–433 (2002)
14. Zhang, W.H., Chen, A., Rasch, M.J., Wu, S.: Decentralized multisensory information integration in neural systems. J. Neurosci. **36**(2), 532–547 (2016)

15. Shams, L., Seitz, A.R.: Benefits of multisensory learning. Trends Cogn. Sci. **12**(11), 411–417 (2008)

16. Gu, Y., Angelaki, D.E., DeAngelis, G.C.: Neural correlates of multisensory cue integration in macaque MSTd. Nat. Neurosci. **11**(10), 1201–1210 (2008)

17. Molholm, S., Ritter, W., Javitt, D.C., Foxe, J.J.: Multisensory visualauditory object recognition in humans: a high-density electrical mapping study. Cereb. Cortex **14**(4), 452–465 (2004)

18. Fetsch, C.R., Pouget, A., DeAngelis, G.C., Angelaki, D.E.: Neural correlates of reliability-based cue weighting during multisensory integration. Nat. Neurosci. **15**(1), 146–154 (2012)

19. Hairston, W.D., Wallace, M.T., Vaughan, J.W., Stein, B.E., Norris, J.L., Schirillo, J.A.: Visual localization ability influences cross-modal bias. J. Cogn. Neurosci. **15**(1), 20–29 (2003)

20. Seitz, A.R., Kim, R., Shams, L.: Sound facilitates visual learning. Curr. Biol. **16**(14), 1422–1427 (2006)

21. Odegaard, B., Wozny, D.R., Shams, L.: The effects of selective and divided attention on sensory precision and integration. Neurosci. Lett. **614**, 24–28 (2016)

22. Carandini, M., Heeger, D.J.: Normalization as a canonical neural computation. Nat. Rev. Neurosci. **13**(1), 51–62 (2012)

23. Stanford, T.R., Quessy, S., Stein, B.E.: Evaluating the operations underlying multisensory integration in the cat superior colliculus. J. Neurosci. **25**(28), 6499–6508 (2005)

24. Driver, J., Noesselt, T.: Multisensory interplay reveals crossmodal influences on sensory-specificbrain regions, neural responses, and judgments. Neuron **57**(1), 11–23 (2008)

25. Kim, R.S., Seitz, A.R., Shams, L.: Benefits of stimulus congruency for multisensory facilitation of visual learning. PLoS One **3**(1), e1532 (2008)

26. Zhang, W.H., Wang, H., Wong, K.T.M., Wu, S.: Congruent and opposite neurons: sisters for multisensory integration and segregation. In: Advances in Neural Information Processing Systems, pp. 3180–3188 (2016)

27. Wang, H., Zhang, W.-H., Wong, K.Y.M., Wu, S.: How the prior information shapes neural networks for optimal multisensory integration. In: Cong, F., Leung, A., Wei, Q. (eds.) ISNN 2017. LNCS, vol. 10262, pp. 128–136. Springer, Cham (2017). doi:10.1007/978-3-319-59081-3_16

Encoding Multisensory Information in Modular Neural Networks

He Wang[1](✉), Wen-Hao Zhang[2], K.Y. Michael Wong[1], and Si Wu[3]

[1] Department of Physics, Hong Kong University of Science and Technology,
Hong Kong, China
{hwangaa,phkywong}@ust.hk
[2] Center for the Neural Basis of Cognition, Carnegie Mellon University,
Pittsburgh, USA
wenhaoz1@andrew.cmu.edu
[3] State Key Laboratory of Cognitive Neuroscience and Learning,
and McGovern Institute for Brain Research, Beijing Normal University,
Beijing, China
wusi@bnu.edu.cn

Abstract. The brain is capable of integrating information in multiple sensory channels in a Bayesian optimal way. Based on a decentralized network model inspired by electrophysiological recordings, we consider the structural pre-requisites for optimal multisensory integration. In this architecture, same-channel feedforward and recurrent links encode the unisensory likelihoods, whereas reciprocal couplings connecting the different modules are shaped by the correlation in the joint prior probabilities. Moreover, the statistical relationship between the difference in the optimal network structures and the difference in the priors and the likelihoods clearly shows that the network can encode multisensory information in a distributed manner. Our results generate testable predictions for future experiments and are likely to be applicable to other artificial systems.

Keywords: Recurrent neural networks · Multisensory Bayesian inference

1 Introduction

Real-life perception and behavior are usually based on comparison and integration of multiple sources of information in diverse modalities (e.g., vision, audition, the vestibular sense of motion, etc.), which provide complementary facets about certain external or internal states of organisms and enable reliable representation of the ambiguous environment. A wide range of psychophysical and neurobiological studies indicate that the brain can integrate sensory cues in an optimal way, as predicted by Bayesian inference [1–3].

In a series of visual-vestibular multisensory tasks, where monkeys make use of visual cues and/or vestibular cues to infer the direction of self-motion (heading

© Springer International Publishing AG 2017
D. Liu et al. (Eds.): ICONIP 2017, Part IV, LNCS 10637, pp. 658–665, 2017.
https://doi.org/10.1007/978-3-319-70093-9_70

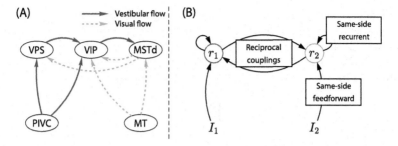

Fig. 1. The neural substrates and our network implementation of multisensory Bayesian inference. (A) A schematic diagram of information flows between candidate cortical areas involved in the visual-vestibular integration. Visual information starts at the middle temporal area (MT), while vestibular information enters the cortex at the parietoinsular vestibular cortex (PIVC). Multisensory neurons are found extensively in the dorsal medial superior temporal area (MSTd), the ventral intraparietal area (VIP), and visual posterior sylvian area (VPS). Adapted from [10]. (B) The bi-modular neural field model in the present study. There are same-side feedforward and same-side recurrent connections within each module. Cross-talks between the two modules are mediated by reciprocal couplings.

direction), Angelaki and coauthors recorded activities of multisensory neurons in several cortical areas involved in visual-vestibular integration [3–5], suggesting that these areas are likely to form parallel and partially redundant pathways for multisensory information processing [6]. A diagram summarizing the information flows among these areas is shown in Fig. 1(A). A decentralized network architecture inspired by these findings has been proposed to elucidate the crucial role played by the extensive reciprocal connections in integrating multisensory information [7–9] (Fig. 1(B)).

In the present study, we take a theoretical approach and elucidate the optimal structure of a decentralized network under the constraint that for a given stimulus prior, the network's output matches the profile of the posterior of the stimulus. We find that the reciprocal couplings are remarkably shaped by the correlation structure in the multisensory prior. In the general case, statistics of the difference in the network structure induced by the difference in the prior and the likelihoods are revealed, suggesting that the decentralized network can encode the prior and the likelihoods in a distributed manner. Predictions generated by these results can be tested in future experiments and may help unveil general computational principles in multisensory information processing.

2   Bi-modality Bayesian Inference for Circular Variables

Suppose there are two external stimuli s_1 and s_2 in different modalities. For convenience, we consider circular variables in the range $|s_i| \leq \pi$ for $i = 1, 2$. They could be the direction of self-motion, or the direction of incoming sound or flash. The corresponding sensory observations are x_1 and x_2, which are presumably represented in different sensory pathways independently. We assume each unisensory likelihood function is a von Mises distribution, which has been widely used in modeling statistics of circular variables [11]. The multisensory likelihood is simply a product of both the unisensory likelihoods,

$$p(x_1, x_2 | s_1 = \theta_1, s_2 = \theta_2) = \prod_{i=1,2} p(x_i | s_i = \theta_i) = \prod_{i=1,2} \frac{e^{\kappa_i \cos(\theta_i - x_i)}}{2\pi \mathcal{I}_0(\kappa_i)}, \quad (1)$$

where κ_i are the reliability of the sensory observation in modality i, and $\mathcal{I}_n(\cdot)$ is the nth order modified Bessel function of the first kind.

Multisensory information processing relies on the prior experience about correlations among sensory cues, which usually benefits organisms in forming a unified and coherent perception of the external world [12,13], yet sometimes evokes interesting illusions [14,15]. In general, the joint prior should be composed of an independent part and a correlated part. Denote the unisensory prior distributions as $p(s_1)$ and $p(s_2)$. The joint prior can be described as $p(s_1, s_2) = (1 - p_c)p(s_1)p(s_2) + p_c q(s_1, s_2)$. Here, $q(s_1, s_2)$ is a correlated distribution and $p_c \in [0, 1]$ the statistical weight that s_1 and s_2 originate from that distribution. Several kinds of the multisensory prior have been formulated to account for different correlation structures [9,16]. According to Bayes' theorem, the marginal posterior distribution for s_1 is given by $p(s_1 | x_1, x_2) \propto \int p(x_1 | s_1)p(x_2 | s_2)p(s_1, s_2) \, ds_2$. Optimal multisensory integration requires that the marginal posterior are represented in the population activity of multisensory brain areas [17].

3   The Decentralized Network Implementation

In order to implement multisensory Bayesian inference, we adopt a bi-modular decentralized network model [7–9]. We seek the optimal couplings between the two modules such that the network activity in each module at the steady state matches the profile of the corresponding marginal posterior, given the likelihood functions as external inputs. The dynamical equation for module i is,

$$\tau_s \dot{U}_i(\theta, t) = -U_i(\theta, t) + \rho \int_{-\pi}^{\pi} d\theta' \left[W_i^{\text{ff}}(\theta, \theta') I_j(\theta') + \sum_{j=1,2} W_{ij}^{\text{rec}}(\theta, \theta') r_j(\theta', t) \right], \quad (2)$$

where U_i is the synaptic input of module i, r_i is the firing rate of module i, I_i is the external input on module i, W_i^{ff} is the same-side feedforward connection in module i, W_{ii}^{rec} is the same-side recurrent connection within module i, and W_{ij}^{rec} ($i \neq j$) is the reciprocal couplings from module j to module i. This network architecture is illustrated in Fig. 1(B).

The global inhibition, a mechanism achieved by divisive normalization, is incorporated into the activation function [18],

$$r_i(\theta, t) = \frac{[U_i(\theta, t)]_+^2}{1 + k_1 \rho \int [U_i(\theta', t)]_+^2 \, d\theta'}, \quad (3)$$

where k_I is the strength of global inhibition, and $[x]_+$ is equal to x when $x \geq 0$, otherwise 0. In the present work, we fix $k_I \rho$ at 0.9, while small changes in $k_I \rho$ does not affect the results significantly.

We define the mean squared error between the stationary firing rate and the corresponding marginal posterior in both modules as the cost function $\overline{L} \equiv \left\langle \sum_i \int d\theta \left[r_i^*(\theta) - p(s_i = \theta | x_1, x_2) \right]^2 \right\rangle_{x_1,x_2}$. Our objective is to find out what kind of couplings can minimize the cost function $\overline{L}$ when the external inputs on the two modules are the corresponding likelihood functions, $I_i(\theta; x_i) = p(x_i | s_i = \theta)$. The cost function is minimized through stochastic gradient descent [19]. In the next section, we will first show typical results for three different kinds of priors, then explore how the priors and the likelihoods are encoded in the couplings in the general case.

4   Results

4.1   Reciprocal Couplings Shaped by Multisensory Prior

To investigate the capability of the decentralized network model of performing optimal multisensory integration for a variety of correlation structure, three kinds of multisensory prior are chosen due to their distinctive profiles [9]:

1. Congruent prior, which describes the environment where the two stimuli are positively correlated when they are originated from the same source (Fig. 2(A)).
2. Opposite prior, which describes the environment where the two stimuli may come in the same or opposite directions (Fig. 2(B)).

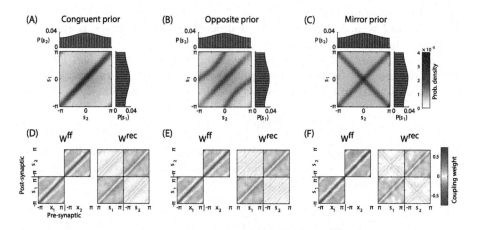

Fig. 2. Three kinds of multisensory prior and the optimal coupling weights. (A) The congruent prior. (B) The opposite prior. (C) The mirror prior. (D–F) The optimal coupling weights for the prior in (A–C). The reliabilities of the likelihoods are the same $\kappa_1 = \kappa_2 = 10.7$ for the three cases.

3. Mirror prior, which describes the environment where the two stimuli may be the same or the mirror image of each other (Fig. 2(C)).

Coupling weights of the networks optimized with the same likelihoods and the three example priors are shown in Fig. 2(D–F). The same-side connection weights (W_1^{ff}, W_2^{ff}, W_{11}^{rec} and W_{22}^{rec}) are almost identical for the three cases. We observe that they are excitatory in the short range and inhibitory in the intermediate range, typical of center-surround antagonism in visual processing of the brain. More interestingly, the pattern of the reciprocal couplings (W_{12}^{rec} and W_{21}^{rec}) resembles that of the corresponding prior, suggesting that the reciprocal couplings can store the correlation structure between sensory stimuli and play a crucial role in multisensory integration.

4.2   Optimal Couplings in Different Environments

In order to reveal the different impact of the noises and the prior distributions on the recurrent neural network model, we compare the coupling weights of networks optimized in different environments. We define the multisensory "environment" by the joint prior distribution $p(s_1, s_2)$ and the sensory reliabilities κ_1 and κ_2. Suppose in environment A, the prior distribution is $p_A(s_1, s_2)$ and the sensory reliabilities are (κ_1^A, κ_2^A). In environment B, the prior and the sensory reliabilities are $p_B(s_1, s_2)$ and (κ_1^B, κ_2^B), respectively. We define the difference in the noise as $L_{\text{noise}} = (\kappa_1^A - \kappa_1^B)^2 + (\kappa_2^A - \kappa_2^B)^2$. We resort to the Jensen-Shannon divergence in defining the difference between priors,

$$L_{\text{prior}} = D_{JS}(p_A(s_1, s_2) \| p_B(s_1, s_2)) \equiv (D_{KL}(p_A \| M) + D_{KL}(p_B \| M))/2, \quad (4)$$

where D_{KL} denotes the Kullback-Leibler divergence and $M \equiv (p_A + p_B)/2$. Note that if p_A and p_B are sufficiently different, which means p_A is non-zero only when p_B is nearly zero, and vice versa, $D_{KL}(p_A \| M) = \sum_{p_A \neq 0} p_A \ln(2p_A/(p_A + p_B)) \to \sum_{p_A \neq 0} p_A \ln 2 = \ln 2$. Therefore, the Jensen-Shannon divergence is symmetric and upper-bounded at $\ln 2$.

Denote the optimal couplings in environment A as $\{W_i^{\text{ff}}\}^A$ and $\{W_{ij}^{\text{rec}}\}^A$. The changes in the feedforward coupling weights are simply measured by the Euclidean distance of coupling weights in difference environments,

$$D_{ij}^{\text{rec}} = \sqrt{\|\{W_{ij}^{\text{rec}}\}^A - \{W_{ij}^{\text{rec}}\}^B\|^2}. \quad (5)$$

D_i^{ff} can be defined similarly. In order to differentiate the crosstalks between the two modules from the interactions within each of themselves, we define the same-side and reciprocal differences as

$$D_{\text{same}} = \left(D_1^{\text{ff}} + D_2^{\text{ff}} + D_{11}^{\text{rec}} + D_{22}^{\text{rec}}\right)/4, \quad (6)$$

$$D_{\text{recip}} = \left(D_{12}^{\text{rec}} + D_{21}^{\text{rec}}\right)/2. \quad (7)$$

Our network model is optimized with 55 different reliabilities and 84 different priors of the three kinds above, which resulted in a total of 4620 different environments and optimal couplings. We compare the differences of the likelihoods

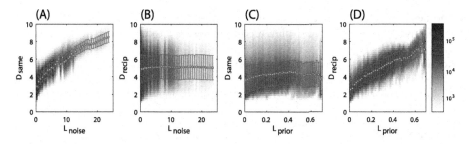

Fig. 3. Differences in the coupling weights due to differences in the sensory reliabilities and priors. The color code denotes number of data points. The dots are the mean differences in the weights given the divergence in the reliabilities or the priors. The error bars are standard deviations. (A) D_{same} increases as L_{noise} increases. (B) D_{recip} is not correlated with L_{noise}. (C) D_{same} is not correlated with L_{prior}. (D) D_{recip} increases as L_{prior} increases. (Color figure online)

and the priors for each pair of the environments, and the differences of the couplings for the corresponding pair of networks ($\sim 10^7$ pairs in total). The results are shown in Fig. 3. We find that the differences in same-side coupling weights are highly correlated with the differences in the sensory reliabilities (Fig. 3(A)), but almost independent of the differences in the joint priors (Fig. 3(C)). However, the differences in reciprocal coupling weights can be largely attributed to the differences in the joint priors (Fig. 3(D)), but not to the differences in the noise (Fig. 3(B)). This result clearly indicates that the independent likelihoods are encoded in each sensory channel respectively, while the multisensory prior is likely to be encoded in the cross-talks between neural pathways in a distributed manner.

5   Conclusion

We have investigated how the priors and the likelihoods are encoded in modular neural networks to achieve optimal multisensory integration. We have explored the optimal network structure in a wide range of environments. We found a remarkable correspondence between the profiles of the reciprocal couplings and the correlation pattern in the multisensory prior. The statistical relationship shows that differences in the multisensory prior are correlated with differences in the reciprocal couplings, whereas differences in the likelihood are correlated with differences in the same-side couplings. Our work developed a theoretical link between the multisensory network structure to the statistical structure of Bayesian inference, demonstrating that the multisensory priors and likelihoods can be encoded in a decentralized fashion. These theoretical insights can be potentially extended to other artificial intelligence systems benefiting from exploiting multiple information sources, such as computer vision and robotics.

Cross-talks between different neural pathways can also be carried out by feedforward cross-links, which directly convey sensory inputs to different modules [13]. Our on-going work suggests that they also contribute to optimal multisensory integration and play a complementary role.

Acknowledgments. This work is supported by the Research Grants Council of Hong Kong (N_HKUST606/12, 605813 and 16322616) and National Basic Research Program of China (2014CB846101) and the Natural Science Foundation of China (31261160495).

References

1. Alais, D., Burr, D.: No direction-specific bimodal facilitation for audiovisual motion detection. Cogn. Brain Res. **19**(2), 185–194 (2004)
2. Ernst, M.O., Banks, M.S.: Humans integrate visual and haptic information in a statistically optimal fashion. Nature **415**(6870), 429–433 (2002)
3. Gu, Y., Angelaki, D.E., DeAngelis, G.C.: Neural correlates of multisensory cue integration in macaque MSTd. Nat. Neurosci. **11**(10), 1201–1210 (2008)
4. Chen, A., DeAngelis, G.C., Angelaki, D.E.: Macaque parieto-insular vestibular cortex: responses to self-motion and optic flow. J. Neurosci. **30**(8), 3022–3042 (2010)
5. Chen, A., DeAngelis, G.C., Angelaki, D.E.: Functional specializations of the ventral intraparietal area for multisensory heading discrimination. J. Neurosci. **33**(8), 3567–3581 (2013)
6. Chen, A., Gu, Y., Liu, S., DeAngelis, G.C., Angelaki, D.E.: Evidence for a causal contribution of macaque vestibular, but not intraparietal, cortex to heading perception. J. Neurosci. **36**(13), 3789–3798 (2016)
7. Zhang, W.H., Chen, A., Rasch, M.J., Wu, S.: Decentralized multisensory information integration in neural systems. J. Neurosci. **36**(2), 532–547 (2016)
8. Zhang, W.H., Wang, H., Wong, K.Y.M., Wu, S.: "Congruent" and "opposite" neurons: sisters for multisensory integration and segregation. In: Lee, D.D., Sugiyama, M., Luxburg, U.V., Guyon, I., Garnett, R. (eds.) Advances in Neural Information Processing Systems 29, pp. 3180–3188. Curran Associates, Inc. (2016)
9. Wang, H., Zhang, W.-H., Wong, K.Y.M., Wu, S.: How the prior information shapes neural networks for optimal multisensory integration. In: Cong, F., Leung, A., Wei, Q. (eds.) ISNN 2017. LNCS, vol. 10262, pp. 128–136. Springer, Cham (2017). doi:10.1007/978-3-319-59081-3_16
10. Chen, A., Gu, Y., Liu, S., DeAngelis, G.C., Angelaki, D.E.: Evidence for a causal contribution of macaque vestibular, but not intraparietal, cortex to heading perception. In: The 16th Japan-China-Korea Joint Workshop on Neurobiology and Neuroinformatics (NBNI 2016), Hong Kong (2016)
11. Murray, R.F., Morgenstern, Y.: Cue combination on the circle and the sphere. J. Vis. **10**(11), 15 (2010)
12. Girshick, A.R., Landy, M.S., Simoncelli, E.P.: Cardinal rules: visual orientation perception reflects knowledge of environmental statistics. Nat. Neurosci. **14**(7), 926–932 (2011)
13. Ghazanfar, A.A., Schroeder, C.E.: Is neocortex essentially multisensory? Trends Cogn. Sci. **10**(6), 278–285 (2006)
14. Sato, Y., Toyoizumi, T., Aihara, K.: Bayesian inference explains perception of unity and ventriloquism aftereffect: identification of common sources of audiovisual stimuli. Neural Comput. **19**(12), 3335–3355 (2007)
15. Shams, L., Ma, W.J., Beierholm, U.: Sound-induced flash illusion as an optimal percept. NeuroReport **16**(17), 1923–1927 (2005)

16. Shams, L., Beierholm, U.R.: Causal inference in perception. Trends Cogn. Sci. **14**(9), 425–432 (2010)
17. Ma, W.J., Beck, J.M., Latham, P.E., Pouget, A.: Bayesian inference with probabilistic population codes. Nat. Neurosci. **9**(11), 1432–1438 (2006)
18. Fung, C.C.A., Wong, K.Y.M., Wu, S.: A moving bump in a continuous manifold: a comprehensive study of the tracking dynamics of continuous attractor neural networks. Neural Comput. **22**(3), 752–792 (2010)
19. Pascanu, R., Mikolov, T., Bengio, Y.: On the difficulty of training recurrent neural networks. In: Proceedings of the 30th International Conference on Machine Learning, vol. 28 (2013)

Biomedical Engineering

Using Transfer Learning with Convolutional Neural Networks to Diagnose Breast Cancer from Histopathological Images

Weiming Zhi[1(✉)], Henry Wing Fung Yueng[2], Zhenghao Chen[2],
Seid Miad Zandavi[2], Zhicheng Lu[2], and Yuk Ying Chung[2]

[1] Department of Engineering Science, University of Auckland,
Auckland 1010, New Zealand
wzhi262@aucklanduni.ac.nz
[2] School of Information Technologies, University of Sydney,
Sydney, NSW 2006, Australia
{hyeu8081,zhenghao.chen,miad.zandavi,zhlu2106,
Vera.chung}@sydney.edu.au

Abstract. Diagnosis from histopathological images is the gold standard in diagnosing breast cancer. This paper investigates using transfer learning with convolutional neural networks to automatically diagnose breast cancer from patches of histopathological images. We compare the performance of using transfer learning with an off-the-shelf deep convolutional neural network architecture, VGGNet, and a shallower custom architecture. Our proposed final ensemble model, which contains three custom convolutional neural network classifiers trained using transfer learning, achieves a significantly higher image classification accuracy on the large public benchmark dataset than the current best results, for all image resolution levels.

Keywords: Histopathological image analysis · Convolutional neural networks · Transfer learning · Deep learning

1 Introduction

Breast cancer is a common cancer with a high mortality rate, relative to other types of cancer. Breast Cancer is the most prevalent type of cancer in women worldwide [1]. Early diagnosis of breast cancer, and selection of its treatment greatly improves survival chances.

Biopsy is the only diagnostic procedure that can definitely determine if a suspected region is cancerous. A biopsy involves the extraction of sample cells or tissues for examination. Psychopathology refers to the examination of the specimens extracted. Diagnosis from analysis of histopathological images is the gold standard in diagnosing a considerable number of diseases, including almost all types of cancers [2].

© Springer International Publishing AG 2017
D. Liu et al. (Eds.): ICONIP 2017, Part IV, LNCS 10637, pp. 669–676, 2017.
https://doi.org/10.1007/978-3-319-70093-9_71

Despite recent advances in computer vision, breast cancer diagnosis continues to rely heavily on visual inspections conducted by experienced pathologists. The diagnosis is subjective in nature, and may vary between observations. This process can also be very tedious and time-consuming. Image processing and machine learning techniques can be utilised to build computer systems to automate parts of this diagnosis process, and therefore boost the efficiency and improve the consistency of human pathologists.

In this paper, we explore the use of using transfer learning and Convolutional Neural Networks (CNNs) to diagnose breast cancer from histopathological images. Transfer learning aims to transfer knowledge between related *source* and *target* domains [3]. Transfer learning allows us to utilise CNN architectures deeper than the previous best performing architectures on our benchmark dataset, *BreakHis* [5]. In particular, we transfer weights from models pre-trained on the large image database, ImageNet [4]. We compare the performance of transfer learning two different models: VGGNet [6], and a custom six layer architecture. The results obtained using transfer learning on the custom model surpass the current best performance, for all of the resolutions in the benchmark dataset.

2   Related Work

2.1   Automation of Breast Cancer Diagnosis

Automatic diagnosis of cancer is an ongoing topic of research. Previous research have been conducted on breast cancer diagnosis. Doyle et al. [7] investigated the classification of breast cancer histopathology from textual, nuclear architectural features and images. Classification of benign and malignant histopathological images based on 10 syntactic structure features, extracted after the Gabor filter had been applied, was used by van Diest et al. [8]. The usage of Support Vector Machines (SVMs) on relevant numerical features extracted from histopathological images in a semi-automatic fashion has also yielded very promising results [9]. George et al. [10] developed an intelligent remote detection and diagnosis system for breast cancer from cytological images, and tested out the results of applying different classification models on the extracted features. In light of the lack of large public datasets of breast cancer histopathological images, Spanhol et al. released *BreakHis* [5], a large dataset of breast cancer histopathological images, acquired from 82 patients, and evaluated the performance of several classifiers on this dataset. More recently, Spanhol et al. used a three layer CNN architecture, combined with different methods of sampling small (32×32 and 64×64 pixels) patches from whole histopathological images, to achieve a high performance on the BreakHis dataset [11].

2.2   Convolutional Neural Networks and Transfer Learning

Convolutional Neural Networks (CNNs) have achieved great success in large-scale image recognition, and have been used to win the large scale visual recognition challenge (ILSVRC) since 2012 [12]. CNNs are high capacity classifiers, with multiple trainable layers and can contain a very large number of parameters, which must be trained by sufficient training data. CNNs with many layers, or *deep* CNNs, are able to extract abstract information from images, without the need to specifically identify and handcraft features. This allows CNNs to be more robust, and less prone to being affected by distorted images. CNNs have been widely adopted in computer vision, and have proved to be very successful at certain medical imaging tasks [13].

However, deep CNNs often have a number of parameters so large that it cannot reasonably be trained without a very large dataset. Medical imaging datasets are often not sufficiently large to train a deep CNN model from scratch adequately. Thus, the usage of transfer learning in medical imaging has been explored [14]. Transfer learning seeks to transfer knowledge between large *source* and small *target* domains [3]. For CNNs, this is often done by pre-training a CNN model with the source dataset, then re-training parts of the model with the target dataset [15]. Yosinski et al. [16] showed that features in deep neural networks, in particular those learnt in at the bottom layers of a CNN are highly transferable, and *fine-tuning* the entire CNN model with the target dataset improves the performance of the model. Shin et al. [14] demonstrated that the performance of deep CNNs on several CADe problems can be consistently improved by transfer learning from the large scale natural image dataset, ImageNet.

3   Approach

3.1   Dataset

The dataset used for training and evaluation is the benchmark *BreakHis* dataset [5]. It contains 7909 breast cancer histopathology images acquired from 82 patients. These images are labeled either "Benign" or "Malignant" and can be of four different resolutions (40X, 100X, 200X, 400X). The dataset is subsequently separated into five folds. Different samples from the same person will only ever be in the same fold. We will use four folds for training, and one fold for evaluating (Figs. 1, 2).

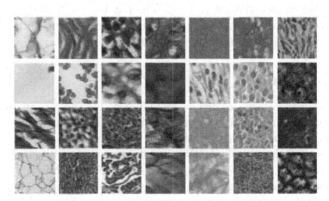

Fig. 1. *Extracted from* [11]. Examples of some of the textures available in the *BreakHis* dataset.

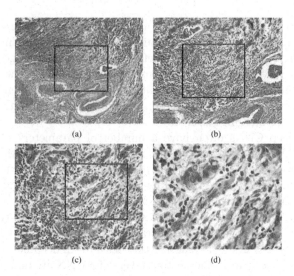

Fig. 2. *Extracted from* [11]. A slide of malignant tumor seen in four magnification factors: (a) 40X (b) 100X (c) 200X (d) 400X. The area of interest to be detailed in the next highest magnification factor is marked out by the rectangle.

3.2   Patch Extraction and Image Augmentation

The image samples on the benchmark dataset are of size 700 × 460 pixels, this is relatively large and will require a model with an impractically large number of parameters. To mitigate this problem, a patch-based approach is used to train the CNN models. Patch-based approaches have been used by Hou et al. [17] on large whole slide image tissue samples. Similar methods were also used by Spanhol et al. [11] to generate thousands of small (32 × 32 and 64 × 64 pixels) patches per image. Our method randomly samples only five patches, each of size 224 × 244 pixels, from every histopathological image used. The size of the patches corresponds to the input size of VGGNet, and can be inputted directly into a VGGNet model without resizing. The patches are further augmented by randomly zooming in by 20%, and performing horizontal and vertical flips. The data augmentation is done on-line and randomly.

3.3   Transfer Learning

The architectures we are exploring are VGGNet [6], and our custom architecture. Our custom architecture is inspired by VGGNet, and contains convolutional layers with parameters derived from the bottom six convolutional layers of VGGNet. VGGNet is a deep CNN architecture with very small convolution filters, and contains 16 or 19 trainable layers. We will use the 16 layer version of VGGNet. Our custom-made model contains six convolutional layers, with configurations similar to the bottom layers of VGGNet. The architecture of our model is shown in Table 1.

Table 1. Architecture of our proposed model

Type	Filter size	Channels
Input ($224 \times 224 \times 3$)		
Convolution	3×3	64
Convolution	3×3	64
Maxpool		
Convolution	3×3	128
Convolution	3×3	128
Maxpool		
Convolution	3×3	256
Convolution	3×3	256
Maxpool		
Fully-connected	256	
Dropout (Probability $= 0.5$)		
Fully-connected	2	
Sigmoid		

After obtaining the weights of both models pre-trained on ImageNet, transfer learning is done in the following steps:

1. **Training fully connected layers with features extracted from fixed convolutional layers:** The fully connected layers in each network have randomly initialised weights. We freeze the convolutional layers of the networks, and only train the fully connected layers using our training data. The top fully connected layers are trained from scratch on the features extracted from the fixed convolutional layers.
2. **Fine-tuning all the layers:** The convolutional layers of each network are then un-frozen, and the entire network is fine-tuned on the training data. This involves re-training the CNN, starting from the retained weights, and using a very small step size.

3.4   Ensemble Models

Due to the random nature of our patch sampling, we can train several slightly different models from the same training images. Each patch extracted from the original image is random, and likely to be different from other patches extracted from the same image. We can train several different CNN classifiers using the different patches, and ensemble the different classifiers. Ensembling is the process of creating multiple models and combining them to produce a desired output, as opposed to creating just one model. The *Max* rule [18] is used to ensemble the individual classifiers.

3.5   Training Details

The training data and test data are split in a roughly 8:2 ratio, with four pre-determined folds as training data and one fold as testing data. Five patches, of size 224 × 224 pixels, are randomly sampled from each inputted training image. We train fully connected layers with features extracted from the frozen convolutional layers for 30 epochs, with a stochastic gradient descent optimiser and a learning rate of 10^{-4}. We then fine-tune the entire model for 30 epochs, with a stochastic gradient descent optimiser and a learning rate of 10^{-5}. The loss function in both stages of the training is cross entropy. During each epoch, all the samples in the training set are used.

4   Results

One patch of size 224 × 224 pixels is extracted from each testing image, and inputted to the trained model for prediction. The percentage accuracy of correctly identifying whether the image is benign or malignant is recorded.

4.1   VGGNet and Custom-Made Model Results

The results of VGGNet and our custom model, on each of the four resolutions, with the first fold held as testing data, are shown in Table 2.

Table 2. Results of transfer learning using VGGNet and custom model

	40X	100X	200X	400X
VGGNet	89.12	89.47	86.69	83.35
Custom model	91.28	91.45	88.57	84.58

Using transfer learning on our six-layer custom model out-performs, applying transfer learning on VGGNet, at every resolution. This indicates that although a deep off-the-shelf model such as VGGNet may perform well on large-scale image recognition, a shallower model such as our custom model may be better suited to specific tasks, especially tasks with a relatively smaller amount of training data.

4.2   Custom Model Results

We can see that the custom model consistently out-performs the VGGNet on classifying histopathological images from our chosen benchmark dataset. We proceed to train three classifiers on each resolution separately, and then ensemble the classifiers using the *Max* [18] rule, to obtain a unified model. Five-fold validation is done using the ensembled model, and compared to the current best results on the benchmark dataset. The percentage accuracies and standard deviations are shown in Table 3.

Table 3. Percentage accuracy after ensembling and coss-validation

	40X	100X	200X	400X
Our results	93.3 ± 2.3	94.6 ± 2.2	94.8 ± 3.2	88.4 ± 4.1
Current best [11]	89.6 ± 6.5	85.0 ± 4.8	84.0 ± 3.2	80.8 ± 3.1

The results from our ensembled transfer learning surpasses the current best results on the BreakHis dataset by significant margins, at every resolution. This increase is particularly visible at the 100X and 200X resolution.

5   Conclusion

We propose using transfer learning to diagnose breast cancer from histopathological images. The public benchmark dataset BreakHis was used to train and evaluate our approach. We have explored using transfer learning on VGGNet, an off-the-shelf model, and using transfer learning on a proposed custom model. The performance of using transfer learning on both VGGNet and the custom model is higher than the current best result on the benchmark dataset, presented by Spanhol et al. [11]. Spanhol et al. trained a three-layer CNN model from scratch using thousands of small patches extracted from each image. Furthermore, our custom model out-performs the off-the-shelf VGGNet model. Relative to the approach used by Spanhol et al., our approach samples less (five per image), but slightly larger patches, reducing the overall training time. Deeper models can be successfully trained using transfer learning. Random sampling allows us to train several distinct CNN classifiers, which can then be ensembled to provide a better estimate.

We can also see that our shallower model out-performs the deeper off-the-shelf VGGNet. This indicates that for certain specific tasks, applying transfer learning on a smaller, shallower model may yield a better performance than applying transfer learning to a deep model. After ensembling, our model performed significantly better than the state-of-the-art results [11] on the BreakHis dataset, for all four resolutions. At some resolutions, our performance surpassed the state-of-the-art results, on the chosen benchmark, by 10%.

References

1. Ferlay, J., Soerjomataram, I., Ervik, M., Dikshit, M., Eser, S., Mathers, C., et al.: GLOBOCAN 2012 v1.0, cancer incidence and mortality worldwide: IARC CancerBase no. 11 (2013)
2. Rubin, R., Strayer, D., Rubin, E., McDonald, J.: Rubin's pathology: Clinicopathologic Foundations of Medicine. Lippincott Williams and Wilkins (2007).
3. Pan, S., Yang, Q.: A survey on transfer learning. IEEE Trans. Knowl. Data Eng. **22**(10), 13451359 (2010)

4. Deng, J., Dong, W., Socher, R., Li, L.-J., Li, K., Fei-Fei, L.: ImageNet: a large-scale hierarchical image database. In: IEEE Computer Vision and Pattern Recognition, CVPR (2009)

5. Spanhol, F.A., Oliveira, L.S., Petitjean, C., Heutte, L.: A Dataset for Breast Cancer Histopathological Image Classification. IEEE Trans. Biomed. Eng. **63**(7), 1455–1462 (2016)

6. Simonyan, K., Zisserman, A.: Very deep convolutional networks for large scale image recognition. In: ICLR 2015 (2015)

7. Doyle, S., Agner, S., Madabhushi, A., Feldman, M., Tomaszewski, J.: Automated grading of breast cancer histopathology using spectral clustering with textural and architectural image features. In: IEEE International Symposium on Biomedical Imaging: From Nano to Macro. IEEE International Symposium on Biomedical Imaging, vol. 29, pp. 496–499 (2008)

8. van Diest, P.J., Fleege, J.C., Baak, J.P.A.: Syntactic structure analysis in invasive breast cancer: analysis of reproducibility, biologic background, and prognostic value. Human Pathol. **23**(8), 876–883 (1991)

9. Sewak, M., Vaidya, P., Chan, C.C., Duan, Z.: SVM approach to breast cancer classification. In: 2nd International Multi-Symposiums on Computer and Computational Sciences, IMSCCS 2007, Iowa City, IA, pp. 32–37 (2007)

10. George, Y.M., Zayed, H.H., Roushdy, M.I., Elbagoury, B.M.: Remote computer-aided breast cancer detection and diagnosis system based on cytological images. IEEE Syst. J. **8**(3), 949–964 (2014)

11. Spanhol, F.A., Oliveira, L.S., Petitjean, C., Heutte, L.: Breast cancer histopathological image classification using convolutional neural networks. In: 2016 International Joint Conference on Neural Networks, IJCNN, Vancouver, BC, pp. 2560–2567 (2016)

12. Krizhevsky, A., Sutskever, I., Hinton, G.E.: Imagenet classification with deep convolutional neural networks. In: Advances in Neural Information Processing Systems, pp. 1097–1105 (2012)

13. Roth, H.R., Lu, L., Liu, J., Yao, J., Seff, A., Cherry, K.M., Kim, L., Summers, R.M.: Improving computer-aided detection using convolutional neural networks and random view aggregation. IEEE Trans. Med. Imaging **35**(5), 11701181 (2016)

14. Shin, H.C., Roth, H.R., Gao, M., Lu, L., Xu, Z., Nogues, I., Yao, J., Mollura, D., Summers, R.M.: Deep convolutional neural networks for computer-aided detection: CNN architectures, dataset characteristics and transfer learning. IEEE Trans. Med. Imaging **35**(5), 1285–1298 (2016)

15. Oquab, M., Bottou, L., Laptev, I., Sivic, J.: Learning and transferring mid-level image representations using convolutional neural networks. In: IEEE Conference on Computer Vision and Pattern Recognition, pp. 1717–1724 (2014)

16. Yosinski, J., Clune, J., Bengio, Y., Lipson, H.: How transferable are features in deep neural networks?. In: Advances in Neural Information Processing Systems, pp. 3320–3328 (2014)

17. Hou, L., Samaras, D., Kurc, T.M., Gao, Y., Davis, J.E., Saltz, J.H.: Patch-based convolutional neural network for whole slide tissue image classification. In: Proceedings of the IEEE Conference on Computer Vision and Pattern Recognition, pp. 2424–2433 (2016)

18. Kittler, J., Hatef, M., Duin, R.P., Matas, J.: On combining classifiers. IEEE Trans. Pattern Anal. Mach. Intell. **20**(3), 226–239 (1998)

Real-Time Prediction of the Unobserved States in Dopamine Neurons on a Reconfigurable FPGA Platform

Shuangming Yang[1], Jiang Wang[1(✉)], Bin Deng[1], Xile Wei[1],
Lihui Cai[1], Huiyan Li[2], and Ruofan Wang[3]

[1] School of Electrical and Information Engineering, Tianjin University,
Tianjin, China
jiangwang@tju.edu.cn
[2] School of Automation and Electrical Engineering,
Tianjin University of Technology and Educations, Tianjin, China
[3] School of Information Technology Engineering,
Tianjin University of Technology and Educations, Tianjin, China

Abstract. Real-time prediction of dynamical characteristics of Dopamine (DA) neurons, including properties in ion channels and membrane potentials, is meaningful and critical for the investigation of the dynamical mechanisms of DA cells and the related psychiatric disorders. However, obtaining the unobserved states of DA neurons is significantly challenging. In this paper, we present a real-time prediction system for DA unobserved states on a reconfigurable field-programmable gate array (FPGA). In the presented system, the unscented Kalman filter (UKF) is implemented into a DA neuron model for dynamics prediction. We present a modular structure to implement the prediction algorithm and a digital topology to compute the roots of matrices in the UKF implementation. Implementation results show that the proposed system provides the real-time computational ability to predict the DA unobserved states with high precision. Although the presented system is aimed at the state prediction of DA cells, it can also be applied into the dynamic-clamping technique in the electrophysiological experiments, the brain-machine interfaces and the neural control engineering works.

Keywords: Dynamical prediction · Field-programmable gate array (FPGA) · Dopamine neuron · Real-time

1 Introduction

The dopamine system located in the ventral midbrain is an elementary component of the neural circuits for the control of cognitive and motor behaviors [1], which has been investigated in the treatment of multiple psychiatric and neurodegenerative disorders [2–4]. Midbrain dopamine (DA) neurons fire in a low frequency, metronomic manner predominantly, and generate occasional but functionally vital burst-like episodes in high frequency [5, 6]. Abnormal DA levels are related to psychiatric disorders from depression to Schizophrenia [7, 8]. The mechanism how changes in firing activities of

© Springer International Publishing AG 2017
D. Liu et al. (Eds.): ICONIP 2017, Part IV, LNCS 10637, pp. 677–684, 2017.
https://doi.org/10.1007/978-3-319-70093-9_72

DA neurons facilitates this biological function is not well understood yet. While network connectivity affects the neural responses to stimuli, intrinsic excitability properties of individual neurons do the same thing that defines synchronization qualities and neural coding strategy [9].

A number of previous studies have suggested that Ca^{2+}-independent currents contributes to oscillations [10, 11]. Since the dynamics of the L-type voltage-gated calcium current cannot be measured directly in the electrophysiological experiment, they are considered as the unobserved states of the DA neurons. At the cellular and microcircuit levels, dynamic-clamping technique can investigate more complicated neuronal dynamical characteristics. However, the dynamic clamp technique uses the measured membrane potential to adjust the amount of injected current into a cell. Thus, the accurate prediction of the unobserved hidden states would be meaningful in the dynamic-clamping experiment. There exist two challenges in the application of the dynamic clamp techniques. The first challenge is that the critical hidden states underlying the neural activity behaviors cannot be observed directly by the dynamic-clamping techniques. Besides, the dynamic clamping technique based on hardware is limited by the low programmability, and software-based technique cannot meet the requirement of real-time computation. Therefore, it is demanded to present a novel technique with the advantages of computational speed and good programmability to solve the challenges in the conventional technique.

Recently, the unscented Kalman filter (UKF) has been increasingly used in the dynamical estimation of nonlinear neural systems [12, 13]. However, the states of the ionic channels of the DA cell evolve rapidly in the process of external stimulation. To the end of real-time online computational performance of the UKF, a hardware design with high computational efficiency is needed. The heavy computational burden is a challenging problem for a fixed-point processor on a hardware chip. Manifesting the advantages of high programmability, low energy expedition and parallel calculation, the field programmable gate array (FPGA) is an ideal candidate for a neural dynamical prediction system [14, 15].

In this study, we present a portable neural dynamical prediction processor implemented by FPGA for the real-time prediction of the dopamine hidden states, which has the advantages of high modularity and high computational efficiency in comparison with the state-of-the-art techniques. The proposed work facilitates the mechanism investigation of the dopamine dynamics, and improves the performance of the current electrophysiological dynamic-clamping technique. The proposed UKF-based scheme of the hidden states predictions can be applied in the studies using the dynamic clamp, and the performance enhancement of the closed-loop neural control and dynamical exploration of the psychiatric disorders from depression to Schizophrenia.

2   Methods

2.1   Platform Setup

The overview of the platform setup is shown in Fig. 1, which is implemented to predict the hidden states of the DA cells in real time. The proposed platform is equipped with an analog-to-digital converter (ADC) device, a data acquisition (DAQ) device, a FPGA platform and a personal computer (PC). The FPGA platform can receive the biological activities of the DA neuron in two approaches. The first is the direct acquisition from the ADC device. The second approach is using a DAQ device to acquire the biological activities of the DA neuron and then transmitted the measured signals from the PC. The PC can monitor and show the prediction results, and conduct further theoretical analysis about the prediction results.

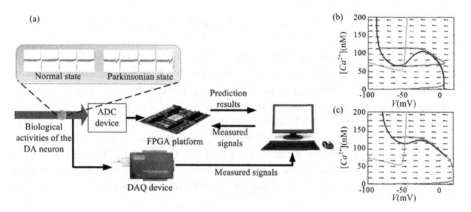

Fig. 1. Overview of the prediction system for FPGA implementation of unobserved states for the DA neurons. (a) The platform setup. (b) Phase portraits of DA model when I_{app} = 2 mA. (c) Phase portraits of DA model when I_{app} = 40 mA.

2.2 The DA Neuron Model

The biological model of DA neuron is a conductance-based neuron model. Their variables in the DA neuron model are governed by the following current-balance equations:

$$\begin{cases} c_m \frac{dv}{dt} = g_{Ca}(v)(E_{Ca} - v) + (g_{KCa}([Ca^{2+}]) + g_K(v))(E_K - v) + g_l(E_l - v) + I_{app} \\ \frac{d[Ca^{2+}]}{dt} = \frac{2\beta}{r}\left(\frac{g_{Ca}(v)}{zF}(E_{Ca} - v) - P_{Ca}[Ca^{2+}]\right) \end{cases}$$

$$(1)$$

where c_m is the membrane capacitance; g_{Ca} is calcium conductance; g_{KCa} is calcium dependent potassium conductance; E_K is potassium reversal potential; g_l is leak conductance; E_l is leak reversal potential. In the calcium equation, β is the calcium buffering coefficient, $i.e.$ the ratio of free to total calcium, r is the radius of the compartment, z is the valence of calcium, and F is Faraday's constant. P_{Ca} represents the maximum rate of calcium removal through the pump. The subgroup of intrinsic ionic currents: I_{Ca}, I_{KCa}, I_K, and I_{leak} work on pacemaking mechanism of the DA neuron. Detailed definitions of neuron variables and parameter values are given in [9]. The dynamical behaviors of the DA model are shown in Fig. 1(b) and (c).

2.3   The Application of the UKF in the Synaptic State Estimation

In order to predict the hidden states of the DA neuron, the UKF is demanded to be implemented in the DA neuron model. For an N-dimensional estimated state x, the sigma points from the mean value and covariance P_{xx} is chosen as:

$$\begin{cases} X_i = \bar{x} + (\sqrt{NP_{xx}})_i^T & (i = 1, 2, \ldots, N) \\ X_i = \bar{x} - (\sqrt{NP_{xx}})_i^T & (i = N+1, \ldots, 2N) \end{cases} \tag{2}$$

where P_{xx} is the prediction covariance matrix and $\sqrt{NP_{xx}}$ is the matrix square root. Sigma points are the sample.

The function G is applied to the sigma points with results $\tilde{X}_i = G(X_i)$ $(i = 1,2\ldots 2N)$. The observation of the new state is described by $\tilde{Y}_i = M(\tilde{X}_i)$. The nonlinear G (X) and $M(X)$ use the DA neuron model to obtain the transformed points and observations, which are implemented in the transformed point's acquirement module. In order to implement the UKF algorithm into the DA model, the augmented state vector x as an $N = p + M$ dimension vector consisting of p parameters and n dynamic variables. In order to predict the external current of the DA neuron, the external applied current I_{app} is regarded as a time-varying parameter and added into the state vector. Therefore, the process equations of the UKF algorithms are described by:

$$\dot{x} = \begin{bmatrix} \dot{I}_{app} \\ V \\ [Ca^{2+}] \end{bmatrix} = \begin{bmatrix} I_{app} \\ (I_{Ca} + I_{KCa} + I_K + I_{leak})/c_m \\ h(v, [Ca^{2+}]) \end{bmatrix} + Q \tag{3}$$

where the function $h(v, [Ca^{2+}])$ is expressed as follows:

$$h(v, [Ca^{2+}]) = \frac{2\beta}{r} \left(\frac{g_{Ca}(v)}{zF} (E_{Ca} - v) - P_{Ca}[Ca^{2+}] \right). \tag{4}$$

Since the only measurable state is the membrane potential V of the DA neuron, the measurement equation is described by $y = Cx + R$, where $C = [0\ 1\ 0]$. Besides, the proposed method can also predict both the unobserved states and system parameters by adding the observed states with unobserved hidden states and system parameters.

The updating module of UKF can assimilate noisy measurable states to update the system states. The mean values are defined as $\tilde{x} = \frac{1}{2N} \sum_{i=1}^{2N} \tilde{X}_i$ and $\tilde{y} = \frac{1}{2N} \sum_{i=1}^{2N} \tilde{Y}_i$, which represents the a priori state prediction and a priori measurement prediction respectively. The a priori covariance of the ensemble members is described as follows:

$$\begin{cases} \tilde{P}_{xx} = \sum_{i=1}^{2N} (\tilde{X}_i - \tilde{x})(\tilde{X}_i - \tilde{x})^T + Q \\ \tilde{P}_{yy} = \sum_{i=1}^{2N} (\tilde{Y}_i - \tilde{y})(\tilde{Y}_i - \tilde{y})^T + R \\ \tilde{P}_{xy} = \sum_{i=1}^{2N} (\tilde{X}_i - \tilde{x})(\tilde{Y}_i - \tilde{y})^T \end{cases} \tag{5}$$

where Q and R are the covariance matrix of the process noise and observation noise. The current state and error covariance are updated by the a posteriori equation $\hat{x} = \tilde{x} + K(y - \tilde{y})$ and $\hat{P}_{xx} = \tilde{P}_{xx} - K\tilde{P}_{yy}$ respectively. The Kalman gain matrix is obtained by $K = \tilde{P}_{xy}\tilde{P}_{yy}^{-1}$ and y represents the measurement state. The observation state is updated by the equation $\hat{y} = M(\hat{x})$, and updated $\hat{x}$ and $\widehat{P_{xx}}$ will be used in the next iteration.

3   Results

3.1   FPGA Implementation of the Prediction System

In order to predict the hidden states of the DA neuron, the UKF is demanded to be implemented in the DA neuron model. The proposed hardware architecture is designed according to the reformed UKF algorithm. The top layer module implements several sub-modules according to the major steps in UKF calculation as illustrated in Fig. 2(a). In aspects of control logic, the top-level module uses a finite-state machine (FSM) to issue data signals in different states based on the calculation order in the UKF algorithm. The proposed system contains both the decoding and encoding parts of the UKF algorithm, which is different from previous works that only implement the decoding part. The Cholesky decomposition is implemented in the "Cholesky decomposition" module. The "X$_i$_calc" module calculates the sigma points X$_i$, and the DA computation module is used to propagate each sigma point from time step t to $t + 1$ to obtain transformed points and observations. The prediction of the new mean values according

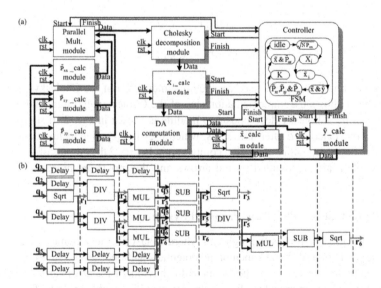

Fig. 2. Overview of the platform setup for FPGA implementation of hidden states prediction for the DA neurons. (a) The top-level description of the prediction system. (b) The digital structure of the Cholesky decomposition algorithm. (Color figure online)

Table 1. The computation procedure of the Cholesky decomposition in the FPGA system.

Step	1	2	3	4	5	6	7	8	9
r_1	$\sqrt{q_1}$								
r_2		q_2/r_1							
r_3			r_2^2	$q_3 - r_2^2$	$\sqrt{q_3 - r_2^2}$				
r_4		q_4/r_1							
r_5				$r_4 r_2$	$q_5 - r_4 r_2$	$(q_5 - r_4 r_2)/r_3$			
r_6					r_4^2	$q_6 - r_4^2$	r_5^2	$q_6 - r_4^2 - r_5^2$	$\sqrt{q_6 - r_4^2 - r_5^2}$

to the transformed points is computed in the "$\widetilde{x}$ _calc" and "$\widetilde{y}$ _calc" modules, and the estimation of the updated covariance is computed in the "$\widetilde{P}_{xx}$ _calc", "$\widetilde{P}_{xy}$ _calc" and "$\widetilde{P}_{yy}$ _calc" modules. The "parallel Mult." module contains 2N multiplication blocks in parallel, which is used to compute the Kalman gain matrix K, updated mean and covariance matrix.

The digital topology of the Cholesky algorithm is shown in Fig. 2(b), which is a parallel structure for high-speed implementation. The Cholesky algorithm is described as follows. Suppose two covariance matrices Q and R as:

$$Q = \begin{bmatrix} q_1 & q_2 & q_4 \\ q_2 & q_3 & q_5 \\ q_4 & q_5 & q_6 \end{bmatrix} = R^T R, \; R = \begin{bmatrix} r_1 & 0 & 0 \\ r_2 & r_3 & 0 \\ r_4 & r_5 & r_6 \end{bmatrix} \tag{6}$$

where the corresponding equations are as follows:

$$r_1 = \sqrt{q_1}; r_2 = q_2/r_1; r_3 = \sqrt{q_3 - r_2^2}; r_4 = q_4/r_1;$$
$$r_5 = (q_5 - r_4 r_2)/r_3; r_6 = \sqrt{q_6 - r_4^2 - r_5^2}. \tag{7}$$

The pipeline technique is used for the implementation of the Cholesky decomposition. Table 1 shows the implementation step of the Cholesky algorithm and the digital structure of the Cholesky decomposition module is shown in Fig. 2. It shows that the results can be calculated after nine steps, which uses the pipeline technique to take the advantage of the capability of parallel computation. In Fig. 2 the outputs of the Cholesky decomposition are represented by red arrows and the inputs are represented by blue arrows.

3.2 Prediction Results of the Proposed System

The prediction system uses the high-end Altera FPGA pro development board for the real-time prediction of the DA membrane potentials and unobserved states. Figure 3(a) shows the prediction results of the dynamical behaviors of the DA neuron. The performance of the UKF dynamical prediction strategy provides credible results. The proposed RTDE system is implemented on a Stratix-III EP3SL150 FPGA. As shown in Fig. 3(a), the required neuronal dynamics can be obtained by the presented system in real time.

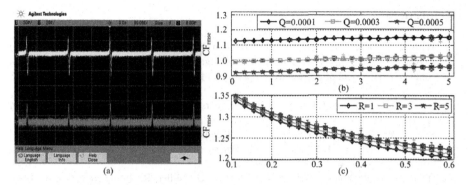

Fig. 3. The real-time prediction results on the proposed reconfigurable platform. (a) The prediction results of the membrane potential and unobserved state of DA neuron shown on the oscilloscope device. (b) Effects of the covariance matrices for process noise Q and observation noise R on the cost function CF_{rmse}.

We use the cost function CF_{rmse} for root mean square error (RMSE) to evaluate the precision performance of the prediction system, which is defined as

$$CF_{rmse} = \sqrt{\frac{\sum_{i=1}^{n} \left(x_{est,i} - x_{tru,i}\right)^2}{n}} \tag{8}$$

where n is the number of estimated points, and $x_{est,i}$ and $x_{tru,i}$ stand for the estimated (*est*) and the true values (*tru*), respectively. The cost function CF_{rmse} is used to evaluate the estimation error of the algorithm including the effects of the process noise (Q) and observation noise (R) covariance matrices on the cost function CF_{rmse} as shown in Fig. 3(b) and (c), which is plotted in a form of boxplot. The boxplots show the points, L-estimators, interquartile range, midhinge, range, mid-range and trimean. The estimation error of the membrane potentials increases with increasing observation noise R, or with the decreasing process noise Q as shown in Fig. 3.

4 Conclusions

In this study, a real-time system for the dynamical prediction of DA cells is developed on a reconfigurable FPGA platform. The system can effectively predict the membrane potentials with noise and unobserved states with high precision. Implementation results show that the proposed system can predict the membrane potentials and unobserved states of DA cells in a high precision. This study opens a pathway for the future works of neural dynamical estimation of DA system by overcoming the high hardware cost, scalability and computational efficiency challenges.

Acknowledgements. This work was supported in part by the National Natural Science Foundation of China under Grant 61374182, in part by the National Natural Science Foundation of China under Grant 61601331, and in part by the National Natural Science Foundation of China under Grant 61471265.

References

1. Wise, R.A.: Dopamine, learning and motivation. Nat. Rev. Neurosci. **5**(6), 483–494 (2004)
2. Braver, T.S., Barch, D.M., Cohen, J.D.: Cognition and control in Schizophrenia: a computational model of dopamine and prefrontal function. Biol. Psychiatry **46**(3), 312–328 (1999)
3. Iversen, S.D., Iversen, L.L.: Dopamine: 50 years in perspective. Trends Neurosci. **30**(5), 188–193 (2007)
4. Schultz, W.: Behavioral dopamine signals. Trends Neurosci. **30**(5), 203–210 (2007)
5. Grace, A.A., Bunney, B.S.: The control of firing pattern in nigral dopamine neurons: burst firing. J. Neurosci. **4**(11), 2877–2890 (1984)
6. Hyland, B.I., Reynolds, J.N.J., Hay, J., Perk, C.G., Miller, R.: Firing modes of midbrain dopamine cells in the freely moving rat. Neuroscience **114**(2), 475–492 (2002)
7. Grace, A.A.: Phasic versus tonic dopamine release and the modulation of dopamine system responsivity: a hypothesis for the etiology of Schizophrenia. Neuroscience **41**(1), 1–24 (1991)
8. Nestler, E.J., Carlezon, W.A.: The mesolimbic dopamine reward circuit in depression. Biol. Psychiatry **59**(12), 1151–1159 (2006)
9. Morozova, E.O., Zakharov, D., Gutkin, B.S., Lapish, C.C., Kuznetsov, A.: Dopamine neurons change the type of excitability in response to stimuli. PLoS Comput. Biol. **12**(12), e1005233 (2016)
10. Guzman, J.N., Sánchez-Padilla, J., Chan, C.S., Surmeier, D.J.: Robust pacemaking in substantia nigra dopaminergic neurons. J. Neurosci. **29**(35), 11011–11019 (2009)
11. Khaliq, Z.M., Bean, B.P.: Pacemaking in dopaminergic ventral tegmental area neurons: depolarizing drive from background and voltage-dependent sodium conductances. J. Neurosci. **30**(21), 7401–7413 (2010)
12. Yang, S., Deng, B., Wang, J., Li, H., Liu, C., Fietkiewicz, C., Loparo, K.A.: Efficient implementation of a real-time estimation system for thalamocortical hidden Parkinsonian properties. Sci. Rep. **7**, 40152 (2017)
13. Julier, S.J., Uhlmann, J.K.: Unscented filtering and nonlinear estimation. Proc. IEEE **92**(3), 401–422 (2004)
14. Yang, S., Wang, J., Li, S., Deng, B., Wei, X., Yu, H., Li, H.: Cost-efficient FPGA implementation of basal ganglia and their Parkinsonian analysis. Neural Netw. **71**, 62–75 (2015)
15. Yang, S., Wang, J., Li, S., Li, H., Wei, X., Yu, H., Deng, B.: Digital implementations of thalamocortical neuron models and its application in thalamocortical control using FPGA for Parkinson's disease. Neurocomputing **177**, 274–289 (2016)

A Subject-Specific EMG-Driven Musculoskeletal Model for the Estimation of Moments in Ankle Plantar-Dorsiflexion Movement

Congsheng Zhang[1,2], Qingsong Ai[1,2(✉)], Wei Meng[1,2], and Jiwei Hu[1,2]

[1] School of Information Engineering, Wuhan University of Technology, Wuhan 430070, China
qingsongai@whut.edu.cn
[2] Key Laboratory of Fiber Optic Sensing Technology and Information Processing, Wuhan University of Technology, Ministry of Education, Wuhan 430070, China

Abstract. In traditional rehabilitation process, ankle movement ability is only qualitatively estimated by its motion performance, however, its movement is actually achieved by the forces acting on the joints produced by muscles contraction. In this paper, the musculoskeletal model is introduced to provide a more physiologic method for quantitative muscle forces and muscle moments estimation during rehabilitation. This paper focuses on the modeling method of musculoskeletal model using electromyography (EMG) and angle signals for ankle plantar-dorsiflexion (P-DF) which is very important in gait rehabilitation and foot prosthesis control. Due to the skeletal morphology differences among people, a subject-specific geometry model is proposed to realize the estimation of muscle lengths and muscle contraction force arms. Based on the principle of forward and inverse dynamics, difference evolutionary (DE) algorithm is used to adjust individual parameters of the whole model, realizing subject-specific parameters optimization. Results from five healthy subjects show the inverse dynamics joint moments are well predicted with an average correlation coefficient of 94.21% and the normalized RMSE of 12.17%. The proposed model provides a good way to estimate muscle moments during movement tasks.

Keywords: EMG signals · Musculoskeletal model · Ankle plantar-dorsiflexion · Joint moment

1 Introduction

With the development of society and technology, the aging problem has become more and more serious. At the same time, patients with limb disability are also increasing. There are nearly 1.4 million people losing their ability to live independently because of stroke in China each year [1]. Health care and rehabilitation for elderly and disabled people are increasing. The ankle joint plays an important role in human standing and walking but easy to be damaged [2]. It is important to estimate the moments for

© Springer International Publishing AG 2017
D. Liu et al. (Eds.): ICONIP 2017, Part IV, LNCS 10637, pp. 685–693, 2017.
https://doi.org/10.1007/978-3-319-70093-9_73

ankle joint during its movement and rehabilitation. There are three degrees of freedom (DOFs) in the ankle joint: dorsiflexion/plantarflexion, abduction/adduction, inversion/eversion. Plantar-dorsiflexion (P-DF) is the most important among them with great significance in gait rehabilitation [3] and foot prosthesis research [4].

The electromyography (EMG) signals are commonly used in rehabilitation to estimate the moments for the physiological significance. There are two main methods for muscle forces and moments estimation based on EMG signals: black-box method and musculoskeletal model [5]. Black-box method includes neural network, support vector regression and other fitting algorithms or related improved algorithms to build the relationship between EMG signals and muscle forces or moments. Predictive model can be built easily through part of the input and output data using black-box method. But this method cannot inform us of muscle changes during body movement and cannot provide good reference for rehabilitation analysis [6].

Musculoskeletal model is able to provide a better understanding in mechanical process of human motion. Hassani et al. estimated muscular activities detection by realistic musculoskeletal models of the muscles actuating the knee joint, realizing an active rehabilitation strategy following the wearer's intention [7]; Manal et al. built a musculoskeletal model for ankle P-DF and proposed an EMG-driven modeling approach and data processing framework that allowed them to predict Achilles tendon force in real-time [8]. Musculoskeletal model provides a better method for muscle forces and moments estimation with physiological significance, realizing quantitative estimation of physical exercise ability.

Many existing researches on ankle P-DF moment only focus on simple applications of mature models without musculoskeletal model optimization. Some researchers use sophisticated equipment such as nuclear magnetic resonance imaging (MRI) [9] or motion capture systems [10] to obtain actual model parameters, which is not suitable for practical applications. In this paper, EMG signals from four muscles related to ankle P-DF are collected along with angle signals. A subject-specific musculoskeletal geometry model is proposed to estimate muscle lengths and muscle force arm lengths. Difference evolutionary (DE) algorithm is applied to adjust the parameters of the whole model in off-line condition.

The contributions of this paper are: (1) Subject-specific musculoskeletal model provides a simple and effective method for moments estimation; (2) Muscle status and rehabilitation information can be easily obtained during the estimation process; (3) The proposed method has the potential in intuitive and continuous robot control. The rest of this paper is arranged as follows: Sect. 2 presents design details of the ankle P-DF musculoskeletal model. The experiment is carried out to verify the performance of the model in Sect. 3. Section 4 draws conclusion of the paper.

2   Musculoskeletal Modeling Methods

The EMG-driven musculoskeletal model in ankle P-DF derives from Hill-based muscle model and joint forward/inverse dynamics. The model in this paper builds direct relationship between EMG signals and joint moments, which consists of three modules: muscle activation dynamics, Hill-type muscle-tendon model and subject-specific musculoskeletal geometry model, as shown in Fig. 1. Raw EMG signals collected from medial gastrocnemius (MG), lateral gastrocnemius (LG), soleus (SO) and tibialis anterior (TA) are pre-processed and then used as input to the muscle activation dynamics to get muscle activations. Angle signals are taken as input to the subject-specific musculoskeletal geometry model to get the muscle lengths and muscle force arm lengths. Hill-type muscle-tendon model is used to get muscle forces of each muscle and finally get the whole moment output.

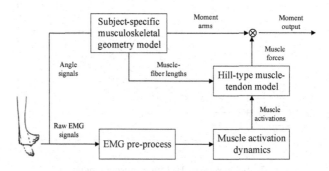

Fig. 1. EMG-driven musculoskeletal model

2.1 EMG Pre-process and Muscle Activation Dynamics

In ankle movement, raw EMG signals contain some low-frequency noise due to EMG acquisition device itself and environment impact. Therefore, EMG signals can be pre-processed by a band-pass filter with a cutoff frequency in the range of 5–30 Hz and then be rectified and normalized. The processed EMG signals are expressed by $e(t)$. Neural activation $u(t)$ is related to its past magnitude and $e(t)$. A two order difference equation can be used to describe the dynamic relationship between them as follows:

$$u(t) = \gamma e(t - d) - \beta_1 u(t - 1) - \beta_2 u(t - 2) \tag{1}$$

where d is time delay. γ, β_1 and β_2 are the scaling coefficients. To realize a positive stable solution, they must satisfy the constraints as shown in Eq. (2).

$$\beta_1 = c_1 + c_2, \beta_2 = c_1 \cdot c_2, \gamma - \beta_1 - \beta_2 = 1 \tag{2}$$

where $|c_1| < 1$, $|c_2| < 1$. Since there is a nonlinear relationship between $u(t)$ and muscle activation $a(t)$, a simple transformation is needed as follows:

$$a(t) = \frac{e^{Au(t)} - 1}{e^A - 1} \tag{3}$$

where A represents nonlinear shape factor with values ranging from -3 to 0.

2.2   Hill-Type Muscle-Tendon Model

When muscles are activated, they contract and produce muscle contraction forces. This section focuses on the process of muscle contraction dynamics from muscle activations to contraction forces. Muscles are mainly composed of muscle fibers and tendons. The Hill model equates muscle fibers with a passive element CE in parallel with the shrink element PE and equates muscle tendon to a non-linear spring element, as shown in Fig. 2.

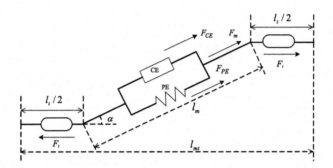

Fig. 2. Schematic of Hill-type muscle-tendon model

The relationship between muscle unit l_{mt}, muscle fibers l_m and muscle tendon l_t can be described by Eq. (4), where α is pennation angle. Muscle fibers contraction force F_m is equal to the sum of the forces from the shrink element and the passive element, as shown in Eq. (5).

$$l_{mt} = l_m \cos \alpha + l_t \tag{4}$$

$$F_m = F_m^{\max}[F_A(l) \cdot F_V(v) \cdot a + F_P(l)] \tag{5}$$

where $F_m^{\max}$ is maximum isometric contraction force, a denotes muscle activation, $F_A(l)$ is active force-length relationship. $F_V(v)$ describes force-velocity relationship and $F_p(l)$ represents passive elastic force–length relationship. And $l = l_m/l_{mopt}$ is normalized muscle fiber length, $v = v_m/(0.5 \cdot v_{\max}(a+1))$ represents the normalized muscle fiber contraction velocity, l_{mopt} is the muscle fiber length when maximum isometric force is produced and $v_{\max}$ is the maximum velocity.

Tendons are rather stiff and the strain is only about 3% of tendon length for maximum muscle force, which can be neglected [11]. Tendon changes during general ankle P-DF in OpenSim (Simtk, Stanford. USA) [12] also illustrate the stiffness of the tendons. Thus, tendon slack length scale s_t can be introduced to adjust personalized tendon length l_t based on tendon slack length l_{tr}, as shown in Eq. (6). In this way, the muscle fiber length can be calculated indirectly by Eq. (4) when tendon length l_t and muscle length l_{mt} are obtained.

$$l_t = s_t \cdot l_{tr} \tag{6}$$

2.3   Subject-Specific Musculoskeletal Geometry Model

Muscle paths need to be determined first to build musculoskeletal geometry model, which can be done by medical device or by anatomical measurement on cadaveric samples. General musculoskeletal geometry model is built based on the mean of data from above methods, which is not suitable for different people. Muscles are attached to skeletons, whose paths are related to skeletal morphology. Muscle path assessment equations [13] are used to determine morphological parameters of the actual subjects, and the paths of main muscle group in ankle joint are obtained. For simplicity, muscles are assumed as straight lines. Suppose the body sitting on a chair with the knee staying 90 degrees, taking SO as an example, as shown in Fig. 3.

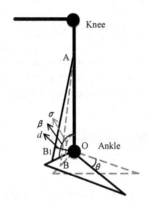

Fig. 3. Musculoskeletal geometry model of SO

A is the origin point, B is the insertion point and O represents the center of ankle joint. The dashed line indicates the foot is in the horizontal position and the solid line indicates the foot moves to a certain position. θ is the angle of ankle motion, l_{AB_1} represents the muscle length and d is arm length. The calculation equation is shown in Eq. (7).

$$\begin{cases} l_{AB_1} = \sqrt{l_{AO}^2 + l_{B_1O}^2 - 2l_{AO}l_{B_1O}\cos\beta} \\ d = l_{AO}l_{B_1O}\sin\beta/l_{AB_1} \end{cases} \tag{7}$$

2.4   Parameters Identification

There are many parameters with subject-specific differences and difficult to be measured directly in musculoskeletal model. To improve the prediction accuracy, subject-specific parameters should be selected for different subjects.

Since pennation angle α is usually small, it can be assumed as consistent to reduce the complexity of parameters identification. Also, it can be assumed that all muscle activation parameters are the same and v_{max} can be replaced by $10l_{mopt}$. Therefore, the parameters needed to be identified are d, c_1, c_2 and A in muscle activation model, F_m^{max}, l_{mopt} and s_t in Hill-type muscle-tendon model. Different people have different muscle lengths and arm lengths even when their ankle joints are at the same angle. Subject-specific height and weight should be measured to determine the muscle lengths and arm lengths of different people in Eq. (7). Then the original data [14] for parameters identification and their physiological limits [15] are selected.

Based on the principle shown in Eq. (8) that the joint moments estimated by EMG-driven model should match the moments calculated by inverse dynamics, a difference evolutionary algorithm is used to tune the parameters, which can minimize the difference between the calculated moments from inverse dynamics and the estimated moments. Thus, the best parameters for different people can be found, realizing subject-specific parameters identification. The left side of the equation is the reference moment calculated according to the inverse dynamics and the right side is the estimated moment of the skeletal muscle model.

$$I\ddot{\theta} + Mgl\cos\theta = -F_{MG}d_{MG} - F_{LG}d_{LG} - F_{SO}d_{SO} + F_{TA}d_{TA} \tag{8}$$

where θ is the angle of ankle P-DF movement, I, M, g, l are the moment of inertia for the foot, the mass of the foot, the gravity acceleration and the gravity arm length of the foot, respectively, F and d represent muscle forces and arm lengths.

3   Experimental Results and Analysis

3.1   Experiment Setup

Five healthy subjects were recruited to collect the signals (right foot, 23.5 ± 1.6 years old). Written informed consent was obtained from them. The subjects were asked to sit on a suitable table to conduct the ankle P-DF movement, with his/her knee joint staying 90 degrees and foot above the ground. Four electrodes were attached to the skin surface on muscle bellies center of MG, LG, SO and TA. Raw EMG signals were recorded by the EMG acquisition equipment (DataLOG MWX8, Biometrics Ltd. UK) and angle signals were recorded by the angle acquisition equipment (MPU6050, InvenSense. USA). To reduce the effects of muscle fatigue, subjects relaxed their ankle muscles for 10 min between every two experiment trials.

3.2   Experimental Results

Without loss of generality, two subjects (Sub.1 and Sub.4) are selected to analyze the results, as shown in Fig. 4. Both the value and trend of the estimated moment (the green and the red dashed line) are in good agreement with the reference moment (the blue solid line) calculated by inverse dynamics illustrated in the section of parameters identification, which shows the EMG-driven model can be used for ankle joint moment estimation with an acceptable accuracy. It can be seen that under the same test conditions, the subject-specific model (the green dashed line) is closer to the reference moment and better than the general model (the red dashed line), which demonstrates different people have different parameters and it is thus of great importance to build their own subject-specific musculoskeletal model.

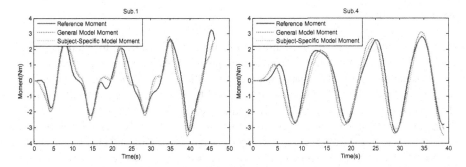

Fig. 4. Comparison of moments (Color figure online)

The same experiments have been performed with five subjects. The normalized RMSE and correlation coefficient are used as the main criteria to show the performance of the model. A high level of correlation coefficient and a low normalized RMSE have been obtained as presented in Table 1.

Table 1. Summary of prediction performance of the model

Subject	Model	Correlation coefficient (%)	Normalized RMSE (%)
Sub.1	G	93.57	14.71
	S	95.08	11.78
Sub.2	G	92.62	13.62
	S	94.44	11.97
Sub.3	G	89.35	17.37
	S	91.16	14.29
Sub.4	G	96.02	12.48
	S	96.78	10.29
Sub.5	G	91.76	14.28
	S	93.57	12.53
Mean	G	92.66	14.49
	S	94.21	12.17

(Note: G and S stands for general model and subject-specific model, respectively.)

It is evident that the estimated moments in G and S are both in good agreement with the reference moment but the result in S is better than that in G. The mean of the correlation coefficient in S (94.21%) is greater than that in G (92.66%) and the mean of the normalized RMSE in S (12.17%) is smaller than that in G (14.49%). In addition, the correlation coefficients in S are greater and the normalized RMSEs in S are smaller across all subjects compared with that in G, indicating the proposed model is capable of providing better representations of the subjects involved in this study. Through the results of the experiment, it can be concluded that the performance of subject-specific model is superior to general model.

4   Conclusion

Researches on ankle P-DF musculoskeletal model are beneficial to evaluate muscle contraction forces and moments from a more physiological point of view, and can thus improve the flexibility of human-computer interaction. In this paper, EMG signals and angle signals in ankle P-DF movement are used to estimate the joint moments off-line through a subject-specific EMG-driven musculoskeletal model with an acceptable accuracy. The average correlation coefficient is high (94.21%) and the normalized RMSE is relatively small (12.17%), which shows the feasibility and effectiveness of the proposed model. Experimental results also show the importance of acquiring more accurate musculoskeletal geometry. Limitations of this paper include the fact that off-line musculoskeletal model is studied but without using it in real-time applications. Therefore, our future work is to verify the on-line effectiveness of EMG-driven model proposed in this study and consider the practical conditions. Further more, the approach proposed in this study will be expanded to other joints that can be employed in various applications.

Acknowledgments. Research supported by The Excellent Dissertation Cultivation Funds of Wuhan University of Technology with No. 2016-YS-062 and National Natural Science Foundation of China under grants Nos. 51475342 and 61401318.

References

1. Zhang, T.: Stroke rehabilitation in China (2011 edition). Chin. J. Rehabil. Theor. Pract. **18**(4), 301–318 (2012). (in Chinese)
2. Meng, W., Xie, S., Liu, Q., et al.: Robust iterative feedback tuning control of a compliant rehabilitation robot for repetitive ankle training. IEEE/ASME Trans. Mechatron. **22**(1), 173–184 (2017)
3. Vivian, M., Tagliapietra, L., Reggiani, M., et al.: Design of a subject-specific EMG model for rehabilitation movement. Biosyst. Biorobotics **7**, 813–822 (2014)
4. Patar, A., Jamlus, N., Makhtar, K., et al.: Development of dynamic ankle foot orthosis for therapeutic application. Procedia Eng. **41**, 1432–1440 (2012)
5. Meng, W., Ding, B., Zhou, Z., et al.: An EMG-based force prediction and control approach for robot-assisted lower limb rehabilitation. In: 45th IEEE International Conference on Systems, Man, and Cybernetics, pp. 2198–2203. Institute of Electrical and Electronics Engineers Inc., San Diego (2014)
6. Ai, Q., Ding, B., Liu, Q., et al.: A subject-specific EMG-driven musculoskeletal model for applications in lower-limb rehabilitation robotics. Int. J. Humanoid Robot. **13**(03), 1650005 (2016)
7. Hassani, W., Mohammed, S., Rifaï, H., et al.: Powered orthosis for lower limb movements assistance and rehabilitation. Control Eng. Pract. **26**(1), 245–253 (2014)
8. Kurt, M., Karin, G., Buchanan, T.: A real-time EMG-driven musculoskeletal model of the ankle. Multibody Sys. Dyn. **28**(1–2), 169–180 (2012)
9. Zhang, M., Meng, W., Davies, T., et al.: A robot-driven computational model for estimating passive ankle torque with subject-specific adaptation. IEEE Trans. Biomed. Eng. **63**(4), 814–821 (2016)

10. Prinold, J., Mazzà, C., Marco, R., et al.: A patient-specific foot model for the estimate of ankle joint forces in patients with juvenile idiopathic arthritis. Ann. Biomed. Eng. **44**(1), 247–257 (2016)
11. Fleischer, C., Hommel, G.: A human-exoskeleton interface utilizing electromyography. IEEE Trans. Robot. **24**(4), 872–882 (2008)
12. Delp, S., Anderson, F., Arnold, A., et al.: OpenSim: open-source software to create and analyze dynamic simulations of movement. IEEE Trans. Bio-med. Eng. **54**(11), 1940–1950 (2007)
13. Zheng, R., Liu, T., Kyoko, S., et al.: In vivo estimation of dynamic muscle-tendon moment arm length using a wearable sensor system. In: 12th IEEE/ASME International Conference on Advanced Intelligent Mechatronics, pp. 647–652. Institute of Electrical and Electronics Engineers Inc., Xi'an (2008)
14. Delp, S., Loan, J., Hoy, M., et al.: An interactive graphics-based model of the lower extremity to study orthopaedic surgical procedures. IEEE Trans. Biomed. Eng. **37**(8), 757–767 (1990)
15. Shao, Q., Bassett, D., Manal, K., et al.: An EMG-driven model to estimate muscle forces and joint moments in stroke patients. Comput. Biol. Med. **39**(12), 1083–1088 (2009)

Real-Time Scalp-Hemodynamics Artifact Reduction Using a Sliding-Window General Linear Model: A Functional Near-Infrared Spectroscopy Study

Yuta Oda[✉], Takanori Sato, Isao Nambu, and Yasuhiro Wada

Graduate School of Engineering, Nagaoka University of Technology,
1603-1 Kamitomioka, Nagaoka, Niigata 940-2188, Japan
yoda@stn.nagaokaut.ac.jp

Abstract. Functional near-infrared spectroscopy (fNIRS) measures temporal hemoglobin changes in gray matter, reflecting brain activity. The primary advantage of fNIRS is real-time estimation of brain activity, with applications such as neurofeedback training. However, task-related scalp-hemodynamics distributed across the whole head are superimposed onto cerebral activity, leading to false estimation of brain activity. To prevent this, we propose a real-time artifact rejection method using short distance probes, by applying a sliding-window general linear model (GLM) with a real-time updated design matrix via a global scalp-hemodynamics model (GSHM). To assess the performance of our proposed method, we performed simulation, assuming that fNIRS signals, consisting of local cerebral blood flow (CBF) and scalp-hemodynamics, had a spatially common pattern. Simulation results were compared with off-line analysis and previous on-line methods, with scalp-hemodynamics excluded from the design matrices. The proposed method showed significantly higher performance for estimating CBF.

Keywords: Functional near-infrared spectroscopy · Sliding window analysis · Scalp-hemodynamics artifact reduction

1 Introduction

Functional near-infrared spectroscopy (fNIRS) is a noninvasive functional neuroimaging technique that can measure concentration changes in oxygenated and deoxygenated hemoglobin (ΔOxy- and ΔDeoxy-Hb) associated with gray matter changes. The advantages of fNIRS include fewer physical constraints for the participant and portability facilitating wider adoption in rehabilitation and brain computer interfaces. Neurofeedback training aims to improve brain functions by offering feedback on brain activity. The relationship between brain activity and behavior has been studied recently [4].

In neurofeedback training process, some stimuli that represent individual brain activity are given to a subject in real-time, for example, showing up a map

© Springer International Publishing AG 2017
D. Liu et al. (Eds.): ICONIP 2017, Part IV, LNCS 10637, pp. 694–701, 2017.
https://doi.org/10.1007/978-3-319-70093-9_74

of brain activity on a screen. In such situations, signal processing is required to estimate actual brain activity with signal artifact reduction.

Due to the nature of fNIRS measurement, the signals show scalp-hemo-dynamics flowing over gray matter layers and cerebral blood flow (CBF). It is known that scalp-hemodynamics increase task sections, but do not reflect task-related brain activity. A previous study approached this problem using a Short PCA-GLM with an advanced design matrix, including a global scalp-hemodynamics model (GSHM) [3]. In that study, it was assumed that the 1st principal component of PCA, estimated from 15 mm wide multi-short channels, would be global scalp-hemodynamics, showing similar temporal patterns over the whole head. Those authors also simulated synthesized fNIRS data, with the scalp-hemodynamics component, to evaluate performance. Significantly higher performance was observed for reduction of estimation error than for the standard design matrix containing no elements for scalp-hemodynamics.

Our study aimed to apply a Short PCA-GLM to real-time data. A real-time general linear model (GLM), with a sliding sample window, was used for real-time estimation in previous studies [1,2]. However, scalp-hemodynamics are potentially superimposed on real-time feedback signals. This leads to over-estimated subject brain activity feedback. Our proposed method to reject the effect of scalp-hemodynamics was designed via a sliding-window GLM with a real-time updated design matrix, with a real-time estimated GSHM sample by sample. To evaluate the advantage of this proposed method, we first explored real-time estimation of a GSHM. We then tested the performance of brain activity estimation by applying the proposed method to fNIRS simulation data.

2   Methods

2.1   Proposed Real-Time Scalp-Hemodynamics Artifact Reduction Method

The proposed method was constructed with three real-time processing steps: (1) pre-processing, (2) GSHM estimation was carried out, and (3) weight parameter estimation was carried out with a GLM. Process diagram is shown in Fig. 1.

Sliding Window Analysis. For the overall proposed method, sliding window analysis (SWA) is important for real-time estimation. SWA sets a certain sample range window, and the most recent data are used for computing statistics. As the sample window is maintained at the same size, it has the advantage of constant sensitivity through a continuous experimental period [2]. Our proposed real-time method is based on SWA.

Pre-processing. To reject periodic artifacts superimposed on fNIRS raw signals (not scalp-hemodynamics), a second-order Butterworth bandpass filter with a cut-off frequency of 0.01–0.3 Hz was applied. During on- and off- line process-ing, the same filter performance was set, without sample window size. In the

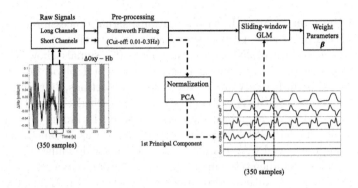

Fig. 1. Real-time signal processing with a 350 sample-wide window. Processing flow of long and short channels are represented by *solid* and *dotted lines*, respectively.

off-line process, all samples were applied at one time. In the on-line process, a 350-sample-wide window was applied, sample by sample. This number, which is related to the sample size of a single trial section (described in a following section), has at least a single task block [1,2].

Real-Time Estimation of the GSHM by Sliding-Window PCA. The previous study assumed that the 1st principal component could explain a spatially common pattern, such as the GSHM [3]. In our proposed method, we estimated it by using a 350-sample-wide sliding-window PCA from pre-processed data. To unify the amplitudes with elements of the design matrix, the filtered signals were normalized before applying PCA, so that the mean and standard deviation (SD) were 0 and 1, respectively.

$$\hat{y}^{\text{Short}} = \frac{y^{\text{Short}} - \mu}{\sigma}, \tag{1}$$

where y^{Short} is measured via signals on short channel applied pre-processing, μ and σ are the mean and SD of each channel, respectively. The calculated newest value is stored in the design matrix for the sliding-window GLM.

Real-Time Estimation of Cerebral Activity by a Sliding-Window GLM. The GLM expresses response variables y in terms of a linear combination of explanatory variables (design matrix) X:

$$y = X\beta + \varepsilon, \tag{2}$$

where β is an unknown weight vector. ε is a normally distributed error term with a mean of zero and variance of σ^2 (i.e., $\varepsilon \sim N(0, \sigma^2)$). Optimized estimation weight vector $\hat{\beta}$ was calculated by least-square,

$$\hat{\beta} = X^{\dagger}y, \tag{3}$$

$$X^{\dagger} = (X^{\mathrm{T}}X)^{-1}X^{\mathrm{T}}. \tag{4}$$

We constructed a design matrix with five elements, same as the off-line ShortPCA-GLM: cerebral hemodynamic model (CHM), with temporal and dispersion derivatives (CHM$^{(1)}$, CHM$^{(2)}$), constants (Const.), and a GSHM estimated in real-time PCA. All elements without the GSHM were normalized, with 1 for the maximum amplitude of averaged signals. Accordingly, β corresponded to the maximum amplitudes.

2.2 Evaluation of the Performance of the Proposed Method for fNIRS Simulation Data

We investigated the performance of a real-time Short PCA-GLM by applying fNIRS simulation data. fNIRS signals were measured and 43-long and 4-short channels were generated (Fig. 2B). Each channel signal consisted of the linear combination of 3 elements: CBF, scalp-hemodynamics, and white noise. As we assume the global scalp-hemodynamics are task-related fluctuations, the generation of template data was based on real fNIRS data measured in the following experiment.

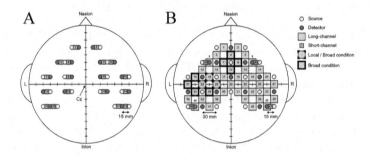

Fig. 2. fNIRS channel arrangements. (A) Arrangement for scalp-hemodynamics measurement. (B) Arrangement for simulation data. *Thick lines* represent channels with assumed brain activation in each condition, i.e. CBF is superimposed.

fNIRS Experiment for Scalp-Hemodynamics Measurement. To generate scalp-hemodynamics templates, fNIRS signals were measured by a multichannel optical imaging system, at wavelengths of 780, 805, 830 nm (FOIRE-3000, Shimadzu Corp., Kyoto, Japan). Channel arrangements are shown in Fig. 2A. The block-designed experiment consisted of 6 trial loops constructed as rest-task-rest (15 s each), and 13 subjects were asked to perform finger tapping or hand grasping with the right hand in task blocks [3]. Sampling periods were set to 100 ms.

Synthesized Simulation fNIRS Signals. CBF was generated by convolving box-car function, following the task sections, with hemodynamic response function. To simplify, we assumed that the temporal pattern of CBF was common in long-channels with brain activity. Scalp-hemodynamics were generated by Gaussian spatial interpolation, with the measurement values of ΔOxy-Hb measured in the fNIRS experiment, by changing the sampling period to 130 ms and channel arrangement to Fig. 2B. The interpolation parameters, mean μ and radius r, were set to 0 and 15 mm. Assuming localization of human brain functions, spatial distribution of CBF had two conditions: local and broad, as shown in Fig. 2B with thick-lined boxes. Furthermore, we selected random CBF amplitudes during trials with normal distributions, $N(0.01, 0.002^2)$. Assuming that amplitude fluctuation during channels were common, the amplitude ratios during channels were always the same. The amplitude of scalp-hemodynamics was selected as equal to CBF. The SD of white noise was randomly selected from the normal distribution $N(0.005, 0.001^2)$. Finally, simulation data consisted of randomly fluctuating linear waves: CBF, scalp-hemodynamics, and white noise.

Evaluation of the Proposed Method

Real-Time Estimation of the GSHM. For our proposed method, it is important to estimate the true GSHM in real-time. Therefore, we assessed the differences between on- and off-line PCA. First, we compared the 1st principal component in the time-domain, estimated by applying sliding (on-line) and all samples (off-line) windows. Next, the contribution ratio was calculated for all components.

Performance Comparison to Design Matrices. To compare performance during methods, we prepared two comparative design matrices: simple (CHM + Const.) and standard (CHM + CHM$^{(1)}$ + CHM$^{(2)}$ + Const.). We calculated the mean absolute error (MAE) from the absolute value of subtraction the theoretical β_{CHM} from estimated $\hat{\beta}_{CHM}$ for assessment of effectiveness of scalp-hemodynamics. $\hat{\beta}_{CHM}$ was calculated by a GLM with each design matrix and theoretical β_{CHM} was calculated by applying a GLM with the standard design matrix to synthesize the fNIRS data, which excluded scalp-hemodynamics (CBF + white noise). We also calculated adjusted R^2 for each design matrix for evaluation of a GLM fitting:

$$\text{Adjusted } R^2 = 1 - \frac{(\boldsymbol{R}_y)^T(\boldsymbol{R}_y)}{(\boldsymbol{y} - \bar{\boldsymbol{y}})^T(\boldsymbol{y} - \bar{\boldsymbol{y}})} \frac{N_T - 1}{N_T - N - 1} \tag{5}$$

where $\boldsymbol{y}$ is the measured signal, $\boldsymbol{R}$ is the residual matrix, N_T and N are the numbers of time-domain samples and elements in the design matrix, respectively. To assess the performance for each design matrix condition, we performed two-tailed paired t-tests between ShortPCA-GLM, and each of the other two design matrix conditions for each CBF condition (with Bonferroni correction, significant level $\alpha = 0.05/2$). We excluded the 1st trial data from statistical analysis, due to potential instability until the full-window sized samples were stored.

Comparison with Off-Line Analysis. Performance comparison of on-line (by applying sliding-window) and off-line (by all samples window) GLM analyses were carried out. The design matrix of off-line GLM was expanded task-by-task (i.e. the number of elements was 20 for ShortPCA-GLM). Evaluation statistics were performed using two-tailed paired t-tests ($\alpha = 0.05$).

3   Results

3.1   Real-Time Estimation of the GSHM

Figure 3A shows a typical temporal waveform of the 1st principal component estimated by on- and off- line Short PCA-GLM. The 1st principal component, estimated from 4 short channels in real-time, shows a similar temporal pattern to off-line estimation, even though substantially fewer samples were used in on-line estimation for each sample. We also calculated the contribution ratio of the 1st-4th principal components (Fig. 3B). The results show that the 1st principal component explained approximately 90% of the data during short channels in both processing methods. Moreover, on-line processing, with a restricted sample size for estimation by sliding window, was sufficient to estimate the GSHM distributed over the whole head, as with off-line processing.

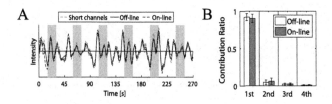

Fig. 3. Comparison of the estimated GSHM by on and off-line PCAs. (A) Typical temporal waveforms of 1st principal component. (B) Contribution ratios of 1st-4th principal components. *Error bars* show SD over samples.

3.2   Performance Comparison with Design Matrices

Typical real-time estimated CBF and scalp-hemodynamics waveforms in the broad condition are shown in Fig. 4. In Ch 19 located ipsilateral side, with right-hand motor execution, the signal is mostly described by scalp-hemodynamics. By

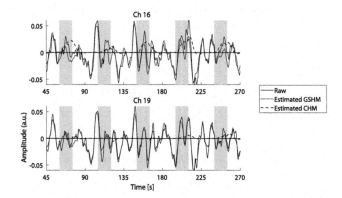

Fig. 4. Typical estimated waveforms on symmetrically located channels by on-line ShortPCA-GLM for a representative sample during a right-hand movement task.

contrast, on the opposite side, it was explained by CBF and scalp-hemodynamics (Ch 16). This result concords with the conditions of spatial distribution of CBF. Differences in MAEs between the theoretical β_{CHM} and estimated $\hat{\beta}_{\text{CHM}}$ broad condition are shown in Fig. 5A. The MAE for the real-time ShortPCA-GLM, was significantly lower than the other design matrices. As the results were the same in both conditions, the proposed method, which adds the GSHM into the design matrix, rejects the potential scalp-hemodynamics artifact in real-time. The result of adjusted R^2 is shown in Fig. 5B. ShortPCA-GLM was significantly higher than the others, and the best way to fit ΔOxy-Hb during motor execution tasks performed in block-designed experiments between comparison methods.

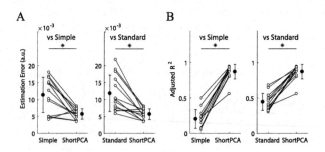

Fig. 5. Comparisons of (A) mean absolute errors of estimated $\hat{\beta}_{\text{CHM}}$ and (B) adjusted R^2 in the broad condition. *Open circles* show each subject sample. *Filled circles* and *error bars* represent the mean and SD of samples, respectively.

3.3 Comparison with Off-Line Analysis

Figure 6 shows MAE and adjusted R^2 for on- and off- line processing results, respectively. The on-line SD of R^2 was larger than the equivalent values off-line; however, the fitting score was still high. By contrast, the on-line MAE was significantly higher than off-line, suggesting that on-line cannot identify the maximum off-line performance. Thus, there is still room for improvement.

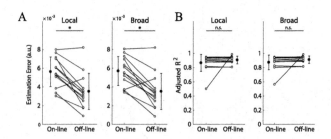

Fig. 6. Comparison between on and off-line ShortPCA-GLM. (A) Mean absolute errors of β_{CHM}. (B) Adjusted R^2.

4   Conclusion

In this study, we aimed to remove task-related scalp-hemodynamics artifacts from fNIRS signals in real-time. The proposed method consisted of a sliding-window GLM with the GSHM element measured over the whole head. Estimation of a GSHM by on and off-line PCA showed similar results for identification of the 1st principal component, with a high contribution ratio. When compared with standard design matrices, a ShortPCA-GLM showed a significantly lower estimation error for β_{CHM} than the other methods, whilst adjusted R^2 was significantly higher. The results suggest that a real-time ShortPCA-GLM performs sufficiently to explain the GSHM and estimate brain activity, with lower estimation error than for previous methods. However, real-time estimation errors were higher than off-line; thus, there is still room for improvement in real-time processing. Our next step will be to assess the performance of real-time ShortPCA-GLM for complex data with different latencies and durations of peak responses.

References

1. Mihara, M., Miyai, I., Hattori, N., Hatakenaka, M., Yagura, H., Kawano, T., Ok-ibayashi, M., Danjo, N., Ishikawa, A., Inoue, Y., Kubota, K.: Neurofeedback using real-time near-infrared spectroscopy enhances motor imagery related cortical activation. PLoS ONE **7**(3), 1–13 (2012)
2. Nakai, T., Bagarinao, E., Matsuo, K., Ohgami, Y., Kato, C.: Dynamic monitoring of brain activation under visual stimulation using fMRI? The advantage of real-time fMRI with sliding window GLM analysis. J. Neurosci. Methods **157**(1), 158–167 (2006)
3. Sato, T., Nambu, I., Takeda, K., Aihara, T., Yamashita, O., Isogaya, Y., Inoue, Y., Otaka, Y., Wada, Y., Kawato, M., Sato, M., Osu, R.: Reduction of global interference of scalp-hemodynamics in functional near-infrared spectroscopy using short distance probes. NeuroImage **141**, 120–132 (2016)
4. Thibault, R.T., Lifshitz, M., Raz, A.: The self-regulating brain and neurofeedback: experimental science and clinical promise. Cortex **74**, 247–261 (2016)

Liver Segmentation and 3D Modeling Based on Multilayer Spiral CT Image

Yanhua Liang[1] and Yongxiong Sun[2(✉)]

[1] College of Software, Jilin University, Changchun 130012, China
[2] College of Computer Science and Technology, Jilin University,
Changchun 130012, China
sunyx@jlu.edu.cn

Abstract. The 3D reconstruction can facilitate the diagnosis of liver disease by making the target easier to identify and revealing the volume and shape much better than 2D imaging. In this paper, in order to realize 3D reconstruction of liver parenchyma, a series of pretreatments are carried out, including windowing conversion, filtering and liver parenchyma extraction. Furthermore, three kinds of modeling methods were researched to reconstruct the liver parenchyma containing surface rending, volume rendering and point rendering. The MC (marching cubes) algorithm based on 3D region growth is proposed to overcome the existence of a large number of voids and long modeling time for the contours of traditional MC algorithms. Simulation results of the three modeling methods show different advantages and disadvantages. The surface rendering can intuitively image on the liver surface modeling, but it cannot reflect the inside information of the liver. The volume rendering can reflect the internal information of the liver, but it requires a higher computer performance. The point rendering modeling speed is quickly compared to the surface rendering and the volume rendering, whereas the modeling effect is rough. Therefore, we can draw a conclusion that different modeling methods should be selected for different requirements.

Keywords: Liver CT image · Liver segmentation · 3D modeling · MC algorithm

1 Introduction

In recent years, the incidence of liver cancer is on the rise, and surgical resection is still the most effective way to treat liver cancer [1]. With the application of CT, MRI and other medical imaging technology, a group of 2D tomographic images of the patient's liver lesions can be obtained. Through these 2D tomographic images, the doctor can analyze the lesion site. However, these medical instruments provide a 2D image of the liver can only reflect the liver of a cross section of the anatomical information [2]. Doctors can only use the experience to estimate the size and shape of the lesion. This method relies heavily on physician's subjective imagination and clinical experience, lacking intuition and accuracy. The 3D modeling of human liver is constructed and displayed by 2D tomographic images, which can make up the deficiency of imaging

© Springer International Publishing AG 2017
D. Liu et al. (Eds.): ICONIP 2017, Part IV, LNCS 10637, pp. 702–712, 2017.
https://doi.org/10.1007/978-3-319-70093-9_75

equipment and improve the veracity and scientific of medical diagnosis and treatment plan [3]. 3D visualization methods are divided into two categories: surface rendering and volume rendering. In 1987, Lorensen and Cline proposed a classical surface rendering method-MC algorithm [4], which is simple and is easy to adapt to large-scale parallel computing. However, there are ambiguities in the connection of the triangular patches in the MC algorithm. Nielson and Hamann proposed a method to determine the "hole" in the MC algorithm, but it leads to the low efficiency of implementation [5]. Levoy in 1988 proposed a ray casting algorithm [6]. The two methods of surface rendering and volume rendering are all displayed on the 3D data field, but there are great differences in drawing effect, modeling time and interactive performance. From the point of imaging quality, the volume rendering is better than the surface rendering. While, regarding interaction performance and algorithm efficiency, the surface rendering is superior to the volume rendering at least on the current hardware platform [7]. Therefore, it is necessary to choose different rendering methods according to actual needs. In this paper, we study and implement three liver modeling methods.

2 Liver Parenchyma Extraction Based on CT Image

At present, the international common medical image format through the 64-slice CT machine is DICOM3.0; this format is not common in Windows system. So DICOM3.0 format of liver CT images should be converted to Windows commonly used BMP format [8]. The default window width and window level (400, 40) cannot guarantee that the liver parenchyma will display the best results with its adjacent tissues in the liver CT images. By contrast, we choose the window width and window level is (200, 50). The dataset is provided by the Jilin Province Tumor Hospital, we selected two abdominal CT images from two different patients and then windowing the images, as illustrated in Fig. 1. The two images on the left are the liver CT images of patient A in different window widths, while the two images on the right are the liver CT images of patient B in different window level.

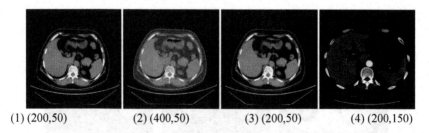

(1) (200,50)   (2) (400,50)   (3) (200,50)   (4) (200,150)

Fig. 1. CT images of the liver in patients with a contrast effect window.

In the process of liver image collection by spiral CT scanner and image conversion, a large amount of noise is generated, which has a great influence on the subsequent liver parenchyma extraction and liver 3D modeling. In this paper, an improved adaptive anisotropic filter algorithm is adopted [9], which is a combination of adaptive median filter and anisotropic filter. This filtering algorithm can dynamically change the template size according to the number of noise, which can not only smooth the image with strong pollution, but also save the edge information well.

Figure 2 is a comparison of liver CT images using a variety of filter algorithms. (a) is the original image. (b) is processed by the adaptive median filter algorithm. (c) is processed by the anisotropic filter algorithm. And (d) is the processed image using the filter algorithm presented in this paper. Compared with adaptive median filter and traditional anisotropic filter, the filtering algorithm used in this paper has the better de-noising effect than both of them.

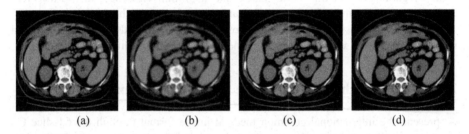

(a)   (b)   (c)   (d)

Fig. 2. Comparison of different filtering methods for liver CT image processing.

To achieve a 3D model based on liver CT images, we used the region growing algorithm based on RBF-CI (RBF-Confidence Interval) [10] to extract the liver parenchyma. And this algorithm was presented by our laboratory task force. The following shows the image using the region growing algorithm mentioned in this paper.

The result show that there are many voids in the extracted liver parenchyma, which will have a great influence on the subsequent 3D reconstruction of the liver. In this paper, the extracted liver pods are treated using the image expansion and erosion algorithm [11] to obtain the ideal liver parenchyma. The last CT image in Fig. 3 is the image that fill the segmented results.

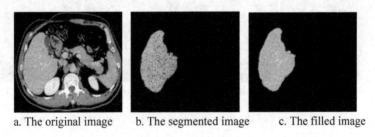

a. The original image   b. The segmented image   c. The filled image

Fig. 3. Patient A 147th liver CT image segmentation results.

3   3D Modeling of Liver Based on CT Image

3.1   Surface Rendering

MC algorithm is a classic method of surface rendering [12]. The traditional MC algorithm each reads two adjacent liver CT images. First, extract the corresponding four vertices in each CT image to form a cube, and assign each vertex to the cube in a certain order. And then make the gray value of each vertex in the cube compare with the threshold set in advance. If the gray value is greater than the threshold, the corresponding vertex is marked as 1. Otherwise, the flag is 0, resulting in an octet binary sequence. According to the eight vertices of the tag is different, we had a total of 256 kinds of circumstances, thus establishing the state table 0–255. As the cube has symmetry and rotational symmetry, the state table eventually corresponds to 15 modes, as shown in Fig. 4.

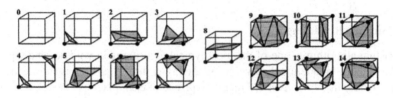

Fig. 4. 15 patterns correspond to the state table.

Each stat corresponds to the establishment of different triangular patches. Because of the ambiguity of the MC algorithm leads to a large number of voids in the extracted isosurface, as shown in Fig. 5. To fill the missing triangles patches in the void, extended eight-modes are proposed to supplement the previous 15 models. As shown in Fig. 6.

Fig. 5. Traditional MC algorithm produces voids.

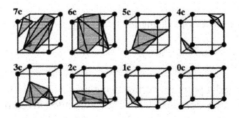

Fig. 6. Supplemented by eight kinds of isosurface extraction patterns.

In addition to generating a large number of voids, the use of traditional MC algorithm for liver 3D modeling process is also time-consuming. To solve this problem, MC algorithm based on 3D region growth is presented. This method uses the coordinates of the midpoint of the cube edge as the coordinates of the vertex of the triangle patch, eliminating the difference calculation, and reducing the traversal of the useless cube, and then improving the speed of 3D reconstruction. How to reduce the traversal of the useless cube in this paper? A cube has six faces, and when we find a cube that can extract the isosurface, there must be an isomorphic surface connected to it in its adjacent cube. Thus, we can find one or a group of cubes intersecting the isosurface as a seed cube, and the seed cube is selected from the middle of the extracted liver parenchyma. And then to the seed cube as the center, outward expansion to find the existence of equatorial cube. So go on until there is no new seed cube. And the number of traversal cube can be further reduced according to the continuity of the isosurface.

Extract all isosurface stored in the STL file. And then through the OpenGL to add light and shadow effects, the three-dimensional display of liver parenchyma.

3.2   Surface Rendering

Ray-casting algorithm is one of the most typical algorithms in volume rendering [13]. The flow chart of modeling the liver using the ray-casting algorithm is shown in Fig. 7. Volume rendering of the liver CT image sequence of each pixel as a voxel, and draw it to the 3D scene, the volume rendering can be a good way to retain the liver internal information.

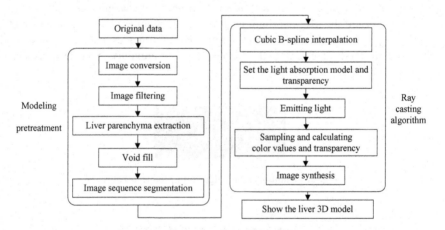

Fig. 7. Flow chart of modeling the liver using the ray-casting algorithm.

3.3   Point Rendering

This paper also achieved a point rendering to the liver 3D modeling. Before performing 3D modeling, extract the contours from the segmented liver parenchyma. Figure 8 shows the effect of contour extraction of two different liver parenchymas.

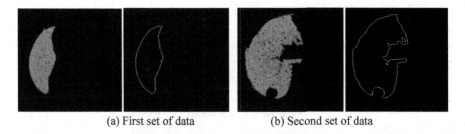

<table>
<tr><td>(a) First set of data</td><td>(b) Second set of data</td></tr>
</table>

Fig. 8. Extract the edge of liver parenchyma.

All the extracted liver edge images were arranged in sequence. Then, we adjust all the pixels to the appropriate size. Ultimately, 3D reconstruction of the liver was performed using OpenGL package. The flow chart of modeling the liver using the fast point rendering is shown in Fig. 9. Fast point rendering does not require equivalent surface extraction in a large number of 3D data, and does not require linear interpolation. So it has the advantages of small computation, less processing data, less resource consumption and fast modeling speed.

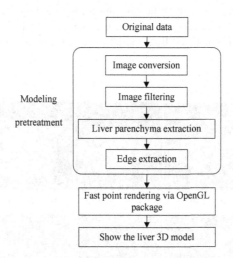

Fig. 9. Flow chart of modeling the liver using the point rendering.

4 Simulation

Three sets of data were selected and reconstructed using the traditional MC algorithm and the MC algorithm based on 3D region growth. Experiments were performed by scanning the number of cubes and the time taken for 3D modeling in the surface rendering process. The experimental results are shown in Table 1.

Table 1. Comparison between traditional MC and improved MC

Number of layers (pcs)	Thickness (mm)	Algorithm	Number of cubes (pcs)	Time (s)
27	5	Traditional MC algorithm	6789146	55.84
		Improved MC algorithm	77606	24.31
26	5	Traditional MC algorithm	6528025	55.43
		Improved MC algorithm	68645	24.12
215	0.625	Traditional MC algorithm	55879894	423.99
		Improved MC algorithm	721781	95.34

As can be seen from Table 1, the number of cubes scanned by the surface drawing is large. Among them, the number of scanning cube and the modeling time of the first group and the second group are closer, while the third set of data modeling time is longer and the number of cube scans is more than the previous two sets of data. This is because the third set of data has 215 CT images, each imaging interval 0.625 mm, while the previous two sets of data CT images less and larger interval. Through the above analysis, we can see that the more the number of CT images, the more the number of cubes scanned, and the longer the time required for 3D modeling. Compared with the traditional MC algorithm, the MC algorithm based on 3D region growth is not only greatly reduces the number of cube scans, but also greatly reduces the modeling time. These three sets of data for liver modeling results and local triangular meshes are shown in Fig. 10.

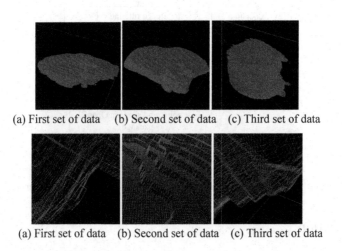

(a) First set of data   (b) Second set of data   (c) Third set of data

(a) First set of data   (b) Second set of data   (c) Third set of data

Fig. 10. Three groups of data on liver modeling effects.

From the modeling effect of Fig. 12, it can be seen that the established 3D model of the liver has obvious ladder shape. This is due to the large spacing between adjacent two CT images. Compared with the modeling results of the previous two sets of data, there is no obvious ladder in the modeling results of the third set of data. Thus, the smaller the spacing between adjacent CT images, the more the number of scanning cube, and the better the effect of 3D modeling. However, the time required for modeling will be longer.

In this paper, two sets of data were selected for 3D modeling of the liver using the ray casting algorithm. As shown in Fig. 11. In both experiments, the image (b) is rotated and scaled by the image (a).

Figure 12 shows the 3D model of liver CT images using the point rendering. In the two sets of liver model images, the size of the point of the image (a) is set to 10 pixels, and the size of the point of the image (b) is set to 3 pixels.

From the experimental results, it can be found that when modeling the liver with point rendering, the effect of the established liver model is different because the pixel values of the set points are different. When the pixel value of the point is set to small, there is a gap between the points, so that the 3D image of the liver produces a perspective effect. When the pixel value of the point is set to 10, it is possible to completely obscure the gap between point and point, but the modeling effect is rough.

In the following thesis, three kinds of modeling methods were used to model the liver CT images of three groups of patients. The first set of data had 33 liver CT images, and the distance between adjacent CT images was 5 mm. The second set of data included 46 CT images, and the spacing between adjacent CT images was 5 mm. The third group of data had 236 CT images, and the distance between adjacent images was 0.625 mm. Table 2 lists the time required to model these three sets of data using the three modeling methods described above. The modeling results of these three sets of data are shown in Figs. 13, 14 and 15.

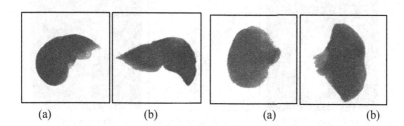

(a)   (b)   (a)   (b)

Fig. 11. Ray-casting algorithm for modeling the liver.

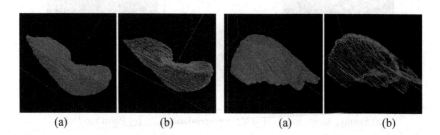

(a)   (b)   (a)   (b)

Fig. 12. Point rendering for modeling the liver.

Table 2. Comparison of modeling time using three modeling methods.

Number of layers (pcs)	Thickness (mm)	Modeling method	Time (s)
33	5	Improved MC algorithm	41.8
		Ray casting algorithm	5.3
		Point rendering	0.9
46	5	Improved MC algorithm	52.6
		Ray casting algorithm	6.7
		Point rendering	1.8
215	0.625	Improved MC algorithm	95.3
		Ray casting algorithm	13.9
		Point rendering	2.4

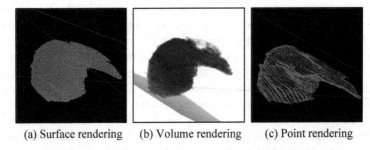

(a) Surface rendering   (b) Volume rendering   (c) Point rendering

Fig. 13. Three 3D modeling of the first set of data.

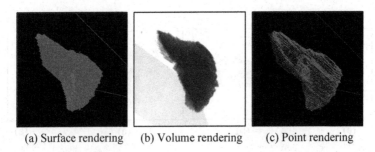

(a) Surface rendering   (b) Volume rendering   (c) Point rendering

Fig. 14. Three 3D modeling of the second set of data.

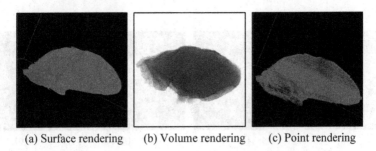

(a) Surface rendering   (b) Volume rendering   (c) Point rendering

Fig. 15. Three 3D modeling of the third set of data.

Through the comparison experiments above, it can be found that the more the number of liver CT images, the longer time spent on modeling. Since the surface drawing requires scanning a large number of cubes to extract all isosurfaces, the modeling takes longer time. The point rendering modeling is significantly faster, but the modeling effect is rough than other two algorithms.

5 Conclusion

In this paper, we research on three modeling methods based on liver CT images. By CT image conversion and liver parenchyma segmentation using region growing algorithm and an improved adaptive anisotropic filter algorithm, we firstly realize the image segmentation of CT series. The 3D modeling algorithms include the growing surface rendering of the MC algorithm, the volume rendering of ray casting algorithm and the point rendering. After the extraction of liver parenchyma, three modeling methods have been studied and realized in the experiment. The experiment results show that the surface rendering method can better show the liver surface information, the volume rendering method can better save the internal liver information, while, the point rendering modeling saves modeling time most. Therefore, it can be concluded that three modeling methods are suitable for different modeling requirements. Whereas, our future research work will focus on the segmentation and 3D modeling of liver lesion area, the rendering technologies and the volume proportion calculation of liver lesion to further facilitate the liver cancer diagnosis.

Acknowledgments. This research is supported by Jilin Province Nature Science Foundation (No. 20130102082JC), Jilin Province Development and Innovation Committee's High and New Technology Projects (No. JF2012C006-6).

References

1. Chen, L., Luo, H., Dong, S., et al.: Safety assessment of hepatectomy for huge hepatocellular carcinoma by three dimensional reconstruction technique. Chin. J. Surg. **54**(9), 669–674 (2016)
2. Huang, Q.: Research of 3D modeling and liver volume calculation method based on liver CT image. Jilin University (2016)
3. Fang, C., Feng, S., Fan, Y., et al.: Study on the application of three-dimensional visualization technique in evaluation of residual liver volume and guidance for hepatectomy. J. Hepato-biliary Surg. **20**(2), 96–98 (2012)
4. Lorensen, W.E., Cline, H.E.: Marching cubes: a high resolution 3D surface construction algorithm. ACM SIGGRAPH Comput. Graph. **21**(4), 163–169 (1987)
5. Nielson, G.M., Hamann, B.: The asymptotic decider: resolving the ambiguity in marching cubes. In: Proceedings of the 2nd Conference on Visualization, pp. 83–91. IEEE Computer Society Press, Los Alamitos (1991)
6. Zhou, J.: 3D-volume reconstruction of medical images based on ray-casting algorithm. Comput. Sci. **43**(11A), 156–160 (2016)

7. Hu, Z., Gui, J., Zhou, Y., et al.: Methods for implementation of CT visualization based on surface and volume rendering. Microcomput. Inf. **25**(12), 107–108 (2009)
8. Liu, L.: Research of liver image segmentation method and 3D modeling method based on multilayer spiral CT. Jilin University (2015)
9. Fu, L., Yao, Y., Fu, Z.: Filtering method for medical images based on median filtering and anisotropic diffusion. J. Comput. Appl. **34**(1), 145–148 (2014)
10. Sun, Y., Liu, L., Huang, Q., et al.: Liver image segmentation algorithm based on RBF confidence interval. J. Inf. Comput. Sci. **12**(5), 1703–1711 (2015)
11. Sun, J., Wu, B., Liu, X.: Cellular neural network applicating manner in pre processing image. Chin. J. Comput. **28**(6), 985–990 (2005)
12. Wang, M., Feng, J., Yang, B.: Comparsion and evaluation of marching cubes and marching tetrahedra. J. Comput.-Aided Des. Comput. Graph. **26**(12), 2099–2106 (2014)
13. Li, L.: Design method of transfer function for 3D medical data volume rendering. University of Science and Technology of China (2016)

Deep Retinal Image Segmentation: A FCN-Based Architecture with Short and Long Skip Connections for Retinal Image Segmentation

Zhongwei Feng[1], Jie Yang[1(✉)], Lixiu Yao[1], Yu Qiao[1], Qi Yu[2], and Xun Xu[2]

[1] Institute of Image Processing and Pattern Recognition,
Shanghai Jiao Tong University, Shanghai, China
{fengzhongwei,jieyang,lxyao,qiaoyu}@sjtu.edu.cn
[2] Shanghai General Hospital, Shanghai Jiao Tong University, Shanghai, China
1683289010@qq.com, 820478428@qq.com

Abstract. This paper presents Deep Retinal Image Segmentation, a unified framework of retinal image analysis that provides both optic disc and exudates segmentation. The paper presents a new formulation of fully Convolutional Neural Networks (FCNs) that allows accurate segmentation of the retinal images. A major modification in these retinal image segmentation tasks are to improve and speed-up the FCNs training by adding short and long skip connections in standard FCNs architecture with class-balancing loss. The proposed method is experimented on the DRIONS-DB dataset for optic disc segmentation and the privately dataset for exudates segmentation, which achieves strong performance and significantly outperforms the-state-of-the-art. It achieves 93.12% sensitivity (Sen), 99.56% specificity (Spe), 89.90% Positive predictive value (PPV) and 90.93% F-score for optic disc segmentation while 81.35% Sen, 98.76% Spe, 81.64% PPV and 81.50% F-score for exudates segmentation respectively.

Keywords: Optic disc segmentation · Exudates segmentation · Convolutional neural network · Skip connections · Class-balancing loss

1 Introduction

Retinal image segmentation is key for the diagnosis of ophthalmological diseases such as diabetes and hypertension. However, manual segmentation of retinal images by ophthalmologist is both time consumed and lack accuracy and. Accordingly, automated segmentation systems can play an important role in improving the segmentation accuracy as well as diagnosis accuracy.

Optic Disc Segmentation: Prior attempts of optic disc segmentation includes handcrafted features [1] and morphology [2]. Morales et al. [3] use morphological

© Springer International Publishing AG 2017
D. Liu et al. (Eds.): ICONIP 2017, Part IV, LNCS 10637, pp. 713–722, 2017.
https://doi.org/10.1007/978-3-319-70093-9_76

operators with Principal Component Analysis to obtain the optic disc in retinal images. In the last ten years, machine learning based methods have gained attentions, which is a powerful tool for feature classification. In [4], the authors propose an automatic method of segmenting OD and optic cup based on a statistical model technique. A superpixel classification method is proposed in [5]. However, handcrafted features has limit to different datasets. Recently, Convolutional Neural Networks (CNNs) is applied in the optic disc segmentation task. Lim et al. [6] describe a comprehensive solution based on applying convolutional neural networks to feature exaggerated inputs emphasizing disc pallor without blood vessel obstruction. Deep Retinal Image Understanding (DRIU) [7] uses FCNs [8] based on VGG-16 [9] with two set of specialized layers to solve both retinal vessels and optic disc segmentation. However, the models have large number of parameters, which are difficult to fit extremely small datasets.

Exudates Segmentation: Previous work on exudates segmentation can be broadly divided into unsupervised methods and supervised methods. Harangi et al. [10] use an active contour based method to extract accurate borders of the candidates and let a boosted Naive Bayes classifier eliminate the false candidates. In [11], the authors propose a svm-based method using the Gabor features and GLCM features to classify retinal images into exudates or nonexudates. Ardiyanto et al. [12] use maximum entropy principle to determines a reasonable threshold value for separating exudates areas. CNN-based methods are also arised in the exudates segmentation task. Prentašić et al. [13] uses CNNs combined with high level knowledge about landmark points, which contains vessels and optic disc, in order to increase the accuracy of exudate detection. However, its performance depends upon landmark points detection, which is a difficult problem factually. Perdomo et al. [14] work on LeNet [15] convolutional neural network architecture to improve the classification of nonexudates and exudates in retinal images to correctly diagnosis the disease. Both existed CNN-based methods on exudates segmentation have been trained by patch-based approach, which consumes a lot of computations and time.

Contribution: In this paper, we propose a modified FCN architecture by adding short and long connections for both optic disc and exudates segmentation, which is much more lightweight in number of parameters. Taking advantage of FCN, our proposed model can achieve end-to-end training, which consumes less time compared with patch-based training. To the best of our knowledge, our work is the first of its kind to bring the combined advantage of FCNs approach for exudates segmentation in retinal images. The rest of the paper is organized as follows: Sect. 2 describes the proposed methodology in detail. In Sect. 3 shows the experimental results. Finally, in Sect. 4 we conclude our paper with a summary.

2   Materials and Methods

Figure 1 presents a pipeline of our method for both optic disc and exudates segmentation. Only Contrast Limited Adaptive Histogram Equalization (CLAHE) is used as the preprocessing for the proposed method.

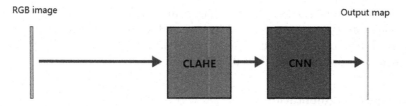

Fig. 1. Pipeline of the proposed method for the task of both optic disc and exudates segmentation.

2.1 Dataset Description

In this paper, results of optic disc segmentation are reported for the publicly available dataset DRIONS-DB [16], which contains groundtruth for optic disc. DRIONS-DB contains 110 full eye fundus images in total, which is divided into two parts for training and testing and each part contains 55 images equally. Figure 2(a) and (b) shows an example from the DRIONS-DB dataset. The results of exudates are tested on the privately dataset with precise groundtruth segmentation, which manually labeled by experienced ophthalmologists in Shanghai

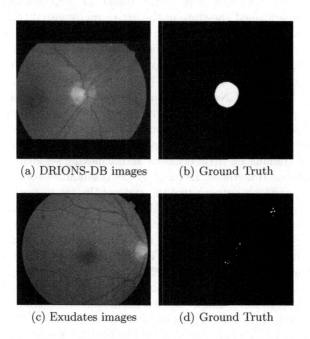

(a) DRIONS-DB images   (b) Ground Truth

(c) Exudates images   (d) Ground Truth

Fig. 2. Examples from datasets. (a), (b) are retinal images and its corresponding groundtruth from DRIONS-DB. (c), (d) are retinal images and its corresponding groundtruth from the EX dataset.

No. 1 Hospital. The exudates dataset has 62 full eye fundus images totally, which is divided into the trainset with 42 images and the rest for testing. Figure 2(c) and (d) shows an example from the exudates dataset. Limited to GPU memory, all the images are resize to $256 * 256$ for training.

2.2 Data Augmentation

As for retinal image segmentation tasks there is very little training images available. On the other hand, CNN models easily have the chance of overfitting the training data. However, data augmentation is used to help prevent this behaviour. Data augmentation is the process to generate new samples by applying transformations to the original images. In this paper, we extend keras [17] ImageDataGenerator to real-time SegImageGenerator for both optic disc and exudates segmentation tasks. In particular, during training phase, we apply same transformations on image and corresponding groundtruth, such as flip, crop & resize, rotation, shift and etc.

2.3 Network Architecture

In this section, a unified method is proposed for segmentation of optic disc and exudates. Our approach is primarily based on deep learning techniques, which have outperformed the-state-of-the-art in many visual tasks. As for the classification task, residual networks [18] with short skip connections prominently performs the-state-of-the-art. Furthermore, CNNs have been applied successfully to a large variety of general recognition tasks such as object detection [19], semantic segmentation [8], and contour detection [20].

The U-net [21] is a FCNs architecture for image segmentation that accepts image as an input and returns softmap as an output, which shows good performance for biomedical image segmentation. The standard U-net has only long skip connections by recovering spatial information lost during downsampling, which has capable of training on little datasets and achieving segmentation competitive with patch-based methods. Inspired by the residual network [18] with short skip connections, which outperforms the-state-of-the-art in many classification tasks. We extend standard U-net by replacing regular convolutional layers with modified residual blocks, which are similar to the ones introduced in residual networks, in order to speed-up training and improve precise segmentation. The proposed architecture is illustrated in Fig. 3(a). Its input to the simple block in the proposed architecture is a $3 \times 256 \times 256$ image from the training set. The simple block, consisting of a 1×1 convolution, followed by a batch normalization [22] layer for speed-up training and a rectified linear unit, is illustrated in Fig. 3(b). The encoding path follows the typical architecture of U-net, but replacing regular convolutional layers with modified residual blocks. Each modified residual block, which is showed in Fig. 3(c), consists of the repeated application 3 times of the basic block. In order to equal the channels of its input and output for element-wise addition, a extra regular convolutional layer on shortcut will be added if

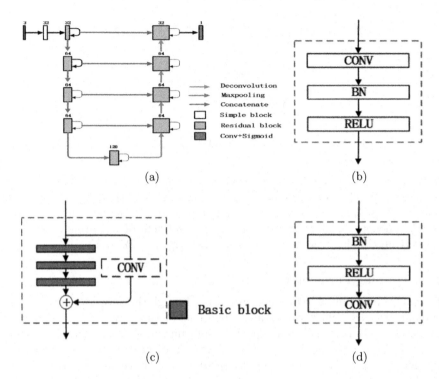

Fig. 3. Proposed CNN architecture for retinal image segmentation.(a) U-net with short skip connections built from residual blocks (b) simple block (c) residual block and (d) basic block

necessary (Fig. 3(c) dotted rectangle on shortcut). The basic block consists of a batch normalization layer, a rectified linear unit and a 3×3 convolution with padding for same size, which is illustrated in Fig. 3(d). Each modified residual block followed by a 2×2 max pooling operation with 2 pixels stride for downsampling to reduce the amount of parameters and computation. In order to concate different-size feature maps through long skip connections, an upsampling of the feature map followed by a 2×2 deconvolution [8] is used. This important architecture is that allow the network to propagate context information to higher resolution layers for more precise segmentation. After a concatenation, each followed by a modified residual block in decoding path. At the final convolutional layer a 1×1 convolution is used to map multi-channel maps to the desired number of classes. Finally, a sigmoid operation is applied on the last convolutional layer to scale the softmap. The output is of dimension $1 \times 256 \times 256$.

For training the network, the segmentation tasks are learnt by class-balancing loss function originally proposed in [23] for contour detection in natural images. The class-balancing loss function is then defined as:

$$L(y_i, \hat{y}_i) = -\beta \sum_{i \in Y_+} (y_i \log \hat{y}_i + (1 - y_i) \log(1 - \hat{y}_i))$$

$$- (1 - \beta) \sum_{i \in Y_-} (y_i \log \hat{y}_i + (1 - y_i) \log(1 - \hat{y}_i))$$

$$= -\beta \sum_{i \in Y_+} \log \hat{y}_i \tag{1}$$

$$- (1 - \beta) \sum_{i \in Y_-} \log(1 - \hat{y}_i)$$

where $\hat{y}_i$ is the predicted segmentation map obtained by the last sigmoid layer while $y_i \in \{0, 1\}$ is the ground truth. The multiplier β is used to achieve the balance of the large number of background compared to foreground pixels. Y_- and Y_+ denote the background and foreground sets of the ground truth Y, respectively. In this case, we use $1 - \beta = |Y_+|/|Y|$, $\beta = |Y_-|/|Y|$.

2.4   Training Parameters

Training consists in an iterative propagation through the network and modification of its weights, which are initialized by xavier method. The cycle of presenting all training examples, called an epoch, is split into smaller units called batches. In this approach, we use mini-batch stochastic gradient descent with momentum at 0.95 and fixed the learning rate to be 10^{-3}. The model, which includes new 400 batches that generated by real-time SegImageGenerator in each epoch, is trained 500 epochs with a batch size of 4. The implementation is based on keras framework [17], which performs all computation on GPUs in single precision arithmetic. The experiments are conducted on Intel Core i7-6700 CPU with a NVIDIA TitanX card. In practical, these initial parameters is same for optic disc and exudates segmentation.

3   Experiment and Discussions

In test phase, we present the performance of networks tested on the test set in terms of Sen, Spe, PPV and F-score, which is obtained on pixel-wise. In this case, positive decision is made when the output of the unit associated with the positive class in the sigmoid output layer is greater than the 0.5; otherwise, negative decision is made. In fact, this threshold can be arbitrarily selected from [0, 1], which leads to different result.

Optic Disc Segmentation: Figure 4 shows exemplary segmentations produced by our proposed model for optic disc segmentation. For the task of optic disc segmentation, we compare our solution with the method from DRIU, which is the best method that we have found for investigated datasets. In Table 1, we present the performance of networks tested on the test set. The results show that our method performs a little better than DRIU comparison on DRIONS-DB dataset, which has strong performance and significantly outperforms the-state-of-the-art for optic disc segmentation.

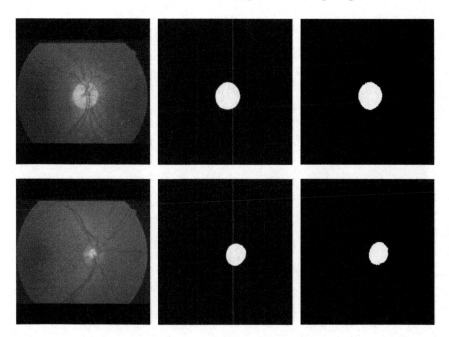

Fig. 4. Output map produced by our proposed method on the DRIONS-DB: first column, original retinal images; middle column, expert annotations; last column, results obtained by our method

Table 1. Performance on optic disc segmentation.

Method	Sen	Spe	PPV	F-score
Yin et al. [4]	.8542	.9826	-	-
Lim et al. [6]	.8742	-	.8535	.8636
Maninis et al. [7]	.9252	**.9972**	.8758	.9082
proposed method	**.9312**	.9956	**.8990**	**.9093**

Exudates Segmentation: Figure 5 shows accuracy segmentations produced by our method for exudates segmentation. In Table 2, we present the performance tested on the test set. The proposed method has strong performance and significantly outperforms the-state-of-the-art. Interestingly, taking a closer look at some true negative of our technique (see Fig. 5 red rectangles on the middle), we observe that almost true negative is very small. Although many convolution operators are applied on, it is hard to learn useful features to classify very small exudates in retinal images. Recently, we extend our proposed architecture to multi-scale inputs, which is expected to solve this problem.

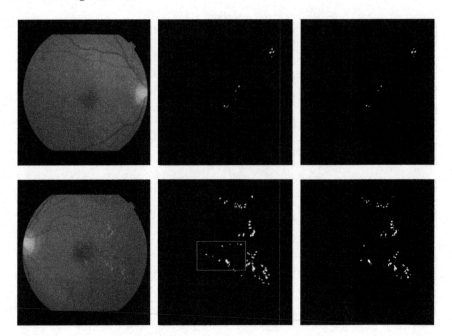

Fig. 5. Output map produced by our proposed method on the EX dataset:first column, original retinal images; middle column, expert annotations; last column, results obtained by our method (Color figure online)

Table 2. Performance on exudates segmentation.

Method	Sen	Spe	PPV	F-score
Ardiyanto et al. [12]	.1690	**.9960**	.6940	-
Harangi et al. [10]	.7123	-	.6613	.6854
Prentaic et al. [13]	.7854	.9954	.7834	.7812
proposed method	**.8135**	.9876	**.8164**	**.8150**

4   Conclusion

Combining with class-balancing loss, we presented a fully convolutional neural network with short and long skip connections for both optic disc and exudates segmentation. The proposed architecture has high-capability to learn hierarchical features and context information from raw pixel data without handcrafted features and any prior domain knowledge.

In this paper, the proposed method has strong performance and significantly outperforms the-state-of-the-art for both retinal optic disc and exudates segmentation. Taking advantage of FCN with short and long skip connections, our method has full potential of carrying out more robust and precise segmentation than traditional methods. So, we currently work on verifying this claim for hemorrhages and microaneurysms segmentation on retinal images.

Acknowledgments. This research is partly supported by NSFC, China (No: 81600776), Committee of Science and Technology, Shanghai, China (No: 16411962100) and (No. 17JC1403000)

References

1. Youssif, A.A.H.A.R., Ghalwash, A.Z., Ghoneim, A.A.S.A.R.: Optic disc detection from normalized digital fundus images by means of a vessels' direction matched filter. IEEE Trans. Med. Imaging **27**(1), 11–18 (2008)
2. Walter, T., Klein, J.-C.: Segmentation of color fundus images of the human retina: detection of the optic disc and the vascular tree using morphological techniques. In: Crespo, J., Maojo, V., Martin, F. (eds.) ISMDA 2001. LNCS, vol. 2199, pp. 282–287. Springer, Heidelberg (2001). doi:10.1007/3-540-45497-7_43
3. Morales, S., Naranjo, V., Angulo, J., Alcaniz, M.: Automatic detection of optic disc based on PCA and mathematical morphology. IEEE Trans. Med. Imaging **32**(4), 786–796 (2013)
4. Yin, F., Liu, J., Wong, D.W.K., Tan, N.M., Cheung, C., Baskaran, M., Aung, T., Wong, T.Y.: Automated segmentation of optic disc and optic cup in fundus images for glaucoma diagnosis. In: 2012 25th International Symposium on Computer-Based Medical Systems (CBMS), pp. 1–6. IEEE (2012)
5. Cheng, J., Liu, J., Xu, Y., Yin, F., Wong, D.W.K., Tan, N.M., Tao, D., Cheng, C.Y., Aung, T., Wong, T.Y.: Superpixel classification based optic disc and optic cup segmentation for glaucoma screening. IEEE Trans. Med. Imaging **32**(6), 1019–1032 (2013)
6. Lim, G., Cheng, Y., Hsu, W., Lee, M.L.: Integrated optic disc and cup segmentation with deep learning. In: 2015 IEEE 27th International Conference on Tools with Artificial Intelligence (ICTAI), pp. 162–169. IEEE (2015)
7. Maninis, K.-K., Pont-Tuset, J., Arbeláez, P., Van Gool, L.: Deep retinal image understanding. In: Ourselin, S., Joskowicz, L., Sabuncu, M.R., Unal, G., Wells, W. (eds.) MICCAI 2016. LNCS, vol. 9901, pp. 140–148. Springer, Cham (2016). doi:10.1007/978-3-319-46723-8_17
8. Long, J., Shelhamer, E., Darrell, T.: Fully convolutional networks for semantic segmentation. In: Proceedings of the IEEE Conference on Computer Vision and Pattern Recognition, pp. 3431–3440 (2015)
9. Simonyan, K., Zisserman, A.: Very deep convolutional networks for large-scale image recognition. arXiv preprint (2014). arXiv:1409.1556
10. Harangi, B., Lazar, I., Hajdu, A.: Automatic exudate detection using active contour model and regionwise classification. In: 2012 Annual International Conference of the IEEE Engineering in Medicine and Biology Society (EMBC), pp. 5951–5954. IEEE (2012)
11. Ruba, T., Ramalakshmi, K.: Identification and segmentation of exudates using SVM classifier. In: 2015 International Conference on Innovations in Information, Embedded and Communication Systems (ICIIECS), pp. 1–6. IEEE (2015)
12. Ardiyanto, I., Nugroho, H.A., Buana, R.L.B.: Maximum entropy principle for exudates segmentation in retinal fundus images. In: 2016 International Conference on Information & Communication Technology and Systems (ICTS), pp. 119–123. IEEE (2016)
13. Prentašić, P., Lončarić, S.: Detection of exudates in fundus photographs using deep neural networks and anatomical landmark detection fusion. Comput. Methods Programs Biomed. **137**, 281–292 (2016)

14. Perdomo, O., Arevalo, J., Gonzalez, F.A.: Convolutional network to detect exudates in eye fundus images of diabetic subjects. In: 12th International Symposium on Medical Information Processing and Analysis, p. 101600T. International Society for Optics and Photonics (2017)
15. Lenet, B., Komorowski, R., Wu, X.Y., Huang, J., Grad, H., Lawrence, H., Friedman, S.: Antimicrobial substantivity of bovine root dentin exposed to different chlorhexidine delivery vehicles. J. Endod. **26**(11), 652–655 (2000)
16. Carmona, E.J., Rincón, M., Garcia-Feijoó, J., Martínez-de-la Casa, J.M.: Identification of the optic nerve head with genetic algorithms. Artifi. Intell. Med. **43**(3), 243–259 (2008)
17. Chollet, F., et al.: Keras (2015). https://github.com/fchollet/keras
18. He, K., Zhang, X., Ren, S., Sun, J.: Deep residual learning for image recognition. In: Proceedings of the IEEE Conference on Computer Vision and Pattern Recognition, pp. 770–778 (2016)
19. Girshick, R., Donahue, J., Darrell, T., Malik, J.: Region-based convolutional networks for accurate object detection and segmentation. IEEE Trans. Pattern Anal. Mach. Intell. **38**(1), 142–158 (2016)
20. Yang, J., Price, B., Cohen, S., Lee, H., Yang, M.H.: Object contour detection with a fully convolutional encoder-decoder network. In: Proceedings of the IEEE Conference on Computer Vision and Pattern Recognition, pp. 193–202 (2016)
21. Ronneberger, O., Fischer, P., Brox, T.: U-Net: convolutional networks for biomedical image segmentation. In: Navab, N., Hornegger, J., Wells, W.M., Frangi, A.F. (eds.) MICCAI 2015. LNCS, vol. 9351, pp. 234–241. Springer, Cham (2015). doi:10. 1007/978-3-319-24574-4_28
22. Ioffe, S., Szegedy, C.: Batch normalization: accelerating deep network training by reducing internal covariate shift. arXiv preprint (2015). arXiv:1502.03167
23. Xie, S., Tu, Z.: Holistically-nested edge detection. In: Proceedings of IEEE International Conference on Computer Vision (2015)

Computer-Aided Diagnosis in Chest Radiography with Deep Multi-Instance Learning

Kang Qu[1], Xiangfei Chai[2], Tianjiao Liu[3], Yadong Zhang[2], Biao Leng[4(✉)], and Zhang Xiong[4]

[1] School of Computer Science and Engineering, Beihang University, Beijing, China
[2] Huiying Medical Technology Inc. (Beijing), Beijing, China
[3] Department of Electronic Engineering, Tsinghua University, Beijing, China
[4] State Key Laboratory of Software Development Environment,
School of Computer Science and Engineering, Beihang University, Beijing, China
lengbiao@buaa.edu.cn

Abstract. The Computer-Aided Diagnosis (CAD) for chest X-ray image has been investigated for many years. However, it has not been widely used since limited accuracy. Deep learning opens a new era for image recognition and classification. We propose a novel framework called Deep Multi-Instance Learning (DMIL) on chest radiographic images diagnosis, which combines deep learning and multi-instance learning. Besides, we preprocess images with the alignment based on the key points. This framework can effectively improve the diagnosis effect in the image level annotation. We quantify the framework on three datasets, respectively with different amounts and different classification tasks. The proposed framework obtained the AUC of 0.986, 0.873, 0.824 respectively in classification tasks of the enlarged heart, the pulmonary nodule, and the abnormal. The experiments we implement demonstrate that the proposed framework outperforms the other methods in various evaluation criteria.

Keywords: Chest radiograph · Deep learning · Multi-Instance Learning · Medical image

1 Introduction

Chest radiographs, as the most common examinations in medical radiographs, have a critical role in the medical image research. With a wide range of information about the lungs, heart, mediastinum and ribs, the chest radiograph is often used to diagnose various diseases, such as pneumonia, effusion and lung cancer. Figure 1 shows some examples of normal and abnormal chest radiographic images. CAD is an essential development direction to improve the quality and efficiency of diagnosis, which is also widely used on chest radiographs. In general, researchers get features from medical images utilizing descriptors such as the histogram of oriented gradients (HOG) [1] and the Local Binary Patterns (LBP) [2]. Recently, with the rapid development of Deep Learning (DL), more

© Springer International Publishing AG 2017
D. Liu et al. (Eds.): ICONIP 2017, Part IV, LNCS 10637, pp. 723–731, 2017.
https://doi.org/10.1007/978-3-319-70093-9_77

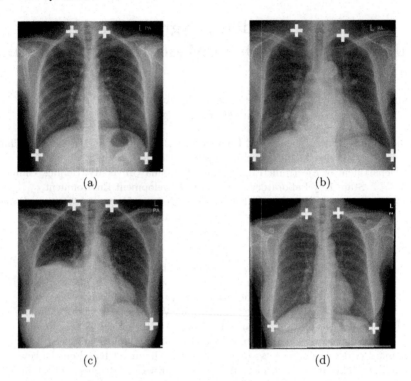

Fig. 1. Chest radiographic images categories are shown as (a) normal (b) enlarged heart (c) pulmonary effusion (d) pulmonary nodule.

and more researches have begun using the deep feature in the fields of medical image with Convolution Neural Network (CNN).

Obtaining leap development in computer version tasks [3], CNN has been proved to be effective in many other domains. For the generality of the CNN feature, transfer learning has been used in various fields, including medical image processing. Roth et al. [4] adapt CNNs to improve three existing CAD system for the detection of colon polyps, rigid spinal metastases, and lymph node. Moreover, Bar et al. [5,6] in two papers use models pre-trained on the non-medical database as feature extractors. They fuse the deep feature with other different normalized features and obtain excellent results in the tasks of various diseases diagnosis.

However, there are still many difficulties in the research over the medical data with CNNs [7]. One important problem is that the lesion may only occupy a small part of the whole image, but the annotation is image-level. It is worth noting that, in most computer vision tasks, CNNs are supervised with the label on the same level. So this is a great challenge to the transfer learning in tasks of the medical image. To deal with this problem, we propose a Deep Multi-Instance Learning framework to improve the diagnosis of chest radiographic images.

Compared with standard supervised learning, Multi-Instance Learning (MIL) deals with the whole bag, which consists of instances. In the task of natural image recognition and automatic annotation, MIL have been adopted and gratifying results have been achieved [8]. In MIL problems, we only get bag-level annotations, and instance annotations are absent. In a standard MIL issue, we label a bag as positive when at least one instance of the bag is positive, which means that all instances of the negative bag are negative [9]. Considering our task, we deal with the whole image as a bag, and the local region is an instance. However, sometimes most patches may be normal (negative) even if the whole image is labeled as abnormal (positive).

To improve the CAD of the chest radiographic image, we propose a novel method using the pre-trained CNNs from different domains combined with MIL method to automatically detect the abnormal images and two specific diseases. Besides, our experiments prove that alignment based on key points has a certain effect on chest radiographic image.

2   Methods

2.1   Image Alignment

Image preprocessing is an extremely important part of CAD. In the face recognition, the alignment based on key point is very common, but the research in medical images is still blank.

Although the data we collected is homologous, there is still a large difference between images, which is shown in not only the tilt angle of the body but also the proportion of lung area occupied the whole picture as well. These differences may have a misleading effect on the learning process. To solve this problem, we utilize image alignment, which is widely used in face recognition [10]. The middle parts of the top ribs and the outer edges of eighth ribs are manually marked as four key points. To perform the alignment, we obtain the four standard points by counting the distribution of the key points of all samples and get a square by the position of these standard points. The coordinates of the upper left and the lower right corner are defined as (0.0, 0.0) and (1.0, 1.0) so that we can get the coordinates of the standard point with function:

$$Pts_{std} = \frac{\sum_{i=1}^{N} Pts_{gti}}{N} \tag{1}$$

where N is the total number of samples, Pts_{std} is the coordinates of standard points, and Pts_{gti} is the coordinates of key points in the ith image. Utilizing the corresponding point coordinates of the real images, similarity transformation is adopted as:

$$T_i = SimilarityTransform\left(Pts_{gti} \to Pts_{std}\right) \tag{2}$$

$$I_i^* = Transform\left(I_i, T_i\right) \tag{3}$$

where I_i^* is the transformed i^{th} image, I_i is the original i^{th} image, and T_i is the transformation matrix. This transformation is to make the real point match the standard position and obtain a corresponding square. We can get the pre-processed image after cropping the transformed image according to the square. In this way, we can reduce background interference to a great extent.

2.2 CNNs for Transfer Learning

Compared training from scratch, transfer learning has been proved to be effective in medical imaging [11]. The weight distribution of pre-trained model can provide a better initialization for our network. Among those multitudinous network models, we choose ResNet [12] as our pre-trained model, mainly because of its excellent performance in computer vision tasks. ResNet, which is about 20 times deeper than AlexNet and 8 times deeper than VGG [13], won the champion of ILSVRC 2015.

2.3 Deep Multi-Instance Learning

According to the standard MIL definition, each instance has its distinct label, which means one patch in an image is either normal or abnormal. However, the truth has not always been the case. In some cases, patches of our images can be described as normal or abnormal, but in other cases, patches could be judged only when combined. For example, the most common criterion of the enlarged heart is the cardiothoracic ratio (CTR), and when CTR is greater than a certain threshold, we can label the bag as abnormal. It is evident that we are unable to judge each patch independently.

So our framework does not supervise the learning of instances alone. Firstly, the framework focuses on the patches of the whole image, which is the instance. Then, our framework combines the information from each instance and learns the representation of the entire image, which is the bag. The DMIL framework is illustrated in Fig. 2.

Instance-Level Learning. To synthesize the representations of instances and further learn in bag-level, we should obtain the proper representation of instance level. We extract the feature map from the last convolution layer of Res50 as the expression of each instance, which is $2048 \times 7 \times 7$. Because of the small amount of dataset, we need to use the pre-trained model to initialize the weights of our network. However, it should be pointed out that our medical images are gray pictures with only one channel, while the natural image dataset the model pre-trained on consists of three channel (RGB). Obviously, there exists a difference between the two datasets. To use the pre-trained model, we copy the data of one image three times and then sent them into three channels. In the instance learning, we resize the size of the preprocessed images to 512×512 pixels, and then random crop nine patches with 224×224 pixels, which is the largest number we can adopt because of the memory limitations in the experimental environment. These patches would adjust the weights of the instance-level network during fine tuning. This instance-level network is set up with several parallel sub-networks, and all these networks share the same weights. To ensure the experimental validation information integrity, we use grid cropping on the validation set, and the whole images are cropped by 3×3 with partial overlapping.

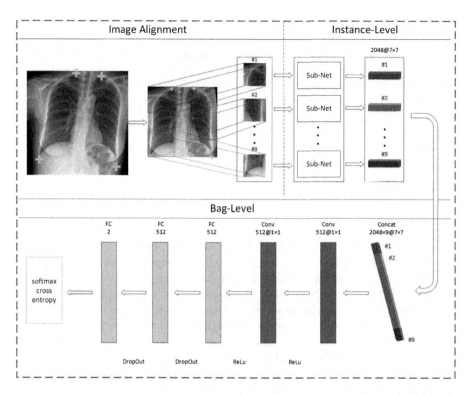

Fig. 2. The Deep Multi-Instance Learning Framework contains three stages. In the first stage, one image is aligned based on the key points and cropped to nine patches (224 × 224 pixels). In the second stage, the networks learn the expression of each instance by transfer learning. In the last stage, the network combines feature maps and is supervised with the bag-level annotation.

Bag-Level Learning. After obtaining feature from the instance-level network, we use a bag-network to combine all the information of instances. We set two convolution layers and two fully connected layers after the concatenation of feature maps coming from one original image. In foremost, there are two convolution layers with filters with 1 × 1, whose primary function is to learn the small range association in different instance feature maps. The number of filters of each layer is 512. Since the dimension in the bag-level network is very high, the 1 × 1 convolution kernel not only can be used to learn the local feature association but also play the role of dimension reduction. After each convolution layers, we utilize ReLu as the activation function, which can increase the sparsity at the start of learning from feature maps. There are also two fully connected layers initialized with the Gauss distribution, and the variance is set to 0.01. Each fully connected layers have 512 nodes. The two fully connected layers are mainly to learn the overall relevance of the expression of different instances. After each fully connected layer, we use dropout layer to reduce overfitting. Following these layers, we choose softmax cross entropy loss function as the loss of the whole network.

2.4   Implementation Details

We built pre-trained models from [12] based on the Caffe framework. We trained networks using the NVIDIA Titan X (Pascal). In each batch, we sent nine of the patches of one chest image. The models were trained with 30 epochs using stochastic gradient descent and we set the initial learning rate to 0.001. We conducted experiments on each dataset using GIST, HOG and the fine tuning ResNet-50 for comparison. These contrast experiments were based on the same environment and training strategy except that the batch size was set to 16.

3   Experiments

3.1   Datasets

Our dataset, obtained from Medical Technology Inc., is originally from tertiary referral hospitals. All the images are labeled by the experts from these hospitals. We extracted the original DICOM format data to images. With the information of the header of DICOM files, we only selected radiographic images with "Chest PA" label. Since the annotation of the dataset is expressed in the form of text, we obtained positive samples through the keywords and then extracted the same number of negative samples of each disease from the samples which are completely normal. Eventually, we got three datasets. The first dataset is for the enlarged heart, including 510 positive samples and 510 negative samples. The second dataset is for the pulmonary nodule, including 1071 positive samples and 1071 negative samples. And the third dataset including normal or abnormal images, in which the 5000 positive samples contain a variety of diseases such as enlarged heart, pulmonary nodule, and pleural effusion, and the rest 5000 negative samples are all completely normal. We validated our approach on these three datasets.

3.2   Evaluation

For each dataset, we selected eighty percent of all data as the training set and the remaining twenty percent as the validation set. We train our model with two ways, one with alignment and the other without alignment. We call them DMIL and DMIL*. Considering the common evaluation criteria in the medical image domain [14,15], we chose AUC, Accuracy, Sensitivity and Specificity as the evaluation measures.

The results of experiments are illustrated in Tables 1, 2, 3 and Fig. 3. From these results, we notice that our proposed method outperforms the other three methods in three tasks. In particular, in the classification of pulmonary nodules, our method obtained a 5% AUC improvement in comparison with Res50. Even in the heart shadow classification task, in which baseline has been very high, there is still 2% improvement. Compared with HOG and GIST, all of the evaluation results of our proposed method have a vast improvement. This increase demonstrates that our framework is much better compared with traditional methods.

Table 1. Results on the classification of the enlarged heart

Method	AUC	Accuracy	Sensitivity	Specificity
HOG	0.963	0.901	0.923	0.880
GIST	0.959	0.901	0.894	0.907
ResNet	0.967	0.915	0.918	0.912
DMIL*	0.983	0.941	0.961	0.921
DMIL	**0.986**	**0.946**	**0.963**	**0.928**

Table 2. Results on the classification of the pulmonary nodule

Method	AUC	Accuracy	Sensitivity	Specificity
HOG	0.720	0.668	0.646	0.689
GIST	0.705	0.653	0.663	0.644
ResNet	0.823	0.757	0.745	0.741
DMIL*	0.844	0.759	0.716	**0.806**
DMIL	**0.873**	**0.792**	**0.830**	0.755

Table 3. Results on the classification of the abnormal

Method	AUC	Accuracy	Sensitivity	Specificity
HOG	0.776	0.713	0.788	0.639
GIST	0.772	0.717	0.717	**0.717**
ResNet	0.790	0.711	0.721	0.702
DMIL*	0.816	0.731	**0.790**	0.671
DMIL	**0.824**	**0.733**	0.751	0.715

Moreover, compared to the result of Res50, our framework also obtained a definite improvement of 2–3%. This increase shows that multi-instance methods have a particular promotion effect on medical image dataset with only the image-level annotation.

We find that the DMIL with alignment has a certain degree of improvement in most evaluation result than which without alignment, especially in the classification of the pulmonary nodule. This increase proves that alignment based on key points has a good effect on chest X-ray. Besides, these experiments in different datasets show that our method has excellent performance in various tasks. It should be noted that our methods not only perform well in tasks with the relatively fixed lesion in shape and position but also be effective in tasks with vague features and uncertain lesion. In general, the experimental results demonstrate that the proposed method has superior generality and robustness.

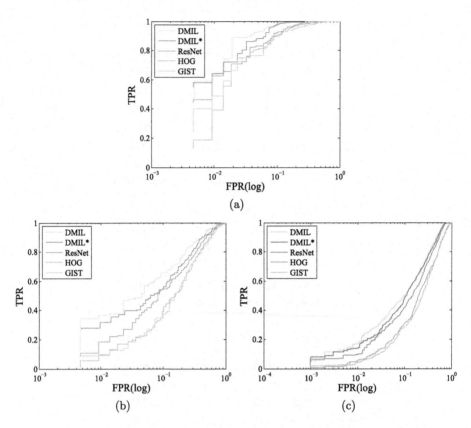

Fig. 3. ROC curves on three datasets and our method outperforms the others. The x-axis represents log value of FPR and The y-axis represents TPR. (a) is ROC of the enlarged heart. (b) is ROC of the pulmonary nodule. (c) is ROC of the abnormal.

4   Conclusion

In this paper, a novel Deep Multi-Instance Learning framework is proposed for the automatic diagnosis of chest radiographic images. This approach combines deep learning and multi-instance learning. The results demonstrate that the proposed framework outperforms fine-tuning models and other traditional methods. Besides, the proposed framework has superior generality and robustness when dealing with different disease and various amount of data. In further work, we will explore the automatic mark of key points in the medical image and integrate it into our framework to get a further improvement.

Acknowledgments. This work is supported by the National Natural Science Foundation of China (No. 61472023) and the State Key Laboratory of Software Development Environment (No. SKLSDE-2016ZX-24).

References

1. Song, Y., Cai, W., Zhou, Y., Feng, D.D.: Feature-based image patch approximation for lung tissue classification. IEEE Trans. Med. Imag. **32**(4), 797–808 (2013)
2. Sorensen, L., Shaker, S.B., De Bruijne, M.: Quantitative analysis of pulmonary Emphysema using local binary patterns. IEEE Trans. Med. Imag. **29**(2), 559–569 (2010)
3. Krizhevsky, A., Sutskever, I., Hinton, G.E.: Imagenet classification with deep convolutional neural networks. In: Advances in Neural Information Processing Systems, pp. 1097–1105 (2012)
4. Roth, H., Lu, L., Liu, J., Yao, J., Seff, A., Cherry, K., Kim, L., Summers, R.: Improving computer-aided detection using convolutional neural networks and random view aggregation. IEEE Trans. Med. Imag. **35**(5), 1170–1181 (2015)
5. Bar, Y., Diamant, I., Wolf, L., Greenspan, H.: Deep learning with non-medical training used for chest pathology identification. In: SPIE Medical Imaging, pp. 94140V. International Society for Optics and Photonics (2015)
6. Bar, Y., Diamant, I., Wolf, L., Lieberman, S., Konen, E., Greenspan, H.: Chest pathology detection using deep learning with non-medical training. In: 2015 IEEE 12th International Symposium on Biomedical Imaging (ISBI), pp. 294–297. IEEE (2015)
7. Albarqouni, S., Baur, C., Achilles, F., Belagiannis, V., Demirci, S., Navab, N.: AggNet: deep learning from crowds for mitosis detection in breast cancer histology images. IEEE Trans. Med. Imag. **35**(5), 1313–1321 (2016)
8. Wu, J., Yu, Y., Huang, C., Yu, K.: Deep multiple instance learning for image classification and auto-annotation, pp. 3460–3469 (2015)
9. Zeng, T., Ji, S.: Deep convolutional neural networks for multi-instance multi-task learning. In: 2015 IEEE International Conference on Data Mining (ICDM), pp. 579–588. IEEE (2015)
10. Ren, S., Cao, X., Wei, Y., Sun, J.: Face alignment at 3000 FPS via regressing local binary features. In: Proceedings of the IEEE Conference on Computer Vision and Pattern Recognition, pp. 1685–1692 (2014)
11. Shin, H.C., Roth, H.R., Gao, M., Lu, L., Xu, Z., Nogues, I., Yao, J., Mollura, D., Summers, R.M.: Deep convolutional neural networks for computer-aided detection: CNN architectures, dataset characteristics and transfer learning. IEEE Trans. Med. Imag. **35**(5), 1285–1298 (2016)
12. He, K., Zhang, X., Ren, S., Sun, J.: Deep residual learning for image recognition. In: Proceedings of the IEEE Conference on Computer Vision and Pattern Recognition, pp. 770–778 (2016)
13. Simonyan, K., Zisserman, A.: Very deep convolutional networks for large-scale image recognition. arXiv preprint arXiv:1409.1556 (2014)
14. Melendez, J., van Ginneken, B., Maduskar, P., Philipsen, R.H., Reither, K., Breuninger, M., Adetifa, I.M., Maane, R., Ayles, H., Sánchez, C.I.: A novel multiple-instance learning-based approach to computer-aided detection of Tuberculosis on chest X-rays. IEEE Trans. Med. Imag. **34**(1), 179–192 (2015)
15. Ciompi, F., de Hoop, B., van Riel, S.J., Chung, K., Scholten, E.T., Oudkerk, M., de Jong, P.A., Prokop, M., van Ginneken, B.: Automatic classification of pulmonary peri-fissural nodules in Computed Tomography using an ensemble of 2D views and a convolutional neural network out-of-the-box. Med. Image Anal. **26**(1), 195–202 (2015)

A Hybrid Model: DGnet-SVM
for the Classification of Pulmonary Nodules

Yixuan Xu[1], Guokai Zhang[1], Yuan Li[1], Ye Luo[1], and Jianwei Lu[1,2(✉)]

[1] School of Software Engineering, Tongji University, Shanghai, China
{xuyixuan,zhang.guokai,ly_foxer,yeluo,jwlu33}@tongji.edu.cn
[2] Institute of Translational Medicine, Tongji University, Shanghai, China

Abstract. We investigate the problem of benign and malignant pulmonary nodules classification for thoracic Computed Tomography (CT) images. Although various methods have been proposed to solve this problem, they have bottlenecks of poor input image quality and subjective or shallow feature extraction. In this paper, we propose a Denoise GoogLeNet model with the classifier of Support Vector Machine (DGnet-SVM) to improve the final classification accuracy. We apply Denoise Network to improve the CT image quality by reducing the noise, and GoogLeNet is utilized to extract high-level features for better generalization of data. Furthermore, SVM is applied to classify the nodules owing to its great classification performance. The experimental results show that our hybrid model outperforms other state-of-the-art methods with the accuracy of 0.89 based on five-fold cross validation and the AUC is 0.95. The advantages of the proposed model and our future work are also discussed.

Keywords: Pulmonary nodules · Denoise network · GoogLeNet · SVM

1 Introduction

Lung cancer has become a tremendous threat to humans' health with high morbidity and mortality. Especially, the number of lung cancer deaths has ranked first in the malignant cancer, which surpasses prostate, colon and breast cancer [1]. According to clinical experience, the malignant pulmonary nodule is a significant omen to lung cancer [2]. If it can be diagnosed on the early-stage, patients' survival rates will be efficiently increased by professional treatments.

Pulmonary nodules are usually round or oval shape tissues in lung, and they can be detected by Computed Tomography (CT) which has higher examination precision and better image quality than other scanning tools [3]. However, it is still challenging for radiologists to discriminate nodules' categories owing to massive numbers of CT images [2] need to be handled in their routine work. And it inevitably brings about misdiagnosed and missed diagnosis cases. Therefore, some Computer-Aided Diagnosis (CAD) systems have been developed to address this problem in pulmonary nodule classification field (e.g., [4,5]), which replace the tradition of completely relying on radiologists.

© Springer International Publishing AG 2017
D. Liu et al. (Eds.): ICONIP 2017, Part IV, LNCS 10637, pp. 732–741, 2017.
https://doi.org/10.1007/978-3-319-70093-9_78

In the last decades, various CAD methods have been proposed to classify pulmonary nodules. Sun et al. [6] used a Mean-Shift algorithm [7] which is a non-parametric estimation method to segment CT images. Song et al. [8] applied a SIFT feature extraction algorithm [9] to compute the labeled locations (foreground or background) of the images. Song et al. [10] proposed SURF [11] and LBP [12] descriptors to present pulmonary nodules textures, in this case, more comprehensive features can be extracted to generate more valuable information. Although previous works have shown favorable performance in pulmonary nodules analysis, extracted features tend to be subjective, which limits the quality of the network model.

With neural networks springing up in computer vision field, relevant researches have been studied in feature extraction stage on more large-scale medical image datasets. Dandil et al. [13] used Artificial Neural Network (ANN) to classify the nodules, which reached an accuracy of 0.906. Chen et al. [14] adopted a Neural Network Ensemble (NNE) scheme, a multi-layer neural network using the back-propagation algorithm, and it achieved the accuracy of 0.787. Hua et al. [15] proposed an unsupervised deep learning model called Deep Belief Nets (DBNs), and it computed the morphology and texture features of pulmonary nodules for classification. Shen et al. [16] presented a Multi-crop Convolutional Neural Network (MC-CNN) by using regions cropped from the convolutional feature maps which shows the accuracy of 0.87.

Though these previous works have devoted great contributions to pulmonary nodules classification, they still bear bottlenecks. For instance, original CT images have unavoidable interferential noise (e.g., system noise) and these noise will deeply affect the final classification accuracy. Meanwhile, the shallow depth of designed network structure can significantly influence the high-level features extraction process. Yet it will greatly increase the computational budget if we expand networks' scales directly.

In view of the previous works limitations, we propose a novel model using Denoise GoogLeNet with Support Vector Machine (DGnet-SVM) to improve the situation. Our major works can be concluded in the following aspects:

(1) We adopt Denoise Network [17] to enhance the quality of CT images by reducing the noise. It minimizes the influence of noise by forcing most stochastic pixels to be blank. Notably, we design a kind of fusion input strategy which mixes the noisy images with denoised ones to improve the robustness of our model.

(2) In order to avoid over-fitting problem caused by limited data in medical image processing field, and extra computational complexity owing to directly increasing the depth and the width (the number of nerve cells) of the network [18]. GoogLeNet [19] is applied to analyze our CT images by extracting deeper and more abstract features in convolutional units.

(3) We substitute the default classifier of softmax function in GoogLeNet with SVM for further improving the final accuracy, considering that SVM [20] is more local objective especially in classification task.

Overall, by adopting this hybrid model, our DGnet-SVM can improve the quality of CT images, and extract more salient information from the data. Furthermore, to the best of our knowledge, our method is unprecedented by the way of integrating the advantages of multiple networks aforementioned in pulmonary nodules classification problem.

The rest of this paper is organized as following: In Sect. 2, we will present our dataset in detail. In Sect. 3, our proposed DGnet-SVM model will be introduced. Then, we validate our models efficacy and conduct contrast experiment in Sect. 4. Finally, we make a conclusion of our work in Sect. 5.

2   Data Description

2.1   LIDC Dataset

We used LIDC (Lung Cancer Data Consortium) [21] - IDRI dataset in our research, which includes 1018 lung cancer screening CT scans with marked-up annotated lesions, and it is an open source dataset which offers standard CT images for analysis. The diameters of these nodules are between 3 mm and 30 mm, and these nodules are labeled according to four radiologists' blinded review with the locations of them listed in an XML file. In addition, radiologists also identify the benign and malignant pulmonary nodules with respective scores. The score below 3 was regarded as benign, on the contrary, the score above 3 was regarded as malignant and we discarded the score of 3. Moreover, we also reexamined the data by professional radiologists to further ensure the accuracy of the data. In the end, we selected 742 as benign samples and 553 as malignant samples.

2.2   Data Augmentation

We aimed to use limited data to generate more information, which will contribute to avoid over-fitting phenomenon in network models. We implemented random augmentation operations assembled with Hinton's research [22] on CT images to augment our dataset. To be specific, we flipped the images, and magnified them by setting the zoom at 0.2. In addition, the rotation and translation were done of $[30°, 60°]$ and in range of $[-6, 6]$ voxels. After these augmentation operations, data images can be more abundant and robust. In the end, the final number of the dataset samples were augmented to 2590.

3   DGnet-SVM Model

In recent years, deep convolutional neural network has been brought to the forefront in images classification field. Apart from seeking for a high-quality network, computation consumption is also a crucial aspect which deserves our attention. And we adopted GoogLeNet [19] which proposed an inception module concept with the structure of concatenated convolutional layers and pooling layers, and it reduced computational complexity to some degree. Additionally, for improving the quality of the input images which is beneficial to the feature extraction stage, Denoise Network was used to reduce the noise effect on CT images. Furthermore, SVM was our better choice in the classification stage instead of default softmax function. We carried out the research by the integration of these methods into our system which will be explained in Sects. 3.1 and 3.2. And the overview of it is presented in Fig. 1.

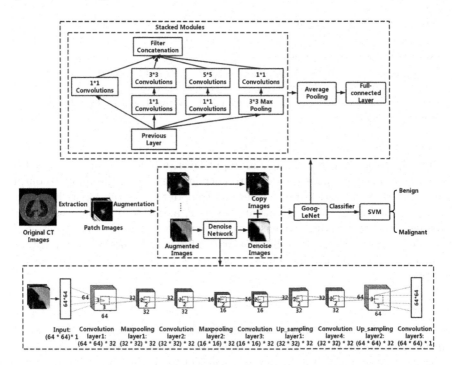

Fig. 1. An overview of our proposed DGnet-SVM. In the upper part, the form $k \times k$ is applied to express convolutional kernel size. In the lower part, the form $(m \times m) \times n$ is used to express the convolutional process, where $(m \times m)$ means the convolutional output size, and n is the number of the output units.

3.1 Denoise Strategy

Considering that noise is a vital aspect in CT image quality, and as the input data of our network, if CT images contain massive noise, it may deeply affect the observation of pulmonary nodules especially for some small ones.

Under this circumstance, we first extracted patches from original CT images based on the radiologists annotations [21], then copy these patches into two parts for different processing. The first part of the noisy data was put into Denoise Network for denosing processing, and the other part was left without any processing. Moreover, we merged these two parts as input data into GoogLeNet, so as to improve robustness of the designed model and generate more comprehensive features information.

Table 1. Denoising algorithm

Input: original samples: $X \in D^N$ /* N: the number of features of the images */
Output: weight matrics: $\{W\}$, bias: $\{b\}$

Step 1: Preprocessing

- Initialization: set weights and bias randomly
- Processing the images: adding stochastic noise by mapping x to $\tilde{x}$
- Network parameters training: input x are mapped to corrupted $\tilde{x}$ by means of a stochastic noise $\tilde{x} \sim q_D(\tilde{x}|x)$

Step 2: Forward propagation

- find parameters $\{W, b\}$ to present hidden state y through $f_\theta(x) = s(W\tilde{x} + b)$

Step 3: Backward propagation

- reconstruct $\tilde{x}$ to vector z through $g_{\theta'}(y) = s(W'y + b')$ and correct the weights and bias by minimizing the cross entropy loss $L(x, z) = \alpha(-\sum_{j \in J(\tilde{x})}[X_j log z_j + (1 - x_j)log(1 - z_j)]) + \beta(-\sum_{j \in J(\tilde{x})}[X_j log z_j + (1 - x_j)log(1 - z_j)])$ between input $\tilde{x}$ and reconstructed z

---

Our Denoise Network contains an input layer, a hidden layer and an output layer. Its structure can be divided into encoder part (from the input to output data) and decoder part (from output to input data). As the algorithm shown in Table 1, we define our original samples as x, and corrupted data as $\tilde{x}$ which processed through forcing stochastic components to be zero by q_D function. In the encoder part, we get non-linear sigmoid function values after our training of hidden units, and the hidden state y can be obtained by the expression as following:

$$f_\theta(x) = s(W\tilde{x} + b), \tag{1}$$

where W is the weight of the units and b is the bias (which can measure the output is good or not), and $\theta = \{W, b\}$. In this case, $f_\theta(x)$ can be used to present the function of the hidden state y. Note also that, during the training stage, these parameters are initialized randomly, and then optimized by adaptive moment estimation algorithm based on gradient descent. In the decoder stage, given the hidden state y, we decode the data to get the reconstruction vector (i.e., z) of x. And the function $g_{\theta'}(y)$ is used to express it as follows:

$$g_{\theta'}(y) = s(W'y + b'); \tag{2}$$

Hereafter, we fine tune the weight and bias by minimizing the errors between the input x and reconstruction vector z so that the output data can approximate the input data as much as possible. And we use binary cross entropy loss to measure the error as $L(x, z)$ shows:

$$L(x, z) = \alpha(-\sum_{j \in J(\tilde{x})}[X_j log z_j + (1 - x_j)log(1 - z_j)])$$
$$+\beta(-\sum_{j \in J(\tilde{x})}[X_j log z_j + (1 - x_j)log(1 - z_j)]), \tag{3}$$

where $J(\tilde{x})$ represents the indexes of the x, and the weight β for the corrupted data. In addition, we add the weight α to the original data in case the error from the corrupted data dominates the result.

Specially, owing to some irrelevant components erased from the original images patch, noise was reduced to some degree by decreasing weights (which refers to β) in the training part, so the result data can be more robust and generalized well. In addition, due to the model is trained by corrupted data, the over-fitting problem is better improved. The main structure of Denoise Network is introduced in Fig. 1, and contrast images before and after processing are shown in Fig. 2.

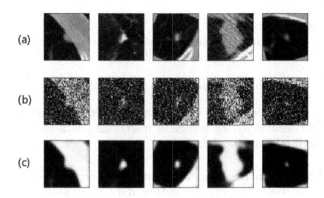

Fig. 2. Illustration of the effect of Denoise Network. (a) Original CT images. (b) Noisy CT images after adding stochastic noise. (c) Noise free CT images after our Denoise Network processing.

3.2 Feature Extraction and Classification

GoogLeNet Design. For extracting deeper and more abstract features, it is crucial to increase the depth and width of the network model. However, if we do that blindly, it is bound to bring the phenomenon of over-fitting due to the lack of massive features. Meanwhile, network complexity will be a heavy burden to the system configuration which consumes much memory. And all these factors pretty affect the efficiency of the system performance.

GoogLeNet is a deep convolutional neural network which was proposed in ILSVRC 2014. It presents a concept of inception modules which stacks convolution kernels and pooling units. And on the basis of sparse learning, which uses non-zero elements to present features as less as possible. GoogLeNet reduces the consumption of computational resources to some degree, which meets our requirements.

In this model structure, inception module is the basic unit of GoogLeNet, and it contains three sizes of convolution kernels $(1 \times 1, 3 \times 3, 5 \times 5)$ which can extract more features information. And the pooling operations make contributions to the current convolutional networks by reducing the dimension of the output vectors and improving the over-fitting problem. However, this mix of pooling layer is still computation-consuming, because 5×5 convolution kernels bring about massive filers at the top of the network, so we use the revised structure to reduce computation further. It applies 1×1 convolution kernel combined with rectified linear activation judiciously to reduce dimension of the input in the next stage. Nevertheless, expensive 3×3 and 5×5 convolutions kernels are still needed to process dense and compressed data of the output of 1×1 convolution kernels. By stacking this kind of inception module, our GoogLeNet is constructed in 22 layers with occasional max-pooling layers with stride 2. In addition, to avoid vanishing gradient phenomenon, in the middle of the processing section, we implement two auxiliary softmax functions to propagate gradients back via layers.

Furthermore, the network is fine tuned by minimizing the cross entropy loss which is used to measure the similarity between two independent multivariate distributions and it can be presented as following:

$$LOSS = -(q \log_2 p_1 + (1 - q) \log_2 p_0), \tag{4}$$

where q is the distribution of our input data predicted by the model, and p_i (i is 0 or 1, which is decided by the benign or malignant characteristics of nodules) is the true distribution of the data.

Classifier Adjustment. As the default final classifier in GoogLeNet, softmax function is a function mapping the output data into the probability of each category. Through additional experiments, we find that SVM can show better classification performance than softmax. Actually, SVM is a supervised machine learning model which was proposed by Cortes et al. [23] in 1995. In recent years, SVM has been applied successfully into medical image processing, such as the research [24] of Alzheimer's Disease diagnosis.

To guarantee good classification result by using SVM, we find a hyper plane to divide our training samples into benign or malignant nodules according to the distance between them. And we use hinge loss function to find the plane which can tolerate local disturbance instead of other alternative planes. Besides, we select radical basis kernel function which is a scalar function with radial symmetry, and it can map the original space into a higher dimensional space for computational convenience.

4   Experimental Results

We selected part of the data that can meet our requirements from the LIDC dataset in our experiments, and after data augmentation, the final number of the total input samples reach to 2590 for classification. Although these CT images contained many slices of lung anatomy, we used the middle slices of them to assure the clearness and completeness of the nodules. Moreover, the ground-truth annotated by radiologists were 2D image patches with the pulmonary nodule in the center. It is also noteworthy that the original size of CT images were 512×512 pixels, but the final patch extracted in our experiments were 64×64 pixels for convenient to process.

In this section, we adopted common evaluation standards such as accuracy, Receiver Operating Characteristic (ROC) curve, sensitivity and specificity to assess the performance of our DGnet-SVM versatilely based on the five-fold cross validation. First, we made a comparison between the state-of-the-art methods and the original DGnet which is shown in Table 2. As a baseline, traditional Convolutional Neural Network (CNN) [25] showed the accuracy of 0.88, and GoogLeNet [19] outperformed a little than it in this task. However, with the processing of Denoise Network for improvement, our original DGnet achieved a relatively higher accuracy of 0.86, which validated its efficacy than other networks. And we also show ROC curve to verify the results in Fig. 3(a).

Table 2. Measures of different methods

Methods	CNN	GoogLeNet	DGnet	DGnet-SVM
ACC	0.81	0.83	0.86	**0.89**
AUC	0.88	0.91	0.94	**0.95**
Sensitivity	0.81	0.81	0.91	**0.92**
Specificity	0.80	0.85	0.82	**0.89**

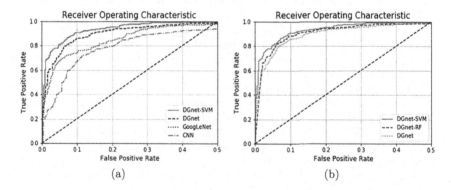

Fig. 3. (a) The ROC curve of four different networks. (b) The ROC curve of classifiers' comparison

Afterwards, we compared SVM with other classifiers such as default classifier softmax function in DGnet and Random Forest (RF) to verify its advantage. By modifying the final classifier to SVM instead of softmax function in original DGnet, our DGnet-SVM had the highest accuracy of 0.89 than the methods aforementioned, which can be ascribed to the high-level feature extraction in DGnet and great classification performance of SVM. Specially, the contrast classification results showed in Fig. 3(b) by ROC curve, and SVM obtained the AUC of 0.95.

5 Conclusions

In this work, we proposed a novel hybrid model named DGnet-SVM which integrated GoogLeNet, Denoise Network, and improved it by using SVM classifier in the classification stage. We focused on the aspect of improving input images quality, extracting deeper salient features and better classification results which based on prior stages. And DGnet-SVM is an efficient and high-powered way to distinguish benign and malignant pulmonary nodules by attaining encouraging results. The superiority of our method over the other state-of-the-art networks was corroborated by additional experiments. In general, our hybrid model is a good choice to solve the problem of classifying pulmonary nodules, and it delivered highly accurate results. A valuable direction of future work will be examining other deep networks to obtain more promising classification results.

References

1. Bach, P.B., Mirkin, J.N., Oliver, T.K., et al.: Benefits and harms of CT screening for lung cancer: a systematic review. JAMA **307**(22), 2418–2429 (2012)
2. Wu, Y., Wang, N., Zhang, H., et al.: Application of artificial neural networks in the diagnosis of lung cancer by Computed Tomography. In: 6th International Conference on Natural Computation, pp. 147–153. IEEE Press (2010)
3. Devinder, K., Alexander, W., David, A.C.: Lung nodule classification using deep features in CT images. In: 14th Conference on Computer and Robot Vision, pp. 133–138. IEEE Press, Edmonton (2015)
4. Matsuki, Y., Nakamura, K., Watanabe, H., et al.: Usefulness of an artificial neural network for differentiating benign from malignant pulmonary nodules on high-resolution CT: evaluation with receiver operating characteristic analysis. AJR Am. J. Roentgenol. **178**(3), 657–663 (2002)
5. Santos, A.M., Filho, A., Silva, A.C., et al.: Automatic detection of small lung nodules in 3D CT data using Gaussian mixture models, tsallis entropy and SVM. Eng. Appl. Artif. Intell. **36**(C), 27–39 (2014)
6. Sun, S.S., Li, H., Hou, X.R., et al.: Automatic segmentation of pulmonary nodules in CT images. In: 1st International Conference on Bioinformatics and Biomedical Engineering, pp. 790–793. IEEE (2007)
7. Cheng, Y.: Mean shift, mode seeking, and clustering. IEEE. Trans. Pattern Anal. Mach. Intell. **17**(8), 790–799 (1995)
8. Song, Y., Cai, W., Wang, Y., et al.: Location classification of lung nodules with optimized graph construction. In: 9th IEEE International Symposium on Biomedical Imaging (ISBI), pp. 1439–1442. IEEE (2012)
9. Lowe, D.G.: Distinctive image features from scale-invariant keypoints. Int. J. Comput. Vis. **60**(2), 91–110 (2004)
10. Song, Y., Cai, W., Zhou, Y., et al.: Feature-based image patch approximation for lung tissue classification. IEEE. Trans. Med. Imaging **32**(4), 797–808 (2013)
11. Bay, H., Tuytelaars, T., Van Gool, L.: SURF: speeded up robust features. Comput. Vis. Image Undert. **110**(3), 346–359 (2008)
12. Ojala, T., Pietikainen, M., Maenpaa, T.: Multiresolution gray-scale and rotation invariant texture classification with local binary patterns. IEEE. Trans. Pattern Anal. Mach. Intell. **24**(7), 971–987 (2002)

13. Dandil, E., Cakiroglu, M., Eksi, Z., et al.: Artificial neural network-based classification system for lung nodules on Computed Tomography scans. In: Soft Computing and Pattern Recognition, pp. 382–386. IEEE Press, Tunis (2014)
14. Chen, H., Wu, W., Xia, H., Du, J., Yang, M., Ma, B.: Classification of pulmonary nodules using neural network ensemble. In: Liu, D., Zhang, H., Polycarpou, M., Alippi, C., He, H. (eds.) ISNN 2011. LNCS, vol. 6677, pp. 460–466. Springer, Heidelberg (2011). doi:10.1007/978-3-642-21111-9_52
15. Hua, K.L., Hsu, C.H., Hidayati, S.C., et al.: Computer-aided classification of lung nodules on Computed Tomography images via deep learning technique. Onco Targets Ther. **8**, 2015–2022 (2015)
16. Shen, W., Zhou, M., Yang, F., et al.: Multi-crop convolutional neural networks for lung nodule malignancy suspiciousness classification. Pattern Recogn. **61**, 663–673 (2017)
17. Vincent, P., Larochelle, H., Lajoie, I., et al.: Stacked denoising autoencoders: learning useful representations in a deep network with a local denoising criterion. J. Mach. Learn. Res. **11**(12), 3371–3408 (2010)
18. Tang, P., Wang, H., Kwong, S.: G-MS2F: GoogLeNet based multi-stage feature fusion of deep CNN for scene recognition. Neurocomputing **225**, 188–197 (2016)
19. Szegedy, C., Liu, W., Jia, Y., et al.: Going deeper with convolutions. In: Proceedings of the IEEE Conference on Computer Vision and Pattern Recognition, pp. 1–9. IEEE Press (2015)
20. Sun, T., Wang, J., Li, X., et al.: Comparative evaluation of support vector machines for computer aided diagnosis of lung cancer in CT based on a multi-dimensional dataset. Comput. Methods Programs Biomed. **111**(2), 519–524 (2013)
21. Armato, S.G., McLennan, G., Bidaut, L., et al.: The lung image database consortium (LIDC) and image database resource initiative (IDRI): a completed reference database of lung nodules on CT scans. Med. Phys. **38**(2), 915–931 (2011)
22. Krizhevsky, A., Sutskever, I., Hinton, G.E.: Imagenet classification with deep convolutional neural networks. In: International Conference on Neural Information Processing Systems, pp. 1097–1105. Curran Associates Inc., Doha (2012)
23. Cortes, C., Vapnik, V.: Support-vector networks. Mach. Learn. **20**(3), 273–297 (1995)
24. Yepescalderon, F., Pedregosa, F., Thirion, B., et al.: Automatic pathology classification using a single feature machine learning support-vector machines. In: Proceedings of SPIE Medical Imaging, p. 187. SPIE (2014)
25. Lecun, Y., Bengio, Y.: Convolutional Networks for Images, Speech, and Time Series: The Handbook of Brain Theory and Neural Networks. MIT Press, Cambridge (1998)

Deep Learning Features for Lung Adenocarcinoma Classification with Tissue Pathology Images

Jia He[1], Lin Shang[1(✉)], Hong Ji[2], and XiuLing Zhang[2]

[1] State Key Laboratory for Novel Software Technology,
Department of Computer Science and Technology,
Nanjing University, Nanjing, China
shanglin@nju.edu.cn
[2] Beijing Computing Center, Beijing, China

Abstract. This paper presents the approach for lung adenocarcinoma diagnosis, using deep convolutional neural networks (CNN) to learn the features from the tissue pathology images. Our multi-stage procedure can detect the lung cancer of adenocarcinoma, in which the preprocessing consists of image enhancement and class imbalance treatment. Then Gradient-weighted Class Activation Mapping (Grad-CAM) and Guided-Backpropagation visualization techniques are employed to produce the visual explanations for decisions from our CNN model. Learned features and details for the specific areas have been generated through the model. Data is collected from 22 different patients with 270 lesion images and 24 normal ones. Experimental result on this data set has achieved F1-score with 0.963. Moreover, the study is not only to pursue precise classification on the tissue pathology images of lung adenocarcinoma, but also learn the specific areas in images which should be more concerned by doctors.

Keywords: Deep learning · Tissue pathology analysis · Visualization

1 Introduction

Deep learning has made great breakthroughs, some remarkable CNNs [3–5] have excellent performance on natural image dataset (i.e. ImageNet). Surveys [1,2] indicate that deep learning techniques, in particular CNN, have been widely used for analyzing medical images. Deep learning methods for solving computer vision tasks, including image classification, object detection, segmentation and others, are correspondingly applied to lesion classification [11,12], landmark and region localization [13,14], organ and substructure segmentation [15,16] as well as other medical images fields.

Inspired by previous works on medical images analysis, we have trained serval CNNs to classify whether lung adenocarcinoma tissue pathology images are normal or not. Lung adenocarcinoma is the most common type of lung cancer

© Springer International Publishing AG 2017
D. Liu et al. (Eds.): ICONIP 2017, Part IV, LNCS 10637, pp. 742–751, 2017.
https://doi.org/10.1007/978-3-319-70093-9_79

which has a high morbidity and mortality, accounting for almost half of all lung cancer. Our CNNs has achieved a good performance of 0.965 F1-score.

Medical images classification not only pursues high accuracy, but also attaches more importance to the medical interpretation of diagnosis results. Thus we use visualization techniques, Guided Backpropagation [10] and Grad-CAM [8], which are developed for CNN, to explore which features and details on images have influenced the lung adenocarcinoma classification model. We attempt to learn the specific features of tissue slices images from our trained CNN model. (i) It makes us understand what factors of image activate neurons and which parts of lung adenocarcinoma images are bases of classification. The combination of interpretable classification result and doctors's experience makes diagnosis of the deep learning model undoubted. (ii) Deep learning model for medical image may learn latent features. The visualization of the model allows us to notice some details that doctors may ignore, as providing more medical reference.

Besides, our visualization method has an ability of weakly location (shown as Fig. 1) to remind us of paying attention to pathologic features. Some works (i.e. [13,14]) on medical image analysis with deep leaning are detection task with using bounding-box, for the dataset having location labels of lesions. Detecting the tissues on medical images is better than just classifying whether this image is normal or not. However, there are little annotated medical image datasets, as data acquisition is difficult, and quality annotation is costly. Even in medical image detection task with deep learning, our visualization method can be used to understand medical characteristics which belong to the marked lesion tissues. The method is still meaningful for more medical image analysis situations.

So, in this paper, we classified tissue pathology images of lung adenocarcinoma and visualized internal representations of the used CNN via Grad-CAM and Guided-Backpropagation. The highlighted regions resulted in the visualization helps in improving interpretability of medical images analysis with deep model. This is an important contribution of this paper, and the other is to localize features that are used to give a prediction.

The rest of this paper as structured as followed. We introduce the related work in Sect. 2. Dataset, training details and visualization methods are described in Sect. 3. The comparison of different CNNs and the analysis of visualization results are described in Sect. 4. We discuss conclusion in Sect. 5.

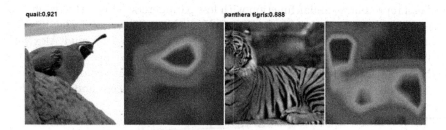

Fig. 1. Grad-CAM has weakly location ability by highlighting the discriminative regions which detected by CNN

2 Related Work

2.1 Visualization

In order to understand black box of CNN, researchers have proposed several approaches to visualize the filters to probe what kinds of patterns are these filters favoring. Krizhevsky et al. [3] directly visualized the filters learned in the first layer. Since filters in high layers receive inputs from their previous layers instead of pixels, there is no direct way to visualize them in pixel space. Zeiler and Fergus [6] in their deconvolution work (Deconvnet) introduced a slight modification to the backward pass of ReLU, to pass only the positive gradients from higher layers. Their methods can visualize filters of all layers by patches with highest activations, together with their reconstructed versions via deconvolution network. Springenberg et al. [10] developed CNN architecture without max pooling layers which Deconvnet would not work. Therefore they developed Guided Backpropagation [10] which enhanced the clarity and sharpness of the projection as an improvement of Deconvnet. Zhou et al. [7] proposed a technique called Class Activation Mapping (CAM) for identifying discriminative regions used by a restricted class of image classification CNNs. But this visualization method do not contain gradients from fully-connected layers. Grad-CAM [8] is a improved method of CAM which can be applicable to a wide variety of CNNs including CNNs with fully-connected layers. Grad-CAM uses the gradients of any target concept flowing into the final convolutional layer to produce a coarse localization map highlighting the important regions in the image for predicting the concept.

2.2 Convolutional Neural Network Architectures

We mainly use three common CNN architectures (AlexNet [3], VGG-16 [4] and GoogLeNet [5]), fine-tuning model pre-trained from natural image dataset to our medical image classification tasks. AlexNet is relatively shallow networks, consisting of five convolution layers, three pooling layers, and two fully-connected layers. VGG-16 is a much deeper architecture, and employed small, fixed size kernels in each layer. GoogLeNet is significantly more complex and deep than other two CNN architectures. It also introduces a new module called "Inception", which concatenates filters of different sizes and dimensions into a single new filter. Stacking smaller kernels instead of a single layer of kernels with a large receptive field can represent a similar function with less parameters. This is the theoretical basis of designing Inception block. GoogLeNet achieved 5.5% top-5 classification error on the ImageNet challenge, compared to VGGNet's 7.3% and AlexNet's 15.3% top-5 classification error. With CNN model performance improving, the CNN architecture is becoming more complex and deeper. Such complexity makes these models hard to interpret. Thus it is important for deep model to explore the spectrum between interpretability and accuracy.

These remarkable works have got good performance on natural images. However, differences of cells or tissues on medical images are much less obvious than that of various objects on natural images. In addition, medical images are difficult to identify even with medical knowledge. Thus it is more significant to insight features and details in CNN model identifying medical images.

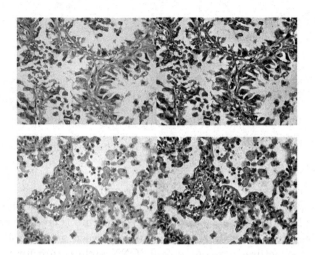

Fig. 2. Original lung adenocarcinoma tissue pathology images and pre-processed images. The top is lesion sample and the bottom is normal sample. Left one is original and right one is pre-processed

3 Method

3.1 Dataset

Beijing Computing Center provides experimental datasets. These lung adenocarcinoma tissue pathology images labeled lesion (positive) or normal (positive) without lesions location label. We then cut each image into four parts and have data-augmentation with random over-sampling and rotation to balance data. Finally, the lung adenocarcinoma tissue pathology images dataset consists of 280 lesion samples and 140 normal ones, and they are split into 340 for training and 80 for testing. Besides, the contrast of these images is low, and the edges of some cells in images are ambiguous. In order to obtain more details, We have made two pre-processing ways for them. CLAHE [9] algorithm is used to enhance the contrast of images, and then bootstrap filter is applied to making cells' edges more distinct (shown as Fig. 2).

3.2 Training

TensorFlow1.0 and Ubuntu14.04 are our default development environment. We have used three CNN architectures to classify lung adenocarcinoma tissue pathology images, AlexNet [3], VGG-16 [4] and GoogLeNet [5]. Owing to a small number of images, which exit in almost deep learning tasks on medical image, using

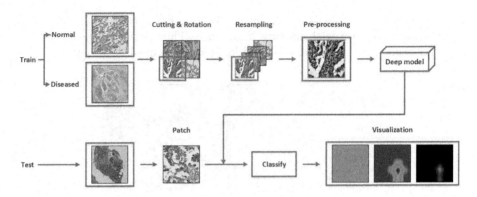

Fig. 3. The framework of lung adenocarcinoma classification and visualization

fine-tuning is a necessary choice. Our training and testing process are shown as Fig. 3. These three trained models performing very well (evaluation index shown as Table 1) and ensure that visualization of the feature is credible enough.

Considering VGG-16 achieving a balance between good performance and relatively neat structure, it is our default CNN architecture for using Guided Backpropagation [10] and Grad-CAM [8] for features analysis and visualization in the remainder of the paper.

3.3 Visualization Approach

Comparing Guided Backpropagation and Deconvnet. In CNNs, backpropagation uses information from lower layers and input images to determine which of the pixel values were important in computing the activations of the higher layers. Deconvnet [6] uses gradient information from the higher layers to determine which of the pixels from the input image had a beneficial effect on the activations of the units of interest. It project the activations from the feature space back to the input space. And Guided Backpropagation [8] combines Deconvnet and backpropagation. It masks out the value of any unit that had either a negative activation during the forward pass or a negative contribution value during the backward pass (schematic illustration of these procedures shown as Fig. 4).

Both Deconvnet and Guided Backpropagation are proposed as a probe of an already trained CNN, and they are not used in any learning capacity. By backward passing gradient information of the last softmax layer and projecting it to pixel space, we get visualization images generated by the two methods. We compare the two visualization methods on natural images and medical images, with trained totally three models (one for natural images, two others for medical images including lung adenocarcinoma tissue pathology images and chest X-ray images). Our experimental results confirm that Guided Backpropagation enhance the clarity and sharpness of the projection as an improvement of

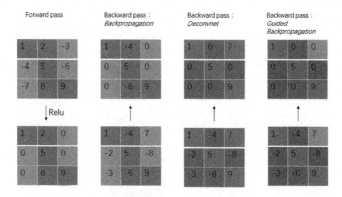

Fig. 4. Backpropagation, Deconvnet, and Guided Backpropagation have different influences on gradient when backward pass through Relu layer. The forward pass through the ReLU layer is shown for comparison.

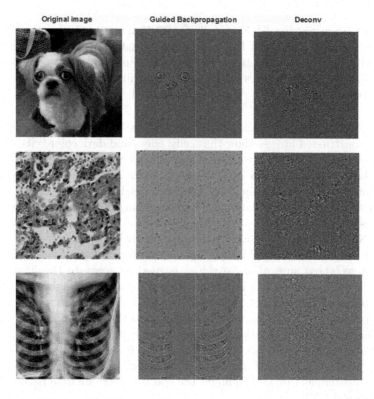

Fig. 5. Guided Backpropagation and Deconvnet are used for natural images and medical images. Both of them reflect which pixels activate the neurons greatly, but the Deconvnet generated images is vague.

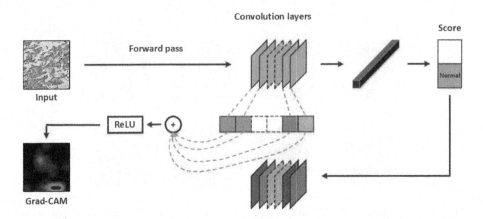

Fig. 6. Grad-CAM method

Deconvnet (shown as Fig. 5). Especially for medical images, for reasons of details in original images are much less obvious than natural images (the difference described in introduction section), the Deconvnet method almost impossible to provide deep learning features which can be recognized by us. We chose Guided Backpropagation as a better visualization method.

Grad-CAM. As shown in the second row of Figs. 5 and 7, Guided Backpropagation reconstructed images show that cells' edges and dark color cells affect our CNN classification results greatly. But it is not specific enough for us to understand which particular parts of lung adenocarcinoma tissue pathology images play a important role. We then use Gradient-weighted Class Activation Mapping (Grad-CAM) [8] to enhance our understanding of the features and details of images.

In CNNs, pooling layer outputs the spatial average of the feature map of each unit at the last convolutional layer when forward pass through trained model. A weighted sum of these values is used to generate the final output. So class activation maps can be obtained by computing a weighted sum of the feature maps of the last convolutional. Grad-CAM uses the gradients of any target concept (i.e. logits of 'tiger'), flowing into the final convolutional layer to produce a coarse localization map highlighting the important regions in the image for predicting the concept (shown as Fig. 6). This method we used has ability of weakly location. It allows us to visualize the predicted class scores on lung adenocarcinoma image, highlight the specific discriminative parts detected by the trained CNN.

We combine Grad-CAM and guided backpropagation reconstructed image for a better visualization (shown as Fig. 7).

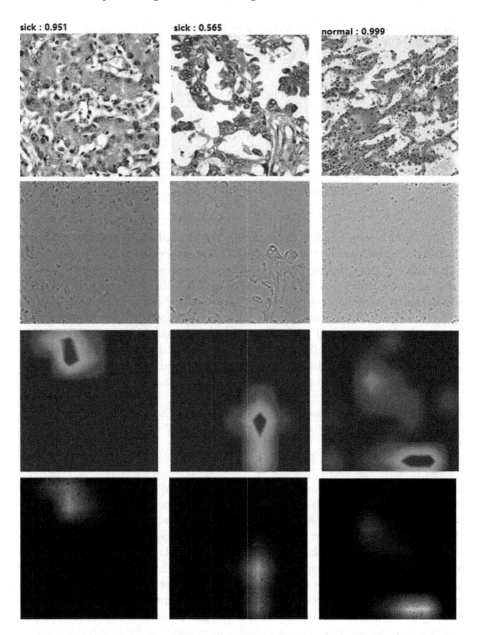

Fig. 7. Examples of visualizing features and details which detected by CNN. These lung adenocarcinoma tissue pathology images are correctly classified.

4   Experimental Results

Performance comparison of three CNN architectures on lung adenocarcinoma tissue pathology images test set is shown as Table 1. Although there is a slight gap in performance among them, all three trained models have shown very good performance in classification. The high-precision classifier ensures that visualizing features detected by the trained CNN are reliable.

Table 1. F1-score and accuracy on lung adenocarcinoma tissue pathology images test set of classification-trained model

Network	Precision	Recall	F1-score	Accuracy
AlexNet	0.902	0.949	0.923	0.925
Vgg-16	0.928	0.975	0.951	0.950
GoogLeNet	0.951	0.975	0.963	0.962

Grad-CAM and Guided Backpropagation visualization results of tissue pathology images classified by trained model are shown as Fig. 7. We select three representative samples which have been correctly classified to present lung adenocarcinoma features and details detected by CNN. Images in first line are the original tissue pathology ones, showing there is a little different in shape and size between cancer cells and normal cells. Images in second line are Guided Backpropagation reconstructed ones. They show that tissue's edges active neurons greatly. And the cancer cells in the lesion images are displayed clearly while the normal cells are not obvious. It means that our trained CNN model does depend on the morphology of cells and tissues, rather than background or other factors. Images in the third line are Grad-CAM reconstructed ones, showing cancer cells which are gathered together are located by CNN. The last line images are the combination of two ways for a better visualization effects.

5   Conclusion and Discussion

In this study, we use convolution neural networks to classify tissue pathology images of lung adenocarcinoma. Besides, we combine two visualization methods, Grad-CAM and Guided-Backpropagation, to understand our trained deep classification model. The reconstructed visualizing images help us to understand what factors on tissue pathology images determine classification and remind us to pay attention to pathologic features or details of the specific areas.

Therefore, these results suggest us that we may get rid of unimportant parts or enhance parts which trained CNN focus on to improve performance of deep model. In the future work, we are adding this pre-processing way to lung adenocarcinoma tissue pathology images classification and visualization based on deep learning. More generally, we attempt to integrate deep learning-based approaches into the work-flow of the diagnostic pathologist could drive improvements in the accuracy and clinical value of pathological diagnoses.

Acknowledgements. This work is supported by the National Natural Science Foundation of China (No. 61672276) and Natural Science Foundation of Jiangsu, China (BK20161406).

References

1. Shen, D., Wu, G., Suk, H.I.: Deep learning in medical image analysis. Ann. Rev. Biomed. Eng. (2017)
2. Litjens, G., Kooi, T., Bejnordi, B.E., Setio, A.A.A., Ciompi, F., Ghafoorian, M., van der Laak, J.A., van Ginneken, B., Sanchez, C.I.: A survey on deep learning in medical image analysis. arxiv preprint arXiv:1702.05747 (2017)
3. Krizhevsky, A., Sutskever, I., Hinton, G.E.: Imagenet classification with deep convolutional neural networks. In: Advances in Neural Information Processing Systems, pp. 1097–1105 (2012)
4. Simonyan, K., Zisserman, A.: Very deep convolutional networks for large-scale image recognition. arxiv preprint arXiv:1409.1556 (2014)
5. Szegedy, C., Liu, W., Jia, Y., Sermanet, P., Reed, S., Anguelov, D., Erhan, D., Vanhoucke, V., Rabinovich, A.: Going deeper with convolutions. In: Proceedings of the IEEE Conference on Computer Vision and Pattern Recognition, pp. 1–9 (2015)
6. Zeiler, M.D., Fergus, R.: Visualizing and understanding convolutional networks. In: Fleet, D., Pajdla, T., Schiele, B., Tuytelaars, T. (eds.) ECCV 2014. LNCS, vol. 8689, pp. 818–833. Springer, Cham (2014). doi:10.1007/978-3-319-10590-1_53
7. Zhou, B., Khosla, A., Lapedriza, A., Oliva, A., Torralba, A.: Learning deep features for discriminative localization. In: Proceedings of the IEEE Conference on Computer Vision and Pattern Recognition, pp. 2921–2929 (2016)
8. Selvaraju, R.R., Das, A., Vedantam, R., Cogswell, M., Parikh, D., Batra, D.: Gradcam: Why did you say that? visual explanations from deep networks via gradient based localization. arxiv preprint. arXiv:1610.02391 (2016)
9. Zuiderveld, K.: Contrast Limited Adaptive Histogram Equalization. Academic Press Professional Inc., Cambridge (1994)
10. Springenberg, J.T., Dosovitskiy, A., Brox, T., Riedmiller, M.: Striving for simplicity: the all convolutional net. arxiv preprint arXiv:1412.6806 (2014)
11. Kawahara, J., Hamarneh, G.: Multi-resolution-Tract CNN with hybrid pretrained and skin-lesion trained layers. In: Wang, L., Adeli, E., Wang, Q., Shi, Y., Suk, H.-I. (eds.) MLMI 2016. LNCS, vol. 10019, pp. 164–171. Springer, Cham (2016). doi:10. 1007/978-3-319-47157-0_20
12. Shen, W., Zhou, M., Yang, F., Yang, C., Tian, J.: Multi-scale convolutional neural networks for lung nodule classification. In: Ourselin, S., Alexander, D.C., Westin, C.-F., Cardoso, M.J. (eds.) IPMI 2015. LNCS, vol. 9123, pp. 588–599. Springer, Cham (2015). doi:10.1007/978-3-319-19992-4_46
13. de Vos, B.D., Wolterink, J.M., de Jong, P.A., Viergever, M.A., Isgum, I.: 2D image classification for 3D anatomy localization: employing deep convolutional neural networks. In: Medical Imaging 2016: Image Processing, vol. 9784 (2016)
14. Yang, D., Zhang, S., Yan, Z., Tan, C., Li, K., Metaxas, D.: Automated anatomical landmark detection ondistal femur surface using convolutional neural network. In: 2015 IEEE 12th International Symposium on Biomedical Imaging (ISBI), pp. 17–21. IEEE (2015)
15. Ronneberger, O., Fischer, P., Brox, T.: U-Net: convolutional networks for biomedical image segmentation. In: Navab, N., Hornegger, J., Wells, W.M., Frangi, A.F. (eds.) MICCAI 2015. LNCS, vol. 9351, pp. 234–241. Springer, Cham (2015). doi:10. 1007/978-3-319-24574-4_28
16. Wells, W.M., Viola, P., Atsumi, H., Nakajima, S., Kikinis, R.: Multi-modal volume registration by maximization of mutual information. Med. Image Anal. $\mathbf{1}$(1), 35–51 (1996)

The Analysis and Classify of Sleep Stage Using Deep Learning Network from Single-Channel EEG Signal

Songyun Xie[(✉)], Yabing Li, Xinzhou Xie, Wei Wang, and Xu Duan

School of Electronics and Information,
Northwestern Polytechnical University, Xi'an 710129, China
syxie@nwpu.edu.cn

Abstract. Electroencephalogram (EEG)-based sleep stage analysis is helpful for diagnosis of sleep disorder. However, the accuracy of previous EEG-based method is still unsatisfactory. In order to improve the classification performance, we proposed an EEG-based automatic sleep stage classification method, which combined convolutional neural network (CNN) and time-frequency decomposition. The time-frequency image (TFI) of EEG signals is obtained by using the smoothed short-time Fourier transform. The features derived from the TFI have been used as an input feature of a CNN for sleep stage classification. The proposed method achieves the best accuracy of 88.83%. The experimental results demonstrate that deep learning method provides better classification performance compared to other methods.

Keywords: Convolutional neural networks (CNN) · Time-frequency decomposition · Sleep analysis

1 Introduction

Sleep stages analysis is not only beneficial for treatment of sleep disorders (such as apnea, insomnia and narcolepsy), but also aids several psychophysiological monitoring (aid physical recovery and enhance the ability for learning and memory) [1]. According to the Rechtschaffen and Kales's (R&K) standard, the state of sleep can be divided into four major stages: awake stage (W); light sleep stages (stage 1 and stage 2); deep sleep stages (stage 3 and stage 4); and rapid eye movement (REM) [2]. In general, study for sleep is visually scored by experts according to two available standards: R&K and American Academy of Sleep Medicine (AASM) [3]. While, this work is expensive and is often tedious and time consuming [4]. Considering these problems, some signal processing techniques along with machine learning algorithms based on electroencephalogram (EEG) analysis are developed. Herrera et al. analyzed sleep state using Self-Organizing Maps (SOM) and mutual information (MI) based variable selection algorithm, and the results provided success rates around 70% [5]. Ronzhina et al. detected the sleep stages by combining Power Spectrum Density (PSS) and Artificial Neural Network (ANN), the performance achieves the best accuracy of 81.42% [6]. Fraiwan et al. also evaluated the sleep stages with an accuracy of 83% by using time-frequency techniques and random forest classifier [7]. Hassan et al. analyzed sleep

© Springer International Publishing AG 2017
D. Liu et al. (Eds.): ICONIP 2017, Part IV, LNCS 10637, pp. 752–758, 2017.
https://doi.org/10.1007/978-3-319-70093-9_80

scoring combined spectral feature (SF) and Adaptive Boosting (AdaBoost), and an accuracy of 82.83% was reported [8]. In summary, the accuracy for automatic sleep classification is still a challenge, especially for identifying sleep stages with a single-channel EEG signal.

In recent years, deep learning approaches have been used successfully in many fields to learn features and classify different types of data (speech recognition, hand-writing font recognition, and visual object recognition) [9–11]. On one hand, CNN has been proved to be very good at discovering the intricate structures in complex data [12]. On the other hand, continuous EEG signals recording for sleep provide a large amount of data that is benefit for training of deep learning. However, the number of studies that employ deep learning on sleep stage classification is very limited compared to the huge applications in other fields.

A new method combined CNN and time-frequency image (TFI) is presented in the paper to classify the sleep stages based on a single-channel EEG signal. The TFI of EEG signals is obtained by using the smoothed short-time Fourier transform. The raw data of the TFI have been used as an input feature of a CNN for sleep stage classification. The experimental results have achieved better classification accuracy of classifying sleep stages from EEG signals.

The remaining sections of this paper are organized as follows. In Sect. 2, the feature extraction is elucidated before describing the classifier. Section 3 presents the performance of proposed method, and also makes the compare between proposed method and some traditional methods. Finally, the conclusions and further work are described in Sect. 4.

2 Methods

The structural diagram of the automatic sleep stages classification method proposed in this study is described in this section, which included data acquisition, signal preprocessing, feature extraction and classification using single-channel EEG signal. First, the sleep–EDF database is obtained from MIT-BIH Database which is publicly available and widely used in the literature [13]. In the stage of preprocessing, the band-pass filters are used for EEG signals. And then, time-frequency decomposition is applied during features extraction session from these signals automatically. Finally, we present the CNN to classify sleep stages. The pro-posed algorithm is shown in Fig. 1.

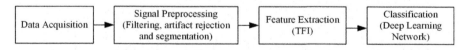

Fig. 1. The structure of the proposed sleep stage classification algorithm

2.1 EEG Feature Extraction

In order to extract TFI, the following steps are applied to EEG signals:

Step 1 (pre-processing): the raw EEG data were band-pass filtered between 1 Hz and 40 Hz. After that, EEG signals were segmented into epochs of 30 s with each epoch corresponding to a single sleep stage. The bad trials, which contained obvious artifacts (eye blink, eye movement and electromyography (EMG)), were rejected.

Step 2 (resample and electrode selected): each dataset resampled at sampling rate of 100 Hz. In this study, EEG data from Pz-Oz electrodes is selected to analyze.

Step 3 (time-frequency-domain feature): spectral features of EEG signals for sleep from all subjects were obtained by using short-time Fourier transform (STFT). And all time smoothing windows were set as Hamming 256-point length windows.

Step 4 (feature normalization): time-frequency feature normalization is used to tackle the remaining mismatch between different epochs.

2.2 Deep Learning Network

The architecture of CNN used in this paper is structured as a series of stages (shown in Fig. 2) [14, 15]. It is composed of 5 layers (2 conventional layers, 2 pooling layers and output layer). The input of TFI is a 20 × 40 pixel image.

Conventional Layer1 means conventional layer with 5 feature map of size 16 × 36. And each neuron is connected to a 5 × 5 neighborhood of the TFI. It can be denoted as:

$$\left(C^k\right)_{i,j} = \left(W_k * input\right)_{i,j} + b_k \tag{1}$$

Here, $k = 1, 2, \ldots, 5$ denotes the index of feature map, $input$ is the TFI for EEG signals, W_k and b_k mean weight and bias for the $k - th$ filter, respectively, and C^k is the $k - th$ output feature map. "*" denotes the 2D spatial convolution.

Pooling Layer1 is a sub-sample layer with 5 feature map of size 8 × 18. Each pixel of feature map in this layer is connected to a 2 × 2 neighborhood of feature map in Conventional Layer1. Therefore, feature maps in this layer have half number of rows and column as feature map in Conventional Layer1. It can be expressed as follows:

$$P_k = \max_{i,j} C_{i,j}^k \tag{2}$$

The $C_{i,j}^k$ denote the elements at location (i, j) of Conventional Layer1 obtained by the $k - th$ feature map. The P_k means new feature map elements in Pooling Layer1.

Conventional Layer2 means conventional layer with 5 feature map of size 4 × 14. And each neuron is connected to a 5 × 5 neighborhood of the Pooling Layer1.

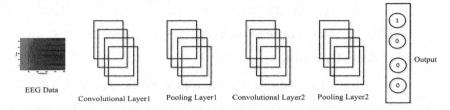

Fig. 2. CNN architecture used in this paper

Pooling Layer2 is a sub-sampling layer with 6 feature maps of size 2 × 7. And each unit in each feature map is connected to a 2 × 2 neighborhood in corresponding feature map in Conventional Layer2. The relationship between Conventional Layer2 and Pooling Layer2 is similar with the relationship between Conventional Layer1 and Pooling Layer1.

Output Layer is composed of sigmoid function contains 84 units. The feature maps of Pooling Layer2 are sent into traditional neural network for further results.

To validate the performance of the classifiers, the "leave-one-subject-out" method was used, which meant that mixed data for each subject was used for the training of classifier, and the rest is used to test the classifier.

3   Results and Discussion

In order to evaluate the performance of method combined CNN and time-frequency feature, the within-class classification metrics (accuracy, sensitivity, specificity and confusion matrix and global classifier performance metric (overall accuracy) were used to assess classification [16]. The proposed method was compared with different classifiers listed in this paper.

Figure 3 shows the sensitivity, specificity, and accuracy of CNN class detection. The best performance of these parameters for all stages are higher than 0.8. And the best performances obtained for epochs among all classes are deep sleep with the best accuracy (for sensitivity, specificity, and accuracy were 0.70029, 0.94475, and 88.363%, respectively). Additionally, specificity was above 0.94 for all stages, except for awake stage. While accuracy was above 84% for all stages except for awake stage. In other words, the sensitivity, specificity and accuracy results are high for all classes, with one exception being the drop in sensitivity for REM stage.

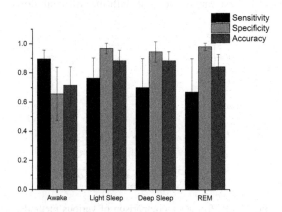

Fig. 3. The sensitivity, specificity, and accuracy for CNN performance

Figure 4 shows the percent of accuracy for each subject and the average for all subjects. For all subjects, the best performance was larger than 88%, which is S3 (88.83%) and S9 (88.37%). And for most of the subjects, the accuracy was above 83%. It is clear that accuracy for all subjects were well above the chance level of 25% (the probability that 1 out of 4 of randomly detecting the right class), which was above 80%, except for S6 (79.82%).

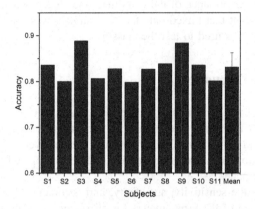

Fig. 4. The performance of recognition for all subjects

The performance of our method was also compared with three other widely classifiers. Figure 5 illustrated the accuracy for various methods, which represented the best performance using the same sleep-EDF database. Compared to other groups, the proposed method can achieve a better performance (best performance was 88.83%, which was 6%–19.03% higher than the other methods). The results indicated that the proposed method is the best capable to deal with the different sleep stages.

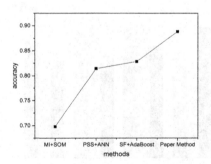

Fig. 5. Performance comparison of various methods

4 Conclusion

In this paper, the method combined of time-frequency feature of EEG data and CNN is proposed to analyze the stages of sleeping. In comparison with several recently available methods on the single channel classification of sleep stages, the present study has certain advantages with 88.83% in terms of the accuracy. The experimental results suggest that the proposed method is effective and has the potential of sleep monitoring.

Acknowledgments. This work was supported in part by National Natural Science Foundation of China (61273250), the Fundamental Research Funds for the Central Universities (No. 3102017jc11002) and the graduate starting seed fund of Northwestern Polytechnical University (Z2017141).

References

1. Pan, S.T., Kuo, C.E., Zeng, J.H., Liang, S.F.: A transition-constrained discrete hidden markov model for automatic sleep staging. BioMed. Eng. Online **11**(1), 1–19 (2012)
2. Rechtschaffen, A.Q., Kales, A.A.: A manual of standardized terminology techniques and scoring system for sleep stages in human subjects. Psychiatry Clin. Neurosci. **26**(6), 644 (1968)
3. Schulz, H.: Phasic or transient? Comment on the terminology of the AASM manual for the scoring of sleep and associated events. J. Clin. Sleep Med. JCSM Official Publ. Am. Acad. Sleep Med. **3**(7), 752 (2007)
4. Aboalayon, K.A.I., Faezipour, M., Almuhammadi, W.S., Moslehpour, S.: Sleep stage classification using EEG signal analysis: a comprehensive survey and new investigation. Entropy **18**(9), 272–303 (2016)
5. Herrera, L.J., Mora, A.M., Fernandes, C., Migotina, D., Guillen, A., Rosa, A.C.: Symbolic representation of the EEG for sleep stage classification. In: International Conference on Intelligent Systems Design and Applications 2011, vol. 60, pp. 253–258. IEEE, Cordoba, Spain (2011)
6. Ronzhina, M., Janoušek, O., Kolářová, J., Nováková, M., Honzík, P., Provazník, I.: Sleep scoring using artificial neural networks. Sleep Med. Rev. **16**(3), 251–263 (2012)
7. Fraiwan, L., Lweesy, K., Khasawneh, N., Wenz, H., Dickhaus, H.: Automated sleep stage identification system based on time-frequency analysis of a single EEG channel and random forest classifier. Comput. Methods Programs Biomed. **108**(1), 10–19 (2012)
8. Hassan, A.R., Bashar, S.K., Bhuiyan, M.I.H.: Automatic classification of sleep stages from single-channel electroencephalogram. In: India Conference 2015, pp. 1–6. IEEE, New Delhi, India (2015)
9. Schmidhuber, J.: Deep learning in neural networks: an overview. Neural Netw. **61**, 85–117 (2014)
10. Hinton, G., Deng, L., Yu, D., Dahl, G.E., Mohamed, A.R., Jaitly, N., Senior, A., Vanhoucke, V., Nguyen, P., Sainath, T.N.: Deep neural networks for acoustic modeling in speech recognition: the shared views of four research groups. IEEE Sig. Process. Mag. **29**(6), 82–97 (2012)
11. Lawrence, S., Giles, C.L., Tsoi, A.C., Back, A.D.: Face recognition: a convolutional neural-network approach. IEEE Trans. Neural Netw. **8**(1), 98–113 (1997)
12. LeCun, Y., Bengio, Y., Hinton, G.: Deep learning. Nature **521**(7553), 436–444 (2015)

13. Kemp, B., Zwinderman, A.H., Tuk, B., Kamphuisen, H.A.C., Oberye, J.J.L.: Analysis of a sleep-dependent neuronal feedback loop: the slow-wave microcontinuity of the EEG. IEEE Trans. Biomed. Eng. **47**(9), 1185–1194 (2000)
14. Lecun, Y., Bottou, L., Bengio, Y., Haffner, P.: Gradient-based learning applied to document recognition. Proc. IEEE **86**(11), 2278–2324 (1998)
15. Kang, L., Ye, P., Li, Y., Doermann, D.: Convolutional neural networks for no-reference image quality assessment. In: Conference on Computer Vision and Pattern Recognition 2014, pp. 1733–1740. IEEE, Columbus, OH, USA (2014)
16. Lajnef, T., Chaibi, S., Ruby, P., Aguera, P.E., Eichenlaub, J.B., Samet, M., Kachouri, A., Jerbi, K.: Learning machines and sleeping brains: automatic sleep stage classification using decision-tree multi-class support vector machines. J. Neurosci. Methods **250**(30), 94–105 (2015)

Thin-Cap Fibroatheroma Detection with Deep Neural Networks

Tae Joon Jun[1], Soo-Jin Kang[2], June-Goo Lee[3], Jihoon Kweon[2], Wonjun Na[2], Daeyoun Kang[1], Dohyeun Kim[1], Daeyoung Kim[1(✉)], and Young-hak Kim[2]

[1] KAIST, Daejeon, Korea
{taejoon89,ikasty,stalker7,kimd}@kaist.ac.kr
[2] University of Ulsan College of Medicine, Asan Medical Center, Seoul, Korea
{sjkang,nwj,mdyhkim}@amc.seoul.kr, kjihoon2@naver.com
[3] Asan Institute for Life Sciences, Asan Medical Center, Seoul, Korea
junegoo.lee@amc.seoul.kr

Abstract. Acute coronary syndromes (ACS) frequently results in unstable angina, acute myocardial infarction, and sudden coronary death. The most of ACS are related to coronary thrombosis that mainly caused by plaque rupture followed by plaque erosion. Thin-cap fibroatheroma (TCFA) is a well-known type of vulnerable plaque which is prone to serious plaque rupture. Intravascular ultrasound (IVUS) is the most common methods for imaging coronary arteries to determine the amount of plaque built up at the epicardial coronary artery. However, since IVUS has relatively lower resolution than that of optical coherence tomography (OCT), TCFA detection with IVUS is considerably difficult. In this paper, we propose a novel method of TCFA detection with IVUS images using machine learning technique. 12,325 IVUS images from 100 different patients were labeled with equivalent frames from OCT images. Deep feed-forward neural network (FFNN) was applied to a different number of selected features based on the Fishers exact test. As a result, IVUS derived TCFA detection achieved 0.87 area under the curve (AUC) with 78.31% specificity and 79.02% sensitivity. Our experimental result indicates a new possibility for detection of TCFA with IVUS images using machine learning technique.

Keywords: Acute coronary syndromes · Vulnerable plaque · Thin-cap fibroatheroma · Intravascular ultrasound · Optical coherence tomography · Machine learning · Deep neural networks

1 Introduction

Acute coronary syndromes (ACS) are the common result of unexpected coronary thrombosis with an associated etiology in 55% to 60% plaque rupture, in 30% to 35% plaque erosion, and in 2% to 7% calcified nodule [1]. Patients with ACS have a high possibility of unstable angina, acute myocardial infarction, and sudden coronary death [1]. The most common type of vulnerable plaque, that

© Springer International Publishing AG 2017
D. Liu et al. (Eds.): ICONIP 2017, Part IV, LNCS 10637, pp. 759–768, 2017.
https://doi.org/10.1007/978-3-319-70093-9_81

is prone to serious plaque rupture, is thin-cap fibroatheroma (TCFA) which is characterized by an existence of underlying necrotic core and by less than 65 μm fibrous cap that contains plenty of macrophage [1,2]. Intravascular ultrasound (IVUS) is the most common method for imaging coronary arteries since it provides a tomographic assessment of lumen area, plaque size, distribution, and composition [3]. In addition, IVUS also provides conditions of the vessel wall. However, the axial and lateral resolutions of IUVS are 150 μm and 250 μm which are insufficient to identify less than 65 μm fibrous caps in TCFA lesion. Therefore, medical experts practically use optical coherence tomography (OCT) to clarify TCFA lesion, since the spatial resolution of OCT is 4 to 16 μm which can clearly visualize lipid-rich plaque and necrotic core [4]. Nevertheless, OCT has a limited length (35 mm) of images from single pull-back and vessel wall may not be visualized if the vessel diameter is large.

In this paper, we propose a novel method of TCFA detection with IVUS images using machine learning technique. 12,325 IVUS images from 100 different patients were labeled with equivalent frames from OCT images. Both images were collected from the patients with either stable or unstable angina who underwent both IVUS and OCT procedures. After the IVUS and OCT co-registration, lumen and external elastic membrane (EEM) are segmented from the IVUS images to differentiate our region of interests (ROI). Then we extracted 105 different features from each IVUS image including a proportion of plaque in the vessel, and a proportion of every 10 pixels from 4 different plaque regions. With these extracted features, n-ranked selected features based on Fishers exact test [5] are used for training using deep feed-forward neural networks (FFNN) [6]. As a result, we achieved 0.87 area under the curve (AUC) with 78.31% specificity and 79.02% sensitivity in the case of maximizing the sum of specificity and sensitivity. In addition, we summarized top 10 important features that highly affects the TCFA classification performance.

Several previous works have been reported which are related to the identification of TCFA. Prospective prediction of future development of TCFA with virtual histology IVUS (VH-IVUS) using support vector machine (SVM) was proposed by Zhang in [7]. Jang, Rodriguez-Granillo, and Sawada reported in vivo characterization of TCFA with VH-IVUS and OCT in [4,8,9]. Association of VH-IVUS and major adverse cardiac events (MACE) on an individual plaque or whole patient analysis is presented by Calvert in [10]. Garcia-Garcia reported characteristics of coronary atherosclerosis including TCFA with IVUS and VH-IVUS in [11]. However, as far as we know, our work is the first computer-aided approach to classifying TCFA from IVUS images.

The paper is organized as follows. Section 2 provides detailed methodologies used in TCFA detection. Section 3 includes experimental setup and evaluation results. Conclusion and future works are described in Sect. 4.

2   Methodology

Methodologies used in TCFA detection includes IVUS and OCT frame co-registration, ROI extraction, feature extraction, feature selection, and

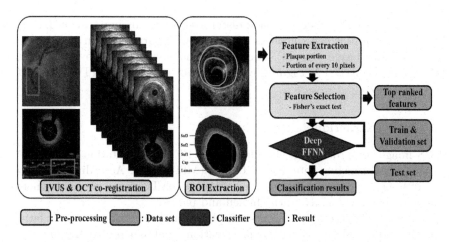

Fig. 1. Overall procedures processed in TCFA classification

classification with deep FFNN classifier. Figure 1 explains overall procedures processed in TCFA classification from data acquisition to deep FFNN classification result.

2.1   IVUS and OCT Co-registration

TCFA classification with machine learning technique requires labeled IVUS image data. Labeling IVUS image from equivalent OCT frame is called IVUS and OCT co-registration. Both IVUS and OCT images are obtained from patients present either stable or unstable angina which has lesions with angiographic 30% to 80% of diameter stenosis.

Within the target segment, every OCT section with 0.2 mm interval was co-registered with its comparable IVUS frame (approximately, every 120th IVUS frame) by using anatomical landmarks such as vessel shape, side branches,

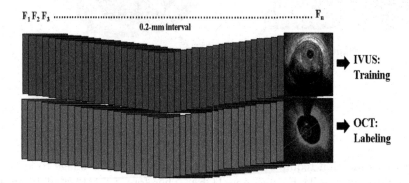

Fig. 2. Overall process of IVUS and OCT co-registration

calcium, perivascular structures, and distances from the ostium. An OCT-defined TCFA had a fibrous cap thickness at the thinnest part less than 65 μm and an angle of lipidic tissue over 90°. Each IVUS image was labeled by OCT-defined TCFA versus non-TCFA. Figure 2 presents the overall process of IVUS and OCT Co-registration.

2.2   ROI Extraction

IVUS image pre-processing with lumen and EEM segmentation were done by using Medical Imaging Interaction Toolkit (MITK) [12]. According to the borders of lumen and external elastic membrane, each IVUS image was segmented into 3 compartments; adventitia, lumen and plaque.

Since the distributions of lipid-rich plaque and necrotic core are relatively more important in superficial plaque location, that is close to the lumen, we additionally divided plaque region into 4 different spaces: Cap, Suf1, Suf2, and Suf3. Cap is a plaque region closest to the lumen that is within 2 pixels from the borders of lumen. Suf1 and Suf2 are superficial plaque regions that are within 2 to 10 pixels and within 10 to 20 pixels from the borders of lumen. Suf3 is a rest of plaque region excluding Cap, Suf1, and Suf2. In conclusion, a single IVUS image is segmented into 6 different regions after the ROI extraction. Figure 3 describes the initial and additional ROI extractions.

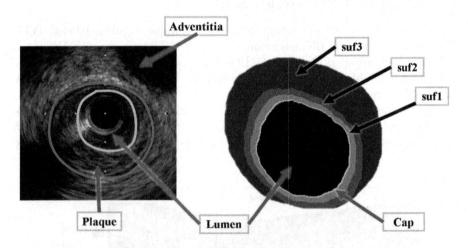

Fig. 3. Initial and additional ROI extraction

2.3   Feature Extraction

TCFA classification requires input features from each IVUS image to learn the differences between TCFA and non-TCFA. Therefore, we extracted 105 different features from every IVUS image (F1–F105). In this research, we intended

Table 1. Extracted feature table

Feature	Description	Region
F1	Plaque/(Plaque + Lumen)	-
F2–F27	Pixels (0–10), (11–20), ..., (251–255)	Cap
F28–F53	Pixels (0–10), (11–20), ..., (251–255)	Suf1
F54–F79	Pixels (0–10), (11–20), ..., (251–255)	Suf2
F80–F105	Pixels (0–10), (11–20), ..., (251–255)	Suf3

to give general information as input features to the classifier since the representative features of TCFA such as lipid-rich plaque, less than $65\,\mu m$ fibrous cap, and necrotic core are difficult to represent as particular values from the IVUS image. Consequently, from the second feature (F2) to the last feature (F105), we extracted proportion of every 10 pixels starting from 0 to 255 in Cap (F2–F27), Suf1 (F28–F53), Suf2 (F54–F79), and Suf3 (F80–F105). The first feature (F1) is a ratio of plaque in the entire vessel. Table 1 presents the summarized descriptions of extracted features from F1 to F105.

2.4 Feature Selection

In general, high-dimensional features may lead classifier fell into the curse of dimensionality. Especially in case of using neural networks for the pattern recognition, it is known that adding irrelevant features beyond a certain point may lead to reduction in the classification performance [13]. Therefore, we processed Fishers exact test to select the most correlated n-ranked features from the 105 features. Selected n-ranked features are used as input features in deep FFNN classifier. We also summarized top 10 ranked features in Sect. 3.

2.5 TCFA/Non-TCFA Classification

Classification between TCFA and non-TCFA is performed using deep FFNN. Deep FFNN is the standard form of an artificial neural network which has multiple hidden layers. Our deep FFNN has 4 hidden layers that each layer has 100, 200, 80 and 40 neurons. Figure 4 shows overall procedures of our deep FFNN classifier. We used rectified linear units (ReLU) as an activation function for the hidden layers, and Adam optimization function for the stochastic gradient-based optimizer [14]. Max-min normalization is processed for the every input data, and the initial learning rate and L2 penalty parameter are set to 0.001 and 0.0001. Learning rate is gradually decreased after the given training step. In addition, to prevent an over-fitting problem in training phase, we processed 10 fold cross validation.

The flow of the classification is followed. The top-ranked feature from feature selection phase is used for the first training. With this single feature, deep FFNN optimizes their neurons to reduce the training cost with the training set. After the

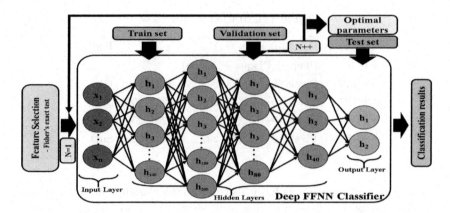

Fig. 4. Overall procedures of proposed deep FFNN classifier

network is optimized, validation set is used to validate the model's performance with AUC. The first average AUC is obtained after 10 fold cross validation. Before increasing the number of ranked features, parameters such as top-ranked feature, the first average AUC, and the optimized deep FFNN model are stored into optimized parameters. The number of ranked features is keep increased until it reaches 105, while optimized parameters are updated only when the average AUC is larger than the previous one. When the training phase is finished, the test set is fit into the optimized model to measure the performance of the classifier.

3 Experiments and Results

3.1 Experimental Setup

For the entry into the study, 100 patients (with 100 native coronary arteries) who required clinically indicated coronary angiography or percutaneous coronary intervention were evaluated. All patients provided written informed consent and the institutional review board of Asan Medical Center approved the study.

OCT images were acquired using a non-occlusive technique with the C7XR? system and DragonFly? catheters (LightLab Imaging, Inc.) at 20 mm/s of pull-back speed. Images with significant signal attenuation that precluded satisfactory evaluation of plaque morphology were excluded from analysis. Grayscale IVUS imaging was performed using motorized transducer pullback (0.5 mm/s) and a commercial scanner (Boston Scientific/SCIMED, Minneapolis, Minnesota), consisting of a rotating, 40-MHz transducer within a 3.2-F imaging sheath. Target segment included the sections with larger than 0.5 mm of maximal plaque thickness.

Within 12,325 IVUS images, 20% of total images were randomly chosen for the test set, and the rest of the data was selected as the training set while preserving similar proportion of TCFA and non-TCFA between two sets. Deep FFNN classifier was implemented using GPU version TensorFlow [15] with NVIDIA

Table 2. AUC interpretation guidelines

AUC	Guidelines
0.5–0.6	No discrimination
0.6–0.7	Pool discrimination
0.7–0.8	Acceptable discrimination
0.8–0.9	Good discrimination
0.9–1	Excellent discrimination

Titan X GPU (3,584 CUDA cores, 12 GB GPU memory). The other parts of classification are implemented using scikit-learn Python module [16].

The evaluation of the TCFA detection considered following four metrics: Area Under the Curve (AUC), Specificity, and Sensitivity. AUC is an area under the Receiver operating Characteristic (ROC) curve. In case of interpreting the AUC result, Hosmer and Lemeshow suggested the general guideline rules in [17]. Table 2 presents the above guideline rules. Specificity is the ratio of negative test results that are correctly classified as non-TCFA. Sensitivity is the probability of positive test results that are correctly identified as TCFA. These two metrics are defined with following four measurements.

- True Positive: Correctly detected as TCFA
- True Negative: Correctly detected as non-TCFA
- False Positive: Incorrectly detected as TCFA
- False Negative: Incorrectly detected as non-TCFA

$$Specificity(\%) = \frac{TN}{FP + TN} \times 100 \tag{1}$$

$$Sensitivity(\%) = \frac{TP}{TP + FN} \times 100 \tag{2}$$

3.2   TCFA Detection Evaluation

To evaluate the TCFA classification, we used the confusion matrix, AUC graph with a different number of ranked features, and the receiver operating characteristic (ROC) curve. Figure 5 describes the AUC graph and Fig. 6 represents the ROC curve. From the Fig. 5 the highest AUC, which is a 0.854, is obtained when top 81 ranked features are used for the validation. With these 81 features and stored parameters from the deep FFNN, AUC for the test set was 0.868. From the Table 2, we can conclude that our TCFA classification result is in "good discrimination" state. Table 3 shows confusion matrix of the classification results when the sum of specificity and sensitivity is maximum. As a result, our TCFA detection achieved 78.31% specificity and 79.02% sensitivity.

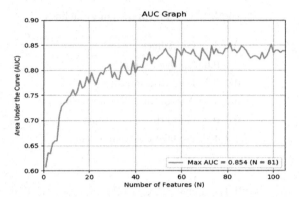

Fig. 5. AUC graph from the validation set

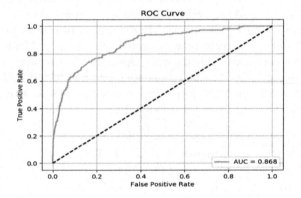

Fig. 6. ROC curve from the test set

Table 3. Confusion matrix (total 2465)

	TCFA	Non-TCFA
Positive	162	490
Negative	43	1770

3.3   Feature Ranking Evaluation

After processing Fishers exact test on the training set, we obtained score list of
105 features based on p-value. Table 4 summarizes the top 10 ranked features
correlated to the TCFA detection. When putting the proportions of F2, F3, and
F4 together, we can get total 25.53%. This implies that dark pixels (0 to 30),
which may be necrotic core, near the lumen (Cap) is the most crucial feature
for classifying TCFA from non-TCFA with machine learning technique. The
second rank feature is F1 which is ratio of the plaque in the vessel. This seems
quite obvious since vessel with TCFA will have more plaque burden than non-
TCFA one in general. From the rank 4 to rank 7 features (F40–43), bright pixels

Table 4. Top 10 ranked features

Rank	Feature	Ratio (%)	Pixel range	Region
1	F2	18.64	0–10	Cap
2	F1	17.52	-	-
3	F3	4.75	11–20	Cap
4	F41	3.69	151–160	Suf1
5	F42	3.48	161–170	Suf1
6	F40	2.62	141–150	Suf1
7	F43	2.17	171–180	Suf1
8	F4	2.14	21–30	Cap
9	F69	1.69	111–120	Suf2
10	F15	1.61	151–160	Cap
etc.	-	41.69	-	-

(141 to 180) in superficial plaque region (Suf1) also affects the TCFA detection in some degree. Although the top 10 ranked features have significant impacts on the TCFA detection, the rest of features enrolls more than 40% of the classification.

4 Conclusion

In this paper, we proposed the first computer-aided TCFA classification with IVUS images labeled by equivalent OCT images. Our TCFA detection includes IVUS and OCT co-registration, ROI extraction, feature extraction, feature selection, and deep FFNN based classifier. As result, we achieved 0.87 AUC with 78.31% specificity and 79.02% sensitivity when top 81 features are used for the training. Also, we investigated several important features when detecting TCFA from IVUS images. We believe our research indicates a new possibility for classifying TCFA with IVUS images using machine learning technique. In future, we plan to apply other well-known machine learning algorithms such as k-nearest neighbor, SVM, and random forest tree.

Acknowledgments. Support of Asan Medical Center providing IVUS images and clinical advices for this research are gratefully acknowledged and this research was supported by International Research & Development Program of the National Research Foundation of Korea (NRF) funded by the Ministry of Science, ICT&Future Planning of Korea (2016K1A3A7A03952054).

References

1. Virmani, R., Burke, A.P., Farb, A., et al.: Pathology of the vulnerable plaque. J. Am. Coll. Cardiol. **47**(8 Suppl.), C13–C18 (2006)

2. Kolodgie, F.D., Burke, A.P., Farb, A., et al.: The thin-cap fibroatheroma: a type of vulnerable plaque: the major precursor lesion to acute coronary syndromes. Curr. Opin. Cardiol. **16**(5), 285–292 (2001)
3. Nissen, S.E., Yock, P.: Intravascular ultrasound novel pathophysiological insights and current clinical applications. Circulation **103**, 604–616 (2001)
4. Jang, I.K., Tearney, G.J., MacNeill, B., et al.: In vivo characterization of coronary atherosclerotic plaque by use of optical coherence tomography. Circulation **111**(12), 1551–1555 (2005)
5. Fisher, R.A.: Statistical Methods for Research Workers. Genesis Publishing Pvt. Ltd., Delhi (1925)
6. Glorot, X., Bengio, Y.: Understanding the difficulty of training deep feedforward neural networks. In: AISTATS, vol. 9, pp. 249–256 (2010)
7. Zhang, L., Wahle, A., Chen, Z., Lopez, J., Kovarnik, T., Sonka, M.: Prospective prediction of thin-cap fibroatheromas from baseline virtual histology intravascular ultrasound data. In: Navab, N., Hornegger, J., Wells, W.M., Frangi, A.F. (eds.) MICCAI 2015. LNCS, vol. 9350, pp. 603–610. Springer, Cham (2015). doi:10.1007/978-3-319-24571-3_72
8. Rodriguez-Granillo, G.A., García-García, H.M., McFadden, E.P., et al.: In vivo intravascular ultrasound-derived thin-cap fibroatheroma detection using ultrasound radiofrequency data analysis. J. Am. Coll. Cardiol. **46**(11), 2038–2042 (2005)
9. Sawada, T., Shite, J., Garcia-Garcia, H.M., et al.: Feasibility of combined use of intravascular ultrasound radiofrequency data analysis and optical coherence tomography for detecting thin-cap fibroatheroma. Eur. Heart J. **29**(9), 1136–1146 (2008)
10. Calvert, P.A., Obaid, D.R., O'Sullivan, M., et al.: Association between IVUS findings and adverse outcomes in patients with coronary artery disease. JACC: Cardiovasc. Imaging **4**(8), 894–901 (2011)
11. Garcia-Garcia, H.M., Costa, M.A., Serruys, P.W.: Imaging of coronary atherosclerosis: intravascular ultrasound. Eur. Heart J. **31**(20), 2456–2469 (2010)
12. Wolf, I., Vetter, M., Wegner, I., et al.: The medical imaging interaction toolkit. Med. Image Anal. **9**(6), 594–604 (2005)
13. Bishop, C.M.: Neural Networks for Pattern Recognition. Oxford University Press, Oxford (1995)
14. Kingma, D., Ba, J.: Adam: a method for stochastic optimization. arXiv preprint arXiv:1412.6980 (2014)
15. Abadi, M., Agarwal, A., Barham, P., et al.: Tensorflow: large-scale machine learning on heterogeneous distributed systems. arXiv preprint arXiv:1603.04467 (2016)
16. Pedregosa, F., Varoquaux, G., Gramfort, A., et al.: Scikit-learn: machine learning in Python. J. Mach. Learn. Res. **12**(Oct), 2825–2830 (2011)
17. Hosmer Jr., D.W., Lemeshow, S., Sturdivant, R.X.: Applied Logistic Regression, vol. 398. Wiley, Hoboken (2013)

Generalization of Local Temporal Correlation Common Spatial Patterns Using Lp-norm (0 < p < 2)

Na Fang and Haixian Wang[✉]

Key Laboratory of Child Development and Learning Science of Ministry
of Education, School of Biological Science and Medical Engineering,
Southeast University, Nanjing 210096, Jiangsu, People's Republic of China
hxwang@seu.edu.cn

Abstract. As one of the effective feature extraction methods, common spatial patterns (CSP) is widely used for classification of multichannel electroencephalogram (EEG) signals in the motor imagery-based brain-compute interface (BCI) system. The formulation of the conventional CSP based on L2-norm, however, implies that it is sensitive to the presence of outliers. Local temporal correlation common spatial patterns (LTCCSP), as an extension of CSP by introducing the local temporal correlation information into the covariance modelling of the classical CSP algorithm, extracts more discriminative features. In order to further improve the robustness of the classification, in this paper, we generalize the LTCCSP algorithm by replacing the L2-norm with Lp-norm (0 < p < 2) in the objective function, called LTCCSP-Lp. An iterative algorithm is designed under the framework of minorization-maximization (MM) optimization algorithm to obtain the optimal spatial filters of LTCCSP-Lp. The iterative solution is justified in theory and the effectiveness of our novel proposed method is verified by experimental results on a toy example and datasets of BCI competitions.

Keywords: Brain-computer interfaces (BCI) · Common spatial patterns (CSP) · Local temporal correlation common spatial patterns (LTCCSP) · LTCCSP-Lp · Lp-norm · Minorization-maximization (MM)

1 Introduction

As a communication system between brain and outside electronic devices without using peripheral nerves and muscle tissues, brain-computer interfaces (BCI) provides a promising pathway of information transfer for human with external world. Recently, electroencephalogram (EEG)-based BCI system is extensively studied and used with several advantages such as simplicity, non-invasiveness and high temporal resolution. However, the EEG signal recordings are often contaminated due to the phenomenon of volume conduction and outliers such as muscle artifacts, ocular artifacts, etc. Therefore, how to extract efficient features becomes the key to EEG-based BCI research.

Among the various machine learning techniques, common spatial patterns (CSP) is a popular approach applying to analyzing multichannel data recordings from two

© Springer International Publishing AG 2017
D. Liu et al. (Eds.): ICONIP 2017, Part IV, LNCS 10637, pp. 769–777, 2017.
https://doi.org/10.1007/978-3-319-70093-9_82

classes. However, its L2-norm-based formulation renders it vulnerable to the presence of outliers. To address this issue, many extensions of CSP have been developed in literature, such as the regularization of CSP [1], local temporal common spatial patterns (LTCSP) [2], local temporal correlation common spatial patterns (LTCCSP) [3], and L1-norm-based CSP (CSP-L1) [4–6].

This paper proposes a generalized LTCCSP algorithm by using Lp-norm rather than L2-norm in the objective function. The proposed approach, named LTCCSP-Lp, is motivated by the basic idea of Lp-norm-based modeling developed recently in the field of subspace leaning [7]. Note that, we limit our attention to the case of $0 < p < 2$ because when $p > 2$ the utilization of large norm would further magnify the effect of outliers and thus makes the obtained filters meaningless.

The remainder of this paper is organized as follows. In Sect. 2, the conventional CSP and LTCCSP methods are briefly reviewed. In Sect. 3, we propose the LTCCSP-Lp method and the iterative solution. The experimental results are reported in Sect. 4. And finally Sect. 5 concludes the paper.

2   Brief Review of CSP and LTCCSP

2.1   Conventional CSP

For the EEG analysis, let $X(i), \ldots, X(t_x) \in \mathbb{R}^{C \times N}$ be the trial segments of EEG signals for one class of imaginary movement and $Y(i), \ldots, Y(t_y) \in \mathbb{R}^{C \times N}$ be the others, where C denotes the number of channels, N is the number of EEG samples per trial, and t_x, t_y denote the different numbers of trials for the two classes respectively.

The CSP algorithm aims to obtain a set of spatial filters $\omega_1, \ldots, \omega_r \in \mathbb{R}^C$, where r is the number of filters, such that these filters can maximize the variance of filtered signals for one class while minimizing it for the other class. Mathematically, the spatial filter of CSP can be formulated by maximizing (or minimizing) the objective function [8]

$$J_{CSP}(\omega) = \frac{\omega^T C^x \omega}{\omega^T C^y \omega} \tag{1}$$

where C^x, C^y are the estimates of the average covariance matrices of two classes.

For the spatial filters of CSP, we can solve it simply by the generalized eigenvalue problem. Technically, the common practice for classification is to select a few eigenvectors from both ends of eigenvalue spectrum as the set of spatial filters. The variances after a log-transformation of the filtered EEG data are used as features for classification.

2.2   LTCCSP

As an extension of CSP, LTCCSP considers the local temporal correlation when computes the covariance matrices for the two classes [2, 3]. Technically, the objective function of LTCCSP can formulate simply as the same form as (1)

$$J_{LTCCSP}(\omega) = \frac{\omega^T C_L^x \omega}{\omega^T C_L^y \omega} \tag{2}$$

where C_L^x, C_L^y are the estimates of the average local temporally correlation covariance

matrices as $C_L^x = \frac{1}{t_x} \sum_{i=1}^{t_x} \frac{1}{2N} \sum_{l=1}^{N} \sum_{m=1}^{N} x_{lm}(i) x_{lm}^T(i) W_{lm}^x(i)$, $C_L^y = \frac{1}{t_y} \sum_{i=1}^{t_y} \frac{1}{2N} \sum_{l=1}^{N} \sum_{m=1}^{N} y_{lm}(i)$

$y_{lm}^T(i) W_{lm}^y(i)$ and $x_{lm}(i) \in \mathbb{R}^C$, $y_{lm}(i) \in \mathbb{R}^C$ denote the differences between pairs of data points per trial of two classes respectively, i.e., $x_{lm}(i) = x_l(i) - x_m(i)$, $y_{lm}(i) = y_l(i) - y_m(i)$. The weight W_{lm}^x is defined by the correlation coefficient as

$$W_{lm}^x = \begin{cases} \exp(corr(x_l, x_m)), & |l - m| < \tau \\ 0, & otherwise \end{cases} \tag{3}$$

where τ is the local temporal range. The weight W_{lm}^y is the similar formulation as the W_{lm}^x. The filters of LTCCSP can be computed as straightforward as CSP with some algebraic operations.

3 Lp-norm-Based LTCCSP

The LTCCSP algorithm models are based on L2-norm. To see this, we rewrite (2) as

$$J_{LTCCSP}(\omega) = \frac{\omega^T C_L^x \omega}{\omega^T C_L^y \omega} = \frac{t_y \sum_{i=1}^{t_x} \sum_{l=1}^{N} \sum_{m=1}^{N} (\omega^T x_{lm}(i))^2 W_{lm}^x(i)}{t_x \sum_{i=1}^{t_y} \sum_{l=1}^{N} \sum_{m=1}^{N} (\omega^T y_{lm}(i))^2 W_{lm}^y(i)} \tag{4}$$

As is well-known that L2-norm is sensitive to outliers because the magnified influence of large deviations by the square operator. Motivated by the basic idea of Lp-norm-based modeling, we propose to replace the L2-norm in the objective function of LTCCSP with arbitrary norm $(0 < p < 2)$ as

$$J_{Lp}(\omega) = \frac{\sum_{i=1}^{t_x} \sum_{l=1}^{N} \sum_{m=1}^{N} |\omega^T x_{lm}(i)|^P W_{lm}^x(i)}{\sum_{i=1}^{t_y} \sum_{l=1}^{N} \sum_{m=1}^{N} |\omega^T y_{lm}(i)|^P W_{lm}^y(i)} = \frac{\sum_{i=1}^{m} |\omega^T x_w(i)|^P}{\sum_{j=1}^{n} |\omega^T y_w(j)|^P} = \frac{\|\omega^T X_w\|_p^p}{\|\omega^T Y_w\|_p^p} \tag{5}$$

where X_w is a new weighted data matrix of one class composed of the weighted signal elements $X_w = (x_w(1), \ldots, x_w(m)) \in \mathbb{R}^{C \times m}$, and each elements is formulated as $W_{lm}^x(i)^{1/P} x_{lm}(i)$. Note that, we reject the elements when $|l - m| >= \tau$ and $l = m$ (i.e., elements would be zeros with nonsense in these points), so m depended on the value of τ, N, and t_x. And the weighted data matrix $Y_w = (y_w(1), \ldots, y_w(n)) \in \mathbb{R}^{C \times n}$ does likewise.

We refer to (5) as LTCCSP-Lp. And the existence of the absolute value operation in the objective function makes the optimization difficult. We propose an elegant iterative algorithm under the framework minorization-maximization (MM) to obtain the optimal filter that maximizes the objective function (5) in Table 1. The minimization of (5) can be likewise solved by exchanging the positions of EEG data signals of two classes. And we can simply extend our iteration to extract multiple spatial filters by applying the same procedure greedily to the updated samples (see [4] for more details).

It is worth noting that if let $\tau = N$ and all of the weights take 1, then LTCCSP-Lp will turn out to be Lp-norm-based CSP in the sense that CSP-Lp can be derived from LTCCSP-Lp under a special case. Obviously, the iterated solution we present in Table 1 can be applied to CSP-Lp as well.

Table 1. Algorithm procedure of LTCCSP-Lp.

Input: the EEG data of two classes as $(X(i),...,X(t_x))$ and $(Y(i),...,Y(t_y))$, the value of p (0<p<2), local temporal range τ ,learning rate parameter η .

Output: optimal spatial filter ω^*

Before the iteration, we transform the EEG data into the weighted data matrix (i.e., X_w, Y_w). Then we use t to denote iteration number, and assume that $\omega \neq 0$.

1) Set t=0. Initialize $\omega(t)$ by any a c-dimensional vector, and rescale it to unit length.

2) Define two polarity function $p_i(t)$ and $q_j(t)$ to implement the absolute value operation in (5): $p_i(t) = \text{sgn}(\omega^T(t)x_w(i))$ and $q_j(t) = \text{sgn}(\omega^T(t)y_w(j))$.

3) Let

$$d(t) = \frac{\sum_{i=1}^{m}|\omega^T(t)x_w(i)|^{p-1} \cdot p_i(t)x_w(i))}{\sum_{i=1}^{m}|\omega^T(t)x_w(i)|^{p}} - \frac{\sum_{j=1}^{n}|\omega^T(t)y_w(j)|^{p-1} \cdot q_j(t)y_w(j))}{\sum_{j=1}^{n}|\omega^T(t)y_w(j)|^{p}}$$

4) Update $\omega(t)$ by $\omega(t+1) = \omega(t) + \eta d(t)$ and rescale the filter to unit length

5) Set $t \leftarrow t+1$.

6) Stop the iteration and set $\omega^* = \omega(t)$ if $J_{Lp}(\omega(t))$ cannot increase significantly. Otherwise, go to Step 2.

3.1   Algorithm Validation

For the iterative algorithm we proposed above, we show the validation in this section. It is well known that the strategies to constructing MM algorithm are the various inequalities derived from the first-order convexity condition [9]. Firstly, before the validation, we introduce a lemma that would be utilized.

Lemma: Let $w = (w_1, w_2, \ldots, w_d) \in \mathbb{R}^d$, $v = (v_1, v_2, \ldots, v_d) \in \mathbb{R}^d$, $v \neq 0$, let $0 < p < 2$, then [7]

$$\sum_{i=1}^{d} |w_i|^p \leq \frac{p}{2} \sum_{i=1}^{d} w_i^2 |v_i|^{p-2} + (1 - \frac{p}{2}) \sum_{i=1}^{d} |v_i|^p \tag{6}$$

holds wherein the inequality becomes equality when $w = v$.

Considering the objective function in (5) at iteration t, we have that

$$
J_{Lp}(\omega(t)) = \frac{\sum_{i=1}^{m} |\omega^T(t) x_w(i)|^P}{\sum_{j=1}^{n} |\omega^T(t) y_w(j)|^P} = \frac{\sum_{i=1}^{m} (p_i(t)\omega^T(t) x_w(i))^p}{\frac{p}{2} \sum_{j=1}^{n} \frac{\omega^T(t) y_w(j))^2}{|\omega^T(t) y_w(j)|^{2-p}} + (1 - \frac{p}{2}) \sum_{j=1}^{n} |\omega^T(t) y_w(j)|^P}
$$

$$
= \frac{\sum_{i=1}^{m} (\omega^T(t) u_i(t))^p}{\frac{p}{2}\omega^T(t) V(t)\omega(t) + (1 - \frac{p}{2})\|z(t)\|_p^p} \tag{7}
$$

where $u_i(t) = p_i(t) \cdot x_w(i)$, $z_j(t) = \omega^T(t) y_w(j)$, $V(t) = \sum_{j=1}^{n} (y_w(j) \cdot y_w^T(j) / |z_j(t)|^{2-p})$ and $z(t)$ is the vector with the entries $\{z_j(t)\}_{j=1,\ldots,n}$.

Now we introduce a surrogate function

$$
Q(\omega(t+1)|\omega(t)) = \frac{\sum_{i=1}^{m} (\omega^T(t+1) u_i(t))^p}{\frac{p}{2}\omega^T(t+1) V(t)\omega(t+1) + (1 - \frac{p}{2})\|z(t)\|_p^p} \tag{8}
$$

We commence with considering the numerator of (8)

$$
0 \leq \sum_{i=1}^{m} (\omega^T(t+1) u_i(t))^p = \sum_{i=1}^{m} (p_i(t)\omega^T(t+1) x_w(i))^p
$$
$$
\leq \sum_{i=1}^{m} (p_i(t+1)\omega^T(t+1) x_w(i))^p = \sum_{i=1}^{m} |\omega^T(t+1) x_w(i)|^P \tag{9}
$$

When $0 < p < 2$, we proceed to consider the denominator of (8) according to Lemma

$$
\frac{p}{2}\omega^T(t+1) V(t)\omega(t+1) + (1 - \frac{p}{2})\|z(t)\|_p^p
$$
$$
= \frac{p}{2} \sum_{j=1}^{n} \frac{\omega^T(t+1) y_w(j))^2}{|\omega^T(t) y_w(j)|^{2-p}} + (1 - \frac{p}{2}) \sum_{j=1}^{n} |\omega^T(t) y_w(j)|^P \geq \sum_{j=1}^{n} |\omega^T(t+1) y_w(j)|^P
$$
$$\tag{10}$$

Combining (9) and (10), we have that $Q(\omega(t+1)|\omega(t)) \leq J_{Lp}(\omega(t+1))$ with the equality holding at $\omega(t+1) = \omega(t)$, satisfying the two key conditions of the MM framework. Therefore, $Q(\omega(t+1)|\omega(t))$ is a feasible surrogate function. And according to the fact, $d(t)$ is the vector that points to the ascending direction of the surrogate function that formulated as the gradient of $Q(\omega(t+1)|\omega(t))$ with respect to $\omega(t+1)$ at the point $\omega(t)$, we have $Q(\omega(t)|\omega(t)) \leq Q(\omega(t+1)|\omega(t))$. So $Q(\omega(t)|\omega(t)) = J_{Lp}(\omega(t)) \leq J_{Lp}(\omega(t+1))$. Above all, we prove that $J_{Lp}(\omega(t))$ is non-decreasing with respect to t, that is to say the iterative algorithm we proposed guarantees the local maximum solution of (5).

3.2 Feature Extraction

By applying the iterative algorithm and greedy solution, we obtain multiple orthonormal spatial filters $\omega_1, \ldots, \omega_r$. For any EEG single trial Z, the feature is extracted as

$$f = (\|\omega_1^T Z_w\|_p^p, \|\omega_2^T Z_w\|_p^p, \ldots, \|\omega_r^T Z_w\|_p^p)^T \tag{11}$$

where Z_w is the weighted data matrix as the same definition as the X_w, f is a r-dimensional feature vector.

4 Experiment

In the experiments, we used two data sets to compare the robustness of the proposed LTCCSP-Lp ($p = 0.5, 1, 1.5$) approach and the conventional CSP methods. The first one was a toy dataset. The second data set was from a publicly available real EEG dataset: dataset IVa of BCI competition III. Note that, we set the initial filtering vectors of LTCCSP-Lp as the solution of CSP in our experiments for more stable results.

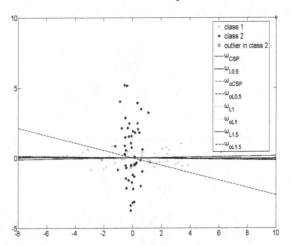

Fig. 1. A toy example of spatial filtering by using CSP and LTCCSP ($p = 0.5, 1, 1.5$) on a 2-D dataset. Data points and first spatial filter obtained by CSP and LTCCSP ($p = 0.5, 1, 1.5$) in the cases with and without outliers were shown.

4.1 Toy Example

A 2-D artificial dataset was randomly generated from two Gaussian classes with 50 data points per class and an additional outlier introduced in as depicted in Fig. 1. For illustration, we applied CSP and LTCCSP-Lp ($p = 0.5, 1, 1.5$) to extract the first spatial filter that maximized the dispersion of class 1 and minimized the dispersion of class 2 in the cases with and without outlier. In using the LTCCSP-Lp, the parameters set as $\tau = 10$, $\eta = 0.01$. And the directions of the filters were shown as the lines in Fig. 1. We could easily observe that, compared with CSP, LTCCSP-Lp alleviated the impact of outlier and with strong robustness.

4.2 Real EEG Dataset

The recorded EEGs were the tasks of right hand vs right foot movement. This dataset contained 280 trials each subject using 118 electrodes. Specifically, 168, 224, 84, 56 and 28 trials were used as training data corresponding to the five subjects: *aa, al, av, aw* and *ay*, respectively. We chose three subjects, *aa, al, aw*, which performed better than the other two subjects of the dataset in previous literatures, for the further analysis.

Preprocessing. The EEG data were band-pass filtered between 8 Hz and 35 Hz by using a fifth order Butterworth filter. And we used the EEG segments recorded from 0.5 s to 3.75 s after the visual cue.

Introducing outliers. In order to simulate the contaminative actual EEG recording, we introduced outliers into the training data as in [4]. The probability of occurrence of the outliers was variable and the time positions we selected to add outliers was randomly distributed in all samples of the training data.

Experiment settings and results. In using the LTCCSP-Lp, η was varied from 1e−6 to 9e−4, and the updated filter with the highest value for the objective function on the train data was adopted. The other parameter settings were shown in Table 2.

The linear discriminant analysis (LDA) classifier was used for classification. Our experiments were repeated 10 times each occurrence frequency of outliers and the average accuracy were recorded in Table 2.

Discussion. From Table 2, it can be observed that LTCCSP-Lp significantly outperformed CSP in almost all experimental trials whether $p = 0.5, 1$ or 1.5. For subject *aa*, CSP achieved a plenty of accuracies less than 50% with various occurrence frequencies of outliers added, which indicated BCI illiteracy phenomenon [10]. This phenomenon may appear because of the large number of outliers, up to half of samples, we introduced in the original noisy dataset. Even so, the proposed methods achieved the superior performance comparatively on the real EEG dataset, especially with outliers.

Although the classification performance was improved for each LTCCSP-Lp, there was slightly different for the various value of p, that is the optimal p-value was determined by the data distribution for each subject. Besides, the different settings of parameters could more or less affect the performance of LTCCSP-Lp, the selections of parameters must be critical to the optimum performance of our methods.

Table 2. Average classification accuracies (%) of the Dataset IVa of BCI competition III for CSP and LTCCSP ($p = 0.5, 1, 1.5$) with increasing occurrence frequencies of outliers.

Sub.	r	Method	Occurrence frequencies of outliers										
		τ	0	0.05	0.1	0.15	0.2	0.25	0.3	0.35	0.4	0.45	0.5
aa	1	CSP	65.2	54.2	48.7	46.3	46.5	46.5	46.6	46.3	46.7	46.3	46.3
	3	L0.5	**72.3**	**61.6**	**63.9**	**61.8**	**60.4**	**57.8**	**57.6**	**54.7**	**51.5**	**49.6**	**48.6**
		L1	66.1	60	60.2	56.0	53.5	54.5	54.0	49.8	49.3	48.1	48.3
		L1.5	65.2	60	56.1	50.1	52.6	50.4	52.0	47.4	47.6	47.5	47.9
	2	CSP	71.4	56.8	50.9	46.7	46.5	46.3	46.5	46.1	46.7	46.4	45.9
	3	L0.5	**72.3**	**65**	**65.2**	**61.4**	**59.5**	**56.6**	**55.4**	**51.5**	**50.5**	**48.8**	**48.7**
		L1	69.6	64.5	59.9	53.8	50.9	51.4	51.3	48.9	48.7	47.5	47.4
		L1.5	71.4	63.1	54.6	50.4	50.2	50.1	49.7	47.1	47.7	46.9	47.1
	3	CSP	69.6	57.0	50.9	46.9	46.6	46.5	46.3	46.3	46.7	46.4	45.8
	3	L0.5	**70.5**	**67.9**	**65.8**	**61.3**	**61.0**	**56.1**	**55.1**	**52.1**	**51.6**	**48.6**	**48.8**
		L1	69.6	66.6	59.6	53.6	53.1	51.6	51.1	50	48.4	47.1	48.4
		L1.5	69.6	65.5	56.3	48.7	50.9	50.8	49.6	48.0	47.8	47.0	48.0
al	1	CSP	91.1	88.2	92.5	77.5	65	61.8	58.9	57.5	54.6	55	54.6
	3	L0.5	**98.2**	**100**	99.3	88.2	79.3	72.1	71.4	68.6	66.4	65.4	62.1
		L1	92.9	98.6	98.9	92.1	82.1	74.0	68.6	69.3	67.9	63.9	61.4
		L1.5	91.1	98.6	**99.6**	**98.2**	**89.6**	**79.3**	**76.8**	**73.6**	**74.0**	**67.9**	**65**
		CSP	91.1	92.7	94.0	80.4	63.6	60.7	60.4	55.7	55	54.3	53.2
	10	L0.5	**92.9**	**98.2**	93.2	78.2	67.5	60	58.2	55.4	55.4	52.9	53.6
		L1	91.1	97.5	**97.5**	83.6	72.5	66.8	64.6	60.7	58.6	57.1	57.1
		L1.5	91.1	97.9	97.1	**92.1**	**81.4**	**72.9**	**69.6**	**66.1**	**62.9**	**59.3**	**59.3**
aw	1	CSP	89.3	74.2	58.8	51.9	57.6	54.2	56.6	58.0	57.5	51.6	56.1
	3	L0.5	89.7	79.5	75.5	**75.1**	75.1	71.6	**74.3**	72.9	**73.8**	71.5	74.8
		L1	91.1	79.3	74.4	73.1	76.7	**73.8**	72.1	**73.5**	69.5	**71.9**	**76.1**
		L1.5	**92.0**	**80.1**	**78.4**	74.0	**77.6**	71.1	69.8	73.1	68.0	69.8	73.4

5   Conclusion

In this paper, we proposed the generalized Lp-norm ($0 < p < 2$) based LTCCSP, called LTCCSP-Lp, for more robust performance. We also designed an iterative algorithm under the framework of MM optimization algorithm for optimizing spatial filters of LTCCSP-Lp. The experimental results confirmed the effectiveness of the newly proposed approach. And further research would be conducted into how to select the parameters analytically.

Acknowledgments. This work was supported in part by the National Basic Research Program of China under Grant 2015CB351704, the Key Research and Development Plan (Industry Foresight and Common Key Technology) - Key Project of Jiangsu Province under Grant BE2017007-3, and the National Natural Science Foundation of China under Grants 61773114 and 61375118.

References

1. Lotte, F., Guan, C.: Regularizing common spatial patterns to improve BCI designs: unified theory and new algorithms. IEEE Trans. Biomed. Eng. **58**(2), 355–362 (2011)
2. Wang, H., Zheng, W.: Local temporal common spatial patterns for robust single-trial EEG classification. IEEE Trans. Neural Syst. Rehabil. Eng. **16**(2), 131–139 (2008)
3. Zhang, R., et al.: Local temporal correlation common spatial patterns for single trial EEG classification during motor imagery. Comput. Math. Methods Med. **2013**, 7 p. (2013). Article ID 591216
4. Wang, H., Tang, Q., Zheng, W.: L1-norm-based common spatial patterns. IEEE Trans. Neural Syst. Rehabil. Eng. **59**(3), 653–662 (2012)
5. Wang, H., Li, X.: Regularized filters for L1-Norm-Based common spatial patterns. IEEE Trans. Neural Syst. Rehabil. Eng. **24**(2), 201–211 (2016)
6. Li, X., Lu, X., Wang, H.: Robust common spatial patterns with sparsity. Biomed. Signal Process. Control **26**, 52–57 (2016). Elsevier
7. Wang, J.: Generalized 2-D principal component analysis by Lp-norm for image analysis. IEEE Trans. Cybern. **46**(3), 792–803 (2015)
8. Blankertz, B., Tomioka, R., Lemm, S., et al.: Optimizing spatial filters for robust EEG single-trial analysis. IEEE Signal Process. Mag. **25**(1), 41–56 (2008)
9. Hunter, D., Lange, K.: A tutorial on MM algorithms. Am. Stat. **58**(1), 30–37 (2004)
10. Vidaurre, C., Blankertz, B.: Towards a cure for BCI illiteracy. Brain Topogr. **23**(2), 194–198 (2010)

fNIRS Approach to Pain Assessment
for Non-verbal Patients

Raul Fernandez Rojas[1]([✉]), Xu Huang[1], Julio Romero[1],
and Keng-Liang Ou[2,3,4]

[1] Human-Centred Technology Research Centre, ESTEM Faculty,
University of Canberra, ACT, Canberra, Australia
raul.fernandezrojas@canberra.edu.au
[2] Department of Dentistry, Taipei Medical University Hospital, Taipei, Taiwan
[3] Department of Dentistry, Taipei Medical University-Shuang Ho Hospital,
New Taipei City, Taiwan
[4] 3D Global Biotech Inc., New Taipei City, Taiwan

Abstract. The absence of verbal communication in some patients (e.g.,
critically ill, suffering from advanced dementia) difficults their pain
assessment due to the impossibility to self-report pain. Functional near-
infrared spectroscopy (fNIRS) is a non-invasive technology that has showed promising results in assessing cortical activity in response to painful
stimulation. In this study, we used fNIRS signals to predict the state of
pain in humans using machine learning methods. Eighteen healthy subjects were stimulated using thermal stimuli with a thermode, while their
cortical activity was recorded using fNIRS. Bag-of-words (BoW) model
was used to represent each fNIRS time series. The effect of different step
sizes, window lengths, and codebook sizes was investigated to improve
computational cost and generalization. In addition, we explored the effect
of choosing different features as neurological biomarkers in three different
domains: time, frequency, and time-frequency (wavelet). Classification on
the histogram representation was performed using K-nearest neighbours
(K-NN). The performance is evaluated by using leave-one-out cross validation and with different nearest neighbours. The results showed that
wavelet-based features produced the highest accuracy (88.33%) to distinguish between heat and cold pain while discriminate between low and
high pain. It is possible to use fNIRS to assess pain in response to four
types of thermal pain. However, future research is needed for the assessment of pain in clinical settings.

Keywords: Haemodynamic · Multiclass · Pain · Time series · Neural ·
Brain

1 Introduction

Functional near-infrared spectroscopy (fNIRS) can be used as a non-invasive neuroimaging method that facilitates the measurement of brain activity by reading

© Springer International Publishing AG 2017
D. Liu et al. (Eds.): ICONIP 2017, Part IV, LNCS 10637, pp. 778–787, 2017.
https://doi.org/10.1007/978-3-319-70093-9_83

cerebral haemodynamics and oxygenation. Specifically, it measures changes in chromophore mobilization, oxygenated haemoglobin (HbO) and deoxygenated haemoglobin (HbR) simultaneously. This technique has been widely used in diverse clinical and experimental settings, offering advantages over other technologies (fMRI, EEG, PET) such as, better temporal and spatial resolution, less exposure to ionising radiation, safe to use over long periods and many times, less expensive, easy to use, and small size (portable).

The absence of verbal (or writing) communication in some patients (also referred as non-verbal) is an obstacle to the evaluation of pain. Patients with impaired communication, unconscious patients, infants, critically ill, persons suffering from advanced dementia, or patients with intellectual disabilities are examples of vulnerable individuals who are unable to speak for themselves [2,5]. Due to the inability to communicate pain status, these populations are in risk to be under- or over-treated while in pain due to inadequate pain control. These conditions create a significant obstacle to evaluate and manage patient's pain experience in a suitable and optimal manner. Therefore, the need for a reliable and objective pain assessment that assists medical practitioners in diagnosis of pain and personalise treatment is critical for this vulnerable population.

Some strategies have been proposed to aid to the objective assessment of pain. In clinical settings, the use of vital signs, metabolic markers, and brain imaging tools have been explored [9]. Even though vital signs, such as heart rate, blood pressure, respiratory rate, and arterial oxygen saturation are commonly used by medical practitioners to measure pain in non-verbal patients, these parameters can be unreliable. In conditions where the patient suffers from physiologic problems, stress, under medications or sedatives, changes in vital signals can happen. In these cases, relying on vital signals can affect the interpretation and assessment of pain in patients.

Neuroimaging methods in conjunction wit machine learning (ML) techniques provide a promising method to assess pain by means of brain activity. ML has shown the advantages of discriminative classifiers in neuroimaging for pain assessment. For example, Brown et al. [1] used fMRI data from eight individuals to train a support vector machines (SVM) to distinguish painful from non-painful thermal stimuli with 81% accuracy. In another fMRI study, Wager et al. [11] predicted pain intensity and identified patterns of painful heat from non-painful warmth using a regression technique with a performance of 93%. Gram et al. [4] used EEG to investigate morphine- and placebo-administered subjects after receiving stimulation using the cold pressor test, the authors used SVM to classify responders with an accuracy of 72%. In a fNIRS study, Pourshoghi et al. [8] made use of SVM to classify between pain and no pain from healthy subjects after a cold pressor test with 94% accuracy. However, these studies focused on two conditions, pain or no-pain using a single type of stimulation (cold or heat). Therefore, the need for robust classification models, which can discriminate multiple pain signatures at different intensities, would be more useful for clinical testing.

In this study, our aim is to advance knowledge towards the establishment of a tool for pain assessment in non-verbal patients using neuroimaging and machine learning techniques. We trained a K-NN classifier to discriminate fNIRS signals according to temperature level (cold or hot), and pain intensity (low or high). We built our classifier based on the bag-of-words (BoW) representation [7] used in text classification, parameter tuning was carried out to produce the best results on our pain database. In addition, we investigated the effect of choosing different features as neurological biomarkers in three different domains: time, frequency, and time-frequency (wavelet). The major contributions of this study are: (1) the use of the quantitative sensory testing (QST) to obtain pain information, which can potentially be used on non-verbal patients; (2) for the first time, heat and cold pain are differentiated as well as their corresponding pain intensity (low or high), which can improve pain control by personalizing medicine to individual needs; and (3) investigating different features as potential functional biomarkers to assess human pain using fNIRS signals. We also included future research to improve our classification task and validate our method for pain assessment.

2   Methods

2.1   Experimental Setup

Subjects. A total of eighteen healthy adults participated in the experiments. All participants were right-handed to avoid any variation in functional response due to lateralisation of brain function. Written consent was obtained from all subjects prior to initiation of the experiments. Subjects with a history of a significant medical disorder, a current unstable medical condition or currently taking any medication were excluded.

Stimulation Paradigm. The pain experiment involved applying cold and heat to the skin (back of the hand) to induce pain. Similar to the quantitative sensory testing (QST) [10], subjects are exposed to gradually increasing or decreasing temperatures with a thermode and they press a button when they experience pain (threshold test) and when they experience highest intensity of pain (tolerance test). All thermal tests were performed using a Pathway CHEPS (Medoc, Israel). We define the thermal pain threshold (low pain) as the temperature of the thermode just as it becomes painful, and the thermal pain tolerance (high pain) as the temperature just as it becomes unbearable. Similarly to Rolke et al. [10], the stimulation paradigm was divided in two tests: thermal pain threshold (low pain) and thermal pain tolerance (high pain), with a two-minute rest between both tests. Thermal pain threshold was first obtained by three consecutive measurements of cold stimuli, followed by sixty-second rest and then, three measurements of heat stimuli. The same procedure for thermal pain tolerance is repeated. Figure 1 summarizes the stimulation paradigm.

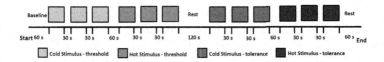

Fig. 1. Stimulation paradigm. Pain threshold (low pain) test was first measured followed by pain tolerance (high pain) test. Cold and heat stimulus were applied on the back of the hand of each subject with a thermode.

fNIRS Setup. Haemodynamic data were collected using an optical topography system ETG-4000 (Hitachi Medical Corporation, Tokyo, Japan). In this experiment, only concentrations of HbO were used for the analysis, due to its higher signal-to-noise ratio [12]. The measurement area was the somatosensory cortex region, as it was expected to obtain haemodynamic response in this cortical region. According to the international EEG 10–20 system, the probes were centred on the C3 and C4 positions. The configuration for this experiment was two probes of 12 channels. Figure 2 shows the 24-channel configuration used in the study. However, the region of interest (ROI) consisted of eight channels: Ch4, Ch6, Ch7, Ch9, on the right hemisphere, and Ch16, Ch18, Ch19, Ch21 on the left hemisphere. The sampling frequency used in the experiments was 10 Hz.

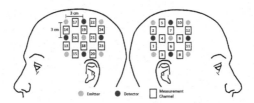

Fig. 2. Probe configuration, right hemisphere (Ch1–Ch12) and left hemisphere (Ch13–Ch24); probes were centred around the C3 and C4 areas (10–20 system). However, for the classification task, the focus was on the primary somatosensory area S1 only, using channels Ch4, Ch6, Ch7, Ch9, Ch16, Ch18, Ch19, Ch21.

2.2   Bag of Words

This study is based on the bag-of-words (BoW) approach for document classification adapted for time series. In the context of fNIRS time series representation, bag-of-words can adopt a similar model to the approach used for image classification. It can be generalized on three main steps: firstly, the detection of keypoints (features) in the fNIRS time series using sliding windows; secondly, keypoints are represented into words (codebook generation) using a quantization technique (e.g. k-means); and finally fNIRS time series are characterized by histograms of number of occurrences for each word observed in each time series. The histogram representation is used as input to a classifier (e.g. K-NN) that finds boundaries between classes. These steps are exhibited in Fig. 3 and described below.

Feature Extraction. The method starts by sliding a window through the fNIRS and obtaining local features from each partition. Consecutive windows are obtained in uniform manner (steps) to divide each fNIRS signal into partitions, the effect of different number of steps and different lengths of sliding windows was studied. Local features are obtained from every partition and represented by their corresponding feature vector. In this study, we also investigated the effect of different features: (1) time-domain features such as mean and variance of the values, slope, area under the activity curve, and kurtosis; (2) frequency-domain features, we were interested in a specific band only, the activation band (0.02–0.08 Hz) [6] since the period of stimulation used in the block design (Fig. 1) corresponds to this band; (3) time-frequency (wavelet), in this case, we were interested in the wavelet coefficients from the activation band during the period of stimulation ($\sim$30 s), we used the continuous wavelet transform (CWT) with the Morlet Wavelet.

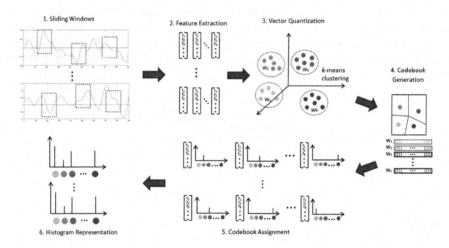

Fig. 3. Bag-of-words (BoW) representation. (1) Sliding window over the entire fNIRS time series. (2) Extract features from each sliding window. (3) Cluster all feature vectors. (4) Each centroid obtained by k-means represent a "word" in the codebook. (5) Assign each feature vector to a "word". (6) Count number of occurrences for each "word" from all feature vectors. For BoW representation of test dataset, it follows steps 1, 2, 5 and 6. Once the codebook is generated is universal for both training and testing datasets.

Codebook Formulation. Clustering of extracted local features and the creation of the codebook is next. Features are quantized using the k-means algorithm to obtain a codebook of k words. The k-means algorithm generates a k set of centroids set of feature vector. Each feature vector is classified with the centroid index closest to it. These centroids are referred as the codebook (or dictionary) and each centroid can be seen as a "word" in the dictionary. The total number (k) of centroids is referred as the codebook size, and codebooks of different size were also investigated.

Word Assignment. Using the extracted feature vectors, we assigned each feature vector to a centroid with the most representative value of a given local pattern. The histogram represents the number of feature vectors associated with each "word" in the codebook. Each bin in the histogram represents a "word" in the dictionary and the magnitude of each bin is the number of occurrences for that specific "word" in the fNIRS signal. It is worth noting that the temporal ordering of the local features in each time series is ignored, and the centroids indexing (i.e. 1, 2, .., k) obtained in the clustering task is used to order each "word" within the histogram. The histograms are the new representation of the fNIRS time series. The codebook generation is only computed once from training data and is generic to be used on test data. Finally, these histograms are then passed to a classifier to learn how to map classes from this BoW representation.

2.3 Classification

The classification model used in our study is the K-nearest neighbour (K-NN) algorithm, which is a robust classifier for time-series classification [3]. This algorithm does not require tuning complex parameters, which makes it ideal to be used by other research groups to validate our results. This algorithm makes no assumptions about the data, apart from a distance measure between any two attributes. The K-NN algorithm determines the K data points that are nearest to the given test point. We used the case of 1-NN for parameter tuning of the BoW representation, and then we investigated different number of nearest neighbours ($K = 1, 3, 5, 7, 9$) using the best BoW parameters. In order to make predictions, we needed to obtain the similarity between the test data and the specified training data, thus Euclidean Distance was used as measure between instances. The best set of parameters was chosen by leave-one-out cross validation.

3 Experimental Results

3.1 Thermal Measurements

Thermal threshold and tolerance of pain perception were obtained following the thermal test of the quantitative sensory testing (QST). By using QST thermal tests we aimed to minimize the subjective nature of self-reported pain scores. In this way, the obtained measurements are not based on self-reported pain rates but on temperature readings to classify a measurement as low pain (threshold) or high pain (tolerance) of cold and heat stimulation. The measured (averaged) values obtained from each experiment are shown in Fig. 4.

The box plots show the threshold and tolerance temperatures for cold and heat thermal tests. The first three measurements in each plot refer to the pain threshold while the last three measurements refer to the pain tolerance. The mean ($\pm$SD) temperatures in which participants first perceived pain to cold (12.6 ± 1.97, 11.8 ± 1.93, 12.6 ± 2.22 °C) and heat (41.8 ± 2.44, 42.16 ± 2.75, 43.25 ± 2.92 °C) clearly presented differences from the maximum pain the participants could take to cold (3.5 ± 2.07, 3.0 ± 2.71, 2.0 ± 2.12 °C) and heat (48.3 ± 1.92, 48.5 ± 2.11, 49.3 ± 1.82 °C).

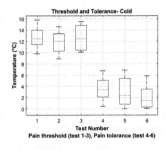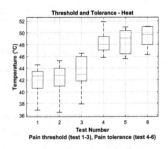

Fig. 4. Thermal threshold and tolerance levels experienced by the participants after cold (left panel) and heat (right panel) stimuli.

3.2   Bag-of-words Parameters

The effect of step size (ss), window length (wl), and codebook size (cz) for the bag-of-words representation was studied to reduce computational time and increase generalization. While testing the BoW parameters, two parameters were fixed and one was studied with different values. The first parameter to test was the step size using $wl = 128$ and $cz = 1000$. Once the best value for the parameter under examination was found, the same procedure was repeated for the other two parameters, updating the search with the best parameter found in the previous step. The parameters that produced the best results with the 1-NN classifier were using a step size of 4, sliding window of 192 samples, and a codebook size of 500 words; Fig. 5 shows these results. Using these parameters, 576 histograms constituted the pain database with 500 bins per histogram.

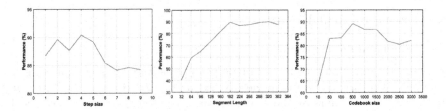

Fig. 5. Parameter search sensitivity of the BoW representation. Using our pain database the optimal values are: $ss = 4$, $sw = 192$, and $cs = 500$.

3.3   Classification of Pain

The main goal of the classification task is to design a classifier that can make predictions using new observations. As mentioned before, we have four types of observations (classes); these observations refer to temperature level (heat and cold) and pain intensity (low and high). Based on these observations, we trained our classifier model to predict new observations to be either: cold threshold (low pain), heat threshold (low pain), cold tolerance (high pain), or heat tolerance (high pain). We first studied the effect of choosing features from different domains (time, frequency, time-frequency), and classification results using the 1-NN classifier are presented in Table 1. In addition, we also investigated the effect of different nearest neighbours on our classifier (with $K = 1, 3, 5, 7, 9$), however, the best performance was obtained with the 1-NN classifier.

Table 1. Performance of the 1-NN using the best results from parameter tuning of the BoW representation. The evaluation includes the three type of feature domains (time, frequency, time-frequency (wavelet)). Minimum (Min) refers to pain threshold while maximum (Max) refers to pain tolerance.

Classifier	Feature domain	Class	Values (%)		
			Accuracy	Sensitivity	Specificity
1-NN	Time	Cold-Min	72.88	67.19	89.04
		Heat-Min	75.86	78.57	90.21
		Cold-Max	74.55	73.21	90.41
		Heat-Max	66.18	70.31	84.77
		Average	**72.08**	**72.32**	**88.61**
1-NN	Frequency	Cold-Min	68.25	67.19	84.50
		Heat-Min	62.75	57.14	86.33
		Cold-Max	60.00	69.64	81.29
		Heat-Max	62.30	59.38	83.21
		Average	**63.33**	**63.34**	**83.83**
1-NN	Wavelet	Cold-Min	88.52	84.38	95.76
		Heat-Min	93.10	96.43	97.53
		Cold-Max	81.54	94.64	92.98
		Heat-Max	91.07	79.69	96.99
		Average	**88.33**	**88.78**	**95.81**

4   Discussions

This study aims to identify cold and heat painful stimuli while discriminating between low pain (threshold) and high pain (tolerance). To that end, we used the bag of words (BoW) representation of fNIRS signals with a K-NN classifier to predict the response of new fNIRS within four categories of painful stimuli: cold threshold (low pain), heat threshold (low pain), cold tolerance (high pain), or heat tolerance (high pain). The classification model exhibited higher sensitivity using features from the time-frequency (wavelet) domain. These results demonstrate the possibility to distinguish between different types of pain and their pain intensity. This study also contributes to the idea of having an objective assessment of pain for non-verbal patients.

In our efforts to find a reliable assessment of pain for non-verbal patients, we used the thermal test following the protocols of the QST test. Our rationale to use the thermal test was to reduce the subjective nature of self-reported pain scores and use the profile of fNIRS signals to represent the type of pain and its intensity. In that context, the classification of fNIRS signals showed promising results to extend the use of QST protocols to non-verbal patients. In this way, by estimating the somatosensory profile (normally done in less than an hour) from patients while recording their fNIRS response, this information can be used to characterize patients with a variety of diseases and medical conditions, know their pain status and possible affected area. Thus, medical practitioners might use this information to treat patients, achieve optimal pain control and further treatment.

In addition, the advantage of obtaining the intensity (low and high) of pain in a patient is to provide a personalised medical treatment. It is well reported that pain judgement is affected by factors such as age, gender, weight, cultural background, etc. Medical treatments that work well for some patients, may not be adequate for others due to physical differences; therefore, tailoring pain control to the individual needs of each patient is not only important for non-verbal patients but also for the whole population. In many cases, medical practitioners prescribe analgesics based only on general information about medical treatments and not on individual patient's characteristics. For this reason, knowing the individual pain sensation will reduce the risks of harmful side effects due to overdose and cut medical treatment costs compared with traditional trial-and-error treatments.

There are studies that successfully discriminated between painful and non-painful signals using ML techniques but they mainly differ in feature extraction (e.g. time mean). However, due to the specific nature and data type of each study, it is recommended to find features that are meaningful for the study (or task dependent). Thus, in our case it was imperative to explore different features from different domain representations that might produce good classification performance, and the results exhibited that wavelet-based features obtained the highest classification (88.33%) on our pain database. These results validated the hypothesis of finding more-relevant neural response in the activation band (0.02–0.08 Hz) [6] than in other bands due to the period of stimulation used in our experiments. This is due to the ability of wavelet analysis to localize frequency content in specific time periods. In our case, the wavelet analysis used only frequency content (activation band) that mostly contributes to the total energy contained within the fNIRS signal at the specific stimulation period.

The findings of the present study represent a step closer to developing a physiologically based diagnosis of human pain that would benefit vulnerable patients who cannot self-report pain. The contributions of this study can be summarized into: (1) the use of the quantitative sensory testing (QST) to obtain pain information, that can potentially be used on non-verbal patients; (2) for the first time, heat and cold pain are differentiated as well as their corresponding pain intensity (low or high), that can improve pain control by personalizing medicine to individual needs; and (3) investigating different features as potential functional biomarkers to assess human pain using fNIRS signals.

However, the development of a bedside monitor for the diagnosis of pain is yet to be fully established. fNIRS has demonstrated to be a method that has potential for the assessment of pain, possesses advantages over PET or fMRI to be used in more realistic clinical settings. Future work should further test different types of noxious stimulation (e.g., pressure, chemical, electrical), use feature selection techniques to obtain a feature set that best represents the three domain representations, increase the number of participants, test different parts of the body, and include subjects with other medical conditions.

References

1. Brown, J.E., Chatterjee, N., Younger, J., Mackey, S.: Towards a physiology-based measure of pain: patterns of human brain activity distinguish painful from non-painful thermal stimulation. PLoS ONE **6**(9), e24124 (2011)
2. Cowen, R., Stasiowska, M.K., Laycock, H., Bantel, C.: Assessing pain objectively: the use of physiological markers. Anaesthesia **70**(7), 828–847 (2015)
3. Ding, H., Trajcevski, G., Scheuermann, P., Wang, X., Keogh, E.: Querying and mining of time series data: experimental comparison of representations and distance measures. Proc. VLDB Endowment **1**(2), 1542–1552 (2008)
4. Gram, M., Graversen, C., Olesen, A.E., Drewes, A.: Machine learning on encephalographic activity may predict opioid analgesia. Eur. J. Pain **19**(10), 1552–1561 (2015)
5. Herr, K., Coyne, P.J., McCaffery, M., Manworren, R., Merkel, S.: Pain assessment in the patient unable to self-report: position statement with clinical practice recommendations. Pain Manag. Nurs. **12**(4), 230–250 (2011)
6. Kirilina, E., Yu, N., Jelzow, A., Wabnitz, H., Jacobs, A.M., Tachtsidis, I.: Identifying and quantifying main components of physiological noise in functional near infrared spectroscopy on the prefrontal cortex. Front. Hum. Neurosci. **7**, 864 (2013)
7. Lin, J., Li, Y.: Finding structural similarity in time series data using bag-of-patterns representation. In: Winslett, M. (ed.) SSDBM 2009. LNCS, vol. 5566, pp. 461–477. Springer, Heidelberg (2009). doi:10.1007/978-3-642-02279-1_33
8. Pourshoghi, A., Zakeri, I., Pourrezaei, K.: Application of functional data analysis in classification and clustering of functional near-infrared spectroscopy signal in response to noxious stimuli. J. Biomed. Opt. **21**(10), 101411 (2016)
9. Ranger, M., Gélinas, C.: Innovating in pain assessment of the critically ill: exploring cerebral near-infrared spectroscopy as a bedside approach. Pain Manag. Nurs. **15**(2), 519–529 (2014)
10. Rolke, R., Baron, R., Maier, C.A., Tölle, T., Treede, R.D., Beyer, A., Binder, A., Birbaumer, N., Birklein, F., Bötefür, I., et al.: Quantitative sensory testing in the german research network on neuropathic pain (DFNS): standardized protocol and reference values. Pain **123**(3), 231–243 (2006)
11. Wager, T.D., Atlas, L.Y., Lindquist, M.A., Roy, M., Woo, C.W., Kross, E.: An fMRI-based neurologic signature of physical pain. New Engl. J. Med. **368**(15), 1388–1397 (2013)
12. Yamamoto, T., Kato, T.: Paradoxical correlation between signal in functional magnetic resonance imaging and deoxygenated haemoglobin content in capillaries: a new theoretical explanation. Phys. Med. Biol. **47**(7), 1121 (2002)

Tinnitus EEG Classification Based on Multi-frequency Bands

Shao-Ju Wang[1], Yue-Xin Cai[2,3], Zhi-Ran Sun[1], Chang-Dong Wang[1(✉)], and Yi-Qing Zheng[2,3]

[1] School of Data and Computer Science, Sun Yat-sen University, Guangzhou, China
{wangshj6,sunzhr3}@mail2.sysu.edu.cn, changdongwang@hotmail.com
[2] Department of Otolaryngology, Sun Yat-sen Memorial Hospital,
Sun Yat-sen University, Guangzhou, China
panada810456@126.com, yiqingzheng@hotmail.com
[3] Institute of Hearing and Speech-Language Science, Sun Yat-sen University,
Guangzhou, China

Abstract. Tinnitus is an auditory phantom percept of chronic high-pitched sound, ringing, or noise. Since the underlying physiological mechanisms of tinnitus are still under study, there is no universally effective treatment to cure tinnitus so far. There is even no method for objectively classifying tinnitus patients from normal people. In this paper, we utilize a Multi-view Intact Space Learning (MISL) method for the analysis and classification of electroencephalogram (EEG) signals using power value of frequency bands. At first, the power values of seven frequency bands are calculated by using Fast Fourier Transform (FFT) so as to obtain seven single views of features. Next, Multi-view Intact Space Learning is applied to integrate the seven single views together to get better classification results. Compared with the single view classification, the Multi-view Intact Space Learning method has achieved significant accuracy improvements by 6.32–23.25%. That is, the best accuracy, precision, recall and F1 of classification performance reach 0.828, 0.811, 0.857 and 0.833 respectively. The proposed method can be applied for auxiliary therapy of tinnitus as well as be extended to assist with the treatment of other diseases.

Keywords: Electroencephalogram · Multi-view Intact Space Learning · Least Squares Support Vector Machine · Tinnitus · Statistical analysis · Classification

1 Introduction

Tinnitus is an auditory phantom percept (ringing of the ears) of chronic high-pitched sound, ringing, or noise without any objective external sound source [2]. Tinnitus can occur at any age, which can affect the quality of life, involving

---

Shao-Ju Wang and Yue-Xin Cai make equal contributions.

D. Liu et al. (Eds.): ICONIP 2017, Part IV, LNCS 10637, pp. 788–797, 2017.
https://doi.org/10.1007/978-3-319-70093-9_84

sleep disturbance, work impairment and psychiatric distress for millions of people around the world [2,11]. Since the underlying physiological mechanisms of tinnitus are still under study [4], there is no universally effective treatment to cure tinnitus to date [8].

The electroencephalography (EEG) is a non-invasive imaging modality that depicts the process of recording electrical activity along the scalp generated from ionic current flow within the neurons of the brain [4,16]. It not only has extremely high temporal resolution but also spatial definition at the level of the scalp [6]. Some researches reveal that EEG study in tinnitus have several tremendous abnormalities in various frequency bands, such as Delta, Alpha, Beta and Gamma [4,8]. Till now, the values corresponding to different frequency bands of the EEG signals have not been applied into the classification distinguishing tinnitus patients and healthy people.

This study introduces a novel but effective method for classifying tinnitus patients from healthy people. At first, we transfer the power values of seven frequency bands using Fast Fourier Transform (FFT) into seven single views of features. The seven frequency bands include Delta (1–4 Hz), Theta (4–8 Hz), Alpha 1 (8–10 Hz), Alpha 2 (10–13 Hz), Beta (14–25 Hz), Gamma 1 (25–45 Hz) and Gamma 2 (55–80 Hz). Next, Multi-view Intact Space Learning (MISL) is applied to integrate the seven single views together to obtain the complete space representation of data so as to express the essence of data [17]. Finally, Least Squares Support Vector Machine (LS-SVM) [14] is used to classify the tinnitus patients and healthy controls on the multi-view intact space.

The main contributions of this study can be summarized as follows:

(1) The experimental results show that there is a significant difference between power values in different frequency bands of tinnitus patients and healthy controls.
(2) The method of Multi-view Intact Space Learning is firstly applied to combine different frequency bands in EEG data to achieve better classification results.
(3) Our study can be applied to tinnitus treatment to provide doctors with objective criteria for better and more accurate diagnosis than the previous research.

The rest of the paper is organized as follows. In Sect. 2, the whole procedure of data description, feature extraction and classification is presented. The experimental results including the experimental settings, statistical analysis, single view result analysis and multi-view result analysis are reported in Sects. 3, and 4 concludes the paper.

2 Method

2.1 Data Description

EEG Data Acquisition. This study adopts the resting-state EEG data of normal people and tinnitus patients collected by EEG analyzer with 128 scalp electrodes from Electrical Geodesics, Inc. in United States. The analyzer is an electrode mesh cap. Before conducting the experiment, in order to ensure the effectiveness of data, every participant has been explicated about the intention and precautions of the experiment. During preparation, participants should wear the electrode mesh cap with the electrode $Vref$ aligning the center of the head. They are also required to be fully relaxed and sit on a chair in a acoustic and electric shielding room. In the process of the experiment, participants should clear their mind and focusing on the tiny object in front of them, while avoiding moving head or hands or blinking that generates disturbances [9]. The sampling rates of EEG is set to 500 Hz, and the resistances of 128 electrodes are all lower than 50 kΩ. The EEG data collected in 6 min continuously would be synchronized by the record system. Finally, the raw EEG data would be preserved by Net Station Analysis system (Electrical Geodesics, Inc.).

Data Preprocessing. Due to the fact that there are more than one records of the raw resting-state EEG data of each participant, in order to guarantee the validity of the experiments, the EEG data with the longest length is chosen to load in EEGLAB. The preprocessing procedure is executed as follows:

(1) Coordinate file containing the data of 128 electrodes is loaded. Sampling rate of EEG data is reduced to 250 Hz.
(2) ERPLAB wave filter is applied to conduct notch filtering, in order to eliminate the interference of 50 Hz power frequency. Bandpass filtering between 0.5–80 Hz is applied [1].
(3) Bilateral mastoid electrodes (E56, E107) are used as reference electrodes. The electrodes (E8, E14, E17, E21, E25, E125, E126, E127, E128) around eyes and nasion are removed. Therefore, EEG data of the remaining 118 electrodes would be applied to the following experiments.
(4) Professionals review the data, remove the EEG data that has a large range of drifts, and replace the bad electrodes by linear interpolation. Independent Component Analysis (ICA) is used to remove the independent components related to artifacts [5].
(5) EEG data is separated into segments, each of which consists of 2 s of the raw data. If amplitude of any electrode exceeds 75 μV, it should be removed (Because the normal data would not exceed this range or it would be considered to have significant artifacts).

Participants. There are 29 volunteers in this study, 15 of which are tinnitus patients (age mean = 38.0667, std. dev. = 14.5429, range between 20 and 63, 9 males and 6 females), 14 of which are healthy controls (age mean = 33.3571, std. dev. = 11.1535, range 20–60, 7 males and 7 females).

2.2   Feature Extraction

Feature Extraction in Different Views. In this study, the STUDY module of the software EEGLAB is used to extract the features of the preprocessed data, i.e., extracting the power values of different frequency bands. 7 different frequency bands are used, including Delta (1–4 Hz), Theta (4–8 Hz), Alpha 1 (8–10 Hz), Alpha 2 (10–13 Hz), Beta (14–25 Hz), Gamma 1 (25–45 Hz) and Gamma 2 (55–80 Hz) [12,15]. Since the EEG signals are divided into 2 s during the data preprocessing, the frequency resolution of the spectrum analysis is 0.5 Hz, and the power of the 0.5:0.5:125 Hz frequency point is obtained. Finally, the power values of 118 electrodes in seven different frequency bands are calculated by using Fast Fourier Transform (FFT). For each electrode, all these power values are normalized according to the whole spectra power (1–80 Hz) of the same electrode. Therefore, each frequency bands would be depicted by values of 118 electrodes. The 7 frequency bands share the same dimensionality, which equals 118.

Multi-view Intact Space Learning. The Multi-view Intact Space Learning (MISL) algorithm can not only avoid leading to insufficiency using each view independently but also integrate the encoded complementary information [7,17]. The MISL algorithm uses the Cauchy loss to measure the reconstruction error from different views for enhancing the robustness of the model. It also develops a novel Iteratively Reweight Residuals (IRR) optimization technique to guarantee convergence. In this study, given the multi-view training data $D = \{z_i^v | 1 \leq i \leq n, 1 \leq i \leq m\}$ where $m = 7$ is view number and $n = 29$ is the sample size. The reconstruction error over the latent intact space $\mathcal{X}$ can be measured using the regularized Cauchy loss as follows [17].

$$\min_{x,W} \frac{1}{mn} \sum_{v=1}^{m} \sum_{i=1}^{n} \log(1 + \frac{\|z_i^v - W_v x_i\|^2}{c^2}) + C_1 \sum_{v=1}^{m} \|W_v\|_F^2 + C_2 \sum_{i=1}^{n} \|x_i\|_2^2, \quad (1)$$

where $x_i \in R^d$ is a data point in the latent intact space $\mathcal{X}$, $W_v \in R^{D_v \times d}$ is the v-th view generation matrix, c is a constant scale parameter, and C_1, C_2 are non-negative constants that can be determined using cross validation. The IRR algorithm is applied to alternately optimize the latent intact space $\mathcal{X}$ and the view generation matrix W.

Valid information for EEG data is contained between 0–80 Hz. The single frequency band power values obtained from the EEG data are insufficient to reflect the essence of the EEG signals [12]. Therefore, the power values of each frequency band are considered as a separate view, which are integrated using a MISL algorithm to obtain the potential intact space representation of the EEG signals.

2.3   Classification

Least Squares Support Vector Machine (LS-SVM) is the least squares version of Support Vector Machine (SVM), which are a set of related supervised learning methods that analyze data and recognize patterns for the classification and regression analysis [14]. LS-SVM is a class of kernel-based learning methods.

Many researches of EEG signals, such as [3,12], show that LS-SVM classifier has very satisfactory performance in the classification of patients and healthy controls. In this study, the LS-SVM classifier with a radial basis function kernel is applied for the classification of tinnitus EEG data. The LS-SVM classifier uses a set of linear equations instead of a convex quadratic programming (QP) problem for classical SVM to find the solution [13]. This helps us to get better experimental results. The LS-SVMlab (version 1.8) toolbox in MATLAB that is available online at http://www.esat.kuleuven.ac.be/sista/lssvmlab/ is applied in the following experiments.

3   Experiment

3.1   Experimental Settings

The experimental data set contains the power values of the 7 frequency bands of 118 electrodes obtained from 29 subjects, which consist of 14 healthy people and 15 tinnitus patients. 14-fold cross validation is used. The test set consists of one healthy person and one tinnitus patient, and the remaining 27 subjects compose the training set. We use accuracy, precision, recall and F1 to measure the classification performance, and calculate the mean and variance of four measurements over 100 runs to evaluate the stability of our classification performance. The formulas for the above measurements are as follows [10]:

$$Accuracy = \frac{TP + TN}{TP + FN + FP + TN}, \quad Recall = \frac{TP}{TP + FN},$$

$$Precision = \frac{TP}{TP + FP}, \quad F1 = \frac{2 \cdot Precision \cdot Recall}{Precision + Recall},$$

where TP is the number of true positive cases, TN is the number of true negative cases, FP is the number of false positive cases and FN is the number of false negative cases.

3.2   Statistical Analysis

We use cluster based permutation tests in statistical analysis to compare FFT data between tinnitus patients and healthy controls. The relative power spectrum maps of Delta (1–4 Hz), Theta (4–8 Hz), Alpha 1 (8–10 Hz), Alpha 2 (10–13 Hz), Beta (14–25 Hz), Gamma 1 (25–45 Hz) and Gamma 2 (55–80 Hz) are shown in Fig. 1. Compared with the normal group, the power values of the parietal and occipital region in Delta, Alpha 1, Beta, Gamma 1 and Gamma 2 bands in tinnitus group are significantly higher. No significant difference between the two groups is found in the two remaining frequency bands, i.e. Theta and Alpha 2.

3.3   Single View Result Analysis

From the statistical analysis, it is shown that there is a significant difference between the healthy controls and tinnitus patients in the 5 frequency bands including Delta, Alpha 1, Beta, Gamma 1 and Gamma 2, indicating that these

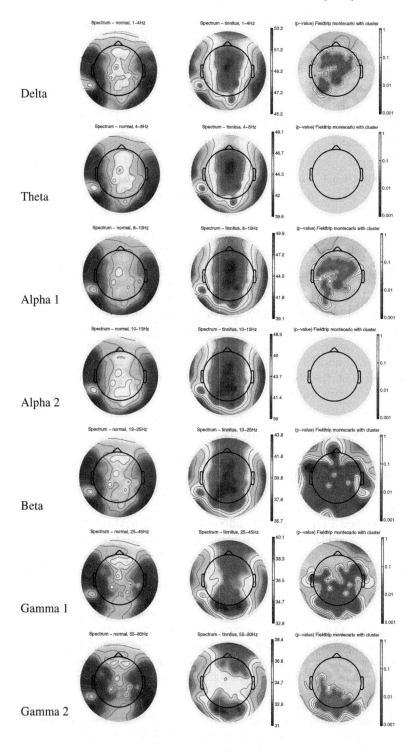

Fig. 1. The relative power spectrum maps of 7 frequency bands.

Table 1. Classification performance comparison by the power value of 7 frequency bands. Mean values and Variances (in parentheses) over 100 runs.

Data	Accuracy	Precision	Recall	F1
Delta band	0.651(0.001)	0.638(0.001)	0.700(0.002)	0.667(0.000)
Theta band	0.723(0.000)	0.725(0.000)	0.719(0.000)	0.722(0.000)
Alpha 1 band	0.683(0.000)	0.705(0.001)	0.632(0.001)	0.666(0.000)
Alpha 2 band	0.596(0.001)	0.633(0.002)	0.461(0.001)	0.533(0.001)
Beta band	0.696(0.001)	0.694(0.001)	0.704(0.001)	0.698(0.000)
Gamma 1 band	0.659(0.002)	0.685(0.002)	0.593(0.003)	0.634(0.002)
Gamma 2 band	**0.765**(0.001)	**0.717**(0.000)	**0.876**(0.003)	**0.788**(0.001)

EEG power values carry the discriminative information between the normal controls and the tinnitus patients. To evaluate how discriminative these 7 frequency bands are, we apply LS-SVM separately on these 7 frequency bands, and the experimental results are shown in Table 1. The classification accuracy of each frequency band is greater than 0.500, indicating that they all carry the partial difference between tinnitus and normal people. Among the frequency bands, the most discriminative frequency band is Gamma 2. The classification accuracy, precision, recall and F1 are 0.765, 0.717, 0.876 and 0.788, respectively. Also, it can be seen that the variance is very small, indicating that the classification performance is very stable.

3.4   Multi-view Result Analysis

The Multi-view Intact Space Learning algorithm is applied to fuse the power values of the 7 frequency bands into a feature space to represent the potential complete features of the EEG data. Therefore, better classification results can be obtained. According to statistical analysis, Alpha 2 and Theta frequency bands have shown no significant difference between healthy controls and tinnitus patients. The classification performance on Alpha 2 frequency band is the worst among the single view classifications. The following three combinations of obtaining multi-view-fusion features and the comparison of different dimensionality of the multi-view-fusion feature space is carried out.

(1) The 7-view-fusion feature space integrates 7 frequency bands.
(2) The 6-view-fusion feature space combines 6 frequency bands except Alpha 2, which is designed according to the effect of single view classification.
(3) The 5-view-fusion feature space incorporates Delta, Beta, Alpha 1, Gamma 1, and Gamma 2 frequency bands, which are designed in conjunction with statistical analysis results.

Their experimental results are shown in Fig. 2, when the dimensionality of the multi-view-fusion feature space is greater than 200, it has little influence on

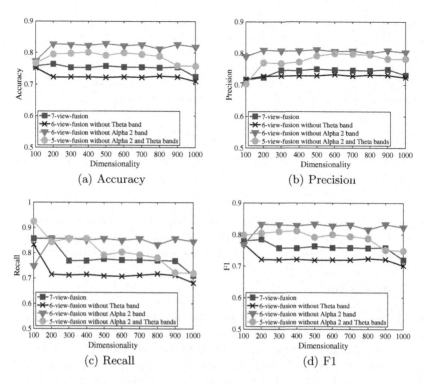

Fig. 2. Classification performance of different combinations of views with different dimensionality of the multi-view intact space features.

the classification performance. Compared with the results of single view classification, the accuracy of multi-view intact space classification results can be improved by 6.32–23.25%. Different combinations of multi-view-fusion features have different classification results. Among them, the best classification performance is obtained from combining 6 views except Alpha 2 frequency band, the accuracy of which is as high as 0.828. The accuracy of combining five views is 0.801, and the worst is obtained by the 7-view-fusion feature, the accuracy of which is only 0.758. As can be seen from the comparison result, the information carried by the Theta frequency band is useful for classifying tinnitus patients and healthy controls. Therefore, the performance of the 6-view-fusion feature is enhanced compared to the 5-view-fusion feature. However, the information carried by the Alpha 2 frequency band is not valid for classifying tinnitus patients and healthy controls, because the response of Alpha 2 frequency band was inhibited during eye-opened condition regardless of tinnitus patients or healthy people, which may become a confounding factor in the classification performance. Therefore, the classification result of the 7-view-fusion feature is worse than that of the 6-view-fusion feature. In order to show that the Alpha 2 band contains more invalid information than valid information from the perspective of classifying tinnitus patients from healthy people, we have designed

Table 2. Classification performance comparison by 4 different combinations of views. Mean values and Variances (in parentheses) over 100 runs.

Data	Accuracy	Precision	Recall	F1
7-view-fusion	0.758(0.000)	0.750(0.001)	0.776(0.001)	0.762(0.000)
6-view-fusion without Theta band	0.752(0.001)	0.718(0.001)	0.833(0.001)	0.771(0.001)
6-view-fusion without Alpha 2 band	**0.828**(0.000)	**0.811**(0.001)	**0.857**(0.000)	**0.833**(0.000)
5-view-fusion without Alpha 2 and Theta bands	0.801(0.000)	0.772(0.001)	0.857(0.000)	0.812(0.000)

a group of 6-view-fusion except the Theta frequency band, and the results are shown in Fig. 2. The best results of the four multi-view-fusion with setting the most suitable intact space dimensionality are summarized in Table 2.

4   Conclusion

The Multi-view Intact Space Learning method is applied in this study to improve the classification performance between the tinnitus patients and healthy controls. The power value of frequency bands of EEG data contains differences between some tinnitus patients and healthy controls. Different partial information of EEG signals exists in different EEG frequency bands. These EEG bands containing valid information are fused by the Multi-view Intact Space Learning algorithm to obtain the potential intact feature space of EEG signals. Applying LS-SVM classifier to the multi-view intact space feature can achieve better classification result than that of single view. In our research, the multi-view-fusion feature classification has achieved an improvement over the single view by 6.32–23.25%. The multi-view classification results can be used to assist doctors in the clinical treatment and improve the treatment efficiency of patients with tinnitus.

Our current research focuses on building intact space representation for classifying EEG data. In the future research, this multi-view method can be extended to model the brain activity of more complete data by fusing the non-invasive functional imaging modalities data, such as magnetoencephalogram (MEG), functional magnetic resonance imaging (fMRI) and EEG for research and treatment of diseases associated with brain dysfunction.

Acknowledgment. This work was supported by NSFC (No. 61502543) and Tip-top Scientific and Technical Innovative Youth Talents of Guangdong special support program (No. 2016TQ03X542).

References

1. Basoeki, A., Rahardjo, E., Hood, J.: PCA-based linear dynamical systems for multichannel EEG classification. In: International Conference on Neural Information Processing (ICONIP), vol. 2, pp. 745–749 (2002)

2. Eggermont, J.J., Roberts, L.E.: The neuroscience of tinnitus. Trends Neurosci. **27**(11), 676–682 (2004)
3. Ghayab, H.R.A., Li, Y., Abdulla, S., Diykh, M., Wan, X.: Classification of epileptic EEG signals based on simple random sampling and sequential feature selection. Brain Inform. **3**(2), 85–91 (2016)
4. Houdayer, E., Teggi, R., Velikova, S., Gonzalez-Rosa, J., Bussi, M., Comi, G., Leocani, L.: Involvement of cortico-subcortical circuits in normoacousic chronic tinnitus: a source localization EEG study. Clin. Neurophysiol. **126**(12), 2356–2365 (2015)
5. Iriarte, J., Urrestarazu, E., Valencia, M., Alegre, M., Malanda, A., Viteri, C., Artieda, J.: Independent component analysis as a tool to eliminate artifacts in EEG: a quantitative study. J. Clin. Neurophysiol. **20**(4), 249 (2003)
6. Li, P.-Z., Li, J.-H., Wang, C.-D.: A SVM-based EEG signal analysis: an auxiliary therapy for tinnitus. In: Liu, C.-L., Hussain, A., Luo, B., Tan, K.C., Zeng, Y., Zhang, Z. (eds.) BICS 2016. LNCS, vol. 10023, pp. 207–219. Springer, Cham (2016). doi:10.1007/978-3-319-49685-6_19
7. Lin, K.-Y., Wang, C.-D., Meng, Y.-Q., Zhao, Z.-L.: Multi-view unit intact space learning. In: Li, G., Ge, Y., Zhang, Z., Jin, Z., Blumenstein, M. (eds.) KSEM 2017. LNCS, vol. 10412, pp. 211–223. Springer, Cham (2017). doi:10.1007/978-3-319-63558-3_18
8. Meyer, M., Luethi, M.S., Neff, P., Langer, N., Büchi, S.: Disentangling tinnitus distress and tinnitus presence by means of EEG power analysis. Neural Plast. **2014** (2014)
9. Perera, H., Shiratuddin, M.F., Wong, K.W.: A review of electroencephalogram-based analysis and classification frameworks for dyslexia. In: Hirose, A., Ozawa, S., Doya, K., Ikeda, K., Lee, M., Liu, D. (eds.) ICONIP 2016. LNCS, vol. 9950, pp. 626–635. Springer, Cham (2016). doi:10.1007/978-3-319-46681-1_74
10. Powers, D.M.: Evaluation: from precision, recall and F-measure to ROC, informedness, markedness and correlation (2011)
11. Roberts, L.E., Eggermont, J.J., Caspary, D.M., Shore, S.E., Melcher, J.R., Kaltenbach, J.A.: Ringing ears: the neuroscience of tinnitus. J. Neurosci. **30**(45), 14972–14979 (2010)
12. Singh, P., Joshi, S., Patney, R., Saha, K.: Fourier-based feature extraction for classification of EEG signals using EEG rhythms. Circ. Syst. Sig. Process. **35**(10), 3700–3715 (2016)
13. Suykens, J.A.K., Gestel, T.V., Brabanter, J.D., Moor, B.D., Vandewalle, J.: Least Squares Support Vector Machines. World Scientific, Singapore (2002)
14. Suykens, J.A.K., Vandewalle, J.: Least squares support vector machine classifiers. Neural Process. Lett. **9**(3), 293–300 (1999)
15. Vanneste, S., De, R.D.: Deafferentation-based pathophysiological differences in phantom sound: tinnitus with and without hearing loss. Neuroimage **129**, 80–94 (2015)
16. Wu, W., Chen, Z., Gao, X., Li, Y., Brown, E.N., Gao, S.: Probabilistic common spatial patterns for multichannel EEG analysis. IEEE Trans. Pattern Anal. Mach. Intell. **37**(3), 639 (2015)
17. Xu, C., Tao, D., Xu, C.: Multi-view intact space learning. IEEE Trans. Pattern Anal. Mach. Intell. **37**(12), 2531–2544 (2015)

Deep Neural Network with l2-Norm Unit for Brain Lesions Detection

Mina Rezaei[✉], Haojin Yang, and Christoph Meinel

Hasso Plattner Institute, Prof.-Dr.-Helmert-Straße 2-3, 14482 Potsdam, Germany
{mina.rezaei,haojin.yang,christoph.meinel}@hpi.de

Abstract. Automated brain lesions detection is an important and very challenging clinical diagnostic task, because the lesions have different sizes, shapes, contrasts and locations. Deep Learning recently shown promising progresses in many application fields, which motivates us to apply this technology for such important problem. In this paper we propose a novel and end-to-end trainable approach for brain lesions classification and detection by using deep Convolutional Neural Network (CNN). In order to investigate the applicability, we applied our approach on several brain diseases including high and low grade glioma tumor, ischemic stroke, Alzheimer diseases, by which the brain Magnetic Resonance Images (MRI) have been applied as input for the analysis. We proposed a new operation unit which receives features from several projections of a subset units of the bottom layer and computes a normalized l2-norm for next layer. We evaluated the proposed approach on two different CNN architectures and number of popular benchmark datasets. The experimental results demonstrate the superior ability of the proposed approach.

Keywords: Multimodal CNN · l2-norm unit · Brain lesion detection and localization

1 Introduction

Annually in the United State alone 24,000 adult and 4,830 children will be diagnosed as new cases of brain cancer. A lot of people have died due to brain tumor, multiple sclerosis, ischemic stroke and Alzheimer diseases[1]. Medical imaging is an important tool for brain diseases diagnosis in case of surgical or chemical planning. Magnetic Resonance Imaging (MRI) can provide rich information for premedication and surgery medication, which is extremely helpful for evaluating the treatment and lesion progress. However the raw data extracted from MR images is hard to be directly applied for diagnosis due to the large amount of the data. An accurate brain lesion detection and classification algorithm based on MR images might be able to improve the prediction accuracy and efficiency, that enables a better treatment planning and optimize the diagnostic progress.

---

[1] http://www.cancer.net/cancer-types/brain-tumor/statistics.

© Springer International Publishing AG 2017
D. Liu et al. (Eds.): ICONIP 2017, Part IV, LNCS 10637, pp. 798–807, 2017.
https://doi.org/10.1007/978-3-319-70093-9_85

As mentioned by Menze et al. [1], the number of clinical study for automatic brain lesion detection has grown significantly in the last several decades. Some brain lesions such as ischemic strokes, or even tumors can appear with different shapes, inappropriate sizes and unpredictable locations within the brain. Furthermore, different types of MRI machines with specific acquisition protocols may provide MR images with a wide variety of gray scale representations on the same lesion cells. Recent research has shown strong ability of Convolutional Neural Network (CNN) for learning hierarchical representation of image data without requiring any effort to design handcrafted features [2–4]. This technology became very popular in computer vision society for image classification [5,6], object detection [7–9], medical image classification [10,11] and segmentation [12,13]. As mentioned by LeCun et al. in [2]: different layers of a network are capable of different levels of abstraction, and capture different amount of structures from the patterns present in the image.

In this work we investigate the applicability of CNN for brain lesions detection. Our goal is to perform localization and classification of single as well as multiple anatomic regions in volumetric clinical images from various image modalities. To this end we propose a novel framework based on CNN with l2-norm unit. A detailed evaluation on parameter variations and network architectures has been provided. We show that l2-norm operation unit is robust to the error variations in the classification task and is able to improve the prediction result. We conducted experiments on a number of brain MRI datasets, which demonstrate the excellent generalization ability of our approach. The contribution of this work can be summarized as following:

- We propose a robust solution for brain lesions classification. We achieved promising results on four different brain diseases (The overall accuracy is over 95%).
- We applied multiple MRI modalities as network input, and this improved the dice coefficient up to 30% on ISLES benchmark.
- We implemented l2-norm unit in Caffe [14] framework for both CPU and GPU computation. The experimental results demonstrate the superior ability of l2-norm in various tasks.

The rest of the paper is organized as follows: Sect. 2 describes the proposed approach, Sect. 3 presents the detailed experimental results. Section 4 concludes the paper and gives an outlook on future work.

2   Methodology

In this chapter we will describe our deep network for classification and detection task in detail. The core techniques applied in our approach are depicted as well. In the recent deep learning context, a deep neural network can be built driven by two principles: Modularity and Residual learning. Modularity is a set of repeatable smaller neural network unit which enables the learning of high-level visual representations. The bottleneck module of the Inception architecture [15] and

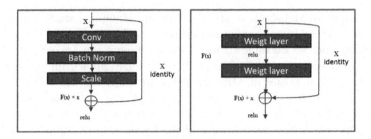

Fig. 1. Exemplary residual building block. The block on the left side shows a vanilla residual block, where the one on the right side is a dense block applied in our classification network.

the corresponding units in VGG-Net [16] can be considered as typical examples. In such networks the wide and depth have been significantly increased. On the other hand residual learning [6] considers new way to each layer. Every consequent layer is responsible for, in effect, fine tuning the output from a previous layer by just adding a learned "residual" connection to the input. This essentially drives the new layer to learn something different from what the input has already encoded. Another important advantage is that such residual connections can help in handling gradient vanishing problem in very deep networks [6]. Figure 1 shows an exemplary residual building block, where $F(x) + x$ denotes the element-wise addition of the original input and the residual connection. The block on the left depicts vanilla residual unit proposed by He et al. [6], where the one on the right side is a dense block that we utilize in our classification network.

2.1   l2-Norm Unit

$$|X_{i,j}| = \begin{bmatrix} x_{1,1} & x_{1,2} & \dots & x_{1,j} \\ x_{2,1} & x_{2,2} & \dots & x_{2,j} \\ \dots & \dots & \dots & \dots \\ x_{i,1} & x_{i,2} & \dots & x_{i,j} \end{bmatrix} \tag{1}$$

$$Forward : |x_{i,j}| = \sqrt[2]{\sum x_{i,j}{}^2} \tag{2}$$

$$Backward : \partial\,|x_{i,j}| = \frac{n\partial(\sum x_{i,j})}{2\sqrt[2]{\sum x_{i,j}{}^2}} \tag{3}$$

In linear algebra, the size of a vector v is called the norm of v. The two-norm (also known as the l2-norm, mean-square norm, or least-squares norm) of a vector v is defined by Eq. 2. Assume we have a 2D matrix $X_{i,j}$ (cf. Eq. 1) which is the output of the specific patch of $a_{i,j}$ from the first convolution layer. Then for each item in feed forward or backward pass we calculate the l2-norm as described by Eqs. 2 and 3. We consider l2-norm operation as a pooling function and apply it to reduce the dimension of the learned representations, which is able to obtain better generalization ability. For example in the classification task an input volume of size $224 \times 224 \times 64$ is pooled by l2-norm operator with filter size 2 and stride 2 into an output volume of size $112 \times 112 \times 64$.

2.2   Brain Abnormality Classification

Recently, ResNet (Deep Residual Network) [6] achieves the state-of-the-art performance in object detection and other vision related tasks. As mentioned above we explored the ResNet architecture with l2-norm unit for brain abnormality classification. Figure 2 depicts the network architecture. Our classification network takes 2D images with three channels, while each channel contains a gray scale copy with the same size and same plane from various MRI modalities with respective class label $l = \{0, 1, \dots, 4\}$. Each gray scale copy extracted from T1, T1c and FLAIR of the same MRI categories has been mapped to the Red, Green and Blue channels of a standard image container, respectively. The proposed network strongly inspired by vanilla ResNet block depicted by Fig. 1.

As shown in Fig. 2, we apply l2-norm operation after the first convolution layer and before the first inner product layer. In the experiments we observed that the l2-norm layer performs a similar effect as a pooling operator, which reduces the spatial size of the feature representations and extracts features that are not covered by standard pooling operators. This allows the network to learn more distinguished feature information such as variance from the data stream, which could improve the overall generalization ability of the model.

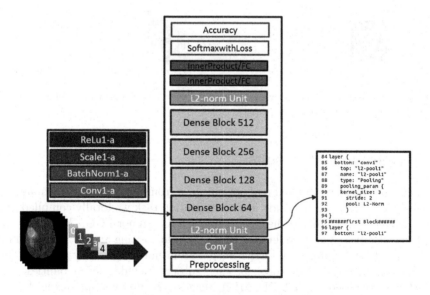

Fig. 2. Brain diseases classification architecture (Color figure online)

2.3   Brain Lesions Detection

Unlike image classification, object detection extracts location and region information of a target object within an image. Figure 3 represents our network for
brain abnormality detection. In our work-flow, we extract and apply multiple
modalities from MRI images, where the images are sampled in 2D slices from
the axial, coronal and sagittal view with various sizes. Inspired by Fast R-CNN
network [17], we build our CNN network based on VGG-16 [16] style architecture
as the feature extractor. Instead of using max-pooling and spatial max-pooling
we place the l2-norm unit after the second convolution (conv1-2) layer and before
the first fully connected (inner product) layer respectively. We utilize selective
search [18] to generate object proposals, which is a set of object bounding boxes.
The proposal sampling process is performed on top of dense feature layer after
layer `conv5-3`. We confirm the suggested solution by Girshick et al. [17] to come
over on heterogeneous collection of computed proposals and divide them into a
pyramid grid of sub-windows. Here three pyramid levels $4 \times 4, 2 \times 2, 1 \times 1$ and
l2-norm "pooling" have been applied in each sub-window to generate the corresponding output grid cell. Subsequently each output feature vector is further fed
into a sequence of fully connected layers, which is followed by two sibling output
layers: the SVM (Support Vector Machine) classifier for object class estimation [19], and the bounding box regression layer to calculate the loss of proposed
object bounding boxes. The overall training is performed in the supervised manner, and the loss of the whole network sums losses from both object classification
and bounding box regression.

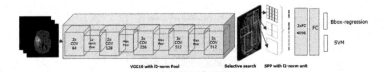

Fig. 3. Proposed architecture with 16 convolutions and l2-norm unit for recognition
and localization of brain lesion

3   Experimental Results

In the experiment we applied real patient data from five popular benchmarks to
evaluate the proposed methods. For classification task we totally compiled 1500
MRI images with label of healthy, tumor-HGG, tumor-LGG, Alzheimer and
multiple sclerosis. We consider 20% of the data for testing and 80% for training.
IXI dataset [20] contains 600 MRI images from normal, healthy subjects. The
MRI image acquisition protocol for each subject includes six modalities, from
which we have used T1, T2, PD, MRA images. The first column of Fig. 4 shows
the healthy brain images from IXI dataset in the sagittal, coronal and axial
sections. The BraTS2016 benchmark [1,21] prepared the data in two part of

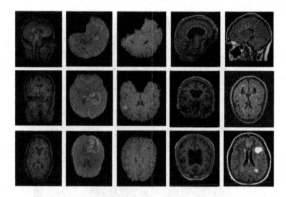

Fig. 4. We trained the proposed network on five different categories of brain MRI. The 1st column shows healthy brain in sagittal, coronal and axial section. The 2nd and 3rd columns show high and low grad glioma, while 4th and 5th columns present some brain images on Alzheimer and multiple sclerosis. Total we have 900k 2D images for the classification task.

Table 1. Brain lesions classification performance of the re-designed ResNet architecture using l2-norm unit. The involved classes include healthy, tumor-HGG, tumor-LGG, Alzheimer and multiple sclerosis. The last two rows show the comparison results to the most recent methods [10,11] that they consider healthy and Alzheimer Diseases only.

	Total MRI	Accuracy	Sensitivity	Specificity	Recall	Kappa
Our method	1500	95.308%	0.91	0.87	87.65	0.92
Paul et al. [10]	191	91.43%	-	-	-	-
El Abbadi et al. [11]	50	94%	0.85	0.87	-	-

High and Low Grade Glioma (HGG/LGG) Tumor. All images have been aligned to the same anatomical template and interpolated to 1 mm, 3 voxel resolution. The training dataset consists of 220 HGG and 108 LGG MRI images which for each patient T1, T1contrast, T2, FLAIR and ground truth labeled by medical experts have been provided. Alzheimer disease dataset[2] comes from Open Access Series of Imaging Studies (OASIS). The dataset consists of a cross-sectional collection of 416 subjects aged from 18 to 96. For each subject, 3 or 4 individual T1-weighted MRI scans were obtained in single scan sessions. 18 MRI images with multiple sclerosis from ISBI challenges 2008 [22] have also been applied in the classification task. ISLES benchmark 2016 [23] (Ischemic Stroke Lesion Segmentation) comes from MICCAI challenge in two part, by which we used only SPES dataset with 30 brain images with 7 modalities in our task. An visual overview of the applied datasets can be found in Fig. 4.

Because the MRI volumes in the BraTS and ISLES datasets do not possess an isotropic resolution, we prepared 2D slices in sagittal, axial and coronal view. As mentioned by Havaei et al. [24], unfortunately brain imaging data are rarely

---

[2] http://www.oasis-brains.org/.

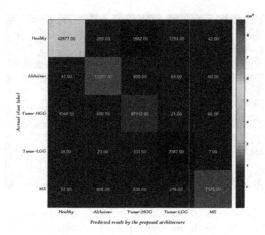

Fig. 5. The confusion matrix of the classification results. X-axis shows predicted results, where Y-axis gives the actual labels.

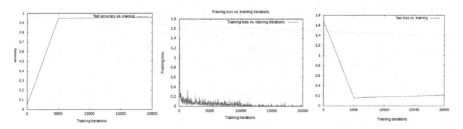

Fig. 6. Learning curves of brain lesions classification

balanced due to the small size of the lesion compared to the rest of the brain. For example the volume of a stroke is rarely more than 1% of the entire brain and a tumor (even large glioblastomas) never occupies more than 4% of the brain. Training a deep network with imbalanced data often leads to very low true positive rate since the system gets to be biased towards the one class that is over represented. To overcome this problem we have chosen volume of MRI with lesions, and augmented training data by using horizontal and ventricle flipping, multiple scaling. By using a re-designed ResNet architecture described in Sect. 2, we achieved over 95% classification accuracy as shown in Table 1, while Fig. 5 demonstrates the confusion matrix of the classification result. We also compared our result with the most recent deep learning based approaches as shown in Table 1, where the reference method also used IXI, OASIS datasets. Figure 6 shows learning curves of testing accuracy, training and testing losses during the training process. For brain lesions detection experiment we applied both BraTS and ISLES datasets. We used 70% of the data for training, 10% for validation and 20% for testing. It is expected that more generalized features could be able to learned from multiple modalities, and the testing accuracy based on more generalized features should be gained. The brain lesions detection results from

Table 2. Dice Similarity Coefficient (DSC) results (for brain lesions detection performance measurement) on the BraTS2016 and ISLES2016 dataset by using incremental modalities. F/D column means the FLAIR modality in BraTS dataset and DWI modality in ISLES dataset.

T1	T1c	T2	F/D	Dice-BraTS16	Dice-ISLES
x	-	-	-	61.8%	42%
-	x	-	-	33.76%	27%
-	-	x	-	36.7%	39.98%
-	-	-	x	73.38%	50.71%
-	x	x	x	81.53%	54.23%
x	x	-	x	82.6%	54.67%
x	-	x	x	83.19%	53.09%
x	x	x	-	82.73%	54.7%
x	x	x	x	83.53%	56.87%

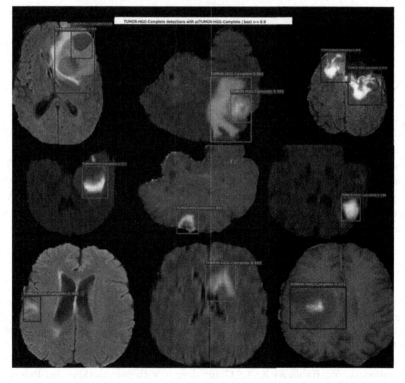

Fig. 7. Visual results of our brain lesions detection approach on Axial, Coronal and Sagittal views. The subjects are selected from the validation set.

Table 2 proved our assumption, where better detection results were achieved by increasing the data modalities in the model training. The detection result can be improved by 20% in BraTS and 30% in ISLES dataset.

Table 3. Evaluation result of the detection network with and without l2-norm unit, which demonstrates the performance gains by using l2-norm unit.

	Dice (without l2-norm unit)	Dice (with l2-norm unit)
BraTS16	72%	83.53%
ISLES16	53.65%	56.87%

From Table 2, we can also infer that the FLAIR modality is the most relevant one for identifying the complete tumor (Dice: 73.38%), However in ISLES benchmark we don't have this modality and it is justified less accuracy on this category. It motivated us to work on generating the missing modalities in the future. The subjects in Fig. 7 are from our testing set, for which the model is not trained on, the detection results from these subjects could give a good estimation of the model performance. Table 3 demonstrates the evaluation results of the detection architectures with and without l2-norm unit. From which we can easily realize the superior ability of the proposed l2-norm operator. We are able to improve the detection performance significantly on both datasets by using this novel operator.

4　Conclusion

In this paper, we explored two important clinical tasks: brain lesions classification and detection. We proposed end-to-end trainable approaches based on state-of-the-art deep convolutional neural networks. We implemented a novel pooling operator: l2-norm unit which can effectively generalize the network, and make the learned model more robust. The applicability, model accuracy and generalization ability have been evaluated by using a set of publicly available datasets. As the future work we will further investigate the automatic segmentation of tumor regions based on the detection results.

References

1. Menze, B., Reyes, M., Van Leemput, K.: The multimodal brain tumor image segmentation benchmark (BRATS). IEEE Trans. Med. Imaging **34**, 1993–2024 (2014)
2. LeCun, Y., Bengio, Y., Hinton, G.: Deep learning. Nature **521**(7553), 436–444 (2015)
3. Szegedy, C., Liu, W., Jia, Y., Sermanet, P., Reed, S., Anguelov, D., Erhan, D., Vanhoucke, V., Rabinovich, A.: Going deeper with convolutions. In: Proceedings of the IEEE Conference on Computer Vision and Pattern Recognition, pp. 1–9 (2015)
4. Gulcehre, C., Cho, K., Pascanu, R., Bengio, Y.: Learned-norm pooling for deep feedforward and recurrent neural networks. In: Calders, T., Esposito, F., Hüllermeier, E., Meo, R. (eds.) ECML PKDD 2014. LNCS, vol. 8724, pp. 530–546. Springer, Heidelberg (2014). doi:10.1007/978-3-662-44848-9_34

5. Krizhevsky, A., Sutskever, I., Hinton, G.E.: Imagenet classification with deep convolutional neural networks. In: Advances in Neural Information Processing Systems, pp. 1097–1105 (2012)
6. He, K., Zhang, X., Ren, S., Sun, J.: Deep residual learning for image recognition. In: 2016 IEEE Conference on Computer Vision and Pattern Recognition (CVPR) (2016)
7. Szegedy, C., Toshev, A., Erhan, D.: Deep neural networks for object detection. In: Advances in Neural Information Processing Systems, pp. 2553–2561 (2013)
8. Girshick, R., Donahue, J., Darrell, T., Malik, J.: Rich feature hierarchies for accurate object detection and semantic segmentation. In: Proceedings of the IEEE Conference on Computer Vision and Pattern Recognition, pp. 580–587 (2014)
9. Ren, S., He, K., Girshick, R.B., Sun, J.: Faster R-CNN: towards real-time object detection with region proposal networks. CoRR abs/1506.01497 (2015)
10. Paul, J.S., Plassard, A.J., Landman, B.A., Fabbri, D.: Deep learning for brain tumor classification, vol. 10137, pp. 1013710-1–1013710-16 (2017)
11. El Abbadi, N.K., Kadhim, N.E.: Brain cancer classification based on features and artificial neural network. Brain 6(1) (2017)
12. Ronneberger, O., Fischer, P., Brox, T.: U-net: convolutional networks for biomedical image segmentation. In: Navab, N., Hornegger, J., Wells, W.M., Frangi, A.F. (eds.) MICCAI 2015. LNCS, vol. 9351, pp. 234–241. Springer, Cham (2015). doi:10.1007/978-3-319-24574-4_28
13. Dai, J., He, K., Sun, J.: Instance-aware semantic segmentation via multi-task network cascades. In: 2016 IEEE Conference on Computer Vision and Pattern Recognition (CVPR) (2016)
14. Jia, Y., Shelhamer, E., Donahue, J., Karayev, S., Long, J., Girshick, R.B., Guadarrama, S., Darrell, T.: Caffe: convolutional architecture for fast feature embedding. CoRR
15. Szegedy, C., Vanhoucke, V., Ioffe, S., Shlens, J., Wojna, Z.: Rethinking the inception architecture for computer vision. CoRR (2015)
16. Simonyan, K., Zisserman, A.: Very deep convolutional networks for large-scale image recognition. arXiv preprint arXiv:1409.1556 (2014)
17. Girshick, R.B.: Fast R-CNN. CoRR abs/1504.08083 (2015)
18. Uijlings, J., van de Sande, K., Gevers, T., Smeulders, A.: Selective search for object recognition. Int. J. Comput. Vis. 104, 154–171 (2013)
19. Liu, G., Zhang, X., Zhou, S.: Multi-class classification of support vector machines based on double binary tree. In: Fourth International Conference on Natural Computation, ICNC 2008, vol. 2, pp. 102–105. IEEE (2008)
20. http://brain-development.org/ixi-dataset/
21. https://www.virtualskeleton.ch/BRATS/Start2016/
22. http://www.medinfo.cs.ucy.ac.cy/index.php/downloads/datasets/
23. http://www.isles-challenge.org/ISLES2016/
24. Havaei, M., Davy, A., Warde-Farley, D., Biard, A., Courville, A., Bengio, Y., Pal, C., Jodoin, P.M., Larochelle, H.: Brain tumor segmentation with deep neural networks. Med. Image Anal. 35, 18–31 (2017)

Emotion and Bayesian Networks

Multimodal Emotion Recognition Using Deep Neural Networks

Hao Tang[1], Wei Liu[1], Wei-Long Zheng[1], and Bao-Liang Lu[1,2,3(✉)]

[1] Department of Computer Science and Engineering,
Center for Brain-like Computing and Machine Intelligence, Shanghai, China
{silent56,liuwei-albert,weilong}@sjtu.edu.cn
[2] Key Laboratory of Shanghai Education Commission for Intelligent
Interaction and Cognitive Engineering, Shanghai, China
[3] Brain Science and Technology Research Center,
Shanghai Jiao Tong University, Shanghai, China
bllu@sjtu.edu.cn

Abstract. The change of emotions is a temporal dependent process. In this paper, a Bimodal-LSTM model is introduced to take temporal information into account for emotion recognition with multimodal signals. We extend the implementation of denoising autoencoders and adopt the Bimodal Deep Denoising AutoEncoder modal. Both models are evaluated on a public dataset, SEED, using EEG features and eye movement features as inputs. Our experimental results indicate that the Bimodal-LSTM model outperforms other state-of-the-art methods with a mean accuracy of 93.97%. The Bimodal-LSTM model is also examined on DEAP dataset with EEG and peripheral physiological signals, and it achieves the state-of-the-art results with a mean accuracy of 83.53%.

Keywords: Multimodal emotion recognition · EEG · Deep neural networks · LSTM

1 Introduction

Automatic emotion recognition has drawn increasing attention due to its potential applications to human computer interaction. There are many modalities that contain emotion information, such as facial expression, voice, electroencephalography (EEG), eletrocardiogram (ECG), pupillary diameter (PD), and so on. However, since emotions are complex and associated with many nonverbal cues, it's difficult to recognize emotions robustly based on a single modality. Saneiro *et al.* detected emotions in educational scenarios from facial expressions and body movements [8]. Koelstra *et al.* built an emotion recognition system based on EEG and peripheral physiological signals [5]. Lu *et al.* used both EEG signals and eye movement signals to recognize three types of emotions and revealed that EEG features and eye movement features were complementary to emotion recognition [7]. Liu *et al.* furthermore used Bimodal Deep AutoEncoder to

© Springer International Publishing AG 2017
D. Liu et al. (Eds.): ICONIP 2017, Part IV, LNCS 10637, pp. 811–819, 2017.
https://doi.org/10.1007/978-3-319-70093-9_86

extract high level representation features and achieved competitive results on both SEED[1] and DEAP[2] datasets [6].

Most of the existing methods treated features at each time step as independent samples, and ignored the temporal dependency property of emotions [11]. Recurrent Neural Networks (RNNs) are powerful tools for modeling sequential data and have the ability to extract temporal information from input signals. Moreover, Long Short Term Memory (LSTM) neural network [4], which is a gated RNN with linear self-loop, has been successfully used to capture temporal dependency property in many fields, such as speech recognition [13] and machine translation [12]. In this paper, to reveal the effect of temporal information in emotion recognition, we introduced a Bimodal-LSTM (Long Short Term Memory) model, which could use both the temporal information and the frequency-domain information to discriminate emotion states. Specifically, the Bimodal-LSTM model consists of two LSTM encoders, for features from EEG and other modalities respectively, and one classification layer. We also extended the implementation of denoising autoencoders in the field of multimodal emotion recognition and introduced the Bimodal Deep Denoising AutoEncoder (BDDAE) model. We evaluated our proposed models on two public multimodal datasets called SEED and DEAP for emotion recognition and achieved the state-of-the-art performance.

2   Bimodal Deep Denoising AutoEncoders

2.1   Denoising Autoencoders

An autoencoder is an unsupervised model, which can be used for dimensionality reduction, data compression, and feature learning [2,3]. Classical autoencoders map the input to its hidden representation with an encoder function and then use a decoder function to map the hidden representation to the reconstruction of input. The reconstruction errors are minimized to train autoencoders.

The denoising autoencoder, which is an extension of the classical autoencoder, reconstructs the input from a corrupted version of it [10]. It can prevent the autoencoder from learning the identity function when the encoder and decoder are given too much capacity. And denoising autoencoders can learn more robust hidden representation.

The Bimodal Deep Denoising AutoEncoder (BDDAE) model consists of two networks, the autoencoder network and the classifier network. The autoencoder network is used to pre-train the encoders' weights. And the classifier network predicts emotion labels using EEG features and other modalities' features.

The autoencoder network of BDDAE, as illustrated in Fig. 1(a), contains one corruption layer, three encoders, and three decoders. The corruption layer randomly sets some of the inputs to zeros according to the dropout probability. The encoders and decoders, whose form is an affine mapping followed by a sigmoid

---

[1] http://bcmi.sjtu.edu.cn/~seed/.

[2] http://www.eecs.qmul.ac.uk/mmv/datasets/deap/.

function, are mirror images of each other. There are two encoders for EEG features and other modalities' features, respectively. The encoded features are then concatenated together, and another encoder is used to extract the combined high-level features. Mean squared error criterion is used to train the network.

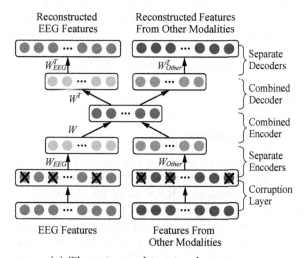

(a) The autoencoder network

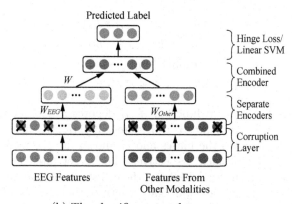

(b) The classifier network

Fig. 1. The structure of BDDAE's autoencoder network and classifier network

The classifier network of BDDAE is depicted in Fig. 1(b). It also contains three encoders, which are the same as the autoencoder network. Moreover, the last layer can be considered as a linear kernel Support Vector Machine (SVM), since the loss function the network uses is L2-regularized hinge loss [9].

2.2 Training

We firstly pre-trained the encoders' weights, W_{EEG}, W_{Other}, W by training the autoencoder network. Moreover, to adapt encoders to the specific task (emotion recognition), we attached the SVM layer to the encoder layers and trained the classifier network. In detail, the pre-trained encoders' weight matrices as in Fig. 1(a) were used to initialize the corresponding encoders' weight matrices in the classifier network, as in Fig. 1(b). The encoder layers' learning rate was set to one percent of the last classification layer's learning rate, so that the classification layer was trained mostly, while the encoder layers were fine-tuned.

3 Bimodal-LSTM

3.1 LSTM Neural Networks

To incorporate the temporal dependency information of features, we introduce Long Short Term Memory (LSTM) neural networks as a temporal encoder. LSTM neural network, which is a RNN using LSTM blocks, can prevent the vanishing (and exploding) gradient problem [1] and has the ability to learn information from long sequences. Each LSTM block contains memory cell states c_t propagated over time. At every time step, the states of memory cells c_t are updated according to the input of current time step x_t and the output of the previous time step h_{t-1} as follows:

$$
\begin{aligned}
f_t &= \sigma(W_f[h_{t-1}, x_t] + b_f) \\
i_t &= \sigma(W_i[h_{t-1}, x_t] + b_i) \\
g_t &= \tanh(W_g[h_{t-1}, x_t] + b_g) \\
c_t &= c_{t-1} * f_t + i_t * g_t,
\end{aligned}
\tag{1}
$$

where σ denotes the sigmoid function, f_t, i_t denotes the forget gate and input gate, g_t denotes the candidate of cell states, W_f, W_i, W_g denotes the weight matrices, b_f, b_i, b_g denotes the biases and c_t and h_t are the memory cell states and the output of LSTM block, respectively. The forget gate controls the process of forgetting information by multiplying the cell states by real numbers between zero and one. Similarly, the input gate controls the process of remembering information. The output of LSTM blocks h_t is a filtered version of memory cell states, as follows:

$$
\begin{aligned}
o_t &= \sigma(W_o[h_{t-1}, x_t] + b_o) \\
h_t &= o_t * \tanh(c_t).
\end{aligned}
\tag{2}
$$

where W_o and b_o denotes the weight matrix and the bias for the output gate o_t. The output gate controls the process of output.

As depicted in Fig. 2, the Bimodal-LSTM network contains one classification layer and two LSTM encoders. Two LSTM encoders are for EEG features and other modalities' features, respectively. The network also uses L2-regularized hinge loss as the objective function to minimize, so the classification layer can be considered as a linear kernel SVM. Dropout is applied to the output of LSTM blocks to obtain more robustness.

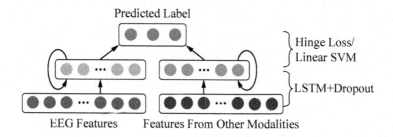

Fig. 2. The structure of Bimodal-LSTM

3.2 Training

We firstly trained the network thoroughly using Adam optimization algorithm. To further minimize the loss of the network after training it thoroughly, we trained a SVM classifier, which minimized the same loss function as the Bimodal-LSTM network. In detail, we used the trained LSTM encoders to extract high-level features from EEG features and features from other modalities at each time step. And then the extracted features were multiplied by the dropout probability to simulate the effect of dropout layer at test time. We used the liblinear package[3] to implement the SVM classifier and trained it using the scaled high-level features.

To minimize the same loss function as in the Bimodal-LSTM network, we optimized the primal problem by setting the option '$-s$' to 2 and set the cost of the SVM classifier equal to $\frac{1}{2\lambda}$, where λ denotes the L2 regularization strength used when training the network thoroughly. After training the SVM classifier, we copied the trained weights back into the last classification layer of the Bimodal-LSTM network to produce the final classifier.

4 Experiment Settings

4.1 The Datasets

The SEED dataset contains EEG signals and eye movement signals of three emotions (positive, neutral, and negative) from 15 subjects . All subjects were watching 15 four-minute-long emotional movie clips while their signals were collected. The EEG signals were recorded with ESI NeuroScan System at a sampling rate of 1000 Hz with a 62-channel electrode cap. The eye movement signals were recorded with SMI ETG eye tracking glasses. To compare with our previous work

---

[3] http://www.csie.ntu.edu.tw/~cjlin/liblinear/.

[6,7], we used the same data, which contained 27 experiments from 9 subjects. Signals recorded while the subject watching the first 9 movie clips were used as training datasets for each experiment and the rest were used as test datasets.

The DEAP dataset contains EEG signals and peripheral physiological signals of 32 participants. Signals were collected while participants were watching one-minute-long emotional music videos. And participants were asked to rate the levels of arousal, valence, like/dislike, dominance and familiarity for each video. We chose 5 as the threshold to divide the trials into two classes according to the rated levels of arousal and valence. Then the tasks can be treated as two binary classification problems, namely high or low arousal and valence. We used 10-fold cross validation to compare with [6,15].

4.2   Feature Extraction

For SEED dataset, we extracted Differential Entropy (DE) features from each EEG signal channel in five frequency bands: δ (1–4 Hz), θ (4–8 Hz), α (8–14 Hz), β (14–31 Hz), and γ (31–50 Hz). The size of Hanning window used when extracting EEG features was 4 s. At each time step, there were totally 310 (5 bands × 62 channels) dimensions for EEG features. As for eye movement data, the same features as in [6] were used. There were totally 41 dimensions including both Power Spectral Density (PSD) and DE features of pupil diameters at each time step. The features were rescaled between 0 and 1 when used as the inputs of the BDDAE model. Before training the BLSTM model, the features were normalized to zero mean and unit variance and then split into small sequences of length 60.

For DEAP dataset, we extracted DE features from EEG signals in four frequency bands: θ (4–8 Hz), α (8–14 Hz), β (14–31 Hz), and γ (31–50 Hz), since a bandpass frequency filter from 4 - 45 Hz was applied during pre-processing. The size of Hanning windows was 2 s. Then there were totally 128 (4 bands × 32 channels) dimensions of extracted 32-channel EEG features. As for peripheral physiological signals, time-domain features were extracted to describe the signals in different perspective, including maximum value, minimum value, mean value, standard deviation, variance and squared sum. So there were totally 48 (6 features × 8 channels) dimensions of extracted peripheral physiological features. Features were rescaled to [0, 1] before fed into the BDDAE model. And we also standardized the features and split them into sequences of length 5 to train the Bimodal-LSTM model.

4.3   Parameter Details

We trained different models for different experiments. For each experiment, we randomly selected some sets of hyper-parameters within a given range to train the model. The hyper-parameters of the BDDAE model include the hidden units' number of three encoders, the dropout probability, the L2 regularization strength, and the learning rate for the autoencoder network and the classifier network. The hyper-parameters and their corresponding range of the Bimodal-LSTM model for SEED and DEAP datasets are shown in Table 1.

Both models were implemented using tensorflow[4]. All weights were initialized from a Gaussian distribution with a mean of 0 and a standard deviation of 0.001. All biases were initialized to zero. And the initial hidden states and cell states of LSTM blocks were set to zero. The Adam optimization algorithm was used to train networks. EarlyStopping was also adopted to stop training when the accuracy had not increased 0.1% in the last 120 epochs.

Table 1. The hyper-parameters and their corresponding range of the Bimodal-LSTM model for SEED and DEAP datasets

Model	Hyper-parameter	SEED range	DEAP range
Bimodal-LSTM	EEG hidden size	16 to 256	32 to 256
	Other modalities' hidden size	8 to 64	16 to 256
	Dropout probability	0.3 to 0.99	0.2 to 0.9
	$\log_{10}$(L2 regularization strength)	-9 to 0	-4.5 to -1
	$\log_{10}$(learning rate)	-4 to -1.5	-2.5 to -0.5

5   Experiment Results

For SEED dataset, we randomly selected 200 sets of hyper-parameters within a given range for each experiment. We compared our models with two other state-of-the-art approaches [6,14] and the baseline method, which uses SVM directly as the classifier. As shown in Fig. 3, Bimodal-LSTM achieves the best accuracy (93.97%), which is about 2% points higher than the state-of-the-art approaches, and the smallest standard deviation (7.03%).

For DEAP dataset, we randomly selected 15 sets of hyper-parameters and tuned the parameters using 10-fold cross validation. The Bimodal-LSTM model was compared with one baseline method and two state-of-the-art approaches [6,15]. Liu *et al.* used Bimodal Deep AutoEncoder to extract high level features and used the preprocessed data as inputs. Yin *et al.* proposed a multiple-fusion-layer based ensemble classifier of stacked autoencoder to recognize emotions, and also estimated the accuracy by 10-fold cross validation. The baseline method used the same features as the Bimodal-LSTM model and used linear kernel SVM as the classifier. As shown in Table 2, Bimodal-LSTM obtains state-of-the-art performance on both arousal and valence classification tasks, with the mean accuracies of 83.23% and 83.83%, respectively.

---

[4] https://www.tensorflow.org/

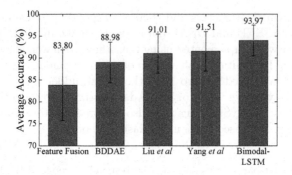

Fig. 3. Results of different models on SEED dataset. Feature Fusion denotes the model using directly concatenated features as inputs and using SVM with a radial basis function (RBF) kernel as the classifier. Liu *et al.* denotes the best result in [6], which uses the Bimodal Deep AutoEncoder model. And Yang *et al.* denotes the best result in [14].

Table 2. Average accuracies (%) and standard deviations of different approaches on DEAP dataset

	Feature fusion	Liu *et al.* [6]	Yin *et al.* [15]	Bimodal-LSTM
Arousal (%)	65.43/7.79	80.5/-	84.18/-	83.23/2.61
Valence (%)	65.29/7.93	85.2/-	83.04/-	83.82/5.01

6   Conclusion

In this paper, we have introduced two models to predict emotions based on EEG features and features from other modalities. The first is an extension of denoising autoencoders, called BDDAE, and the second is the Bimodal-LSTM model, which can use both the temporal information and frequency-domain information of features. Compared with other existing methods, the Bimodal-LSTM model has achieved the best performance with a mean accuracy of 93.97% on SEED dataset. For DEAP dataset, the Bimodal-LSTM model has achieved the state-of-the-art results with mean accuracies of 83.23% and 83.83% for arousal and valence classification tasks, respectively.

Acknowledgments. This work was supported in part by grants from the National Key Research and Development Program of China (Grant No. 2017YFB1002501), the National Natural Science Foundation of China (Grant No. 61673266), the Major Basic Research Program of Shanghai Science and Technology Committee (Grant No. 15JC1400103), ZBYY-MOE Joint Funding (Grant No. 6141A02022604), and the Technology Research and Development Program of China Railway Corporation (Grant No. 2016Z003-B).

References

1. Bengio, Y., Simard, P.Y., Frasconi, P.: Learning long-term dependencies with gradient descent is difficult. IEEE Trans. Neural Netw. **5**(2), 157–166 (1994)
2. Hinton, G.E., Salakhutdinov, R.R.: Reducing the dimensionality of data with neural networks. Science **313**(5786), 504–507 (2006)
3. Hinton, G.E., Zemel, R.S.: Autoencoders, minimum description length and helmholtz free energy. In: NIPS, pp. 3–10 (1994)
4. Hochreiter, S., Schmidhuber, J.: Long short-term memory. Neural Comput. **9**(8), 1735–1780 (1997)
5. Koelstra, S., Yazdani, A., Soleymani, M., Mühl, C., Lee, J., Nijholt, A., Pun, T., Ebrahimi, T., Patras, I.: Single trial classification of EEG and peripheral physiological signals for recognition of emotions induced by music videos. In: Yao, Y., Sun, R., Poggio, T., Liu, J., Zhong, N., Huang, J. (eds.) BI 2010. LNCS, vol. 6334, pp. 89–100. Springer, Heidelberg (2010). doi:10.1007/978-3-642-15314-3_9
6. Liu, W., Zheng, W.L., Lu, B.L.: Emotion recognition using multimodal deep learning. In: Hirose, A., Ozawa, S., Doya, K., Ikeda, K., Lee, M., Liu, D. (eds.) ICONIP 2016. LNCS, vol. 9948, pp. 521–529. Springer, Cham (2016). doi:10.1007/978-3-319-46672-9_58
7. Lu, Y., Zheng, W.L., Li, B., Lu, B.L.: Combining eye movements and EEG to enhance emotion recognition. In: IJCAI, pp. 1170–1176 (2015)
8. Saneiro, M., Santos, O.C., Salmeronmajadas, S., Boticario, J.G.: Towards emotion detection in educational scenarios from facial expressions and body movements through multimodal approaches. Sci. World J. **2014**, 484873 (2014)
9. Tang, Y.: Deep learning using linear support vector machines. Workshop on Representational Learning, ICML (2013)
10. Vincent, P., Larochelle, H., Bengio, Y., Manzagol, P.: Extracting and composing robust features with denoising autoencoders. In: ICML, pp. 1096–1103 (2008)
11. Wang, X.W., Nie, D., Lu, B.L.: Emotional state classification from eeg data using machine learning approach. Neurocomputing **129**, 94–106 (2014)
12. Wu, Y., Schuster, M., Chen, Z., Le, Q.V., Norouzi, M., Macherey, W., Krikun, M., Cao, Y., Gao, Q., Macherey, K., et al.: Google's neural machine translation system: bridging the gap between human and machine translation. arXiv preprint (2016). arXiv:1609.08144
13. Xiong, W., Droppo, J., Huang, X., Seide, F., Seltzer, M., Stolcke, A., Yu, D., Zweig, G.: The microsoft 2016 conversational speech recognition system. In: 2017 IEEE International Conference on Acoustics, Speech and Signal Processing (ICASSP), pp. 5255–5259. IEEE (2017)
14. Yang, Y., Wu, Q.J., Zheng, W.L., Lu, B.L.: EEG-based emotion recognition using hierarchical network with subnetwork nodes. IEEE Trans. Cogn. Dev. Syst. (2017). doi:10.1109/TCDS.2017.2685338
15. Yin, Z., Zhao, M., Wang, Y., Yang, J., Zhang, J.: Recognition of emotions using multimodal physiological signals and an ensemble deep learning model. Comput. Methods Prog. Biomed. **140**, 93–110 (2017)

Investigating Gender Differences of Brain Areas in Emotion Recognition Using LSTM Neural Network

Xue Yan[1], Wei-Long Zheng[1], Wei Liu[1], and Bao-Liang Lu[1,2,3(✉)]

[1] Department of Computer Science and Engineering,
Center for Brain-like Computing and Machine Intelligence,
Shanghai Jiao Tong University, 800 Dong Chuan Road, Shanghai 200240, China
{yanxue_10085,weilong,liuwei-albert,bllu}@sjtu.edu.cn
[2] Key Laboratory of Shanghai Education Commission for Intelligent Interaction
and Cognitive Engineering, Shanghai Jiao Tong University,
800 Dong Chuan Road, Shanghai 200240, China
[3] Brain Science and Technology Research Center, Shanghai Jiao Tong University,
800 Dong Chuan Road, Shanghai 200240, China

Abstract. In this paper, we investigate key brain areas of men and women using electroencephalography (EEG) data on recognising three emotions, namely happy, sad and neutral. Considering that emotion changes over time, Long Short-Term Memory (LSTM) neural network is adopted with its capacity of capturing time dependency. Our experimental results indicate that the neural patterns of different emotions have specific key brain areas for males and females, with females showing right lateralization and males being more left lateralized. Accordingly, two non-overlapping brain regions are selected for two genders. The classification accuracy for females (79.14%) using the right lateralized region is significantly higher than that for males (67.61%), and the left lateralized area educes a significantly higher classification accuracy for males (82.54%) than females (73.51%), especially for happy and sad emotions.

Keywords: Electroencephalography · Emotion · Long Short-Term Memory neural network · Gender differences · Brain areas

1 Introduction

Gender differences can be observed in both behaviour and character. In the field of emotion and psychopathology, gender differences are largely considered, since studying differences in gender of normal subjects, psychiatric or brain-damaged patients is instructive for clinical studies such as depression and obsessive-compulsive disorder [5]. On language processing and visuospatial tasks, gender differences are also widely discussed, with men outperform women on visuospatial tasks especially on mental rotation and spatial perception, and women perform better than men on language tasks particularly on verbal fluency [1]. In this paper, we intend to study gender differences of brain areas in EEG-based emotion recognition using Long Short-Term Memory (LSTM) neural network.

© Springer International Publishing AG 2017
D. Liu et al. (Eds.): ICONIP 2017, Part IV, LNCS 10637, pp. 820–829, 2017.
https://doi.org/10.1007/978-3-319-70093-9_87

The existing study in neuropsychological processes has found that men and women performed more activation in the parietal area and the right frontal region, respectively [11]. Accordingly, emotional processing in men and women is supposed to occur in different brain areas. Moreover, an overall observation that emotion is lateralized on the right hemisphere has been reported. However, there is evidence that the hemispheric asymmetry and lateralization of emotional activities are more complicated and region-specific than the previous observation [9]. Besides, differences on brain asymmetries are influenced not only by tasks but also by its details like stimuli and the presentation methods.

Therefore, in this paper, we focus on investigating the key brain areas for men and women while their emotions are evoked by using film clips as stimuli and analysing with EEG data. In the previous study, little work has been done on investigating gender-related differences of key brain areas on emotions using neural networks [5,7,9]. Since LSTM possesses a great characteristic on incorporating information over a long period of time, which accords with the fact that emotions are developed and changed over time, LSTM is an appropriate method for emotion recognition.

In our previous work, we have demonstrated that there are gender differences using EEG data to recognise three different emotions. We indicated that males are more individually different in emotions while females have more similar emotion patterns in EEG data [13]. In this study, we extend our previous study in two ways. First, we enlarge the dataset from 45 sets to 70 sets to confirm the efficiency of the experimental results, and apart from SVM, LSTM-based method is applied for emotion recognition. Second, we try to find out gender-related key brain areas for different emotions, which has not been investigated yet to our best knowledge. Besides, the influence of the key brain areas on each emotion is studied.

2 Methodology

2.1 Feature Extraction and Feature Smoothing

The EEG data is filtered between 0.05 Hz and 50 Hz and preprocessed by the notch filter to eliminate power-line interference. The Short-Term Fourier Transform (STFT) with a 1 s no overlapping Hanning window is employed to calculate the Differential Entropy (DE) features of EEG data [2]. 30 channels are used in the electrode cap, and the EEG data in each channel is filtered into five frequency bands (δ: 1–3 Hz, θ: 4–7 Hz, α: 8–13 Hz, β: 14–30 Hz and γ: 31–50 Hz), which means the total dimension of EEG features is 150.

There are some rapid fluctuations in EEG data that do not contribute to emotional states, so we further perform feature smoothing. The Linear dynamic system is an efficient method to remove noises and employed in this paper [8,10].

2.2 Feature Selection

In order to obtain the key regions of emotions for men and women, Minimal-Redundancy-Maximal-Relevance (MRMR) [6] is applied as feature selection method. We use selected features with different dimensions to classify emotions. The higher accuracy rates indicate that the more effective features are used, which suggests that areas where selected features are located are more likely to be key areas of emotion.

2.3   Classification

Linear Support Vector Machine (SVM) with soft margin and LSTM are applied as classifiers in the paper. LSTM neural networks provide a solution of vanishing gradient by replacing the summation units of a standard Recurrent Neural Networks (RNN) in the hidden layer with memory blocks [3]. Each block with one cell consists of the input gate, output gate and forget gate that can write, read and reset the information, which ensures the usage of data over a long period of time. The update rule of the block is shown below:
The input gate:

$$i_t = \sigma \left(W_{ix} X_t + W_{ih} h_{t-1} + b_i \right). \tag{1}$$

The forget gate:

$$f_t = \sigma \left(W_{fx} X_t + W_{fh} h_{t-1} + b_f \right). \tag{2}$$

The cell state:

$$\tilde{C}_t = tanh \left(W_{cx} X_t + W_{ch} h_{t-1} + b_c \right). \tag{3}$$

$$C_t = i_t * \tilde{C}_t + f_t * C_{t-1}. \tag{4}$$

The output gate:

$$o_t = \sigma \left(W_{ox} X_t + W_{oh} h_{t-1} + b_o \right). \tag{5}$$

$$h_t = o_t * tanh \left(C_t \right). \tag{6}$$

where σ is the logistic sigmoid function.

Before applying LSTM, EEG features are first normalised to zero mean and unit variance and then divided into 64-s data sequences, which is determined by the length of movie clips. The structure of LSTM model used in the paper is shown in Fig. 1. The processed EEG features are encoded by a single layer of LSTM and decoded by Multi-Layer Perception (MLP) using ReLU as the activation function. The encoder and decoder layer share the same number of neurons, and two dropout layers with a percentage of 0.5 are added between the LSTM decoder and the MLP decoder and before the output layer to improve the generalization ability.

The model is implemented based on Keras.[1] The weights of the linear transformation of the recurrent state and the inputs are initialised by a random orthogonal matrix and Glorot uniform initializer, respectively. The bias of the forget gate is initialised to one, and hyperparameters are determined by cross-validation. During training, the RMSProp method is employed to optimize the loss function and the early stopping strategy is adopted when there is no improvement on the validation set after 10 epochs.

---

[1] https://keras.io/.

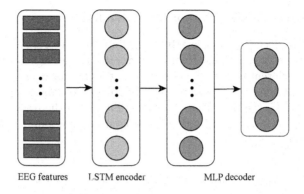

EEG features   LSTM encoder   MLP decoder

Fig. 1. The structure of LSTM nerual networks.

3   Experimental Results

3.1   Experiment Design

Each experiment consisted of 12 trials (4 trials per emotion). At the beginning of each trial, the textual guidance was presented on the screen for 20 s to ask subjects to relax and to calm their emotions. Then, the clip evoking one single emotion was presented for 1.5 to 4 min. During this time, subjects were asked to watch the film with emotional involvement. Finally, the self-assessment stage would last for 10 s for subjects to assess whether and what emotions were evoked and the extent to which their emotions were evoked.

Film clips used as stimuli were evaluated before experiments by 50 college students (25 females) aged from 19 to 26. They were asked to score the clips according to the degree their emotions were evoked and in the end, 60 clips (12 for each of the five experiments) were chosen based on ratings. In order to avoid effects of the former or similar clip, all clips were shuffled randomly and used only once during whole experiments for one subject.

During the experiment, the EEG data was recorded by the ESI NeuroScan System with a 32-channel electrode cap with 1000 Hz sampling rate.[2] The experiments were carried out in a clean and comfortable environment when subjects were in good mental states. Before each experiment, each subject was informed of details and precautions of the procedure, and feedback forms collected after experiments indicated that certain emotions were successfully evoked.

16 healthy subjects (8 females) aged between 18 and 28 were recruited from the university. Each of them participated the experiment for 5 times to avoid individual deviations, generating a total of 70 sets of valid data, half of which are from females. All the subjects were right-handed with normal or corrected-to-normal vision and were told the harmlessness and the goal of the experiment.

---

[2] http://compumedicsneuroscan.com/product/32-channels-quik-cap/.

3.2   Key Features Selection for Two Genders

As mentioned in Sect. 2.2, we use MRMR to find the key brain areas of different emotions for males and females. Figure 2 illustrates the average accuracies of two genders under different feature dimensions using SVM. For males, the accuracy is highest when the feature dimension equals to 20, and for females, the model with 18 features acquires the best accuracy. Generally, when the feature dimension is around 20, the performance of emotion recognition is among the best for both men and women.

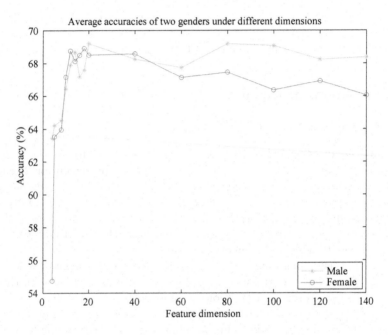

Fig. 2. The accuracy of emotion recognition using features selected by MRMR, with feature dimensions varying from 4 to 140.

To confirm the key brain areas, we investigate the discriminative neural patterns of three emotions using topography. Figure 3 shows the mean square deviation of average DE features across five frequency bands in each electrode for three emotions, and that the electrode location of 20 features based on the feature selection for females and males. The features selected from two methods are consistent with each other on the location, with key electrodes mainly concentrating on the right hemisphere in the α and β bands for females, and on the right part of the left hemisphere in the α and β band and the right temporal lobe in γ band for males. Therefore, the selected 20-dimension features are effective for emotion recognition.

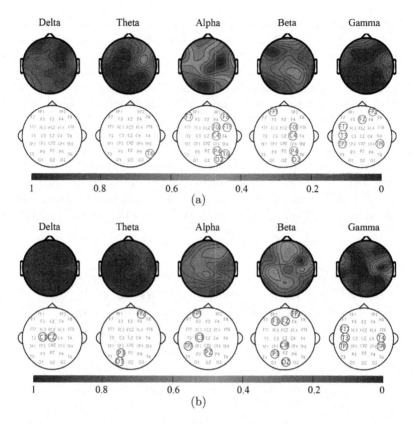

Fig. 3. The mean square deviation of the average energy of three emotions in brain topography, and features selected by MRMR: (a) females, and (b) males. From top to bottom and left to right, the electrodes selected are: (a) T6 in θ, F7, F8, FC4, FT8, C4, P4, T6 and O2 in α, FP1, FC4, C4, P4 and O2 in β, and FP2, FZ, FT7, T3, TP7 and TP8 in γ band; (b) C3 and CZ in δ, FP2, P3 and O1 in θ, FP1, C3, TP2 and PZ in α, FP2, F3, FZ, CPZ, P3 and OZ in β, and FT7, T3, T4, TP7 and TP8 in γ band.

3.3   Key Brain Areas for Two Genders

To investigate gender differences, the common electrode locations of females and males from the 20 features selected by MRMR are removed, and 11-dimension features that are critical only to females (FFs) and males (MFs) are left, respectively. The remaining features are T6 in θ band, F7, F8, FC4, P4, T6 and O2 in α band and FC4, C4, P4 and O2 in β band for females, and C3 and CZ in δ band, P3 and O1 in θ band, C3 and PZ in α band, F3, CPZ, P3 and OZ in β band and T4 in γ band for males. The detailed electrode location is presented in Fig. 4.

In order to explore the impact of FFs and MFs on men and women, respectively, four experiments named the cross-gender training are designed, with FFs and MFs classifying men and women's emotions. The result shown in Fig. 5 illustrates that with LSTM, accuracies of using FFs to recognise emotions of females and males are 79.14% and 67.61%, respectively, and accuracies of using MFs to recognise emotions of females and males are 73.51% and 82.54%, respectively, which are all higher than that of SVM. The one-way analysis of variance (ANOVA) shows that the performance of FFs for females is significantly better than that for males for both SVM ($p = 0.0024$) and LSTM ($p = 0.0022$) models. Also, the performance of using MFs to recognise emotions of males is significantly better than that of females for both SVM ($p = 0.0474$) and LSTM ($p = 0.0213$). Besides, the classification accuracy of the LSTM model is higher than that of the SVM model by 12.46% to 18.32% according to different training tasks. The experimental results indicate that the time-dependent property of LSTM make it perform better than SVM in terms of emotion recognition.

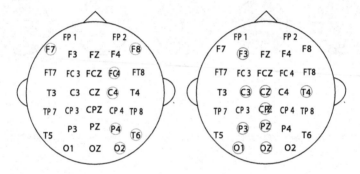

Fig. 4. The electrodes chosen for females and males after removing duplicate electrodes from the 20 features. The features chosen for females are shown on the left.

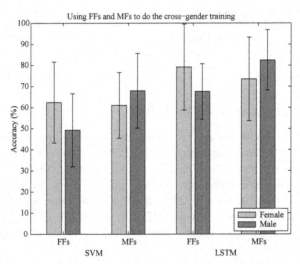

Fig. 5. The cross-gender training on emotion recognition by SVM and LSTM.

In conclusion, the right hemisphere where FFs are located is significantly important to females' emotions than males', and the left hemisphere where MFs are located is significantly critical to males' emotions than females'. The existing study [12] has shown that in terms of lateralisation, gender differences may be caused by using different cognitive strategies in the same task. Besides, in the functional magnetic neuroimaging (fMRI) study [1], males have been found to be more left lateralised in phonological processing tasks and females are more right lateralised in visuospatial processing tasks. These results imply that in our task, the lateralisation can be explained by men being more sensitive to phonological information while women being more sensitive to visuospatial information of movie clips during experiments.

3.4 Influence of Key Brain Areas on Each Emotion

The confusion graphs of cross-gender training strategy trained by LSTM are shown in Fig. 6, which gives insights into the influence of key brain areas on three emotions. Figure 6 demonstrates that as for the right hemisphere where FFs located, all three emotions are recognised with a higher accuracy and a lower misclassification probability for females than for males. Similar observation generally exists for the left hemisphere where MFs located, with which males are recognised with higher accuracy and lower misclassification probability than females, except for a few slight deviation such like the neutral emotion. Studies of attentional processes on EEG have shown that alpha activity can be found on overall parietal lobe when tasks not requiring attention [7]. In our experiment, subjects tended to be relaxed and pay less attention under neutral emotion, which evoked alpha activity. While for two key brain areas, FFs contain many features of the alpha band but MFs only contain few, which explains the

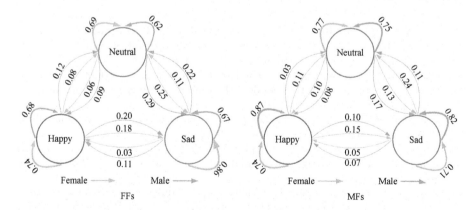

Fig. 6. The confusion graph of the cross-gender training strategy using LSTM models. The left one shows the confusion graph of using FFs to recognise emotions for females and males. The numbers are the proportion of samples in class (arrow tail) that is classified as class (arrow head), and the thickness of the line represents the value.

deviation result on neutral emotion. Furthermore, sad and neutral emotions are more confused with each other, which is consistent with our previous study [4].

In conclusion, key brain regions for females and males primarily have effects on happy and sad emotions. Due to the influence of alpha activity, the critical brain areas for two genders are affected in the neutral emotion.

4   Conclusions

In this paper, we have investigated the gender differences of key brain areas in emotion recognition from EEG data using LSTM neural network. From our experimental results, we have obtained the following observations: (1) Females and males have different key brain areas on emotions, with females showing the right lateralisation and males showing a trend of left lateralisation. (2) As for the overall emotion recognition, the performance of men's critical brain area on men significantly outperforms the performance on women, and the key brain area of women also performed significantly better for women than men. (3) The effect of two key brain areas is mainly on happy and sad emotions among three emotions.

Acknowledgments. This work was supported in part by grants from the National Key Research and Development Program of China (Grant No. 2017YFB1002501), the National Natural Science Foundation of China (Grant No. 61673266), the Major Basic Research Program of Shanghai Science and Technology Committee (Grant No. 15JC1400103), ZBYY-MOE Joint Funding (Grant No. 6141A02022604), and the Technology Research and Development Program of China Railway Corporation (Grant No. 2016Z003-B).

References

1. Clements, A.M., Rimrodt, S.L., Abel, J.R., Blankner, J.G., Mostofsky, S.H., Pekar, J.J., Denckla, M.B., Cutting, L.E.: Sex differences in cerebral laterality of language and visuospatial processing. Brain Lang. **98**(2), 150–158 (2006)
2. Duan, R.N., Zhu, J.Y., Lu, B.L.: Differential entropy feature for EEG-based emotion classification. In: 6th International IEEE/EMBS Conference on Neural Engineering, vol. 8588, pp. 81–84 (2013)
3. Hochreiter, S., Schmidhuber, J.: Long short-term memory. Neural Comput. **9**(8), 1735–1780 (1997)
4. Lu, Y., Zheng, W.L., Li, B., Lu, B.L.: Combining eye movements and EEG to enhance emotion recognition. In: International Conference on Artificial Intelligence, pp. 1170–1176 (2015)
5. Nolen-Hoeksema, S., Aldao, A.: Gender and age differences in emotion regulation strategies and their relationship to depressive symptoms. Pers. Individ. Differ. **51**(6), 704–708 (2011)
6. Peng, H., Long, F., Ding, C.: Feature selection based on mutual information: criteria of max-dependency, max-relevance, and min-redundancy. IEEE Trans. Pattern Anal. Mach. Intell. **27**(8), 1226–1238 (2005)

7. Rowland, N., Meile, M., Nicolaidis, S., et al.: EEG alpha activity reflects attentional demands, and beta activity reflects emotional and cognitive processes. Science **228**(4700), 750–752 (1985)

8. Shi, L.C., Lu, B.L.: Off-line and on-line vigilance estimation based on linear dynamical system and manifold learning. In: Engineering in Medicine and Biology Society, pp. 6587–6590. IEEE (2010)

9. Wager, T.D., Phan, K.L., Liberzon, I., Taylor, S.F.: Valence, gender, and lateralization of functional brain anatomy in emotion: a meta-analysis of findings from neuroimaging. Neuroimage **19**(3), 513–531 (2003)

10. Wang, X.W., Nie, D., Lu, B.L.: Emotional state classification from EEG data using machine learning approach. Neurocomputing **129**(4), 94–106 (2014)

11. Weiss, E., Siedentopf, C.M., Hofer, A., Deisenhammer, E.A., Hoptman, M.J., Kremser, C., Golaszewski, S., Felber, S., Fleischhacker, W.W., Delazer, M.: Sex differences in brain activation pattern during a visuospatial cognitive task: a functional magnetic resonance imaging study in healthy volunteers. Neurosci. Lett. **344**(3), 169–172 (2003)

12. Welsh, T.N., Elliott, D.: Gender differences in a dichotic listening and movement task: lateralization or strategy? Neuropsychologia **39**(1), 25–35 (2001)

13. Zhu, J.-Y., Zheng, W.-L., Lu, B.-L.: Cross-subject and cross-gender emotion classification from EEG. In: Jaffray, D.A. (ed.) World Congress on Medical Physics and Biomedical Engineering. IP, vol. 51, pp. 1188–1191. Springer, Cham (2015). doi:10.1007/978-3-319-19387-8_288

Can Eye Movement Improve Prediction Performance on Human Emotions Toward Images Classification?

Kitsuchart Pasupa[1]([✉]), Wisuwat Sunhem[1], Chu Kiong Loo[2],
and Yoshimitsu Kuroki[3]

[1] Faculty of Information Technology,
King Mongkut's Institute of Technology Ladkrabang,
Bangkok 10520, Thailand
kitsuchart@it.kmitl.ac.th, wisuwat.sun@gmail.com
[2] Faculty of Computer Science and Information Technology,
University of Malaya, 50603 Kuala Lumpur, Malaysia
ckloo.um@um.edu.my
[3] Department of Control and Information Systems Engineering,
Kurume National College of Technology, Fukuoka 830-8555, Japan
kuroki@kurume.kosen-ac.jp

Abstract. Recently, image sentiment analysis has become more and more attractive to many researchers due to an increasing number of applications developed to understand images e.g. image retrieval systems and social networks. Many studies aim to improve the performance of the classifier by many approaches. This work aims to predict the emotional response of a person who is exposed to images. The prediction model makes use of eye movement data captured while users are looking at images to enhance the prediction performance. An image can stimulate different emotions in different users depending on where and how their eyes move on the image. Two image datasets were used, i.e. abstract images and images with context information, by using leave-one-user-out and leave-one-image-out cross-validation techniques. It was found that eye movement data is useful and able to improve the prediction performance only in leave-one-image-out cross-validation.

Keywords: Emotion classification · Eye movement · Abstract image · Image with context information

1 Introduction

Many researchers have focused on understanding contents in digital images. One of the reasons is to build a content-based image retrieval (CBIR) system which ranks images in a database according to similarities between the content in query image and the content in database images rather than by using keywords. Each image is represented as a feature vector which can contain low-level image

© Springer International Publishing AG 2017
D. Liu et al. (Eds.): ICONIP 2017, Part IV, LNCS 10637, pp. 830–838, 2017.
https://doi.org/10.1007/978-3-319-70093-9_88

features such as colour, shape, and texture. It is undeniable that CBIR can dramatically reduce human effort in labelling or tagging images. The ability to utilise low-level feature vectors on CBIR has been well-documented in the literature [8,11,13]. Later, Emotion Semantic Image Retrieval (ESIR)–an extension of CBIR was introduced [16]. It focuses on the conception of image representation based on emotions or moods of humans. Similarly to CBIR, this approach aims to construct a distance metric between query image and images in the database, but with degrees of emotion. The process is broken down into three major phases: emotional semantic representation; image feature extraction; and emotion recognition. A survey of ESIR can be found at [15]. Recently, [17] attempted to cope with this task with one kind of deep neural network called "Convolutional Neural Network (CNN)". They focused on the noisy labelled dataset. However, deep neural networks always lead to poor performance when they are trained by small-sample-size datasets. Thus, in their work, a pre-trained deep model was also re-trained and learned more knowledge, through the transfer learning method utilised to solve the problem.

It is known that a learning to cooperate model with multiple data sources can enhance overall performance of the system. One of the sources which can be easily obtained is user feedback to the CBIR system. The feedback can be both implicit and explicit. Here, we focus on implicit feedback which is involuntarily given by a human i.e. eye movements while the user is looking at the screen. Many works demonstrated that eye movement data can be used to enhance overall performance [1,6,7,10]. Recently, Pasupa et al. introduced an approach to predict human emotions stimulated by abstract art using low-level features with eye movement data [9]. The work evaluated both the user-model and global model based on the leave-one-out technique. In this paper, we built on top of the previous work by investigating on images with context information in order to determine whether eye movement can be useful in ESIR.

The paper is structured as follows: related theories to our approach are shown in Sect. 2. Section 3 describes the two datasets used in this paper which are (i) images with non-context information, and (ii) images with context information. This includes how image features and eye movement information are extracted and combined. Then, our experimental framework is explained in Sect. 4. Experimental results are discussed in Sect. 5. And finally, this work is concluded in Sect. 6.

2 Image and Eye Tracking

2.1 Eye Tracking

Eye tracking techniques relate to the process of monitoring the movement of eyes gazing on something. This can give some informative patterns of implicit user feedback which can benefit wide areas of commercial IT production and research such as Human Computer Interaction (HCI), and marketing psychology [3,14]. The eye tracking process requires an eye tracking device that is controlled by dedicated software in order to locate the coordinates at which user is gazing on the screen. The eye tracking device used in this work is called the "Eye Tribe" (the same device as used in Pasupa et al. [9]). This device can collect 30 coordinates per second. It is the smallest portable eye tracking device with the size of $20 \times 1.9 \times 1.9$ cm. Basically, the Eye Tribe consists of two elementary modules– a camera and infrared LEDs. The tracking process starts with locating user's pupils through the camera in a conjunction with computer vision techniques. Then, it will return a list of the coordinates. All these points are combined with image features in experiments. The experiment is described in detail in Sect. 4. A review of eye tracking methodology can be found at [4].

2.2 Abstract Image

Abstract art is a type of picture that is painted without meaningful objects or forms. Artists also express their feeling and background experiences in forms of different shapes and colours, usually not containing imitated real-world objects [12]. Furthermore, these non-contextual images can stimulate a humans emotion whenever they are looking at them. However, the difference in emotion experienced by people of different backgrounds is greater with abstract art than with representational art [5]. This is because this type of art requires the viewers imagination which is related to their past experiences in order to understand and feel the art. Moreover, in abstract art, each part of the picture can be painted in different shapes and colours with different brush stroke techniques. This can lead to different feeling depending on the viewers. We are of the belief that it would be more effective to predict human emotion if there was information about precisely where viewer was gazing.

2.3 Image with Context Information

An image with context information is a kind of picture with meaningful content inspired by real world objects contained in the image. It is related to figurative art defined as the opposite of abstract art. Figurative works are sometimes created to illustrate information about human life in the past, or evolutions of any specimen [2]. It can be perceived that the content of figurative pictures are formed by realistic representation including objects and their interaction. We presume that figurative art is more sophisticated than abstract images to predict human emotions with low-level image features and eye movement information. This is because not only should low-level representation be taken into consideration, but also understandable content in the image can affect the emotions of a person looking at the image. In this paper, we evaluate both types of image using the same framework to discover how effective and useful the eye movement information is.

3 Data Preparation

The datasets were prepared according to Pasupa et al. [9]. Two sets of images mentioned in the previous section were shown to 20 users. Each image dataset

(a) Abstract image dataset

(b) Image with context information dataset

Fig. 1. A set of example images for dataset.

contains 100 images. Examples of images in the dataset are shown in Fig. 1. The users were asked to label images according to how they felt, i.e. happy, sad, angry, and afraid. In order to show the distribution of emotions labelled by the user, the stacked histograms of labels given by 20 volunteers for each image are shown in Fig. 2. This figure shows an agreement of users' preference regarding the four emotions to be labelled on each image. It can be seen that, while there is only a 57.8% chance that users give the same label on each abstract image, there is a 76.3% chance for images with context information. Therefore, this can confirm that figurative images more often lead people to the same emotion, even though the viewers have had different background experiences, than do abstract images.

After that, we generated three types of image sets: (i) image only, (ii) combining image with eye movement data with Gaussian blur, and (iii) image with

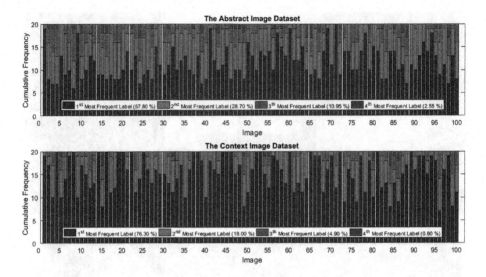

Fig. 2. The stacked histograms show the number of each label defined by 20 volunteers.

eye movement data without Gaussian blur. The size of the Gaussian eye locating area was set to be 100, 125, and 150. The variance parameter of the Gaussian function was also set to be the size of the Gaussian mask divided by six [9]. The images (both with and without pre-processing) were feature extracted by low-level-image feature extraction, i.e. histogram of RGB colour, Sobel shape, and Gabor texture features. Each type of feature was discretised by equal-width-binning discretisation to be a histogram with size of 8, 16, 32 and 64-bin. After that each image was represented as a feature vector by concatenating these histograms. All the samples were then normalised to unit norm.

4   Experiment Framework

We propose two frameworks for the validation model as follows: (i) Leave-one-user-out cross validation–used to test how well the model can perform when there is no information about each user's behaviour in the model. The model learns all the information about all the images in the database and other users' eye movement data on the images too; and (ii) Leave-one-image-out cross validation–the case where we have all the users' behaviour information but have no information about the new images at all.

The first framework contained 1,900 instances as a training set (19 users) and 100 instances as a test set (hold out one user). The experiment was run 20 times as we held out each user for one test set. For the second frame work, the training set consisted of 1,980 images and 20 images for the test set. The experiment was run 100 times as there were 100 images. We evaluated all three pre-processed image sets, as explained in the previous section, on both frameworks and on two datasets (Abstract image and image with context information). A Support Vector Machine with Linear kernel was utilised as a classifier in all the experiments. All the model parameters were tuned to achieve the optimal model. The parameter ranges are as follows: the size of histogram bin feature is $\{8, 16, 32,$ and $64\}$-bin; SVM regularisation parameter (C) is $\{10^{-6}, 10^{-6}, 10^{-5}, 10^{5}, 10^{6}\}$; the size of the Gaussian mask is varied by 100, 125 and 150 for the eye movement dataset. The optimal model was selected based on the highest Area Under the Receiver Operating Characteristic (AUROC).

5   Results and Discussion

5.1   Leave-One-User-Out Cross Validation

This framework is similar to [9]. It should be noted that only 10 users were evaluated in [9] and it was found that eye movement can improve the prediction performance. In this paper, we evaluated all 20 users. The results are shown in Table 1. Performance using image feature and image with eye movement data are at the same level on both accuracy and AUROC. It is clearly seen that using eye movement data is not very useful in the abstract image dataset ($p = 0.9763$). Moreover, applying the Gaussian function to the eye movement data can lead to poor performance at 65.8% of AUROC.

Table 1. The performances from leave-one-user-out cross validation.

Dataset	Feature	Accuracy $(\mu \pm \sigma)$	AUROC $(\mu \pm \sigma)$
Abstract image	Baseline	$51.3 \pm 6\%$	$68.4 \pm 4\%$
	Image	$53.3 \pm 6\%$	$69.1 \pm 4\%$
	Image + Eye movement	$53.3 \pm 5\%$	$68.9 \pm 4\%$
	Image + Eye movement with Gaussian	$49.8 \pm 4\%$	$65.8 \pm 3\%$
Image with context Information	Baseline	$74.0 \pm 6\%$	$82.6 \pm 4\%$
	Image	$76.2 \pm 6\%$	$83.7 \pm 4\%$
	Image + Eye movement	$72.6 \pm 5\%$	$81.1 \pm 3\%$
	Image + Eye movement with Gaussian	$63.0 \pm 4\%$	$72.9 \pm 2\%$

In the case of image with context information, eye movement information does not enhance the performance, but degrades the performance from 83.7% of AUROC to 81.1% for image fused with eye movements, and to 72.6% for the case of image fused with eye movements with Gaussian blur.

In the case of leave-one-user-out and using image features alone, the model had been trained with all the information on images with their labels. The model is likely to predict the test image (which has been already seen/trained by the model) to the label most frequently selected by other users. However, using image features to create models is still useful because it is better than baseline method. The baseline method uses the label for each image most frequently selected by other users as the predicted label. It achieved 51.3% and 74.0% of accuracy in abstract image and in image with context information, respectively. Using image features can yield better performance at 53.3% and 76.2% accuracy in non-figurative and figurative images, respectively.

5.2   Leave-One-Image-Out Cross Validation

This is another way to evaluate the global model by allowing the model to demonstrate an ability to predict an unseen image through knowing all the other users' behaviour. It should be noted that unseen image means unseen data (with or without eye movement) by the model. Taking eye movement information into account can enhance the accuracy of the prediction from 53.6% and 55.2% in non-figurative and figurative images, respectively, to 59.7% and 64.0% as shown in Table 2. The results are significant at $p = 0.045$ for both abstract and figurative images.

Table 2. The performances from leave-one-image-out cross dation.

Dataset	Feature	Accuracy $(\mu \pm \sigma)$	AUROC $(\mu \pm \sigma)$
Abstract image	Baseline	33.9 ± 10%	57.5 ± 9%
	Image	53.6 ± 20%	50.0 ± 0%
	Image + Eye movement	58.7 ± 16%	64.4 ± 6%
	Image + Eye movement with Gaussian	56.2 ± 17%	63.5 ± 8%
Image with Context Information	Baseline	34.9 ± 28%	50.0 ± 8%
	Image	55.2 ± 34%	50.0 ± 0%
	Image + Eye movement	64.0 ± 27%	66.6 ± 9%
	Image + Eye movement with Gaussian	56.7 ± 28%	61.1 ± 7%

We further compared our proposed features to a baseline method which predicts unseen images to a label which is a majority class given by the user. Without image features, the result dropped to 33.9% and 34.9% accuracy in abstract and figurative images, respectively.

6   Conclusion

In this work, two global models–leave-one-user-out and leave-one-image-out– of emotion classification were investigated. The experiments showed that eye movement was not useful and could degrade the average performance compared with using image features or the baseline method on the leave-one-user-out framework. Hence, some users behaviour information might not be useful to another user. In this case, using image features alone is the best option, the same as using the majority vote of each image. In the other framework, eye movement can improve prediction performance of human emotions toward unseen images when the model has learned user's behaviour in both abstract image and image with context information. Apart from these, this experiment confirmed that using Gaussian blur on eye movement data can degrade the overall performance, which is similar recent work.

Acknowledgments. This work was supported by the Faculty of Information Technology, King Mongkut's Institute of Technology Ladkrabang under grant agreement number 2560-06-002.

References

1. Auer, P., Hussain, Z., Kaski, S., Klami, A., Kujala, J., Laaksonen, J., Leung, A.P., Pasupa, K., Shawe-Taylor, J.: Pinview: implicit feedback in content-based image retrieval. In: Proceeding of the Workshop on Applications of Pattern Analysis, WAPA 2010, Windsor, UK, pp. 51–57 (2010)
2. Avital, T.: Art versus Nonart: Art Out of Mind. Cambridge University Press Cambridge, UK (2003)
3. Chandon, P., Hutchinson, J., Bradlow, E., Young, S.H.: Measuring the value of point-of-purchase marketing with commercial eye-tracking data. INSEAD Working Paper Collection, vol. 22, pp. 1 (2007)
4. Duchowski, A.T.: Eye Tracking Methodology: Theory and Practice, no. 328. Springer, Cham (2017). doi:10.1007/978-3-319-57883-5
5. Feist, G.J., Brady, T.R.: Openness to experience, non-conformity, and the preference for abstract art. Empirical Stud. Arts **22**(1), 77–89 (2004)
6. Hardoon, D.R., Pasupa, K., Shawe-Taylor, J.: Image ranking with implicit feedback from eye movements. In: Proceeding of the 6th Biennial Symposium on Eye Tracking Research and Applications, ETRA 2010, Austin, USA, pp. 291–298 (2010)
7. Hussain, Z., Leung, A.P., Pasupa, K., Hardoon, D.R., Auer, P., Shawe-Taylor, J.: Exploration-exploitation of eye movement enriched multiple feature spaces for content-based image retrieval. In: Balcázar, J.L., Bonchi, F., Gionis, A., Sebag, M. (eds.) ECML PKDD 2010. LNCS, vol. 6321, pp. 554–569. Springer, Heidelberg (2010). doi:10.1007/978-3-642-15880-3_41
8. Müller, H., Michoux, N., Bandon, D., Geissbuhler, A.: A review of content-based image retrieval systems in medical applications-clinical benefits and future directions. Int. J. Med. Inform. **73**(1), 1–23 (2004)
9. Pasupa, K., Chatkamjuncharoen, P., Wuttilertdeshar, C., Sugimoto, M.: Using image features and eye tracking device to predict human emotions towards abstract images. In: Bräunl, T., McCane, B., Rivera, M., Yu, X. (eds.) PSIVT 2015. LNCS, vol. 9431, pp. 419–430. Springer, Cham (2016). doi:10.1007/978-3-319-29451-3_34
10. Pasupa, K., Szedmak, S.: Utilising Kronecker decomposition and tensor-based multi-view learning to predict where people are looking in images. Neurocomputing **248**, 80–93 (2017)

11. Rui, Y., Huang, T.S., Chang, S.F.: Image retrieval: Current techniques, promising directions, and open issues. J. Vis. Commun. Image Representation **10**(1), 39–62 (1999)

12. Schapiro, M.: Nature of Abstract Art. American Marxist Association (1937)

13. Smeulders, A.W., Worring, M., Santini, S., Gupta, A., Jain, R.: Content-based image retrieval at the end of the early years. IEEE Trans. Pattern Anal. Mach. Intell. **22**(12), 1349–1380 (2000)

14. Strandvall, T.: Eye Tracking in Human-Computer Interaction and Usability Research. In: Gross, T., Gulliksen, J., Kotzé, P., Oestreicher, L., Palanque, P., Prates, R.O., Winckler, M. (eds.) INTERACT 2009. LNCS, vol. 5727, pp. 936–937. Springer, Heidelberg (2009). doi:10.1007/978-3-642-03658-3_119

15. Wang, W., He, Q.: A survey on emotional semantic image retrieval. In: Proceeding of 15th IEEE International Conference on Image Processing, ICIP 2008, CA, USA, pp. 117–120 (2008)

16. Wei-ning, W., Ying-lin, Y., Sheng-ming, J.: Image retrieval by emotional semantics: a study of emotional space and feature extraction. In: Proceeding of the IEEE International Conference on Systems, Man, and Cybernetics, SMC 2006, Taipei, Taiwan, pp. 3534–3539 (2006)

17. You, Q., Luo, J., Jin, H., Yang, J.: Robust image sentiment analysis using progressively trained and domain transferred deep networks. In: Proceeding of the 29th AAAI Conference on Artificial Intelligence, AAAI 2015, Austin, Texas, USA, pp. 381–388 (2015)

Effect of Parameter Tuning at Distinguishing Between Real and Posed Smiles from Observers' Physiological Features

Md Zakir Hossain and Tom Gedeon[✉]

Research School of Computer Science,
Australian National University, Canberra, Australia
zakir.hossain@anu.edu.au, tom@cs.anu.edu.au

Abstract. To find the genuineness of a human behavior/emotion is an important research topic in affective and human centered computing. This paper uses a feature level fusion technique of three peripheral physiological features from observers, namely pupillary response (PR), blood volume pulse (BVP), and galvanic skin response (GSR). The observers' task is to distinguish between real and posed smiles when watching twenty smilers' videos (half being real smiles and half are posed smiles). A number of temporal features are extracted from the recorded physiological signals after a few processing steps and fused before computing classification performance by k-nearest neighbor (KNN), support vector machine (SVM), and neural network (NN) classifiers. Many factors can affect the results of smile classification, and depend upon the architecture of the classifiers. In this study, we varied the K values of KNN, the scaling factors of SVM, and the numbers of hidden nodes of NN with other parameters unchanged. Our final experimental results from a robust leave-one-everything-out process indicate that parameter tuning is a vital factor to find a high classification accuracy, and that feature level fusion can indicate when more parameter tuning is needed.

Keywords: Physiological features · Real smile · Posed smile · Observers · Parameter tuning · k-nearest neighbor · Support vector machine · Neural network

1 Introduction

A smile is a multifaceted and multi-functional facial display that generally conveys positive feelings, such as enjoyment, warmth, appreciation, satisfaction, and so on. We refer to these as real smiles, and refer to conscious attempts to faithfully mimic these as posed smiles. We also know that people can smile in negative or neutral situations too, such as arrogance, sarcasm, acted, appeasement, or smile to hide something [1] including frustration and puzzlement. We refer to these negative and non-happy smiles as fake smiles, and refer to smiles that signify happiness as happy smiles. With the growing interest in emotion recognition and human centered computing, intelligent machines are being developed to determine people's affective behavior. The recognition of emotion from others' facial expressions is a vital and universal skill for social

© Springer International Publishing AG 2017
D. Liu et al. (Eds.): ICONIP 2017, Part IV, LNCS 10637, pp. 839–850, 2017.
https://doi.org/10.1007/978-3-319-70093-9_89

interaction [2], and smiles attract more attention than any other regions of faces [3]. Thus, developing a system that understands a smiler's affective state could be used in many situations, such as customer service quality evaluation, interactive tutoring systems, video conferencing, patient monitoring, verifying truthfulness during interrogation or hearings, border control or customs, and so on.

Previously researchers focused on smilers' faces and/or observers' verbal responses to distinguish between real and posed smiles. A computational technique is used to recognize real smile from smiler's facial features in [4] and reports 92.9% correctness. Ambadar et al. [5] scrutinize the characteristics of fake smiles along with their perceived meanings, and find that perceived meanings are related to specific characteristics. Calvo et al. [1] inspect perceptual, categorical, affective, and morphological characteristics to recognize happy smiles, and suggest that happy smiles are more likely to occur with congruent happy eyes and smiling mouth. Frank et al. [6] considers observers' self-judgments of happy smiles and notes a 56.0% average response rate. Hoque et al. [7] execute two experiments for classifying happy and fake smiles from smilers' facial features, and report 93.0% and 69.0% accuracies by classifiers and verbal responses respectively. Although observers may have certain impressions or feelings during face to face interaction, watching video clips or listening to music [8], it is not an easy task to distinguish happy and fake smiles from observers' verbal responses [6].

On the other hand, physiological signals have the advantage of immediately being affected by observing facial changes that cannot be posed voluntarily or assessed well visually [8, 9]. In this regard, many studies [10, 11, 12] have considered peripheral physiology to recognize facial behaviors. In this paper, we analyzed three physiological signals from observers – pupillary response (PR), galvanic skin response (GSR), and blood volume pulse (BVP) – to classify smilers' affective states into real or posed classes. We believe 'real' to be a more accurate characterization for our work than 'happy' as the datasets elicited smiles, so they are real but did not consistently test the subjects' emotional state so we cannot confidently say they were happy. The situation is similar for the smiles which are not real: the subjects were asked to smile, which is a posed smile, and not fake in the many possible reasons for which people can do a fake smile.

Pupillary responses can change due to memory load, stress, pain, light intensity, content of face to face interactions and so on. It has the advantage of needing no sensors to be attached to the observer, and certainly not to the smiler [10]. GSR is considered to be a strong physiological signal in emotion detection that measures electrical changes of human skin automatically [11], and has been used for this task previously [12]. BVP measures blood volume changes using infrared light through the tissues, and used as an indicator of emotional changes and affective processing that affects heart rate and pulse amplitude [13]. The temporal patterns of these physiological signals are useful while classifying real and posed smile observations. Six statistical features are extracted from each peripheral physiological signal in each observation and used to compute classification accuracy using a robust leave-one-out process. By robust we mean that in each run, we leave out all the samples collected from a particular subject, along with all other subjects' responses to that particular stimulus (a short video in this case). That is, we go beyond leave-one-subject-out, and use leave-one-subject-and-one-stimulus-out. Thus, our results are not only subject independent, but also stimulus independent. Our aim is to detect smilers' affective states by classifying smiler videos into real and posed, from observers' peripheral physiology based temporal features, using three classifiers, namely k-nearest neighbor (KNN), support vector machine (SVM), and neural network (NN). In this paper we thoroughly examine parameter settings.

2   The Method

We collected twenty smilers' videos chosen at random from four benchmark database. We processed them using MATLAB to convert into grey scale, mp4 format with each video lasting 5 s. The videos are presented in a balanced order, and physiological signals are recorded from each observer while watching these videos. The signals are smoothed, filtered, and normalized. Six statistical time domain features are extracted from each signal. Binary classifiers are employed to classify between real and posed smiles. The overview of this method is shown in Fig. 1.

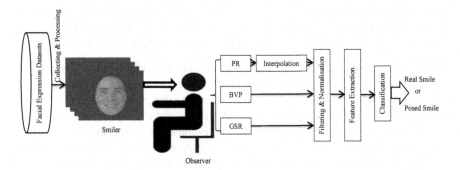

Fig. 1. Outline of the research.

2.1   Smile Videos

Ten videos were collected from UvA-NEMO [14] and MAHNOB [15] databases (five from each) where real smiles were elicited from smilers, by showing a sequence of funny or otherwise pleasant video clips. Ten videos were collected from MMI [16] and CK+ [17] databases (five from each) where posed smiles were created by smilers with experimenters asking or instructing them to display a smile. We used 4 databases to increase the variation in our stimuli. The collected videos were processed using MATLAB R2014b to make them uniform in size, format and duration. These are: fixed aspect ratio of 4:3 (Height = 336 pixels, Width = 448 pixels) with resolution of 72 dpi, grey scale, mp4 format, and lasting 5 s each. Frame rates are adjusted slightly to make each video's length be 5 s long. Luminance (128 ALU – Arbitrary Linear Unit) and contrast (32 ALU) of smilers are also adjusted and kept similar using the MATLAB SHINE toolbox [18]. Only faces of smilers are shown to the observers (see *Smiler* in Fig. 1), to avoid the side effect of light backgrounds on pupil dilation.

2.2 Observers

Twenty-four right-handed healthy volunteers (15 males, 9 females) participated as observers in the study, with mean age of 30.7 ± 6.0 (mean ± SD). All observers had normal or corrected to normal vision. They signed an informed consent form prior to their participation. Ethics approval was received from our Australian University's Human Research Ethics Committee, prior to performing the study.

2.3 Peripheral Physiology

Pupillary Response (PR): This physiological signal is measured from the eyes' responses that vary the size of the pupil. The primary function of this response is to control the amount of light reaching the retina. When luminance is controlled, then other influences on pupil size can be detected. Pupil dilation is the widening of the pupil and is controlled by the sympathetic nervous system and may be happening due to e.g. high curiosity. Constriction means narrowing the pupil, and that is controlled by the parasympathetic nervous system and may be happening due to less curiosity [10]. The magnitude of the pupillary response appears to be a function of the attention and curiosity required to perform a task. This is an involuntary signal that changes with any event observed by the subject, and recorded here using The Eye Tribe remote eye-tracker system (https://theeyetribe.com/), with a sampling rate of 60 Hz.

Blood Volume Pulse (BVP): A photoplethysmographic (PPG) sensor measures the change of blood volume that passes through the tissues over a given period of time, this is the BVP. A light-emitting diode is used to pass an infrared light through the tissues, and the returned light is proportional to the volume of the blood in the tissue. BVP signals are used as indicators of emotional response by measuring the change in peripheral physiology. It is possible to identify observers' or subjects' mental states by analyzing the variation of BVP amplitudes [13]. In this study, each observer's BVP signals are recorded from the wrist of the left hand at a sampling rate of 64 Hz using an Empatica E4 device (https://www.empatica.com/).

Galvanic Skin Response (GSR): Skin conductance, also known as electro-dermal response or psychogalvanic reflex, measures the electrical changes in human skin that varies with changes in skin moisture level (sweating). This is an automatic reaction that cannot be controlled voluntarily and reflects changes in the sympathetic nervous system, and is an indication of psychological or physiological arousal that can be used for affect detection or mental state recognition [11]. When an observer is more curious, skin conductance is increased; conversely, the skin conductance is reduced when an observer finds it easier (less stressful), e.g. to identify a smile video. The same Empatica E4 sensor as used in BVP recording is used here to record the GSR signals from the observer's left wrist, at the maximum sampling rate of 4 Hz.

2.4   Conduct of the Experiment

Each observer (subject) fills in a consent form for their voluntary participation. A 15.6" ASUS laptop and a normal computer mouse are peripherals for interaction between the observer and a laptop running the web-based tool showing the smile videos. The chair of the observer is moved forward or backwards to adjust the distance between the observer and eye tracker. Observers are asked to track a spot displayed in the laptop for calibrating the eye tracker and starting the experiment. Observers are instructed to limit their body movements in order to reduce undesired artifacts in the signals. The videos are presented in a balanced way to the observer to avoid order effects. Thus the positioning of each smiler's video, near the beginning/middle/end of the experiment, is different for each observer. Each video is followed by questions to identify the smile's real or posed nature, implying the smiler's affective state. For brevity of discussion, henceforth we will discuss e.g. 'real smile' as an affective state. Due to poor signal quality and unfinished data collection from two observers, the results are reported by analyzing data from twenty-four observers.

2.5   Signal Processing

Due to the nature of human bodies, physical movements and other effects, the recorded peripheral physiology is affected by noise like small signal fluctuations, eye blinking etc. To reduce this latter effect, eye blinking points are considered to be zero in pupillary responses. Then, cubic spline interpolation and 10-point Hann moving window average are employed to reconstruct and smooth the pupil data respectively [10]. This procedure is applied to the left and right eyes' pupillary response separately, and then averaged to find a single pupillary response signal for a specific observer. A low-pass Butterworth filter (order = 6, normalized cut-off frequency = 0.5) is used to smooth the GSR and BVP signals [11]. Then, maximum value normalization is applied to keep the signals in the range between 0 and 1.

2.6   Feature Extraction

The following six different time domain features are computed for each video related peripheral physiological signal. Let $X(n)$ represents the value of the nth sample of the processed peripheral physiological signals, $n = 1, 2, \ldots \ldots, N$.

1. Mean

$$\mu_x = 1/N \sum_{n=1}^{N} X(n) \tag{1}$$

2. Maximum

$$M_x = Max(X(n)) \tag{2}$$

3. Minimum

$$m_x = Min(X(n)) \qquad (3)$$

4. Standard Deviations

$$\sigma_x = \sqrt{1/(N-1)\sum_{n=1}^{N}(X(n)-\mu_x)^2} \qquad (4)$$

5. Means of the absolute values of the first differences

$$\delta_x = 1/(N-1)\sum_{n=1}^{N-1}|X(n+1)-X(n)| \qquad (5)$$

6. Means of the absolute values of the second differences

$$\gamma_x = 1/(N-2)\sum_{n=1}^{N-2}|X(n+2)-X(n)| \qquad (6)$$

These statistical features convey information such as typical range, gradient, and variation of the signals [19]. Then, we employed a feature level fusion (merge all features from BVP, GSR, and PR) technique before computing classification accuracies. In this case, there are 360 extracted features (20 videos × 6 features × 3 peripheral physiological signals) for an observer (half for posed smile videos and other half for real smile videos) and a total of 8,640 features for all 24 observers. We did not consider any features in the training set of a test observer that is related to that test observer. For example, suppose we consider the test data of observer 1 (O1) when watching the smiler 1, then the data of O1 while watching the other smilers' videos is not used to either train or test the classifier. Thus there are 18 (1 observer × 6 features × 1 smiler × 3 peripheral physiological signals) testing features and 7,986 (23 observers × 6 features × 19 smilers × 3 peripheral physiological signals) training features in each execution. Finally, average accuracies over all possible executions are reported. This leave-one-out process means that our classifiers have seen no physiological signals from training observers nor from training smilers in the test set, our results is thus completely independent.

3 Experimental Outcomes

Smilers' affective states are classified into two classes, namely real smiles and posed smiles from observers' peripheral physiological features. The analysis is executed on an Intel(R) Core™ i7-4790 CPU with 3.60 GHz, 16.00 GB of RAM, Operating System 64-bit computer using MATLAB R2014b. Three different types of classifier are used to compute classification performances, namely k-nearest neighbor (KNN), support vector machine (SVM), and neural network (NN). The feature sets are divided according to the test observer identifications, such as O1, O2 all the way to O24. When the test observer is O1, and other observers' (O2 to O24) features are used to train the classifiers, we call it O1 and so on. In a similar fashion, test smilers are identified by S1, S2 all the way to S20. According to our robust leave-one-out process, the final outcome of O1 is the average value over 20 executions (S1 to S20) for each physiological feature set.

Parameter tuning is found to be vital factor in determining classification accuracies using all three of these classifiers. We choose the default Euclidian distance metric for the KNN classifier and check the variation of classification accuracies with different K values. The results are depicted in Fig. 2, where error bars indicate standard deviations. It is clear from Fig. 2 that the classification accuracies decrease with increasing K. It is also seen that higher accuracies are found for even values of K compared to odd values of K. A similar result was found in the case of a Parkinson dataset [20], possibly this is due to some properties of human data in producing decision regions with unusual topologies.

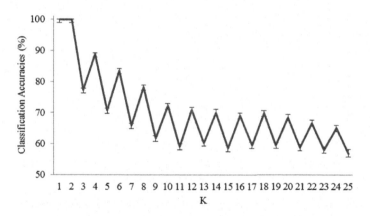

Fig. 2. Variation of accuracies for 'K' values of KNN, (Maximum value is 99.8%).

For SVM, the Gaussian radial basis kernel function is used with various scaling factors to compute classification accuracies. The variation of average accuracies with scaling factors is noticeable and explored in Fig. 3, where error bars indicate standard deviations. The classification accuracies gradually decrease with increasing scaling factors. The rate of decrease is higher for the low values of the scaling factor (from 1 to 5), and then this rate diminishes. Some research has focused on empirical analysis to find the best fitted scaling factors to report best performances from SVM classifier [21].

In NN, Levenberg-Marquardt training function with various numbers of hidden nodes are considered to compute classification accuracies. The variation of average classification accuracies with the various numbers of hidden nodes of NN classifier are shown in Fig. 4, again error bars indicate standard deviations.

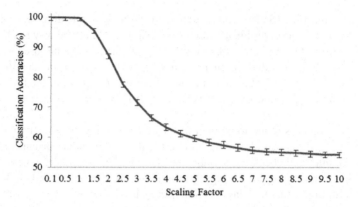

Fig. 3. Variation of accuracies to scaling factors of SVM, (Maximum value is 99.7%).

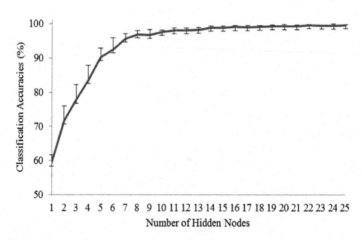

Fig. 4. Variation of accuracies to the no. of hidden nodes of NN, (Maximum value is 99.6%).

The average classification accuracies are improved with increasing number of hidden nodes as shown in Fig. 4. The errors are also seen to decrease with the increasing number of hidden nodes according to standard deviation computation. It has been necessary to focus on empirical analysis to find the best number of hidden nodes, as we expected and did find that the number of hidden nodes affects classification accuracies [22]. We note that our results are robust, yet the very high accuracies do require some discussion. We believe that the results are explained by results [23] showing highly conserved shared structure in neural activity across individuals in a consistent observation task. Our work in removing backgrounds, adjusting luminance and contrast, and order balancing of tasks has led to a consistent task. The results are also robust as they are leave-one-out of both subject and each test video.

4   Parameter Fusion

We investigate fusion of parameters using a simple ensemble over the decision of the above three classifiers (KNN, SVM, and NN) from the three techniques, in four ways. Firstly, we examine the effect of fusing the three best values, representing the situation where a thorough investigation of parameter values was done, to attempt to further improve the results. Secondly, we examine the most perverse setting where we use the worst results and fuse them, representing a very naïve user making particularly bad choices. Thirdly, we fuse midrange values, representing a naïve user who has expended some effort. Finally, we examine some combinations of best/midrange/worst results (see Table 1). We note that our robust leave-one-of-everything-out process is not able to overfit, as each observer-stimulus pair is used as a test set in different runs. This is admittedly a computationally expensive process for robustness.

It is noticeable from Figs. 2, 3, and 4 and Table 1 that the best classification accuracies are found for k = 1 (99.8%), s = 0.1 (99.7%), and n = 25 (99.6%) and the worst accuracies are found for k = 25 (56.7%), s = 10 (54.3%), and n = 1 (59.3%). For the midrange, we choose accuracies between 75–80%. Specifically, we choose k = 8 (78.2%), s = 2.5 (77.5%), and n = 3 (77.7%). Finally, we fuse the physiological features at different combinations of these parameter settings. We find that the ensemble classifier can improve the performance (highlighted in Table 1) for the worst results, and for combinations involving midrange results. We find that our feature level fusion does not improve the results when one or more of the best case results are included.

Table 1. Accuracies of ensemble classification with different combinations of parameters (where K, S, and N are the K values of KNN, scaling of SVM, and Nodes of NN)

	k	s	n	KNN	SVM	NN	Ensemble
All best	1	0.1	25	99.8	99.7	99.6	99.6
All worst	25	10	1	56.7	54.3	59.3	**73.7**
All mid	8	2.5	3	78.2	77.5	77.7	**87.6**
KNN best	1	10	1	99.8	54.3	59.3	87.1
KNN mid	8	10	1	78.2	54.3	59.3	**80.0**
KNN worst	25	0.1	25	56.7	99.7	99.6	94.6
SVM best	25	0.1	1	56.7	99.7	59.3	88.1
SVM mid	25	2.5	1	56.7	77.5	59.3	**80.1**
SVM worst	1	10	25	99.8	54.3	99.6	94.5
NN best	25	10	25	56.7	54.3	99.6	84.6
NN mid	25	10	3	56.7	54.3	77.7	**80.0**
NN worst	1	0.1	1	99.8	99.7	59.3	97.4

5   Conclusion

This paper investigates the effects of K values, scaling factors, and the number of hidden nodes, for KNN, SVM, and NN respectively, to distinguish between real and posed smiles from observers' peripheral physiological features while other factors remain unchanged. From the results of our robust leave-one-out process, we found substantial effects from the parameters we considered on the smile classification as real or posed, from observers' physiological features. We saw that lower K values and scaling factors, and higher number of hidden nodes were needed to find higher classification accuracies according to the architecture of each classifier. We noted that fusing results when we had optimized parameter values for each technique led to no improvement, strongly indicating that the errors made by each classifier must be quite similar. Fusing results from cases with less good parameter values led to improved classification results. We believe this would be a practical test for naïve users of these techniques to indicate that further parameter tuning should be done. In the future, we will consider other parameters of each classifier to check the variability of classification performance and to tune the parameters to design a robust system to distinguish between real and posed smiles in this context. We will also consider aggregation approaches designed for complex structured data such as physiological signals [24, 25], alternative artificial intelligence approaches [26–28], and the use of virtual or synthesised faces [29, 30].

References

1. Calvo, M.G., Gutiérrez-García, A., Del Líbano, M.: What makes a smiling face look happy? Visual saliency, distinctiveness, and affect. Psychol. Res. 1–14 (2016)
2. Libralon, G.L., Romero, R.A.F.: Investigating facial features for identification of emotions. In: Lee, M., Hirose, A., Hou, Z.-G., Kil, R.M. (eds.) ICONIP 2013. LNCS, vol. 8227, pp. 409–416. Springer, Heidelberg (2013). doi:10.1007/978-3-642-42042-9_51
3. Beaudry, O., Roy-Charland, A., Perron, M., Cormier, I., Tapp, R.: Featural processing in recognition of emotional facial expressions. Cogn. Emot. 28(3), 416–432 (2014)
4. Dibeklioğlu, H., Salah, A.A., Gevers, T.: Recognition of genuine smiles. Trans. Multimedia 17(3), 279–294 (2015)
5. Ambadar, Z., Cohn, J.F., Reed, L.I.: All smiles are not created equal: morphology and timing of smiles perceived as amused, polite, and embarrassed/nervous. J. Nonverbal Behav. 33(1), 17–34 (2009)
6. Frank, M.G., Ekman, P., Friesen, W.V.: Behavioral markers and recognizability of the smile of enjoyment. J. Pers. Soc. Psychol. 64(1), 83–93 (1993)
7. Hoque, M.E., McDuff, D.J., Picard, R.W.: Exploring temporal patterns in classifying frustrated and delighted smiles. Trans. Affect. Comput. 3(3), 323–334 (2012)
8. Kim, J., Andre, E.: Emotion recognition based on physiological changes in music listening. Trans. Pattern Anal. Mach. Intell. 30(12), 2067–2083 (2008)
9. Gong, P., Ma, H.T., Wang, Y.: Emotion recognition based on the multiple physiological signals. In: International Conference on Real-Time Computing and Robotics, pp. 140–143. IEEE, Angkor Wat (2016)
10. Hossain, M.Z., Gedeon, T., Sankaranarayana, R., Apthorp, D., Dawel, A.: Pupillary responses of Asian observers in discriminating real from fake smiles: a preliminary study. In: 10th International Conference on Methods and Techniques in Behavioral Research, pp. 170–176. Measuring Behavior, Dublin (2016)
11. Xia, V., Jaques, N., Taylor, S., Fedor, S., Picard, R.: Active learning for electrodermal activity classification. In: Signal Processing in Medicine and Biology Symposium, pp. 1–6. IEEE (2015)

12. Hossain, M.Z., Gedeon, T., Sankaranarayana, R.: Observer's galvanic skin response for discriminating real from fake smiles. In: 27th Australian Conference on Information Systems, pp. 1–8. University of Wollongong Faculty of Business, Wollongong (2016)
13. Peper, E., Harvey, R., Lin, I., Tylova, H., Moss, D.: Is there more to blood volume pulse than heart rate variability, respiratory sinus arrhythmia, and cardiorespiratory synchrony? Biofeedback 35(2), 54–61 (2007)
14. Dibeklioğlu, H., Salah, A.A., Gevers, T.: Are you really smiling at me? Spontaneous versus posed enjoyment smiles. In: Fitzgibbon, A., Lazebnik, S., Perona, P., Sato, Y., Schmid, C. (eds.) ECCV 2012. LNCS, vol. 7574, pp. 525–538. Springer, Heidelberg (2012). doi:10. 1007/978-3-642-33712-3_38
15. Soleymani, M., Lichtenauer, J., Pun, T., Pantic, M.: A multimodal database for affect recognition and implicit tagging. IEEE Trans. Affect. Comput. 3(1), 42–55 (2012)
16. Pantic, M., Valstar, M., Rademaker, R., Maat, L.: Web-based database for facial expression analysis. In: International Conference on Multimedia and Expo, p. 5. IEEE, Amsterdam (2005)
17. Lucey, P., Cohn, J.F., Kanade, T., Saragih, J., Ambadar, Z., Matthews, I.: The extended Cohn-Kanade dataset (CK+): a complete expression dataset for action unit and emotion-specified expression. In: Conference on Computer Vision and Pattern Recognition, pp. 94–101. IEEE, San Francisco (2010)
18. Willenbockel, V., Sadr, J., Fiset, D., Horne, G.O., Gosselin, F., Tanaka, J.W.: Controlling low-level image properties: the SHINE toolbox. Behav. Res. Methods 42(3), 671–684 (2010)
19. Picard, R.W., Vyzas, E., Healey, J.: Toward machine emotional intelligence: analysis of affective physiological state. Trans. Pattern Anal. Mach. Intell. 23(10), 1175–1191 (2001)
20. Chih-Min, M., Wei-Shui, Y., Bor-Wen, C.: How the parameters of k-nearest neighbor algorithm impact on the best classification accuracy: in case of parkinson dataset. J. Appl. Sci. 14, 171–176 (2014)
21. Romero, R., Iglesias, E.L., Borrajo, L.: A linear-RBF multikernel SVM to classify big text corpora. BioMed Res. Int. 1–14 (2015)
22. Zou, W., Li, Y., Tang, A.: Effects of the number of hidden nodes used in a structured-based neural network on the reliability of image classification. Neural Comput. Appl. 18(3), 249–260 (2009)
23. Chen, J., Leong, Y.C., Honey, C.J., Yong, C.H., Norman, K.A., Hasson, U.: Shared memories reveal shared structure in neural activity across individuals. Nat. Neurosci. 20(1), 115–125 (2017)
24. Mendis, B.S.U., Gedeon, T.D., Kóczy, L.T.: Investigation of aggregation in fuzzy signatures. In: 3rd International Conference on Computational Intelligence, Robotics and Autonomous Systems, pp. 17–31. CIRAS and FIRA Organising Committee, Singapore (2005)
25. Mendis, B.S.U., Gedeon, T.D., Koczy, L.T.: On the issue of learning weights from observations for fuzzy signatures. In: World Automation Congress, pp. 1–6. IEEE Press (2006)
26. Treadgold, N.K., Gedeon, T.D.: A cascade network algorithm employing progressive RPROP. In: Mira, J., Moreno-Díaz, R., Cabestany, J. (eds.) IWANN 1997. LNCS, vol. 1240, pp. 733–742. Springer, Heidelberg (1997). doi:10.1007/BFb0032532
27. Khan, M.S., Chong, A., Gedeon, T.D.: A methodology for developing adaptive fuzzy cognitive maps for decision support. JACIII 4(6), 403–407 (2000)
28. Tikk, D., Bíró, G., Gedeon, T.D., Kóczy, L.T., Yang, J.D.: Improvements and critique on Sugeno's and Yasukawa's qualitative modeling. IEEE Trans. Fuzzy Syst. 10(5), 596–606 (2002)

29. Asthana, A., Gedeon, T., Goecke, R., Sanderson, C.: Learning-based face synthesis for pose-robust recognition from single image. In: British Machine Vision Conference, pp. 1–10. British Machine Vision Association and Society for Pattern Recognition (2009)
30. Asthana, A., Goecke, R., Quadrianto, N., Gedeon, T.: Learning based automatic face annotation for arbitrary poses and expressions from frontal images only. In: Computer Vision and Pattern Recognition CVPR, pp. 1635–1642. IEEE Press (2009)

Brain Effective Connectivity Analysis from EEG for Positive and Negative Emotion

Jianhai Zhang[(✉)], Shaokai Zhao, Wenhao Huang, and Sanqing Hu

College of Computer Science, Hangzhou Dianzi University,
Hangzhou 310018, China
{jhzhang, sqhu}@hdu.edu.cn, lnkzsk@126.com,
hwhzzly@gmail.com

Abstract. Recently, there have been increasing evidence which supports that multiple brain regions are involved in emotion processing. Therefore, research on emotion from the perspective of brain network is becoming popular. In this study, based on the Granger causal analysis method, we constructed brain effective connectivity network from DEAP emotional EEG data to investigate how emotion affects the patterns of effective connectivity. According to our results, prefrontal region plays the most important role in emotion processing with interactions to almost all other regions. More interactions are found under negative emotion than positive one. Parietal region in charge of human's alert mechanism is more active under negative emotions. These results are consistent with the previous findings obtained in neuroscience, which illustrate the effectiveness of our methods. Furthermore, the brain effective connectivity network shows significant differences to different emotional states, so it can be used to recognize different emotional states with EEG.

Keywords: Emotion processing · Granger causality · Brain effective network · EEG

1 Introduction

The topic on how the brain processes emotional stimuli has attracted increasing attention worldwide for a long time. The recent researches have shown that different emotional states are related to a specific neural response pattern, thus, the investigation on the neural correlations to emotion probably can be a better way to understand its internal mechanisms [1, 2]. So far, a variety of models have been proposed for this aim [3, 4]. These models can be roughly divided into two categories. Some studies suggest that each kind of emotion corresponds to distinct single brain regions, while others argue a set of interacting brain regions are involved for any kind of emotion experience. Recently, more and more evidences strongly support the latter one [5] and brain connectivity analysis becomes predominant for this research.

Effective connectivity based on causal theory has been widely used to construct brain network because it can quantify not only the intensity but also the direction of information flow between the interesting brain regions. So far, most studies are based on fMRI data due to its high spatial resolution and clean signal. But fMRI has crucial

© Springer International Publishing AG 2017
D. Liu et al. (Eds.): ICONIP 2017, Part IV, LNCS 10637, pp. 851–857, 2017.
https://doi.org/10.1007/978-3-319-70093-9_90

drawbacks of low temporal resolution, high-cost and inconvenient to use. EEG signals has advantages such as high temporal resolution, low cost, higher availability and feasibility of applying different cognitive tests during recording. The present study aims at finding how the positive and negative emotional state affects brain effective connectivity based on Granger causality theory and EEG recordings from multiple electrodes.

2 Materials and Methods

2.1 DEAP Database

This study was performed on the publicly available database, namely DEAP [6]. DEAP is a multimodal dataset for the analysis of human emotional states. The EEG signals (32 channels) and peripheral physiological signals (8 channel) of 32 subjects (aged between 19 and 37) were recorded during watching music video. The forty 1-minute-long videos were carefully selected to elicit different emotional states according to the dimensional valence-arousal emotion model. In DEAP, each video clip is rated on a scale of 1 to 9 for arousal and valence by each subject after the viewing.

In this paper, for simplicity, the preprocessed DEAP dataset in MATLAB format is used. In the preprocessing procedure, the sampling rate of EEG signal was down sampled from 512 Hz to 128 Hz and a band pass frequency filter from 4.0–45.0 Hz was applied. In addition, EOG artifacts have been removed from the EEG signal. The processes of detrending and ensemble demeaning are also implemented in order to avoid the interference of neutral data and ensure adequate data for our research, we screened the subjects and trials. The trials with valence rated larger than 7 (positive) or less than 3 (negative) were selected, and only the subjects both of whose positive and negative trials are more than 9 were selected. After screening, sub1, sub12, sub13, sub14, sub22, sub25, sub26, sub28 were left. The following table lists the number of remaining trials which belongs to the subjects mentioned above (Table. 1).

Table 1. Trials for positive and negative emotion of selected subjects

Subjects	Positive trials (valence > 7)	Negative trials (valence < 3)
Subject 01	16	12
Subject 12	10	11
Subject 13	11	11
Subject 14	12	12
Subject 22	13	15
Subject 25	12	11
Subject 26	16	10
Subject 28	16	11

2.2 Granger Causality Calculating

In our study, a standardized toolbox GCCA [7] was adopted to calculate the Granger causality value. Considering the characteristics of emotion eliciting process and the stationary of the signal, we chose the middle 30 s (from 15 s to 45 s) data of each trial. In addition, since our research goal is the effective connectivity between brain regions with interesting under different emotional states, the corresponding 8 channels located in the frontal lobe, temporal lobe, parietal lobe, occipital lobes were chosen as shown in Fig. 1.

The channel number that appears in the left panel in Fig. 1 according to typical 10–20 system. To facilitate the subsequent study and help understanding, we renumbered the channels as in the right panel. Detailed information is listed in Table 2.

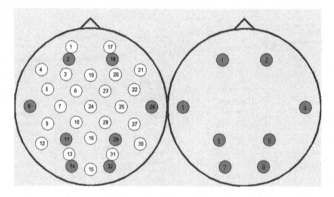

Fig. 1. The channels which appear in the left panel come from the DEAP database and the selected channels are marked yellow. The selected channels with new number are shown in right panel. (Color figure online)

Table 2. Detailed information of selected channels

Channel name	DEAP channel number	New number	Brain region
AF3	2	1	Left prefrontal
AF4	18	2	Right prefrontal
T7	8	3	Left temporal
T8	26	4	Right temporal
P3	11	5	Left parietal
P4	29	6	Right parietal
O1	14	7	Left occipital
O2	32	8	Right occipital

The Granger causality was calculated on the assumption that the signal must be stationary. Although all trials were preprocessed with detrending and ensemble demeaning, previous studies have shown that the stationary of signal can't be ensured after the above processing. Therefore, the signal covariance stationarity should be tested by the Augmented Dickey Fuller (ADF) test [8] or the Kwiatkowski-Phillips-Schmidt-Shin (KPSS) test [9]. In this study, the KPSS test was used, and all signals passed the test. Later, the optimal model order 20 was selected in terms of the Akaike information criterion [10] for MVAR model. The Durbin-Watson test [11] was employed to test whether the residual between raw curves and fitting curves was "white". The consistency test [12] assesses the consistency between fitting signals and original signals. When the consistency is less than a certain percentage (usually 80%), the obtained MVAR model is in a state of incomplete, then these data will be discarded and can't been used for analysis. In our experiments, all data consistency values are more than 80%, so no data need to be discarded. The GC values of each trial were calculated under a significance threshold p = 0.05 (Bonferroni corrected) (Fig. 2).

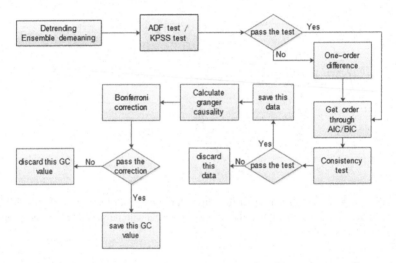

Fig. 2. Standard steps for calculating the time-domain granger causality.

3   Results

After the above process, the Granger causality (GC) matrix (8 × 8) of each trial was obtained. Then the significant interaction matrix corresponding to the GC matrix of each trial was obtained by Bonferroni's method. The significant interaction matrix is composed of 0 and 1 and have the same size with GC matrix. If the value of an element in the significant interaction matrix is 1, it means that the causal value of the same position in the causality connection matrix passes the Bonferroni corrected. The performance of all the connectivity between channels can be obtained by averaging the significant interaction matrix of the same emotions across all the trials from different subjects. We choose 0.8 as the threshold and plot the selected channel-pairs (Fig. 3). A threshold of 0.8 means that the GC value of this channel pair has passed the Bonferroni correction in 80% trials. Table 3 gives the channel pairs which passed the

Bonferroni correction under the two emotional states. The arrow '→' indicates the direction of information flow. The green numbers represent the different causal flow between positive and negative emotions.

Positive   Negative

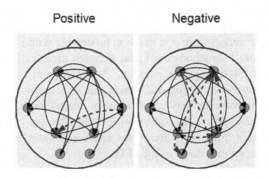

Fig. 3. Brain mapping of Table 3. The black arrows indicate the information flow existing in both emotional states. The green arrows indicate the information flow only existing in one of emotional states. (Color figure online)

Table 3. Significant interaction of two emotional states

	Positive	Negative
	1->2/3/4/5/6/7	1->2/3/4/5/6/7/8
	2->1/3/4/5/6/7/8	2->1/3/4/5/6/7/8
significant interaction	4->5	5->2/3/4/6/7
	5->3/4	6->2/3/4/5
	6->3/4/5	

From the Table 3, the amount of the significant interactions under negative emotions is more than that under positive state, which indicates that there are more interactions among brain regions under negative emotions. This result is same with Wang et al. [13]. Previous studies [14–16] have shown that prefrontal lobe plays an important role in the emotional processing, which also can be seen from our results where prefrontal region (AF3 and AF4) has significant connectivity with almost all other selected regions. And under both positive and negative emotions, information loops exist between the left prefrontal and right prefrontal lobes. In comparison with positive emotion, the frontal lobe under negative emotion is more active. The information flows of T7→AF4, T8→AF4, AF3→O2 were found under negative emotion. In our results, the biggest difference between positive and negative emotion is in the bilateral parietal lobe. Under negative emotion, more interregional interactions (P3→AF4, P4→AF4, P3→O1) and intraregional interaction (P3→P4) are found. However, information flow from right temporal lobe to left parietal lobe (T8→P3) is found under positive emotion. Previous studies [17] showed that the parietal lobe has a negligible effect in processing of negative emotions. This is mainly because the negative emotion can trigger the parietal lobe which controls the human's alert mechanism.

4 Conclusion

In this paper, based on the Granger causality analysis methods, we constructed brain effective connectivity network from emotional EEG data. With the brain network, we can investigate the changes of interaction and information flow among different brain regions under positive and negative emotions. According to our results, prefrontal region plays the most important role in emotion processing with interactions to almost all other regions. More interactions are found between prefrontal region and parietal and occipital region under negative emotion. Parietal region is more active under negative emotions because it controls the human's alert mechanism. The above results are consistent with the previous findings obtained in neuroscience, which illustrate the effectiveness of our methods. From our results, brain effective connectivity network demonstrates significant differences under two emotional states, so it can be used to recognize different emotional states using EEG.

Acknowledgments. This work was supported in part by the National Natural Science Foundation of China under Grant 61100102 and Grant 61473110 and Grant 61633010, in part by the International Science and Technology Cooperation Program of China under Grant 2014DFG12570.

References

1. Bono, V., Biswas, D., Das, S., et al.: Classifying human emotional states using wireless EEG based ERP and functional connectivity measures. In: IEEE International Conference on Biomedical and Health Informatics. IEEE (2016)
2. Shahabi, H., Moghimi, S.: Toward automatic detection of brain responses to emotional music through analysis of EEG effective connectivity. Comput. Hum. Behav. **58**, 231–239 (2016)
3. Hamann, S.: Mapping discrete and dimensional emotions onto the brain: controversies and consensus. Trends Cogn. Sci. **9**, 458–466 (2012)
4. Tettamanti, M., Rognoni, E., Cafiero, R., et al.: Distinct pathways of neural coupling for different basic emotions. NeuroImage **59**, 1804–1817 (2012)
5. Lindquist, K., Wager, T., Kober, H., et al.: The brain basis of emotion: a meta-analytic review. Behav. Brain Sci. **35**(3), 121–143 (2015)
6. Koelstra, S., Muhl, C., Soleymani, M., et al.: DEAP: a database for emotion analysis. Using Physiol. Signals **3**, 18–31 (2016). IEEE
7. Seth, A.K.: A MATLAB toolbox for Granger causal connectivity analysis. J. Neurosci. Methods **186**, 262–273 (2010)
8. James, D.H.: Time Series Analysis. Princeton University Press (2007)
9. Kwiatkowski, D., Phillips, P.C.B., Schmidt, P., et al.: Testing the null hypothesis of stationarity against the alternative of a unit root ☆: how sure are we that economic time series have a unit root? J. Pap. **54**(1–3), 159–178 (1990)
10. Akaike, H.: A new loot at the statistical model identification. IEEE Trans. J. Autom. Control **19**(6), 716–723 (1974)
11. Durbin, J., Watson, G.S.: Testing for serial correlation in least squares regression. I. In: Kotz, S., Johnson, N.L. (eds.) Breakthroughs in Statistics. Springer Series in Statistics (Perspectives in Statistics), pp. 237–259. Springer, New York (1992). doi:10.1007/978-1-4612-4380-9_20

12. Ding, M., Bressler, S.L., Yang, W., et al.: Short-window spectral analysis of cortical event-related potentials by adaptive multivariate autoregressive modeling: data preprocessing, model validation, and variability assessment. J. Biol. Cybern. **83**(1), 35–45 (2000)

13. Wang, N., Wang, Y., Li, Y., Tang, Y., Wang, J.: Gamma oscillation in brain connectivity in emotion recognition by Granger causality. In.: International Conference on Biomedical Engineering and Informatics, vol. 2, pp. 762–766. IEEE (2011)

14. Friedman, D., Shapira, S., Jacobson, L., Gruberger, M.: A data-driven validation of frontal EEG asymmetry using a consumer device. In: International Conference on Affective Computing and Intelligent Interaction, pp. 930–937. IEEE (2015)

15. Kim, M.K., Kim, M., Oh, E., Kim, S.P.: A review on the computational methods for emotional state estimation from the human EEG. J. Comput. Math. Methods Med. **2013**, 13 p. (2013). Article ID 573734. doi:10.1155/2013/573734

16. Heller, W.: Neuropsychological mechanisms of individual differences in emotion, personality, and arousal. J. Neuropsychol. **7**, 476–489 (1993)

17. Lin, Y.P., Wang, C.H., Jung, T.P., Wu, T.L., Jeng, S.K., Duann, J.R., et al.: EEG-based emotion recognition in music listening. J. IEEE Trans. Biomed. Eng. **57**, 1798–1806 (2010)

Efficient Human Stress Detection System Based on Frontal Alpha Asymmetry

Asma Baghdadi[✉], Yassine Aribi, and Adel M. Alimi

REGIM: REsearch Groups in Intelligent Machines, National School of Engineers,
University of Sfax, BP 1173, 3038 Sfax, Tunisia
{baghdadi.asma.tn,yassine.aribi,adel.alimi}@ieee.org

Abstract. EEG signals reflect the inner emotional state of a person and regarding its wealth in temporal resolution, it can be used profitably to measure mental stress. Emotional states recognition is a growing research field inasmuch to its importance in Human-machine applications in all domains, in particular psychology and psychiatry. The main goal of this study is to provide a simple method for stress detection based on Frontal Alpha Asymmetry for trials selection and time, time-frequency domain features. This approach was tested on prevalent DEAP database, and provided us with two subdatasets to be processed and classified thereafter. From the variety of features produced in the literature we chose to test Hjorth parameters and Band Power as a time-frequency feature. To enhance the classification performance, we tested the SVM classifier, K-NN and Fuzzy K-NN.

Keywords: Human stress · EEG · Frontal Asymmetry · Hjorth · Band Power · FK-NN

1 Introduction

Biometrics are becoming highly important in Human-machine interaction, and are used for the recognition of persons' identity, mental and physical health [37–43]. Biometric identifiers are distinctive, examples include, but are not limited to: handwriting [20,22,36], face recognition, speech and physiological signals (EMG, EEG, etc.).

Nowadays people suffer from mental stress during their daily life. Although there is a contiguous link between mental health and stress, psychological stress (and correlated emotions such as depression, anxiety, and anger), can also have negative effects on physical health. Actually, incessant psychological stress can modify the broad-mindedness of the central-peripheral regulatory systems [21], conceivably rendering them less adaptive or efficient in terms of health supporting. Conditions such as incessant anxiety, depression, and stress have been found to be correlated with abnormal autonomic nervous system (ANS) functioning [23]. Appropriately, stress is one of the main factors leading to lifelong disorders [24,25]. The desire to work can be influenced by stress, work performance, company achievements, and one's general mood facing life [26–28].

© Springer International Publishing AG 2017
D. Liu et al. (Eds.): ICONIP 2017, Part IV, LNCS 10637, pp. 858–867, 2017.
https://doi.org/10.1007/978-3-319-70093-9_91

Research demonstrates and indicates the robust relationship that exists between stress and brain activity [13,14] while another research is required for more intuition in order to establish stress detection systems based on brain activity analysis.

During negative emotions, the right cerebral hemisphere activities govern the activities in the left cerebral hemisphere [16], this suggests a stress detection field. Fast Beta wave frequencies from decline in Alpha wave frequencies are the major characteristics announcing stress [15–19]. Researches explore EEG for various issues like stress levels recognition in computer game players [13] and biofeedback games [14].

Based on these scientific claims and conclusions, we propose in this work to investigate the Frontal Alpha Asymmetry obtained from the frontal right and left cerebral hemispheres as an annotation characteristic in order to label all Deap trials into two classes: Stress and NoStress. We then process this data to extract relevant features and perform classification task. This paper is organized as follows. Section 2, reviews researches conducted with the intention to detect stress from EEG. The proposed method is presented in Sect. 3. We discuss in Sect. 4 the results of classification. Finally, in Sect. 5, a conclusion is presented with some perspectives for our approach.

2   Review of Stress Detection Systems

Researches conducted for stress/anxiety detection based on EEG signals analysis are few compared to those done for emotion recognition surveyed in our previous paper [24].

Giannakakis *et al.* [2] defined a set of conditions for both stress and relax states. By defining thresholds for valence and arousal, the author extracted trials from the convenient "DEAP dataset" and constructed two subdatasets (trials labeled with stress and trials labeled with relax). These levels are summarized by stress and relaxed states and subjects that conform to adequate norm of stress/anxiety scale were named outstanding to a subset of 18 subjects. In order to represent accurately the states under investigation, spectral, temporal and non linear EEG features were presented, but classification wasn't performed.

A review of three EEG signals feature extraction techniques is presented by Bastos-Filho *et al.* in their paper [32]. Their system was validated through the stress and calm emotional states classification using the K-NN classifier from convenient "DEAP dataset".

Other authors in [33] have proposed a new method to classify emotional stress in the two major fields of the valence and arousal space by adopting bio-signals. They defined two specific fields of valence and arousal emotional stress space, corresponding to two states, calm and negatively excited. Qualitative and quantitative psychophysiological signals evaluation have been used to choose relevant EEG signal segments for improving performance and efficiency of emotional stress recognition systems.

In the work of Vanita and Krishnan [3], the authors proposed an EEG-based stress detection system for students. The purpose of the study is to determine the stress level for students in higher academic institution. An experimentation was conducted in order to activate stress and record EEG signals which are preprocessed thenceforth in order to remove noise and ocular artefact. A time-frequency analysis was applied to extract useful information from EEG and hierarchical SVM was implemented as classifier and obtained accuracy of 89.07%.

To detect stress in healthy subjects, Sulaiman et al. [1] used in their study k-NN classifier and some parameters such as Shannon Entropy (SE), Relative Spectral Centroid (SC) and Energy Ratio (RER). Their study employed 185 EEG data from different experiments.

Lahane et al. [4] proposed a real-time System to Detect human Stress using EEG signals. Data in this work was gathered using an android application then processed in order to extract different EEG frequency bands. Relative Energy Ratio was calculated for each frequency band as feature.

As shown, several researches are based on "DEAP dataset" giving it is one of the few available emotional datasets which is characterised by the number of participant and the used neuro cap that contains most important channels.

3   The Proposed System

The common point of researches elaborated recently for emotional states recognition and which used "DEAP dataset" is the way data is prepared for the classification. Dependably of the emotion to detect, most works were based on the Valence-Arousal dimension in order to map data into emotion. While there are claims that prove the strong relation between Frontal Asymmetry and emotional states, we propose to use this relationship to annotate/labelize the DEAP trials in order to detect stress and nostress states.

Our system contains three steps, presented in details in the Fig. 1:

1. Data annotation: 2 subdatasets are constructed from whole DEAP trials based on alpha Frontal Asymmetry.
2. Features extraction: extracting useful information from the signal (Hjorth parameters as Time Domain feature and Power bands and Time-Frequency Domain feature).
3. Classification: Detect from the prepared data and based on the extracted features the two emotional states: Stress/Nostress).

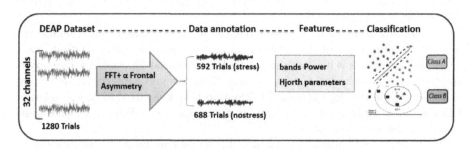

Fig. 1. The framework of our method

3.1   Frontal Alpha Asymmetry and Mental States

Increased alpha activity has been conceptualized into a psychophysiological trait-like characteristic distinguishable in current and remitted depressed participants from healthy participants [5,6]. Notably, frontal alpha EEG asymmetry, defined as the difference in alpha activity over right vs. left cerebral hemisphere, that presents higher scores, indicate greater relative left activation (i.e., increased right alpha activity). More specifically, data has shown that current and remitted depression is associated with increased left (versus right) alpha activity, which corresponds to decreased left (versus right) cerebral hemispheric activation.

Alpha activity has been associated with approach and withdrawal related motivation measured by the Behavioral Activation System and Behavioral Inhibition System Scales [7]. Specifically, less relative left frontal alpha activity (greater left activation) has been associated with heightened behavioral activation sensitivity, or an increased motivation to approach when goal-directed action is indicated [8,9].

The consistency and strength of association between EEG asymmetry and behavioral activation and inhibition sensitivity suggests that they strongly relate to an individual's tendency to approach or avoid. Minimal research, however, has linked frontal EEG asymmetry to avoidance behavior in depressed individuals before and after a treatment designed to modify behavioral avoidance.

Finally, as frontal EEG asymmetry has been linked with affect, data have indicated an association between greater relative left prefrontal activation (i.e., less alpha activity) with positive affect measured by the Positive and Negative Affect Schedule [10]. Conversely, greater relative right prefrontal activation (i.e., more alpha activity) has been associated with negative personality factors [9] and more intense reactions to negative films [12], in addition to anxiety and negative emotions trait [8]. This differential association between frontal EEG asymmetry and negative/positive affect may indicate a biological substrate of affect [9].

3.2   Data Description

Our approach is conducted on the available multimodal dataset for the analysis of the stress states of human affective "DEAP dataset" [44]. It contains EEG signals of 32 Healthy participants aged between 19 and 37 (mean age 26.9) while watching 1 min music video. 40 trials per subject, after each trial the participant noted a Self-Assesement Manikan SAM to rate Arousal, Valence, Dominance and Familiarity. "DEAP dataset" contains RAW data and preprocessed Data. Since we need to test our stress detection approach we are interested in using the preprocessed data.

3.3   Data Annotation

The EEG signals issued from both right and left Hemisphere of the human brain channels were analysed in an off-line way. We calculated The EEG frequency band power by using Fast Fourier Transform (FFT) with Hamming window. The window was set to 256 with 50% extending. Then EEG Asymmetry Index formula is used in order to classify the power for each into 2 groups as shown below in Eq. 1.

$$Asymmetry\ Index = \ln(\alpha)\Big|_{LChannel} - \ln(\alpha)\Big|_{RChannel} \qquad (1)$$

The resulting sub-datasets are: 592 trials labeled with stress and 688 trials labeled with nostress.

3.4   Feature Extraction

In this section, we review relevant features for EEG emotion recognition, while the stress detection is considered as an emotion recognition problem. We tested on our approach two different types of features: Time Domain feature which is Hjorth parameters and Time-Frequency Domain feature which is Band Power.

3.4.1   Time Domain Feature

Hjorth parameters [30] are: Activity, Mobility, and Complexity. The variance of a time series represents the activity parameter. The mobility parameter is represented by the mean frequency, or the standard deviation proportion of the power spectrum. Finally The complexity parameter represents the variation in frequency. It besides indicates the deviation of the slope.

Assume that $dx_i = x_{i+1} - x_{i'} (i = 1, .., n-1), ddx = dx_{i+1} - d_{i'} (i = 1, .., n-1)$, then calculate the variances includes $\sigma_1 = \frac{1}{n} \sum_{i=1}^{n} x_i^2, \sigma_2 = \frac{1}{n-1} \sum_{i=1}^{n-1} dx_i^2$ and $\sigma_3 = \frac{1}{n-2} \sum_{i=1}^{n-2} ddx_i^2$.

The expressions of Hjorth parameters are: $activity = \sigma_1, mobility = \sigma_2/\sigma_1$ and $complexity = \sqrt{\sigma_3/\sigma_2 - mobility}$

Hjorth were used in many EEG studies such in [29–31]. In our work, we calculated Hjorth parameters for all EEG channels, that produce a size Feature Vector of 96×1 for each trial.

3.4.2   Time-Frequency Domain Feature

Power bands features are the most popular features in the context of EEG-based emotion recognition. This assumes stationarity of the signal for the duration of a trial. The Definition of EEG frequency bands differs slightly between studies. Commonly they are defined as given in the first two columns of Table 1. In order to extract frequency bands, we have applied the wavelet decomposition technique with the function 'db5' for 5 levels decomposition as shown in the last column of Table 1. Power was calculated thereafter for each frequency band and for all channels to construct the Feature Vector with a size of 128×1 per trial.

3.5 Classification

In our study, the classification task is more challenging since the number of samples is more important, up to 1280.

After extracting the relevant features, we still have to find the related emotional stress states in the EEG signals. In this research, we have used three classifiers which are K-NN, Fuzzy K-NN and SVM with a scheme of 5-folds cross validation that consists of training 4 folds and test 1 fold, five iterations.

The performance of the stress detection system is the average of all measured accuracy for each iteration; and it was evaluated through Classification Accuracy computed as:

$$Accuracy = \frac{Number\ of\ correctly\ classified\ trials\ in\ test\ set}{Total\ number\ of\ trials\ in\ test\ set} \qquad (2)$$

4 Results and Discussion

In Table 1, our method is compared to the most recent methods in terms of features, classifiers and accuracy. The results of the proposed method are compared with those of some other systems using the same emotional benchmark. In fact, according to Table 1, we confirm that the proposed method showed promising results. Power band and Hjorth features giving an accuracy rate of 86.42% and 87.27% respectively based on Fuzzy K-NN classifier. The K-NN classifier also gave a good result, very close to FK-NN (85.63%) and (85.39%) respectively for Hjorth and Band power features. Nevertheless, the SVM classifier with RBF kernel [11], C = 1 did'nt give good results only 65.35% (Band Power) and 68.26% (Hjorth parameters). While K-NN and FK-NN are not significantly influenced over bands changes, accuracies are quite close. SVM shows much weaker performance especially in terms of classification accuracy with Band Power features as shown in Fig. 2. In this study, we have tested only simple features to show the efficiency annotation method proposed and explained previously, a variety of features can be applied and higher performances can be reached. So as future work, we plan to maximize the number of features based on this dataset. Investigation of features selection techniques is required for huge features amount. A deep work is planed as an extension to this one (Table 2).

Table 1. EEG signal frequency bands and decomposition levels at fs = 128 Hz

Bandwidth (Hz)	Frequency band	Decomposition level
1–4 Hz	Delta δ	A5
4–8 Hz	Theta θ	D5
8–13 Hz	Alpha α	D4
13–32 Hz	Beta β	D3
32–64 Hz	Gamma Γ	D2

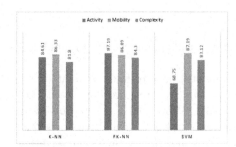 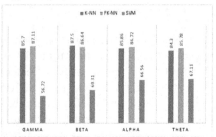

Fig. 2. Classification accuracy according to different bands and parameters, from left to right according features are: Hjorth and Band Power

Table 2. Quantitative comparison of results of the most pertinent studies dealing with the stress identification systems from the EEG recording

Study	Features	Classifier	Accuracy
Hosseini et al. [33]	FD, CD and wavelet entropy	Linear Discriminant Analysis (LDA) and SVM	LDA: 80.1% SVM: 84.9%
Bastos-Filho et al. [32]	Statistical features, PSD and HOC	K-nearest neighbor (K-NN)	Stat.: 66.25% PSD: 70.1% HOC: 69.6%
Garca-Martnez et al. [34]	SEn, QSEn and DEn	Decision tree	DT: 75.29%
Garca-Martnez et al. [35]	QSEn, PEn and AAPEn	SVM	SVM: 81.31%
Our study	Asymmetry Index, Power band and Hjorth	SVM (RBF Kernel) K-NN (5 NN) Fuzzy K-NN (FK-NN)	SVM: 68.26% K-NN: 85.63% FK-NN: 87.27%

5 Conclusion

In this paper, we proposed a Stress Detection System using a new data annotation technique based on Frontal Alpha Asymmetry of "DEAP dataset" trials. The results of our study indicate that our Dataset subdivision approach gives us an acceptable classification rate honestly compared with other previous works using the same database but based on the Valence-Arousal approach for the emotional mapping. We plan as future work to deepen our research in term of features variety and feature selection investigation, in order to select the more relevant channel/band combination for stress/anxiety detection system.

Acknowledgments. The research leading to these results has received funding from the Ministry of Higher Education and Scientific Research of Tunisia under the grant agreement number LR11ES48.

References

1. Sulaiman, N., Taib, M.N., Lias, S., Murat, Z.H., Aris, S.A.M., Hamid, N.H.A.: Novel methods for stress features identification using EEG signals. Int. J. Simul.: Syst. Sci. Technol. **12**(1), 27–33 (2011)
2. Giannakakis, G., Grigoriadis, D., Tsiknakis, M.: Detection of stress/anxiety state from EEG features during video watching. In: Conference of the IEEE Engineering in Medicine and Biology Society (2015)
3. Vanitha, V., Krishnan, P.: Real time stress detection system based on EEG signals. Biomed. Res. **27**, 271–275 (2016). Special Issue
4. Lahane, P., Vaidya, A., Umale, C., Shirude, S., Raut, A.: Real time system to detect human stress using EEG signals. Int. J. Innovative Res. Comput. Commun. Eng. **4**(4) (2016)
5. Brenner, R.P., Ulrich, R.F., Spiker, D.G., Sclabassi, R.J., Reynolds, C.F., Marin, R.S., Boller, F.: Computerized EEG spectral analysis in elderly normal, demented and depressed subjects. Electroencephalogr. Clin. Neurophysiol. **64**(6), 483–492 (1986)
6. Pollock, V.E., Schneider, L.S.: Topographic electroencephalographic alpha in recovered depressed elderly. J. Abnorm. Psychol. **98**(3), 268–273 (1989)
7. Gray, J.A.: The psychophysiological basis of introversion-extraversion. Behav. Res. Ther. **8**(3), 249–266 (1970)
8. Coan, J.A., Allen, J.J.: Frontal EEG asymmetry and the behavioral activation and inhibition systems. Psychophysiology **40**(1), 106–114 (2003)
9. Sutton, S.K., Davidson, R.J.: Prefrontal brain asymmetry: a biological substrate of the behavioral approach and inhibition systems. Psychol. Sci. **8**(3), 204–210 (1997)
10. Tomarken, A.J., Davidson, R.J., Wheeler, R.E., Doss, R.C.: Individual differences in anterior brain asymmetry and fundamental dimensions of emotion. J. Pers. Soc. Psychol. **62**(4), 676–687 (1992)
11. Dhahri, H., Alimi, A.M.: The modified differential evolution and the RBF (MDE-RBF) neural network for time series prediction. In: IEEE International Conference on Neural Networks - Conference Proceedings, p. 2938 (2006)
12. Tomarken, A.J., Davidson, R.J., Henriques, J.B.: Resting frontal brain asymmetry predicts affective responses to films. J. Pers. Soc. Psychol. **59**(4), 791–801 (1990)
13. Dharmawan, Z.: Analysis of computer games player stress level using EEG data. Master of Science Thesis report, Faculty of Electrical Engineering, Mathematics and Computer Science, Delft University of Technology, Netherlands (2007)
14. Interactive Productline IP AB-Mindball. http://www.mindball.se/index.html
15. Novák, D.: EEG and VEP signal processing. Technical report. Czech Technical University in Prague, Department of Cybernetics (2004)
16. Horlings, R.: Emotion recognition using brain activity. In: Proceedings of the 9th International Conference on Computer Systems and Technologies and Workshop for Ph.D. Students in Computing, Gabrovo, Bulgaria, p. II.1-1 (2008)
17. Morilak, D.A.: Role of brain norepinephrine in the behavioral response to stress. Prog. Neuro-psychopharmacol. Biol. Psychiatry **29**(8), 1214–1224 (2005)
18. Hoffmann, E.: Brain training against stress: theory methods and results from an outcome study. Stress Rep. **4** (2005)
19. Lin, T., John, L.: Quantifying mental relaxation with EEG for use in computer games. In: International Conference on Internet Computing, Las Vegas, Nevada, USA, pp. 409–415 (2006)

20. Alimi, A.M.: Evolutionary computation for the recognition of on-line cursive handwriting. IETE J. Res. **48**(5), 385–396 (2002). SPEC

21. Fuchs, E., Uno, H., Fluegge, G.: Chronic psychosocial stress induces morphological alterations in hippocampal pyramidal neurons of the tree shrew. Brain Res. **673**, 275–282 (1995)

22. Bezine, H., Alimi, A.M., Derbel, N.: Handwriting trajectory movements controlled by a beta-elliptic model. In: Proceedings of the International Conference on Document Analysis and Recognition, ICDAR, p. 1228 (2003)

23. Hughes, J.W., Stoney, C.M.: Depressed mood is related to high-frequency heart rate variability during stressors. Psychosom. Med. **62**, 796–803 (2000)

24. Baghdadi, A., Aribi, Y., Alimi, A.M.: A survey of methods and performances for EEG-based emotion recognition. In: Abraham, A., Haqiq, A., Alimi, A.M., Mezzour, G., Rokbani, N., Muda, A.K. (eds.) HIS 2016. AISC, vol. 552, pp. 164–174. Springer, Cham (2017). doi:10.1007/978-3-319-52941-7_17

25. Lawrence, D.A., Kim, D.: Central/peripheral nervous system and immune responses. Toxicology **142**, 189–201 (2000)

26. NIOSH, Stress at Work, NIOSH Publication Number 99-101 (1999)

27. Cooper, C.: Stress in the workplace. Br. J. Hosp. Med. **55**, 559–563 (1996)

28. Manning, M., Jackson, C., Fusilier, M.: Occupational stress, social support, and the costs of health care. Acad. Manag. J. **39**, 738–750 (1996)

29. Ansari-asl, K., Chanel, G., Pun, T.: A channel selection method for EEG classification in emotion assessment based on synchronization likelihood. In: Proceedings of 15th European Signal Processing Conference, pp. 1241–1245 (2007)

30. Hjorth, B.: EEG analysis based on time domain properties. Electroencephalogr. Clin. Neurophysiol. **29**(3), 306–310 (1970)

31. Horlings, R., Datcu, D., Rothkrantz, L.: Emotion recognition using brain activity. In: Proceedings of International Conference on Computer Systems and Technologies, p. II.116 (2008)

32. Bastos-Filho, T.F., Ferreira, A., Atencio, A.C.: Evaluation of feature extraction techniques in emotional state recognition. In: IEEE Proceedings of 4th International Conference on Intelligent Human Computer Interaction, Kharagpur, India, 27–29 December 2012

33. Hosseini, S.A., Khalilzadeh, M., Changiz, S.: Emotional stress recognition system for affective computer based on bio-signals. J. Biol. Syst. **18**, 101–114 (2010). Special Issue

34. García-Martínez, B., Martínez-Rodrigo, A., Cantabrana, R.Z., García, J.M.P., Martínez, R.A.: Application of entropy-based metrics to identify emotional distress from electroencephalographic recordings. Entropy **18**, 221 (2016)

35. García-Martínez, B., Martínez-Rodrigo, A., Zangróniz, R., García, J.M.P., Alcaraz, R.: Symbolic analysis of brain dynamics detects negative stress. Entropy **18**, 221 (2017)

36. Elbaati, A., Boubaker, H., Kherallah, M., Alimi, A.M., Ennaji, A., Abed, H.E.: Arabic handwriting recognition using restored stroke chronology. In: Proceedings of the International Conference on Document Analysis and Recognition, ICDAR, p. 411 (2009)

37. Aribi, Y., Wali, A., Alimi, A.M.: Automated fast marching method for segmentation and tracking of region of interest in scintigraphic images sequences. In: Azzopardi, G., Petkov, N. (eds.) CAIP 2015. LNCS, vol. 9257, pp. 725–736. Springer, Cham (2015). doi:10.1007/978-3-319-23117-4_62

38. Aribi, Y., Wali, A., Hamza, F., Alimi, A.M., Guermazi, F.: ARG: a semiautomatic system for ROI detection on Renal Scintigraphic images. In: Proceedings of the 14th International Conference on Hybrid Intelligent Systems (HIS 2014), Kuwait, December 2014

39. Aribi, Y., Wali, A., Alimi, A.M.: An intelligent system for renal segmentation. In: Proceedings of the 15th International Conference on e-Health Networking - Healthcom 2013, Lisbon, Portugal, pp. 1–6, October 2013

40. Aribi, Y., Wali, A., Chakroun, M., Alimi, A.M.: Automatic definition of regions of interest on renal scintigraphic images. In: Proceedings of the Conference on Intelligent Systems and Control, Vancouver, Canada, AASRI Procedia, vol. 4, pp. 37–42 (2013)

41. Aribi, Y., Wali, A., Alimi, A.M.: A system based on the fast marching method for analysis and processing DICOM images: the case of renal scintigraphy dynamic. In: Proceedings of the International Conference on Computer Medical Applications (ICCMA 2013), Sousse, Tunisia, pp. 1–6, January 2013

42. Aribi, Y., Wali, A., Hamza, F., Alimi, A.M., Guermazi, F.: Analysis of scintigraphic renal dynamic studies: an image processing tool for the clinician and researcher. In: Hassanien, A.E., Salem, A.-B.M., Ramadan, R., Kim, T. (eds.) AMLTA 2012. CCIS, vol. 322, pp. 267–275. Springer, Heidelberg (2012). doi:10.1007/978-3-642-35326-0_27

43. Aribi, Y., Hamza, F., Wali, A., Alimi, A.M., Guermazi, F.: An automated system for the segmentation of dynamic scintigraphic images. Appl. Med. Inform. 34(2), 1–12 (2014)

44. DEAP dataset, a dataset for emotion analysis using EEG, physiological and video signals. http://www.eecs.qmul.ac.uk/mmv/datasets/deap/

A Pattern-Based Bayesian Classifier
for Data Stream

Jidong Yuan[✉], Zhihai Wang, Yange Sun, Wei Zhang, and Jingjing Jiang

School of Computer and Information Technology, Beijing Jiaotong University,
Beijing, China
{yuanjd,zhhwang}@bjtu.edu.cn

Abstract. An advanced approach to Bayesian classification is based on
exploited patterns. However, traditional pattern-based Bayesian classifiers cannot adapt to the evolving data stream environment. For that,
an effective Pattern-based Bayesian classifier for Data Stream (PBDS)
is proposed. First, a data-driven lazy learning strategy is employed to
discover local frequent patterns for each test record. Furthermore, we
propose a summary data structure for compact representation of data,
and to find patterns more efficiently for each class. Greedy search and
minimum description length combined with Bayesian network are applied
to evaluating extracted patterns. Experimental studies on real-world and
synthetic data streams show that PBDS outperforms most state-of-the-
art data stream classifiers.

Keywords: Data stream · Frequent pattern · Bayesian · Lazy learning

1 Introduction

Classification based on patterns has attracted significant attention and research
effort in recent years [1,14]. Pattern is a subset of data or a set of items, where an
item refers to a pair of attribute-value. Frequent patterns (or itemsets) are generated w.r.t. minimum support. Since frequent pattern combines the set of single
features non-linearly and indicates more underlying semantics of data [4], except
associative classifier [14], it has also been successfully employed to approximate
the joint probability of Bayesian classifier [1]. However, those previous pattern-
based classifiers built on traditional datasets cannot adapt to the highly dynamic
and complex data stream environment.

For mining patterns over data stream, state-of-the-art approaches mainly
focus on discovering frequent pattern [10] or its variants. Most of the exiting
classifiers for non-stationary data stream pay more attention to adaptive tree
models [9], decision rules [7], ensemble algorithms [8,12] and kNN [3]. However,
the above algorithms ignore the potential of incorporating pattern in graphical
models.

© Springer International Publishing AG 2017
D. Liu et al. (Eds.): ICONIP 2017, Part IV, LNCS 10637, pp. 868–877, 2017.
https://doi.org/10.1007/978-3-319-70093-9_92

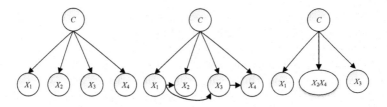

Fig. 1. (left) The structure of Naive Bayes classifier; (middle) The structure of Bayesian network classifier; (right) The structure of pattern-based Bayesian classifier.

As is well known, Bayesian classification is a graphical model based on Bayes theorem[1]. The major challenge of Bayesian classification lies in the computation of joint probability $P(\mathbf{x}, y)$. The simplest way to address this issue is the Naive Bayes model (as shown in Fig. 1 *left*). Although Naive Bayes classifier could be applied on evolving data stream, it cannot adapt to concept drift directly. To alleviate the strong assumption, researchers proposed the conditional dependence model, *e.g.* Bayesian networks [6]. Besides the low-order dependencies among variables, some higher order dependency models, *e.g.* frequent pattern based Bayesian [1,11], have also been proposed. Long and overlapped patterns are used to build Bayesian model previously [11], while independent pattern based Bayesian classifier presents its effectiveness and flexibility recently [1]. For pattern-based model, the joint probability is (as shown in Fig. 1 *right*):

$$P(\mathbf{x}, y_i) = P(x_1, x_2, x_3, x_4, y_i) = P(y_i) \cdot P(x_1|y_i) \cdot P(x_2 x_4|y_i) \cdot P(x_3|y_i) \quad (1)$$

Previous works on streaming data mining either focus on finding interesting patterns efficiently [10] or creating classifier effectively [3]. However, none of them combines pattern with classifier practically. Unlike these listed methods, our objective is to learn an efficient and effective **P**attern-based **B**ayesian classifier for **D**ata **S**tream (PBDS) that adapts to concept drift. Meanwhile, several challenges remains.

First, traditional pattern mining methods may generate excessive number of patterns that are useless in the classification. However, the computing of lazy classifier is performed on a demand driven basis, only the "useful" portion of the training data is mined for generating patterns applicable to the test case. Second, since frequent pattern mining in data stream can just check each instance for a single time, a simple but effective summary data structure for each class based on the sliding window model is proposed, which means the probability approximation for PBDS is separately tailored to each class. Third, for pattern-based Bayesian classifier, a set of long and not overlapped patterns that fully covers the given test case should be found, but this problem is NP-hard in general. Therefore, a heuristic pattern extraction mechanism is adopted, which is based on greedy search and the minimum description length (MDL) for Bayesian

---

[1] $P(y_i|\mathbf{x}) = \frac{P(\mathbf{x}, y_i)}{P(\mathbf{x})} = \frac{P(y_i) \cdot P(\mathbf{x}|y_i)}{P(\mathbf{x})}$.

classifier to reduce the generation of candidate itemsets, and to ensure the fitness between extracted patterns and original data records.

The remaining of this paper is organised as follows: Sect. 2 introduces MDL for PBDS. Section 3 describes our summary data structure, establishing the principle of extracting and updating local frequent patterns. In Sect. 4, experimental evaluations of the proposed approach is studied. Section 5 concludes our work.

2 MDL for Pattern-Based Bayesian Classifier

A data stream is an infinite sequence of training records: $\mathbf{U} = \{\mathbf{x}_1, \mathbf{x}_2, \ldots, \mathbf{x}_t, \ldots\}$, where $\mathbf{x}_t$ is the most recent record arriving at time stamp t. A record $\mathbf{x} = \{x_1, x_2, \ldots, x_m, y\}$ is represented by a set of m items, where y is the class label of $\mathbf{x}$. Let X_i be an attribute that describes a data feature. Usually an item (X_i, x_i) is represented by a single character x_i. An itemset (or a pattern) $\mathbf{z} = \{x_1, x_2, \ldots, x_k\}$ is a set of k items, which is also called k-itemset. In order to mine frequent patterns for building lazy Bayesian classifier, we should find frequent patterns for each possible class $(sup_{y_i}(\mathbf{z}) \geq min\_sup)$. The support of itemset $\mathbf{z}$ in sliding window for each class is: $sup(\mathbf{z}) = sup_{y_i}(\mathbf{z}) = \frac{|\mathbf{z}|_{y_i}}{N}$.

The MDL is closely related to Bayesian classifiers [6]. It is used to select optimal model and to avoid overfitting. Let B be a Bayesian network built on sliding window D, The MDL scoring function of B given a training dataset (or a sliding window) D, is written as: $MDL(B|D) = DL(B) + DL(D|B)$, where $DL(B)$ represents the length of the description of the model B, and $DL(D|B)$ the length of the description of the data D when encoded with the help of the model B. According to MDL, the best model of a given dataset is the one that minimizes the sum of $DL(D|B)$ and $DL(B)$.

Suppose for each node X_i, there are k_i parents for it. Let r_i $(1 \leq i \leq m)$ be the cardinality of X_i, $pa(X_i)$ the parents of X_i. The MDL for a Bayesian model is

$$MDL(B|D) = \sum_{i=1}^{m} (k_i \log m + \tfrac{\log N}{2}(r_i - 1) \prod_{j \in pa(X_i)} r_j) + N \sum_{i=1}^{m} (H(X_i) - I(X_i; pa(X_i))) \ (2)$$

where $H(X_i)$ is the entropy of X_i, $I(X_i; pa(X_i))$ the mutual information of X and $pa(X)$.

For our pattern-based Bayesian classifier, if the dependency among items of pattern are not considered, the parent node will be the only class attribute. In addition, a lazy method is used to find patterns for each possible class label $(r_j = 1)$, so the mutual information $I(X_i; pa(X_i))$ will be 0 because the value of parent node is constant. The MDL for our pattern-based Bayesian model is

$$MDL_{PBDS}(B|D) = \sum_{i=1}^{m} \frac{\log N}{2} \times (r_i - 1) + N \sum_{i=1}^{m} H(X_i) \qquad (3)$$

where m represents the number of patterns, r_i the cardinality of pattern X_i, N the number of records for one special class.

3   Pattern-Based Bayesian Classification

In this section, the construction of Candidate Frequent Itemsets forest (CFI-forest) that only scan the data stream once is introduced at first, then details about how to select patterns for characterizing each test record will be given.

3.1   Construction of CFI-Forest

In order to construct CFI-forest, for a basic sliding window $D = \{\mathbf{x}_1, \mathbf{x}_2, \ldots, \mathbf{x}_N\}$, all the records are read and split to $|C|$ folds according to their class label y_i, $|C|$ is the number of classes in D. At the same time, CFI-forest CFI_i for each class fold, $\mathbf{CFI} = \{CFI_1, CFI_2, \ldots, CFI_{|C|}\}$ are built and updated.

Suppose the current sliding window D_1 contains four records, $D_1 = \{T_1, T_2, T_3, T_4\}$, where $T_1 = \{a_1, b_1, c_1, d_2, y_2\}$, $T_2 = \{a_1, b_1, c_1, d_1, y_2\}$, $T_3 = \{a_2, b_1, c_1, d_2, y_1\}$ and $T_4 = \{a_3, b_2, c_1, d_2, y_1\}$. The CFI-forest structure consists of three parts: a list of Candidate Frequent Items (CFI_list), a list of Candidate Frequent Item Trees (CFIT) and a list of Candidate Suffix Frequent Items (CSFI_list). Accordingly we split records into two folds, $\{T_3, T_4\} \in y_1$ and $\{T_1, T_2\} \in y_2$ at first step. Here we take the fold $\{T_1, T_2\} \in y_2$ as an example to illustrate the construction of CFI-forest structure CFI_2.

As shown in Fig. 2(a), for the first record $T_1 = \{a_1, b_1, c_1, d_2\}$, a set of sub-records $\{a_1, b_1, c_1, d_2\}$, $\{b_1, c_1, d_2\}$, $\{c_1, d_2\}$, $\{d_2\}$ are generated and inserted to x_i. CFIT separately. A similar situation is applied on T_2. CFI_list includes all the items and their corresponding support number, while x_i. CSFI_list gives the suffix items and their support number of x_i (note that item d_2 and d_1 have no suffix item). After sliding the window, the previous D_1 is updated to $D_1' = \{T_2, T_3, T_4, T_5\}$. Now the distribution of records is also changed to $\{T_3, T_4, T_5\} \in y_1$ and $\{T_2\} \in y_2$, so we need to remove $T_1 = \{a_1, b_1, c_1, d_2\}$ from CFI_2, the latest CFI_2 is shown in Fig. 2(b).

3.2   Lazy Pattern Mining

The previous section mainly illustrates the process of creating and updating CFI-forest for each possible class, here we will describe how to find effective patterns for Bayesian Classifier. This problem is not straightforward and some issues need to be addressed. For example, which pattern or set of patterns should be selected for classifying? If the length and support of the selected patterns are different from each other, which one(s) should be chose?

Definition. Orders for patterns

Given two frequent patterns $\mathbf{z}_1$ and $\mathbf{z}_2$, $\mathbf{z}_1 \succ \mathbf{z}_2$ (also called $\mathbf{z}_1$ precedes $\mathbf{z}_2$ or $\mathbf{z}_1$ has a higher precedence than $\mathbf{z}_2$) if

(1) The length of $\mathbf{z}_1$ is longer than that of $\mathbf{z}_2$, or
(2) Their length are the same, but the support of $\mathbf{z}_1$ is greater than that of $\mathbf{z}_2$, or

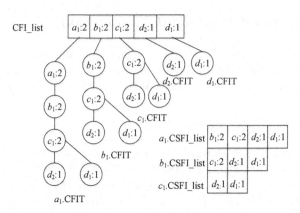

(a) CFI-forest after processing T_1 and T_2 of class y_2

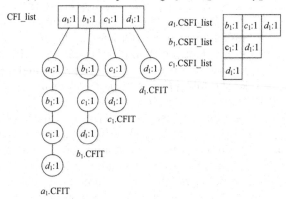

(b) CFI-forest of class y_2 after sliding the window to D'_1

Fig. 2. CFI-forest construction

(3) Both the length and supports of $\mathbf{z}_1$ and $\mathbf{z}_2$ are the same, but the MDL_{PBDS} of $\mathbf{z}_1$ based Bayesian is shorter than the $\mathbf{z}_2$ based one.

Our aim is to find patterns that cover the test record lazily, since set covering is a NP-hard problem, a heuristic pattern selection method is adopted to find patterns greedily. In another words, our goal is to find the longest disjointed patterns, by using min_sup and MDL_{PBDS} to break ties. Note, we cannot ensure that each item or pattern occurs together with a given class. In other words, sometimes the discovered patterns of CFI_i are not enough to cover test record since some items are infrequent. Hence, it arise the zero frequency count or non-frequency problem. For Bayesian classifier, multiplied by zero probability will ignore the contributions of other patterns, so Laplacian smoothing is used to address this issue. The pseudo code of PBDS is shown in Algorithm 1.

Algorithm 1. PBDS($\mathbf{x}_{test}$, **CFI**, min_sup)

Input: $\mathbf{x}_{test}$, test record to be classified; **CFI**, a list of CFI-forests; min_sup, minimum class support
Output: predicted class label y_{test}
1: **for** each CFI_i of **CFI do**
2:    $finalPattern = \phi$
3:    **repeat**
4:      $bestPattern = $ selectPattern($\mathbf{x}_{test}$, CFI_i, min_sup)
5:      $\mathbf{x}_{test} = \mathbf{x}_{test} - bestPattern$
6:      $finalPattern = finalPattern \cup bestPattern$;
7:    **until** $\mathbf{x}_{test} = \phi$
8:    $P(\mathbf{x}, y_i) = P(y_i) \prod_{\mathbf{z} \in finalPattern} P(\mathbf{z}|y_i)$ {$\mathbf{z}$ is the selected pattern}
9: **end for**
10: **return** $y_{test} = max(P(\mathbf{x}, y_i))$

---

4 Experiment and Evaluation

In order to evaluate the performance of our PBDS approach on Massive Online Analysis (MOA) platform [2], synthetic and real datasets that have relatively large records (for example, $\geq$10,000) are selected[2]. Since continuous attribute values cannot be directly employed in classification by means of patterns, here we are only interested in the case where all the variables are discrete. The entropy based discretization approach is performed as a preprocessing step for continuous variables [5].

4.1 Classification Accuracy

Classification accuracy measures the ability of a classifier to correctly predict the class label of unknown record. Here we compare PBDS with a list of classifiers that belong to specific categories, such as Bayesian classifier, rule based classifier, instance based classifier, tree model and ensemble method. In particular, we consider the MOA default settings of Nave Bayes (NB), Naive Bayes Multinomial (NBM), kNN, kNNwithPAW (PAW) [3], RuleClassifierNBayes (RCNB) [7], HoeffdingTree (HT) [9], and Accuracy Weighted Ensemble classifier [13](AWE). For the proposed PBDS, the min_sup that decides the quantity of patterns, and the size of sliding window N are learned by a standard 5-fold cross validation on a 10% sampling of the original dataset, where $min\_sup \in \{0, 0.05, 0.1 : 0.1 : 1.0\}$, and $N \in \{1, 10, 100 : 100 : 1000\}$. The prequential scheme that interleaves training and test records is used for evaluating above selected classification algorithms.

---

[2] Datasets Chess, Connect-4, EEG, MAGIC, PokerHand and CoverType are downloaded from UCI Machine Learning Repository http://archive.ics.uci.edu/ml/; Others are generated by the classical data generators separately with 1,000,000 records via MOA.

As shown in Table 1, PBDS performs better than other classical methods, especially on datasets LED, SEA, Chess and EEG. The average rank of PBDS on 12 synthetic and real data streams is 2.75, it wins all other classifiers. In addition, it is notable that the accuracy of PBDS on dataset Chess is 99.80%, while NB (or NBM) just classify 5.00% (or 2.10%) of instances accurately.

Table 1. Accuracy comparison with state-of-the-art classifiers (%)

Data	NB	NBM	RCNB	kNN	PAW	HT	AWE	PBDS
RBF	73.90 (5.5)	68.70 (8)	75.10 (3)	74.70 (4)	76.50 (2)	**85.40 (1)**	73.30 (7)	73.90 (5.5)
HyperPlane	**88.50 (1)**	79.70 (6)	85.50 (4)	77.50 (7)	75.80 (8)	86.80 (3)	80.50 (5)	87.90 (2)
LED	71.90 (7)	70.50 (8)	73.10 (2)	72.60 (3)	72.20 (4)	72.00 (5.5)	72.00 (5.5)	**73.40 (1)**
LEDdrift	28.80 (5)	27.80 (6)	33.40 (2)	23.10 (8)	24.80 (7)	**33.60 (1)**	29.70 (4)	30.80 (3)
SEA	81.00 (5)	65.10 (8)	84.90 (2)	80.70 (6.5)	80.70 (6.5)	84.60 (3)	84.50 (4)	**85.70 (1)**
STAGGER	100.00 (4)	93.33 (8)	100.00 (4)	100.00 (4)	100.00 (4)	100.00 (4)	100.00 (4)	100.00 (4)
Chess	5.00 (7)	2.10 (8)	40.60 (4)	87.70 (2)	73.10 (3)	22.30 (5)	5.60 (6)	**99.80 (1)**
Connect-4	48.10 (8)	53.40 (7)	64.50 (3)	**66.50 (1)**	65.40 (2)	61.50 (4)	59.50 (6)	60.20 (5)
EEG	83.60 (6)	82.30 (7)	90.40 (4)	91.20 (2)	91.10 (3)	79.60 (8)	87.00 (5)	**92.90 (1)**
MAGIC	47.50 (8)	68.50 (7)	100.00 (3.5)	100.00 (3.5)	100.00 (3.5)	100.00 (3.5)	100.00 (3.5)	100.00 (3.5)
PokerHand	46.20 (6.5)	46.20 (6.5)	62.50 (2)	49.40 (5)	49.60 (4)	**79.70 (1)**	44.60 (8)	58.30 (3)
CoverType	80.10 (6)	64.80 (8)	96.90 (4)	**98.00 (1)**	97.60 (2)	71.80 (7)	88.10 (5)	97.50 (3)
Average	5.75	7.29	3.13	3.92	4.08	3.83	5.25	**2.75**

Figure 3 shows the evolving of classification accuracy on the SEA dataset. When a sudden concept drift appears, the accurate rates of all tested algorithms show a dip downwards except for our method. In addition, PBDS maintains a higher, more stable accuracy, preserving the smallest descending. It might be

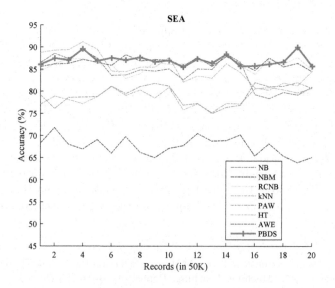

Fig. 3. Classification accuracy on the SEA dataset

contributed by the pattern-based model which could adapt to or even capture sudden concept drifts instantly, building classifiers lazily in a timely manner to cope with this type of drift.

4.2 Time and Memory Usage

In terms of classification time, as shown in Table 2, it is straightforward that NB or NBM is one of the most efficient classifiers. Compared with NB or NBM, PBDS needs to update CFI-forest, searching local frequent patterns for each record, so it is no doubt that PBDS is slower than NB or NBM. However, in comparison with instance based classifiers (*e.g.* kNN and PAW), PBDS is faster when items of the datasets are relatively small (*e.g.* HyperPlane, LED, LEDdrift and so on).

Similarly, simple but not so accurate classifiers (*e.g.* NB and NBM) achieved minimal memory consumption, as shown in Table 3. In most cases, especially when the number of items or class attributes are not large, the memory usage of PBDS is relatively less than instance based classifiers (*e.g.* kNN and PAW), ensemble method (*e.g.* AWE), and even RCNB (*e.g.* HyperPlane, SEA and so on). This is partly because our method employs a compressed tree structure to store each record efficiently.

In a nutshell, our approach performs better than other algorithms in the following three aspects: (1) it is more accurate than classical classifiers; (2) it is more suitable for scenarios with concept drift; (3) When the number of items or class attributes are relatively small, our algorithm is more efficient than other instance based approaches in terms of classification time and memory usage.

Table 2. Time comparison with state-of-the-art classifiers (second)

Data	NB	NBM	RCNB	kNN	PAW	HT	AWE	PBDS
RBF	5.59	**4.84**	177.63	435.05	674.53	16.39	209.03	1816.89
HyperPlane	4.69	**4.63**	502.33	345.53	547.09	14.67	250.27	122.07
LED	**3.73**	6.38	864.58	270.03	420.08	6.33	120.41	218.69
LEDdrift	**9.41**	15.23	620.97	1198.89	1574.38	11.59	227.19	1058.93
SEA	**1.86**	4.80	579.13	105.98	191.84	3.91	56.91	71.43
STAGGER	**1.61**	1.95	314.06	129.39	217.53	1.83	23.02	65.43
Chess	**0.27**	0.39	3.33	6.69	10.53	0.50	2.61	3.06
Connect-4	**1.20**	1.38	4.41	130.02	176.67	1.94	19.70	289.34
EEG	0.23	**0.22**	0.45	6.73	9.78	0.44	1.92	6.71
MAGIC	**0.23**	0.25	0.36	6.97	9.89	0.41	1.78	6.19
PokerHand	**5.42**	7.77	233.94	534.72	757.53	10.14	147.77	186.39
CoverType	**13.89**	17.81	62.77	958.78	1226.58	25.64	943.64	2121.18

Table 3. Memory usage comparison with state-of-the-art classifiers (KB/Hour)

Data	NB	NBM	RCNB	kNN	PAW	HT	AWE	PBDS
RBF	0.089	**0.061**	99.730	87.134	190.549	163.168	35.668	439.353
HyperPlane	0.019	**0.015**	260.583	66.008	149.858	67.012	82.706	3.316
LED	**0.009**	0.010	36.497	42.290	94.340	0.341	11.058	38.692
LEDdrift	0.071	**0.063**	56.618	473.934	894.272	1.226	42.572	174.814
SEA	0.003	**0.002**	4947.741	12.011	31.215	1.051	4.080	1.721
STAGGER	**0.001**	**0.001**	0.880	14.556	35.127	0.002	1.391	0.947
Chess	**0.001**	**0.001**	0.029	0.977	2.191	0.009	0.207	0.033
Connect-4	0.011	**0.009**	0.209	77.171	148.864	0.161	4.577	415.236
EEG	**0.001**	**0.001**	0.005	1.689	3.477	0.008	0.410	0.075
MAGIC	**0.001**	**0.001**	0.002	1.321	2.627	0.004	0.138	0.138
PokerHand	0.034	**0.028**	12.630	102.161	207.466	2.295	17.065	3.077
CoverType	0.275	**0.195**	6.050	763.954	1410.557	81.086	1505.897	2123.484

5   Conclusion and Future Work

In this paper, we propose PBDS, an effective pattern-based Bayesian classifier for evolving data stream. PBDS exploits a lazy learning strategy to find local frequent patterns when a classification request occurs. A sliding window based tree structure is presented to process streaming data. To ensure the quality of extracted patterns, MDL based principle is used for pattern selection. The experimental results on real and synthetic datasets show the potential of PBDS. Future direction of our work could involve investigating how to create faster and more accurate PBDS based on adaptive sliding windows, and to learn numerical attributes based pattern for classification directly.

Acknowledgments. This work is supported by National Natural Science Foundation of China (Nos. 61672086 and 61702030) and the Fundamental Research Funds for the Central Universities (Nos. 2016RC048 and 2016YJS036).

References

1. Baralis, E., Cagliero, L., Garza, P.: Enbay: a novel pattern-based bayesian classifier. IEEE Trans. Knowl. Data Eng. **25**(12), 2780–2795 (2013)
2. Bifet, A., Holmes, G., Kirkby, R., Pfahringer, B.: MOA: massive online analysis. J. Mach. Learn. Res. **11**(May), 1601–1604 (2010)
3. Bifet, A., Pfahringer, B., Read, J., Holmes, G.: Efficient data stream classification via probabilistic adaptive windows. In: Proceedings of the 28th Annual ACM Symposium on Applied Computing, pp. 801–806. ACM (2013)
4. Cheng, H., Yan, X., Han, J., Hsu, C.W.: Discriminative frequent pattern analysis for effective classification. In: 2007 IEEE 23rd International Conference on Data Engineering, pp. 716–725. IEEE (2007)

5. Fayyad, U.M., Irani, K.B.: Multi-interval discretization of continuous-valued attributes for classification learning. In: Machine Learning, pp. 1022–1027 (1993)
6. Friedman, N., Geiger, D., Goldszmidt, M.: Bayesian network classifiers. Mach. Learn. **29**(2–3), 131–163 (1997)
7. Gama, J., Kosina, P., et al.: Learning decision rules from data streams. In: IJCAI Proceedings-International Joint Conference on Artificial Intelligence, vol. 22, p. 1255 (2011)
8. Gomes, H.M., Barddal, J.P., Enembreck, F., Bifet, A.: A survey on ensemble learning for data stream classification. ACM Comput. Surv. (CSUR) **50**(2), 23 (2017)
9. Hulten, G., Spencer, L., Domingos, P.: Mining time-changing data streams. In: Proceedings of the Seventh ACM SIGKDD International Conference on Knowledge Discovery and Data Mining, pp. 97–106. ACM (2001)
10. Li, H.F., Shan, M.K., Lee, S.Y.: DSM-FI: an efficient algorithm for mining frequent itemsets in data streams. Knowl. Inf. Syst. **17**(1), 79–97 (2008)
11. Meretakis, D., Wüthrich, B.: Extending Naive Bayes classifiers using long itemsets. In: Proceedings of the Fifth ACM SIGKDD International Conference on Knowledge Discovery and Data Mining, pp. 165–174. ACM (1999)
12. Sun, Y., Wang, Z., Liu, H., Du, C., Yuan, J.: Online ensemble using adaptive windowing for data streams with concept drift. Int. J. Distrib. Sens. Netw. **12**, 4218973 (2016)
13. Wang, H., Fan, W., Yu, P.S., Han, J.: Mining concept-drifting data streams using ensemble classifiers. In: Proceedings of the Ninth ACM SIGKDD International Conference on Knowledge Discovery and Data Mining, pp. 226–235. ACM (2003)
14. Yuan, J., Wang, Z., Han, M., Sun, Y.: A lazy associative classifier for time series. Intell. Data Anal. **19**(5), 983–1002 (2015)

A Hierarchical Mixture Density Network

Fan Yang[✉], Jaymar Soriano, Takatomi Kubo, and Kazushi Ikeda

Graduate School of Information Science, Nara Institute of Science and Technology, Ikoma, Nara, Japan
{yang.fan.xv6,jaymar.soriano.jk2,takatomi-k,kazushi}@is.naist.jp

Abstract. The relationship among three correlated variables could be very sophisticated, as a result, we may not be able to find their hidden causality and model their relationship explicitly. However, we still can make our best guess for possible mappings among these variables, based on the observed relationship. One of the complicated relationships among three correlated variables could be a two-layer hierarchical many-to-many mapping. In this paper, we proposed a Hierarchical Mixture Density Network (HMDN) to model the two-layer hierarchical many-to-many mapping. We apply HMDN on an indoor positioning problem and show its benefit.

Keywords: Mixture Density Network · Hierarchical many-to-many mappings

1 Introduction

In real problems, it is common to find that the same variable could be generated under different conditions, resulting in different values. Different variables, conversely, could have the same value. Such a relationship between two variables can be denoted as a many-to-many mapping. Although it is difficult to model a many-to-many mapping directly in the continuous space, it can be simplified to a one-to-many mapping when conditioning on one variable.

Generally, Mixture Density Network (MDN) is used for one-to-many mapping in a continuous space, as its output is a multi-modal distribution, which is suitable to approximate multiple targets [3]. Many applications have indicated the successes of applying MDN to model one-to-many mapping between two variables. Inspired by these works, we consider whether MDN could be used for modeling a two-layer hierarchical many-to-many mapping among three variables (see Fig. 1).

In this paper, we suppose a two-layer hierarchical many-to-many mapping can be approximated by two many-to-many mappings, connected by a sampling method. Therefore, we proposed a Hierarchical Mixture Density Network (HMDN), by integrating two pre-trained MDNs together. In the indoor positioning experiment, we show that our HMDN can take both WLAN fingerprint and illumination intensity into account directly, and make better predictions.

© Springer International Publishing AG 2017
D. Liu et al. (Eds.): ICONIP 2017, Part IV, LNCS 10637, pp. 878–885, 2017.
https://doi.org/10.1007/978-3-319-70093-9_93

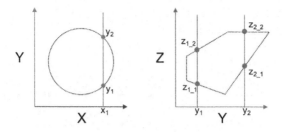

Fig. 1. A hierarchical many-to-many mappings. x_1 can be mapped to y_1 and y_2, while y_1 can be mapped to z_{1_1} and z_{1_2}, and y_2 can be mapped to z_{2_1} and z_{2_2}. If x_1 and z_{1_1} are observed features, the target is most likely to be y_1

2 Mixture Density Network

Typical neural networks only consider an unimodal distribution for one output. However, it may not be sufficient to represent the statistics of a complex output. Mixture Density Network (MDN) is a neural network that was designed to model an arbitrary distribution of output, by replacing the unimodal distribution to a linear combination of kernel functions [3]. In other words, MDN still models the mapping between the input x and the output y, but instead of directly giving y, it provides the parameters of a mixture density from which we can sample y.

For most applications, the kernel function is simply chosen as Gaussian, whose probability density is represented in the form

$$p(y \mid x; w) = \sum_{k=1}^{K} \pi_k(x; w) \mathcal{N}\left(\mu_k(x; w), \sigma_k^2(x; w)\right), \tag{1}$$

with the constraints

$$\sum_{k=1}^{K} \pi_k(x) = 1, \ 0 \leq \pi_k(x; w) \leq 1, \tag{2}$$

where K is the number of mixture components, w is the neural network parameters, $\pi_k(x; w)$, $\mu_k(x; w)$ and $\sigma_k(x; w)$ are the mixing coefficients, the means, and the variances of Gaussian mixtures.

What MDN output layer generates are activation units, which can be divided into three types as a^π, a^σ and a^μ (see Fig. 2). Hence, $\pi_k(x; w)$, $\mu_k(x; w)$ and $\sigma_k(x; w)$ can be derived as follows:

$$
\begin{aligned}
\pi_k(x) &= \frac{exp(a_k^\pi)}{\sum_{l=1}^{K} exp(a_l^\pi)}, \\
\sigma_k(x) &= exp(a_k^\sigma), \\
\mu_{ki}(x) &= a_{ki}^\mu, \\
k &\in \{1, 2, \ldots, K\}, \\
i &\in \{1, 2, \ldots, D\},
\end{aligned}
\tag{3}
$$

where D is the dimension of y. Hence, the number of elements for a^π, a^σ and a^μ are K, K and $D \times K$, respectively.

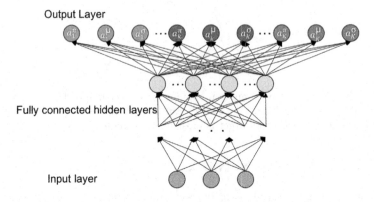

Fig. 2. Representation of the Mixture Density Network

In MDN, what we want to get is the conditional density function $p(y \mid x)$, for each pair (x, y). Therefore, a MDN is trained by maximizing the conditional density likelihood over all variables, yielding the following cost function

$$E(w) = -\sum_{n=1}^{N} ln \left\{ \sum_{k=1}^{K} \pi_k(x_n; w) \mathcal{N}\left(y_n \mid \mu_k(x_n; w), \sigma_k^2(x_n; w)\right) \right\} \qquad (4)$$

Since the derivatives of $E(w)$ and output activations (i.e. a^π, a^σ and a^μ) can be calculated [3], MDN training can be performed using general gradient descent algorithms.

3   Extensions of MDN

In this section, we review several neural network structures integrated with MDN. Once we get the idea about how they are constructed, HMDN can be treated as a similar construction.

Although MDN is popularly used in a range of applications, there is little difference in the structure of how it integrates with other networks. Since it is difficult to back-propagate the gradient from other networks to an MDN, the common integration is done by putting MDN at the final output layer, modeling the one-to-many mapping between extracted features and final targets.

Denoting the neural networks before MDN as g_1, while MDN as g_2, and their corresponding parameters are w_1 and w_2, respectively, we can represent aforementioned structure in a general form:

$$g_1 = f(x; w_1),$$

$$g_2 = \sum_{k=1}^{K} \pi_k(g_1; w_2) \mathcal{N}\left(\mu_k(g_1; w_2), \sigma_k^2(g_1; w_2)\right), \qquad (5)$$

where f could be a convolutional neural network (CNN) [8], a recurrent neural network (RNN) [5], or a combination of CNN and RNN. To some extent, an variational autoencoder [7] can be seen as putting two MDNs together symmetrically, in which the mixture density layer is shared. Extensions of MDN are summarized in Fig. 3.

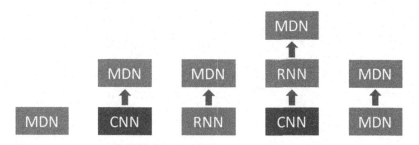

Fig. 3. Extensions of MDN. A single MDN was used in [4,12]. CNN+MDN is used in [6,9]. CNN + RNN is used in [2,11,14]. CNN + RNN + MDN is used in [1]. MDN + MDN is similar to [7].

4   Hierarchical Mixture Density Network

Referring to Fig. 3, our proposed HMDN structure can be interpreted as MDN + MDN. Nevertheless, in contrast to a variational autoencoder, the output of the first MDN is the input of the second MDN in HMDN (see Fig. 4). To concrete the idea about HMDN, following example is used.

Suppose we have datasets X and Z as features, while Y are the targets. Herein, Y is the bridge to link X and Z together. In general, we train a model using both X and Z as inputs and Y as the output. However, it may not work when the relations from X to Y and Y to Z are both many-to-many mappings in the continuous space.

To tackle this issue, we use one MDN (g_1) to model the mapping from X to Y, conditioning on each variable x; while using another MDN (g_2) to model the mapping from Y to Z, conditioning on each variable y. The formulas can be represented as

$$g_1 = p(y \mid x; w_1) = \sum_{k=1}^{K} \pi_k(x; w_1)\mathcal{N}\Big(\mu_k(x; w_1), \sigma_k^2(x; w_1)\Big),$$

$$g_2 = p(z \mid y; w_2) = \sum_{k=1}^{K} \pi_k(y; w_2)\mathcal{N}\Big(\mu_k(y; w_2), \sigma_k^2(y; w_2)\Big). \tag{6}$$

Given X, we can sample $Y_{sample} = \{y_1, y_2, \ldots, y_m\}$ from $g_1(X)$. We further select samples from Y_{sample}, depending on which sample can give higher $p(z \mid y)$.

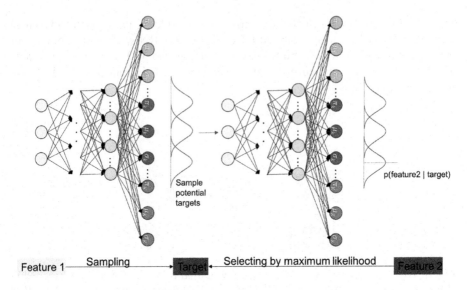

Fig. 4. The model structure of HMDN. Sampling potential targets using feature 1, and further selecting targets using feature 2.

Despite $p(y \mid z)$ is what we truly aim at, based on following Bayesian formula, $p(y \mid z)$ is proportional to $p(z \mid y)$.

$$p(y \mid z) = \frac{p(z \mid y)p(y)}{p(z)},$$

$$p(y \mid z) \propto p(z \mid y). \tag{7}$$

Since Z are given, after Y_{sample} were sampled, both $p(y)$ and $p(z)$ are constants in above formula.

5   Experiment

To apply HMDN in a real situation, we utilize the UJIIndoorLoc Data Set [13], whose features are WLAN intensity received from mobile phones, while the targets are the two-dimensional coordinates with respect to each WLAN fingerprint. Due to the signal reflection or limited Wireless Access Points (WAPs), the same WLAN fingerprint could be mapped to several potential positions. Conversely, one position could have several fingerprints at the different time, as the signal is unstable.

In order to make indoor positioning more accurate, besides WLAN fingerprint, we hope to use other location related signals. Since illumination intensity could be location dependent indoor, it may support the indoor positioning [10]. However, even in the same room, illumination intensity does not remain constant. For instance, the illumination intensity could vary in three conditions:

sunny outside, cloudy outside and night with light opening. Therefore, using the light meter at a fix position, the illumination intensity could be three levels in the previous assumption. Moreover, several positions could have the same illumination intensity.

Putting WLAN fingerprint and illumination intensity together, there are two many-to-many mappings. Our proposed HMDN is pertinent in such case.

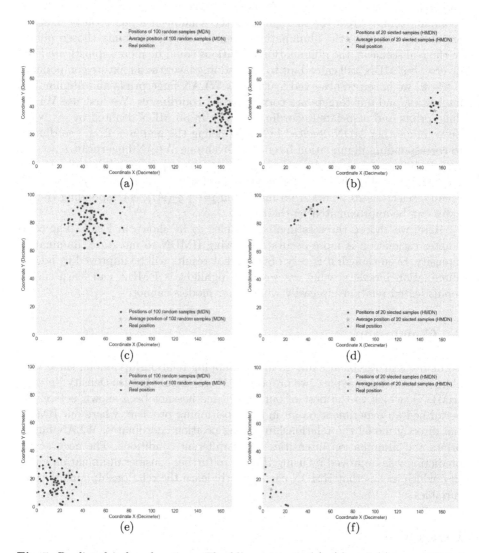

Fig. 5. Predicted indoor locations. The blue points in (a), (c) and (e) are predicted locations that only depend on WLAN fingerprints. Each of them contains 100 predicted locations. The green points in (b), (d) and (f) are prediction locations that are further selected using illumination intensities. Each of them contains 20 predicted locations (Color figure online).

In the UJIIndoorLoc Data Set, there is no illumination information. For this reason, we customize this dataset by adding simulated illumination intensity under three aforementioned conditions. As our main focus is for demonstration, a simple location dependent luminous field was utilized. Through inspecting the coordinate data of one room in UJIIndoorLoc Data Set, we assume the size of a room could be 17×10 m, the average height of holding a phone could be 1.5 m, and ceiling height could be 4 m. Without considering any light reflection in the room, and simply treat the sunlight from the window as a point source light, we are able to calculate the illumination intensity at 100 randomly chosen points. In the real scenario, the illumination conditions could be more sophisticated and diverse, but MDN still can adapt to it by adding layers and mixture components.

Now, we have two observed features as WLAN fingerprints and illumination intensities, and our targets are corresponding coordinates. We first use WLAN fingerprints and associated coordinates to train an MDN denoted by g_1. Next, we train another MDN denoted by g_2, modeling the mapping from coordinates to corresponding illumination intensity. Given one WLAN fingerprint s, we can sample 100 positions as $A = \{pos_1, pos_2, \ldots, pos_{100}\}$ from $g_1(s)$ and calculate the conditional likelihood of illumination intensity I given A, i.e., $p(I \mid g_2(A))$. Finally, selecting 20 points from maximum $p(I \mid g_2(A))$, corresponding coordinates can be approximated by their mean.

Here, we choose three estimated samples to be shown in Fig. 5. The coordinates estimation is more accurate by using HMDN to integrate illumination intensity. Even though it is very obvious that results will be improved by adding illumination intensity, what we want to highlight is HMDN can learn such a sophisticated relationship easily while other models cannot.

6   Conclusion

To model a hierarchical many-to-many mapping relationship among three variables in a continuous space, we proposed a Hierarchical Mixture Density Network (HMDN), which to the best of our knowledge has not been shown before. We performed an experiment on an indoor positioning problem, where our HMDN can directly model the relationship among position coordinates, WLAN fingerprints and illumination intensities under different conditions. The accuracy of prediction was improved by using HMDN to further consider illumination intensity, which means that HMDN can correctly learn the relationship among three variables.

References

1. Bazzani, L., Larochelle, H., Torresani, L.: Recurrent mixture density network for spatiotemporal visual attention. arXiv preprint (2016). arXiv:1603.08199
2. Berio, D., Akten, M., Leymarie, F.F., Grierson, M., Plamondon, R.: Sequence generation with a physiologically plausible model of handwriting and recurrent mixture density networks (2016)

3. Bishop, C.M.: Mixture density networks (1994)
4. Herzallah, R., Lowe, D.: A mixture density network approach to modelling and exploiting uncertainty in nonlinear control problems. Eng. Appl. Artif. Intell. **17**(2), 145–158 (2004)
5. Hochreiter, S., Schmidhuber, J.: Long short-term memory. Neural Comput. **9**(8), 1735–1780 (1997)
6. Iso, H., Wakamiya, S., Aramaki, E.: Density estimation for geolocation via convolutional mixture density network. arXiv preprint (2017). arXiv:1705.02750
7. Kingma, D.P., Welling, M.: Auto-encoding variational bayes. arXiv preprint (2013). arXiv:1312.6114
8. Krizhevsky, A., Sutskever, I., Hinton, G.E.: Imagenet classification with deep convolutional neural networks. In: Advances in Neural Information Processing Systems, pp. 1097–1105 (2012)
9. Moon, S., Park, Y., Suh, I.H.: Predicting multiple pregrasping poses by combining deep convolutional neural networks with mixture density networks. In: Hirose, A., Ozawa, S., Doya, K., Ikeda, K., Lee, M., Liu, D. (eds.) ICONIP 2016. LNCS, vol. 9949, pp. 581–590. Springer, Cham (2016). doi:10.1007/978-3-319-46675-0_64
10. Randall, J., Amft, O., Bohn, J., Burri, M.: LuxTrace: indoor positioning using building illumination. Pers. Ubiquit. Comput. **11**(6), 417–428 (2007)
11. Rehder, E., Wirth, F., Lauer, M., Stiller, C.: Pedestrian prediction by planning using deep neural networks. arXiv preprint (2017). arXiv:1706.05904
12. Richmond, K.: A trajectory mixture density network for the acoustic-articulatory inversion mapping. In: Ninth International Conference on Spoken Language Processing (2006)
13. Torres-Sospedra, J., Montoliu, R., Martínez-Usó, A., Avariento, J.P., Arnau, T.J., Benedito-Bordonau, M., Huerta, J.: UJIIndoorLoc: a new multi-building and multi-floor database for wlan fingerprint-based indoor localization problems. In: 2014 International Conference on Indoor Positioning and Indoor Navigation (IPIN), pp. 261–270. IEEE (2014)
14. Wang, W., Xu, S., Xu, B.: Gating recurrent mixture density networks for acoustic modeling in statistical parametric speech synthesis. In: 2016 IEEE International Conference on Acoustics, Speech and Signal Processing (ICASSP), pp. 5520–5524. IEEE (2016)

A New Bayesian Method for Jointly Sparse Signal Recovery

Haiyan Yang[1], Xiaolin Huang[1($\boxtimes$)], Cheng Peng[1], Jie Yang[1], and Li Li[2]

[1] Institute of Image Processing and Pattern Recognition,
Shanghai Jiao Tong University, Shanghai, China
umiiwa.y@outlook.com, {xiaolinhuang,jieyang}@sjtu.edu.cn,
pynchon1899@gmail.com
[2] Department of Automation, Tsinghua University, Beijing, China
li-li@mail.tsinghua.edu.cn

Abstract. In this paper, we address the recovery of a set of jointly sparse vectors from incomplete measurements. We provide a Bayesian inference scheme for the multiple measurement vector model and develop a novel method to carry out maximum a posteriori estimation for the Bayesian inference based on the prior information on the sparsity structure. Instead of implementing Bayesian variables estimation, we establish the corresponding minimization algorithms for all of the sparse vectors by applying block coordinate descent techniques, and then solve them iteratively and sequently through a re-weighted method. Numerical experiments demonstrate the enhancement of joint sparsity via the new method and its robust recovery performance in the case of a low sampling ratio.

Keywords: Compressive sensing · Multiple measurement vectors · Sparse signal recovery · Joint sparsity

1 Introduction

Compressive sensing (CS) theory aims to recover unknown sparse signals accurately from undetermined linear measurements. Over the past few years, it becomes increasingly attractive and important in many fields, such as medical image [1,2], data compression [3], sensor network [4] and so on. At first, single measurement vector (SMV) model was widely used to solve the sparse signal recovery problem. However, when the dimension of the measurement is small, the restricted isometry property condition for the sensing matrix becomes poor, thereby most likely leading to incorrect recovery.

In some applications, such as Magnetoencephalography (MEG) [5], nonparametric spectrum analysis of time series [6] and so on, a group of signals share the same jointly sparse structure, which could be further exploited if we turn to solve the multiple measurement vector (MMV) problems. Quite a few studies [7–9] have already shown that compared to the SMV cases, the MMV model can

© Springer International Publishing AG 2017
D. Liu et al. (Eds.): ICONIP 2017, Part IV, LNCS 10637, pp. 886–894, 2017.
https://doi.org/10.1007/978-3-319-70093-9_94

improve the successful recovery rate greatly due to the common sparse support. The standard form of this problem is given by

$$\mathbf{y}_l = \Phi \mathbf{x}_l + \epsilon_l \tag{1}$$

where $\Phi \in \mathbb{R}^{m \times n} (m < n)$ is the compressive sensing matrix, $\mathbf{y}_l \in \mathbb{R}^m$ is the measurement vector, ϵ_l is a m-dimensional vector of noise, $\mathbf{x}_l \in \mathbb{R}^n$ is the corresponding sparse solution to $\mathbf{y}_l$, $l = 1, \ldots, L$, and L is the number of measurements. The additional information to this problem is that all $\mathbf{x}_l$ have the same sparsity structure, that is, share the same support set. The support set of $\mathbf{x}$ can be described as $supp(\mathbf{x}) = \{i : x_i \neq 0\}$.

Motivated by various applications mentioned above, how to solve MMV problems has been extensively studied over the past few years. For example, in [10–13], the solution vectors and measurement vectors are combined as $X = [\mathbf{x}_1, \ldots, \mathbf{x}_L]$ and $Y = [\mathbf{y}_1, \ldots, \mathbf{y}_L]$ respectively. Then the jointly sparse solutions to MMV problems turn to be an X with a low rank. Kim *et al.* [14] develop an algorithm called compressive MUSIC which identifies the parts of support using CS and then estimates the remaining parts by using a generalized MUSIC criterion. Vila and Schniter propose a Bayesian approach with a Gaussian-mixture signal prior in [15], namely, EM-GM-AMP, which combines expectation maximization with approximate message passing. In [16], Li *et al.* propose an iterative re-weighted algorithm based on [17] to recover a set of jointly sparse vectors by applying block coordinate decent and Majorization-Minimization techniques.

In this paper, we are going to provide a Bayesian inference scheme for the MMV model. Specifically speaking, we carry out maximum a posteriori (MAP) estimation for the sparse solutions based on the Bayesian inference and the prior information on the sparsity structure. Then we use the block coordinate decent technique and the re-weighted method in [16] to recover the set of jointly sparse vectors sequently and iteratively. In each iteration, every vector is calculated based on all the obtained solutions which could provide the prior information on the sparsity structure and help modify the values of weights. Thus, the original problem turns to be a minimization problem with a mixture penalty of weighted ℓ_1 and squared ℓ_2 norms. In addition, every element of the vector is assigned with the spike and slab prior weighted by Bernoulli variables.

The paper is organized as follows. Section 2 focuses on the Bayesian formulation for the recovery of two vectors. Section 3 proposes the algorithms for multiple sets of measurements in both noisy cases and noiseless cases. Section 4 presents some numerical tests to show the efficiency of the proposed method. Finally, Sect. 5 concludes the paper.

2   The Bayesian Formulation

Let us first consider the recovery of two jointly sparse vectors. Suppose $\mathbf{y}_1, \mathbf{y}_2 \in \mathbb{R}^m$ are two measurement vectors corresponding to two sparse vectors $\mathbf{x}_1, \mathbf{x}_2 \in \mathbb{R}^n$ via compressive sensing matrix $\Phi \in \mathbb{R}^{m \times n}$ respectively, that is,

$$\mathbf{y}_1 = \Phi \mathbf{x}_1 + \boldsymbol{\epsilon}_1, \quad \mathbf{y}_2 = \Phi \mathbf{x}_2 + \boldsymbol{\epsilon}_2, \tag{2}$$

where $\boldsymbol{\epsilon}_1, \boldsymbol{\epsilon}_2 \in \mathbb{R}^m$ are two vectors of zero-mean Gaussian noise with variances $1/\lambda_1$, $1/\lambda_2$ respectively. Note that the compressive sensing matrices for two measurements are not necessarily the same in this paper, which is similar to [16,18]. $\mathbf{x}_1, \mathbf{x}_2$ are assumed to share the same sparsity structure. To represent the sparsity structure of a vector, we construct a diagonal matrix $Z(\boldsymbol{\omega}, \tau)$ of which the diagonal entries $z_{ii}(i = 1, 2, \dots, n)$ are defined as

$$z_{ii}(\boldsymbol{\omega}, \tau) = \begin{cases} 0 & |\omega_i| > \tau, \\ 1 & \text{otherwise,} \end{cases} \tag{3}$$

where τ is a positive threshold parameter which is expected to be reasonably small. When $|\omega_i| \le \tau$, ω_i will be considered approximately as zero. Thus, 1 indicates that ω_i is a zero entry, 0 indicates that ω_i is a nonzero entry.

In this paper, $\mathbf{x}_1$ and $\mathbf{x}_2$ are assumed to obey Gaussian distributions with variances υ_1, υ_2 respectively when they are non-zero. Moreover, the elements in any of them are unrelated with each other, that is, x_{1i} is irrelevant with $x_{1j}(i \ne j)$ and so is $\mathbf{x}_2$. Under the above assumptions, the MAP solutions of problem (2) are formulated as

$$\begin{aligned} (\mathbf{x}_1, \mathbf{x}_2)_{\text{MAP}} &= \arg \max_{\mathbf{x}_1, \mathbf{x}_2} p(\mathbf{x}_1, \mathbf{x}_2 | \mathbf{y}_1, \mathbf{y}_2) \\ &= \arg \min_{\mathbf{x}_1, \mathbf{x}_2} - \log[p(\mathbf{x}_1 | \mathbf{x}_2) p(\mathbf{x}_2) p(\mathbf{y}_1, \mathbf{y}_2 | \mathbf{x}_1, \mathbf{x}_2)]. \end{aligned} \tag{4}$$

Once $\mathbf{x}_2$ is estimated, its sparsity structure $Z(\mathbf{x}_2, \tau)$ will be estimated at the same time. Since $\mathbf{x}_1, \mathbf{x}_2$ are jointly sparse, the diagonal entries of $Z(\mathbf{x}_2, \tau)$ can be seen as the sparsity indicators of $\mathbf{x}_1$ as well. Like the spike and slab prior described in [19], when $z_{ii}(\mathbf{x}_2, \tau) = 0$, x_{1i} is nonzero and obeys a Gaussian distribution (the slab). Otherwise x_{1i} is zero and its probability density could be denoted as an origin-center point probability mass $\delta(x_{1i})$ (the spike). Thus, the posterior of $\mathbf{x}_1$ is explicitly expressed as

$$\begin{aligned} p(\mathbf{x}_1 | \mathbf{x}_2) &= \prod_{i=1}^{n} p(x_{1i} | x_{2i}) = \prod_{\substack{i=1, \\ x_{2i} \ne 0}}^{n} \mathcal{N}(x_{1i} | 0, \upsilon_1) \prod_{\substack{i=1, \\ x_{2i} = 0}}^{n} \delta(x_{1i}) \\ &= \prod_{i=1}^{n} \mathcal{N}(x_{1i} | 0, \upsilon_1)^{1 - z_{ii}(\mathbf{x}_2, \tau)} \delta(x_{1i})^{z_{ii}(\mathbf{x}_2, \tau)}. \end{aligned} \tag{5}$$

Similarly, the prior of $\mathbf{x}_2$ could be expressed as

$$\begin{aligned} p(\mathbf{x}_2) &= \prod_{i=1}^{n} p(x_{1i} = 0) p(x_{2i} | x_{1i} = 0) + p(x_{1i} \ne 0) p(x_{2i} | x_{1i} \ne 0) \\ &= \prod_{i=1}^{n} p(x_{1i} = 0) \delta(x_{2i}) + p(x_{1i} \ne 0) \mathcal{N}(x_{2i} | 0, \upsilon_2). \end{aligned} \tag{6}$$

$$p(\mathbf{y}_1, \mathbf{y}_2 | \mathbf{x}_1, \mathbf{x}_2)] = \lambda_1 ||\mathbf{y}_1 - \Phi \mathbf{x}_1||^2 + \lambda_2 ||\mathbf{y}_2 - \Phi \mathbf{x}_2||^2. \tag{7}$$

From the former three equations, we can draw a conclusion that (4) is non-convex and difficult to be solved directly. Considering $\mathbf{x}_1, \mathbf{x}_2$ are interlaced in (4), we choose the block coordinate descent algorithm to update $\mathbf{x}_1, \mathbf{x}_2$ sequently and iteratively.

3   Iterative Algorithm Using Block Coordinate Descent

Block coordinate descent algorithm is used to solve problems with multiple variables which could be split into several blocks. In each iteration, a single block will be updated while the others will be fixed. To solve the minimization problem in (4), we regard $\mathbf{x}_1, \mathbf{x}_2$ as two blocks of variables and update one of them while keep the other one fixed alternately.

Firstly, $\mathbf{x}_2$ is initialized as a nonzero vector denoted by $\mathbf{x}_2^0$. Based on this estimation of $\mathbf{x}_2$, we can further estimate $\mathbf{x}_1$ as

$$\mathbf{x}_1^1 = \arg\min_{\mathbf{x}_1} \ \frac{1}{\upsilon_1}||[\mathbf{I} - Z(\mathbf{x}_2^0, \tau)]\mathbf{x}_1||^2 + \sum_{i=1}^{n} -z_{ii}(\mathbf{x}_2^0, \tau)\log \delta(x_{1i})$$
$$+ \lambda_1||\mathbf{y}_1 - \varPhi\mathbf{x}_1||^2, \tag{8}$$

at the first iteration. To avoid calculating $\log \delta(x_{1i})$ when $\delta(x_{1i}) = 0$, we replace this term with $\log(\delta(x_{1i}) + \varepsilon)$ where ε is a small positive constant. Note that $z_{ii}(\mathbf{x}_2^0, \tau) = 0$ implies $x_{2i}^0 \neq 0$ which will imply $x_{1i} \neq 0$. In turn, $-z_{ii}(\mathbf{x}_2^0, \tau) \cdot \log(\delta(x_{1i}) + \varepsilon) = 0 \cdot \log \varepsilon = 0$. On the other hand, $z_{ii}(\mathbf{x}_2^0, \tau) = 1$ implies $x_{2i}^0 = 0$ which will further imply $x_{1i} = 0$ and in turn, $-z_{ii}(\mathbf{x}_2^0, \tau)\log(\delta(x_{1i}) + \varepsilon) = -1 \cdot \log(1 + \varepsilon)$. Thus, the second term in (8) can be expressed as $-\log(1 + \varepsilon)$ multiplying the number of zero entries in x_1, that is, $-\log(1 + \varepsilon)(n - ||\mathbf{x}_1||_0)$. Inspired by [16,17], we then replace $||\mathbf{x}_1||_0$ with $||W_1\mathbf{x}_1||_1$ where $W_1 \in \mathbb{R}^{n \times n}$ is a weight matrix. At the j-th iteration, the objective function is given as

$$\mathbf{x}_1^j = \arg\min_{\mathbf{x}_1} \ \frac{1}{\upsilon_1}||[I_n - Z(\mathbf{x}_2^{j-1}, \tau)]\mathbf{x}_1||^2 + \beta||Z(\mathbf{x}_2^{j-1}, \tau)W_1^{j-1}\mathbf{x}_1||_1$$
$$+ \lambda_1||\mathbf{y}_1 - \varPhi\mathbf{x}_1||^2, \tag{9}$$

where I_n is a n-dimensional identity matrix and $\beta = \log(1 + \varepsilon)$. Then $\mathbf{x}_2$ will be estimated by calculating

$$\mathbf{x}_2^j = \arg\min_{\mathbf{x}_2} \ \frac{1}{\upsilon_2}||[I_n - Z(\mathbf{x}_1^j, \tau)]\mathbf{x}_2||^2 + \beta||Z(\mathbf{x}_1^j, \tau)W_2^j\mathbf{x}_2||_1$$
$$+ \lambda_2||\mathbf{y}_2 - \varPhi\mathbf{x}_2||^2. \tag{10}$$

In each loop iteration, $\mathbf{x}_1, \mathbf{x}_2, W_1$ and W_2 will be updated and then fed to the next loop. The loop will terminate either if a sparse optimal solution is found or both $\mathbf{x}_1$ and $\mathbf{x}_2$ converge, or we reach the maximum iterations. In summary, the proposed algorithm for the recovery of two vectors is written as Algorithm 1. Similarly, we propose an algorithm for multiple (more than two) vectors as Algorithm 2.

Algorithm 1. Re-weighted Bayesian approach for two vectors

Input: $\mathbf{y}_1, \mathbf{y}_2, \Phi, \upsilon_1, \upsilon_2, \tau, \epsilon, \beta, \lambda_1, \lambda_2, j_{max}$
Output: $\mathbf{x}_1, \mathbf{x}_2$
1: initialize $j \leftarrow 0$, $W_1^0 \leftarrow I_n$, $W_2^0 \leftarrow I_n$, $\mathbf{x}_1^0 \leftarrow [1, 1, \ldots, 1]^T$, $\mathbf{x}_2^0 \leftarrow [1, 1, \ldots, 1]^T$
2: **repeat**
3:　　$j \leftarrow j + 1$
4:　　Update $\mathbf{x}_1$ via (9)
5:　　$W_{2,ii}^j \leftarrow \frac{1}{|x_{2i}^{j-1}|+\epsilon} + \frac{z_{ii}(\mathbf{x}_1^j, \tau)}{z_{ii}(\mathbf{x}_1^j, \tau) \cdot |x_{2i}^{j-1}|+\epsilon}$
6:　　Update $\mathbf{x}_2$ via (10)
7:　　$W_{1,ii}^j \leftarrow \frac{1}{|x_{1i}^j|+\epsilon} + \frac{z_{ii}(\mathbf{x}_2^j, \tau)}{z_{ii}(\mathbf{x}_2^j, \tau) \cdot |x_{1i}^j|+\epsilon}$
8: **until** both $\mathbf{x}_1$ and $\mathbf{x}_2$ converge or $j = j_{max}$ or successful recovery

---

Algorithm 2. Re-weighted Bayesian approach for multiple vectors

Input: $\mathbf{y}_1, \mathbf{y}_2, \ldots, y_L, \Phi, \upsilon_1, \upsilon_2, \ldots, \upsilon_L, \tau, \epsilon, \beta, \lambda_1, \lambda_2, \ldots, \lambda_L, j_{max}$
Output: $\mathbf{x}_1, \mathbf{x}_2, \ldots, x_L$
1: Sequentially initialize $j \leftarrow 0$, $W_l^0 \leftarrow I_n$, $\mathbf{x}_l^0 \leftarrow [1, 1, \ldots, 1]^T$, $l = 1, 2, \ldots, L$
2: **repeat**
3:　　$j \leftarrow j + 1$
4:　　Sequentially update $\mathbf{x}_l, l = 1, 2, \ldots, L - 1$ via

$$\mathbf{x}_l^j \leftarrow \arg\min_{\mathbf{x}_l} \frac{1}{\upsilon_l} ||[I_n - Z(\mathbf{x}_{l-1}^j, \tau)]\mathbf{x}_l||^2 + \beta||Z(\mathbf{x}_{l-1}^j, \tau)W_l^{j-1}\mathbf{x}_l||_1 + \lambda_l||\mathbf{y}_l - \Phi\mathbf{x}_l||^2$$

5:　　Update the weighting matrix W_{l+1} via $W_{l+1,ii}^j \leftarrow \frac{1}{|x_{l+1,i}^{j-1}|+\epsilon} + \frac{z_{ii}(\mathbf{x}_1^j, \tau)}{z_{ii}(\mathbf{x}_1^j, \tau) \cdot |x_{l+1,i}^{j-1}|+\epsilon}$
6:　　Update $\mathbf{x}_L$ via

$$\mathbf{x}_L^j \leftarrow \arg\min_{\mathbf{x}_L} \frac{1}{\upsilon_L} ||[I_n - Z(\mathbf{x}_{L-1}^j, \tau)]\mathbf{x}_L||^2 + \beta||Z(\mathbf{x}_{L-1}^j, \tau)W_L^{j-1}\mathbf{x}_L||_1$$
$$+ \lambda_L||\mathbf{y}_L - \Phi\mathbf{x}_L||^2$$

7:　　Update the weighting matrix W_1 via $W_{1,ii}^j \leftarrow \frac{1}{|x_{1,i}^j|+\epsilon} + \frac{z_{ii}(\mathbf{x}_L^j, \tau)}{z_{ii}(\mathbf{x}_L^j, \tau) \cdot |x_{1,i}^j|+\epsilon}$
8: **until** both $\mathbf{x}_1$ and $\mathbf{x}_2$ converge or $j = j_{max}$ or successful recovery

---

In noiseless cases, the error term in (1) is equal to zero. Thus, the original squared l_2 norm of noise will vanish. In addition, the objective function for any vector, e.g. $\mathbf{x}_l$, will be constrained by $\mathbf{y}_l = \Phi\mathbf{x}_l$, which could be written as

$$\mathbf{x}_l^j = \arg\min_{\mathbf{x}_l} \frac{1}{\upsilon_l} ||[\mathbf{I_n} - Z(\mathbf{x}_{l-1}^j, \tau)]\mathbf{x}_l||^2 + \beta||Z(\mathbf{x}_{l-1}^j, \tau)W_l^{j-1}\mathbf{x}_l||_1 \tag{11}$$
$$s.t. \quad \mathbf{y}_l = \Phi\mathbf{x}_l.$$

Then the algorithm for multiple vectors in noiseless cases will be easily obtained by replacing the objective functions in Algorithm 2 with (11).

4   Numerical Experiments

In this section, we evaluate the proposed algorithms by two numerical experiments. The first experiment aims to recover two jointly sparse vectors and evaluate the performance of our algorithm by showing the root mean squared error (RMSE) with respect to the signal-to-noise ratio (SNR). RMSE and SNR are defined as

$$\text{RMSE} = \sqrt{\frac{1}{n}\sum_{i=1}^{n}||\mathbf{x} - \hat{\mathbf{x}}||^2}, \quad \text{SNR} = 10\log\frac{\upsilon}{1/\lambda},$$

where $\upsilon, 1/\lambda$ denote the variances of the signal and the noise respectively. The setup for each trial is as follows. We construct two sparse signals $\mathbf{x_1}, \mathbf{x_2}$ of length $n = 256$ with $||\mathbf{x_1}||_0 = ||\mathbf{x_2}||_0 = K$. The positions of the K nonzero entries in $\mathbf{x_1}, \mathbf{x_2}$ are the same and selected randomly. Moreover, the values of them are generated following a Gaussian distribution $\mathcal{N}(0, 1)$. We set $m = 32$ and select the elements of the compressive sensing matrix Φ randomly from a Gaussian distribution $\mathcal{N}(0, 1)$. We also construct two vectors of Gaussian noise of which the variances $1/\lambda_1, 1/\lambda_2$ depend on the signals and the SNR. Then $\mathbf{y_1}, \mathbf{y_2}$ are derived from $\mathbf{y} = \Phi\mathbf{x} + \epsilon$. In this experiment, we are going to recover the two 256-dimensional signals given Φ and $\mathbf{y_1}, \mathbf{y_2}$. For each SNR, we conduct 50 trails. The average RMSE for each SNR is calculated as $\frac{1}{50}\sum_{i=1}^{50}(\text{RMSE}_{\mathbf{x_1}} + \text{RMSE}_{\mathbf{x_2}})$. We choose three MMV algorithms, $i.e.$, CS-MUSIC, EM-BG-AMP, SOMP, which have been introduced in Sect. 1 for comparison and set their parameters as the corresponding originals suggest so as to provide a fair comparison. The result is shown in Fig. 1, from which we can see that our algorithm exhibits the uniform superiority over SOMP and CS-MUSIC, and exceed all the other three algorithms especially when SNR > 15 dB.

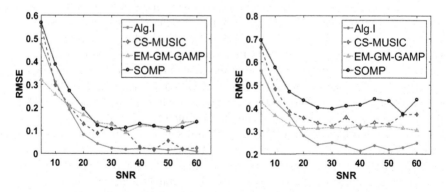

Fig. 1. RMSE versus SNR for various MMV algorithms with (left) $K = 9$, (right) $K = 15$.

An Electrocardiogram (ECG) is a record of the electrical voltage in the heart during a period of time. This kind of signals is not sparse, whereas by applying

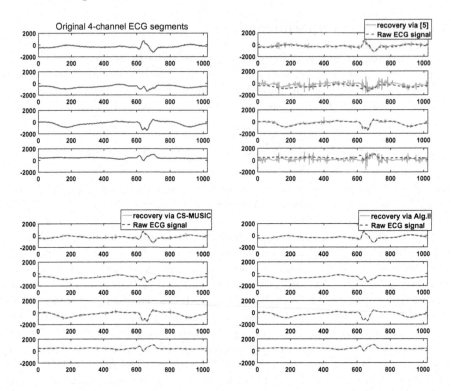

Fig. 2. (Top left) Raw 4-channel ECG signals. Reconstruction of 4-channel ECG signal segments via (top right) the SMV algorithm in [17]; (bottom left) CS-MUSIC; (bottom right) Algorithm 2 proposed in Sect. 3.

a Daubechies wavelet transformation, the transformed signals turn to be sparse. The second experiment aims to reconstruct the multi-channel ECG signals by recovering the transformed sparse signals, of which the process is as follows. To show the reconstruction results more clearly, we cut out four segments with length $n = 1024$, each of which contains a complete heart beat as shown in Fig. 2(a), from a raw complete 4-channel ECG signal with length 38400. Firstly, we can easily obtain a Daubechies wavelet transformation matrix $\Psi \in \mathbb{R}^{n \times n}$. We then generate the elements of the measurement matrix $A \in \mathbb{R}^{m \times n}$ randomly following a Gaussian distribution $\mathcal{N}(0, 1)$ and set $m = 150$. Next the compressive sensing matrix is obtained as $\Phi = A\Psi$. By multiplying the compressive sensing matrix by each transformed ECG segment vector respectively, the measurements are derived. We observed that the transformed signal segments are jointly sparse, which means that we could use MMV algorithms to recover them from the measurements. The original signals are then obtained through a inverse transformation. By splitting the complete signal into quite a few segments and recovering the segments sequently, finally we can realize the recovery of the whole signal.

In this experiment, we choose the SMV algorithm in [17] and the MMV algorithm CS-MUSIC for comparison. Furthermore, to provide a fair comparison, we adjust the parameters of the three algorithms so that the recoveries of the entire 4-channel transformed signals have similar sparsity levels as shown in Table 1. The sparsity level is defined as the percentage of the non-zero entries in the entire vector. Table 1 also demonstrates that our method can achieve a fairly smaller error. Figure 2(d) shows the reconstruction result of the ECG segments via our algorithm after 15 iterations. By comparing Fig. 2(b), (c) and (d), we can see that our method performs better than the SMV algorithm when recovering multiple jointly sparse signals, which means that our method could take full use of the prior information on sparsity structure and enhance joint sparsity. In addition, our method exceeds CS-MUSIC as well.

Table 1. The sparsity levels of the recoveries of the entire transformed signals and the RMSEs of the reconstructions of the entire ECG signals.

Method	Sparsity level				RMSE			
	$ch1$	$ch2$	$ch3$	$ch4$	$ch1$	$ch2$	$ch3$	$ch4$
Algorithm 2	12.5	13.5	13.8	13.3	**83.19**	**93.74**	**100.82**	**73.94**
CS-MUSIC	12.3	13.3	13.4	13.1	89.86	101.17	107.11	75.66
SMV method in [17]	12.9	12.9	13.1	12.6	127.63	152.73	154.50	137.85

5 Conclusions

In this paper, we address the sparse signal recovery problem of the MMV model and provide a Bayesian inference scheme for this model. By exploiting the prior information on the sparsity structure, we develop a novel method to carry out MAP estimation for this Bayesian inference. By means of the block coordinate descent techniques, we establish the corresponding minimization algorithms for all of the sparse vectors. Different from the single penalty of ℓ_1 norm in the objective function which has been proposed in many classical studies, a mixture penalty of weighted ℓ_1 and squared ℓ_2 norms is developed. Then we use a re-weighted method to solve the new minimization problems iteratively and sequently. Both simulation and application experiments show the enhancement of joint sparsity via the new method and its robust recovery performance in the case of a low sampling ratio.

References

1. Candès, E.J., Romberg, J., Tao, T.: Robust uncertainty principles: exact signal reconstruction from highly incomplete frequency information. IEEE Trans. Inf. Theory **52**(2), 489–509 (2006)

2. Lustig, M., Donoho, D., Pauly, J.M.: Sparse MRI: the application of compressed sensing for rapid MR imaging. Magn. Reson. Med. **58**(6), 1182–1195 (2007)

3. Candès, E.J., Tao, T.: Near-optimal signal recovery from random projections: universal encoding strategies? IEEE Trans. Inf. Theory **52**(12), 5406–5425 (2006)

4. Bajwa, W., Haupt, J., Sayeed, A., Nowak, R.: Compressive wireless sensing. In: Proceedings of the 5th International Conference on Information Processing in Sensor Networks, pp. 134–142. ACM (2006)

5. Gorodnitsky, I.F., Rao, B.D.: Sparse signal reconstruction from limited data using focuss: a re-weighted minimum norm algorithm. IEEE Trans. Signal Process. **45**(3), 600–616 (2002)

6. Petre, S., Randolph, M.: Spectral analysis of signals (POD). Leber Magen Darm **13**(2), 57–63 (2005)

7. Cotter, S.F., Rao, B.D., Engan, K., Kreutz-Delgado, K.: Sparse solutions to linear inverse problems with multiple measurement vectors. IEEE Trans. Signal Process. **53**(7), 2477–2488 (2005)

8. Eldar, Y.C., Mishali, M.: Robust recovery of signals from a structured union of subspaces. IEEE Trans. Inf. Theory **55**(11), 5302–5316 (2009)

9. Eldar, Y.C., Rauhut, H.: Average case analysis of multichannel sparse recovery using convex relaxation. IEEE Trans. Inf. Theory **56**(1), 505–519 (2006)

10. Davies, M.E., Eldar, Y.C.: Rank awareness in joint sparse recovery. IEEE Trans. Inf. Theory **58**(2), 1135–1146 (2010)

11. Lee, K., Bresler, Y., Junge, M.: Subspace methods for joint sparse recovery. IEEE Trans. Inf. Theory **58**(6), 3613–3641 (2012)

12. Blanchard, J.D., Davies, M.E.: Recovery guarantees for rank aware pursuits. IEEE Signal Process. Lett. **19**(7), 427–430 (2012)

13. Gogna, A., Shukla, A., Agarwal, H.K., Majumdar, A.: Split Bregman algorithms for sparse/joint-sparse and low-rank signal recovery: application in compressive hyperspectral imaging. In: IEEE International Conference on Image Processing, pp. 1302–1306 (2015)

14. Kim, J.M., Lee, O.K., Ye, J.C.: Compressive MUSIC: revisiting the link between compressive sensing and array signal processing. IEEE Trans. Inf. Theory **58**(1), 278–301 (2012)

15. Vila, J.P., Schniter, P.: Expectation-maximization Gaussian-mixture approximate message passing. IEEE Trans. Signal Process. **61**(19), 4658–4672 (2013)

16. Li, L., Huang, X., Suykens, J.A.K.: Signal recovery for jointly sparse vectors with different sensing matrices. Signal Process. **108**(C), 451–458 (2015)

17. Candès, E.J., Wakin, M.B., Boyd, S.P.: Enhancing sparsity by reweighted ℓ_1 minimization. J. Fourier Anal. Appl. **14**(5), 877–905 (2008)

18. Ji, S., Dunson, D., Carin, L.: Multitask compressive sensing. IEEE Trans. Signal Process. **57**(1), 92–106 (2009)

19. George, E.I., Mcculloch, R.E.: Approaches for Bayesian variable selection. Stat. Sin. **7**(2), 339–373 (1997)

Author Index

Printed in the United States
By Bookmasters